The Magnificent Bastards in Ramadi and A Father's Journey There

The 2004 Battle of Ramadi and a 2008 Gold Star Father's Memorial Service There

Gregory Janney

Copyright © 2024 Gregory Janney

M. LiClar Publishing Co., LLC
Monroe City, MO

All rights reserved, including the right to reproduce this book or portions thereof in any form whatsoever.

ISBN-13: 979-8-9902673-0-5

CONTENTS

PROLOGUE

2ND BATTALION, 4TH MARINES - RAMADI DEPLOYMENT OVERVIEW — 1
- MAJOR J.D. HARRILL III, BATTALION OPSO (NOW LTCOL)

6 APRIL 2004 - ROUTE NOVA FIREFIGHT AND QRF — 16
- ROMEO SANTIAGO, SGT, SCOUT SNIPER, HEADHUNTER II TEAM LEADER
- ANONYMOUS SCOUT SNIPER - KNOWN AS AM IN INTERVIEW AND (REDACTED) IN BOOK
- BRANDON LUND, LCPL
- RAMON BARRON, CPL
- KYLE KATZ, PFC
- ROY THOMAS, LCPL

6 APRIL 2004 - ROUTE GYPSUM AMBUSH AND QRF — 80
- DESHON OTEY, LCPL - AAR (AFTER ACTION REPORT)
- EVAN NULL, CPL
- MATT SCOTT, PFC
- AARON VERGARA, LCPL
- ERIC SMITH, CPL
- RYAN MILLER, PFC
- JOSE VALERIO, SGT
- DOC ADAM CLAYTON, HMC
- GREGORIO CIENFUEGOS, PFC
- FRANK GUTIERREZ, CPL AND VICTOR MADRILLEJOS, PFC (PROMOTED TO LCPL)
- SHAWN SKAGGS, CPL
- "DOC" KEITH GRIMES, HM3

6 APRIL 2004 GOLF COMPANY — 248
- GLENN FORD, SGT
- DAMIEN COAN, SGTMAJ
- WINSTON JAUGAN, GYSGT

PHOTOS OF A FATHER'S JOURNEY TO IRAQ MEMORIAL MISSION — 271

6 APRIL 2004 GOLF COMPANY QRF — 283
- NICK KELLY, CPT
- MARVIN ENDITO, SGT (HOG)
- KRISTOPHER PRIVITAR, LCPL
- JUSTIN WEAVER, PFC (PROMOTED TO LCPL DURING DEPLOYMENT)
- REAGAN HODGES, CPL

6 APRIL 2004 COMBAT OUTPOST — 324
- PATRICK LEBLANC, CPL
- RYAN SAVAGE, LCPL

PIGS (Professionally Instructed Gunmen) and HOGS (Hunters of Gunmen)   335
- Jesse Longoria, Sgt (HOG)
- Jonathan Wood, Cpl (PIG)
- Shawn Spitzer, Sgt (HOG)

7 – 10 April 2004 including "Operation Bug Hunt"   354
- Marc Coiner, Sgt
- Ben Musser, LCpl
- Logan Degenheart, Cpl
- Omar Enrique Morel, LCpl

HQ, H&S, Intel Marines   394
- Charles Lauersdorf, Sgt., HQ-Intel
- Pete Rosado, Sgt., H&S Co
- Carlos Segovia, PFC, H&S Co
- Jamie Bunette, SSgt., HQ-Intel
- Ed Hines, Cpl, H&S Co

Combat Replacements and Delayed Deployment Marines   431
- Jason Adams, Cpl, Weapons Co
- Mike Martinez, Cpl
- James (Buttrey) Baum, Sgt

Battalion SgtMaj James Booker - Deployment Overview   444

Gold Star Family Interviews   468
- Sheila Cobb (PFC Christopher Cobb)
- Mark Crowley (LCpl Kyle Crowley)
- Margaret Kellum (LCpl John "J.T. Sims)
- Dianne Layfield (LCpl Travis Layfield)
- Frances Mabry (PFC Christopher Mabry)
- John and Shawn Wroblewski (2ndLt John "J.T." Wroblewski)

A Father's Journey To Iraq   531
– Gold Star Dad John Wroblewski's Memorial Service in Ramadi
- 2007 First Iraq Embed and Memorial Mission Attempt
- 27 Feb – 13 Mar 2008 Our 2nd Journey to Iraq

Disclaimer

The author is not responsible for any errors or omissions in this book. As many Marines and Corpsmen as possible were interviewed via telephone and social media messaging. Numerous posts were made on the 2nd Battalion, 4th Marines, and Ramadi veterans' social media sites requesting participation in the interviews, as well as efforts by 2/4 veteran Robert Gibson to get 2/4 Marines Ramadi veterans to complete interviews. I also messaged and texted numerous Marines directly whose names I had been given by other 2/4 veterans or other Marines that expressed an interest in completing an interview, but not everyone responded. The difficulties of transcription of telephone interviews may also have caused inadvertent errors. The fog of war, along with the passage of up to 20 years since the 2004 Battle of Ramadi may also have contributed to errors and omissions. The fog of war was evident as several Marines who literally fought side by side had different perspectives of the same firefight even though they were within feet of one another. It was my sincere hope to interview everyone deployed to Ramadi about their personal experiences and remembrances, but space limitations also limited the book to no more than 800 pages. Since the author was not present during the deployment, there is no first-person perspective other than that of the Marines and Corpsmen who completed interviews.

Organization of Chapters

I have organized the interviews based primarily on what the interviewees were doing on 6 April 2004 since that was the first day of the Battle of Ramadi. Some interviewees experienced combat prior to 6 April, so their interview may contain that information prior to what they experienced on 6 April. For example, LCpl Vergara was involved in a 25 March action against a mortar team, but was also involved in the 6 April Route Gypsum action, so his 25 March action is described before his 6 April actions, even though his interview is in the Route Gypsum ambush 6 April section.

Other men had duty the night of 5 April, so they had to rest on 6 April. Since they didn't have a direct role in the events of 6 April, they are organized by what they were doing in the following days. For example, LCpl Ben Musser had duty all night 5 April, so was only involved in the cleanup activities on Route Gypsum on 6 April. Musser's actions are documented in the book's 7-10 April section of the book, even though he participated in some activities on 6 April.

If someone isn't listed in the 6 April section, it's not to say that they did nothing on 6 April. They just may not have had a direct combat role on 6 April, or if they did, they did not describe it to me in their interview.

My suggestion is "just read the book." Everyone has their own piece of a giant jigsaw puzzle, and through reading, as much of the puzzle as possible will be completed. The fog of war prevents a complete picture of everything that occurred those days, but this is as complete as I was able to make the puzzle via the 46 interviews I conducted with the Marines and Corpsmen of 2/4 Marines.

Prologue

What were you doing on 6 April 2004? While most of us went about our daily lives, the men of the 2nd Battalion, 4th Marines, also known as "The Magnificent Bastards," were fighting and dying in Ar-Ramadi, Iraq, and twelve families' lives were forever changed. On a day that few of us remember, 11 Marines and a Navy Corpsman were killed in fighting across the city 8000 miles from home, including a daylight ambush in a narrow marketplace on the east side of Ramadi. This book is about the Battle of Ramadi and of two long journeys I made to Iraq with Gold Star father John Wroblewski in an attempt to visit the site where his son, USMC 2ndLt J.T. "Ski" Wroblewski, fell in battle that terrible day in 2004. Our mission goal was to honor the sacrifices of these men and their families by performing a memorial service at the location where they lost their lives in service to our Nation and their brothers-in-arms. "The Magnificent Bastards" of 2nd Battalion, 4th Marines, 1st Marine Division lost 34 men in combat during their 2004 deployment in Ramadi.

Most of the book details the Battle of Ramadi via the combat experiences of forty-six men of the 2nd Battalion, 4th Marines in the Marines' and Corpmen's own words, followed by interviews with some of the Gold Star parents. The final chapters of the book are my diary of the two trips I made to Iraq with Gold Star dad John Wroblewski in a quest to get him to Ramadi to perform a memorial service where his son, USMC 2ndLt J.T. "Ski" Wroblewski and 11 other men of 2/4 Marines were killed in action (KIA) during savage Route Gypsum and Route Nova ambushes on April 6, 2004 and dozens of other firefights across the city April 6-10. Our first embed in January 2007 was unsuccessful in getting to Ramadi, but I promised John Wroblewski that we would try again. John and I made it to Route Gypsum in Ramadi during our second embed in March 2008. On March 6, 2008, John Wroblewski, USMC Major General John Kelly, and General Kelly's PSD (Personal Security Detachment) performed a memorial service for the fallen heroes of 2/4 Marines near the spot where 2ndLt Wroblewski was mortally wounded. The memorial service was filmed by USMC Cpl Angel and aired on Fox News on Memorial Day 2008 in a piece with journalist Jennifer Griffin, and BYU-TV filmed a story later. Our 2008 mission was the only successful visit by a Gold Star parent to the actual site in Iraq where their son or daughter was killed in combat. Profits from the sale of this book will be used to establish a scholarship fund for the survivors of the 34 men of 2/4 Marines that were KIA during the 2004 Ramadi deployment.

2ND BATTALION, 4TH MARINES - 2004 RAMADI DEPLOYMENT OVERVIEW

Major J.D. Harrill, III (now LtCol) – Battalion OPSO (aka Bastard 3)

Janney: It's 15 December 2020. Tell me when and why you decided to enlist in the Marine Corps.

Harrill: As crazy as it sounds, my dad was a Marine and I grew up wanting to be a pro football player for the Washington Redskins or wrestle in the Olympics, one of the two. In high school, I realized that it probably wasn't happening on the Washington Redskins, so I decided that I would wrestle in college. Went to boot camp between my junior and senior year. Back then, you could do that for the Navy so that I could get an appointment; make myself more attractive for an appointment to Annapolis and go wrestle with them and become a Marine officer because I thought that would probably be the next best thing to being a pro football player. Went there for two years; I didn't do too well because I wasn't focused on what I should have been focused on, so I got kicked out. Went to Auburn. On my way down to Auburn from leaving the Naval Academy, I drove by the officer selection office and said, "Hey, I want to sign up." I wouldn't have signed me up, but they said, "All right. You must be dedicated." So, eventually I got commissioned through the OC program. The big reason I did it was I always wanted to be part of a team. All my dad's stories from Vietnam and all the books that I read. I read tons of books on the Marines growing up and Vietnam and everything my dad could put me onto. Hemingway, and all his writings on "For Whom the Bell Tolls" and battle; it just seemed like something I wanted to do. To be a part of something bigger than myself and lead Marines. You can't beat being around your buddies and doing things outside with them. So, that's what kind of attracted me.

Janney: Out of curiosity, what unit was your dad with in the Marine Corps in Vietnam?

Harrill: He was a Company commander for India 3 / 4 in Cam Lo, Con Thien and a little bit in Da Nang. He was a COVAN, which is an adviser to the Vietnamese; that was in '66 – '67 in Con Thien, Cam Lo and then he went back in '71 – '72 as an adviser for the Vietnamese Marines COVAN.

Janney: I'm sure he had some incredible stories to tell you as well.

Harrill: Yeah, he sure does. My grandpa was a Marine too, so three generations with Purple Hearts in battle. So, it's pretty cool.

Janney: I really thank you for your service and your family's service and their sacrifices. That's an incredible legacy to continue.

Harrill: Yeah, thank you.

Janney: What year was it that you enlisted in the Marine Corps?

Harrill: Well, I enlisted in the Navy in '86. I had no desire to be in the Navy. Just so I could, I talked to a Navy recruiter, and he put me in an interview with the Reserve unit Commander and he said, "Hey, if you enlist in the Navy and drill your junior and senior year of high school and you get honor grad at boot camp, then I'll give you a nomination to the Naval Academy. You don't have to worry about a Senator or Congressman." So, I thought that sounds good, so I got honor grad at boot camp, and he gave me a nomination to the Naval Academy prep school. Went there for a year and then went to the Navy for a year. Then blew a beautiful opportunity. I think that may be the person who I am.

Janney: Laughs. Well, we all have some stories like that. I was pre-med at Emory at one point and blew my last semester at Oxford, so yeah, we all have stories like that to tell. Once you got in the Marine Corps, what was your career path at that point?

Harrill: I got in and went to TBS; wanted infantry and infantry only, got infantry and went to 3rd Battalion, 6th Marines as a rifle platoon commander. I was lucky enough to stay there for about, I think, four and a half years as a rifle platoon commander, rifle Company executive officer, scout sniper platoon commander, mobile assault platoon commander or the Weapons Company. Weapons Company XO, fire support coordinator, HNS Company commander. Made Captain there so I had a great long first lead tour. Most people only get about a year and a half or two years sometimes so I was lucky to get about 4 deployments there. Fun ones there. Went to Haiti, Cuba, on my first deployment and did the stuff down in Haiti. Did some migrant operations in Cuba. Went on a couple of Med floats, Okinawa deployment, so I got some good experience. Then from there I went to a TOW unit in 23rd Marines in Broussard, Louisiana and my next HQ was in San Francisco. So I had a great time there learning how to plan training, execute training. I had great reservists who had all been in the first Gulf War as co-gunners; machine gunners so learned a lot about machine guns and missiles and employment there. Then went from there to the Army Captains Career Course and got to go to Cav leader course there and do a bunch of mechanized stuff. Then they told me, "You better graduate honor grad there" so I had to do that. Then I went and picked up the MEC Company, Golf Company, 2/4 after that and got ready for a MEU deployment. Okinawa, 31st MEU and we got extended there on the 31st MEU.

Once the invasion of Iraq kicked off, we were trapped there for another six months. We all thought we missed the fight, and the fight was over. We got to do some stuff around the Philippines and my Company went out on a boat on my own with the tracks and did a MEDCAP humanitarian mission in East Timor and then we got back from that year long deployment. Our battalion CO was fired while we were on that deployment. I got chosen to be the OPSO (Operations Officer) while we were there, but I was competing with a bunch of high-class guys - I wouldn't have chosen me that's for sure. Much smarter guys than me. Then Kennedy flew over while we were still on Oki. He had come back from the war and said he was taking over the battalion. He took over the battalion there in Oki and said, "Boys, get ready to go back to war." No one really believed it then, but we prepped for war. I think we got back in August – September and then we went back immediately after a yearlong deployment.

Immediately redeployed to Iraq. I think I left in January. So, we only had about three or four months to get the battalion up to speed and go into Ramadi. Then we went and Ramadi went down. General Dunford, Colonel Dunford at the time, called me about a week before. I had gotten selected on the RSCO Board, the Recruiting Station Commander Board but I didn't have to go, thank goodness, because I was going to Ramadi instead and one of my mentors, Colonel Frenese, now General Frenese said, "Hey, if you miss the train to Auschwitz, don't go to the station. So, you lucked up, JD." About a week before we left Ramadi, Colonel Dunford pulled me into the division CP and said, "Hey, we just fired a recruiting station commander in Virginia. You're going to come back to Pendleton and fly immediately a week later to take over RS Richmond, Recruiting Station in Richmond, Virginia." So, I called my wife. She packed our shit, and I flew to Virginia, and she moved us on out. I took over RS Richmond which was a very beautiful time in my development because that's some rough duty, especially during that time on the quotas and the mission. I did pretty well out there. Or my Marines did pretty well out there. They hadn't made mission in a year. We turned that thing around. They were some real fighters and made mission and were lucky enough to get out of there alive. Went to commanding staff there. Went to SAW, School of Advanced Warfighting which was the most beautiful learning experience of my life. It's kind of a school where you read a book a day, you seminar with some doctors for a couple of days and then about every three months you fly to Europe or Asia or Africa and walk the battles of the books that you read and talk about the campaign plans and OP orders and terrain and enemy. Beautiful experience prepared me to be a planner, but I never really had to go into the division planning ground. Went straight to work for Kennedy as his Operations Officer again, this time in 2nd Marine Regiment and got ready to go to Afghanistan and was selected as the battalion commander. Spent half a year with Kennedy and then took over 2/8 and then quickly deployed to Afghanistan.

Went to Afghanistan a couple of times when I was in limbo between SAW and battalion command to see the ground and do a little operation or a collection, but then got over there with the battalion. We were in Marjah and we fought over there for about eight months. I was wounded there. Marines did a great job there. Killed a lot of bad guys. Helped a lot of people. We had turned that AO around. It was pretty darn safe by the time we left and headed in a super direction. Then I went to be a regimental XO for 2nd Marines and then 8th Marines. I decided when I was in Afghanistan that I didn't want to go to top level school. My daughter was going into high school and my son was going into middle school, so I wanted to keep them stable. I decided to get out and not go to top level school and end the chapter of the Marine Corps book. So, that was a long story, but that was it.

Janney: That's an incredible career, sir. Let me step back just a little bit though. What year and month was it that you linked up with 2/4? In that you had mentioned Kennedy and 2/4 at some point.

Harrill: Yeah, I got to 2/4 in June of 2001.

Janney: So, you had been with them quite a while before all the casualty recruit drops prior to deployment in February in 2004. Is that what I'm understanding?

Harrill: Yeah, I got there in 2001. 9/11 went down; we deployed to Okinawa on a MEU, so I got to take Bronzi's Golf Company on a MEU for a year in Okinawa. So, that was June of 2002 to July of 2003 that we went on the MEU. Eventually Lieutenant "Big Bad" Ben Kaler got his arm blown off and had to leave us in Ramadi. He was Echo Company's XO. I had spent two or three years by that time in the battalion before we deployed with all those Lieutenants as a Company commander and OPSO so I had a pretty intimate knowledge of a lot of the enlisted Marines in the battalion and of a lot of the staff NCOs and obviously a lot of the officers. I had partied with them in Okinawa, being trapped on Okinawa for a year. Laughs.

Janney: Laughs. Yeah, for a year. I know this is a sidetrack, but you mentioned 2/8. That's who I was embedded with when I got to Ramadi in early March 2008.

Harrill: Oh, wow! That's awesome! I would have liked to seen it then. No, no, I wouldn't.

Janney: Laughs. Yeah, I'm sure.

Harrill: Because I can imagine what it's like now. There was a point in time where I would have liked to have seen it though. Laughs.

Janney: Yes, in early 2007 it was pretty hairy, but by 2008 it had calmed down quite a lot. It was just IEDs and an occasional sniper attack. It was nothing compared to what 2/4 experienced in 2004.

Harrill: No, some great men did a lot of great stuff after we left. That's for sure.
Janney: I mean, you all did. It was incredible. I read your thesis[i] and it almost seemed like a magic trick, but you guys pulled it off somehow. It's impressive and hard to believe.

Harrill: Yeah, I think we all knew what was going to happen in the end though back in 2004. That was kind of the sad part, you know.

Janney: One question I really, that's really key in my mind was when Operation Vigilant Resolve kicked off on 4th April 2004 in Fallujah. I know you guys took over for the National Guard unit in Ramadi in February, and from what I understand those guys were pretty much just cruising around the city, weren't really doing any dismounted patrols. When the Marines got there it was quite a different story. You said, "Get out of the vehicles; walk the streets, say hello to the people."

Harrill: Yeah.

Janney: Tell me what your perspective was when you guys first got there as compared to what the National Guard unit was doing before you took over from them.

Harrill: It's been a long time, but here's the deal. Before we even went it was all unknown. And all the smart people had pretty good ideas about what was going to happen, or what could happen. To paint the background; the division before we left wasn't quite sure. And there were some smart people, right? You had Renforth, Kelly and Dunford, the smartest guys in the world, and it was an unknown because it was currently going pretty good with the National Guard and the 82nd Airborne on deck. But the original mission was painted as SASO, Stability and Support Operations, this new term that was kind of squishy. We'd seen the doctrine on staff, and we understood what they meant by it but squishy. We had ideas, we were getting intel both ways on the 1st and the 124th and the 82nd and what everybody was doing on deck. I think that was mainly due to, it was a pregnant pause, nobody COULD know what was going to happen.

Janney: I read that in your thesis.[1] That is a great description.

Harrill: You know, the ship went down in 2003 and it went down hard, and I think the bad guys knew that the ship could go down hard, and they probably had a pretty good plan. So, it went down hard. The National Guard, whether they were doing, I'm sure they were doing their job. I mean every professional does.

Janney: You know, I'm not trying to say that they weren't. I'm just curious about the difference.

Harrill: Maybe they were in the pregnant pause. They were probably out and about. I don't think that anybody was ready to decisively engage anybody at that point in time. They were ready to feel some stuff out, but I think the bad guys were much smarter than that as we saw when they created the Islamic State.

---

[1] "Phased Insurgency Theory: Ramadi" - Thesis by Major J.D. Harrill: chrome-extension://efaidnbmnnnibpcajpcglclefindmkaj/https://apps.dtic.mil/sti/pdfs/ADA491315.pdf

Those are the same guys we fought, for the most part. So, they were a little smarter than that and I think that was just the pregnant pause. The National Guard was out and about. I'll tell you what, they definitely made some great relationships. The National Guard had all these Miami policemen. They knew the tribes. They knew the people. They had the police force; although corrupt, like everything, you can't expect anything different. They had everything up and starting to roll. So, there was a thought process that it could be a walk on easy street, or it could not be. Kennedy flew over there about a month before we left and his take was, "Boys, it's gonna be hot." He was right, obviously. So, I don't think it was because of their actions. I think that was the natural pregnant pause. Let's rearm, refit on the bad guy's perspective. Let's get a coherent plan and let's make it all go down. Then it all went down a little early. I don't know if you got this, but it all went down a little early on April 6th. It shouldn't have gone down on April 6th. From a bad guy perspective.

Janney: I heard there was a lot of action in late March. The mortar attacks and sporadic attacks.

Harrill: There was action. Like compared to most places in the world, most fights going on throughout Iraq, there was action, but I think the bad guys had a plan from what we gathered after April 6th went down to launch a major operation in conjunction with Fallujah because Fallujah was cordoned too tight. So, it drove many more fighters into our area. But they didn't want to do it on the 6th. It was meant for a week or two later. But what happened, as best as we can tell, on the night of April 5th/morning of April 6th we conducted Operation Wild Bunch which rounded up a HVT and 10 of his best friends who was responsible, in a Zarqawi-type role at the time, for the big offensive that was going to happen in Ramadi. And by taking him down, it forced relatively incoherent actions all around the town on the morning of April 6th. I think it could have been much more coordinated. I kind of went off a little bit from your original question.

Janney: No, that leads me to a lot of other questions too, because those are on my list. You've kind of thrown me for a loop here a little bit about the HVT. I know there had been ongoing mortar attacks on Combat Outpost and the FOB, but on the 25th of March your Marines got some telemetry information from the Army that allowed your 2/4 guys to go out and actually assault, kill most of the mortar team that was involved in that and capture one guy. So, was that, like, the initial kickoff to what started in early April? Or had there been other initiating actions?

Harrill: If I read to you the SIG acts from the time we hit the ground, the significant actions, it would boggle your mind. It boggles my mind because I hadn't looked at it in 15 years. Every single day and every single night, I mean, this is as early as, say, February.

Janney: So, when you guys got there, it started kicking off?

Harrill: Yeah, it kicked off in small actions here, small actions, four enemy KIA here, cache of 25 Katyusha rockets in February, 7 IEDs go off simultaneously in the vicinity of Route Nova. I mean, I could just go on and on and on, every day. Hurricane Point was hit by 8 Katyusha rockets on February 16th. What you mentioned is one little SIG act of over a hundred between February and March.

Janney: That's why I wanted to speak to you, because these other guys that I spoke with really don't have that overall knowledge of what was going on. Whatever they were personally affected by is what they could tell me. So, that leads me to kind of discount what I was concerned with was Operation Vigilant Resolve, the foreign fighters and the other fighters moving west from Fallujah to Ramadi. They were already there or at least the local insurgents, the former Ba'athists and officers and operations guys were already trying to test you guys and see what was going on.

Harrill: Yeah. They were definitely already there. They probably weren't already there in force. Let me read you something from Operation Wild Bunch that I remember: "The rise in attacks in Ar Ramadi over the last month", so this is dated like April 1st, "are attributed to the return and relocation of foreign fighters and fighters from Fallujah." So, we've known this for a month, and this is what initially gave us the Intel to launch that operation on April 5th/morning of April 6th. It goes into great detail, I remember, on all the foreign fighters and the people and "motivated Jihadist ideology is mounting in our area of operations over February and March. Offensive to regain the initiative by the enemy is expected very soon. Creating chaos in Iraq to disrupt the transition to sovereignty." So, there was a lot going on. But you got to know this, every day we would get an Intel report. This is no exaggeration. Every single day. Intel report: "Tomorrow morning at 0600, there'll be 2000 foreign fighters and they're going to destroy the government center and take over the Combat Outpost."

Janney: I guess at some point you have to take that with a grain of salt if the intel didn't pan out.

Harrill: You do, but we had a very highly developed source network very quickly. In part due to the work the 1st and the 124th and the National Guard did. 116, who was on the ground before us who we RIPed (Relief in Place) with. Other government agencies were in the area, so we had a pretty well-developed source network. You can't believe everything anybody says but we had some sources that we could tell, with a reasonable amount of certainty, what could be true and what could not be true.

Janney: Sure. It's incredible that they developed that.

Harrill: Yeah, and we had a Special Forces ODA team on the ground who had done a lot of good work. Obviously OGA and ODA are all limited by their ability to interact with the populace like we could because we had numbers, so we could interact a little better, but there was a good source network to at least point you in the right direction. But, every day there was something coming from a pretty decent source or somebody saying that the shit was getting ready to go down. So, it's hard to say, but from everything I know, I really believe it was a pregnant pause. Everybody was getting their stuff together and they (insurgents) wanted about mid-April or a little later, they wanted to make some stuff happen. That they didn't get to launch on their terms. Not because of any greatness because we happened upon it, you know, almost, in Operation Wild Bunch.

Janney: So, going back to your thesis, maybe I'm getting off your train of thought, but how did you guys achieve the alliance with Sheik Sattar? Were those friendships and alliances cultivated by the National Guard unit and your sources that you are talking about?

Harrill: Yeah, you know, the first time I met Sheik Sattar was before the main body got on deck. I think I flew in on my birthday, maybe February 8th, or something. And I don't think the Battalion got there until March, if I remember right. Or I could be a month late. I did a high five, literally, first with the 1st and 124th's XO. I had twenty minutes with him and then he left.

Janney: Wow.

Harrill: First, there was the 116th, who I love to this day because they stayed with us the whole time. They had gotten there two weeks before us, so they kind of did our battle hand over with 1st and 124th. I would patrol every night as the OPSO with the 1st and 116th. I would go out on their raids, their named operations, on their assaults. I would go on their key leader engagements so I could learn the AO and know how to write something that made some sense. One of the key leader engagements was with Sheik Sattar. He pulled me aside at that meeting over at his house and he said, "I think we can make some stuff happen if you want to." He told me a story about how his father and his grandfather, whether this is true or not I doubt, but his father, his grandfather and his great grandfather had worked with the British during their Battle of Ramadi. Have you looked that up?

Janney: Yes, back in 1917, I think?

Harrill: Yes. He had been working with the Intelligence community since then and he thought he had something to offer us. Obviously, he did, and he made some stuff happen, mostly driven by economic gain, I think, but who really cares, you know? Obviously, everybody cares, but…

Janney: I understand. I saw a briefcase full of cash handed over to the mayor of Al Faris, south of Ramadi when I was there on a mission with RCT1.

Harrill: I mean, cash is great. Cash makes things happen and rightly so, small venture capital and all that. It gets it going but you've got to sustain it and if you don't it doesn't work.

Janney: Yeah, I understand that. I read your thesis with great interest about how that would work, and then other tribes would sabotage those efforts. Or the insurgents might sabotage those efforts or whatever. Cash is king, really.

Harrill: Yeah, but you gotta sustain it. Like you can't sell interest in a place for ten years and leave. When you commit to a counterinsurgency, I don't know that there has ever been a counterinsurgency that's been won. Someone could argue with me on that, but when you commit to a counterinsurgency, you commit your nation and your treasure for a very, very long time. But, most people don't have the stomach for that.

Janney: Your dad is a perfect testament to that as far as him being in Vietnam then.

Harrill: Yeah.

Janney: But in any case, it sounds like AQI and the insurgents were already heavily invested in Ramadi even before 4th April when Vigilant Resolve kicked off.

Harrill: Halfway, because Ramadi was very embedded with the Fedayeen. It was a retirement community for the Saddam Fedayeen. Even Saddam was hesitant to roll into Ramadi because he understood the capacity of the former Fedayeen, the Intel network people with their early warning systems. This was a big community of former Saddam loyalists, Generals, Special Forces guys, Fedayeen. I think Ramadi, as much as Fallujah or more, and probably more in a lot of ways, was the retirement community or the sanctuary for, and I mean sanctuary in the brightest of terms, sanctuary for the people who wanted to see Iraq succeed. One of the biggest things that really tipped that to the bad was, we had some inroads with a lot of those guys, so when Bremer divorced ourselves from former Saddam and Ba'ath Party loyalists, that gave them no hope in life. So, then it became war.

Janney: So, were they basically afraid that you were going to turn everything over to the Shias?

Harrill: I don't know about that, yeah probably, and they (Anbar Sunnis) hated the Shias, I'll tell you that. I thought there was some hate in the south, but they hated the Shias. But no, they felt like, according to former Ba'athist generals who desired to take their tribe and fight alongside us, when they were disenfranchised from any hope of ever obtaining a status again, they had no option but to go to the Moog as we called it at the time, or ISIS or the Islamic State, or whatever you want to call it. Their recourse was taken away and they had no more upward social mobility and no hope of power. If we wanted to make them a police chief they couldn't do it. If we wanted to make them a governor; they couldn't do it. Fire chief can't do it. Make them in charge of the local defense force; they can't do it. So, there was no recourse for them to take but to side with the other side.

Janney: They felt like they couldn't do it because of what? Because of no upward mobility?

Harrill: Yeah, because. All right, this is me talking way out of my lane, but I saw it play down to my lane. Do you remember Bremer who said that no former Ba'athist parties will have a stake in our government, and they are all fired as generals? They didn't have positions anymore.

Janney: Yeah. In my opinion that was the whole cause of the destabilization of Iraq was because they fired everybody right off the bat. All the military police forces.

Harrill: That steam rolled down to our level. We couldn't get them vetted to be in charge of our ICDC, Iraqi Self Defense Force, or whatever it's called. Iraqi Civil Defense Corps. I'm getting Afghanistan and Iraq confused, so Iraqi Civil Defense Corps. We couldn't get them vetted to take over the jobs that they wanted so they became, naturally, I would too, became disillusioned with the American involvement. So, you look at that during the pregnant pause, all through February and March and into April, and then you start to see the tide start to change in a line with their only hope of survival which is to go to the Islamic state or the mujahideen at the time or start their own networks in order to outlast us.

Janney: Because they didn't think, at that point, that you would prevail, I guess is what you are saying? They didn't think the US forces would win, so they thought, "Well, this is going to be just a flash in the pan like the British were, and so we're going to kick them out and win."

Harrill: Maybe that, but here's the thing that I had a hard time understanding for a little bit. It's that their mind was wired completely differently than ours. We think about retirement and we think, "What am I going to do today in order to set myself up for a nice retirement where I have an RV when I'm 60 and I can then spend the next 20 years traveling around. I have all the money in the world, medical is taken care of and I have a great life. I'm going to sit on the beach in St. Petersburg and do whatever I want." That's what we would base our decisions off of with them is that kind of mentality, what will benefit me in the future. But their mentality is, "What do I do right now in this very minute to survive until tomorrow? To gain as much power and money that I can until tomorrow to keep myself and my people alive." And when you look at life in those two different spectrums, your sense of morality and ethics becomes completely and utterly different. I saw it play out many times. I saw, one day with Golf Company, a RPG fired into a mass of schoolgirls, and Lance Corporal Bolding was killed instantly, and I thought, "This is a tipping point." These villagers just thought we opened the school; the Iraqi police gave them backpacks. They went to the school the first day and in this crowd of girls, they've got bodies stacked and limbs torn apart. You ask the family, "Hey, who did this?" and they would say, "Inshallah." My daughter just died, but inshallah (God willing.

Because He willed it and that's just the way it happened - it's a different logic. Until you understand that, there's not going to be a tipping point, that the logic is different, then you start really setting yourselves up for failures. Quickly, you realize that there's not going to be a tipping point - that I need to start thinking in their logic and not my logic of retirement. When you do that, the iterative learning process about your enemy and what's going to happen doesn't set you up for heartbreak. It sets you up to be more adaptive.

Janney: It sounds like you guys adapted to that. I mean, you were successful.

Harrill: I don't know if we were or not. We killed a lot of insurgents.

Janney: Yes, and we'll address that in a few minutes. I think your understanding of the difference between their value of life versus ours and your analogy of retirement versus their retirement makes perfect sense to me. I do remember that story about Bolding, and I can't tell you enough how much I grieve to the extent that I can about the loss of any of your Marines. Like I said, this is why I'm doing this for their families, so I really appreciate that you are doing this interview.

Harrill: Bolding was my main man. He was in Golf Company with me. I mean, they were all my main men, but Bolding was my main man. He went UA and we got him back in the fold before we deployed to Japan and he was just a great Marine. Always full of smiles. I held him the day that he got hit. That one hits me hard all the time. He's a great one.

Janney: I'm so sorry, Lt. Colonel. Like I said, I got so close to John that I feel the loss of his son and therefore, to some extent, feel the loss of all those guys, so I really appreciate what you did for them, and I really appreciate your sacrifices as well.

Harrill: I don't know if you know, I'm sure someone has told you this but John hit everybody hard. What I mean by everybody is that John was the first officer that people saw a casualty call. In officer country, people started to see casualty calls come around. So now you have officer country, the vehicles coming from door to door and that's starting to go down. And John's, to this day, I know that John's casualty call has affected more women than you can count. Be they Colonels, Generals, Lieutenant's wives, 2ndLt John Wroblewski's death had an effect on us all.

Janney: As you know, it was just a terrible day with twelve guys that were lost on 6 April.
I understood a lot of things from your thesis, but I just want to clarify a little bit. How did you accomplish what Galula talked about as far as the tipping point and Kilcullen talked about the evolving phase of the insurgency? I know you mentioned in your thesis that it was alliances, money and security, but was it all based on relationships? It seems that way to me, but you tell me.

Harrill: So, I don't know that we hit a tipping point when I was there. Obviously, a tip line.
Janney: The Awakening that happened after you guys left, of course, was the real sea change in what happened and what was happening in Anbar. But it seems to me that what you're saying is the development of these relationships first started by 124th was really a key factor in you guys developing those relationships that helped save Anbar.

Harrill: Yes, you're right. Relationships are everything. One of the greatest men that I've ever met in my life was a Company commander in 2/4. Two of them. And he asked me, "Sir, what are we doing?" I said, "What do you mean?" He said, "I don't understand what we're doing. Like I got our mission, but what are we really doing?" and the only thing I could think of was, "We're saving and helping as many Iraqi people as we can. At our level, so that we can get the best Intelligence as we can to kill as many of the right people as we can so that we can accomplish our mission and bring as many Marines home as we can. And take care of them as best as we can." And at our level, I don't know that there is any better way to say it. I saw it work in Helmand province again and again. I saw it start to work in Ramadi and eventually work. I saw it throughout history and it's that either you go and kill everybody, which is a technique. Or you have some humanity, and you befriend as many people as you can, help them with their farm work, you get to know them, you get to know what makes them tick, you get to know what motivates them, you court them, you see what their grievances are, or what their indifferences are, and you understand that. And then you become human to them and once you become human to them then they start to confide in you. And once they start to confide in you then they start to trust you and once they start to trust you then they tell you who the real bad guys are if you do it right. And then you start to kill the real bad guys. And then you create a psychological fence. So, there's no such thing as a real fence, right? But you create a psychological fence on the battlefield where the enemy is scared to operate in your area because they don't know who to trust anymore. That was their strength. They could blend in the population. But now you have a psychological fence on the enemy, and they can no longer operate within your area. And then once that starts to happen, beauty starts to happen. And then you start having offices open up, you start having infrastructure created, you start having the beginnings of rule of law, you start having the beginnings of governments. And then the trick after that point is, by doing all that you take care of your Marines because you don't inflict as many casualties. You don't kill the wrong people that cause vendetta killings, you don't do all the things that cause people to hate you. When you do that, a beautiful thing happens where people start to side with you. The enemy becomes uncomfortable operating in your area because of the psychological fence and they don't know who to trust anymore when that was their center of gravity. Then you start reaping all the benefits of security and peace. But, what we don't do well, is we don't sustain that. So, that's what we tried to create there. I think all of us were smart enough. We had all read enough to understand that, but obviously by 2010, we didn't continue to invest in that.

Janney: Yes, my personal opinion is that it was terribly unfortunate that the Bush administration chose to withdraw without leaving a security force in place since they didn't negotiate a status of forces agreement, and Obama's administration didn't want one either.

Harrill: Yeah, but you know how I go to sleep at night? I say Clausewitz told me when I was 14 years old that war is an extension of politics and if you believe anything else then you are setting yourself up for a disaster. So, that's how I go to bed. I figured that out about Clausewitz when I was little. So, I don't expect it to work out. If any of us believed anything different then we're in the wrong business. Our business was acting like Americans, treating people right, befriending people, trying our best to take care of them so that we could get the intelligence to kill enough of the right bad guys to cripple their network. Build a psychological fence, create mission success, take our Marines home in case the nation wants to play it further. If you look deeper into it than that, then you're not reading Mr. Clausewitz.

Janney: Yes, your citations of Clausewitz and Kilcullen in your thesis were spot on. You guys executed that to the best of your ability in spite of the logistical difficulties. It seemed like a magic trick.

Harrill: We didn't do shit well. If you do anything in the book, those Marines did everything well. The Condis and the Oteys and the Dave Dobbses and the Ben Kalers and the Aperts and the Dukewells, and especially the Bronzis and the Weilers, those guys were incredible, along with the platoon commanders JD Stevens and Valdez. I think one of the reasons our time with the kinetic was we didn't want it to be kinetic. I truly say that we really didn't want it to be kinetic. We came there hoping that it wouldn't be kinetic, laying the groundwork that it wouldn't be kinetic. But our guys, when it turned kinetic, I have never in my life since, in any battle that I've ever been in, seen people when contact happened, when IEDs happened, when firefights happened, I've never seen anyone with more of a bias for action. They charged into the enemy every time and I don't think the enemy wanted to meet them. Those young Marines and those Lieutenants, I don't think I could have done what they did. Those Captains, they were some brave dudes. They didn't stand their ground and fight; they stood their ground and then they charged.

Janney: Exactly. As I've said, I've heard that from so many of the Marines that I have interviewed. I appreciate you confirming that.

Harrill: Yeah. One of Wroblewski's big buddies, Machine Gun Tommy Hogan, that dude. Wroblewski and Hogan, they were two animals. Lieutenant Hogan was an animal. When his Company Commander could not do what a Company Commander should do, Hogan rallied the day and charged enemy machine gun nests, won the Silver Star and decimated the enemy. I mean, this guy was incredible. I don't know how many tens of people he took out, but he single handedly turned the tide of the battle. An amazing man. Look up Tommy Hogan's Silver Star citation.

Janney: Yes sir, I will. There are so many people that I haven't spoken to. I want to talk to everyone that I can. It's just incredible what everyone went through and the bravery that they exhibited.

Harrill: It's hard though, I know. I respect what you're doing.

Janney: As far as the foreign fighters, at the point at which you guys were in Iraq in 2004, was there much Iranian influence? I know there were EFPs and other Iranian weapons in theater, but I think that was a later development than when you 2/4 was there in 2004.

Harrill: We saw the introduction of that, yes, and the introduction of technology. The foreign fighter thing was weird because there was a disconnect between the taking of prisoners and the releasing of information down the chain of command. So, I'll give a couple of examples. I do know there were a lot of people we took prisoner who weren't from the Middle East. They were either from Sudan or Gamon or more African origin. Whatever happened to them, I have no idea.

Janney: Any Chechnyans?

Harrill: Yes, Chechnyans, too. So, we had white people, though I never knew who they were, and I call them white people.

Janney: Yeah, they weren't Arabs.

Harrill: Then we had a lot of, which was surprising, on at least 15 different occasions, Asian people. Some of them said they were from China and other places. You can look it up and it will pop up. They were there, probably Uyghurs. Then, we had white dudes who wouldn't speak and were pretty evil looking. We had black guys who obviously weren't from Iraq. One time, July the 13th, we used to call it Wicked Wednesdays. Every Wednesday we would go down and we would get into this huge fight. For about a month it was pretty brutal. One time, I think we killed about 40 or 50 dudes. We were ambushed right in the middle of Ramadi by the Saddam Mosque. There's some good videos of it. Car bombs going off everywhere. It was a good fight. Probably lasted six hours. Car bombs coming from every direction. Dudes everywhere. Fire coming from at least six or seven buildings. Golf Company maneuvered on one of the buildings that we were taking heavy fire from, and we caught like 40 guys with dreadlocks. They were Arab looking guys, but they had dreadlocks and we never heard who they were. Of the 40, I think we killed 15. The arms cache that we got from them was truly amazing - I haven't seen much like it since. It was thousands of weapons and 155 rounds or whatever the Soviet equivalent is, you name it.

Janney: Was that from the Mosque or just that general AO?

Harrill: There was a building out from the Mosque, not the Mosque. It was right by the Saddam Mosque, right by checkpoint 297.

Janney: Let me ask you this. This is an unlearned theory that I have regarding the coordination of the 6th April attacks. First off, there was an IED that had been placed on Route Nova. Santiago had a scout sniper team; they're doing over watch on Nova. That four man team, because of a kid walking down near the Euphrates spied them and blew their cover, got swarmed by thirty or so fighters that tried to envelope them, to push them into the river and kill them. That, all in conjunction with the Route Gypsum ambush which was pretty well coordinated. They had a DShK (Soviet 12.7mm heavy machine gun – equivalent to our .50 caliber M2 machine gun) set there at Nova and Gypsum.

Harrill: So, they had a DShK on the approach, they had mortars laid in; they had IEDs on the road, which was beautiful, BEAUTIFUL. They had IEDs in the buildings as the Marines maneuvered into them. So yeah, beautiful ambush.

Janney: I want you to expound on that a lot more but my question about that is I know you talked in your thesis about a lack of coordination and control (C & C) by the insurgents but, it seems to me either they spent a lot of time in placing those things and knew that at some point those Marines were going to come up Route Gypsum because that was the closest point from the Combat Outpost, or do you think there was a little bit more coordination in that whole day? Does that question make sense at all?

Harrill: No, it does. I think that everything was starting to get in place, from all the Intelligence that we got afterwards, everything was starting to get into place for an attack after April 10th if you look at the Intel reports that we were getting at the time. Let me give you the op order for that week to give you some context if I can find it.

Janney: You know what I am saying? Gypsum was the quickest route from MSR Michigan from Combat Outpost to Nova so if they…

Harrill: You have to know that Gypsum is irrelevant.

Janney: Okay. But they spent a lot of time on that ambush. Some of the guys that I interviewed said that there were piles of cigarette butts at the DShK site, so they sat there and waited all day or for whatever period of time until that ambush took place. My question is how well planned was that or was it that they just had a blanket battle plan?

Harrill: I think it was well planned obviously. It was a beautiful ambush.

Janney: They had that IED placed east of Santiago's position that one of the 2/4 squads found, and they called EOD and then took cover in the house waiting for EOD and then that's when those Echo Company guys got hit, along with the insurgents swarming Santiago's Headhunter team.

Harrill: I can give you the minute that happened. If you read the SIG act report for the month following all the way up to that moment in time, there's a SIG act every hour of some direct fire contact or IED. Then, when that happened, that moment, say 12:00 on April 6th, the SIG act report by the S2 is that the entire city just erupted in violence. At that moment, what you just described to me, happened literally in thirteen different places. It happened in 8 different places in Golf Company AO. Golf Company had a squad separated, they had people that they could not find. They had squads pinned down here and there. Fox Company had two squads pinned down. Weapons Company was in fire fights here and there. Echo Company was at Nova, but they all kicked off at noon on April 6th. So, was that laid in advance, yes, but it was laid in thirteen other places in advance.

Janney: Putting together a timeline by piecing everything together from the interviews is confusing.

Harrill: Yes, unfortunately it's horribly confusing. That week, we never told Echo to do a route sweep because we would never do that. We told them to ambush here and there. But, on 6th April, we had thirteen squads, at the same time at noon, we had thirteen squads or elements pinned down in Fox Company AO, 7 in Golf Company AO separated between the cemetery - they couldn't find two or three of their Marines. We had Echo Company that couldn't find an entire squad. We had guys bleeding out in a pump house. We had guys bleeding out in a palm grove. We had a sniper platoon isolated and being attacked. Golf Company was in so much contact, they had so many WIAs and KIAs that mobile Company (Weapons) had to launch and start evacuating their injuries. They were engaged en route. Colonel Kennedy and I went out. We were isolated. We were trapped with a barrage of hand grenades and machine gun fire. Charged a machine gun nest. At that moment in time, there was shit going on everywhere.

Janney: I know they initiated an ambush on the 5th, but they really kicked it off on the 6th?

Harrill: They kicked it off, but what we got was that they didn't want to kick it off until mid-April. So, they weren't quite ready.

Janney: Because, as you told me earlier, because of the capture of the HVT on the 5th, they were kind of unprepared. They meant to do it later, but they felt like they had to take immediate action?

Harrill: That's what we think. I won't say it was because of that. He was the head guy. He was the HVT at that point in time, of the MNFI level or the theater level HVT. Our capture of him and 10 others on the 5th probably caused them to initiate their mid-April battle plan early on 6 April. A whirlwind on the 6th, but not so much that we brought the battalion command element down to Golf Company. We aggregated forces with Golf and Weapons and then we moved those forces to aggregate with Echo. Then on the 7th, we started launching battalion sized attacks into Sofia, and Echo Company AO at Gypsum and Nova. We started launching battalion sized attacks into there and then we started launching battalion sized attacks into Easy and 20th Street. Then, we eventually planned a battalion size action because we got Intel after the 7th that everyone fled. General Dunford said we had killed many insurgents – there were 500-800 people in the morgue. Unclaimed bodies, which generally in the Muslim culture at that time, meant that these guys weren't taken up by their buddies or anyone else. So, there were 500 – 800 people in the morgue. Foreign fighters unclaimed. We attacked into Sofia, 20th Street, Easy Street, and down to the soccer stadium and then we regrouped. On the 10th, we launched Operation Bug Hunt which was off of actionable Intel. All of the fighters had gone into the Fishhook. On April 10th, we got Intelligence that all the bad guys had moved out of Ramadi into Sofia and the Fishhook and prepared defensive positions. So, we launched Operation Bug Hunt on the 10th, and we did a battalion sized attack into Sofia and the Fishhook. Under the cover of darkness, we started with ten simultaneous raids on non-illuminated objective areas. Once we cordoned those objective areas off, we trapped the battlefield with 155 artillery illumination, 81s and 60s and we just laid fire over the entire battle area in non-populated areas to take the enemy off balance. Then we went into our target areas. On April 10th, we started killing bad guys as they were trying to get out of the AO and our fire.

Janney: I've heard some stories about that from the enlisted and NCOs. It sounds like it was an incredibly successful operation.

Harrill: Yeah, after we never encountered a big offensive like that anymore.

Janney: No, they never stood and fought after that is what I understood. Is that true?

Harrill: Yeah. They stood and fought in the Company and battalion size, but never stood and fought again in regimental or divisional size.

Janney: I heard an anecdotal story about the 11th of April that there were dump trucks full of enemy KIA leaving the AO headed to wherever. Have you ever heard that?

Harrill: I witnessed that, yes. I think there were hundreds of people. That was not a small uprising. It was a pretty good attack on a little battalion. I would estimate it at a regimental size if not more. Like 5,000 fighters. So, yeah, there were a lot of bodies.

Janney: It sounds like you guys got some pay back for what happened on the 6th.

Harrill: Yeah, but you know pay back in our war is, I don't know, it's kind of hard to say.

Janney: You talked about developing relationships, but from my limited understanding of Iraqi tribal relationships is that they have a long memory and definitely have kind of a vengeance system going on. I know some of these guys were foreign fighters but how did that affect your relationship or the battalion's relationship with the locals as far as what happened after that. I know some of these were hangers on and some were just jumping on the bandwagon.

Harrill: The bulk of the tribes, I think, understood who we were. We were pretty transparent, and they understood who we were and what we stood for. They understood what we stood for was them and whatever destiny they wanted to create. Surprisingly, we had guys that came up and talked about Thomas Jefferson and Thomas Paine and the Founding Fathers. I thought, "Wow." We understood our operating environment. There were two, three, four tribes that had hardcore alliances to AQI, ISIS, or mujahideen - I don't like that word, but whatever term you choose. They stayed true, I'm sure. When Wroblewski died, it was evident with a lot of people. You interrogate somebody from where Wroblewski died at Gypsum and Nova and the guy didn't have a soul. We could look him in the eye and knew that he was a bad dude. So, there were tribes who wanted no part and had no part of us. It was pretty apparent in the end. So, we won the tribes we won. We understood the tribes we didn't. Only time would cure that, but time can't heal all wounds.

Janney: JD, what you just said struck a chord. John and I did foot patrols with 2/8 almost every day that we were there, almost a month. We walked Nova and Gypsum and, particularly at that intersection, I can't tell you how many times I said salaam alaikum to everybody that we met trying to be friendly. I do remember specifically the people at that intersection just stared holes through us and did not acknowledge our greetings. They definitely seemed a lot less friendly than people on the southern end of Gypsum.

Harrill: Yeah, that intersection is electric. I think it goes back to what we talked about. So, my great grandpa was a moonshiner, and he hid moonshine under the babies. And my grandma was in charge of taking the moonshine, throwing it in the well, putting it under the babies, and hiding it from the law. That's the way you do it, right? Even though you and I probably think the law is right. So, my grandma, when she told me the stories, I was compelled by family, culture, tradition, and the matriarch, and what I believed to be true didn't really seem to be true anymore. So, I think that if you look at Route Gypsum and Nova, it's very similar in that they believe the Islamic culture and the tribal values usurp everything that we want to introduce. And although foreign to us, I think they acted out of what they perceived to be honor. They will continue to do so. Because when I looked into their eyes, they were some hard men. You can look into the eyes of some of the Iraqis, you interrogate them and look into their eyes, and you know they signed up for a gig they weren't ready for. But, you look into some dudes from Gypsum and Nova and you know that you just met some dude that is much more evil than you ever wanted to meet in your life.

Janney: Yes. Where we did the 2/4 memorial service with General Kelly is about three hundred yards south of the intersection of Gypsum and Nova. There was a cinder block lined alley that turned off to the west, and then arched back down south with some big houses kind of out of the norm for the area. There was a vacant lot there at the intersection. That's where General Kelly chose to do the memorial site for security reasons even though we weren't really at the actual spot where JT was mortally wounded. Once we completed the memorial service and came out of the alleyway, the people in that house came out and spoke to General Kelly and John Wroblewski through his interpreters. They said they had lost sons in that same fight, so it was just kind of a surreal circumstance knowing that some of your Marines probably killed some of his sons. He acknowledged that he was saddened by all of the loss and death and was congenial enough to come out and give his condolences to John Wroblewski.

Harrill: Yeah, you know what I saw from the Iraqi people and the Afghan people is that they do that. I hope that continues to be true. We had some scenarios very much like that and they respected the dead in a lot of ways. Frazier and Green were killed at the Euphrates and we recovered their bodies. I think in some ways they are more like us than we want to believe based off the circumstances we encountered. They will allow you to grieve. I'm so happy John Wroblewski got to see his son's intersection.

6 April 2004 – Route Nova Firefight and QRF

Headhunter II Sniper Team

Team members: Santiago – Team Leader, Name Withheld by Request, Lund, Barron, Katz, Thomas,

Janney: Today is 11 April 2020. First off, tell me why did you decide to enlist in the Marine Corps?

Santiago: Well, when I was in high school, I just had to enlist in the Marine Corps because I wanted to make sure that I was able to fend for myself basically. I was initially wanting to go to DeVry to be a computer aided drafter, but my junior year I was thinking, "How am I going to pay for that?" I don't really have a decent job. I was a Kentucky Fried Chicken delivery driver at the time in the Phoenix area. Going through school, I had to live with my uncle and my aunt and I don't want to be a burden to them. So, I decided to enlist in the Marine Corps and see what that was all about.

Janney: You were living in Phoenix at the time?

Santiago: Yeah, yeah.

Janney: You went to boot camp in California or Parris Island?

Santiago: I went to San Diego. I was an 0311 for three and a half years with Golf Company 2/4.

Janney: How did you go from Infantry to being a scout sniper? I mean, that's a pretty big leap.

Santiago: I guess to give you a good understanding of how I became a scout sniper. I almost kind of fell into it. Let me tell you the backstory behind it. When I enlisted in the Marine Corps, I talked to my recruiter Staff Sgt. Witte. So, I was talking to him and he was actually trying to get me to go into the air wing. At the time I was a junior in high school, I think I was seventeen and everything that I knew about the Marine Corps was like "we get to go out and blow stuff up and shoot things" and it was kind of cool so my whole idea of the Marine Corps was basically infantry. He kept trying to convince me to go to the air wing and we left an open contract from then and then the guy that was in charge of the recruiting station walked out of his office one day when I was meeting with my recruiter and he had asked me to go into his office. And this guy was like maybe 5'5" old crusty Marine. Had the raspy voice; had the coffee cup in his hand. I don't even remember what kind of rank he was, but all I remember was he had a lot of stripes and a lot of medals on his chest. I go inside his office and he starts asking me what I wanted to do in the Marine Corps. I told him what my idea of the Marine Corps was, so he started telling me that he was actually recon and he started showing me some videos of 1980's or early 1990's Marine Corps scout snipers and recon guys. I distinctly remember that video and I think if you go on YouTube you can probably still find that video. It looks like it was filmed in the late '80s or early '90s and this guy was in a ghillie suit and he had an old school M40A1 and he was using the Vietnam era grease stick paint to camouflage the barrel.

I thought, wow, that is pretty interesting. I want to do that. He started telling me the requirements to become a scout sniper or a recon guy. I said, "Well, I'm not very athletic, I'm like 5'3" on a good day. I never played any sports." He was telling me all these requirements and said, "I guess the route to that is being an 0311, being an infantry guy" so fast forward, I went through boot camp. I went through SOI; did my stint as a regular grunt; went through the ranks.

At the time, before my enlistment was up, I think I was three and a half years in, my enlistment was going to be ending in a few months. I had no money saved up, no education, didn't even take any college courses or anything like that. So, I had nothing set up. At this point, Marine Corps scout sniper or recon was in the back of my head just because of things that happened in the Marine Corps. I had gotten to the point that I was physically capable enough to start keeping up with folks. One of my good buddies, Jesse Longoria actually, it was Jesse who just out of the blue comes up to me and asks, "Hey man, you want to take the sniper indoc?" I said, "Uh, yeah, sure." So, we actually had to do a pre-screening before we could even do the indoc. Then, we did the indoc: me, Jesse (Longoria), Marvin (Endito) and a couple of other guys that I'm still friends with and we got lucky enough to get selected to be in the platoon.

Janney: How hard was Scout sniper school? I imagine it was incredibly tough.

Santiago: Yeah, yeah, definitely physically tough and mentally tough. Like I told you at the time when I was going through sniper school or even the Marine Corps, I was like 5'3" and the heaviest that I got was 135 lbs. So, all the equipment that you have to carry could equal up to 50 lbs. They include your pack, your gear, your rifle, and your pistol. So, pretty heavy stuff. Actually, at the time even in school you have to carry a 35 lb. pig egg. It's basically a sandbag weighing 35 lbs. and you put that in your ruck, and you ruck everywhere with that. You go to chow, anywhere you go, it goes. It was definitely mentally and physically demanding for me and I actually failed out the first time I went through school. I failed out during the stalking portion of it.

Janney: So, they spotted you at some point.

Santiago: Yeah. I don't know if you want to give you a step by step of what it was to give you an idea of what was the requirement regarding the stalking or you just want me to give you a, basically, a synopsis of it.

Janney: I don't know much about that. I know you have to do some kind of mission where they try to identify you and pick you out while you're wearing your ghillie suit.

Santiago: So, the stalking phase of Marine Corps sniper school is a series of ten stalks and you have to score 90% if I remember right. You get scored at different portions of the stalk. They drop you off at a certain distance, typically around 1,000 to 1,500 yards away from the observers. You have a bunch of instructors on an elevated position looking through binoculars trying to detect you. Your task is to basically get within 200 yards of the observers, set up a position where you can take a shot from it, take a shot, and if you're able to complete your two shots without being seen, you get 100 points. During each part of that, you'll get points deducted if you fail. Let's say you actually made it to 200 yards; within 200 yards of the instructors, but you get busted; you get seen and they walk on you, I think that was 70 points.

If you get to a position where you can actually set up and take your first shot, I think that's 80 points. If you complete both shots without being seen, you score 100 points. It's all cumulative scoring, so you have ten stalks with the same grading scale and you have to score within 90% on everything.

Janney: That's like some crazy Gunny Hancock stuff.

Santiago: Laughs. Almost. You just gotta crawl without being seen just to clarify the two shots. You set up your position within 200 yards of the instructors, you set up and you take a shot and the shot is, of course, they're blanks. Once you take that first shot an instructor will come within 10 feet of you and the instructors who are observing will start looking around within that area where that instructor is standing. So, it makes it very difficult. If you didn't set up your position correctly, they're going to be able to see you. If they can't see you after that instructor is within 10 feet of you guiding that instructor from the crews observing, then you're allowed to take your second shot.

Janney: That sounds incredibly difficult.

Santiago: Yeah, so anything that you mess up on like you have some piece of vegetation in front of you that you didn't clear out. I mean, that blast along from the blank reveals your position because the bush would have moved, right? It incorporates all the things that they teach you about camouflage and being hidden and stuff like that. Yeah, it was pretty difficult and I failed out and I think I missed it by 10 points. It came down to my last stalk. I didn't score enough points and I ended up getting kicked out of school.

Janney: But you still had the gumption or drive to go back and do it all over again even though that meant a lot more pain and suffering.

Santiago: I was extremely lucky to get another shot to go into sniper school. At the time it was pretty, not rare, but it wasn't very often that you got a second shot so most of the time it's only one shot, one kill and that's it. If you don't make it through school, you're not going to get your certification.

Janney: Did Jesse (Longoria) make it through his first time?

Santiago: No, no, Jesse ended up going to school numerous times. He was one of the lucky ones where he just got opportunity after opportunity to go to school. I got a second chance within a couple of weeks of failing out of school. I think within the next month I went back and if I remember correctly, it was blind luck. They had an extra slot and I happened to get that slot, so I went back to school, went through Day 1, Phase 1 of sniper school; started from scratch. Eventually graduated; passed my stalking phase with flying colors because now I know exactly what to do. Just to give you some idea, I went to sniper school weighing 135 lbs., but after two strings of sniper school when I graduated, I was about 110 lbs. I lost so much weight. It took me several months to recover from that I was pretty beat up. Plus, old school ways of doing shooting calculations. We didn't have ballistic calculators; we didn't have computers at the time, so it's all about using theoretical, using the calculator and using the old school mathematical formulas by hand. Yeah, a lot of math involved.

Janney: Knowing wind speeds and what quadrant it's coming from so half value or full value?

Santiago: Yes, all of the wind calling. Unfortunately, in my opinion, it's becoming a lost art just because a lot of it was based on your experience and how much you trained and how much you listened to your instructors. Nowadays, you can get away with a lot just by using a Kestrel wind meter, a ballistic calculator, and plugging numbers in or letting the computer plug numbers in and then just take that shot. Back in the day, it was all by hand.

Janney: Yeah, electronics fail, batteries die. That's not a good way to do things, even now.

Santiago: Yeah, unfortunately, nowadays with modern technology they occasionally shit in the bed on you not very often nowadays they're pretty reliable so but then again you always gotta the old school way just like driving a car. I mean, if you can learn to drive a stick shift before you can drive an automatic, I mean it would be great, you know what I mean?

Janney: Yeah, you're even that much better, I guess. You probably didn't have all those electronics when you were in Iraq?

Santiago: No, we didn't. It was still the old school way. I understand where a lot of the electronic stuff comes from because that definitely would have helped us out getting ranges. If we had laser range finders that would have been great. The Marine Corps way is to make do with what you've got.

Janney: So, you guys made range cards, so you knew what your distances were anyway, right?

Santiago: Yep, yep you start using the old school tricks like under light posts or typical distances between blocks and stuff like that so you start pacing it out, you look at your map study and get distances that way or GPS coordinates and just plugging it away. Of course, eventually once you're in an area for such a long time you start realizing the distances and stuff like that because you've been there before. Of course, you have a built-in rangefinder in your reticle. You can just use your formula and measure out a known standard like a door or a tire of a car and then use your mil reticle to measure that out, get into the formula and get an estimated range so you can plug that in.

Janney: Did you actually deploy to Iraq with Jesse or tell me how that all came about.

Santiago: I deployed with Jesse with our entire platoon. I was tasked to support Echo Company and Jesse was tasked to support Golf Company. Golf Company and Echo Company were stationed out of Combat Outpost which is near the Sofia district of Ramadi in the eastern portion of Ramadi. I had three other dudes that I was in command of in the beginning and then things happened.

Janney: What was your rank at the point you deployed?

Santiago: I was a Sergeant, E-5. I was the team leader for Headhunter II.

Janney: Okay. You guys went over there before Echo even deployed, I think Jesse said and so you were there what, probably early March or mid-March?

Santiago: Yeah, around then. We were there maybe a week or so before the rest of the companies got there. Just checking stuff out. We were actually hanging out with the 82nd Airborne, the Army guys, and they were showing us key pieces of terrain around the area. Then we actually went on a couple of patrols with their guys so they could just show us a few things around the area.

Janney: If Jesse was with Golf and you were with Echo, you guys weren't operating together, so your story is going to be a little different than his. Tell me a little bit about what you were doing after the 82nd guys cut you loose.

Santiago: Basically, we were tasked to support Echo Company, so the main threat of Ramadi at the time was IEDs. Improvised Explosive Devices. One of our main missions was to locate and prevent these insurgents from planting IEDs. It was pretty difficult for us just because the way the training that we were doing prior to the point where we deployed to Ramadi was all tailored toward open terrain whereas before that we were training up to go to Afghanistan because word was we were going to go to Afghanistan versus Ramadi. It wasn't until maybe a month or two before that we got word that we were actually going to go to Ramadi which is a built-up urban area. So, a lot of the stuff that we had to do was learn on the fly. Our main mission was to find insurgents who are planning IEDs and take care of that threat.

Janney: The Echo guys, the grunts, were not doing the same things that you were doing. You guys were doing kind of an overwatch mission as compared to kicking piles of trash looking for IEDs like some of the other guys were doing.

Santiago: Well, it depends, like I said, everything was a learning experience, so we were just trying to figure out. 2003-2004 was kind of a difficult time for everybody I think because nobody really knew how to fight that war right. It was insurgent warfare, it was IED prep, that was everywhere so at the time we were just still trying to figure out what would be the best way for us to support the mission whether that's going out and going a traditional sniper mission where we're going out on our own or whether we're attaching ourselves to the live platoon of regular grunts and supporting them in overwatch positions. We kind of did both. And there was even a time where we attached ourselves to the Intel guys who were driving around and getting to know the locals. We would drive around with them and become their security so we could learn the lay of the land. It was different kinds of stuff. We were also doing cordon searches. We would do over watch. If there's no cordon searches at the time, we would do IED missions.

Janney: So, watching MSR Michigan and the other main roads? Looking for guys digging holes.

Santiago: Yeah. I guess, let me just ask you if you have already heard the story about my team, Headhunter II and the melee that we got into on April 6th and April 10th? Have you already heard that story from some other guys that you have interviewed?

Janney: I know a little bit about it, but I want to hear what you have to tell me about that.

Santiago: Okay. Again, I just don't know how much detail you want me to get into or what you actually want to know. If you have specific questions, please just ask me.

Janney: I'd rather you tell me what happened on the 6th and then that way if I've got specific questions related to that I'll ask them.

Santiago: What I don't want to do is just get into answering your questions and then I'm leaving stuff out that would have been something that you wanted to know. Because I don't know how much information you already have, whether you've read other books or listened to other people tell the story. If that makes sense.

Janney: Yes, and that's why I'd rather you tell me what happened, what your team's experience on the 6th rather than me lead the questions and then taint your version of it, you know what I mean?

Santiago: Yeah, so, to answer your question of what our mission was primarily was to find and prevent insurgents from placing IEDs. Whether we do that through your traditional sniper mission where the four of us would go out in darkness and set up a position where we believed that the next IED was being placed and watching MSRs that are always IED heavy and wait for insurgents to plant IEDs. As a secondary mission, of course, we were always supposed to support Echo Company so once in a while we would go on cordon searches and be over watch for the entire Company or certain platoons. We've even had a particular platoon going out on a patrol, and we would support them by just being additional guns basically. If they need long range position shots or support from high ground or over watch position, they can call us up with that. So, basically outright supporting Echo Company.

Janney: Tell me what you were doing on the 6th of April.

Santiago: Okay. The 6th of April. So, Route Nova, which is the road closest to the Euphrates River, was always getting hit with IEDs. Our forces would drive through there and they would always get hit with IEDs. For a few weeks, a few days we were going out and doing our typical IED missions. We would go there; we would watch one area that was hit with an IED or an IED was found and we would sit there and do overnight missions. We would go to one spot and then an IED would blow up on the spot that we couldn't watch on Route Nova. Or we didn't have eyes on where we were the night before so this would happen for several days. It would just flip; we would be on one side and then an IED would be in place on the other side and hit our forces. We would go to that side and then on the other side it would be an IED so we figured that they were watching us. They knew exactly where we were going, the insurgents were watching us. We would go to the other side where we couldn't watch them and then place IEDs. Before April 6th we came up with an idea to do a 24-hour mission to stay out there for 24 hours during the day and the night; one 24-hour mission. A couple of days after we came up with that plan, we briefed the Company Commander, and it was approved. They thought it was a good idea, so we're going to do this 24-hour mission. At the time, we were planning to do it every 24 hours. 24 hours on, 24 hours off, 24 hours on. We got the first mission, it was planned, we had routes, we got the gear, and because we wanted to make sure when we exited the gate, the eastern gate, we knew that we were gonna be seen, so it was late afternoon right before the sunset. Because our plan was for whoever was watching us exit the gate, they would know it's a specialized team when there's only four of you guys stepping out of the gate. It's not a regular infantry squad and it's not a regular mission. They know.

So, we made a point to leave the gate right before, so whoever's watching us and reporting back to the bad guys were going to see us walk out and know that our mission is on. So, we went out, everything was just fine, and this would have been the evening of the 5th. The night before, we leave the gate, we walk the neighborhood, kids are trying to follow us, we're giving candy to kids, we patrol out. Once we got to the date palms and the agricultural field, it was pretty dark at that point or it was getting dark. Our idea was we were going to turn west again for whoever is watching us, to think that we're going to head west. As night fell, we hunkered down behind some pretty vegetative stuff and sat there for a couple of hours until late in the night. Then, we got up from that position and headed east to our planned observation point which would have been near the tank graveyard.

Janney: So, your position was near the tank graveyard off of Nova.

Santiago: Yes, off of Nova and it's right at the edge of the Euphrates River and about 60 to 75 yards away from Route Nova. So, we set up there, we probably got in position pretty late, I think it was maybe around 1:00 or 2:00 a.m. when we finally got into position. During the time of darkness, we were on top of the two pump houses. We found two pump houses that were basically within 10 yards of each other and we split up into two-man teams. That was Richard Stayskal, my assistant (redacted) was with Cameron Ferguson. So, two of them went up on top of one pump house and me and Richard Stayskal were on the other pump house and we had pretty good views of Route Nova and the surrounding area. We sat there late throughout the night. We stayed there on top of these pump houses. Nothing was going on. We had our night vision glasses on. Late at night, you start thinking about different things like how I came to that point of being on top of this pump house thinking about my family and thinking about things I've done in the Marine Corps so far, but just nothing was going on. We're in 50-50 meaning half of my team was asleep; the other half was sitting security and watching the road.

The sun started coming up, of course we were going to be pretty exposed up on top of these pump houses so we decided to go down to the base of the pump houses and have the rest of the team on the base of the river. Right at the edge of the river where it's heavily vegetative, we were going to hunker down right there. We could still see pretty far down Route Nova. Since we were there and we ended up setting up our position at night we wanted to make sure we got the lay of the land and see what's around us so I decided to do a quick foot patrol. Just me and my teammate Richard Stayskal, so we just did a quick patrol down to the orange grove which was further to the east. We were patrolling and sneaking around until we realized that there were some locals that were tending to their crop. So, they were pretty far away and they're not going to come over here, so we ended up going back to our observation position behind these two pump houses. I know my team was pretty tired at that point, so we decided again to go 50-50 just watching the road waiting for something to happen. I forget now, I think we started hearing the gunshots in Golf Company's area. I'm trying to remember to give you a good chronological order of the things that happened that day. I'm watching the road and we start seeing orange and white taxi cabs going down the road and they were full of men. We see one drive by at a high rate of speed and I ask my assistant, "Did you see that?" He said, "Yeah, I'm seeing the same thing that you see." But, we weren't seeing any weapons at the time. We couldn't determine whether or not they were hostile. We're assuming that something strange is going on; a taxicab full of males headed westward on Route Nova wasn't normal. So, we started seeing that. I don't know if we got compromised by the kid first or heard the gunshots first. I think we started hearing some gunshots in Golf Company's area. And this would have been around 9:00 or 10:00 a.m. I believe. We started hearing some gunshots in Golf Company's area and we're like, "Okay, they're in a firefight." And at the time, we didn't think it was pretty heavy gunfire at first.

Janney: And Golf is west of you. They're west of Gypsum, is that correct?

Santiago: Yeah, way west. They're pretty much in the downtown area. Several kilometers away. We can hear gunshots and stuff like that.

Janney: The taxicabs that you saw earlier were headed west, not east toward Gypsum?

Santiago: Yeah, they were headed westward on Route Nova. So, we're already assuming that they had something to do with it, but again we didn't see any weapons. I believe we did send out a radio call; communicated to our command center that we were seeing people; parties of males in taxicabs headed west on Nova. We did communicate that to our operations center. And then, at some point we were just sitting there 50-50 and it was my turn to rest, so I was laying down trying to get some sleep and one of my teammates, Cameron, started telling me that there was a boy walking towards us or was walking in the field in front of us. I'm like, "A boy?" I got up and I started looking and yeah, I saw this boy walking towards us. It didn't look like he knew exactly where we were at; he was just walking but then he happened to walk right on top of our position. And at the time, I'm a big movie buff, and I don't know if you've ever seen the movie "Bravo Two Zero." Have you ever seen that?

Janney: No sir, I have not seen that.

Santiago: Well, "Bravo Two Zero" is based on the first Gulf War, right? There's this British Special Forces team that was compromised by a little boy. They ended up getting into this big melee where they had to basically fight their way through Iraq and make it to Syria and only a couple of them survived. I think there was an eight-man team, or a six-man team and only two or three of them survived. They all got captured or killed. At the time, to me, it was kind of comical in my head that I was thinking about that movie. I couldn't believe that we were getting compromised by a boy that must have been no more than ten or eleven years old. He looked at us and we looked at him like, "Aww, crap, now he's gonna tell somebody that we're here." Of course, we didn't know what to do. We didn't know what to do with this boy, whether to hold on to him, so we were just looking at him. I think he was dumbfounded and, of course, we were dumbfounded, and confused about what to do and this boy ended up running away. I'm assuming that he went back to his parents and now we're compromised. Is the mission over at this point? I mean, we were trying to make a decision to head back to base, scrub the mission and go back to base or stay put. Of course, at the time we're communicating this to our command, that we had just been compromised by a boy. Now understand that we were in the middle of the Sofia district. In order for us to get back into base we would have had to walk several kilometers through open fields of date palms and through the residential neighborhood before we could get back to base. Definitely a good recipe for an ambush.

Of course, we had a mission to accomplish. So, at the time, I decided to stay put, gamble and continue on with the mission. Hunker down, and of course, at this point everybody is 100% up and ready, weapons next to them. We're on the radio trying to make sure if somebody does hit us, we're prepared. So, we're sitting there, and I don't know if a few minutes or an hour go by or some time go by and now I see a grown man and a woman walking towards us. I'm like, "Great, this is probably his parents or some adult that the little boy told that there's these crazy Americans, four of them next to the river." So, this man and woman walk up to us and I don't speak Arabic and I don't speak the language they speak. I don't speak their language, but based on their body language and the tone of their voice, and the loudness of their voice, I could tell that they weren't happy that we were there. And at this point I'm thinking that we were in great trouble. Of course, we didn't see any weapons. He wasn't getting too aggressive walking towards us that would warrant us acting upon what he was doing. We had weapons at the ready just in case he had a concealed weapon on him and started blasting at us. After a few minutes of him and her yelling at us and motioning things that weren't very good, they walk away and go back towards the neighborhood. Again, at this point, this is an adult, two adults that have seen where our position is at. They knew we were a specialized unit because there were only four of us. We have weird, different gear than everybody else. At the time, we were wearing green camis different from the desert camis that everybody else was wearing because we were in a vegetative area and we wanted to blend in. So, they disappear. At this point, of course, we communicated that to our command post telling them what's going on and again I make a decision to stay put because if they are setting up an ambush it'll be silly of us to walk back towards base because all they are going to be doing is setting up that ambush at the likely avenue or road that we're going to take back to base and they're just going to sit there and lay in wait and wipe us out as we are walking towards them.

Janney: Yeah, you would have walked right into it.

Santiago: Yup, we would have walked right into an ambush if the worst happened. At this point, I'm assuming that they are. So, we hunker down and at least we had set up a defensive position and we'd already communicated that we had been compromised to the command center so we should be good. The decision was once nightfall hits, we were going to move. At least we can have an advantage of moving under the cover of darkness. But, like I said, this was 9:00 or 10:00 in the morning. We have several hours of pucker factor ahead of us.

Janney: Yes, that's a long time to sit there and wait for something to happen.

Santiago: Yeah, exactly. So, we're sitting there and of course at this time we're at 100% just waiting to get hit. I made a decision because I could not see the curve on the west end portion on Nova. Route Nova starts curving around as you head more toward the west and I couldn't see what was on the other side. So, what I decided to do, I told my team, "Hey, stay here, you guys are at 100%, just make sure you guys are keeping watch and be on alert. I'm going to punch out about 100 to 200 yards to the west to a different pump house and see if there's anything going on towards the west." I grab my M16, a couple of magazines and a handheld radio so I could still communicate with my team, and I punch out. I do a solo patrol to the next pump house about 100 to 150 yards out. I find a pump house and I go inside this pump house. I'm sitting there, and I had good views of the western part of Nova and all the surrounding areas to our west. So, I'm sitting there looking and a few minutes go by, probably a good thirty minutes go by now. I'm super tired, I've been up all night, and I'm on patrol and observing. Adrenaline dump and now the adrenaline dump is wearing off so I'm getting tired. Then a little voice in my head kept telling me, "Hey, you should look on the other side of Route Nova" because you've been there, and Route Nova is on a berm. A three- or four-foot berm and you can't see over on the other side.

So, I wanted to see what's on the other side just in case insurgents or that couple were insurgents and wanted to make trouble. I wanted to make sure that if they were planning an attack they were not on the other side. Whatever, I just wanted to see what's on the other side, right? So, being short and compact that I am sometimes has an advantage, but it makes it easier for me to hide. So, I see this little ditch. Now understand, it's an agricultural field and between farm fields or between crop fields are like little ditches, and the ditch is probably a good foot deep and it's surrounded by short grass. So, you could actually crawl inside that ditch and just low crawl. I saw one of those ditches and it led to this big bush right on the side of Route Nova where I can get onto the edge of Route Nova or the edge of that berm and look over the other side. I started low crawling; it must have been a good 50 – 60 yards to get to this bush. So, I low crawled inside this ditch and got into this bush. Every so often, I would pop up and scan the side of Route Nova with my three-power ACOG scope. I had good views. I could see 100 to 200 yards. I could see the residential neighborhood where we came from; I could see the tank graveyard. I could see the palm grove, everything, I had good views. So, I pop up one time, I scan; a quick scan, I sit back down. I scan again, nothing, all clear.

I think it must have been my third time that I popped up and slowly scanned and then I saw two guys in traditional dress, pajama looking attire and their headdresses. The head wrappings and it looked like, based on their gestures, it looked like they were arguing. I can see them maybe 150 -200 yards away. I can see them arguing and one of the guys that was there that was talking takes off running westbound. I'm like, "Okay, that was kind of odd." And then the guy that he was talking to, arguing with, starts motioning to somebody in an alleyway or roadway. Then one more guy comes out and starts running west and that guy had something in his hand. So, I'm like, "Why is that guy running? That doesn't seem normal." I followed the second guy running with my ACOG and I followed him, and I couldn't make out what he had in his hand until…you've been to the tank graveyard and the palm groves, right? You can visualize how the sun would shine in between the date palm trees, correct?

Janney: I've been there early in the morning and have seen them, so definitely.

Santiago: So, he happens to run underneath where the sun was shining down on the ground between the palm trees and him being lit up like that, I could actually make out that what he had in his hand was an AK47. I'm like, "Well, that's not good." I traversed my rifle and my scope back to where I first saw him coming out of the roadway and I saw one more guy come out. Then another guy comes out and I start counting them. I stopped counting at around 12-15 and I thought, "That's not good." Each one of them was carrying something in their hand. Now, I'm assuming at this point it was weapons. I'm looking at them and they're getting in a traditional infantry-based tactics of getting online if you can kind of picture that. They're basically side by side and facing towards us. At that point, I grab my radio and I start talking to my teammate Stayskal and say, "Hey, get your stuff on, we're in big trouble." But every time I would click on the mike and try to transmit, he would just tell me, "Hey, you're coming in broken. I can't understand what you're saying." I just kept telling him, "Hey, get your stuff on and get moving. We've got bad guys moving towards us." So, he kept saying, "Hey, I can't understand you." The transmission was broken, so I just stuffed that radio in my pocket and I haul ass running back to my teammates and I come sliding in like I was sliding into home in baseball, right. I slide in and I start telling my guys what's going on, what I saw and we started grabbing our gear, putting our body armor back on, and stage magazines and grenades in front of us. Then, we just sit there and wait. We sat there and waited for these guys to see if they were actually going to find us, right. Last thing I saw was them getting online and moving towards us. So, of course, that's when I briefed my team. I grab the radio and start communicating that to our command center. I kept telling them that we have 12-15 dudes moving towards us all armed and for them to send a QRF – a quick reaction force.

Janney: SOP for your four-man team is to try to escape and evade or sit tight? You guys had your back to the river, so you're kind of in a bad spot there.

Santiago: Yeah, you bring up a good point. So, we're in a bad spot. I mean, at this point I'm trying to determine the best course of action. We could have easily fled, but we don't know if the bad guys have set up on the east side. We couldn't go west because we don't know if bad guys have already set up on the west side. And both east and west of us are nothing but open fields; there's nothing there for us to take cover on if we get hit. In front of us, in between us and Combat Outpost, safe haven, as far as I know, is 12-15 dudes, all armed. So, the best course of action for us was to fight it out. Wait for QRF. We were going to stand our ground there and try to fight it out with these 12-15 dudes until we can get some reinforcement with the QRF. Making sense?

Janney: Yes, it makes sense. Horrifying, but it makes sense.

Santiago: Oh yeah. Tell me about it. I mean, just sitting there waiting, knowing we were about to get hit. It was pretty terrifying. So, a few minutes go by, when we're sitting there prepping for this potentially massive gunfight that we're about to get into. So, one of the main standard operating procedures for scout sniper teams is you never let the bad guys get ahold of your sniper rifle. That is your baby; that is the one thing that you do not want the enemy to get a hold of. The common thing that we were supposed to do was render that rifle useless to the bad guy. If we knew that we were going to get wacked, render that rifle useless. So, standard operating procedure was to smash the scope and take the bolt out. I didn't want to smash my scope because there was still the possibility, I mean, I wasn't going to give up. I wasn't going to say, "We're not going to make it out of this alive." But I ended up taking the bolt out of the gun because we're in a bad spot; this was going to be bad. So, I took the bolt out of the sniper rifle and stuck it in my pocket. I wasn't going to smash the scope just yet if we were going to fight it out and see what happens next. You bring up the point that our backs were to the river, so I'm sitting there, and I mean our most likely avenue of escape is across that river. So, that was one of the things that we discussed, and of course, Cameron Ferguson, he wasn't a very good swimmer and I'm telling him, "Dude, if this gets bad, we're going to have to jump up and cross that river." He was just cussing at me, "Fuck you, I ain't crossing that river. I can't swim." One of the moments of comedy in this whole thing. So, we take the bolts out, place the magazines in front of us for easy access and we're waiting there, waiting, waiting. A few minutes go by and I'm scanning the road and looking for bad guys. Then I see this head pop up. This head is wearing that red shemagh around his head. He pops up his head and he's looking around, scanning left to right. I can vividly remember his little beady eyes looking through that shemagh. I have my ACOG right on his noggin and I'm like, "Should I squeeze one round off? Should I not?" At the time, I'm thinking maybe they don't know exactly where we're at and I'm trying to buy time. As far as I know, the command center has received our transmission requesting a QRF and I just need time for them to get here, to get to where we're at. So, I didn't want to squeeze off a round and alert them to exactly where we were at and start this gunfight. I didn't squeeze the round, but I told my team, "Hey, I see them. I see a head." He ducks back down behind Route Nova right behind the berm and I'm like, "Oh boy, here we go" and he pops his head back up. Again, I had my crosshairs right on his noggin thinking that I should just pop him in the head, and then he ducks back down.

Then the third time that he pops up, my reticle goes right on his noggin again and I hear the distinct sound of an RPG (Rocket Propelled Grenade) being fired. You know that whizzing sound. It just whizzes and hits, and at this point I don't even know where it hits, whether it hits a palm tree right above us, whether it hit in front of us or next to us, but I can distinctly remember the concussion and the heat from it. We get hit by an RPG and the fire fight is on. It was just so loud and so deafening. All four of us were firing in all directions. You can hear the snaps of the bullets going over us, going next to us. Richard is to my left, Cameron is to my right, and (redacted) is to his right, and all four of us were right behind the pump houses that we had been using as an observation position or a defensive position. We're just shooting, shooting, shooting to try to fend these guys off. I grab my radio and start broadcasting for command that we're being hit, requesting QRF and requesting ETA for the QRF and I just keep hearing whoever is on the other end of the radio, "Say again, say again." I guess my transmission was being garbled. He asked me on the radio, "Are you requesting permission to fire?" and I'm like, "No, I'm not requesting permission to fire. We're being engaged. We're in a gunfight." He just kept saying, "Say again, say again" and by then the transmission was broken. So, what I ended up doing instead of talking on the radio is I just keyed the mic and just held the mic for a few seconds so whoever was on the other end of the radio could hear all the shooting that was going on. All the shooting and yelling that was going on.

After a few seconds, I released the mic and stuck that bad boy in my pocket and continued to direct my team and get on the fire fight. So, we're in this firefight already and I guess we should back up a little bit to give you an understanding of what was happening. A few hours before this firefight, one of the squads led by Corporal Barron, passed our position because they were patrolling Route Nova for IEDs. That typical patrol that we used to do with Combat Outpost and they're sweeping the road for IEDs. They happened to pass us up a few hours before this gunfight.

Janney: They were moving west on Nova or east towards Gypsum?

Santiago: They were moving east towards Gypsum. During our gunfight, they actually got hit as well. So, they were dealing with their own gunfight and we were dealing with our own gunfight and once they got done and everything was pretty much under control in their area, Corporal Barron and his guys heard our call for help. These guys, man, I mean, to this day, if I can just go up to them one by one and just thank them for saving us, because without them we would have gotten overwhelmed, surrounded and killed. These guys mounted up on their Humvees and just sped towards us. During this gunfight Stayskal tells me, "Hey, there's a Humvee on the road." I looked over on the side of the pump house to see where this Humvee was at and I could just see the guy on the turret of this open back Humvee just laying waste to whoever was on the other side. A machine gun, you're only supposed to shoot six to eight round bursts, right, just because if you just keep spraying and praying, you're going to melt the barrel and render that gun useless. I guess he wasn't following that six to eight round burst because that guy just had his finger on that trigger and was just sweeping, sweeping back on the other side where the bad guys were.

Janney: So, he's opening up with a 240 Golf I'm guessing or a SAW?

Santiago: Yeah, it was either a 240 Golf or a SAW. I mean definitely a belt fed; he's using it and laying waste to whoever was on the other side of Nova that we were fighting with. Now understand, like I said, this was only 60 – 75 yards away. The only thing that separated us from those 12-15 dudes that I saw was 60 – 75 yards of open ground. And we were just doing everything that we could to pin them down and to keep them from crossing that open ground and we were stuck behind these two pump houses. As soon as I see the Humvee, Stayskal asks me, "Hey, should I pop smoke to mark our location?" I said, "Yeah, go ahead pop smoke," so he popped smoke. He pops white smoke to cover our position and mark our position to let the good guys know where we're at. So, as the smoke billowed, we were going to join in this fight. They start dismounting the Humvees and getting into this fight, so we're going to join them in this fight. We're going to get on online. I directed my team to get online and go to the other side of the berm. So, we just haul ass to the other side of the berm. Pop over the berm expecting to get mowed down by whoever was on the other side of the berm expecting all kinds of bad guys on the other side of the berm. But when we popped over all we saw was AK47's left on the ground; some sandals; blood trails, a couple of RPG rounds left on the ground and we didn't see anybody. At a distance, we could see people running. This squad basically made the enemy move away from us. At this point, we just got into a gunfight. We're still getting shot at. There's still enemy out in the tank graveyard shooting at us. We just started doing your infantry tactics of bounding forward and taking the fight to the bad guys. We're bounding forward, shooting up bad guys, bad guys shooting at us. There's some machine gun fire being shot towards us. Some mortar fire that was being shot towards us and during this crazy melee, Marcus Cherry gets hit. I'm no more than a few yards from Marcus Cherry and I could see him. He's rocking and rolling, shooting up bad guys and he just spins around. I literally see him spin around and drop to the ground. We're receiving some heavy machine gun fire, so everybody was pinned down. That squad that had reinforced us was pinned down. Forward momentum was halted at this point, so basic infantry tactics, I decided that the best course of action was for my team to keep moving forward and to try to flank the bad guys who are shooting at us. So, I directed my team to start fighting forward and do a flanking maneuver to wipe out the bad guys that were shooting at us. We're moving, bounding forward shooting at bad guys, bounding forward, cover, move, basic infantry tactics, and I see to my left my teammate Stayskal yelling. I heard him yelling and pointing and he was like, "Look at that" and I'm like, "What is he yelling at me?" He's yelling and pointing in the direction of the tank hulk and all of a sudden, I see him spin and drop to the ground. Assuming that he just got hit. Again, to give you a background, one of the SOP that we established within my team is that if one of us gets hit, you will not go to the down teammate right away because you have to deal with the threat before you can render aid just because you can become a casualty yourself. If you don't deal with the threat, you're no good if you get shot as well.

So, I'm sitting there trying to find who shot my teammate and I can kind of see them. To this day, I still can, I don't know if I can hear them, but in my head, I can hear them screaming in my head. I don't know whether or not I did hear them scream. My ears are ringing from all the gunfire at this point anyway, so maybe it was just in my head that I can hear them scream. Occasionally, while I'm shooting trying to move forward, I see him kick his leg like he's moving and squirming. A few times I look over to him and he stopped moving and I'm like, "Man, I think he's dead or he's either dead or dying at this point". So, I break my own rule that we established within our own team and I run towards him. When I run towards him and he's just complaining that his back hurts and I'm trying to assess where his wound was and I didn't see massive amounts of bleeding, I didn't see a gunshot wound or none of that. He's complaining about his back and he's asking me to cut off his gear. So, I get my knife and I start cutting off his gear, I open his flak jacket and all I see is some blood on his right shoulder. So, I tell him, "You're good, it's just a little bit of blood on your right shoulder. It looks like your good dude, you're good."

I'm about to start cutting his cami top off to see where the wound was and then I happened to look up. When I look up, I see this guy maybe about 75 yards away either sitting or kneeling behind this burnt up tank hulk. I see him and I'm looking at him and I can distinctly see him swinging and aiming a rifle at me. I'm like, "Fuck, I just broke my own goddang rule and now I am going to be that casualty." So, I grab my rifle and I tell Stayskal, "I'll be right back." I grabbed my rifle and I stood up and I started walking towards this guy who's aiming a rifle at me and I just started popping off rounds trying to keep his head down. I'm walking, walking, walking, a fast walk trying to pin his head down and shooting at him. Occasionally I would stop and try to get a good shooting platform and shoot at him and I got to maybe 40 – 50 yards of this guy and take a shot. I assume that I hit him because I saw him flinch, but he gets his rifle up and tries to aim at me again. I tried to take another well aimed shot as he tried to aim his rifle at me again. I took one final shot and he fell backwards. As soon as I got done with that, Cameron ran up to me and we ran up to this guy I had just shot and I stood over him and I saw the rifle that he was pointing at me and trying to shoot me with was an SVD Dragunov rifle which is basically a Soviet sniper rifle. I'm looking at his face and I'm looking at him and yeah, he's dead. Amidst all this craziness going on, rounds popping off everywhere, snapping around us, people yelling, people screaming, people shooting, mortar rounds going off, I just stood next to this guy looking at his face. I grab his rifle and I sling it behind my back.

Me and Cameron proceeded. I got on my radio and Ferguson told me that (redacted) was with Stayskal, taking care of Stayskal and calling in a medevac for him. So, me and (redacted) were getting in different cover positions. We would look for bad guys to shoot at a distance and he would tell me where the bad guys were and I would just shoot at bad guys here and there. Just fight it out until we all got consolidated back with an Echo Company platoon. We joined them; cleared a couple of houses. Tanks (Bradleys) eventually showed up. Army guys showed up. We continued the fire fight for another hour or so and clearing houses and that stuff and taking prisoners and hauling prisoners away. Then, we ended up driving through Route Gypsum seeing all the melee that happened there, seeing the aftermath of Route Gypsum and then we headed back to Combat Outpost late that afternoon. I remember it was getting dark by the time we got back to Combat Outpost. So, that's basically the gist of April 6th on my part.

Janney: Wow. That's incredible. I'm glad you guys made it through that. I can't even imagine.

Santiago: Yeah, and I guess, after everything was said and done, just talking to the guys that were there and who helped support us, and then after the action reports, guys were saying that estimated numbers behind that berm that were trying to kill us were 30+ dudes. It wasn't the 12-15 dudes that I initially thought. I counted 12 -15 dudes initially, but I knew there were more guys coming out of the woodwork, but I wasn't going to sit there and wait to see how many dudes were going to come out. But, based on the debrief it seemed like there were 30+ dudes that were actually there that were trying to kill everybody.

Janney: That's incredible. Just Rich (Stayskal) got hit at this point?

Santiago: Yeah, Rich got hit. Again, after the debrief, we find out that a round entered his left shoulder. Shattered his shoulder, shattered his shoulder bone, collapsed his lung and exited the top of his back, almost at the base of his neck and that bullet got lodged in his body armor. Then looking at the bullet, it was a round from an SVD Dragunov rifle. It was a big caliber round, so I guess you can just put two and two together. That guy that was trying to shoot me while I was rendering aid to Stayskal was the same guy that shot Stayskal.

Janney: Wow. Man, you guys were blessed to survive that.

Santiago: Well, I don't know about blessed, but there were definitely some great dudes there that, without them, we would have been dead - my team: Rich, (redacted), and Cameron. Barron's squad saved us. But I don't know about blessed.

Janney: Was Roy Thomas part of Barron's squad?

Santiago: I'm not sure. I'm not too familiar with who exactly was in Barron's squad. But I know Marcus Cherry was in Barron's squad and that was it. I don't know their exact names. It wasn't even his full squad because part of the squad was hit during their gunfight. It was like a hodgepodge of dudes that just got in a Humvee and helped us out.

Janney: Cherry was on the 240 Golf or the SAW in the Humvee when he got hit?

Santiago: No, I don't know who was on that 240. I don't know who it was, but Marcus Cherry was actually on the ground with us when we were bounding forward. When we were online with the rest of the squad and all of us were bounding forward, he was actually on the ground with us within just a couple of yards to my left. That's basically it. I'm sure you can get other details. If there's other details that you want me to cover regarding that firefight that we got into just ask away. There are so many things that happened that I could in this particular gunfight that I could talk your ear off for hours. Again, there's other people that have different perspectives of it. There's a couple of books that have been written about that particular firefight. As a matter of fact, I actually talked about it in a podcast that we had, and you can get a good additional detail off of that.

Janney: I'm not familiar with any other books written about that except "Joker One", so who made the podcast?

Santiago: Yeah, there's a book written by Milo Afong. It's called "Hogs in the Shadows." It's basically a group of chapters and stories about certain firefights that recon and scout sniper dudes had gotten into. And that particular firefight, the April 6th firefight that my team got into was one of the chapters, so it went into a lot of detail of exactly what happened that day. There's another book by Bing West called "No True Glory," there's a chapter in there about that particular firefight.

Janney: At some point, (redacted) got wounded. Was that that same day or was that a different day?

Santiago: Yeah, he actually got hit that same day. It was during that same firefight. As we were trying to do that flank maneuver, Stayskal gets hit and I'm not sure whether (redacted) gets hit before Stayskal or after Stayskal got hit. They got hit almost at the exact same time. So, with (redacted) being hit, just talking to him afterwards, talking to him and Cameron, Cameron tells me that he gets hit and he gets knocked to the ground. (Redacted) says that it felt like somebody hit him with a baseball bat and it knocked him down to the ground. Funny thing was Cameron runs up to him and tells him to stop being a pussy, "I don't see any blood, so get up." So, (redacted) gets up and they both continue with the firefight and (redacted) being wounded already still renders aid to Stayskal. He was the one taking care of him, making sure that he gets bandaged correctly, and then he calls the medivac. So, (redacted) and a Corpsman started rendering aid to Stayskal.

Janney: Was that Doc Menares?

Santiago: Probably was. I'm pretty sure people from Echo Company would be able to tell you. I don't remember which Corpsman that was.

Janney: I was just curious if it was Doc Clayton or Menares.

Santiago: I don't think it was Doc Clayton. It probably was Menares.

Janney: No, no, I was just curious what you remembered. So, you guys made it through hell and got back to Combat Outpost, so tell me what happened in the days after that.

Santiago: That evening, we bed down for the night. We pack our gear and make sure we resupply a pack full of ammo and at this point we are pretty beat. We just got through a pretty crazy gunfight.

Janney: Yeah, almost 24 hours awake.

Santiago: First thing in the morning, I wasn't woken up, I was already awake. We get hit. The Combat Outpost gets hit so there's this big crazy melee firefight. All the little posts inside Combat Outpost were rocking and rolling. You can hear the staff that were roused coming toward Combat Outpost, a couple of mortar rounds and stuff like that. I know for a fact that there were several squads that were infantry units that were being hit throughout the city at the time. But my team was inside Combat Outpost. I'm walking around and going to the command post to see what they needed us to do or whether they need us to reinforce a team. Get on top of the roof, whatever, right. Trying to get some Intel as to what's going on and I get told that two guys from a Hawaii sniper team were there to reinforce us. I'm like, "Hawaii? What are they doing here." So, I guess these two guys were just being sent out. Their unit, their platoon was being sent out to different locations within Fallujah and Ramadi to augment and support teams. To up the numbers. So, there were these two guys from Hawaii that I was introduced to and I told them, "Hey listen man, I got other stuff to do. I've got to figure out what Echo Company needs me to do, so go grab your gear, go grab your rifles, grab your body armor, hook up with one of the observation posts inside Combat Outpost and engage hostile targets." They just do that.

Janney: So, were these guys Marines or were they with 82nd?

Santiago: No, they were Marines. A platoon from Hawaii.

Janney: But they're still 2/4, right?

Santiago: No, no, they weren't 2/4. They were from a different unit.

Janney: I was wondering if they were with 3/5 because they were in Fallujah at that time.

Santiago: No, it wasn't Fifth Marines. It was a completely different unit whatever that Battalion is that is stationed in Hawaii. It was that Battalion. Their sniper team apparently was tasked to support other sniper teams throughout Ramadi and Fallujah.

Janney: And you've got two guys down at this point. Rich and (redacted.)

Santiago: Yeah, so they were tasked to do that, so they sent out two to help out 2/4 and they happened to direct them to me because I was down one teammate. (Redacted) actually never even got evac'd. He didn't want to get evac'd; a doctor saw his wound. His wound was actually to his left shoulder also and it just entered and grazed him pretty deeply and the bullet cauterized his wound. So, he had a nasty gash on his shoulder that was cauterized. He was still able to fight. He was a hell of a tough bastard. He was fighting with that crazy wound and he just wanted to put a bandage on it and continue on with the fight. He never went to any aid station or anything. He stayed with me. Stayskal ended up getting flown out to Germany because he was in bad shape; it collapsed his lung; he almost died just trying to get to Germany.

These Hawaii guys just attached themselves to us. I mean, they were up and helped us secure the base and now we're in a planning phase because apparently, they were going to do this big operation in the Sofia district where we were going to take the fight to the bad guys. So, they know, Intel is saying at this point that all the bad guys were in Sofia district. All these insurgents and foreign fighters that wanted to take it to us that started this jihad on April 6th were all hiding out in the Sofia district. So, Echo Company planned this big operation to do a sweep and just get some. I joined the planning phase, and we determined that we were going to support Echo Company in this sweep by getting on top of the highest place in this area which was a power plant. Getting on top of this power plant we had good views of the entire area. They started off this operation pretty early in the morning the following day, I think it's either April 9th or the 10th. I think it was April 10th.

Janney: Was that Operation Bug Hunt?

Santiago: Probably was. If that's what the rest of Echo Company guys were talking about where we went to Sofia district, it probably was Bug Hunt. Is that what people are telling you?

Janney: What I've heard, Santiago, was Bug Hunt. Where they were just basically pushing through the city trying to do whatever they could to get as many of the bad guys as they could possibly get.

Santiago: Yeah, Echo Company did this operation in Sofia district. Golf Company, I'm sure they're doing their own operation in their sector. It was everybody. It was the entire battalion. We were just going to take it to them. So, we get there and as soon as daylight comes, and they start the operation we start hearing some sporadic gunfire. We were on an elevated position and we were thinking to ourselves, "Yeah, this is a true sniper mission." Now we're in this elevated position and we can see everything and we're going to go do what snipers do. Over watch and just bagging bad guys with bolt action rifles and laying hate and discontent.

Janney: You still had your M40s? You weren't using a Barrett at this point?

Santiago: No Barrett – we had our M40s. So, we're up there in that power plant for a few minutes and we're like, "You know, we're not seeing anybody". Everybody that was shooting at us was all hidden within the residential neighborhood. We couldn't get a good bead on them. We could see people popping in and out shooting but it just happened so quickly. By the time you get your scope up they're already gone. We decided to link up with one of the squads, or one of the platoons in this operation. I think it was the first platoon. So, we radio 1st platoon and we get down from our position there and run toward the first platoon and catch up with them and we get online, just like regular infantry dudes. We joined this big firefight that everybody got into on April 10th. There were a couple of funny things that happened on April 10th. Me and (redacted), we're getting shot at and we happen to roll into one of the outhouses to take cover. We went in there and it stunk so bad and there were so many flies everywhere that me and him almost threw up on each other. We just looked at each other and said, "Dude, I would rather get shot than be stuck in here".

Janney: Laughs.

Santiago: Yeah, so we roll out of there and we get back online and get in the firefight and Cameron gets rushed by a cow and he kills a cow. What I think happened was there was this wooden fence that they were all trying to get behind to conceal them from the bad guys and Cameron and a machine gunner goes around this fence and all we hear is yelling and cussing and a burst from an M249 rifle or the belt fed and Cameron's M203 going off. And me and (redacted) at this point are thinking, "Oh, crap, they must have run into some bad guys on the other side of this fence and Cameron is dead." So, we go around this fence and we see Cameron standing over this cow that was still moving. "Dude, are you serious? You just shot a cow, bro?" He's looking at me with this look in his eyes like me being disappointed in him. Dude, he looked at me and I said, "Okay, whatever."

Yeah, kind of funny thing. We continue the fire fight. That was the day, I don't know if you heard about that, that portion where a Huey and a Cobra was shooting at us.

Yeah, so during this firefight, it's a big melee. Units started getting separated. They're doing their own little firefights everywhere but because of how confusing everything is right. We end up taking in two or three regular infantry dudes from Echo Company that got separated from their platoon. And it's just us, me, (redacted), Parker at this point to replace Stayskal, and Cameron and we're just bounding trying to link up with dudes and we get pinned down by this machine gunner and we end up flanking this guy and shooting and killing him. And as soon as we dealt with that machine gunner, we started receiving fire from about 150 yards away from this house and as soon as we looked over to the left there were three or four dudes on top of this rooftop just shooting at us. So, we directed our fire towards that; we communicated to a squad we would clear the house and we cleared the house and got a bunch of bad guys through that. So, after all this thing was happening somebody had called in close air support so a Huey gunship with a Gatling gun on the side of the door and that Cobra gunship showed up and started circling around Echo's area of operations. Me, my team and the rest of these guys who we linked up with are bounding through a series of palm groves trying to link back up with an infantry platoon, so we weren't cut off anymore. We're just bounding, bounding, and next thing you know, we just hear massive amounts of fire, machine gun fire. Just nasty loud gunfire and palm trees are splitting on top of us and just craziness. (Redacted) starts yelling at me, "Dude, it's the gunships that are shooting at us." (Redacted) had the radio at the time, so I started telling him, "Hey, tell Echo Company that they're shooting at us." He starts relaying that information and within a few minutes they pop red smoke signaling that they are shooting at friendlies and these two gunships fly off someplace else. We got shot at by our own dudes.

Janney: That's horrifying.

Santiago: Yeah. That was pretty crazy, I mean, it was pretty wild. We literally had to dive into this irrigation ditch to get away from the fire. It was impacting the palm trees next to us. It was pretty crazy. Good thing somebody had red smoke.

Janney: Thank God somebody had red smoke, so you guys could pop smoke and they realized that they were shooting at friendlies.

Santiago: Yup. So, they fly off and that's that. We ended up linking up with the rest of the platoons. Again, we started clearing houses and taking prisoners and taking them back to Combat Outpost for interrogation of the prisoners. That gunfight alone lasted for hours. Multiple, different gunfights throughout the city, throughout the area. We would get shot at by a couple of insurgents and while we were dealing with that gunfight, another group of individuals would shoot at us from a different direction. It was just crazy. A big confusing time, you know.

Janney: Truly, the fog of war.

Santiago: Yeah, that's definitely fog of war. That was basically the day, April 10th. That was the first battle of Ramadi, I mean, four or five days of crazy melee.

Janney: So, after the 10th, what were your operations like? I mean, you went back to doing more of the same or?

Santiago: Yeah, more of the same, you know. At this point we were running 24 hours on, 24 hours off missions. We went back to IED missions, hunting down these guys. You know, at this point, man, we were pretty spread thin. My four-man team had the biggest area to cover, which is Echo Company's area. It's huge. There's no way; we were running ragged at this point. I mean, it was months and months on end of the same mission. Going to houses, going late at night, 24 hours on and 24 hours off and most of the time you don't get 24 hours off because you're either getting hit or things are going on so you may get 3 or 4 hours of sleep a night or when you're off and back on to another mission. That's pretty much it for the rest of the time. Small fire fights here and there but nothing as crazy as the first five days during April 6th through the 10th. You know, we'd get hit a couple of times with IEDs; a couple of rockets, bad guys actually shot an RPG at one of the houses that we liked using as an observation post. Then, towards the end of our deployment I ended up staying for another week or so to do the turnover with 2/5. Got involved, me and actually one of the Hawaii guys who stayed behind, got involved in a crazy gun fight towards the end of our deployment with 2/5. I mean it was just wild.

Janney: So, that was sometime in September?

Santiago: Yeah, yeah. Towards September right before we went back. I think it was a week before we went back. We got involved in this crazy gun fight.

Janney: What happened that day?

Santiago: So, me and Sergeant Braddock, I forget what Battalion he was in, but he was from Hawaii. We're just sitting there in our squad bay. They had already stripped us of all of our equipment trying to turn it over to 2/5. So, at this point we just had our M16s with no optics, maybe three or four magazines. We had no grenades, no nothing. No sniper rifle, of course. So, we're pretty bare. At this point we are just sitting there just acting like consultants for 2/5 STA Platoon. We started hearing all this gunfire. Gunfire everywhere, I mean. 2/5 and all their units were in this crazy, crazy gunfight. Of course, I'm not going to just sit there and stay away from the fight regardless of whether or not we were about to go home. So, Braddock was just sitting there looking at me. He asked me, "Do you want me to go see what's going on?" and I said, "Yeah man, go to the TOC and see what's up." He comes back and tells me, "Yeah man, multiple units are in a gunfight." He asked, "Well, what do you want to do?" I said, "Let's get in a Humvee." So, we load up, we gear up and we find the next unit that is going to go out. There was this one squad that was going to go out. We asked the Lieutenant, "Hey Lt., can we go with you? We're with 2/4." He said, "Yeah man, hop on in. We need some extra bodies." So, we get in the 7-ton, and we go into the actual Golf Company area near downtown. We exited the 7-ton and started bumping from house to house trying to get shots into the bad guys. Sporadic gunfire through alleyways and stuff like that.

We end up going to this two-story house. At this time, we don't have sniper rifles or anything like that, we're just there supporting this infantry squad. So, me and Braddock just decide, since the squad is on top of the roof doing their thing, we told the Lt., "Hey Lt., we're just going to be down here on the bottom floor securing the bottom floor for you guys and make sure that no bad guys come up." He said, "Roger that." So, we're on the bottom floor just securing it and I can hear all the gunfire and everything. The Lt. said, "Hey, snipers, come up here." So, I get up there and he starts explaining to me what he has.

He's saying that he has a couple of guys who are taking shots at Marines from about 200-300 hundred yards away behind this building and he wanted me to go take a look and possibly take a shot. I explained to him, "I don't have a sniper rifle. I don't even have a scope on my rifle anymore." He's looking at me, "Oh, okay, I didn't realize that." He looks at one of his squad leaders who had the ACOG in his M16 and he tells that squad leader, "Hey, give him your rifle." So, he gives me his rifle with the ACOG and I plop it down on the low wall of the roof. As soon as I plop it down, I see this car, this grey Mercedes Benz looking car, screeching away from the location that the LT was explaining to me they were seeing fire from. So, as soon as I see that I automatically switch that bad boy to fire just based on the Intel that I got and their explanation of what was going on and the amount of fire and I deemed that car to be part of the bad guys. As soon as I saw that, I clicked that bad boy to fire, and I emptied a magazine into the car and that car ended up crashing on a low wall. Then I just look at the Lt. and the Lt. is looking at me. I grab the rifle and I give it back to the Lt. and say, "I think I got him, sir." He just tells me, "Hey, next time you see something, let me know first before you start shooting." I said, "Okay, sounds good, sir." Then, Braddock and I just went back on the steps that lead to the bottom floor and I just sat there. I sat there on the steps thinking to myself, "That was way too easy without even thinking about it." I clicked that bad boy to fire and emptied a magazine into that car. So, yeah, that was my final firefight.

Janney: Wow. That's a hell of a way to end a grueling deployment.

Santiago: Yeah, yeah, that was pretty weird, you know. That moment when I'm sitting there at the steps, holding my rifle. I thought, "Man, that was way too easy." It's time to go home because, at the time, I was actually thinking about requesting my command in 2/4 to let me stay with 2/5. Let me stay with 2/5 STA platoon or whoever would need me. Thinking maybe I could stay for their deployment. So, at that point I decided, "Yeah, it's time to go. That was way too easy."

Janney: Let me circle back around and ask you a question. The fighters that you guys captured, did you determine through any of their paperwork, that they were foreign fighters or just locals?
Did you have any intel on that?

Santiago: No. One of the groups of prisoners that we helped detain, we were clearing this house and, of course, they didn't have papers, and they had no good explanation of what was going on and why they were in that particular house. I mean, there's a group of fighting age males inside this house that had no relationship with the homeowners. They're all bundled up in one room. I distinctly remember two of the guys in this group of five or six dudes inside this house, two of them had the old school, Vietnam era, gun belt. He didn't have weapons on him though. I mean he had that belt on and I was like, "DUDE!" He was dressed like that; he was in this fight. So, what we started learning was these guys would stage weapons in different areas. They would run around because they know we're not going to shoot at them because there were rules of engagement. We're not going to shoot at them if they are not armed. So, what these guys would do is stash weapons, get those weapons, shoot from that position where they found the weapon, empty that magazine and then run away or walk away to another position where they know there's weapons. Shoot again and then run away and hide in different places.

So, these guys were dressed in that type of military gear. A couple of the guys that I interrogated had blood on their shirt and on their hands. Whether they were fighting or not, they were definitely aiding the injured combatants. So, there's no reason for them to have blood on their hands or on their clothing and there's no family member that's happened to have been injured. If there's no family member, or brother or sister or friends that they were lending aid to, and they are running around with blood all over their hands you could only assume that they were there dragging bodies away and aiding casualties or injured combatants. They're just trying to blend back in. Of course, we're not going to engage and kill these guys. I mean, we're going to detain them so interrogators can then determine whether or not they're fighters or not.

Janney: During your entire deployment you guys didn't have any armored Humvees. I heard that the 82nd had some, but I guess they didn't leave those for you. They took those with them when they left?

Santiago: Yeah. Our armor consisted of our Motor T guys welding on steel plates on the side of the Humvees and the 7 tons. Then, putting sandbags on the floor for explosions. That was our armored Humvees for the longest time. It wasn't until maybe towards the end of the deployment; the last quarter of our deployment time that we started receiving these up armored Humvees.

Janney: I heard, I don't know if you know this guy, Sgt. Valerio, he was a Motor T guy and was welding hillbilly armor on for everyone. Did you know Valerio? He seems like a good man.

Santiago: I haven't talked to him since we were in the Marine Corps, but he was Motor T. He was actually a pretty good guy. Once you get to know him and realize how he works, he's a good dude.

Janney: Did you know any of the other guys that got hit during your deployment? I know you were with Echo, but did you know any of the other guys like Crowley, Layfield or Staff Sgt. Walker?

Santiago: No, I did not know them. So, my three guys from my team, they were all Echo Company guys. They came from Echo Company. So, when they took ENDOC, I specifically requested Echo Company dudes and picked those dudes because of some of their good traits that they had that I knew was going to be valuable to a team environment. Aside from that, I made sure that they were going to be Echo Company guys because I already knew that I was going to support Echo Company. I was going to be the Echo Company sniper team. Those guys knew everybody. They knew everybody in Echo Company. As far as personally knowing them, I didn't. I didn't personally know them. I may have run into them from time to time because we worked together during our training to deploy. But personally, no, I didn't know them.

Janney: Were you with Jesse's patrol when he got hit?

Santiago: No, I was not. I was actually doing my rest time at Combat Outpost. My 24 hours off before a mission when I heard the big IED explosion, them rolling in, and I heard Jesse screaming as the Humvee rolled in. Yeah, actually, as soon as that thing calmed down, my team geared up and we were going to go get some. We were going to continue our mission and I swear I was so pissed, I just wanted to get some payback for Jesse. I could see the blood all over his camis. His hand was all wrapped up and I could hear him screaming. You know, Jesse was one of my good friends and I just wanted some payback. I got on behind that gun and as soon as the sun started coming up 'til almost when the sun was going down, I was pretty much on that rifle. You know, on my sniper rifle trying to wait and see if I could just send a 30 caliber round into some insurgent's noggin.

Janney: Did you know Tommy Parker?

Santiago: Yeah, yeah, he was on my team.

Janney: You and Jesse were the only guys that I've talked to that really knew him. Tell me a little bit about Tommy. What was he like?

Santiago: So, Tommy was a good dude, man. He's definitely a highly religious dude from what I can remember. I don't remember where he grew up. I knew he had a wife, a newborn baby at the time. He was actually in another sniper team. I think he was Sal's, Sgt. Lopez's sniper team. He ended up coming into my team when I lost Stayskal. Stayskal ended up being evacuated. Parker was the guy that replaced Stayskal, so he was in my team. He was a good dude from what I could tell. I mean, I remember one thing about him. He had that Miata. This was at the time when The Fast and the Furious movies were little rice burner cars were really popular. He had that Mazda Miata that he tricked out. Everything was stripped on it. I think it only had one seat to lighten it up and make it really fast.

Janney: Did he have nitrous on it? Was he street racing it?

Santiago: I don't know if he was street racing with it but definitely had that exhaust that everybody else had. It was maybe, totally 50 horsepower, you know. But yeah, he had that Mazda Miata. For our training before our combat deployment, again, he was attached to another team. He was part of another team, so my involvement with him was really very little. I wasn't as close to him as let's say Stayskal, (redacted), and Cameron just because those guys were on my team. Every time we would go out and train we would be training together.

Janney: Yeah, I heard he loved that car. You know, Jesse went to see his dad and went to visit his grave and his dad still has that Miata in a garage just the way Tommy left it.

Santiago: Yeah, that's what I heard. I met his wife and then teenage daughter at the last reunion. The ten-year anniversary, I think, of the Battle of Ramadi. They were there, and I got to meet his wife and his daughter.

Janney: That's cool. What about Deshon Otey? Did you know him at all?

Santiago: From what I know, he ended up coming into my team to help me out. Because at the time, since we were down bodies, they came up with this great idea to use infantry platoons or certain infantry guys to augment sniper teams for security. So, I knew of Otey. I heard of Otey. He had a great reputation with Echo Company. So, I didn't necessarily request him to come in and help us out, but he was one of the guys that was slotted, and I was super stoked about that because he had that good reputation. I had heard about the gunfight that he had already been in on Route Gypsum.

Janney: Yeah, he was the only guy that survived out of that first vehicle that got hit.

Santiago: Yeah.

Janney: And ironically, of course, he got killed on that rooftop with Parker, Lopez, and Contreras. What about Contreras or Lopez? Did you know those guys?

Santiago: Contreras, I kind of knew him a little bit. I never really hung out with him, but I just knew him from working with Echo Company. I remember that he was a pretty wild dude. I liked the fact that he was what you would consider a rambunctious Marine. Like a real Marine. I remember he used to carry throwing knives. I'm like, "Dude, throwing knives? What are you going to do? Are you going to throw those at the bad guys?" I knew Lopez. I didn't know him personally. I didn't know much about him. As far as working with him, I never really worked with him. But from what his squad leader and squad mates told me about him, he was really good. He was a reliable kind of dude. So, these guys ended up being a part of my team. I remember I told you, initially we were only a four-man team. But since they came up with this plan to augment us, these guys became a part of my team. So, at the time I was running an eight-man team. So, it was four infantry dudes and four snipers. Then we would divide them up. So, I would split my original sniper team up to Cameron and (redacted) with two infantry guys and then two infantry guys with me to flip flop missions. So, we were running constant missions. So, my team, Headhunter 2 Alpha would go out on a mission while Headhunter 2 Bravo would be at rest and planning their next mission. Once we would come in, they would go out. So, in theory we would always have a sniper team out. I don't know if you really want to get into the nitty gritty. I don't really want to get into the nitty gritty of how those guys got hit on top of the rooftop. I could just give you a quick synopsis of that.

Janney: I would like to hear what you know about it because I actually read the NCIS report. You know, Ben Musser was part of the other four-man team that was sitting on that house.

Santiago: So, without really getting into the nitty gritty of it, I can only tell you my part of it. I read that NCIS report as well, so you have a pretty good idea of that side of the story of what happened on that rooftop and how they got killed. But I can tell you why they ended up on that rooftop in the first place, if that makes sense. Remember I told you that we were being spread thin and we had to cover a large area of operations because Echo Company's area of operations was pretty big. At the time we were still tasked with the IED missions. During this time, we had these mortar men launching the mortars towards Combat Outpost. So, we get tasked with finding out where they're launching those mortars from and to find them. Essentially, it's a hunting mission. We were gonna hunt them down and kill them. So, from what we knew these guys were launching these mortars or rockets pretty deep in the Sofia district and in the badlands where we know a lot of the bad guys were at and where we were potentially were going to go was a pretty bad spot. Of course, we still have one IED mission, so at the time I decided to split my team up to accomplish two different missions. I left Parker in charge with Otey, Contreras and Lopez to take care of the IED mission. And at the time, we already had pre-established houses that we were going to and one of the houses that we were going to watch Route Michigan from was only about 700-800 yards away from Combat Outpost. So, relatively safe, correct? Their whole mission was to get inside this abandoned house and observe the road and make sure that nobody is planting IEDs while me, Cameron, (redacted) and Gonzales who was from Echo Company, one of the other infantry guys, go out and go on this hunting mission to go find these mortar men. We went out one day, pretty uneventful, and then we reset and were going to go out the next day. So, they go out, we go out. I think they left before us to establish their position and then we went out. They go out to the main gate – the northern gate, and we go out the southern gate and go towards Sofia district, and nothing is going on. We're all communicating and then sometime during the morning, Headhunter 2 Bravo lost communication. We didn't have communication with them. We kept trying to reach them on the radio. Command Center, the COC, was trying to get a hold of them, too. At the time, standard operating procedure is if you miss one radio check you have another hour to establish another radio check. If you still can't establish communication with anybody on the two-hour mark, you're supposed to scrap the mission and go home. So, at the 2-hour mark with still no communication with Headhunter 2 Bravo, the decision was made that they were going to send a QRF to check up on Parker's team. At the COC, it's marked; they knew they were there. They go up in that position and we get the radio transmission that you never want to hear as a team leader. We have casualties on this rooftop and then we hear the news that they are all dead. They were all shot on this rooftop. So, that's pretty much how that goes down.

Janney: Yeah, that was a terrible day, too.

Santiago: Yeah, and you know, we get that word over the radio and then, of course, my team goes to 100% and we get in a little mini gunfight where we were at. Somebody, a sniper, shot at us through one of the windows. Yeah, it was a terrible day.

Janney: Jesse said that he was carrying a day/night scope. Did you use the same type of scope or did you use something different?

Santiago: Yeah, the same day/night scope.

6 April 2004 – Route Nova Firefight and QRF

Headhunter II Sniper Team

Cpl Anonymous Scout Sniper (AM) - name withheld by request
Interview conducted via Facebook Messenger

Janney: When and why did you enlist in the USMC? As we discussed previously, your identity will be held strictly confidential.

AM: I enlisted in the delayed entry program early 2001 and went to boot camp in June. I'm not sure I can give you the exact reason. I believe in service to others and have dedicated my life and career following that belief in one way or another since enlisting. I suppose at the time, being 18, I figured that was the best way for me to serve. I had tried university and it wasn't for me. I enjoyed physical activity and a good rush, so the path seemed to line up. Might have been some subconscious rebellion. My parents were supportive but more supportive of humanitarian work. Dad working in genocide prevention and mom in international maternal health.

Janney: Once you finished basic, what was your "job"? Were you Infantry?

AM: I was 0311. Infantry. I was initially supposed to go to Intel, but apparently my minor use of weed was an issue for the background check. Although maybe the recruiter was trying to hit a quota.

Janney: Were you always assigned to 2/4 or did that come later?

AM: I did boot camp and school of infantry on the east coast, but I was sent to 2/4 after SOI.

Janney: So, you deployed with 2/4 to Okinawa in 2003 and then with them to Iraq in Feb 2004?

AM: Yes.

Janney: Can you tell me your rank when you deployed to Iraq?

AM: Corporal.

Janney: When you arrived in Ramadi, what Company were you in and where were you stationed?

AM: I was with the Scout Snipers. So technically with HQ. I was briefly with HQ at Hurricane Point. But I was tasked out to support Echo Company at Combat Outpost. Echo was my original Company before going to SS.

Janney: Tell me your impression of the security situation in Ramadi when you first arrived.

AM: In what way? Could you clarify?

Janney: What was your average day like? What were you doing most days? I know some of the Echo Marines were patrolling and it was more of a "hearts and minds" situation at first. In other words, I know what they were generally doing, but with you being a Scout Sniper, I don't know if you were doing over watch, patrolling with them, or what your responsibilities were at that time.

AM: Yeah. The original notion was the "hearts and minds" thing. I think maybe the more official term would be stability and security operations (SASO). I initially started out going on short patrols and set up observation points with Army snipers to familiarize ourselves with the area of operation. Obviously I can't speak for the situation before arrival but it seemed pretty calm. And again obviously it picked up. I provided over watch during the "cordon and knocks." We set up observation points along the main convoy routes to try to spot planting of IEDs. We did a few security details/overwatch missions for the human interrogation teams. Sometimes we would set up to try to catch insurgents firing mortars at the camp.

Janney: Yes, I heard that at first, Combat Outpost would get mortared around supper some nights, but initially it was mainly indirect fire and some pop shots being taken. When did you notice the change from this type of conflict to more direct engagement?

AM: From what I remember it was after April 4 that things really picked up. It also blends together for me at this point. I remember helping the infantry guys fill sandbags to protect the porta shitters from mortars. So that was an issue apparently. But from what I remember, it wasn't until April that things picked up beyond the mortars and maybe occasional rockets (which might have been in April). I know a rocket went through a wall and ruined a poker game or something. Luckily it didn't explode. I actually had an oath with a friend that if I died shitting in there, he had to wipe my ass, pull my pants up, shoot me, and drag me into the fight so that it didn't get out that I died in the shitter. It went both ways, of course. I would have done it for him, too.

Janney: Trying not to laugh at this point! That's a great story! Were you involved in the 25 March action in which Echo engaged the mortar team?

AM: I have a bunch of dumb stories.

Janney: I'm just as interested in those (if not more so) than any combat details, but we do have to get through to 10 April at least. Share at will.

AM: No. I didn't patrol with the infantry beyond the cordon and knocks. Not that I remember at least. I remember hearing about it. I've heard some funny stories about it but I don't want to put them out there since I wasn't there.

Janney: What were you doing on 6 April? Tell me about your day.

AM: We were on a 2–3-day op. I can't remember exactly. I forgot the road name. Gypsum maybe. Wherever the "tank graveyard" was (actually Route Nova.)

Janney: Nova. So, were you in the sniper team that was engaged near the tank graveyard?

AM: Insurgents had been planting IEDs over there. We (4-man sniper team) spent the night on a couple of pump shacks by the Euphrates River.

Janney: I've heard second hand that you guys were engaged with the enemy, but tell me what really occurred?

AM: How many details do you want?

Janney: Whatever you want to share.

AM: Before daylight, we pushed down by the river for concealment. A squad on patrol went by at some point in the morning. And later we heard that one of the other companies (Fox or Golf) was under engagement in the city. I remember seeing helicopters flying overhead to support. One of us commented that we never get any action. Shortly after, a kid came down by the river. We assume just playing around. I was watching our six down by the river bank so I don't think he saw me. Possibly he only saw 2 of the team. We weren't going to kill the kid or anything, so we decided we needed to move positions.

Janney: This was around dawn? Then what happened?

AM: While we prepared to move, one of the guys moved up to a berm about 30-40 meters away (the road on the berm). When he got to the berm he looked over and saw (and I think he saw the kid pointing where we were) an adult male guiding others to get online to push us. He stopped counting around 15, I think. We estimated around a platoon sized element. They were maybe another 50-100 meters away from the berm.

Janney: So, your back was to the river? In other words, you were just below Route Nova?

AM: Beyond dawn. I'm guessing the patrol went by in the morning. Around 7 or 8. But that's a complete guess and I don't remember what squad that was. I can quickly Google maps for you if you want.

Janney: Sure. Whatever info you can offer, as what happened to your team is a huge missing piece of the day for me. The patrol was headed west on Nova, away from Gypsum?

AM: Eh. I thought I could. But I can't really find the location and I don't want to give bad info. Do you have a map of it? I mean, I know the approximate location. They would have been heading east. The squad. Which was Gypsum? I forget.

Janney: Yes, I do – it's okay. Also, refresh my memory – how far is the tank graveyard (your approximate position) from the Nova/Gypsum intersection?

AM: If you can show me Gypsum on the map it might help me. Hahaha.

Janney: Gypsum is the north-south road through the marketplace that intersects with Nova at the Euphrates River on the north end and MSR Michigan on the south end. Gypsum intersects MSR Michigan near the arches on the east side of Ramadi.

AM: We had our backs to the river. SOP was to run. But for various reasons, that wasn't an option. We couldn't really push east or west because at least back then it was pretty open terrain.

Janney: I understand. In any case, what happened after the kid directed the insurgents toward your position? How many people were in your team?

AM: They pushed. AK 47s. RPKs. And RPGs. We moved behind the pump houses. 4 in the team. You can see some on the map. They are close together. We each took a corner. And waited hoping they didn't actually know where we were.

Janney: Guessing that 2 of you had rifles and your spotters had M4's?

AM: Prior to that we reported it and we were told we know our ROEs. Engage. Not sure they understood our situation. But whatever. It started with an RPG that hit the trees above us. From there we engaged trying to conserve ammo. We mostly worked on preventing them from getting over to our side of the berm. They slowly tried to envelop our flanks. But we prevented them from crossing the berm.

AM: We had only one M40 for that mission. Three M16A2. One with the grenade launcher. M4? Come on man. We were Marines. That shit is expensive.

Janney: Haha. Sorry, I should have known. Approximately what time was first contact?

AM: Prior to the engagement, we removed the bolt from the rifle. To toss in the river if we were overrun. And we would have broken the rifle stock and tossed a grenade in the radio bag. Destroy the rifle so nobody could use it against others even if we didn't make it out.

Janney: That's pretty grim. Obviously, you were concerned that you might not make it out.

AM: Shit. I have no idea at this point. You could probably match up timelines. I don't really think about this stuff. It's part of who I am. And I talk when I need to because it can help. But it's part of my past and I don't let it dictate my future. So, some of this is lost to me. Late morning maybe? Not sure we thought that. Just doing the right thing. Be prepared. We weren't going to use the rifle at that range and in that situation. Eventually, we started getting lower on ammo. We heard that reinforcements were on the way. I think some of the insurgents pushed off to engage them. Not sure the squad that came to help out was prepared for as many as there were. We didn't give the most accurate number probably. Didn't have time.

Janney: So, you knew that a QRF was coming to help you or did you call for it? I don't know SOP.

AM: We decided to just rush them. So, we pushed the berm. We pushed them back and they ran into the other squad. We requested it (QRF.) But all hell had broken loose. So, there were delays. I don't think this was QRF. It might have been the squad on patrol. QRF was weapons platoon that day, maybe.

Janney: Was that an Echo Company squad that was on patrol?

AM: Actually, I think it was the squad on patrol. They found an IED down the road and were waiting for EOD when they heard the fight. Friend said it sounded like a Company engagement. They left some to watch the bomb and some came to help us. Couple guys were dropped off at the wrong place and had to run to us. Yeah. The ones that walked by in the morning because I remember talking to some while they walked by.

Janney: After you rushed them and pushed them back, what happened next?

AM: When we pushed them back, that's when some of the squad from Echo were killed. They got overrun from what I heard. There was an insurgent behind an old tank. He shot one of our snipers. We maneuvered to and killed the insurgent. At some point during the maneuver, I was taking a knee and something hit my back. Felt for blood and didn't feel anything and assumed it was nothing. Later someone asked if I was ok. I said, "Yeah. Why?" "Because you have a hole in your shoulder and you're bleeding." Anyway, insurgent down. I stayed with our team member to start first aid. The other two guys moved on to help the squad.

Janney: From what I know, the insurgents planned an ambush at the Nova/Gypsum intersection because they had a DShK 12.7 mm machine gun trained on that intersection and guys on both sides of Gypsum waiting. They hit the lead Humvee before it could turn left (west) on Nova and killed everyone in that vehicle except DeShawn Otey.

AM: While I was trying to stop the bleed, we got surrounded again. Held them off until 2 Echo guys found me. I sent them to get a Corpsmen and when they got back with the Doc, we held off some insurgents until QRF finally made it and we got the other guy out. On the radio I was essentially told that it was and they were trying to get to us. Took a while. But our guy survived. So maybe it was a different group. I don't know. I know they ran into something. We found imprints in the tall grass. They were waiting in L shape ambush formations.

Janney: I had no idea you were wounded, too. So glad you both made it. What happened next? You said you linked up with a QRF or who?

AM: It was a makeshift QRF, I think. Everyone was out doing something.

Janney: With yourself and another Marine wounded, did you go to Combat Outpost then? Was that Doc Clayton or who?

AM: After I left the guy with them, I grabbed ammo and some Army guys (part of the makeshift QRF.) Some higher ranking enlisted and some young kids. I actually started running off and they asked where I was going. As if I was going to fuck off. I said, "Going to find my team." They said they will come with me. Cleared a few buildings and at some point, the Army guys disappeared. No idea where. But I found some Echo guys and the rest of my team. I think it was Doc Menares. From there, we pushed following some tanks. Our sniper team pushed ahead of the tanks to engage. And I remember being told we needed to get back to help the squad that had been waiting for EOD. Roy Thomas was one of those guys. He can tell you more.

Janney: Do you remember the names of any other of the Echo guys you linked up with? I know Roy very well, but he doesn't want to do an interview. What was your approximate location?

AM: By the time we got there it was mostly over. Roy asked me to take a pic of him (head wound from a grenade) because he wanted to know if he was still sexy. I took the pic and said, "No, but you never were."

Janney: Laughs. Is this when he was wounded?

AM: Yeah. Brian Telinda was there.

Janney: Where was this? Gypsum or closer to your original position on Nova? Were the IED and Marines waiting for EOD west or east of your original position?

AM: Don't really know what happened after that. I remember waiting on the side of the road providing security. Pedro Contreras was one of the guys that came to help me when I was doing first aid. He and I saw some guy with an old war belt poke his head around a corner and then run. So, we chased him down. He went into a house and we kicked in the door. There were a bunch of bodies inside. The people claimed they were just innocent bystanders that were hit. But they had mag pouches and old military belts on. So, it was obviously bullshit. Not sure. I just got in the Humvee and let them take me there. Don't know, sorry. After that at some point we went back to Combat Outpost. Apparently, it had been reported that 3 of us were dead and I was one of them. So, people were surprised to see me.

Janney: From what I understand, the IED discovery and the attack on your team basically kicked off the action that day. It's a miracle that you guys survived that.

AM: I guess maybe we were the first hit from the outpost. But shit was definitely going down inside the city before that. But that would have been Golf or Fox Company

Janney: Did the action slow down after the 6th through Operation Bug Hunt on the 10th?

AM: Sporadic shit from what I remember. I don't remember personally being involved in anything until the 10th. I remember engaging insurgents from outside the walls of the outpost when they were being attacked. But I don't remember when that was.

Janney: What do you remember about the 10th? Tell me about your day.

AM: I'm starting to forget that one. We had a 4-man team again. Working over watch.

Janney: Was your OP the house that the 2 Iraqi brothers were working on that was used 24/7 for days on end, or was it another location?

AM: Yeah. It was all fucked. Nowhere to go. CO demanded we cover the road. Somehow thought we could see 5 km or some stupid shit in each direction.

Janney: I think I know why he chose that house (projection of force), but it seems like that wasn't the way Scout Snipers were typically deployed.

AM: I was the TL (team leader) for Headhunter 2B. I went with Alpha because Alpha was going further out. And wanted people they could trust for sure. Based on how long it took QRF last time. For sure not how we should be deployed. I can't speak for what happened the day those guys were killed. Wish I knew more. April 10. Was that a cordon and knock?

Janney: Yes, sir. That was the day that everyone made a big push to engage whatever insurgents were in the city.

AM: Did it start before sunrise with a raid on a building? With psyops playing music. And mortar fire in the fields to create some chaos.

Janney: I believe so, yes. I hadn't heard about the mortar fire, but one of your terps called the insurgents "out to play" over a mosque loudspeaker. Payback for the 6th.

AM: One day, psyops was playing music. Talking shit to the insurgents. At one point, they played, "Who Let the Dogs Out." I was on a roof and I looked at the other guy and said, "If they don't try to kill us over this shit, they aren't coming at all." They attacked shortly after...

I don't blame them. That song is horrible.

Janney: Yes, it is. Was Ben Musser one of the guys on your SS team?

AM: No.

Janney: I thought he was one of the four guys rotating out with Parker, Lopez, Contreras, and Otey on the house OP. Maybe I've got the names mixed up.

AM: He might have been a supplement. Because they mandated long term OPs. They sent infantry with us to provide more security when we did those OPs. Oh, sure. He could have been one of the infantry guys that did it. Parker was the only SS guy.

Janney: Generally speaking, what was happening after the 10th? I understand that the insurgents never really tried to stand and fight after that.

AM: April 10 I know we were on over watch. At some point when shit hit the fan, we ended up in a building getting surrounded again. Apparently, that was our thing. An Army tank was sent to help us and it somehow went passed us and fucked off. There was a grove of palm trees between us and another building with insurgents. Actually, before going into that building, I was on another one. I saw a couple guys hiding behind trees. Apparently, a machine gunner was shooting at them. I didn't have a shot, so we moved down from the building to engage. Ended up playing Rock Paper Scissors to see who got to rush the machine gun. I lost.

Janney: That's pretty deadly stakes...what happened then?

AM: Team was covering me and I remember thinking, "Why is there so much gunfire coming at me from the machine gun?" I turned and one of the guy's guns had jammed and he was smacking the shit out of it. Got it up again and I covered him until he could get to me. Apparently, one of us killed the machine gunner. The infantry flanked, but he was dead when they got there. My little green notebook was in my cargo pocket. Had some nice bullet holes in the book after that one.

Janney: Damn, son. Apparently, you have 9 lives.

AM: After moving into the building I mentioned before, we engaged some insurgents in another building while a squad pushed up. They threw grenades on the roof to chase them into our line of sight. We got one or two when they ran from the grenades. When we moved to meet up with the squad, apparently Captain Royer called an attack helicopter in on us. He thought we were insurgents. We hid in a ditch until they stopped. When we looked around there was red smoke signifying friendly fire. Guys had popped smoke everywhere. They all knew, but Royer was a moron.

When we got to the building, the squad had cleared it. One of the guys was shot in the legs, I think. Can't remember who. It's a blur from there. Running around engaging. Ran into gunfire at some point and two of us hid in an Iraqi outhouse. Shit everywhere. Flies landing on our faces. We decided fuck this, better to get shot and ran back through the gunfire. Jumping back, I remember running by Marcus Cherry's body on April 6. Good kid. I liked him a lot. But that means his squad was in the area.

Janney: Speaking of Marcus Cherry, I'd like to ask you to share your remembrances of those guys – any funny stories you remember or just telling me what the guys were like that you knew.

AM: Side note: Not sure what the format of your book will be. If you're just recounting the events. If you could just tell the story without any names on our end. I'm doing this to help out and because the money is going to a good cause. But, I've spent most of my life not trying to be recognized and I really don't want anyone to know what I've done. None of their business. I want the story as accurate as I can provide, but recognition should be for those more deserving.

Janney: Yes, sir. I'll do exactly that.

AM: My best friend was Bum Rok Lee. He used to draw up apartment designs for where we would live someday as roommates. Much PTSD comes from survivor's guilt. I've learned to cope, but I moved the team one day and an IED was planted right where I would have been had I not moved. I know I did the right thing, but I still feel shit about it. Bum and I used to set up LAN cables running throughout the ship berthing areas and had little counter strike tournaments on the ship. We also never really took pics together. I have maybe one or two of him. We started taking some on a disposable camera but it was destroyed in the explosion. I think I met him because we were one of the few Asians in 2/4. He was a funny guy. I remember one night driving he must have messed up directions at least 5 times. We just keep making U turns at the same intersections looking for a restaurant.

Janney: That's funny. You said he wanted to be an architect one day?

AM: At Bridgeport Mountain Warfare, I remember talking to him. I was there with the Scout Snipers. His squad moved on. I think we were doing a ropes course. He ran off and suddenly I see him kind of bounding forward back towards me, almost skipping in a really obvious manner expressing, "I forgot something" and the look on his face didn't help. Dumbass forgot his rifle. Probably not that funny a story, but it's stuck in my head.

Janney: Funny, but wow. Hope no NCOs saw that.

AM: Don't know about that. Just him being a fucking weirdo deciding on our living arrangements and closet space. As if my trashy ass has more than 4 pairs of shoes. Deshon and I used to hang out a lot in LA almost every weekend. Me, Deshon, Ian McCall, and Carrol (he didn't deploy with us.)

Janney: Tell me about Deshon. I heard he was quiet until you got to know him.

AM: We spend the weekends in LA with my family. Or hanging out in Crenshaw at block parties. Yeah, for sure. He would definitely open up. We had a lot of good times. Lots of cruising in LA. Drinking. And random shit. After the military, Ian and I were roommates. Ian names his dog Kilo, which was Otey's dog's name at one point if I remember correctly. I've never had the courage to contact his family. I feel bad about it. I keep wanting to. I did visit Bums family once. His mom wanted me to sleep in his bed and cook me his favorite meals. I couldn't do it. That shit fucking sucked. Deshon couldn't run for shit though. Strong motherfucker. But running was not his thing.

Janney: It's never too late if you still want to talk to them. Do it! That would be tough in Lee's case. I can't imagine losing a son. I've interviewed some of the Gold Star parents, but it's all I can do to get through them without breaking down.

AM: He was struggling on a run in Okinawa once. I dropped back to run with him. An NCO came and yelled at us. And we "foolishly" asked permission to speak freely. Of course, it was granted and we proceeded to call out half our seniors for being asses that drop out of runs on a regular basis and then claim that they are just encouraging the stragglers. We got in trouble for that one...

Janney: I could only imagine!

AM: I have a lot of stories with Deshon. He was one of my closest friends as well.

Janney: What other stories can you share with me about your brothers? I'd love to hear whatever you want to share about Deshon or any of the other guys.

AM: Before Reynosa was killed, I remember talking with him about hoping to be home before Christmas because his son was going to be born around then or shortly before. I'd have to get back to you on a lot of them. A lot are of us being stupid and having fun. Not sure I should put them out there. Have to think of some that won't make us sound too ridiculous. Pedro Contreras was hilarious. Let's be honest – not the best-looking guy, but he had a way with women. His personality. Very loud but fun.

Janney: That's tough about Reynosa. I'd really like you to consider writing down whatever stories you want to share and emailing them to me. I can identify (or not) you in the remembrance stories separately from the other things you shared so no one connects the two. Just a thought...

I heard Pedro showed up for a hike wearing a fireman's hat. Were you there for that?

AM: We took him to Crenshaw once for a party. He just got in a random car with some girls. Said something along the lines of, "Where are we going tonight, bitches?" And left with them. Not really sure where he went off to, but he was out with them the whole night from what I heard.

Janney: Someone told me Contreras showed up for a hike wearing a fireman's hat.

AM: Sounds familiar. Something he would do. I didn't really know Parker that well. He wasn't on my team initially. He was always willing to help though. One thing I do remember.

6 April 2004 – Route Nova Firefight and QRF

Echo Patrol Comes To Aid Headhunter II

LCpl Brandon Lund Interview

Janney: Today is 10th April 2022. When and why did you decide to join the Marine Corps?

Lund: Man, I knew I was going to join the Marine Corps probably roughly around middle school. My father was in the military himself. He was in the Air Force. And I wanted to join the military ever since then, even as a little kid. But I always wanted to be a pilot, but my eyesight sucks. So being a pilot was out. And so I wanted to do something exciting. And it was kind of a middle school type thing, where I was like, "Marines, Marines, Marines!" Once I went into high school, it started becoming much more of a reality. I was a garbage person in high school. I didn't do a lot of good things. I wasn't a good student by any means. I have a younger brother, and he looked up to me a lot. It was kind of one of those things that I knew I was going to do anyway. But then I was like, now I have to set a good example for my brother, because he's smart. He's always been a smarter kid than I am. He always had a lot more opportunity, probably as far as college and other things lined up for him. And I was like, I have to set a good example somehow. And so that was the main, final nail in the coffin as far as definitely doing it. And so when I was 17, I talked to my mom, who was, my parents divorced when I was five, but I talked to my mom and they're letting me enlist in a delayed entry program. I talked my dad, who was in the Air Force, fully disapproved of me going into the Marines, into signing my delayed entry program paperwork under the lie that I was going into signals and Intelligence because it would have nothing to do with a combat unit. I told them I was doing that, but I didn't. I signed up for infantry with a guaranteed spot as a, what they call a jazz Marine, security forces with FAST (Fleet Anti-Terrorism Security Team.) I guess that's how that became a whole thing.

Janney: So, what month of the year was that? Do you remember, Brandon?

Lund: I enlisted into the delayed entry program in 2000. I shipped out for boot camp in June, so it was nine days after graduation I graduated. June 11, 2001. I graduated from boot camp on September 7, 2001.

Janney: Okay, and then where did you get stationed after you got out of boot camp?

Lund: After boot camp, obviously I did School of Infantry. There were quite a few guys, actually, that were in boot camp and School of Infantry that ended up in 2/4 right off the bat. And after School of Infantry, we were all, some of us all went to boot camp together and met back up several years later. But after School of Infantry, I had to do Security Forces School for a couple months in Chesapeake, Virginia which is basically nothing too crazy. You get pistol qualification, even as a PFC or a LCpl, which is something the normal infantry does not do because we carry small carbines, and we also carry pistols. Then you also get qualified in non-lethal stuff, too, because you guard things. And so, when you're graduating from security forces school, you do a couple of other things. You can either get a guard post, such as Rota, Spain, or Bangor, Washington, where you're basically guarding ammo dumps, and other things like that. And then you also have FAST, or Fleet Anti-Terrorism Security Team. And I remember, I think I got picked for Rota Spain, which is the most desired post, because it's Spain. And I traded a dude, because you can do that, for FAST because I wanted that. I wanted FAST because they were supposedly high speed. I mean they kind of are.

Janney: Kind of like a MARSOC Light then, basically?

Lund: I wouldn't even go that far. We thought of ourselves as kind of bad asses. We called ourselves the fake ass seal team. Because we trained sometimes on some of the SEAL compounds. We cross trained with some SEALs. We cross trained with some of the police units and things like that. The FBI came to the Chesapeake and did some stuff with us. And we cross trained with the Singapore Special Forces. Other things like that. So, I mean, we did a little bit above and beyond the normal infantry, but we didn't do a lot of the normal pursuit. So, we still would go out and train patrolling, land navigation, and other things like that, because we knew that the post at Bath was only for two years. Out of your four-year tour, you only spent two years there, and then you rotated back out of the fleet unless they chose to keep you. So, we still did a lot of patrolling, but a lot of the things we did above and beyond was urban warfare. We did a lot of heavily advanced urban warfare tactics, tons of CQB tactics, which is SWAT style stuff, urban patrolling. Our class in 2002, I believe it was, was the very first class in the entire Marine Corps to go through advanced urban warfare school.

Janney: Oh, wow. That's a big leg up from a lot of the boot drop guys that came in.

Lund: So I mean, before that there was just CQB School, which was fairly new itself. They would just pick dudes out of FAST platoons, like a couple of dudes here and there, usually like high shooters or things like that and would send them to CQB School. Back then it was very small. Then they'd come back and they would retrain all their tactics they learned at CQB school. Finally, they developed this with a whole new platoon. Now a normal infantry platoon has roughly 33 dudes, usually three squads of three fire teams. A FAST platoon is four squads of three to four fire teams per squad. It's roughly 55 to 60 dudes. The reason why we did that is because we deployed only as a platoon, not as a Company, not as a regiment, not as a battalion. So, it was much more highly mobile. Literally any platoon was on call. We do like what is called Alpha 1. We rotated roughly monthly or so what the platoon was on Alpha 1 React and it was the people in Alpha 1 React. If you're on Alpha-1 React, you are not allowed to get too far away from your compound, your base, because anywhere in the world, you have to be there within 24 hours, period. It was usually something like being activated to secure and evacuate US embassies overseas, which is one of the main things that FAST actually trained to do is evacuate personnel, destroy information. Intelligence and things like that, evacuate personnel, and secure US embassies in less than 24 hours if they came under attack. So that was one of the main focuses. You had to get there, clear it, secure it, evacuate it, all as a platoon inside that one. But, there's obviously way higher tier units out there that do way cooler stuff.

Janney: That's much more additional training over what the 2/4 boot drop guys got. They came out of SOI and basically deployed to Ramadi. You had a big leg up on those guys.

Lund: I mean, they have a high view. I'll tell you a good story about why they thought this. So, after two years, we all started finding out what units we're going to, there's a couple of us, you're told what unit you're going to. And so we're all eager to like, Hey, when we go to our unit, we're going to do some shit. And there's a couple of us that thought we were going to 2/4 and they cannot physically say anything, but our higher demand row is like 2/4. They're getting ready to go to some places. And we're like, for real? And they're like, yeah. They're like, like soon? They're like, yeah. So, we're like, all right. So, we're jazzed because I mean, the big thing is Marines want to go out there and do the shit they've been training for the longest time. When I checked in after a PCS leave at 2/4, that's when I met Sergeant Major Booker for the first time. I looked like a bag of ass because I had to check in and I had to stay in like some sort of check-in facility. I stayed there for two nights straight and I had nothing but my military uniforms and my Alphas. I basically slept on a friggin bench in my Alphas where you're supposed to check in and they look like shit because you slept in them. They don't look good when you check in the next day. And you're freaking out. I had to meet SgtMaj Booker who was just tearing us a new asshole, "More of you FAST guys. You're supposed to be high speed. You guys look like a bag of dicks." He was just ripping us apart. It was kind of a little thing out of our control. We checked in during the middle of the night and they had us sleep on a bench. So, but it was one of those things where you did that. But, I also feel bad because I did have a higher view of us coming in there. Normal fleet units do different things on missions that are different. I was really put into a team leader position which is kind of a shit thing to do because I came in and actually displaced Chris McIntosh as a team leader. It put me in a bad position, especially with McIntosh having been a team leader. I was there just a couple weeks as compared to Mac who had been there for a long time. I only had about 2 weeks to get to know some people and build rapport with my teammates.

Janney: Yeah, that's pretty rough. That is pretty rough being thrown from the frying pan into the fire so to speak. So what rank were you at this point, Brandon?

Lund: I was just a Lance Corporal.

Janney: So, you got thrown in with 2/4, had 2 weeks to get to know everybody and kind of build a team relationship, and then you guys are heading out. So, what happened next?

Lund: So, well, I did meet some good friends there. My biggest friend who attached to me right off the bat was Pedro Contreras. He and I got along right off the bat, and I started hanging out, you know, all the time. And he became like my best friend. And right as he took me in there, and he didn't give a crap where I came from. He didn't give a shit about literally everything. And I instantly clicked and we got on. Contreras didn't give a crap. Pedro said, "I don't care. I've been demoted 4 times. I don't give a shit." One of the greatest things is, again, I only had 2 weeks, so he and I drove up the I-5 at about 120 miles an hour for shits and giggles in his old little black Honda Civic that has these Mickey Mouse gloves in the dash. And shit, he was, you know, I was literally driving up and down the 5 for hours listening to Cannon's/R. Kelly's "Gigolo" over and over because we loved that song. It was just great, just dumb shit like that we did all the time. Again, he was one of those guys that had no fucks given at all. But, he was such a lovable person, that yes, he got demoted or office hours a lot of times, but he was such a lovable person. To me, he was the typical of what Chesty Puller would call a Marine's Marine. Puller said something like, "I don't want the freaking guys who have pressed uniforms. Show me to the brig to meet the true Marines." Contreras would say, "I don't give a shit if I ever get out past PFC again." He doesn't care. But he was a true Marine. That guy made everybody laugh all the time. He taught me how to criss cross in Iraq. That guy, he was literally the life of 2/4, everybody. I love that dude. Now we've spooled up, we're flying over. There were 3 of us - we called ourselves the 3 Amigos. There was Kevin Gaeden, Pedro Contreras and me. We were pretty much inseparable at all times. Contreras was older than I am, and I'm older than Kevin Gaeden who was like the baby brother. We got in trouble for all sorts of shit. We got in trouble for putting cat scratches in our eyebrows - you're not supposed to do that shit, but we had 3 cat scratches in our eyebrows for each other. Every picture we took together, we were throwing up what looked like a gang sign, but it was just 3 with our hands. Everywhere we went, we got pictures of us together like that. We were pretty much inseparable. Kevin Gaeden and I, even though we lost Contreras, hee and I were still like big brother - baby brother. All the way up until I got out.

Janney: Oh man, that's great. Are you guys still in touch?

Lund: Yeah, he and I were like we were all inseparable.

Janney: So how did you guys get to Ramadi? Did you fly in or convoy in?

Lund: OK, so there were 2 different elements that went in. There was one small element with a lot of our equipment. Then the main force stayed in Camp Victory in Kuwait and then we pushed across and flew. I can't remember the base we flew into – TQ, I think, and met up with our main element and then we convoyed with all of our shit into Ramadi. It was a huge convoy. I remember us driving in and these kids were flipping us off and showing us the bottom of their feet (author note: Most Muslims consider the soles of feet/shoes unclean, so showing the bottom of their feet/shoes to someone is an insult.) Some were waving, some were giving us the evil eye. There were just lots of people watching us roll in. When we left Camp Victory, we convoyed on jingle trucks, buses that had the little jingle bells in the front, that were driven to the airport. They had given us one magazine per person (author note: 30 rounds of ammo per magazine) and that's it. Obviously, we had to leave our ammo to get on the plane, flew into TQ (Al Taqqadam), and then we were reissued our ammo and thrown in the back of a 7-ton trucks that were not armored at all. Nobody took it too seriously at first because we were rolling in the middle of the day. It wasn't until nighttime when we stopped in Ramadi that the Iraqis wanted to test us. I remember we were all sitting in the first platoon and we were going over SOPs, then all of a sudden, we got hit. You could hear the 50 cal machine gun over the top of where the first platoon was just rockin' and rollin'. They were hitting our compound with rockets and other things to let us know that, "We saw you and we're gonna flex on you." I don't remember who said it, but I remember it specifically being said, "Shut the fuck up, you hear that shit? Shit is goddamn real." It kind of put a little bit of fear in me.

That night, we took the 82nd Airborne leftsy-rightsy (author note: orientation patrols) on a foot patrol with all the squad leaders and team leaders of the first platoon. I was a team leader, so this was our first patrol out. It was after the curfew, so technically after curfew we'd only see police because nobody else is supposed to be out there. It was one of the creepiest moments of my entire life because the entire time, the hair on the back of your neck is standing right up. You have the feeling you're being watched the whole time through a set of cross hairs. But it's nighttime and we're patrolling around. I don't remember the route we took or anything like that because it was the first night I'd been there. They just took us out on a short foot patrol, no more than three clicks in a short circle to kind of get a feel of the immediate area outside of Combat Outpost. We did see people out quite a bit, just walking around when they shouldn't be. It was just an eerie situation where the city is dead quiet, except for a few things like dogs barking, the hum of those fluorescent lights, a few cars on the road, and police here and there.

Janney: Do you remember what day that was?

Lund: I don't know what day that was. It was the same day we arrived in February. So that patrol, although unnerving, went okay. It was completely uneventful. But this is real. This is no longer training. You feel it deep down in your chest. This is a deep down in your gut freaking nervousness. The first time we actually got hit with anything, there were two of us there when we shot the first person and the first time we almost got hit was the first week we were there. We were out on a patrol at night and I believe it was Brian Telinda and I, we were there and got past our compound. There's a cloverleaf area that was kind of right outside the compound. It kind of splits and it would go towards Fox Company up north, the actual heart of Ramadi, and then the others split into the more rural area, like Route Nova. We were kind of roughly around there, and we saw a convoluted line going across this small field. It was going into a drain pipe, or an irrigation drain thing in the street, and we said, "That's not normal." When we followed it and looked back, there was a dude staring at us, just looking. But we saw him first and we said, "Oh shit, he's a cop." He saw us and just took off. We called EOD, and sure enough, he had dropped an IED into that little irrigation pipe and he was seconds away from completing the command wires on a car battery to blow us up. When EOD arrived, Telinda and I went around the corner to be safe and heard, "Fire in the hole, fire in the hole, fire in the hole." The whole alleyway that we were in was lit up from the controlled detonation. We said, "Oh shit. Whoa."

We were patrolling going through a weird irrigation field that was very lumpy that paralleled Route Michigan towards the arches. I don't know how the gunfight started, but can tell you what happened at the end. We were patrolling in two different squads. My squad was patrolling across the field while the other squad was more on the road. A white sedan pulled up and was commanded to stop, but didn't. Nate Apple started lighting up the windshield. I remember running towards contact with our squad through that field and tripped, did a faceplant, got up and started running again. By the time we got to the gunfight, it was literally over. It was super-fast. Nate Apple shot a dude in the thigh. We made fun of him because that's the first dude he ever shot – didn't kill him, just hit him in the thigh. When I got there, I looked at my muzzle device. The flash hider and the birdcage was full of mud. So, I would have been pretty useless if I'd had to fire. Benjamin Carmen was laughing at me hard and said, "Dude, I saw you fucking disappear in a dirt cloud, and then you were back up. You did the craziest combat roll." I said, "I don't remember any of that shit. I just remember falling." So we're just laughing about that shit. Then we were also laughing because I believe that was the same gunfight where Lieutenant Valdez jumped into an irrigation ditch. Lieutenant Valdez was a short guy in the first place and submerged himself in shit water. It was a lot deeper than he expected.

Janney: What do you remember in the subsequent days? It sounds like it was already getting hairy.

Lund: They were testing us. We were supposed to be SASO operations or stability and stabilization operations, winning hearts and minds. But Ramadi quickly turned into an absolute shit storm because they didn't want SASO operations. When we were first going out there, we were trying to talk to people and shake hands. We didn't realize that Ramadi was the birthplace of the Fedayeen. It was ripe to be a hotbed of jihad, and they weren't having any of our SASO. It's not what they wanted in the first place. There were obviously some people who did, but there was a large group of people that didn't want anything to do with us.

Janney: It seems like most people didn't appreciate the up close and personal aspect of USMC foot patrols versus 82nd Airborne's mounted patrols.

Lund: The 82nd Airborne was there and had a different mission set, but we wanted to not only establish a presence as far as SASO operations, but, if necessary, we're not going to back down from a fight. If you shoot at us, we're going to fucking annihilate some stuff. We were trying to build a hospital there, schools, and some other things. But everything quickly devolved into ridiculous gunfights. Also, Fallujah and Ramadi both kind of kicked off at the same time. Fallujah got a lot more press coverage, but it's actually a smaller, denser city, and it had four battalions of Marines and one Army battalion surrounding the city. Ramadi had one Army unit further outside the city and one Battalion of Marines. Ramadi is a larger city, population-wise, and also covers a larger area. Fallujah was just getting way more press coverage even though Ramadi happened then, too.

Janney: You're exactly right. The encirclement of Fallujah by those five units basically caused a lot of the insurgents to squirt out west toward you guys. Ramadi got all the other bad guys that didn't want to stand and fight in Fallujah, so I'm sure that added to it.

Lund: A couple weeks after we got there, we were patrolling Nova. Our XO Lt. Kaler went out with us. Brian Telinda and I were what were called the guardian angels on this patrol, walking way out front. We had both walked right over five IEDs that were daisy-chained together and buried under the road berm. Telinda turned to me and said, "I still can't believe we came and got..." He didn't even finish the sentence when the IEDs blew up. The Humvee was way behind us and I think that the trigger man thought someone had spotted the IED, got nervous and clacked it off early. The XO was hit and lost most of his left arm. The blast blew him off the road. The weird thing is that Marcus Cherry was his radio operator that day, standing five feet away from him and didn't get a scratch. It just blew him into the ditch. Lt. Kaler was a straight up fucking boss, laying in a bush with his arm blown off, calling out orders, "Form a 360 perimeter, find the triggerman" and giving directions to call in his own medevac. We never found the triggerman, but we got the XO out of there. Lt. Kaler survived and later pinned my second Purple Heart on me.

So, April 6th was my 21st birthday. We were going to set out on a foot patrol that traveled Route Nova and then go down Route Gypsum and end up patrolling back towards Combat Outpost. It was going to be a long foot patrol, almost a whole day. We've done it a couple of times before, but usually we go the opposite direction up Gypsum and then back on Route Nova. Either way, the intersection of Gypsum and Nova was known as a hot area. We'd been hit there before in the past and several people were wounded there. It's almost guaranteed when you go to Gypsum and Nova, something would happen. There's another short, narrow road called Route Apple, and nine times out of ten, we'd have contact on it.

So, when we set out on April 6th, we had two squads, and we were reinforced with a couple guys from weapons, some machine gunners. Two squads on foot. We had one Humvee with a mounted 240 machine gun on the back of it with just basically that plate armor on the side. It was open back. We had our platoon's commander, Lieutenant Bell, who was with us. I believe I was in the second squad at that time. I don't remember the other squad. It might have been first or third. But I switched from third squad into second squad by that point and took over in that. So, we set early in the morning around 0600. I had talked to all the dudes and said, "It's my 21st birthday. We're all gonna have a near beer when we get back."

So, we set out on foot patrol, basically going through that cloverleaf area which is about 300 meters outside of the gate. We're doing satellite patrols with the Humvee driving down the road with a fire team size element and then other fire teams are pushed out to the sides, patrolling farther and farther out in kind of a satellite patrol, kind of almost online, but offset online. I was on the far-left side of it and we got a call for a halt because I think it was Eric Akey who found the IED in the road. It was two 107 rocket warheads with a bunch of sheet metal and stuff piled on top. So, we didn't even get 300 meters out of the gates before we found the first IED. We called a halt. We set a large 360-degree perimeter until EOD gets there to detonate the IED. I remember sitting on a berm like a kid, kind of swinging my feet. I thought this was going to be cool, but I forgot my camera. Everything cool happened when I forgot my camera. Then EOD said, "Fire in the hole" and I said, "Cool!" It was a big explosion. We packed up our stuff and continued the foot patrol. In the background, we heard a lot of gunfire, something not out of the norm, but it was a higher volume of fire than we normally heard. It was coming from the arches area, so in the opposite direction we were headed at that point. It also sounded like heavier fire like we were returning fire with Bradley 25-millimeter cannons.

Because we're on Nova, parallel to the Euphrates River, it kind of swoops up and makes a bell shape and then curves back through a rural farm area that has a lot of pump houses and little rural houses. Where Nova meets Gypsum is a little suburb area. Gypsum travels generally north to south back down to Route Michigan. You pop right out after a little bend on Route Michigan and then we're going back over to Combat Outpost – a long patrol that was roughly going to be about 11 or 12 clicks (kilometers.) Right where Nova starts to do the bell curve is the tank graveyard. There was a four-man sniper team (Headhunter II) doing a 24-hour IED over watch there consisting of Rich Stayskal, Romeo Santiago, (redacted), and Cameron Ferguson. We're on our little PIRs (small radios) and back then, only team leaders and squad leaders had PIRs. They had just disseminated that information to your team leaders. We got a radio call, "We're going to come up on Sniper element. Don't look at them. Don't do anything like that. Don't pay attention to them. You'll compromise the position as we were passing through that tank graveyard." So, we ignored them and walked past them.

Uneventful gunfire in the background, still rocking and rolling here and there, sporadic, but nothing's going on with us. We got about to that little farm area and found another IED. We set up a large security perimeter and we're just kind of waiting for EOD and then we heard a lot of gunfire closer. I heard Lt. Valdez on the PIR saying, "Sniper team is under attack. We need to go help them out." My squad and I all piled on the back of this Humvee while leaving the other squad in place waiting for EOD. So, we piled 12 dudes in the back of this frickin Humvee and we just shot down the road, not knowing what we were getting into. I was sitting on the passenger side of the Humvee itself. I don't remember who was on the machine gun at that time. It might have been Martinez on the machine gun at that time. There's a palm tree grove that was pretty dense and then it cleared out for the tank graveyard which was a bunch of old tank hulls, BMPs and Russian tanks. Once we broke into that palm tree grove, we were up on an elevated berm along the river, so we're just a perfect silhouetted target. There was a bunch of insurgents set up in that tank graveyard and also an element trying to assault forward on foot toward the four-man sniper team which they were engaging. The insurgent assault force attacking the snipers was 12 or more men, plus the guys in the tank graveyard.

When we broke into the palm tree grove, we just absorbed all the fire because we're not a fixed target. It was an insane amount of gunfire and I've never experienced anything like that in my life, even on a firing range. The Humvee came to a quick pause. David Quetglas was driving and two Marines dove out of the back of the Humvee. Quetglas decided to drive a little bit further, so the 2 Marines were running behind it. I saw everybody in the back of that Humvee lighting everyone up so that we could possibly shoot at about 50 yards or less away. The entire gunfight happened within 50 to 100 yards of us on the road. It was terrifyingly close and there was a shitload of gunfire. We came to a dead stop.

The element that was trying to flank or push towards the sniper team turned tail and started running back towards their fixed positions. We were just dropping dudes as they were running back. We all wanted to get out of the Humvee and push towards the element. Martinez and I both got hit at the exact same time because they were shooting at the mounted 240 machine gun. He took probably a ricocheted round off of the 240 in the hand, basically disabled his hand. I took a similar round off the 240 in my right hand between the webbing of my thumb and my index finger and also took a round in my left shoulder blade, although I did not know about the shoulder wound at that time. The bullet had slowed down enough that it hit my shoulder blade and kind of popped back out the same way it came back in, just created a good-sized hole.

As we were dismounting, I turned away from all the gunfire, grabbed the side of the armor, and felt this kind of "thunk" and a Charlie horse feeling in my hand. I looked at it and there was a lot of skin sticking straight up in the air and blood was oozing out. All these thoughts came to mind and I remember looking at that hole in my hand, just dumbfounded, because this doesn't happen to me. This happens to other people. The hot sun was coming through and reflecting off of my blood just perfectly and it almost looked like brick red, like an old 1960s movie where they try and make blood, but they really can't get the hue right. A split second later, the thought went through my head that it was real. I took my left hand and just smothered the wound and squeezed it as hard as I possibly could. I yell over my shoulder, "Guys, I think I'm shot!" That's when I hear Quetglas yelling at me, "Lund, are you OK?"

Benjamin Carmen tried to stand up and pop out of the Humvee, and he sat back down instantly, just hard, sitting straight up, but wasn't moving. We didn't know what happened, but obviously it was bad. Doc Tyrone was still in the Humvee. Martinez had kind of fallen on top of me when I was holding my hand which is why I thought I had bumped into something with my shoulder and that's why it hurt. People were still bailing out of the Humvee. I remember Lieutenant Valdez jumped on the 240, and was trying to fire. He got 20 rounds off at most before that thing jammed, and he just abandoned it and bailed out. I believe Martinez was still in the vehicle, lying on the deck, but trying to get out. Carmen, who just got shot, Doc Tyrone, myself, and David Quetglas who at that point was trying to bandage my hand. David whipped my M16 over my head and laid it across his lap, and then took off my helmet to get my medevac number. He had the radio and began calling in a request for a medevac for Carmen, Martinez and me. I was watching Doc Tyrone peeling armor off Carmen trying to find his wound. Doc Tyrone was yelling at him, "Carmen, keep talking to me, keep talking to me," but he couldn't talk. Doc found the entry hole, but he never found the exit hole.

Doc Tyrone, I believe, knew that Carmen was gone at that point. We heard screams from the main assault element, yelling, "Doc, Doc." So, Doc just bailed out and ran headlong into the gunfire where people were calling for him. He got up and he went face first into the gunfight where they were yelling for him. I believe Doc ran to Marcus Cherry, not knowing that Cherry was already gone, having been shot by a sniper with a SVD Dragunov rifle. There was a guy in the tank graveyard that had one, and we believe he's the one that also shot Stayskal and (redacted.) The good news is Santiago killed that guy. I believe that Doc Tyrone, because of his heroic action of running towards insurgent fire to help Cherry, got a combat V (valor) in his Combat Action Ribbon for that.

As soon as he took off, I was sitting there and realized that I was the only one alive in Humvee. Everybody else had bailed. I thought, "I need to get the fuck out of this thing." Because again, it's still drawing large amounts of fire. You can hear all the rounds pinging off of that side of the armor. You can hear rounds snapping over the top, too. So, I quickly popped up, and I just shot out the back of that thing. You can hear bullets zipping by, sounding like bees going Mach 5 past you. Now, this is a weird memory, maybe my brain made it up or maybe I'm the only one who saw this – I'm on top of the berm and see an Iraqi wearing a man dress riding by on his bicycle down the middle of the road. I screamed at him, "Get out of here!" He stops pedaling, puts his feet on the ground, tippy-toes his feet around in the opposite direction, and starts pedaling in the other direction.

There's a couple of people on the back side of the berm and I just run towards them. I slid down the side of the berm on the back side, which is closer to the Euphrates River. On that side, there was Martinez, David Quetglas, and me. I have no helmet and no weapon. I just have my interceptor vest, and that's it. Martinez, he has literally got an M16 with a 203 on it chicken winged under his arm because his other hand's all fucked up. Quetglas has my M16. He doesn't have his SAW because that's left up in the Humvee. Now, we started receiving rounds through an apple orchard. We couldn't see exactly what was going on, but we started receiving some rounds from it.

The main element of insurgents broke off and were trying to perform an L to flank us and the snipers. So, Martinez is over there chicken wing firing an M16, and Quetglas is firing has my M16. Because my right hand is all jacked up and I'm right-handed, all I could do was keep feeding them ammo. I'm scared out of my goddamn mind because I don't have a weapon. I am half-armored, and I can't do anything but just kind of lie on my back on the berm to try and make myself as thin as possible. These guys, one of them is injured and the other one's shooting a weapon system that's not his. But, they were fighting off this flanking element, just Quetglas and Martinez by themselves.

Then David Quetglas did straight-up one of the most heroic things I've ever seen in my life. He realized that he left the radio inside the Humvee and he couldn't finish calling in nine lines and the QRF. David just went for it, running up the side of that berm under fire the entire time, exposing himself to fire to get the radio as he was reaching into the door to open it, grabbed the radio, and then ran back to us. It was incredible. Unreal. When I wrote my after-action reports, I explained his bravery. David Quetglas was awarded a Bronze Star because of his actions that day. Here was this 19-year-old kid from Miami, Florida who ran through a hail of bullets twice to retrieve a radio and return back to us to call for support.

So, after that, the element was still firing pop shots through the orchard. We kind of pushed back a little bit to a dropped road. I tried to push myself into a small ditch because we're still being flanked. Massive amounts of gunfire that I can no longer see at all, but that's when they started firing RPGs. I remember one goes over the top of us and impacted on the other side of the Euphrates and scared the shit out of me because I've never heard one of those things zip over the top of me before. So, we kind of pushed back a little bit further to drop back from the flanking force. I tried again to make myself as small as possible. I don't have a weapon. I can't do anything. I'm still relying on Martinez and Quetglas to fight off this element while I'm throwing them ammo. It was incredibly terrifying because I can't do anything other than keep feeding these guys ammo, and I'm starting to run out of it. I carried a lot of ammo, but they were burning through it. Because again, Martinez was the machine gunner, Quetglas was the SAW gunner. Neither one of them really carried M16 magazines, so they're using all of mine. When I run out of M16 magazines, they run out of ammo.

It's a whole different story of the QRF when Gypsum and Nova happened with the DShK that hit the lead vehicle. That ambush was for us, but they ended up hitting our QRF forces. At this point, our main assault element had actually started overrunning the Nova ambush, and stuff started exploding. They were trying to drop mortars on us to cover their retreat. When stuff started exploding, the Army pulled up from one direction and the Marines pulled up in the other direction. I remember I was never so happy to see the Army in my entire life. They rolled up with APCs and a couple of Bradleys and they just started pulling us in. I just remember running up the side of the berm towards that APC because they're yelling at everybody who's wounded to get in the back of this one APC. I crawled in with my hand all sausaged up and they had Stayskal laying on the ground and were throwing an IV in him. I'm holding Stayskal's IV over my head with my left hand as we drove to the Army hospital. Since it was my birthday, they let me call my mom who also was celebrating her birthday. My mom and I share our birthday. I got voicemail and left a quick message, "Hey, Mom. Happy Birthday. I got shot, but I'm okay." I went to the hospital for my hand injury.

About a month later, Hernandez asked me, "Hey, you want to go back out?" "I guess." So, I started patrolling again. Then, I got hit a couple months later by a mortar on a rooftop on 2nd July and that took me out for good because it blew my left elbow apart and riddled my body with shrapnel. I was on post two with Kevin Gaeden, and he and I both got hit at the same time by a mortar. Kevin was laying on the ground when we received accurate mortar fire which landed in front of the chow hall. I was yelling at people to take cover, and that's when one detonated on the rooftop behind our compound. I remember the explosion and smoke. Kevin was laying on the ground and he looks up at me and he says, "I can't feel my legs." I fucking lost my mind because Kevin's like my baby brother. We were inseparable. I remember just trying to rip his armor off him. I said, "I don't see any holes." All of a sudden, I started seeing some blood. Kevin rolls over, looks at me and says, "Dude, that's you." I didn't feel it, but my left sleeve was soaked in blood

That's when everybody came up on top of the post to start pulling us off. I was about to go down the ladder and said, "Rifle, rifle." Someone went and grabbed it. It just so happens that day that all of our heads were there – our Battalion Commander, Battalion Sergeant Major Booker, and some General was touring that day. As I came down the ladder, Sergeant Major Booker stood right there, and I said, "That's two, Sergeant Major!" SgtMaj Booker said, "You're a fucking stud, Lund." I'm laughing my ass off. They brought us down into the little med bay that we had there. Staff Sergeant Craig came in here and he was trying to keep it light, messing around with me. A real doctor, a Lieutenant Colonel or Colonel, came in and was asking me questions and said, "We got to fly you out of here for surgery." Apparently, I said, "Fuck you, I'm not going anywhere." But, they flew me to Baghdad for surgery and then they flew me out the next day to Landstuhl Army Hospital in Germany. Funny story, I was born in Landstuhl, the same hospital where I was medevaced. I spent my time there trying to make people laugh to lighten the mood, just like my buddy Pedro Contreras.

Contreras was magic, a different breed of human being. He was magic. And not only different breed of human being, brave as shit, too. Every gunfight he ever got into, he would just disappear and then come back hours later with three other AK-47s slung over him. We're like, "Where the fuck were you?" Pedro replied, "I was with SgtMaj Booker and we were kicking in doors by ourselves." I said, "Holy shit dude." He had been with SgtMaj Booker the whole time, who was kicking in doors with a goddamn M14. Like, that guy was badass. Sergeant Major Booker would go out with that goddamn M14. Contreras had a unique talent for making everybody laugh when they were feeling pretty low. But he was also a good Marine. He was a really good shot, too. My heart sank when I heard that Contreras had been killed. He was my best friend.

I went to boot camp with Tommy Parker. Tommy Parker, and I, got his very first tattoo with me in San Diego when we were in the School of Infantry, which were his meat tags on his rib cage. So, I've known Tommy Parker since boot camp. Tommy Parker, great dude. Super funny. Goofy, kind of like me. He had a little bit of a Southern accent. I didn't know Lopez and Otey as well, but Lopez was in my platoon and he was a quiet, funny guy. Loved him. He's just a straight-shooting guy with a dry sense of humor. He always made fun of me because I was loud and goofy. He told me, "Dude, you need to calm down, bro." He was such a nice human being, just a great dude.

6 April 2004 – Route Nova Firefight and QRF

Headhunter II Sniper Team

Cpl Ramon Barron

Janney: This is 29 April 2022. So, you told me that you did early enlistment and then graduated from high school in May of 2001. You graduated from boot camp on September the 7th, 2001. You said that you had wanted to enlist as a way of paying back since your family had immigrated here, and that was something that you felt like you wanted to do. What was your recollection of what your mom told you on September 11, 2001 when the planes hit the World Trade Center towers?

Barron: She came into the room and said, "Hey, wake up. Something horrible is happening. Come watch the news. That's when I saw the second plane coming in. I said, "We're being attacked. We're going to war." Mom said, "Oh my God, I can't believe it." When I got back from leave, the training was much more real because you were training for war. Just knowing that we had just been attacked heightened the instructors and the students. My MOS was 0311 Infantry and I knew we were going overseas to fight.

When we deployed in February 2004, we'd been told that we were going to go to Fallujah because that's where all the craziness was happening, but then at the last minute they told us we were going to Ramadi. They said, "It's really quiet there. Not too much is going on." I think at one point, everybody was kind of bummed out because this is what we joined for, this is what we trained for, grunts, infantry. It turned out it was totally opposite. Fallujah was rough, but I think nobody knew the gravity of Ramadi. We didn't know that a lot of Saddam's regime had retired there, that generals and lieutenants and officers in his army were there. Special Forces guys that served under Saddam retired there. We just thought we were going to go in there and maintain the peace. But we had no idea what we were in store for.

When I deployed, I was a Corporal in Echo Company and we were at Combat Outpost the whole time. My first patrol there was with the Army, leftsy-rightsy, so we would ride with the Army and they were showing us the way they did business, the dangerous areas, the hot areas, and the quote unquote "friendlies." They would approach people and just kind of check them, doing field interviews with them. We had a curfew in place or were about to put one in place, and the people couldn't be out after a certain time. We started doing patrols 24 hours a day.

Janney: Were the people generally friendly or were they kind of standoffish?

Barron: We would get both. Some people would smile and wave, others would just turn their backs. We would get people that would point to the bottom of their shoe which was a form of being disrespectful.

They would wave with the left hand or flip us off. At the same time, we had people that would smile, wave, and then throw a grenade. It was area specific, too. Gypsum and Nova were really weird, man. We would come into some areas and it would just be cleared out. Like they were told, the Americans are coming, close everything. You wouldn't even see kids or dogs out there. We would be out on some patrols where the kids are playing and would continue to play. But then, you looked up into the second story windows and people on the phones looking at us. Then they would close the blinds or close the doors. If nobody was around that would be when we would get attacked.

If I'm not mistaken, the first actual contact was after the Army took off. I was a squad leader, and Nathan Apple was in a different squad. We started taking pop shots. I think Nate was the first one who actually shot somebody. I don't know if he hit him in the leg or in the torso, but it got real very fast. Nathan and I went to boot camp together, were at SOI together, and went to Japan together. We were in the same platoon in Echo Company. He's my brother. We were together all the time. My buddy just shot somebody, so it became very real to me. I was thinking that people in the Bible walked these streets, still speak the same language, they still dress the same, the same river's here. It was just very surreal being there.

We would get mortared all the time. Every Wednesday for sure, during a certain time, that was their thing. I think that the mortar system that the bad guys had set up was on a vehicle, a truck. They were just trying different areas to mortar us from because eventually it would hit and they would know exactly the distance. It was very difficult to pinpoint because they were on a truck, it wasn't fixed. They would launch a couple mortars and then take off. It was very difficult to track them, but I think eventually the snipers were able to kind of do a reverse azimuth or back azimuth and pinpoint where they were at, and I think they ended up getting them but it was a while before they got them.

My squad was really active, doing round-the-clock missions, whether it was sitting in an LPOP (listening post observation post) to see if they're putting in IEDs in the road, and IED sweeps. We would do raids on different houses to look for bomb makers, weapons caches, or bombs. We did a variety of missions, a few ambushes, and a lot of different types of missions. It seemed like any time that we would go out, my squad specifically, would receive indirect fire followed by either a drive-by, mortars, or we would find an IED. We would set up the perimeter, call in EOD for them to come and blow the IEDs because we couldn't leave. We would set up this big circle around the IED and just were sitting ducks. SOP was to set up a perimeter, sweep for secondary IDs, and wait for EOD, but there was only one EOD team for a whole entire area. Sometimes we're sitting there for two, three hours, just waiting for EOD to come in.

If an IED blew before EOD came, it was almost always followed by a small arms attack. They'd set off the IED and when chaos happened, then small arms fire would follow and we'd start getting attacked. It was crazy because then we started getting machine gun teams attached to us because they knew our own squad was always getting contacts. Around the end of March, I remember our XO, Lieutenant Kaler, had been inside the wire a lot, and he kept asking the CO, "Hey, let me go out. I want to go out. I want to go on patrol." The CO said, "You're going to go out with Barron's squad." I remember the morning he was going to go with us. I did the brief and I said, "Sir, you know we're going to get hit. Right?" I started laughing. Kaler looked at me and I said, "Sir, we're going to get contact. We're going to get some today." I kind of laughed it off and said, "Nah, I'm kidding. We'll be good." It's a huge deal because Lieutenant Kaler was our XO, number two in command. I knew that the guys and I had to have our shit together because if we do something wrong, that's going to be my ass.

So, we went out and everything was perfect. My guys dispersed perfectly the whole patrol, the truck in the back, it was textbook. I'm just giving him a familiarization throughout the area where we were patrolling, explaining things he needed to know. We came to an area where there were a bunch of dirt mounds that were staggered. I said, "Sir, this wasn't here a couple of days ago. This is really weird." I know they might be using this for construction, but this is just weird. So Kaler and I and Cherry, we take a knee next to one of the dirt mounds. I'm pointing and explaining we need to be careful and debrief this when we get back. Half of the dirt mounds were on the road and it would force our vehicles to zigzag around it, so this is a perfect way to blow up our trucks. I said, "Sir, I think we've been here too long. I think we should go ahead and start moving out. We'll do the debrief when we get back." Kaler replied, "Roger that." I turn around and give the hand and arm signal to move out. Cherry and I are on the right side of the road at this point, and we take four or five steps to get back in position.

Then, right where we were taking a knee, an IED exploded, setting off a daisy chain of two more IEDs that blew straight up. It threw me and Cherry over to the opposite side of the road. As soon as they exploded and all the dust cleared, I saw a huge hole and Lieutenant Kaler's missing. So, I started yelling, "Where's Lieutenant Kaler? Where's the XO?" We were also waiting for the small arms fire that usually follows an IED. I'm looking at everybody. Everybody's on the ground. Cherry's fine, I'm alive. We're looking at each other, just still startled. Finally, my SAW gunner in the back said, "He's in the ditch!" I run over and Lt. Kaler's on his belly. I roll him over, grabbing his left shoulder and his left pant pocket and see that half his arm is gone, just evaporated. Kaler was a lefty and that may have saved his life. His rifle was slung on his left shoulder and he had his pistol on his left hip before the blast. His rifle was destroyed, broken into 2 pieces, and his pistol was also damaged. Kaler's leg was also jacked up and he's bleeding from his nose, ears, and eyes.

I started yelling, "Doc, Doc, Corpsman, Corpsman!" Doc ran up and slapped a tourniquet on him and started patching him up. I grabbed a stick I found there and we started splinting up his arm. Lt. Kaler said, "How are the Marines? How's everybody? What happened?" I said, "There was an explosion and you fell. But you're fine. You're good. You're going to be fine. Everybody's good, sir." Kaler would start to doze off and I would kind of smack him in his face, just trying to keep him awake. But, once he would come back, he kept asking, "How are the Marines? How are the Marines?" I said, "Everybody's good. You're going to be fine. Everybody's good." Then he kind of goes out again. We formed a security perimeter and started looking for the triggerman. There was no follow up small arms fire in that event. We were able to get Lt. Kaler medevaced out of there and he made it out alive. I remember we finally got back to the CP and got back to the Combat Outpost. Captain Royer and all the command staff were waiting for us pretty much at the gate. I just broke down. I started crying. I said, "I'm so sorry. I'm so sorry. It's not my fault." They just embraced me. They hugged me and they said, "It's not your fault. Don't worry about it. You did a great job of keeping your guys dispersed. If you had been bunched up, then a lot more people would have got hurt. So, you did a good job."

Activity began to pick up and there were Marines being killed in Anbar and Fallujah. The news reported that a bunch of Marines had been killed, but they weren't putting out any names because they hadn't notified the families. Communication with the families was difficult because we had one satellite phone for the entire Company. We would be able to make a five-minute phone call home about every 10 days. It had been a little bit longer than that since I'd spoken to my parents. I think that week prior to April 6th, we were doing a raid. It was super early in the morning, probably 3 or 4 o'clock in the morning, and we had a photographer with us. During this raid, the reporter David Swanson ended up capturing a photo of me. The photo that David Swanson took of me ended up being printed in my hometown newspaper in Arizona. So, that's how my parents knew that I was alive.

April 6th started off as a normal patrol and IED sweep. We came up through the tank graveyard to Route Nova. We passed up a sniper team, Headhunter II. Headhunter II had already been out there for a couple days. They were right up on the Euphrates River. So, when we ran into them along the river, we just passed them up. We're like, okay, you guys are good. We're good. Yep. They were there because the insurgents were putting in IEDs all over the roads and we could never catch these guys. We didn't know how they were getting the IEDs in. So that's when they started doing the LP mission for the snipers. Hopefully if they'd been out there a couple of days, they could pinpoint where they were putting these IEDs because the tank graveyard was pretty hot all the time. We pass these guys up, maybe a click, two, three clicks or something like that. We got to an area where one of the squads found an IED and we were spreading out in a huge circle.

By the time that happened, we heard Headhunter II on the radio, "We have a squad of enemies online heading towards us." I grabbed Lieutenant Valdez and my whole squad and loaded up into one or two trucks and left the other squad to keep the perimeter around the IED. I think it might've been Hernandez's squad with Thomas, Woodall, all those guys. We drove back to help the snipers and as we're driving back through the houses to get to the tank graveyard where they were at, we start receiving shots from the houses. They're dropping mortars and all this shit and we're getting pinged. We're getting hit. I'm telling everybody, "Hold on. We'll shoot on my command." As we're coming out of the houses and getting shot at, you can see the firefight between the snipers and a bunch of people, and they were starting to pull back as the snipers were shooting back and forth. I take the first shot and hit a dude that's running. I think I hit him in the leg or something.

Right then, everyone opened up. Everybody's cramped in. Quetglas was driving, Lieutenant Valdez was inside. Carmen was in the front left standing up. Martinez was right in the middle on the machine gun and the barrel of his machine gun was over my head. We were all crammed in there. Lund was to my left and Contreras was right behind me. When I took the first shot that was the signal for them to start shooting. Martinez opened up with the machine gun. The pressure and the freaking noise of the machine gun scared the shit out of me. As a natural reaction, I kind of ducked my head. When that happened, rounds came in and hit Martinez in his hands and his fingers. He immediately starts screaming, "I'm hit, I'm hit!" His fingers are just hanging. Lund starts screaming, "I'm hit, I'm hit!" There's just blood all over the back. I yelled, "Stop the vehicle!"

We all jump out. We started getting surrounded. Hajis just started surrounding the vehicle, shooting. I'm trying to get the radio. I'm trying to attend to Martinez and Martinez is trying to grab the radio. Lund is screaming. Everybody's just shooting in every direction at this point. At one point, I think the hajis got as close as 10 feet away from us. They're trying to surround the vehicle. I remember Martinez, his fingers still hanging, holding up the rifle and shooting. I'm holding up my rifle with one hand, just shooting and trying to grab the radio so we can call for help, call for air, call for whatever. As we're trying to get cover behind the vehicle on the river side, a mortar landed right where me and Martinez and those guys were, and sunk in the ground. When it hit, it didn't detonate, but sunk in the mud before it blew. The mud absorbed the majority of the explosion. When that happened, I got shrapnel in my vest and it threw me and knocked me out for a second.

I came to and we all got online with the snipers and started pushing through into the tank graveyard to try to kill these guys. As we're running across and shooting, Cherry and I got separated, Cherry was my radio operator. I was all the way to the far left and Cherry was to my right. We were separated by a fence. I don't know if it was just a natural fence or a wooden fence, I kept telling Cherry, "Get on this side, get on this side" because I didn't want him to get over by himself. We finally stopped and Cherry got to where I was at and they're dropping mortars. We're shooting and throwing grenades. I sent Cherry back to the truck for the radio because he left it in the truck. He got the radio and came back. When he came back, there was a haji that was hiding underneath one of the tanks and he was taking pop shots at us. He just put his rifle out and took a shot. I was about an arm's length away from Cherry when he got hit and dropped immediately. I rolled Cherry over and he was just gone. I remember grabbing his dog tag, a map, and a picture of his girlfriend.

We kept pushing. Stayskal, one of the snipers, got shot through his side and it collapsed his lung. Ferguson was there, (redacted) was there. They're trying to tend to Stayskal. Again, we started pushing. Contreras started throwing grenades, we started throwing grenades and kept shooting. Finally, the firefight was over. I came back to grab Cherry so we could load him in the truck, but he was gone. The first thing that came to my mind was, "Holy shit, they fucking took him." I start freaking out. We had a haji there that was all shot up, but was still alive. I start losing my shit and yell, "What did you guys do with him?" Lieutenant Valdez says, "He's in the truck." We got back to the truck and Cherry was in the truck and was already covered up. They had also gotten Carmen. I didn't even know, but Carmen was shot through this side as well, and I think he was killed almost instantly. When the truck had first pulled up, I fired and Martinez started shooting, I think some of the rounds that hit Lund and Martinez also hit Carmen because we were all on the same side of the Humvee. Carmen was already in the back of a different truck.

The Army was there at that point to help us with our medevacs. It was just a complete mess. Quetglas, the driver that was there, ended up calling for air. I think he actually received a Bronze Star because of his actions that day. Once that firefight finished, we heard from all those guys that had stayed with the IED, "We need help!" We got back in the trucks and started hauling ass back the other way to go help them out, but by the time we got there, they'd already received a QRF. They were pulling out Thomas who was injured by a grenade, and Carter was missing a chunk of his heel from the grenade. So, it was just chaos, and that was just that one day.

The next few days were just a bunch of fighting all over town. We'd lost a bunch of guys and were trying to regroup. The next day that I remember was the 10th. I think we received intel that there was a bomb maker and we went to go check the house and ended up getting bad intel. We just started tracking all the nearby houses to see if we can find either a weapons cache, bomb making material, or IEDs. As we're coming out of one of the houses, we had split up. Hernandez' squad was a couple clicks away doing the same thing. We were on the other side around Route Apple. We ended up getting pop shots from a second story building. My whole squad was pinned down. All of us were on our hands and knees, with the exception of Telinda. He had a SAW and was trying to stand up and see where they were shooting from. Staff Sergeant Craig was with us at the time as we're trying to advance and we kept getting pop shots. We couldn't really do anything.

Spencer, Craig and myself were on the far-left side of the little retaining wall separating different houses. At this point we're in people's backyards. I started noticing the incoming rounds start shifting more and more to the left. I told Sgt. Craig, "Hey, I think these guys are trying to surround us. They're trying to flank us." Craig said, "Grab somebody and get to higher ground. See what you can see." I grabbed Gaeden and we run back to one of the houses and there was nobody there. I told Gaeden, "Hey, you point your rifle straight ahead and I'll aim high." At this point, we're looking at an area behind the house with a wall. I can touch the back of the house and the wall that separates the other property with both my arms, so probably like six feet between the wall and the house. I'm pointing up to make sure nobody's on the roof and looking at the wall, and then I see an AK47 fly over the wall. So, we stopped and he's right on my hip, pointing straight ahead in case anybody comes around the corner. Then, all of a sudden, I see a sandal. I see a leg. I see a man dress. As the dude is straddling the wall, he sees us. He jumps over and is trying to go for the rifle. I just let him have it. We took him out. He had some pictures and a map.

We grabbed his rifle, ammo that he had, and the stuff in his pocket. We ran back and said,

"They're flanking us. They're coming around." I grabbed one of the AT4s and tried to put a rocket inside the building where they were shooting us from, but the rocket wouldn't work. At this point, Telinda and the other guy are running low on rounds already. I said, "I'll run back to the truck and get some ammo." I don't know how far it was, maybe 300-500 meters. I started running back towards the truck and I got underneath and there was an agricultural fence. I laid down on my back to make myself a smaller target, using a big palm tree as cover for my back. I ended up getting stuck on the wire and I was trying to get loose.

While this was happening, I got shot right in my knee. It felt like you've ever landed on a rock with your knee playing soccer. I thought, "That hurt!" Then I looked down and saw there was a hole in my pants, right where my knee is at. "Damn it. I ripped my pants." When I touched it, it was hot and then it started soaking with blood. I freaked out. "I'm shot, and there's nobody around." Everybody was pinned down somewhere else. I'm in the middle of freaking nowhere at this point and I need to get out of here. Nobody knew where I'm at and they already shot me once and I'm fucked if I stay here. So, I ended up using my cravat (like a large bandana), put it around my knee and got through. I eventually ended up getting back to the truck to get some ammo. At this point, Sergeant Major Booker and his whole QRF had arrived on scene already. I said, "Sergeant Major, we need help. I've got my whole squad pinned down over there. We need your guns." SgtMaj said, "We're on it." I ended up getting the cans of ammo and took it back to my guys. Sergeant Major Booker was there with his QRF, pulled all my guys back and put me on a five-ton or a seven-ton. We started driving around picking up all the wounded so we could take them to the medevac site.

I think they sent me to Charlie Med or somewhere in Kuwait. I do remember arriving there because they were trying to take my rifle away and I wouldn't give it to them. They said, "Hey, you're in the Green Zone. You're fine. We've got to get you checked out." I said, "You ain't taking my rifle. I don't give a shit." They ended up putting two IVs in me and I woke up naked with a space blanket and my rifle was gone. Apparently when I got shot, it ripped my patella. I was taken to Germany for 10 days and then I was sent back to the States. Rehab in the States took several months. I went back to the unit to help tracking appointments, keeping track of the troops, where everybody was at, everybody that was coming back to the States. When somebody would be killed, I would have to go pick up the SRB so we can start making notifications to the families. I helped with a couple of funerals for some of the guys. I ended up reenlisting and went to SOI as an instructor.

I was promoted to Sergeant, instructor at the School of Infantry, and I ended up promoting again to Staff Sergeant while I was there. I had about 11 months left on contract. I was sent to 1/5 after my tour at the schoolhouse. So, because I was going to get out and not be able to make that deployment, I was deployed to Colombia almost as soon as I got to 1/5 in October of 2008. I ended up deploying to Colombia as part of a planning assistance training team out of the embassy in Bogota. It was pretty much an Army unit, training the military and police. I was attached to a Seventh Group, Army Special Forces. Most of them were all Special Forces units, Colombian Special Forces, both Army and Marines. It was that and also the PSDs (personal security details) for the Colombian generals. We would run them through infantry basics. I was the chief instructor over that portion of the course and the reconnaissance portion. Then, the Seventh Group guys would run them through sniper school. The training would be 2-3 weeks. Once we were done, we would fly to another remote base and then train another group of guys, and repeat. We did that throughout the deployment. I did 3 years at SOI and my last deployment was to Columbia. I came back for a couple of months, did my terminal leave, and I ended up getting out in September of 2009.

You know that Lopez, Contreras, Otey, and Tommy Parker were the sniper team that were all killed. I was back because I was wounded in April and we ended up doing Contreras' funeral. His car and all his belongings were left at my sister's house here in California. He would go to Arizona with me and stay at my parents'. When he was killed, his dad flew over here to California and had to get the car from my family's house and take it back to Texas. That was really tough because Pedro and I were super close. So again, I ended up doing his funeral. I ended up doing Lopez's funeral. I was the only Spanish speaker in our detail. I, Nathan Apple, and Lt. were part of the funeral, plus a couple other guys that were wounded from Golf and Fox Company were part of the detail.

That was a whole fiasco in itself. We requested to bring weapons into Mexico to do the full military honors. Lopez was buried in Mexico, in Guanajuato, which is a couple hours away from Mexico City. I think it's called San Luis de la Paz. They told us no weapons because of the Treaty of Mexico with the U.S. We arrived in Mexico City, got with the U.S. Embassy, and they provided a color guard for us. The color guard rifles weren't even real. It was just for ceremony. We ended up arriving at Juan's family's house. That's where he was and they had the viewing. It's more traditional in Mexico to have the viewing at a house than an actual location. There were a lot of family and guests. It was a huge deal because Lopez was a Mexican citizen who was in the United States Marine Corps, fighting for America, died in Iraq, and was buried in Mexico. There was a lot of press. Half the casket was draped with the Mexican flag and the other half of the casket was draped with the American flag. Beautiful setup. It was really somber, really sad. They did a ceremony where they gave up his citizenship posthumously and it was really cool to see.

His wife was pregnant at the time he was killed, so it was tough. We ended up putting him in the hearse and the little town that he was from did a parade for him. The school band and everybody followed the hearse, all of us, all the way to the church. We were marching behind once we loaded him. They did the ceremony at the church, put him back in the hearse, and then we marched all the way to the cemetery. Once we're in the cemetery, again I'm the only Spanish speaker, so I'm going to present the flag to his wife and to his mom. The color guard is playing taps. We folded the flag already at this point, the ambassador is across from me. As I'm holding the flag and everybody's at present arms, taps is being played. I see a bunch of Mexican soldiers come in from opposite sides of the cemetery and a couple of Mexican soldiers put their hands on the rifles of the Marines while they're at present arms. They're trying to take the rifles away from the color guard. I look at the ambassador - I'm in shock, like what the heck do I do? We're right in front of all the family and friends. Everything's being televised on the cable channels - everything's live. He kind of gives me the nod to present the flag, so I present the flag to his wife in Spanish and then I give a secondary flag to his mom. While all this is going on, a couple other soldiers are still trying to take the rifles away from the Marines. We got in the van we had there. We're sitting there and they're trying to take us into custody because we have rifles. Finally, after an hour or so, they finally let us go. You can find this on YouTube. It's pretty crazy. So, that's what happened at Lopez's funeral.

One of my relatives that lives in Mexico called my mom. She said, "Where's Ramon?" Mom said, "He's in Mexico City right now at a friend's funeral. My relative said, "He was all over the news right now." Mom said, "What are you talking about?" She explained that they saw us on TV and there was a huge international incident because the Mexican Army was trying to interrupt the funeral. Mexico's President Fox ended up writing a letter to the family apologizing for what happened.

Janney: It sounds like you were pretty close to Contreras.

Barron: He was one of my squad leaders, I have a lot of crazy stories about Pedro just from living together in the barracks. Always taking care of his ass because he would always get all of us in trouble. He could always make everybody laugh even when it was the darkest times. The thing was that he was older than all of us, so he could drink. He was a bad drunk. When he would drink, we'd end up babysitting him and taking care of him. But, if anybody picked on us, he was like the big brother. He'd say, "Oh, ain't nobody gonna mess with you guys. This is what I joined for, and I'm gonna take care of you guys, and they're gonna have to kill me to get to you guys." He always said, "If God wants me, God's gonna take me." That was just his mentality. He was just a warrior, a great friend, an awesome brother. We'd go to the mall and he'd go up to a girl, get on his knees and say, "Marry me, girl. Marry me." That dude didn't have a care in the world. He was just a happy person. He'd say, "This is me. You're going to love me or you're going to hate me, but this is me." And, Pedro was very, very loved.

6 April 2004 – Route Nova Firefight and QRF

Headhunter II Sniper Team

PFC Kyle Katz

Janney: It's 24 April, 2020. Why did you decide to enlist in the Marine Corps?

Katz: When I was younger, there were two things that I always wanted to do, and my number one choice that I didn't get was to play professional baseball. My second choice was to be a police officer, and I thought my best way of making it was to join the Marine Corps and join the infantry.

Janney: What was your MOS or your job in layman's terms for the civilians?

Katz: My MOS in the Marine Corps was 0341 Mortarman. That was my main job, but when I was deployed, it was mostly 0311.

Janney: Rocha and Weaver said that you guys weren't really allowed to use mortars in the city.

Katz: If we did use mortars, we would use the lume (flare) rounds.

Janney: Once you get out of boot camp and SOI, what unit were you assigned to?

Katz: I was assigned to Echo Company, 2nd Battalion, 4th Marines in San Mateo, California. I went to boot camp June 15th of 2003 and got out of SOI on October 20th. I went to 2/4 that same day.

Janney: When you deployed to Iraq, did you go with the main element in mid-February?

Katz: Yes, I went with the main unit. Our Company flew in. Some of the Weapons Company guys convoyed in because of the gun trucks. From there, we convoyed into Combat Outpost.

Janney: What were you guys doing on a day-to-day basis?

Katz: Basically, let's show our presence, let's walk around, let's patrol, let's see if we can find IEDs, find who's planting IEDs. Every day we were running and gunning, getting barely any sleep.

Janney: Did you do a leftsy-rightsy (familiarization) patrol with 82nd Airborne or the Guard unit? there. Did you guys have any of that? Or they just basically said, look, you know, here's the routes and you need to start patrolling.

Katz: I know some of the other guys got to go ride around and get shown the city and all that kind of stuff and drive around with the Commander, but I didn't get to do that because of other issues.

Janney: I know things started to get really hairy in March. I heard you guys got mortared pretty regularly at chow time and a squad took out that mortarmen and captured one guy.

Katz: Yeah, one mortar literally hit right next to the door that I was sleeping at. I was in the top bunk, right next to the door. We were in the building, but the buildings were paper thin with little aluminum doors. So basically, nothing we were in would stop their 82 mm mortars.

Janney: Did anybody get wounded when that happened, or did you guys escape that?

Katz: I believe Sergeant Hernandez got hit with shrapnel during one mortar attack.

Janney: What do you remember that happened next? We're towards the tail end of March now.

Katz: Towards the tail end of March, we would do IED sweeps, and cordon and knocks. Every time we did a sweep and found an IED, we would have to wait for EOD to show up, or we would just shoot the IEDs ourselves if EOD couldn't show up. Just my platoon would average two to four IEDs a day every time we did a sweep. More sometimes.

Janney: What platoon were you in in Echo Company and what else were you doing?

Katz: I was originally with Weapons platoon, but then I got tasked out to First platoon. I remember we did set up a lot of OPs. In late March or April, we had set up a 24-hour OP just right down the street from the outpost to catch guys planting IEDs because literally guys were planting IEDs in the same spot on Michigan. I remember getting hit by an IED in the same exact spot and getting knocked out for five to ten seconds. The guys were screaming at me to wake up. I know that there was a massive crater from the day before from one blowing up. I think that what they said was that there were four or five shells there. Then, literally when we drove by it later on that day, they had planted another one in that same exact hole. McIntosh had his goggles cut off of his helmet.

Janney: What did you notice about April that was different?

Katz: April was probably our worst time because April 6th through April 10th was called the Battle of Ramadi. All of the Syrian fighters, all the fighters coming in from Saudi Arabia, everybody and anybody started coming from those different areas. I guess they just decided, "These new guys are here. Let's see if we can kick them in the ass, take a few of these Marines out, and see if we can take a big chunk out of them." But, we definitely are the ones who took a chunk out of them. During my first two weeks, I wasn't able to go outside the wire.

April 6th was the day of my first patrol over on Gypsum and Nova, and we found an IED. I was scared out of my mind, and all of a sudden, we're walking up to the house, trying to find out anything, just talk to the people around the area. We walked over to a bunch of houses, me and one of the engineers. My platoon went off to go over to the tank graveyard and deal with a firefight over there. They were already shooting left and right over there. They told me to just stay with the engineers.

I was over a mile up the road from my platoon. Me and another engineer walked towards the houses and all of a sudden, we saw the whole town start running inside. I hear a round just fly in between me and the engineer. We jumped behind the berm. That's when everything started popping off. At first, we were still behind the berm. I'm trying to shoot in front of me, trying to shoot behind me, and the engineer's doing the same thing. Woodall and the Sergeant had run about 3 quarters of a mile up the road, and it was literally just the two of them stuck by themselves. They had to take cover in some building because they had four or five Iraqis on them, trying to get in to get them. I had a 203. Because I see that they got their four or five guys on them, I'm trying to shoot over there to help. I wasn't trying to hit Woodall, but I almost accidentally hit him, trying to help them.

Then, we had to deal with guys shooting behind us, shooting in front of us. I'd say there were eight of us against maybe 50. We ran into the house and then another team came inside the house. They were guarding the front door of this house. I was trying to keep the family calm that was inside. Since it was my first time in combat, I was scared. I did not want to leave the inside of the house. I will not deny that. I was scared out of my mind. I did not want to leave that living room.

Janney: I can completely understand that. I was scared most of the time I was there, and we weren't even getting shot at, so I can't even imagine being in your position.

Katz: I got Nava-Castro, McIntosh, and Thomas on the roof of the house, and Carter, the radio operator. All of a sudden, I hear, "Grenade!" They all jumped to the side, but that's how Thomas lost his eye. Then, Thomas and them come downstairs. I had to take the radio from Carter, and I had to help patch up Thomas. Me and McIntosh patched up Thomas. As soon as I got back, I was so tired after that adrenaline dump. I heard on the radio that the sniper team on Nova was calling for a QRF. They were about a mile up the road from us. Someone started organizing guys to go help them. I was inside trying to call over the radio saying, "We got three guys wounded."

My squad leader at the time was Corporal Barron. Corporal Barron said, "Get in." Then, all of a sudden, he said, "No, get out. Stay here with them." My whole squad went to go help the snipers. I think I was the only one not up there with them. I stayed with another squad and the engineers. That's when Carmen and Cherry got hit. I think Carmen was on the machine gun. I heard Carmen just unloading on 240. I guess Carmen was the first one to get hit because they saw Carmen just shooting at everybody with the 240. If you knew Carmen like I did, from what I saw from Carmen when shit hit the fan, Carmen didn't give a fuck, "Die motherfucker, die!" He'd say, "Let me just hold this motherfucker down until this belt is gone. Give me a new belt!"

Janney: Absolutely. That 240 broke the insurgents' attack. Santiago said the same thing. He said, normally you fire a 240 in short bursts, but he just had the hammer down and was laying waste to those guys. He said, "Thank God he did, because otherwise we would have been flanked and overrun." It was your squad that saved the snipers and kept them from getting overrun. Headhunter II team leader Sgt. Santiago told me that if your squad hadn't come to rescue them when they did, their whole team would have been killed. They already had 2 men wounded at that point, with Stayskal very seriously injured.

Katz: Yes, my squad literally hopped in one of those paper-thin unarmored Humvees. Ten or eleven guys hopped in the back of that and hauled ass up to that tank graveyard. Later on, I found something that scared me, too. After we got back, I looked at my vest and there's a bullet hole coming from one side of my vest to the other side. It literally went under my pleat and out the other side, but it didn't even nick me or anything. Also, it was also the most praise I've gotten from my chain of command in a long time when they said, "Katz, we didn't expect you to be so calm over the radio when you took the radio from Carter to give a sitrep."

Janney: Somebody was watching out after you that day.

Katz: Yeah, I didn't even feel the bullet pass through my vest. I didn't even feel it go through. Like I said, it went under my plate and out the other side.

Janney: Somebody was definitely watching out for you that day. Santiago said that he wasn't sure who all else was in that squad other than Barron, Carmen and Cherry.

Katz: I don't know if he realized I was in that squad since I wasn't in the truck that went to the tank graveyard. Like I said, I stayed with the engineers and the other guys over a mile away from there.

Janney: I understand that, but without you helping the wounded, and being so calm and organized over the radio to get the QRF call and call in a sitrep, a bad situation would have been much worse. That's outstanding. I understand you were back at the house, but the rest of your team pushed through to Santiago and those guys. If they hadn't taken that initiative, those snipers would have been dead. Okay, so tell me what you remember about Benjamin Carmen and Marcus Cherry.

Katz: All right, so what I knew about Carmen is that Carmen was really quiet. And I'd say my first two to three months in the unit, I was, let's just say, a shitbag. I'm 18, 19 years old, and I don't know what's going on. I'm this little kid that thinks, I'm a Marine now. I'm invincible. You can't fucking talk to me this way. You can't fucking try to treat me like a piece of shit. I worked with Carmen for a little bit. He was a machine gunner. I was dealing with all these issues. Carmen helped me out a little bit. He tried to teach me a lot of stuff. Carmen was pretty knowledgeable with that 240, 50 cal, and Mark 19. Carmen knew what he was doing. If I ever asked Carmen for help, yeah, he may have been pissed off, he may have been in a bad mood at times, but he still was there to help you out if you needed it. He would have your back even if he liked you or not. I still talk to Carmen's sister every now and then about Benjamin.

Cherry is one of those guys, if he knew that you were down, and he knew that you were going through a lot of stuff, and you needed somebody to talk to, Cherry's just one of those guys that was always there to talk to you. He was just always there for you. He was not a hard person to talk to. Cherry was mostly the life of the group. Cherry was just all around a good person. I cannot say one bad thing about him, from what I knew from the first few months of meeting Cherry until he passed. There was not one bad thing I could say about Cherry.

Janney: Did he ever talk about his family or things that he liked to do?

Katz: All the time. Cherry was engaged. He always talked about his brother. They were pretty close. I mean, I heard him talk about his family, his brothers. Cherry was probably one of the most likable guys you could ever meet. Cherry was just friends with anybody and everything. He would talk to anyone about everything, no matter who you are.

Barron loved Cherry. When Cherry got hit, it probably hit Barron the hardest. When Cherry got hit, the first thing he said to me was, "Katz, go grab Cherry's dog tags." I said, "Corporal, I don't wanna do that." He said, "Katz, go get it now." I really did not want to touch Cherry. I got that's just how it worked. I really did not want to touch him. But he made me grab his dog tags off his body and that was like one of the hardest things to do was just touch Cherry and grab his dog tags.

Janney: Let me ask you another question. We'll shift gears a little bit. When all this happened, you were moving east on Nova. That's where the IED was in between the sniper team and gypsum. When you spotted the IED, was it well hidden or was it out in the open and meant to be found?

Katz: I believe that it was in the road. They literally dug up the concrete road and put it in the road. The only way we knew it was there was by the wires.

Janney: I want your opinion on this because this is just a theory that I have. My theory is that they planted that IED because they knew that the quickest way to get to that point was for a team to come up Gypsum because they had that ambush set up at Nova for hours and hours before those guys got there. They planted it there in the hopes that either it would be found or that they could hit somebody and that the QRF or the patrol responding to that IED was going to come up Gypsum and walk right into their ambush.

Katz: So, I'm guessing that they planted it just to start something. They knew that we were going to stop and be there for a while. So, they knew. It wasn't just that area that was hit. I think it was all over that area that was hit that day. We found another IED further west down Nova, you know, as it curves down into the city. I think the only thing that literally slowed it all down and made it short was when the Army brought in their tanks - when the Bradleys showed up. Because, all of a sudden, I just see tanks coming out of nowhere and I thought, "Oh, thank God." That had a major effect of turning the tide of the battle. I don't know where those Bradleys were coming from or if they were coming from Junction City or Hurricane Point. I don't know if that's where they were coming from, but it took them forever to get to us. I think our firefight lasted four or five hours.

Janney: So, tell me, when you guys got out of the house, what happened?

Katz: So, when the Bradleys showed up, we walked out, we started searching the area, and we finally took care of the IED. I don't even remember where we went to or how long we were gone for, but then we did another patrol. We went back to that house that we were pinned down in, and I tried to kick down the door. It took me two times to kick down the door because the first time I kicked the door, it closed back on me. Kicked it again. The family wasn't in there, so we went and searched the house, then searched the area. I think we came back with eight detainees just from that firefight.

Janney: Where did they take those guys?

Katz: I think they took them to the Combat Outpost at first, and then I don't know where they took them from there. I didn't ask where they usually took them.

Janney: Did you guys search them or did they get searched when they got back to Combat Outpost?

Katz: We searched them right before we put them in the truck. They don't really have too many pockets. We searched for devices on them and if they had anything else on them.

Janney: Did you find foreign passports or other documents identifying these guys as foreign fighters as compared to locals?

Katz: Yes, that was a big thing. A lot of these guys weren't even from Iraq. A lot of them were coming from Saudi Arabia. A lot of them were just coming in from other Middle Eastern countries.

Janney: What happened in the days after that? What can you tell me?

Katz: We got four more days of bullshit. Four more days of basically the same exact thing. More poppin' off, more gunfights, shootin' at other mujahideen left and right. I remember running down the road thinking, "How the fuck am I running down the road right now? I am so fucking tired." I see Corporal Telinda and Lieutenant Valdez sitting in this little trap door thing in the road. It didn't have that much cover, but we're sitting over close to them. We're sitting behind these little walls, and I got Corporal Dexter in a truck on a .50 cal. I saw guys run across the road. I said, "Hey, over there, down the road." Dexter yells, "Katz, where?" I'm shooting in that direction and I'm just pointing and it was non-stop for the next four days. The whole Battalion had to do cordon and knocks. We took apart that whole town and got into a big firefight that day, too. We had to put our cooks on the OPs on Combat Outpost and everywhere else. Basically, if you weren't an infantry MOS, you were standing OPs that day.

Janney: Tell me some stories about some of the other fellows.

Katz: Sometimes we would have to go stand OPs at Fishhook. McIntosh would go back there and take a shit. Instead of him just taking a shit back there, sometimes you would see him straight full-on naked, fucking running around back there, just doing the fucking most random shit. If you were in a bad mood, McIntosh would put a smile on your face.

If you were in a bad mood, Lund and Gaeden were usually the life of the party and were pretty good at putting you in a better mood. Even though Lund and Gaeden were Purple Heart magnets. Raynor got shrapnel in the ass while he was in a port-a-john. There were times Raynor and I got into arguments, but even after we got into arguments, Raynor was still a pretty cool guy. I liked Raynor. He would put you in a good mood if you needed to be cheered up. He's a good guy. If there was one person you want on your fucking side when you're getting shot at, Gardner was all about it. Gardner knew anything about everything when it came to a machine gun. He was probably one of the most knowledgeable guys I've ever met when it came to a machine gun. Nava-Castro and Mercado were two more great guys. I went through S.O.I. with them. Nava-Castro, that kid. He's a character. I love Fredi. We talk shit to each other, but I love Fredi.

My chain of command, Sergeant Major Coleman, who was Gunny Coleman at the time, First Sergeant Winfrey, and Lieutenant Schickel, who took over for Captain Royer, were great. When I got promoted to Lance Corporal, they gave me the most hope after I got promoted. They said, "Katz, you've been doing so much better since you got in trouble the first two months." If you've ever met Sergeant Major Coleman, talk about Drill Instructor all the time, Sergeant Major Coleman will not be afraid to tell you how it is. He is probably the most straightforward guy you'll ever meet in your life. He wants you to succeed and would tell you that you can succeed, and he will not stop showing you and telling you what you can be and showing you how you can be until you get there. Sergeant Major Coleman, probably the best Gunnery Sergeant I ever had. First Sergeant Winfrey was one of the other most outstanding guys you'll meet. If you had problems, you needed somebody to talk to, he would be there. His door was always open, no matter what.

Sergeant Major Booker was another cool guy. So, two weeks into being at the unit, I get a call, "Hey, Katz, Sergeant Major wants to see you." I'm like, "Fuck." I walked in and SgtMaj said, "Oh, you're Katz. I just wanted to see your face." Sergeant Major Booker got to know me very well. Sergeant Major Booker was one man in the Marine Corps you did not want to cross paths with, especially if you were not on his side.

I knew Ryan Jerabek very well. We went through SOI and were roommates. He'd say, "Katz, what's going on? Want to talk?" Ryan was such a good person. Ryan was another shit hot Marine. I think I cried the hardest when he died. He was brave, too. All I know is that he hopped up on the 240 and he kept shooting. Through a hail of lead, he just kept shooting guys with that 240. I could see Ryan if he was still around today staying in the Marines for a while. I could have seen him make it in a career. Ryan was a really good Marine. I love Ryan's mom. I talk to her all the time when I get the chance to. I call her Mama Jerabek.

Demarcus Reed was definitely another guy I was glad to have on my side. You wanna talk about fucking big ass dude, this motherfucker was huge. They gave him a SAW, and that SAW was nothing to him. He would carry that thing around like it was an M16. Reed wasn't afraid to unload that 249 either.

Then, I had Tor. I didn't like him at first. But, as I got to know him, he just started to grow on me. He was cool. I hung out with him when we were in Japan. Tor was always about his breakdancing. He always had good intentions. If he were to yell at you, he would come by later and say, "Katz, I wasn't trying to be a dick to you. I was just trying to get some motivation in you." Just trying to be the senior Marine that he was. Tor was a great mentor, too.

Sergeant Jeffries, oh man, he's a character. So, at first, Sergeant Jeffries was one I didn't like because he gave me shit. But, he's another one that grew on me. Sergeant Jeffries always talked shit, but always had good intentions. He's a good man. Sgt. Jeffries was the first person that came up to me when I saw him at the 2014 reunion and put his arm around my shoulder and said, "Katz, is that you? What the fuck happened to you? You're not that scrawny little fuck anymore." I said, "I started working out." Sergeant Jeffries was my section leader at first when I was with the mortars.

6 April 2004 – Route Nova Firefight and QRF

Headhunter II Sniper Team

LCpl Roy Thomas

Janney: Tell me why and when did you decide to enlist in the Marine Corps?

Thomas: Well, I enlisted in my junior year of high school. I enlisted simply because I just wasn't a fan of school. I wanted to be done with school. As a young teenage kid, my mindset and thoughts were to go to the military and bypass college. When I got out of the military, I would have the training, the discipline to lead into a career. I always wanted to be like the FBI SWAT team. I'll go to the Marine Corps, get the best training, and when I get out, it'll translate. But I was young and life had different plans.

Janney: I'm so sorry about you getting injured, but I'm so thankful you survived.

Thomas: It's all good. I'm here and it's a part of life that made me who I am today. You know, each and every day that I wake up, I'm thankful that I was blessed to live. I try to live a life worth their sacrifice. In my instance, as close as I was to the hand grenade, I was well within the kill radius. So, each and every day that I'm here is a blessing and I don't take it for granted.

Janney: After SOI, did you go to 2/4 or were you with another unit or what happened?

Thomas: Nope, I went straight to 2/4. So, the night I graduated high school, I was on a plane to San Diego to boot camp. I went to boot camp in June of 2001. I graduated from boot camp August 31st, 2001. I flew back to California the night of September 10th. Went and partied with a few buddies, you know, down there in San Diego and we woke up that morning. I had to check into SOI on September 11, 2001. The SOI instructors were running around saying, "You better get the best training you can. We're going to war. You could be taken tomorrow. You're a United States Marine. At any moment, you can go." Lo and behold, it was another three and a half years before we got the chance.

After we came back from 31st MEU in Okinawa, they gave us 30 days to go home. When we got back, I think it was September of 03, we went to Bridgeport, CA for mountain warfare training because at that point, we were starting to get word that we were heading to north Afghanistan. Northern Afghanistan. Maybe mid-January, I think we found out that we were heading to Iraq, and then Valentine's Day of 04, we left. Flew to Kuwait and convoyed to Ramadi.

Once we got to Ramadi, the 82nd and the Army unit were there, but had very different parameters for what they were doing versus what we would do. We got to the Combat Outpost and it was very desolate, very minimal. We got there, got unloaded, got situated. We immediately started filling sandbags, setting up jersey barriers, and immediately began fortifying. We built sandbag fortifications up on the roofs for the 50 cals. It was just a whole bunch of work because there was just no protection.

Even in the midst of continuing the fortification, we started foot patrols immediately. Royer wanted the Marines out there for their presence in the city. We got there and we hit the ground running. Initially, the war was supposed to be over, so it was SASO, stability and security ops. We are going over there to win the hearts and minds of the people. It wasn't poised to be the violent operation it became. I personally think we were way undermanned for the size of the city that we were in. But we rose to the task and we hammered down. We did the impossible. Adapted, improvised, and overcame.

Our first couple patrols, we didn't receive any contact. It was very weird and very eerie walking in a foreign country and all these people are staring at you and just not knowing. We had briefed like crazy with the rules of engagement and knew we can't shoot unless fired upon and escalation of force. But, I'm not sure we were ready for the hailstorm we started enduring. At first, we heard that they were getting mortared or shot at. We went over for support and by then, the people were gone. There were a couple of small firefights. I think the big one that really kind of opened our eyes was the roadside IED around Michigan that killed LCpl Wiscowiche, who was an engineer.

At that point, it seemed like an everyday occurrence. It was, "Oh shit, here we go. Shit's getting real." Every time we went out, there were pop shots or IEDs that we found. Colby got blown up. Very quickly that led to the morning of 6 April. The sad thing is our government, to whatever levels that it extends, sent us over there with soft top canvas Hummers and unprotected 7-tons. We put a sheet of OSB on the outside and a sheet of OSB on the inside with sandbags in the middle, with a ratchet strap over the OSB holding in the sandbags. That was protection for the troops in the back of the 7-tons. I think it was mid-deployment when they finally got some up-armored Humvees, while the unit we replaced already had armored vehicles. We got sent over with Humvees straight from CA - the whole downfall, the unprotectedness. I don't know how far up the chain of command that goes. Do I feel that you go as high as Mattis, the division? I definitely feel they could have fought harder for us and said, "Hey, these guys are going to hostile zones. Let's protect them." But I don't know if they did, if it fell on deaf ears, or if they thought, "We're Marines. They'll figure it out."

Janney: Tell me what you're doing on the morning of 6th April 2004?

Thomas: We were tasked with the morning patrol on the 6th. We had 24-7 constant patrols out there. Between the first platoon and second platoon and third platoon, two platoons were out, one platoon was back on QRF, weapons maintenance, sleeping, eating, cleaning, whatever. But they were the platoon back on QRF, too. If needed, they were available to go out and help on the other two platoons or help covering their AOs. We left out the west side of the compound. We skirted through the back country roads and fields. Because the thing over there, it was a cat and mouse game with them. There was a little gate we went out of, because we had never, not once did you do the same route a couple of times because that's what they were waiting for you to do.

Obviously, we had to go north to get to Route Nova. So, through our journey northwest there, through the back country land to get to Nova, we went northwest and traveled out and we met at Nova where Nova Y'd off. We were west of the tank graveyard where we picked up Route Nova. When we got onto Route Nova, we started heading back east because Nova ran east to west along the Euphrates River. Ideally, our route was to go all the way down to Gypsum and go home. Generally, we came back on Gypsum and went back through. We were about a hundred yards down Nova and PFC or LCpl Eric Akey was the point man for the squad. He stopped, turned around, and I mean, the kid was whiter than a ghost.

He was standing on an IED, two 88s daisy-chained together. So, obviously, the IED never blew up. We followed those wires because it was still early on, and everything was wired to the battery and detonated that way.

As their technology advanced through our deployment, they were able to use cell phones and garage door openers and other remote devices to blow them. But early on, there were still wires to a battery. We followed those wires back and found a battery and we found fresh human feces. We're pretty sure what happened is we skirted through that countryside, getting up to Nova. We walked right by that guy, and we must have scared him off. Because essentially, we kind of did a 180 right around him. We passed him, went to Nova, and started coming back down Nova. So, he was right in that 180 of our U-turn onto Nova. But he was long gone. So, we got into our 360-security perimeter that the squad does when we find an IED.

We got on the radio, called EOD per SOP (standard operating procedure), EOD would come blow it up, and we'd move on. But when we called that morning, they said, "EOD's busy. We can't get to it. If you can blow it up, blow it and move on." At that point, we could hear a firefight very faintly off in the distance. This was early, between 0800 and 0900. The gunfire would have been in Golf's AO. Golf was obviously already in the mix, taking fire and hammering down.

We cleared the guys out, detonated it, and moved on. I always took pictures of the IEDs for Intel. Then, we moved on. We approached the tank graveyard and made radio contact with the sniper team there, "Eyes on you. We see you. Carry on." We couldn't see Headhunter II, but knew they were in concealment viewing us. We kept going. We got about a half a mile, not quite a full mile, and we found another IED. We put the guys down in a 360-security perimeter again.

It was right then when stuff just started going haywire. The snipers called for help over the radio. They had been compromised. They had a line of fighters coming right for them. Lieutenant Valdez was the XO or the platoon commander and he took off with the first squad in the Hummer, and they shot back to go assist the snipers. As soon as they got within that vicinity, you just heard gunfire, machine guns, grenades, RPGs. Everything just went to shit. We were just sitting there holding 360 and then all of a sudden, we started hearing "pop, pop, pop" and looking at puffs of sand from impacts of rounds. We were taking fire.

Myself and the combat engineer that was with us that day jumped up on the road. We found a couple guys, 700 meters down, shooting. We started shooting at them, providing cover fire. I ordered the rest of our squad to run across the street. The rest of the squad ran across the street to the nearest house. When we first found the IED, Sergeant Conner put McIntosh and Akey in a forward observation, like a lookout post on the south side of Nova, and he chose to put himself and Woodall out on another lookout post on the northside of Nova. I still had the radio operator Carter with me. I jumped up on the road, we shot, we provided cover fire, and we went to the house.

We get to the house and we knock on the door. A guy cracks it open. We said, "We need in!" Guy shuts it like no, sorry, man. I said, "I'm sorry, dude. We're coming in!" He had a family there, and we took them to the very back room to protect and safeguard them as much as possible. Then, we went up to the roof to start defending the house and the position. This whole time, I'm on our PIR, our little personal radio and I'm calling for McIntosh, I'm calling for Connor, because I knew those guys were out there, "Hey, where are you guys at?" I was finally able to get through to McIntosh, who said, "Where are you at? I said, "I'm at a house 700 yards back to the west of you. Where are you at? I'll come get you." He explained to me where they were, and he said, "I'm trying to reach Connor, I can't reach him or Woodall." I said, "Me either." So, this whole time, I kept trying to make contact with Sergeant Connor on the PIR -

nothing.

I grabbed a few guys from the squad and we left the house we were in. We skirted the perimeter wall (because you know most of the houses over there had perimeter walls) for a while, and then we ended up against this wall. There was a little Z-shaped jig jog in the wall near a small creek. We kind of huddled in that little jig jog there for protection. At that point, I told Chris McIntosh, "Listen, we're a hundred yards away. I got eyes on your house right now. You know where you're at. I'm going to order these guys to throw down some cover fire. You guys need to get here. You better effing run because we're wasting rounds if you don't come." I count, "One, two, three!" and we fired.

We started providing cover fire across that field. It was like slow motion, like the lifeguards on the beach at Baywatch, him and Akey - like you just pressed the slow-mo' button, watching those guys do that hundred-yard dash to us. It was just so surreal. McIntosh got there back with us and then Akey made this last ditch jump and he landed in that little creek. I reached out, grabbed him by the back of his flak jacket, tugged him and threw him over to us. As I did that, a bullet hit the sole of my boot and I jumped a mile high. So, we got those two. We skirt back along this wall. This whole time there's rounds ricocheting off that wall and you can feel that concrete flying off, pieces of those cinder blocks chipping off from the impact of the rounds. We could feel those pieces of block landing on the back of our necks and going down our flak jackets.

We got back to the house and went back up to the roof. That's when I told those guys, everyone that was up there, "Hey, if they're outside right now, shoot them." There were grenades, machine gun fire, heavy machine gun fire, small arms fire, RPGs going off - it was just complete chaos. At that point, it was a full-fledged gunfight. Those guys were running all around the house. We could see them. There was a road, a field, and then the river, and we could also see them in the tree line. I told those guys, "Just start shooting. Shoot every one of them mother effers you see. They shouldn't be out right now."

We started defending the house and I remember the radio operator Carter was next to me. He started hearing the KIA names coming across the radio. They were Cherry, Carmen; they were his era of Marines that he'd been part of with the boot drop. Those were his guys, you know. He said, "Nooo!" I grabbed him and I said, "Stop or else it's going to be us. Fight, shoot!" He said, "I can't see anything." I dragged him over to me and said, "There's a whole field of them. Shoot!" He and I started shooting and defending the house.

It didn't feel like it was long after that when I heard that distinct "ting, ting, ting, ting, ting, ting, ting." Then, we heard Macintosh yell, "FRAG!" As I looked down to my right, two feet away from my right foot was a hand grenade. McIntosh was at the door that led to the entrance to the room off the roof. He tucked back in the little hallway to get to the room. Carter was obviously next to me, so I looked at Carter, and I said, "RUN!" I turned, grabbed him right by the back of his pack, and I just gave the biggest shove I could, and then I just tucked my M16 down the left side of my body, and I tucked my Kevlar to the left, and then the grenade went off. The next thing I remember, I come to, and I'm on my hands and knees, and blood would just pour out of my face. At that point, I thought, "I'm dead. I'm done. That much blood pouring like a faucet out of my head, that's not okay. Well, this is it." And it's crazy how things slow down and everything that I thought of. I thought of my family, I thought of friends, stuff back home. It was just like that pause and I just thought of everything in my life. But I remember I jumped up screaming.

The very first person I saw was Nava-Castro. He just had a look on his face like, "Oh man. What just happened?" I was able to get over to McIntosh and I remember telling him, "Help me, help me." He helped me down the stairs and we went to the left to the living room of the house. So, on the west side of the house, which is where we assumed that the people snuck up on, there was a tall, probably 10-foot wall on that side of the house. There was an alleyway right there that led to Nova. We're assuming they snuck up on the inside of that wall and just lobbed grenades up over the wall onto the roof. But, we got down there in the living room and that whole wall was glass. I said, "Gosh, this isn't safe. Let's go to another room." Then, we went to the kitchen and that's where that photo of me over there was taken. He leaned me up against that counter and I said, "Hey, take a picture." Mac said, "No, man." I pulled the camera out of my cargo pocket and he took that picture.

    I think at that point, the adrenaline was over. McIntosh laid me down on the floor and started putting pressure dressings on my face, on my wrist, and on my legs to help mitigate the bleeding. I remember lying on that floor, man, and obviously started feeling a lot of pain. I was tired, and wanted to go to sleep. I think if I would've gone to sleep or went unconscious, I wouldn't be here. McIntosh sat there, poured water on my face, and he slapped me to keep me awake. I'd start dozing off and he'd smack me. I'm alive today because of him. You speak to anybody from our deployment and everybody loves Chris. He's the Caucasian version of Contreras.

Janney: When you guys got Headhunter II's call for help at that house, how far away do you think you were from the intersection of Nova and Gypsum?

Thomas: Miles for sure. We were a long way away.

Janney: I have a theory that the IED was there to lure someone up Gypsum to Nova into the ambush.

Thomas: Not that I want to give those guys any credit, but I'll give credit where credit's due. They put together a very well-coordinated citywide attack against us. They had that Nova/Gypsum ambush planned and they got us. They got us at the Tank Graveyard. They got Golf over on the other side of the city. I think it was in the works. They watched us enough. They knew some of our routes. They scouted our AO and set up multiple attack points, and were destined to fight us.

    Unfortunately for me, I was injured early on. I was a priority casualty and I was gone. They took me to the Army hospital by Hurricane Point, got me stable, flew me to Baghdad, did my surgery there, and then flew me to Balad. I spent the night in Balad, and then I was on a plane to Germany.

Janney: I'm just thankful you survived and made it home. If your patrol had come up Gypsum to Nova instead of the other way, most of you would have been killed in that ambush there.

Thomas: I'm very thankful that we endured in that battle. We were in a fight for our lives and started running low on ammo. It was getting to the point of dispersing grenades. I'm very thankful that we didn't suffer any more casualties at the house even though we lost 2 men at the tank graveyard.

6 April 2004 Route Gypsum Ambush And QRF

6 April 2004 After Action Report (AAR) by LCpl Deshon Otey

(This AAR is transcribed word for word from Otey's handwritten AAR. Otey was in the lead Humvee that was ambushed at the intersection of Gypsum and Nova. Otey was the only Marine in the lead Humvee that survived the 6 April 2004 ambush. Staff Sgt. Allan Walker, LCpl Travis Layfield, PFC Kyle Crowley, PFC Christopher Cobb, and PFC Ryan Jerabek in that lead Humvee were KIA. Ironically, LCpl Otey would be assassinated on the roof of a sniper observation post on 21 June 2004, along with his three teammates USMC Scout Sniper Tommy Parker, LCpl Pedro Contreras, and LCpl Juan Lopez.)

6 April 2004
At 1200 till 1600

3rd Sqd conducted a patrol. We patrolled east to go to Gypsum (338) but the route we ended up taking was too long so time wasn't permitting, and we ended up crossing MSR Michigan where we caught contact. We seen muzzle flashes coming from some of the houses on Frontage Road. Cpl Null – TM flanked the enemy after they stopped shooting so we could punch across the street. Now the enemy wasn't shooting anymore so we conducted a Raid on one of the houses we see muzzle flashes coming from. We ended up not finding anything so we called for QRF to start cordon and knock the area. We ended up with 3 detainees after we were finished. Instead of coming back to the Firmbase, there was a message over the radio to help 1-3, they had just found a I.E.D. Now myself and LCpl Tate were watching the detainees and Cpl Waechter hit me on my P.I.R. to go to the Humvee & bring the detainees. SSgt Walker told me I was on the first Humvee along with himself, LCpl Layfield, Doc Mendez, PFC Cobb, PFC Roberts, PFC Jerabek and driver LCpl Crowley and one detainee. As we headed up Gypsum, we got ambushed. I heard rounds hit the Humvee and told everyone to get down, but then I looked around and there was no armor on the vehicle and told everyone to get out. I hopped out quick and got what I thought was cover next to this wall that was about 10 to 20 feet from the Humvee. No one followed me. Now there was a shit load of fire like it was raining lead. Now this wall I was using for cover didn't work. Once I looked around an RPK and one enemy personnel raised up from the berm so I double tapped his ass, looked to my left and another one popped up but he opened up along with 5 more RPK and a .50 Cal. I took off to the next Humvee to see if everyone was alright and the driver Clyne snatch me up and told me to get right next to him. PFC Shores was still in the second Humvee shooting the (SAW) killed 4 of them motherfuckers. I looked to my right and seen 3 guys jump down off the berm – shot one in the head & twice in the back and the other two double tapped twice in the head. Now split second after that Shores jumps out of the vehicle a RPG hits the second Humvee and I hear Shores and

Downing say they are hit. Cpl Waechter and LCpl Tate cleared this room. We had to get cover in. So in the process of getting shot at while we were in this room, Cpl Waechter want to go back out to lay down some fire. I told him that wasn't a good idea. Soon as we step out we would have been shot up either from the left or right with the .50 Cal that was by the market. We were pinned down for about 3 hours. We caught a couple enemy motherfuckers walking by the window & opened up on them and killed them. We tried to get comm over the P.I.R.s, we had all switched to different channels until Waechter got in touch with Cpl Smith from second PLT to link up. They were pinned down too now at the same time. It was still raining lead. They kept shooting for about 2 ½ hours. Finally we got some help from the Army and 2nd PLT and got out of there.

6 April 2004 Route Gypsum Ambush And QRF

Cpl Evan Null, Echo Co, 3rd Platoon

Janney: First off, tell me why you decided to enlist in the Marine Corps.

Null: To go to school.

Janney: Just basically to go to college?

Null: Yes.

Janney: So, did you have any other family members that served?

Null: Yeah. I had an uncle. He was active Army for a bit and then National Guard after that.

Janney: So, his experience didn't turn you off to the military? You saw it as a good way to get some money for school and see the world and all of that. Does that sound about right?

Null: Yes.

Janney: So, tell me Evan, when did you enlist?

Null: I was in the pool program for about a year before I actually went to boot camp. Between 2001 and 2002.

Janney: So, the pool program is kind of like the Marine Corps Junior ROTC?

Null: I guess, something like that. You know, I was trying to do things to keep you motivated and involved before you actually leave.

Janney: Right. When you signed up, what did you decide your MOS was?

Null: Well, at first, I was going to go in as a reservist military police. Then I changed my mind and went into active-duty infantry.

Janney: All right, so you were an 0311?

Null: Yes.

Janney: So, once you got out of boot camp and SOI, where were you stationed?

Null: Camp Pendleton, California. San Mateo to be exact, with 2nd Battalion, 4th Marines.

Janney: Okay, so you were with them all along. A lot of the guys that I've talked to were what they call boot drops. They came in right before you guys deployed.

Null: Yes.

Janney: All right. So, since you went in before 9/11, how did that terrorist attack change your mind about anything or how did it affect the things that were going on in the Marine Corps at that time?

Null: I was too young to know any better. How things were going to change in the future. But, I guess the base went on lockdown for a bit and had like a QRF set up or something, but I wasn't really a part of that. But things did change, obviously, in the world.

Janney: Oh, yes sir. Drastically so. You were with 2/4 all along, so did you deploy prior to Iraq, or did you go to Okinawa with 2/4 when they went and did that deployment?

Null: Yeah, I went to Okinawa with them. That is to the point that we got extended for about six months, roughly. I think that was 2002.

Janney: You guys got back from Okinawa sometime during 2003 and a lot of the guys took Christmas leave in 2003. Did you take leave then, or did you stay on the post, or what did you do?

Null: No, I didn't get to take Christmas leave. So, I stayed around and did whatever.

Janney: Right. I know Staff Sgt. Allan Walker was there and Sgt. Major Booker was there. I don't know if you ran into any of those guys.

Null: Yeah. Staff Sgt. Walker was my platoon Sgt. when we went on our deployment to Iraq.

Janney: Tell me a little bit about Staff Sgt. What was your rank?

Null: I was a Lance Corporal at the time. When we went to Kuwait, we stayed for about a month to get acclimatized and that's when I actually picked up Corporal. Staff Sgt. Walker had just come from the drill field. He had his pet peeves like anybody else does.

Janney: Now, you tell me, based on your experience, I've talked to this First Sergeant that were NCOs together, so they were friends, and I heard that Staff Sgt. Walker really cared about his Marines and that he wanted you guys to come home safely. Did you pick up on any of that when he was your platoon Sergeant?

Null: Yeah. We could tell that he constantly wanted to be out there with us. But it didn't always work out like that. You could tell he did care about his platoon and the people he was over and tried to look out over it as best he could. But he expected a lot out of you, too.

Janney: I think that comes from being a DI. You know how boot camp was and I'm sure a lot of that carried over and I know that would be tough having him as a platoon Sergeant when he just came from the drill field. After you guys climatized in Kuwait, did you convoy into Iraq in Ramadi, or did you fly in?

Null: We flew in. I think most of the rest of the Company convoyed in, but we ended up flying in later.

Janney: When you flew in, did you fly in directly to the airport at Ramadi or did you fly in to TQ?

Null: I'm not sure what spot we flew into. There were two different airfields we had dealings with. Once when we got there and once when we left. I don't remember which was which. I think Al Taqqadam is one of them.

Janney: Now, when you got to Iraq with 2/4, what Company were you in?

Null: Echo.

Janney: Okay, you were always in Echo Company?

Null: Yes, in the Third platoon.

Janney: I know you guys took over from 82nd, so some of the guys got to do some joint patrols with the soldiers. Did you get any of that experience at all?

Null: I wasn't with them. We just saw them around from time to time.

Janney: So, you guys just basically had to learn the city on your own?

Null: Yes.

Janney: So, you were at Combat Outpost with Echo?

Null: Yes.

Janney: What was your first impression of Ramadi? I mean, did the people seem friendly or indifferent or hostile? What was your feeling on that?

Null: You know, you're always on your toes regardless but when we got there it was kind of neutral, I guess. We still didn't have the crowds of kids that come up and want chocolate and stuff like that. We also had those people showing you the bottoms of their feet and scowling at you.

Janney: Yeah, that's a bad sign. That's a major insult to them. When you guys started patrolling were you basically looking for IEDs or just doing general patrols through the city to get the lay of the land, or what was happening?

Null: No, we did a lot of IED sweeps and night patrols. It just varied. Our operation tempo when we first got there was insane. We were going probably eight-to-twelve-hour foot patrols, go on back, get a couple of hours down and back at it again.

Janney: I've heard you guys worked a lot harder than some of the other companies. I don't know if you knew that or not.

Null: No, I didn't. We were beat down there pretty good about two months in. We were just dragging.

Janney: I heard that you guys got mortared every day around chow time. Did you experience some of that?

Null: Whenever chow time was, I mean, we'd pretty much eat when we could eat. I lived for a long time off of warm milk and cereal.

Janney: Warm milk and cereal. That sounds terrible. Yeah, I don't envy you that. I guess what I'm getting at is, and I don't know if they were Weapons Company guys, or Echo Company guys, but they said that toward the end of March there was, they had gotten mortared and they sent out a patrol to try to capture or kill the mortar team on the 25th of March.

Null: Yeah. I think it was at night.

Janney: Yes, it was at night. So, you were involved in that or knew about that?

Null: No, I was on the outpost during that. I believe that was the first actual firefight that our Company had been in at that point.

Janney: After the 25th of March, this is the point at which Fallujah was kicking off, I want to say a few days later, this is when the contractors got killed and the bodies were hung from the bridge and all that. There were five battalions attacking inward in the city in an encirclement of Fallujah, and my theory was that a lot of those fighters escaped from the Marines in Fallujah west to the next town down the road which was Ramadi. Sgt. Major Booker said that his perception was, "Yeah, that might have been the case, but they were also getting a lot of fighters from Syria."

Null: Yeah, that's what I heard at the time.

Janney: Okay. Booker said it was probably more foreign fighters coming in from Syria than it was people coming from Fallujah.

Null: Yeah.

Janney: From what I understand, things started picking up a little bit with more IED attacks and ambushes then. Tell me what your experience was in those last days of March.

Null: The last days of March, to me, were typical. The point where it started ramping up, in my eyes, that I was likely involved in was April 6th. We were on a squad patrol on April 6th, and we were going out and we crossed a four-lane going under a railroad trestle and a few guys up on some houses across from us started shooting at us. It wasn't any big deal, it was only a few of them, you know. Anyway, we called a QRF out at that time. So, QRF came out, peeked around; looked for these dudes for a while, but didn't find anything. That's when we started loading up onto our vehicles. The Humvee that we hopped in originally had armor on it and it had a SAW. Staff Sgt. Walker wanted the 240 Golf to be first in the convoy, but the one with the SAW on it actually had armor on it, so that's the one my fire team hopped out of and got into.

So, the other part of my fire team now is in the vehicle that I was originally in. It didn't have any armor or anything on it. We started heading up Gypsum and we had five-tons behind us, we had a little convoy heading up there. The only communication we had were these inter-squad radios. The five-tons started staying back a little. I said, "Hey, we're losing contact with the five-tons back here. We need to let them catch up." Then we heard gunfire back there with the five-tons. I said, "Hey, we got to stop." The two Humvees there got stopped and Staff Sgt. Walker was in the Humvee in front of me.

Janney: Okay, so you were in the second Humvee in the convoy?

Null: Yes.

Janney: Walker and Kyle Crowley and Jerabek were in the lead Humvee?

Null: Yes. Once we stopped, we started getting lit up. Luckily, we had the iron plates around us, so we heard the bullets dinging off the iron plates. I was completely unaware of what was happening to the Humvee in front of us at this point. I said, "We have to get out of this Humvee. Let's go." I'm at the tail gate of the Humvee and I hop out and shoot straight across the road where I see a wall and I start shooting at insurgents in this field. When I ran up, Ryan Downing and Justin Tate and Shores were saying this. I hopped out and went to the wall and Shores was shooting at people on the rooftops up there. I guess an RPG hit the back of the Hummer and threw Tate out. Then when I was sitting there shooting, Deshon Otey was the only survivor of the first Humvee. He came to where I was, took a knee beside me, and started shooting too.

Janney: Evan, could you actually see the lead Humvee from where you were?

Null: No. It was completely tunnel vision at that time.

Janney: I know you guys were receiving crazy amounts of fire. Zach Shores was on the SAW, right?

Null: Yeah. I think him and Ryan Downing were watching what was happening to the first Humvee. We started hearing footsteps behind us. It was Shores and Ryan and Tate, and I started pulling them in where we see this shack right here. Well, the wall was surrounding this shack. I said, "Okay, get in here." Downing was complaining about his arm, saying he was hit. So, we went in there and I set up the angles of fire and tended to Downing. It wasn't a big deal with him, so we brought him back into the fight. We had no radio at that point because Layfield was in the first Hummer and was the radio man. He didn't make it. So, we were sitting there for what seemed like a long time without any comms with anybody trying to hold off this insurgent attack.

Janney: So, it was you, Downing, Otey, Shores, and Matt Scott was in one of those houses. I don't know if Scott was with you or not.

Null: He was. I didn't know who he was. Scott was the guy that I set at the door apparently, but I didn't know his name.

Janney: Right. You didn't have any comms. Otey described it as a "wall of lead" that was directed at you guys.

Null: That's probably the best description.

Janney: You're fighting and trying to hold off these insurgents. Are they acutely aware that you guys are in that house and they are attacking you specifically?

Null: I don't know if they were aware or not. This one dude was not aware. He walked by our window and ended up getting hit by us. He didn't know we were there. I don't know. I don't think they did. We were trying to take people from the exits the best way we could. We were already low on ammo from the little skirmish we had before, so we had to pick our shots carefully.

Janney: Yeah, that's the last thing you want to do in a firefight is run out of ammo, but I know they only gave you so much walking out the gate anyway.

Null: Yes.

Janney: What happened next? Did the fire eventually die down where you guys could get out of there or did QRF come or what happened?

Null: Nobody knew we were out there, I don't think. Nobody knew what was going on with us. We didn't know what was going on with the rest of the people out there. Anyway, after a couple of hours, I heard vehicles coming. So, I looked out and there were soldiers in two Army Bradleys. They started clearing up whatever was down the road from us. Then I heard them coming back, so they didn't know we were there. They didn't know what was going on. When they were coming back, I threw a rock at them to get their attention. They waited for us to come out. So, we came out and started taking up positions and clearing up what was left and then loading up the bodies.

Janney: So, let me back up just a little bit Evan. The house where you guys were at, was it on the left side of the road or the right side of the road going toward Nova, toward the river?

Null: Left side.

Janney: Okay, so it was on the west side. Then you had the unenviable task of loading your brothers up at that point. I can't even imagine that. I cannot imagine having to do that. So, the Bradleys waited for you. At that point, had any other Marines come out to join you or was it just the guys in the Bradleys?

Null: I think it was, forgot what platoon it was, and they came out and we kind of got together with them and started cleaning up.

Janney: I don't know if you knew this or not, but there was an Echo Company sniper team, Romeo Santiago, Stayskal, (redacted), and Ferguson pinned down on Route Nova.

Null: Yes.

Janney: They were pinned down at the Euphrates River with the river to their back. Another Echo squad further east of Santiago and those guys found an IED and then they got hit with some small arms fire. That was when Roy Thomas got injured. He was on the roof, and somebody tossed a grenade up. They sent Ramon Barron, Marcus Cherry, Benjamin Carmen, David Quetglas, and Brandon Lund in a Humvee to try to help Santiago and his team because they were about to get overrun.

Null: Yes.

Janney: Did you ever link back up with those guys? Because at some point, they made their way back toward Gypsum.

Null: I don't remember seeing them. But we were there. It was in the same area.

Janney: So, you guys are with the 82nd guys in the Bradleys and you said another squad pushed up? Was Cpl Eric Smith with those guys? Do you remember him?

Null: Yeah, I remember him. I thought he was in one of the five-tons that we were breaking off from. I think that was where Lt. Ski was.

Janney: Yeah, that's right. So, they were in the trucks that kind of fell behind your patrol?

Null: Yes.

Janney: At some point did you ever link back up? I know Cpl Smith somehow facilitated sending Lt. Ski back to be evac'd, but at some point, did they link back up with you? Or did you link up with another group of Marines?

Null: I remember linking back up with the squad that Apple and Cherry were in.

Janney: Like I said, I've heard this story from several different people, so I'm just trying to put all the parts together. The guys that weren't injured or killed out of Apple's squad linked back up with you at that point?

Null: Yeah.

Janney: Tell me what happened next.

Null: After we loaded the bodies and stuff up on the truck and everything, we hopped in the five-tons and rolled back to the Combat Outpost.

Janney: Did they send you guys back out again immediately or what happened?

Null: We were walking through an open field because there really wasn't anywhere else to go at that point. When we hit the field, we started getting fire opening up on us. I said, "We can't stay in this field. We gotta go." We got up and we started sprinting across this field and we got back into another house to take cover. They started setting up security in the house while most of us went up to the roof to get high ground to see what was going on. I had one member of my team watching the rear door and I was keeping eyes on the family we had gathered from the house. I put them all in one room while the rest of the squad went up to the roof to see what was going on. That was the day that Sims was also on a rooftop and got hit.

Janney: All right. Sims wasn't clearing houses? He got hit while he was on the roof?

Null: I believe so. That was what I've always understood.

Janney: I know they wanted to evacuate him, but they said that the LZ was too hot. There was too much incoming fire, so that didn't happen.

Null: Yeah.

Janney: This house that you were in, was Captain Royer with you that day?

Null: No, Royer was never with us. He was somewhere out there, but not with us.

Janney: Because I know there was another house and Royer was there with Ben Musser.

Null: Yeah, David Swanson with the Philadelphia Inquirer was with Royer.

Janney: Okay, so that was a different bunch of guys. So, once you made entry into the house and got the people squared away, you guys got on the roof of the house. Tell me what happened after that.

Null: I never went up to the roof of the house. I was down there watching the door and watching this family at the same time making sure they didn't cause any problems.

Janney: Did you continue to get fire or did things calm down?

Null: Fire went on for a little bit. Then it died down and I went up to the roof to check things out and everything was dead and quiet. There was livestock and everything was just laying around. We continued our sweep, and we went into this house and pulled weapons out. There was brain matter all over the windowsill and nothing but women in this house. They were telling us, "No Ali Baba, no bad guys here." We said, "There's brain matter here on this windowsill, so obviously some bad dude was here." There's also a blood trail of a body being drug through the house. Anyway, they said, "No, no, no." We left the house and I thought it was a sniper's view, so I said, "These women down here, they're restocking this house and we have to go back." The women are still saying, "No, no, no" but this time there were men with them. We couldn't really get much out of them. That's what I remember from the 10th.

Janney: Did you guys have interpreters at that point or just one?

Null: I think they had one Company interpreter.

Janney: Okay, so not down to squad level, just one for the Company.

Null: Yeah.

Janney: It would have been very difficult to get anything out of them unless one of y'all spoke Arabic. After the 10th what happened? Did things calm down or just more hit and run type attacks?

Null: Then they started resorting more toward mortar fire, IEDs, and they started shooting these long green rockets at us. It was sporadic.

Janney: Now, did you guys suffer any casualties from the mortar attacks or the rocket attacks?

Null: There were some people from the OPs on the Outpost that took shrapnel from them. There was a couple of times that they had been so close that they were throwing debris on us.

Janney: Were you still doing IED sweep patrols?

Null: I guess when our major IED sweep patrols kind of slowed or stopped. We weren't out there with people with metal detectors anymore. We just started doing more foot patrols throughout the city and doing a lot of raids at night then.

Janney: Now, I was told that every Company had an Intel node, somebody that would pass intel down from headquarters Company and that would kind of direct which houses to include in the raids. Did you experience any of that or does that sound reasonable?

Null: That sounds reasonable. We would get a briefing; this is what we are doing, and we'd just go do it.

Janney: Was the rest of the deployment pretty much like that? You just had on again off again attacks and IEDs?

Null: Yeah. Besides, toward the end of the deployment, we started kind of buckling down in place; not being "out there" so much.

Janney: I can completely understand that. 2/5 Marines replaced you, right?

Null: Yes.

Janney: What other aspects of that deployment stood out to you? We talked a little bit about what most people consider to be the Battle of Ramadi which took place the first 10 days or two weeks of April. What other aspects of your deployment really stood out?

Null: The big city-wide sweeps we would do. We would do every door, every lock, everything got searched. I think the Army, the Big Red One, the Brigade we were attached to, would actually do outer cordons of Ramadi and 2/4 would actually do sweeping through the city. I remember a couple of those. I remember capturing prisoners and taking them to Hurricane Point. There were just so many people that we would round up on these sweeps. We would have to label their clothes with sharpies, notes, or something.

Janney: I know that they interrogated those folks back there and then they sent them on to some other place. As a Lance Corporal, did you ever find any papers on anybody that indicated that they were definitely foreign fighters or is that something that was out of your pay grade?

Null: No, nothing like that.

Janney: The main thing I'm interested in besides the story you've shared with me are stories about your brothers. I don't know how well you knew these guys, like Travis Layfield, and Kyle Crowley and Ryan Jerabek. Tell me some things that you remember about those guys.

Null: Jerabek and them were in a different platoon than me. The people in my platoon are the ones that you're going to be closest with. Layfield was on my fire team. Christopher Cobb was in my squad.

Janney: You probably knew Lt. Ski.

Null: I knew of him. He wasn't my platoon commander. Staff Sgt. Walker was mine.

Janney: You shared a little bit about Walker, but tell me what you remember about Travis Layfield.

Null: I was always fond of Layfield. You could tell the first time you met him he was a pretty sharp guy. He always had something smart to say and was a very uplifting person, too. He was always very positive, very sharp, quick witted.

Janney: You said he was your radio operator or a radio operator?

Null: Layfield had the radio that day.

Janney: What about Christopher Cobb? What do you remember about Christopher?

Null: Cobb was, I don't remember where he was from, but he was generally pretty positive, too. He was a good dude. We had a good group of guys. I remember him falling into one of the irrigation canals on one patrol that we were on. He just fell over and dropped in a canal of shit. I couldn't help but laugh at him because he was covered in this black crap.

Janney: Yeah, that's pretty horrible. Tell me what you remember about Deshon Otey. I've heard a few stories from other people, but not everybody was super close to him. He was pretty quiet unless he really had a point to make or something to say.

Null: He was pretty quiet. He was from Louisville, Kentucky, I believe. He was always laid back, relaxed and always talking about what he wanted to do in the future, "I want to do this. I want to do that." He was pretty ambitious when it came to that. He always wanted to be some type of Force Recon or SEAL.

Janney: Do you think that his goal to do Recon was the reason behind being on the roof with Tommy Parker and those guys?

Null: I don't know exactly why he wanted to go. He volunteered to go to the sniper team.

Janney: Circling back to Travis Layfield, did Travis ever tell you he had gotten a car right before he got into the Marine Corps. Did he ever tell you about his car or anything personal?

Null: No. I was a Corporal at the time. We would be out on ops and I wouldn't holler at them. We just got them not long before we deployed. I was trying to jam as much infantry knowledge into them as I could. So, I would be quizzing them constantly.

Janney: That's a sign of a good NCO because I know you were trying to get these guys up to speed,
They were part of your boot drop that 2/4 got before you deployed, so I certainly understand that. Did you know John Sims very well?

Null: Yeah, I did. Yeah, Sims was always a jokester. He was surprisingly agile too. He would knock off back flips, back-to-back, like you wouldn't believe.

Janney: Yeah, I heard he was a short, muscular guy. He was a Southern boy and had a pretty crazy sense of humor.

Janney: All right Evan, can you think of some other stories about some of the other guys that we haven't talked about?

Null: I remember a few stories about Zach Shores. I don't know if you've heard his name or not.

Janney: I've actually met Zach on a couple of occasions, but I don't know anything about him when he was a Marine in combat. We were just hanging out at a couple of reunions so tell me a couple of stories about Zach.

Null: Shores was the SAW gunner in our squad and there was this mean ass dog we saw on a patrol. This dog was showing Zach his teeth, barking at him and getting aggressive. This dog ran up and started trying to bite his feet and it caused Shores to back up and fall down on his ass. I started yelling at Shores, "You have a damn light machine gun and this dog has messed you up."

Janney: Laughs. I know that Lieutenants and Corporals didn't necessarily hang out, but do you have any remembrances about Lt. Wroblewski?

Null: What I'd seen of him, he was the cool, calm, collected type of guy.

Janney: Right. Yeah, he just didn't get much of a chance to prove it unfortunately.

Null: Yes. It was really a surprise when we heard he passed because we heard he had got shot in the jaw area and he seemed okay, he was fine. I don't know if complications happened with him later, but we were surprised when we heard he passed.

Janney: You probably know that I took his dad to Route Gypsum to do that memorial service for your brothers that fell there on the 6th. The Marine interviews will talk about the Battle of Ramadi.

Null: I can't understand how you did that. That was quite a feat getting that done.
It's kind of disappointing that such a large battle or event could have been swept under the rug because we had 20% Battalion casualties or something? It was ridiculous.

Janney: Yes, you had higher casualties than any other unit since Vietnam. So, 2/4 lost 34 KIA and 255 wounded. That was much higher casualty figures than any other unit because you were trying to control a city of that size with one battalion. And you know the kind of guys that you were facing. It's just amazing that you guys were able to control that city with just a battalion. If you look at it on paper, it's nearly impossible, but you did it. That's sheer testimony to the bravery and fortitude of you and your brothers. They had five battalions around Fallujah, which is a city much smaller and they couldn't get the job done. Actually, they made them stop in early 2004 and pulled them back for political reasons. But, look at what you guys accomplished with just 2/4 in the City of Ramadi. It's an incredible story.

Null: I think it's an awesome story in the way that you put it, too. I would like for people to be aware of what happened there. You know the people who orchestrated that large ambush which killed half of my squad? We got intel of who orchestrated the ambush and where they were at. That was one of the night raids and I don't know whose decision it was, it may have been Captain Royer's, I don't know. He let what was the remnant of our squad actually be first through the door to get them.

Janney: That's pretty cool! Did you actually get those guys?

Null: Yeah.

Null: Yeah. I don't even know if you're aware of who was in this squad. Third platoon, third squad: Marcus Waechter, Deshon Otey, Travis Layfield, Ryan Downing, Christopher Cobb, Zach Shores. I don't remember this dude's first name, something Noble. I never even heard anything about him. I asked around about him, "Hey, whatever happened to him?" Nobody knows.

Janney: Zach is on Facebook.

Null: I went down there to North Carolina to see Zach because we're only about two hours apart. He was the first person that I've ever seen outside of when I was discharged. He's the first person that I've seen since I got out. I met his wife and kids then.

6 April 2004 Route Gypsum Ambush And QRF

PFC Matt Scott

Janney: This is 10th May 2019, and this is my interview with Matt Scott of 2/4 Marines. Matt, just to get started, tell me a little bit about yourself. Where did you grow up?

Scott: Originally from Barton, Louisiana, we moved around, and I ended up growing up in Darrington, Washington. Went to school there, played basketball and graduated from there. I went right into the Marine Corps.

Janney: What led you to enlist in the Marines?

Scott: I did Young Marines early on, so they always had a portion of my heart. I was actually going to move to college, and serve time in the military. Mom said no, so I went and enlisted. Brought the paperwork back because I enlisted when I was 17. I just signed up, so yeah, I always admired them.

Janney: When and where did you go through boot camp?

Scott: So, June, 15th of '03 in San Diego.

Janney: What did you choose as far as your MOS?

Scott: It's kind of a funny story. I signed up as 03, infantry, UV option, so I was going to go to security forces for my first two years and from there I would get bumped to the police. You do boot camp and from there you go to the School of Infantry (SOI.) There they designate you. I was designated 0311 UV option for security forces. A couple of weeks out of graduation, they started calling off names. My name doesn't get picked and I don't know why. In my contract, you can actually see it's whited out and it's UH, Infantry option. That's how I went to 2/4. My designation was whited out and I was thinking they made a mistake, but it was what it was.

Janney: When were you assigned to 2/4?

Scott: I was not initially. I was actually assigned to 1/5, First Battalion, 5th Marines. I'm not sure what happened. Basically, your Battalion Corps Major and other folks come to pick everyone up. We're waiting on the bleachers and thought we're going to 1/5 to Camp Warnock. These Marines say, "When we call your names off, you need to go over there." They start calling our names off and mine gets picked. They say, "Everyone we just picked, you're going up to 2nd Battalion, 4th Marines." I asked where they were at. "They're in San Mateo." I said, "Where is that?" Sgt. Major Booker was there to take us up the hill. That summarizes how everyone from SOI got designated. There was basically a lotto. We just got picked, and that's where we went.

Janney: When you got assigned to 2/4, were those the guys you ended up deploying with in 2004?

Scott: About 40+ of us got selected for 2/4, then we dropped to the unit (Company.) SgtMaj Booker took care of that as well. I got selected for Echo Company. It was right before Thanksgiving, and we got a pass for a four-day weekend.

Janney: What did you do on your 4-day pass?

Scott: I got picked up where I grew up by some folks whose house I had worked on that had actually moved from Washington to California. They found out I was there, so they picked me up for Thanksgiving, and I stayed with them for a couple days.

Janney: When you got back from leave, how long was it before you guys deployed to Iraq?

Scott: We came back from that 96 and started doing some unit training. You get dropped, you find out what platoon you're going to, what team you're going to. We did training from the end of November through Christmas. At Christmas, we had two weeks of leave. We got back, did some more training, and went to March Air Force base from the beginning of January through February. Then we deployed February 15, 2004.

Janney: Did you actually go on the ship or did they fly you over there or how did that go?

Scott: We flew out of March AFB on a C5, stopped over in Germany and then flew into Kuwait and then Al Qadim. First off, we were going to drive in with the fuel Company, but ended up driving from Al Qadim to Ramadi. I want to say Golf Company ended up driving from Kuwait.

Janney: They drove all the way from Kuwait to Ramadi?

Scott: Yeah.

Janney: Wow, at 30 miles an hour that would have been a fun trip.

Scott: Laughs, yeah.

Janney: So, when you got to Ramadi, from what I understand, the FOB wasn't really built yet.

Scott: We drive in from Al Qadim and the first thing we notice as we're getting off the trucks is that the trucks are not armored. There was a piece of plywood with sandbags in between the plywood and the exterior of the truck. I thought, "Well, that's kind of strange." We drive into Ramadi and get to Combat Outpost. The Army had established it, but they had not fortified it. They had some posts, flat posts, set up at the corners, but it was not fortified at all. So, the first week or so was just filling sandbags and fortifying everything. I was a PFC, so we were just doing what our Sergeant ordered.

Janney: So, were they using Hesco's then or did they have T walls or what?

Scott: We sandbagged up everything. We did not get Hesco barriers, I would say, for a month to two. The first thing we did was build these walls in front of the doors with sandbags to protect from snipers. I was actually on the SAT phone with Mom and ended up getting shot at, so had to drop the phone and run behind the sandbags.

Janney: Were you getting any indirect fire at that time or just sniper fire?

Scott: We were told that every Thursday, we'd get mortared at 9 o'clock at night. On the dot, you're going to get indirect fire. The only thing was the Army really would not do anything. So, the more we prepped and worked the city, the more indirect fire we would get. So, we'd get hit about two, three times per week, something like that. Sometimes more, some weeks less. I was at Combat Outpost. Echo Company and Golf Company were at Combat Outpost. Fox Company, Headquarters and Service Company (H&S Co), and Weapons Company were at Hurricane Point.

Janney: Once you guys got there and you got your basic fortifications done, did you immediately start pushing out into the city doing foot patrols or vehicle patrols?

Scott: Every day. The way it worked was that a platoon would be on base security. Next would be on the Main Supply Route (MSR.) We would bump out early in the morning, go walk and look for IEDs. Head down Route Apple, head back up to Route Gypsum and back down MSR Michigan to Combat Outpost. Every morning there would be a platoon out there walking that route, looking for IEDs, looking for wires. That's on top of your day-to-day patrols and raids.

Janney: Were you guys encountering a lot of IEDs at that point?

Scott: Yeah, there were quite a few buried.

Janney: Were you doing patrols at a set time every day or were you guys switching it up?

Scott: No, it was all different times of day, sitting in ambushes, setting up listening posts, whatever.

Janney: What was the general mood of the civilians that you were encountering?

Scott: Ramadi was where a lot of your Saddam's Republican Guard went to retire. When we did training at March AFB, we trained for Stability and Security Operations (SASO), mainly a police force. We were initially handing out soccer balls. We really did not train to be in full combat as much as what it was. The people's mood, overall, was half and half. Some people hated us; some people loved us. The big thing was, as we found out later, was that if there were kids around, you were good. If there ain't, you're fucked.

Janney: What do you remember about the day of 6 April 2004?

Scott: Our squad was Quick Reaction Force (QRF) Charlie. We were given sleep time. I forget what we were doing April 5th, but it was a long day (author note: 5th was a long patrol with Lt. Wroblewski), so we were resting, doing laundry, and chilling out. I would say about 0800, they woke us up, "Hey, you need to get up." We go running out back and you hear a bunch of shots. Golf Company was in contact and had casualties. Within Echo Company, you have 2nd platoon, 2nd squad that was out all night. We go load the truck up with ammo and water. As we're loading that stuff up, we're told, "Hey, stop that shit." I'm thinking, "OOOOkay." We're told 3rd platoon, Echo Company, has had contact at Gypsum and MSR. So, we started offloading that truck. We get in two Humvees in the QRF. Lieutenant Ski is with us. We go ahead and run out there to the 3rd platoon. We showed up on MSR and they're facing north. We roll up Gypsum in the direction of Nova. I'm driving the lead Humvee. We pull up, and Lt. Ski hops out and is figuring out what's going on and where to go.

Then 1st platoon got in contact (author note: on Route Nova during the makeshift QRF to help Headhunter II sniper team by the Echo Company squad that found the IED on Nova to help.) That's when Lance Corporal Cherry went down. They're on the bank at Nova at the tank graveyard and needed some help. That's what led to us pushing there because the 3rd platoon was not in contact after that – the bad guys ran off. Captain Royer was there, but I couldn't tell you where. Lt. Ski is going to take this convoy and we're going to push. So, I've got to credit Lt. Ski, I hope you don't mind if I call him that.

Janney: No, that's fine. That's how I know him, too.

Scott: We called him Lt. Ski - it was just so much easier. So, I have to give him credit; he's why I'm alive now. He decided right before we pushed out that because I had the SAW (squad automatic weapon) sitting on top of my truck and Lance Corporal Crowley's truck that was actually second in line with Jerabek on a 240 Golf sitting up there with a few ports in a deflecting shield, "Hey, we're going to push that gun up front, and yours will be second". So, the order was Crowley's truck, my truck, the 7-ton, Lt. Ski, and I'm not sure the configuration after that. We're heading out – mind you, comms were terrible. Each squad had one radio, but no comms with one another. Nobody could talk to anybody. So, we're rolling, we get on Gypsum, we're signaling the other vehicles. Somebody starts banging on the top steel door on top of the Humvee. We didn't have armor on our Humvees. We tied on shit and sandbagged it. People start yelling at me; banging on the top of the Humvee. I'm like, "I don't know what the hell you're yelling about, but we're going." Come to find out later, they were banging on the top because the middle of the convoy had started receiving fire.

Janney: About how far south from Nova were you at that point?

Scott: When they started banging on the roof, probably a couple hundred meters, 300 meters. Roll up, lead Humvee entered the market, you've walked the area, so you know. I'm about to turn to get to Nova, you know how it banks up, there's a big old berm?

Janney: Right.

Scott: Before that last turn, the lead Humvee went ahead and entered the turn. As I'm approaching, that's when all the firing started. I had Corporal Waechter, I didn't have anyone from the 2nd squad in my truck. I had a squad from 3rd platoon in my truck. I really cannot tell you why he did it; trying to push through; but they blew the son of a bitch apart. I turned to Corporal Waechter and said, "Get the fuck out." So, we both hop out. Right then, a RPG hit the front of the truck. This whole time everything is going off. Everybody in the back of the truck bails out, starts scanning, trying to figure out what's what. I saw Otey running back from the first truck. I started covering him as he ran back. We all ran into a little building for cover.

Janney: Where you guys were in that building, could you actually see the intersection of Gypsum and Nova?

Scott: No. You had a courtyard, so where my truck was, before the turn, you have a brick wall on your right side. On your left side, you have a brick wall and then a door that goes into a little courtyard and then you go in. The room that we were in had one window; one or two windows, I want to say one window, the backside was a wall. That's when I took a knee and then got in the prone.

Janney: So, let me backup a little bit. I don't know if this is true or not, but I had heard at the intersection of Nova and Gypsum, they had a Soviet DShK 12.7s machine gun set up.

Scott: They had one set up by the northeast corner of Gypsum and Nova. They had that set up and then from there, they had dudes lined up along the rooftops. They had RPKs, AKs, RPGs, they had dudes there walking along, whatever it was that they were doing. Otey was able to get a couple of them. Basically, they were surrounding everything. I honestly don't think they knew where we were.

Janney: So, you personally, in that building?

Scott: Yeah. I don't think they knew that that's where we had gone. I think they were running around everywhere else trying to figure it out, because had they known, that would have been pretty easy for them.

Janney: Well, thank God that they didn't know where you were at.

Scott: Yeah.

Janney: So, at this point, they've hit the lead vehicle and your vehicle. Then you guys piled out of the second vehicle, ran into this little building on the left (west) side of Gypsum.

Scott: Yeah, pretty much, we were lined up against this wall. That's when I was covering Otey as he's running back. I'm like, "Oh shit, nobody's coming back." That's when the call was made that let's get the gear and regroup because everyone's bell was rung from the truck blowing up. Some of the bad guys were close because they were fragging the truck. So, I don't know where those dudes were, but they were within grenade throwing reach, or they could have been on the other side of that wall on the right-side tossing frags over. I know that one frag ended up blowing my door off.

Janney: Once you guys started running low on ammo, you went ahead and went back to the COP to get some more?

Scott: No, the Army had sent out some of their Bradleys. By then, Weapons Company had pushed out the big guns. Nobody knew where we were. We didn't have comm with anyone, so that whole time we were talking to no one.

Janney: So, you were just basically cut off out there?

Scott: Yeah, so they didn't know where we were until they started rolling up and we started hearing big machinery. That's when we got out and we got picked up, got driven back, our dead got taken care of, and I went in and started reloading and headed back on out.

Janney: So, prior to the 6th of April, you guys weren't getting ambushed on a regular basis. From what I understand, and you tell me if I'm wrong, the battle of Fallujah had started a few days earlier, and the fighters were being pushed out. I think they just naturally went west to Ramadi if they didn't want to stay in Fallujah which was encircled by five U.S. Battalions.

Scott: Yeah. What happened was the battle of Fallujah began, so however many units started pushing in there, but then, they pulled back out. They figured if you don't want to be here, then you need to leave because they had the whole city surrounded. Pretty much your smart people said let's get on out of here, go somewhere else. So, they ended up bumping down to Ramadi. The catalyst obviously around that time was March 31, 2004 when the four Blackwater guys got strung up on that bridge. That kind of set the tone. Prior to, I want to say, that there were a couple skirmishes; none that I was a part of; skirmishes here and there. We did go out and find the dude that ordered the mortaring at 2100 every Thursday. So, then there was that big, 3rd platoon that did go out. We took a couple of casualties with that one, Cousins and Lloyd. Lloyd got hit in the leg. Cousins got hit on the side. Prior to 6th April, there was no prolonged fighting.

Janney: Later that day when you guys pushed back out, was there still sporadic fighting or had those guys dispersed?

Scott: No, it was done. It was over with. They even cleaned up their brass pretty good. When I got back out there, I found a dude, and thinking back now, I should have just shot him. He was hidden underneath a stack of stuff. I jumped down there and snatched him up. The reason I should have shot him because he could have been laced with a S-vest (suicide vest.) He got whisked off. The bodies of the enemy – three, four, or five were up there at that intersection.

Janney: At the Nova and Gypsum intersection? They were scattered around because I've seen Swanson's photos. Is that what you're talking about?

Scott: Yeah, yeah.

Janney: There's a taxicab there, too. Do you remember seeing that?

Scott: Yeah. We had to go pick them up. One of the dudes had a grenade on him so we had to wait on EOD before we could move him. We reloaded and we bumped back out to pump up security. We did that for a while waiting until the trucks came out so they can tow off all of our vehicles that are down out of there. We ended up getting back; our platoon; two of our squads were back at base when this stuff was all going down. So, when we came back, our platoon rejoined with them since our platoon had taken the least casualties.

Janney: So then, the events of that day are pretty much done. You guys go back and crash. I'm sure you were exhausted.

Scott: No, we went out on patrol that night.

Janney: What? You went out again that night?

Scott: We ended up pushing out pretty soon after getting back on a pretty long patrol. We went just around Sofia, out in the tar fields; patrolling, giving our presence, "Hey, you might have beaten our asses, but we're still here." Went around Fishhook Lake; just making our presence known and patrolling. We went out until the morning. We ended up out all night. They wanted to give a presence patrol, like I said, you kicked our asses the first time but we're still here. So basically, we just walked around giving our presence until we got back in the morning. Then we crashed. Laughs.

Janney: Then the following day, the 7th of April; I think Sims was killed that day, right?

Scott: No, Sims was killed on 10th of April.

Janney: On the 10th. What happened on the 7th after you had been out all night on patrol?

Scott: I don't remember the 7th, 8th or 9th. The 7th I'm assuming we got the day off to sleep. I don't remember what we did on the 8th or 9th though. The 10th was Operation Bughunt. No, that was a different one. I don't know what this operation was but, maybe it was Bughunt, I don't remember. That's when we went out and started rooting out specific targets and looking for fighters. Now we're coming out and we're on the offensive. We're taking the fight to you. We ended up getting in pretty much an all-day fight that day. Essentially, we were trying to root out all these fighters. We were fighting IEDs. They had all the ammo in the middle of the fields, guns everywhere and they had themselves set up. Basically, it was time to go out and find them.

Janney: They had caches set up everywhere and you guys were looking for them and for weapons?

Scott: No, we didn't know that they did have caches set up. We were just out looking for them. What they had was fighting positions all over the place. But, we went out and caught them off guard pretty good.

Janney: Let me ask a question about something that I don't know that much about. There were a couple of Marine Corps snipers that were doing overwatch that were tragically killed. When did that take place, do you remember?

Janney: Yeah, yeah. I went and found them. So, you're talking about Corporal Parker, Otey, Contreras and Lopez, right?

Janney: Yes.

Scott: All right. That would have been July 21st, '04. I think it was the 21st.

Janney: So, they were on top of a building doing over watch and what do you guys think happened? Somebody snuck up on them or what happened?

Scott: So, the sniper teams were short on what's called a HOG, school trained snipers. In a platoon, you're also going to have what's called a PIG. They're trained to be snipers, but hadn't passed all the tests. The sniper team, which was part of H & S Company, was short on people. What they were doing was they were pulling people from the line companies to fill-in, so they could go out in 4-man teams. What they were doing was going out for 24-hour shifts. I still have the Navy Times from that. They kind of botched that story up. So, they go out, whatever time it was, they went out. The house that they were setting up on, Captain Royer would not let them deviate their routes – you're going to go there and you're going to come back. They sat up on the roof for 24 hours. Well, at the house that they set up on all the time, that house was under construction. During the day, they're going to have workers down below doing their thing.
As far as how they (the assassins) got up there? I don't know what time workers show up for work, but I believe their last radio check in was about 0700ish. I want to say at about 1000, we got a radio call

because we were at the graveyard doing overwatch position right there at Gypsum and MSR, "Hey, go check them out." So, we went up there and that's when we found them. So, I can speculate all day long, but to tell the truth, nobody knows that the hell happened up there. Workers might have shown up to work and one came up and they executed them.

Janney: From what I understand, their rifles were taken, but later, one was recovered by 3/5.

Scott: Yeah, 2nd unit, 3rd Battalion, 5th Marine Corps. They were fighting, and they recovered that rifle. They shot the shooter, or not the shooter, but they shot the guy with that rifle. When they went down, they grabbed the rifle and went, "Oh crap, that's ours." They traced it back to us.

Janney: I kind of got ahead of myself. I didn't mean to jump subjects. I just didn't understand the timeline about when those guys were killed, but going back to April, you guys were continuing to push out into the city looking for the insurgents and quite a lot of things went on in April. What's the next day that you want to share with me? Let me ask you a question again about the 6th, so I have some clarification. Cpl Smith was with Lt. Ski, and you guys got separated in the ambush? Wasn't Eric Smith, Cpl (later meritoriously promoted to Sgt.) Smith, kind of in the middle?

Scott: I believe Eric Smith was in the third truck with Lt. Ski. He was behind us.

Janney: Was Doc Urena with them or was he in a different truck?

Scott: So, the lead vehicle, you had Doc Mendez-Aceves, Staff Sgt. Walker, Lance Corporal Crowley, PFC Jerabek on the 240, Christopher Cobb, Otey, and Layfield. That's who was on the first truck.

Janney: You were with Waechter. Who else was in your vehicle; in the second vehicle?

Scott: So, we had Waechter, Apple, Shores, Downing; I believe that's how you spell it. Don't quote me on that one.

Janney: You guys were pretty much, the names that, yourself, and the other guys that you just named, you were in that building on the west side of Gypsum? Ski and Staff Sgt. Smith were in the third vehicle. Do you remember who else was with Ski and Smith in the third vehicle?

Scott: You had Burmaline. You'd have Roger Wales. You'd have Danks and past that, I can't tell you.

Janney: I was just curious. I didn't expect you to know that

Scott: Right. It's been so long, I have to go, "Oh, wait a minute."

Janney: Anyway, let's go ahead and move to the 10th. Can you tell me, were you involved in the action where John Sims was killed? Or where he was mortally wounded?

Scott: Yes. So, you had Burmaline, Ortiz, no Ortiz his would have been on the 6th too. Backtrack a little bit, once Crowley got hit, I got bumped up to team leader. When Sims got hit, they ended up carrying him back to where everyone had consolidated. That's where we performed CPR – that's where we found out that he was hit in the armpit through the lung.

Janney: Right, yeah. I knew that it passed through an opening in his body armor.

Scott: Yeah. So, they ended up coming on back and what we were going to do, Eric Smith, he was trying to do, he was trying to tell me to take the team, and push across that field and secure an LZ. We couldn't do it; there were too many fighters on the other side. The birds wouldn't land. There was no way we would be able to push across there with a 3-man team to be able to secure that.

Janney: Yeah, I know, it would have been a suicide mission.

Scott: So, pretty much, we sat there.

Janney: Do you remember the journalist Dave Swanson that was there? On the 6th he was with Capt. Royer, but on the 5th of April, he patrolled with Ski?

Scott: On the 5th, I was with Lt. Ski and Swanson. We did MSR. We went out in the morning and went around, hit Apple, hit Nova, and we found one or two IEDs and waited for EOD to come blow them in place. I was actually driving the truck that day. You know that picture that was taken of him and Derek Callaway? I was driving that truck behind him. On our way, that's when we found those two claymores with trip wires sitting up in the palm trees.

Janney: Yeah, I heard about that – that's pretty chilling. Thank God you found the claymores on the 5th. Were they US military issued claymores or were they a Soviet version?

Scott: I can't remember that one. I want to say they were Soviet.

Janney: So, you guys had a pretty decent day on the 5th. I mean, relatively speaking.

Scott: It was just a long day. Found a bunch of stuff, that's when we started noticing, especially the shops on Gypsum and Nova, they were closed. Kids weren't coming out, so it was like, "Ah, something is going on." So, they were given prior warning like, "Hey, don't come out because we are going to kill a few Americans." It was a ghost town. One thing I didn't know is right at the corner, when you make that sharp left, if you go straight, there's a road back straight through there, and it will come all the way back to MSR. So, hindsight is 20/20; I would have been able to punch back through there and get on over there and start making my way back, but I didn't know it, nobody knew it was there. So, that shop on the left was the last building before you made that left.
In the grand scheme of things, I ended up back in Ramadi in '06. A bunch of us had been there before, so we knew the area. We knew where to go, so we didn't need to go walking around for no reason. We knew the problem areas.

Janney: After the 10th and throughout the month of April, it was much more of the same. I guess you guys were just trying to take back the city from those guys.

Scott: Well, so the 10th happened and they got their dicks beat to the turf. There was a lull for a little while. They figured out that they could not come at us conventionally because now we're not in those SASO operations anymore; we're on the offensive. We're out there to fight and we know you're here. So, they turned to more IEDs, shot a few rounds, later on, a few firefights here and there, but nothing to that magnitude.

Janney: Let's shift gears here a little bit, can you tell me any funny stories about Lt. Ski or any of the other guys that you knew that didn't make it back?

Scott: So, Lt. Ski, I didn't know him that well. He got dropped to us in December of '03. When he showed up, he was quiet. He would drill PT, stuff like that but he kind of knew his role as Lieutenant, as far as not wanting to get in everyone's business hardcore, which is a very good trait for him. I was brand new and so I didn't know anything. By the same concept, a good Lieutenant should keep their mouth shut and listen and he did that. He wanted to learn from people which made him a great leader. I remember him at March AFB, always working. I don't think I ever saw the guy play. In Iraq, we didn't have many Humvee drivers. They just kind of gave us this hodgepodge of vehicles and picked out who you wanted. They said, "You're going to drive, too" and you said, "Okay" and they gave you a little course. We'd drive a route around Combat Outpost and they'd say, "You're good." So, on the 5th of April, they brought up trucks and Lt. Ski asked, "You're driving?" I replied, "Yes, sir." He asked, "How old are you?" I said, "19, sir." He was with us a short time, but he was always on the ball; he was always working. Always moving, figuring it out.

Janney: What do you remember about Layfield, Jerabek, Cobb, and Crowley?

Scott: I did not know Jerabek all that well. I think I saw him a couple of times. He was with the weapons platoon. We were kind of spread out, everyone was doing their own thing over there. He would get tasked with, "Okay, you're going to go with these people today." Crowley, hard worker. Crowley and I were fighting each other for the team lead position, and he ended up picking up Lance Corporal. We were good friends, always just like the rest of us, figuring it out.
Layfield, I ended up fighting with him on the 6th. To show the kind of person he was, before we deployed, they needed a few guys on a Saturday. I think they had to drive to San Diego and pick something up. But Layfield knew I had been out pretty late the night before working. They kept banging on my door, "Hey, grab your shit. We've got to go do this." Layfield said, "You know what, I got this" so he ended up letting me go back to sleep. He was a good dude.

Janney: That's a neat story. Do you have any other stories that you can share about Sims or Otey?

Scott: Sims was a senior Marine, he was Corporal. I don't think he would hang out with us. Otey, great dude! Same as Lopez. I get to 2/4 and one day we play this game called the Echo Challenge. The team that won got a 72 (hour pass) or something like that. On our team, I forgot the 4th member, he ended up not deploying with us, but it was myself, him, Lopez and Otey. Otey had a strong personality. He wanted to lead. He took charge and led our asses through that thing, and it was a lot of fun. I think it's called the Balling game or Molly game. They have a bunch of stuff that they're going to lay out, a bunch of different items, and you're going to have about thirty seconds to look at it, and then you're going to have to complete a task. Then you come back and the group needs to tell them what you saw, what you had memorized. It was a memorization game. We were sitting there and Otey looked over and said, "You can speak up anytime." I said, "I saw, this, this, this and this" and he said, "Thanks, finally" kind of giving me that boot prompt, "Hey, open your mouth."

    Thinking back, I was actually incorrect on when our sniper team got killed. That was 21 June. 21st of July is when Calavan died. He got hit by a big IED and he took shrapnel to his temple. Calavan and I were out at the recruiting station out of Marysville, Washington, so we went to boot camp together. We were in different platoons, but we ended up going through SOI together. He was in the Weapons Company, and then we got dropped to Echo Company together. It was kind of cool knowing someone from home. Contreras, good dude. If you could put a megaphone on someone he would be about that loud.

Every morning, at 0530, whenever we were to get up, he'd be the first to wake up, "WAKE UP ECHO!" It was loud. He would wake your ass up.

6 April 2004 Route Gypsum Ambush And QRF

LCpl Aaron Vergara

Janney: This is August 31, 2019. Tell me when and why you decided to join the Marine Corps?

Vergara: It had to be 1991 watching the Gulf War on TV and I was sitting there, and my mom was on the couch and I was on the floor. I was about 9 years old and I was watching the Gulf War because it was on TV. I know the Vietnam war was televised and everything, but this was the first one that was actual new age everything and I turned around and told my mom that I'm going to be there one day. I'm going to Iraq one day. I remember her getting up and smacking me across the mouth, telling me to shut up and not to say that. Ever since then it was, you know, the Marine Corps. I wanted to be a Marine and then we got cable because back then there were only like four TV stations. Then we got the cable box with the History channel and I was just addicted to the History channel and WWII, Korean, and Vietnam. When I became 17, I went and did the delayed entry program into the Marine Corps. My mom moved down here. I was in Pennsylvania living up there towards Hershey my entire life and my mom was a jockey for a horse racing track and they were trying to get slot machines and gaming put in up there, but it didn't pass. So down here in West Virginia, it passed, so she moved down here. I was in my senior year and she gave me the choice to either move down there with me or stay up here and finish up your senior year. So, I told her that I was going to stay up in PA and finish my senior year. Well, between me being on my own at 17; I just went buck wild and I was failing out of school. Failing twelfth grade and then I got to the point where I was, you can put this on the record, I already did my time in the Marine Corps so they can't do anything to me. I was tripping on acid one night with a bunch of friends and we were partying and when I went to sleep, I had a dream of seeing myself in dress blues and sitting there staring at my dress blues, my reflection told me if you don't get your shit together you will never be a Marine; you'll never wear this uniform. That next day, I dropped out of school, got my GED. By this time, I was 18, and I put myself in the Marine Corps and within two weeks, I was in Parris Island.

Janney: Wow. That's a pretty powerful dream, too.

Vergara: Yeah. I told myself, I haven't touched drugs since then, but I had that trip and that vision came to me and I thought, you know what, if I don't get my shit together, I'm not going to be anything. I got out of that situation up there and I'm glad I did because if I hadn't I would have probably died because four of my good friends died of drug overdoses throughout the years. If I wouldn't have gotten myself out, I probably would have been in the exact same state as they were.

Janney: I completely understand. Because that's a pretty eye-opening moment and it's certainly a visionary type dream that caused you to change your life, so I'd like to leave that in if it's okay.

Vergara: Yeah, that's fine. There's nothing they can do to me now. I already got my honorable discharge.

Janney: All right, so, what year was it that you joined the Marine Corps?

Vergara: It was in March of 2001. So, 9/11 hasn't happened yet.

Janney: So, when that did happen, were you already with the unit?

Vergara: Yeah, I was with Echo Company, 4th Marines, and we were actually in machine gunner school down on the main side of the base there. I forget what that camp was called, but it's where they have all the specialty schooling. The Recon Battalion is in there, but we were doing our machine gun school. I remember Stayskal waking me up because he was a machine gunner, too. He woke me up in the morning and said, "Hey let's go to the chow hall." I just heard the rumors that a plane hit the World Trade Center, and I was just thinking back during WWII, a B-25 hit the Empire State building because of fog. So, that's what I was thinking - maybe it's foggy or the plane had problems. I was thinking maybe a Cessna or a small passenger plane hit the World Trade Center. I got up and dressed and we went over to the chow hall. We got our food and were sitting down, and they had the news on. They were showing the World Trade Center, and as we're sitting there watching it, the second plane hit and fucking everybody in that chow hall said, "We're going to war." The base put out the alarm and locked everything down. They rolled Abrams tanks up to the main gate and had them sitting there. Helicopters were starting to fire up and there were patrols around the base and it was all hell. We got pulled back into the school. They sat us down and said, "Listen boys, this is an attack. It's not over yet. There are still planes missing and we're going to war, so now is the time to get your shit together because it's happening." I just got that weird sickening feeling that here we go.

Janney: When you guys deployed to Ramadi in February 2004, what was your rank?

Vergara: I was a Lance Corporal, E3 and was a machine gunner.

Janney: And were you on a 240G, a SAW, or a M2 "Ma Deuce" 50 caliber machine gun?

Vergara: I was on a 240. The line companies like Echo, Fox, all them, we were all 240s. Weapons Company had the heavy guns, the 50 cals and everything. Then, the line companies started taking 50 cals for their Hummers. Since we were machine gunners, we were trained on all machine guns, the Mark 19, the 50 cal, and the 240. Our Company and the machine gunners would rotate out between the 50 cals and the Mark 19 and the 240s, but I always carried the 240.

Janney: What was your general assessment of the situation when you arrived in Ramadi?

Vergara: They came in on trucks. The first thing was the rest of the Battalion went in on trucks from Kuwait. Somehow our armor plates got messed up and were in the shipping containers instead of with us, so we couldn't drive in because we didn't have armor plates for our flak jackets. So, they took us down to a Kuwait airbase and we got on C130s. We all fit on one or two C130s. When we were coming in on the C130, they said, "Hey we're dropping the door, but we are not stopping. You guys got to go. We'll hit the runway until we start taking off, and you get off. Everybody's got to move." They weren't lying. I mean that ramp door was a good two feet off the ground there. The wheels touched and the door was up a little bit. It was pitch black. I got to the end of the ramp and stepped out to touch ground, but didn't touch ground. I went face first to the deck. By the time I got back up, that plane was taking back off.

They were afraid they were going to get a mortar attack or rocket attack or so that was kind of setting the tone. I thought we were doing security operations and everything was peaceful. Everybody was out there buying DVDs and stuff at stores. What's changed? What's going on now? We got on trucks and got to Combat Outpost that night and offloaded the trucks. Golf Company was already there, so it was Echo and Golf in Combat Outpost. Golf Company was manning the positions and next thing you know, within 10 minutes of us getting there and getting our stuff off the trucks, all hell broke loose. All the posts started opening up. There were tracers everywhere. I thought, "Jesus Christ, what the hell is going on?" We got all our stuff situated into our rooms and then the Staff Sgt. came in and said there was an attachment of the Army that was still there. I think they were a Puerto Rican National Guard unit. They were American, but I guess they were Puerto Rican.

They came in and were asking for volunteers, "Who wants to go out on patrol? We'll give you guys a tour of the city" and I volunteered. Ryan Miller volunteered, too. It was a hodgepodge, not my whole platoon - just a couple of us got together. Miller was carrying the 240 and I carried my M16 that night. Cpl Frank Gutierrez was there, too. It was basically the Army doing a foot patrol around the city and it was just really weird, kind of creepy. In the middle of the night going out there. People were not really supposed to be out there because of the curfew, but you're seeing Iraqis on the rooftops. We got back into the base and that was my first little taste of Ramadi. Just got in there that night and went out and came back in. It was pretty eye opening.

Janney: After that, were you just patrolling and handing out soccer balls?

Vergara: The lead up to April, third platoon was out there doing IED searches which kind of got me. The Army is out there in their armored vehicles with IED specialized people going out. We're out there kicking coffee cans, picking up bags, and finding IEDs that way, hoping they don't go off on us. We were handing out soccer balls, going to schools. At first, the kids were all happy and nice to us. Then, one time, we were out there on patrol and stopped by a school. All the kids came out and started throwing rocks at us. It got to the point that we said, "Let's get out of here before a hand grenade comes in." We don't know it's a hand grenade and think it's a rock. We got out of there.

The 25th of March was the first time that I saw actual combat. I actually got an award for that action. Every Thursday night, the Iraqis would mortar us. The insurgents would start dropping mortars on the base. That night, Royer told Lt. Cogan, "Go get your boys together, go out, find them and take care of them. We're tired of it. We've got to let them know that we're not going to sit here. We're not the Army, and we're not going to just sit here and be mortared and not do anything about it." So, Lt. Cogan got a few of us together. Half of us were on post, the other half that were not on post was the QRF (quick reaction force.) So, they got a couple of us together.

My squad leader Corporal Medina got PFC Lloyd, who was an 0311, but we were so undermanned that they brought him over to the machine gun, so I trained him on using a machine gun. So, Lloyd was my gunner and we took out a group of QRF from third platoon and we went out there looking around with a Humvee, a 7-ton truck that my gunner was in, and I was in the back with Doc Medina, and a couple of other guys from the 11s from third platoon were in the rear Humvee. We were out there on Route Michigan. From what the Army told us because the Army had sensors up everywhere, they could see where the round would reach its peak and impact. They could backtrack it. So, we had this vicinity area where they were mortaring us from. It was the middle of the night and we couldn't see. We made a turn and Lt. said, "Okay, we'll make a turn over here" because these guys had GPS in the front Humvee. We stopped because we were on this dirt road and came to the middle of a field and it was a dead-end. The road wasn't on the map. We're all sitting there in the back of the 7-ton tripping around like we were a bunch of rednecks on the back of a pickup truck, basically like we're going out hunting and people are yelling out, "Sooey!" I just got this sick butterfly feeling in my stomach. For some reason, it's kind of like somebody spoke right into my ear and said, "Get ready" and as soon as that happened, all hell broke loose. It was like Star Wars, you know. We had red tracers or orange tracers and then we had green tracers and all hell's breaking loose. I remember my gunner up in the top of the turret on the 7-ton started screaming and screaming. I started climbing my way up to him. He turned around and his hands were full of blood and he was screaming. My buddy Rogers was sitting shotgun in the 7-ton and I told him, "Shut him up" because he was just yelling. The way the lights were on the road; they were backlighting us and we couldn't see in front of us. People don't understand here, but out there, when it gets dark out, it gets dark. You can't see a foot in front of your face because there is no other light. The road lights out on Route Michigan were lighting us up, but we couldn't see in front of us.

Janney: Yeah, I know exactly what you mean. You guys were silhouetted, and they were in the dark out of the cone of light and so you couldn't see them, but they could see you perfectly.

Vergara: Yeah, and we just saw tracers flying towards us. I remember Rogers grabbing Lloyd and pulling him down in the truck and telling him to shut the fuck up and the driver, Bayonne was his last name (he got killed over there), but Bayonne opened the door up, jumped out and pulled Lloyd out of the truck. Doc was in back of the 7-ton with us; he jumped down, grabbed Lloyd and they drug him back to the rear Humvee out of the incoming rounds; that was basically where the casualties were being taken care of while we were trying to figure out what the hell was going on. It felt like it was coming from every direction.

I was on the top of the truck because I've got to get to the gun. I've got to lay down some fire because right now, we're getting our asses handed to us because no one knows where it's coming from. Tracers were flying in both directions. I remember getting up towards the gun and then I was laying on my stomach, my legs hanging down the back of the cab. I grabbed the gun, took it off safety, pulled the trigger, and nothing happened. I thought, "What the fuck?" I charged it, the bolt went to the back, pulled the trigger, and the bolt slammed forward. I'm like, what the hell. I'm looking around. I guess when I pushed Lloyd down and Rogers pulled him down into the truck, Lloyd ripped the belt of ammo out of the machine gun.

So, I looked down the hole in the turret and I'm looking at Rogers and I yelled to him, "Look for the rounds. The belt fell down." He's down there and he's trying to cover up the flashlight beam and trying to look around and next thing you know, a RPG hits the front of the truck. It picked that truck up, I want to say, at least a foot off the ground. It picked it up and slammed it back down and I thought, "Fuck!" My squad leader grabbed me by my ankles and said, "What are you doing, fool? You ain't dying today" and he pulled me down. My whole thing was I needed to find the ammo.

So, I got pulled down from the top of the truck and jumped down and jumped in the cab with Rogers. We were in there just digging around looking for the rounds. Then, we started laughing. Him and I, he had a sick sense of humor like I did because the 7-ton driver had a CD player on the dash with speakers plugged into it and it was playing Sublime. We were laughing at that. Everyone else was panicking, but Rogers and I were laughing. We got along really well because we had the same sick sense of humor. We were laughing and then the second RPG hit the tire of the truck and I told Rogers, "We need to get out of here. We're the biggest target. They can see us."

So, Rogers and I got out and there was what we called a shit trench; it was a trench between Route Michigan and this back road. It was nothing but oil, waste, and shit, and Rogers and I started telling everybody, "Get to the shit trench. We're lit up. They can see us." I went to the back of the 7-ton and I remember seeing people back there. PFC Nelson said, "I got hit" and he's looking at his hand and his hand was a little mangled up. I told him, "Get down off of the truck; go to the rear of the Humvee. Doc is back there." He jumped down and that's when everybody in the truck jumped off, "They can see us. Get off the truck." Everybody started getting off the truck and going to the trench. There was a Marine there that had a SAW, so I gave him my M16 and took his SAW. I went over to the passenger side of the back cab of the 7-ton and I was looking out because these rounds were coming in that direction. I could see that there was a dirt berm. That's where they're at, behind that little bit of a wall, still firing at us, but it wasn't like it was before.

I see PFC Mendoza running out in between that front Humvee and our 7-ton; he was out in the kill zone. I told him, "I'm going to lay down fire," because he was stopped out in front of the 7-ton truck. I said, "You're out in the open. I'm going to lay down fire. Haul ass back to the rear Humvee or get in the shit trench." He said, "Okay." I said, "On the count of three - one, two, three" and I just held that trigger down on that SAW and I dumped rounds into that dirt berm. After that, it was a little shot here and there. Basically, that was the end of the combat. 2nd platoon came out to reinforce us. By that time, there were maybe 7 Iraqis that did attack us that night and there was only one left alive. I think one of them got away, and the rest of them were dead. But there was one alive. It was a high school aged kid, and he was telling them, "I'm sorry. I'll never do it again. Don't kill me. I have my finals tomorrow." That was our squad's first and my first time in combat.

Janney: Aaron, when you said it was Doc, was it Clayton or Urena or which Doc was it? Mendes?

Vergara: Doc Escalera. Doc was down there in the middle of the shit taking care of these men. When you're in the Marine Corps and you've got the Navy doctors and you're always laughing and playing jokes on them because they're not Marines, but they wear the uniform. Docs are always falling out of the hikes and getting yelled at for not doing what they're supposed to do, but when shit hit the fan, I was surprised. Here was this Doc, not returning fire or anything and he is worried about patching these Marines up. That was what changed my mind about the Navy Corpsmen. These guys are badass. They go out there with us. We gave them rifles after that because all they had was pistols at the time. After we were in combat, we gave them rifles to protect themselves. Doc Escalera really got my respect that day. After then, our little group that we had within the platoon was my squad leader Medina, Escalera, and I. We were always together. That really got to me that as much as we make fun of Docs, they risk everything to take care of these Marines.

Janney: I know you love your Docs. What happened in the ten days leading up to the 6th of April?

Vergara: Everything got quiet. It was really quiet to the point where you felt like everybody in that town was watching you. Because they would watch us. We would always change our stuff up, but being in the Marine Corps is all about routine. We would get into a routine and they would see that and then they would attack. There was like a lull. Nothing happened; no IEDs; no nothing.

All of a sudden, the IEDs started picking up. The night before the 6th, I believe it was, we went out;

3rd platoon, on an observation post up there towards the graveyard in the city coming into Ramadi. That graveyard up on the hill; it was like the highest point in Ramadi and we went up there and we're watching the streets for IEDs, for people to come out and plant IEDs. It got really fucking cold and that's the night that I was carrying that 240 up there. Because when Lloyd my gunner got hit, he went out to Berlin because he was done for. He got hit right in the knee with a machine gun and it tore his knee all up.

It was really hot during the day and all of a sudden, it started getting really cold at night. Because it was hot during the day, no one would pack anything for cold weather. I knew it was getting cold out, so I wore my pack and I brought my big sleeping bag with me. We were out there that night and 50% of us were up watching the road while the others slept. We would switch on and off throughout the night. When it was time for us to lay down, I got my sleeping bag and everybody else was freezing and Doc Mendes said, "I don't have sleeping bags, but I have 4 or 5 body bags that you guys can use for sleeping bags." Most of us were like, "No, no, no. That is bad juju." Staff Sgt. Walker, all those guys that died the next day, slept in a body bag that night. Staff Sgt. Walker and all those guys that were in that Humvee that got shot up slept in body bags that night. From that day on, if you want to call it superstition or whatever, but that was a bad thing.

Janney: Yeah, that's ironic.

Vergara: Yeah, back to that night. The other part of 3rd platoon was at base, so in the morning when we got up after our OP, we got up and hiked back down to Combat Outpost. We got in and we were passing our other squad from the third platoon; they were going out to do a patrol. We were getting back there and it was our time off to wash clothes, get breakfast, whatever, and within an hour, I remember them coming over the radio saying, "Hey, we're being ambushed. People are out here acting weird; there's barriers up." The next thing you know, they got ambushed. Golf Company got ambushed at the same time that our Echo Company squad got ambushed and just all hell broke loose. The whole Combat Outpost just emptied within an hour. I remember getting on the 7-ton, putting the 240 up in the turret and Gunny Coleman came out there and said, "Take that off there" and I asked him, "Why, what's going on?" and he said, "I got you something" and next thing I know I see them taking the Mark 19 down from the top of the post and they were bringing it up to me. They put the Mark 19 up there. I gave the 240 to my squad leader Medina and our 7-ton was loaded up with guys from 2nd platoon.

Lt. Wroblewski was in front of us in a Humvee, the guys from 2nd platoon on the truck, and there was a Humvee behind us that had a couple guys from 2nd in there, too. We went out Route Michigan and got up there toward Gypsum and Nova and we got hit. The Humvee stopped that had Lt. Wroblewski in it, our 7-ton stopped, and the Humvee behind us stopped, but the rest of the vehicle convoy kept on going. That convoy, as soon as they made it up there, got to the turn up there where the shops were at up by Gypsum and Nova, they got ambushed pretty hard. The insurgents had a DShK, a Russian 50 cal machine gun sitting right up on the intersection and just totally wasted everybody there. What stopped us dead in our tracks and why Walker's guys kept on going was we got ambushed, too. We started getting hit by fire and we stopped. We got out and started returning fire. The Mark 19 was so close to targets that it wasn't arming itself, the rounds weren't arming and it was just punching holes into stuff and rounds weren't going off, so I ceased firing because it was doing no good. Then the firing in our area stopped. Lt. Wroblewski had a headset up to his head and he was talking, and he had the map out. It was a single shot. One single shot. So, it had to be an insurgent marksman or a sniper. I remember Lt. went down. I don't know if Doc was there, but I remember they were putting stuff on his face and they threw him in the back of the Humvee. We started getting sporadic fire here and there, and we were getting everybody together, trying and trying to get a hold of headquarters. Captain Royer was out there too trying to get ahold of them, "Hey we're going back. We've got a cas evac. We've got to get back."

They loaded Wroblewski up into the Humvee. We were in the 7-ton, and Corporal Gutierrez had Lt.'s weapon, gear, and his rifle and I said, "Hey I don't have any other weapon. I have a 9 mil and the Mark 19 is useless right now. We're too close for anything." So, he gave me Lt. Wroblewski's gear and his rifle and I carried that throughout that day. When we got there, there were cars on the sides of the streets and I remember that 7-ton pushing cars out of the way, running over cars to clear a path for the casevac. We got back in the Combat Outpost and the Army was just getting in there. They're coming in with their tanks and their armored personnel carriers and they were setting up, getting organized to come out and reinforce us. The Humvee went over to the medical bay.

I ran to the shitter because I had to shit really bad because that happens when the adrenaline starts kicking in. I got ammo and ran back to the truck and we went back out again. By the time we got back out there, the fighting was almost at a halt. It was basically clean up. That day we got our asses handed to us. We lost a lot of Marines that day and it was just disorganized. I knew that pissed Captain Royer off that the Iraqis did this to us.

I believe it was the 10th, when we went back out there. When we went back out there, this time we were ready. We went out there early in the morning. They got our translator onto the speakers of one of the mosques and told them, "We're out here, we're waiting, come out. Let's play. We're ready for you, let's go." As soon as that happened, all hell broke loose that day. But that day, from now on, you Marines are off your leash. The Army came in and gave us permission – do what you guys do. Because up until then, we had our hands tied, and that pissed us off. We barely had any ammo previously. They had given us a can of ammo which is 400 rounds for the 240 machine gun. Throughout the little fire fights and shit leading up to that point, we were running out of ammo. That's what happened on the 25th. All of us ran out of ammo and it was to the point of, "Are we going to have to fix bayonets?" This was kind of ridiculous.

On the 10th though, the Army came down to us and said do what Marines do. We were told by our Staff Sgt. and by Lt. Token, "We're going out there and we're doing what we do. We're going out there for a fight. You guys are off your leashes. You guys do what you do." That's what we did. We mopped up that day. Between us, we had three or four 7-tons full of prisoners. My machine gun squad was in a 7-ton going up and down the road, just taking fire, returning fire, and picking up wounded. Brains were scattered throughout the thing. We were out there with the Army too, so Marines were fighting side by side with the Army. That got to me. That was kind of cool. Marines laying down fire with Army guys beside them. I thought this has to be the first time since WWII that the Marines and Army have been shoulder to shoulder fighting. That day was a good day. We did take a few casualties that day, but we handed the insurgents their ass.

After that, they never fought us face to face again. It was always IEDs or rocket attacks by made up positions that were foot patrolled. The rockets were set off that way and the IEDs were set off that way at the time. From that day on, that city was locked down; even the IEDs stopped. We had the town locked down. I would say June, July, there were little fire fights here and there that they would try to attack us, but I mean it was moot. We just handed them their ass every single time after that once we were off our leash, we could get the intel and do what we had to do. Afterwards, sitting back here in the civilian world and watching ISIS come in and take over Ramadi, it pisses you off that we had that city locked down. People started returning back and resuming their regular lives without insurgencies and without the insurgents there. Years later, that town is basically destroyed now because of what ISIS came in there and did.

Janney: You know, when I was there in 2007 and '08, I interviewed random people that I came across and I said, "Look, there's folks back home that say bring you guys back. How do you feel actually being here?" Every one of them except one guy said, "Look, let us stay and finish our job, otherwise our sons and daughters will be here in 10 years doing this same shit all over again." So, I totally understand. It's the politics of it. Bush didn't seek a status of forces agreement and the Obama administration sure didn't want to stay there. We withdrew because of the lack of a status of forces agreement. Let me go back to the 6th just a little bit because I have some questions. Were you with Sgt. Valerio and Cpl. Eric Smith or was he in Wroblewski's Humvee? I'm trying to clarify in my mind who was where.

Vergara: Eric Smith came up to us because the 2nd platoon got scattered. When we got hit, it was all-out, it was crazy. We had everybody from 1st platoon, 2nd platoon, and some guys with 3rd platoon with us. When we went there with the QRF, everybody on base who was able bodied got together, got gear, and got in trucks and went. We got out there and I remember after Lt. Wroblewski got hit, Eric Smith was up there trying to get everything organized because it was chaos at first. That day was nothing but chaos. We got our asses handed to us because we weren't organized. Smith came up to the 7-ton driver saying, "Hey we've got to get the wounded back to base and I can't call it in. The headset was hit, and we don't have any other radio. We don't have another headset."

Janney: And the radio in the lead Humvee with Staff Sgt. Walker was destroyed too, so you guys had no comms.

Vergara: Yes, Staff Sgt. Walker and that first Humvee kept on going. There was a bend there and as soon as the first lead Humvee got around the bend, it was weird like the Matrix when they show the bullets flying in slow mo' - that haze behind them, that's what it was like. It was like a wall of haze. It doesn't register in your brain right away until you realize that they are bullets and they are shooting at you. It was just like this weird haze. The Humvee in front of the 7-ton where Lt. Wroblewski was stopped, our 7-ton stopped, and the Humvee behind us stopped because they had nowhere to go. Those other Humvees with Staff Sgt. Walker kept on going around the bend and they weren't looking behind them.

We were just trying to get up Route Gypsum to Nova. That's where we wanted to get; that was our meeting point. We were going to create a line there, but that was where all the insurgents were, and we didn't know it. Our guys from 3rd platoon that got hit were trapped in a house, a small little shed and they didn't have any radio contact with us at all either because they were basically pinned down. Any time they moved their heads or whatever, the insurgents opened up on that house. It was complete chaos that day. After Lt. Wroblewski got hit, the handset that was with the radio went down and we had no comm. I remember Marines running up to us coming out from the right side because where we were at, to the left of us was an alleyway and then there was a house, directly to our, I want to say, our 10:00 and that's where we thought the sniper shot came from that hit Lt. Wroblewski. I remember opening up on that thing with the Mark 19, but it wasn't doing anything. After we all fired back, their firing stopped and that's when Lt. Wroblewski got out of the Humvee. He was talking on the headset and had the map out and that's when that single sniper shot rang out. We opened back up again. By this time, no one else got hit, so we all thought that maybe we got him. This guy had to be a trained sniper because you don't fire two shots from the same position. So, he knew that he wasn't going to fire again. That's when they were taking care of Lt. Wroblewski. I remember them putting him back in the Humvee and we said, "We've gotta get him back to the cas evac area." When we were leaving, a helicopter was coming in to take him away. I remember the next day, they said that he died. I was thinking to myself, "How did he die?" It shot through one side of his mouth and came out the other side of his mouth; straight in and straight out. I have no idea what could have happened. That always blew my mind.

Janney: Yeah, John Wroblewski heard that he actually died in the CSH (Casualty Support Hospital) back in Baghdad and there is a nurse there that he met that said that she was with him when he passed away. I don't know if it was just blood loss or what exactly happened, but it was just terrible that you guys lost so many brothers that day.

I was just trying to clarify in my mind where Sgt. Valerio was versus where Cpl Eric Smith and Lt. Ski were. Valerio said that Staff Sgt. Walker and that lead vehicle kind of pushed ahead when they started up Gypsum. Valerio and Walker had this conversation, "Let me go, I've got it" and, "No, no, I've got it." Walker basically told Valerio to stand down, that he was going to be the lead vehicle. That was probably the only thing that saved their lives was that Walker pushed ahead and then they got hit at Nova and Gypsum.

Vergara: Yeah.

Janney: Can you tell me about the men that didn't make it back? How well did you know them?

Vergara: Pretty good. The day before we went out on that night patrol, we were on a working party filling sandbags. Talking to them about what we were going to do when we got out. All the money we were going to make here because it was tax free. Cherry wanted to get a crotch rocket. And then the next day, none of those guys made it.

Janney: What about Chris Mabry, Jerabek, or Layfield? Did you know those guys at all?

Vergara: Yeah. Jerabek, I believe he was a 0341. He came over and trained as an 0331 with us because the mortars were just going to stay back there at base. They didn't need too many people to man those, so any able body got switched around from positions where we needed them. Jerabek was really quiet and then he started opening up. He was really shy, but he started opening up. He was a good kid. I really liked him. I've got a picture with him and one of the Marines that came in with me was Ellwell. Him and Ellwell looked identical; they both wore those, we called them BC glasses, the birth control glasses that the Marine Corps issues, the Drew Carey ones. I've got a picture of those two guys with them wearing the same glasses. After Jerabek got hit that day on the machine gun, that picture of those glasses on the ground with blood on them and crushed really hit home pretty hard because he always had them on.

Janney: Yeah. David Swanson took that picture. That's a pretty tough picture to look at.

Vergara: Yeah. Every machine gunner that came in after me when we got back from Japan, the new set of Marines, they all got hit: Calavan got hit, Jerabek got hit, Carmen got hit, they never made it back. I just wish that they had more experience, more training, because it was just tough. You know when you come back and you're getting trained up to go over there and those guys just got dumped right in there from SOI. Right into us, maybe 2 months before we deployed. We tried to train them up, but never had time. They just didn't have the experience. I felt really bad.

Janney: What about Layfield or Cherry? Did you know those guys at all?

Vergara: Somewhat. They were in the platoon. They were 0311s. When we did our training, I was in the weapons platoon with Echo. We always did our own stuff training on the machine guns and the 11's did their thing. Then every once in a while, we would go out to the field all together and match up what we knew with what they knew and do maneuvers. But those guys got hit kind of early on. When we went over there, they told the platoon, "3rd squad machine guns, you're going to be with 3rd platoon," the 11s you're attached to after that, getting to know those guys, seeing them, but it wasn't like you really know them.

Third platoon was in two or three different berthing areas or huts in one building. The higher ups like the Lance Corporals that were team leaders and squad leaders were all in together. It's the Marine Corps mentality - the lower grunts like the 0311s were all together and as you went up the ranks, you got put in with higher ranks. The squad leaders basically got their own room. I was a Lance Corporal and I was in there with the other senior Lance Corporals that I came in with. There was a Sgt. in our berthing area with us. I didn't really know the newer guys. I knew our newer guys in Weapons, but I didn't really know some of the other new guys, the 0311s, you know. I saw them and we went on patrols together, but when we went out on patrols, I always knew the guys that I came in with and who I would talk to and everything.

Janney: Right. What about Staff Sgt. Walker? Did you know him at all?

Vergara: Yeah. I remember the first time I met Staff Sgt. Walker. We knew that 3rd platoon got a new Staff Sgt. A couple of the senior 11s that I came in with - before we deployed, we were all getting together one night to go out and go drink and party and the Weapons platoon was on the top deck of our area. It's basically like hotel rooms, but not hotel rooms. It's a three-story building and Weapons platoon was on top. Mixed in on the bottom was 3rd platoon, 2nd platoon, and then 1st platoon was on the ground level and that was where my buddy and I were one night. We were down there with Chris McIntosh and waiting for him to get ready because we were going out. I remember this big guy coming up to us. He was sitting out there talking and saying hey. He was wearing either a Rancid or a Drop-dead Murphy shirt, had baggy pants on with the wallet chain, and he had a flat-billed hat. Like the California style - they wear their hats that way. He was down there talking to us and I looked over at him and I said, "Who the fuck are you?" I thought maybe he was one of the new 0311s or something. He said, "I'm Staff Sgt. Walker, muthafucker" and I was like, "OOH!"

Janney: Laughs.

Vergara: Then when we got over to Iraq, I was with 3rd platoon, so Staff Sgt. Walker was their Staff Sgt. A really down to earth, good guy. But that night that we were going out, he gave us his number, wrote it down and gave it to us and said, "You guys have got it. If you get drunk, it doesn't matter what time it is, call me and I will come pick you guys up. I don't want you guys getting NJPed or getting in trouble before we deploy. Whatever you guys get into; whatever is out there, my lips are sealed." I was like, "Damn." So, when we got to Iraq and he was our Staff Sgt., I was kind of happy because he was one of the Staff Sgts. that weren't too proud that they were a Staff Sgt. Some you can't talk to or whatever because you're a lower rank. The only time you could talk to any of them was about business. Staff Sgt. Melana wasn't that way. You could come up and talk to him about anything and he would talk to you man to man. Not like, I'm a Staff Sgt. and you're a Lance Corporal or PFC; don't talk to me. Staff Sgt. Walker was the same way. Staff Sgt. Walker would sit down and bullshit with you and everything, but you always gave him respect. If he said something, you did it. You guys were friends, but there was that rank thing there that some of the Marines coming in did not understand. That yes, we're friends, and you can talk to me, but at the end of the day, I'm a higher rank than you and if I tell you to do something, don't back talk. That's how Staff Sgt. Walker was. You always did what he told you to do and if you asked anything of him, he would make sure that it would get done. He would do it to the best of his ability or he would see if he could get it done for you. He was what I thought a Staff Sgt. should be. Because the Weapons platoon went through a bunch of platoon Sgts. because we were rowdy. There was a couple of platoon Sgts. that quit and transferred because they didn't want to fucking deal with us. We got Staff Sgt. Melana and we took a liking to him and he took a liking to us. Same way with Staff Sgt. Walker. I met him and it was the same thing. I really liked him, I really did. That day after he got hit, that was pretty hard. After that, we got Staff Sgt. Martin. Staff Sgt. Martin was the same way and I felt very fortunate and lucky that we got another Staff Sgt. that was about his men; about his boys. You could talk to and bullshit with, but they knew their shit.

Janney: What about Lt. Ski? Did you have much interaction with him?

Vergara: I'll give you a little example of what our Lieutenants were before. Our Lts. before were the Lieutenants, "We've been through OCS and college, so we're smarter than you and we know what we're talking about." And I hated that, I fucking hated it. He's a guy who is two or three years older than me. I understand that you're an officer and giving me orders. They were like, "I don't respect you guys, but you guys have to respect me before I respect you." We had Lt. Cogan when we deployed with 3rd platoon and he and Lt. Wroblewski were good friends because they went through OCS together. I imagine Lt. Wroblewski was a lot like Lt. Cogan because he was the same way - you could talk to him. It came to the point where Lt. Cogan knew that Vergara fired first and asked questions later, so I was always in his Humvee; I was always the gunner in his Humvee. I knew Lt. Cogan really well and whenever we'd get back to base, him and Lt. Wroblewski would always bullshit and joke around together, so that's how I knew Lt. Ski. He seemed like a really great guy. You would say hi to him and salute him and he would nod and say, "Hi and how is your day going Vergara?" The officers that we got when we went over there were great officers.

Janney: Yeah. What about Lance Corporal John Sims? Did you know Sims?

Vergara: Yes. Sims was really good. He was one of the 0311 Marines that came in with me. I think he came in a month after I did. Real loud and rowdy. He was a Southern boy, so he had that Southern twang. Obnoxious, but really good hearted, you know?

Janney: Did you ever hear the story about partying in the back of the U-Haul and Sims jumping off the mom's attic onto the snowboard?

Vergara: Yeah, I rented that U-Haul.

Janney: Did you really? Yeah, that was a great story!

Vergara: I was driving the U-Haul because it was a stick shift. I went out and rented it and it was all of us that went in there and partied. We raided the Company lounge upstairs; chairs, sofas, and shit and put them back later. They had that damn snowboard and a half pipe ramp set up in the back. I remember I was driving. At the time, I think it was a Corporal riding shotgun and Rogers and all of them were in the back, and we had walkie talkies. We came to a stop sign where this mall was at and that U-Haul was rocking back and forth and those guys were back there just partying. We had a walkie talkie to talk to them back there and say, "Settle down because everybody is looking at us" because that U-Haul was just rocking back and forth.

Janney: That's hilarious, man.

Vergara: When I took the U-Haul back, I opened the back of it up and it just reeked of alcohol. The U-Haul guy started laughing and asked, "You guys have a good time?" I said, "Yeah, we had a good time".

Janney: That's great! That's a neat story. Can you think of any other guys that you knew?

Vergara: Otey was…man, that guy. He was a good Marine. He was a Marine through and through and he loved his job. I remember the day before he died. My thing with the Marine Corps, I was what they call a gear queer. I had everything and anything people could ever want. I had underground gear that people would come look for. Otey came up and asked me if I had a drop holster for a 9 mil pistol that went around your waist and clipped to your leg. I rummaged through my bag and gave it to him. That's when he was going over to the snipers and was going out that next day. That's the last time that I spoke to him, but man, I loved Otey. He was a good guy. Contreras was a wild man. We hung out a bunch of times. Lopez was really quiet.

Janney: I heard Contreras was quite the practical joker and heard about him showing up in a fireman's hat for a march or PT. Did you know anything about that?

Vergara: No, I don't know anything about that, but I remember one night we came back from partying. We had a formation out there Monday morning. Contreras was still fucking drunk and they were trying to get him out of his bed and they finally got him out there. He was in formation yelling, fucking screaming, and being a wild man, and they had to carry him away.

Janney: Oh, man. And you said Lopez was a pretty quiet guy?

Vergara: Yeah, he was really quiet. Everybody had their cliques. The Mexicans had their clique and so on. Myself, Stayskal, and McIntosh were the ones that went in between the cliques and partied with everybody and that's how we met Contreras.

Janney: Chris McIntosh, he's a wild man and hilarious. I got a chance to meet him at a reunion.

Vergara: Yeah, and he's a good kid.

Janney: He had a kilt on one of those days and had to show everybody that Scots really don't wear anything under their kilts. I said, "Chris, that's a side of you I wish I'd never seen."

Vergara: Alegre was the baby of weapons platoon. We all took care of Alegre because when he came in, he was only 17. We watched over him. He felt that people were watching over him because of his situation, but we always took care of Alegre. But they got hit that day and we lost Lee and Reynosa Suarez. Reynosa was a good friend of mine. He was deferential to Medina, he was my squad leader. That day hurt really bad because, you know, the 11s; we knew the 0311s we worked with them here and there, but like the weapons platoon guys, we'd been pretty much together for three and a half years straight so Lee, Reynosa Suarez, Alegre; when they all got hit; Calavan, that just really hurt.

Janney: Do you want to tell me about what happened that day? I don't know that story.

Vergara: There was a car parked out on Route Michigan and we would always go around it. It was just suspicious. That place was a junkyard with cars everywhere. But there was a car on Route Michigan that was sitting there, and we knew about it. We had reported it, but no one had done anything about it. The next day, that afternoon, I believe they were coming back from patrol. Quick reaction forces sometimes would go out and pick up patrols that were at the end of the time limit to get back to base. They would run out the Humvees to go pick up those guys and drive them back. That's what Alegre and all of them were doing – they were the quick reaction force that ran out. I think it was the 1st platoon that was out there doing a patrol. Alegre and all of them went to pick those guys up and on the way back they went by that car. Instead of going around it on the other side of the road, they were just trying to get back to Combat Outpost. A couple of the Humvees passed that car and when that last Humvee got past that car, it blew; that car was basically a VBIED; a vehicle borne explosive device. They blew that car and they took out the Humvee and took out everybody in back of it. I think the only people that got out alive were Alegre and Doc Melana, and Lee; they all got hit with shrapnel and everything. That was a bad day. That is the only day that I actually broke down; that day almost broke me. You know, because we thought Alegre was dead. We knew Lee died; we knew Suarez died. I remember my squad leader Medina; he was really good friends with Reynosa. He came up to me and he said, "Hey bro, don't do this right now. Don't break on me right now. I need you." I pulled my shit together and went through the shit.

Janney: I am so sorry, Aaron.

Vergara: Hmm. It's all right. I mean, it happens, it's war.

Janney: I know. I know it hurts though. I'm just sorry that you lost so many friends.

Vergara: That's the thing that got to me over there sometimes versus back here in the world. Over there, I was to the point that, "If I die today, then I die today. Ain't nothing I can do about it. Not my choice, I have no choice in it." What's written is written. It's a pretty cold and heartless way to live, but all of us over there were like that, you know? If it's your day, it's your day. Nothing you can do about it.

Janney: Hey, and this can be off the record if you want, have you sought any treatment for any of the issues you brought back with you?

Vergara: You can put it in there. The VA is really disheartening. I've gone there a bunch of times and their thing is medication, medication, and I told them I don't want medication. I don't want to be drugged. I don't want to lose who I am and become some kind of zombie. Then, we'll have sessions and go in there and talk about PTSD. The thing about it that really pissed me the fuck off is I love the armed forces, I love all the other branches, but you got Army guys in there who were cooks and say, "Oh, we got mortar attacked" and they talk about this shit, and I'm like how the fuck do you guys have PTSD from that. They come around to you and you tell them I saw so and so get their leg blown off and picked them up and throw them in the back of a Humvee with the leg hanging off with a piece of tendon. It gets me kind of pissed off that there are some people in the military here at the VA just trying to get benefits saying they've got PTSD. That's why I stopped going because I got sick and tired of hearing bullshit stories.

Janney: I understand. That's completely understandable. My wife is a psychiatric mental health nurse practitioner and she was telling me that there's this magnetic treatment where they stimulate the cerebral cortex, the front part of your brain, which has shown a lot of promise as far as dealing with and treating PTSD, plus EMDR. I don't know if you've heard of eye movement desensitization and reprocessing. Some people have had success with that. I don't know if you've tried any of those things or would consider doing it if you need to or want to.

Vergara: My thing is my wife has been helping me a lot because at first when we got together, she understood why I had anger issues. Why I would blow up at the littlest things. I told her it's from being in combat, loud noises. I remember one day she dropped a bunch of pots and pans and I mean that set me off and I started yelling at her, but it was that, I guess, being scared. That scared me; it startled me and being in combat when something startles you, your first reaction is anger.

Janney: Yeah, I know, you gotta react, right?

Vergara: You've got to push through it. Yeah.

Janney: So, I mean, that adrenaline rush; that fight or flight response.

Vergara: Yeah, and with us, it's always fight, not flight.

Vergara: That's what I had to try to break myself out of. I told her no when we talked and I'm like I'm going to be really open about everything about combat and everything and she asked me why am I like this. I said, "Most veterans don't talk about stuff, but if I don't talk about it and keep it in, I'll go insane. I'll shut down. It's better to get the stuff off your chest and take that weight off you than try holding it in."

Janney: Well, as I said, you've certainly told me some things that I didn't know, and I appreciate you taking the time to share that stuff. I mean, I'm hoping that it's not going to ruin the rest of your day.

Vergara: No, no, no. I don't mind talking about it. You know, I get a little moody and testy during the end of March-April time period. I didn't know why. My wife just asked me what's wrong with you today; why are you quiet? I want to know why, but then on Facebook and everything people will put up there, remembrances, and I'm like, "Oh that's why." My body's subconscious knows that right around this time this is what happened, and my wife just knows to give me my space. The PTSD thing, I don't have nightmares. I don't have flashbacks or any of that stuff. I just have the anger. After we pulled out of there and to see what happened because all my friends that died over there basically died for nothing. That's what pissed me off after the Obama administration and everything; we just pulled out; we just left. So, what we did over there was for nothing. Fallujah overshadows us and really makes the guys from Ramadi feel really bad because we don't get any mention. Everything is Fallujah.

Janney: That's something that I found out through these interviews. I really didn't realize that there's not a lot out there about the Battle of Ramadi and through my research, if you Google it there is very little online other than stuff that was written by the few journalists that were embedded with you guys. David Swanson, his "Echoes of War" video and then the news reports that he was writing for the Philadelphia Enquirer during that time. There's not a lot out there and I hope I can just shed a little light on that with the book and give you guys the honor for all the sacrifices that you made.

Vergara: We know there's that book out, "Echoes of Ramadi."

Janney: Yeah, that's Scott Huesing from 2005-06. I haven't read that. I guess I need to though.

Vergara: He was out there the second go round that they went out there…I'm not talking down on those guys that were out there for the second round but I don't think it was as bad as when we were there and that's why some of the Marines. He's always pushing his book and putting on 2/4's Facebook page: get your copy of this book. They were all out there after us; years after the big shit went down and that's why the group that I was there with wants somebody to tell our side. Tell our story because it was shit. That's where you come in, Greg. Fallujah was bad, but they had a couple of battalions of Marines out there taking care of Fallujah. It was just our battalion in Ramadi and it was the Wild West for a couple of months until they took us off our leash - until the Army came in and the top brass said, "Go out there and fuck shit up. Do what Marines are trained to do." We got guys trained to be security, basically MPs, handing out soccer balls. No more of that. Now it's time to do what Marines do and that is basically to fuck shit up. When they told us that we were happy. Chains are off; leashes are off; here's all the ammo you guys can carry; go out there and clean fucking house. And that's what we did.

Janney: One of the guys, I can't remember if it was Musser or Miller; after the 10th or on the 10th; that there was two 18-wheel dump trucks that went down Route Michigan that were piled up with insurgent bodies, piled up to the top after you guys did Operation Bug Hunt on the 10th. So, I mean, you definitely laid waste to a bunch of those guys.

Vergara: Yeah. We had 7-tons full of prisoners. We had dump trucks full of bodies. They came out and fought us because they thought after the 6th that these Marines are easy. We'll take care of them and when they came out it was another story because basically when we got over there, our hands were tied. Smile, wave, hand out candy, hand out soccer balls, hand out medicine, you know.

Janney: Yes, salaam alaikum, and here's some chocolate.

Vergara: Yep. But after the 10th when we got cut loose and could do what we could do; after that they were scared. They never really faced us ever again. It was always a drive by here; there at the base or an IED here or there, but IEDs stopped after we started picking them off at nighttime.

Janney: Were you with Lt. Ski's patrol on the 5th? Did you guys go out with him?

Vergara: No. It was the 6th that we went out.

Janney: You know he did a foot patrol with Swanson and some of the other guys on the 5th and found a claymore strung in some palm trees which was kind of an eerie foreboding of what was going to happen the next day.

Vergara: Yeah, we went and we picked them up. They were out there moving towards Gypsum and Nova out towards another spot. I remember when they found those IEDs. I went out there because Corporal Lenz, might have been Sergeant Lenz at the time. I remember I was worried about him. Lenz was in 2nd platoon; he was a machine gunner in weapons and he and I had a rivalry. We were good friends, but we had a rivalry about who was going to be the best machine gunner. When they said that Lenz got hit - it was just a little piece of shrapnel, but I remember running back towards him to make sure that he was all right. We went out the back gate of Combat Outpost. We went out that way and they got hit. Lt. Ski and his patrol did find the claymores that day that were in the trees. I remember that because that was where we were going to pick them up. That was an open field towards the end of a business section or intersection and that's where we were going to pick them up, but they found the claymores in those trees. So, we picked them up right on the road. I guess there was an IED set over there, too. because that's where that field was if we were going to come out on the road. That's where they were going to hit us with the claymore I guess and also hit us right there at the intersection with the IED and since we were right by the IED it went off. It threw shrapnel, but it didn't really mangle anybody. People got hit with shrapnel, but that was it. I was in the rear Humvee that day picking up rear security and stopping traffic. It was a Y intersection.

Janney: Did you get a chance to meet David Swanson when he was embedded with you guys?

Vergara: Yeah, I met him at Combat Outpost. He was mostly sticking around Royer and he was mostly with 2nd and everything. The area of operations with the platoons of 2nd was basically in the outskirts in the city area. My platoon was more Gypsum and Nova. That was our area of operation basically. With 3rd platoon, it was Gypsum, Nova, the graveyard, back to the railroad tracks and that levee back to Combat Outpost. That was our area of operation. So, I know Gypsum and Nova really well. But, there were times too we went out and picked up the other squads or we did a night patrol, or a night raid and I was lost in the sauce where we were at. I didn't know the names of the routes like the other guys did in that area.

Janney: John Wroblewski and I did foot patrols with 2/8 up and down Nova and Gypsum for several days during the time that we were there, so I know that area pretty well. Where we did the memorial service, I think was right there at that alleyway that you were talking about. There's an alleyway that went off to the West or left of Gypsum and kind of curved back around. At the time that we did it, it was cinder block lined, and there was a big open lot and behind that was a pretty big brown house that was set back along the left side of that alleyway that was also on the left or the West side of Gypsum. We probably weren't in the exact spot Ski got hit even though Eric Smith gave us GPS coordinates for that, but I'm sure that had a lot to do with the fact that General Kelly was with us. He thought it would be more prudent to do the memorial service off in that alley rather than right there in the middle of the marketplace along Gypsum.

Vergara: It would be safer.

Janney: Exactly. Because that was my biggest fear when I was with John. How ironic and terrible would it be if John got hurt? It would have been my fault since I planned the embed and brought him with me. So, I was terrified that something was going to happen to him.

Vergara: Yeah.

Janney: Before we left, I said, "Look, we've got to talk to somebody and get some body armor." When he showed up at the airport when I picked him up before we flew out a couple of days later, he had a Vietnam era flak. I'm thinking this is freaking terrifying because he didn't have IBA. But at that point, there was nothing that I could do. It was too late to get anything else, but he walked up and down Gypsum and Nova with that Vietnam era flak. I was just praying that nothing would happen to him.

Vergara: Yeah, that's all on Facebook. A couple of my friends were friends with you on Facebook but I didn't know you at the time but I did some Facebook snooping and I thought, "Oh, wow. He's got some balls taking him out there."

Janney: It was an honor to do it.

Vergara: Yeah. I'm glad you guys did it. I'm glad you got to go out there. I always want to go back, but now it's just a giant shithole.

6 April 2004 Route Gypsum Ambush And QRF

Cpl Eric Smith interview (meritoriously promoted to Sgt during deployment)

Janney: Just for the record, it is 1st September 2019. When and why did you join the Marine Corps?

Smith: Good question. I joined the Marine Corps spring break time frame my senior year which was in 2001. So, mid-March 2001. As far as why, I come from a big family, so my parents were going to make every attempt to send me to college, but I felt like it was going to put them in a financial burden to do so. I made the decision at that time, pre-9/11, that I wanted to go in the Marine Corps. If I was going to go into the service, I wanted to be amongst the best. I was young and eighteen, so I decided that it would be awesome to wear that Marine Corp uniform and blow stuff up. So, thinking with that eighteen-year-old mind, I went and joined. Actually, it's kind of a funny story. My parents were out of town when I did it. Not that I needed to sneak around them, but I thought it would be a little lot easier if I did it without a whole lot of push back. So, they went out of town for spring break. I stayed home and I went and joined the Marine Corps. So yeah, spring break 2001. Mainly it was for college money, but I also come from a long line of civil servants. My Dad, my stepdad, and my Mom were all in public service. Whether it was the police department or whatever. My Dad was a game warden, so I felt like there was a call to serve for a long time. I knew I wanted to, but the college pushed it over the top for me.

Janney: Once you finished with basic, what was your job? For the civilians out there, not your MOS, but your actual job in the Marine Corps?

Smith: I was Marine Corps infantry. I chose that job during my recruitment process and went through basic training in MCRD San Diego. I graduated in August 2001. I came home for my ten-day leave, went back to Camp Pendleton for SOI (school of infantry) in early September 2001. Actually, 9/11 was my first day of infantry school. First official day; we were checking out gas masks there on Camp Pendleton when we got the call that something had happened. We didn't know; we could tell that the base was buzzing. There were people moving around a lot from place to place. It was different. We were isolated. We didn't have television or anything like that. So, a lot of people that I talked to throughout the years; their recollection of 9/11 was sitting at home watching the news stories. Mine's a little different than that. We got called back from the supply area to our main barracks where we slept and a Captain, who was our CO at the time, who none of us had even met, came in and told us about 9/11 and what had happened. They posted one picture, and it was the gentleman that had jumped out of the building. It was a pretty famous photo of one of the towers. That was posted on a cork board. Basically, the message that the CO gave us was, "Pay attention to what you are going to learn here the next month or two because you're probably going to need it." So, it was pretty sobering - having joined during peacetime and that quickly changed before my peers were even done with their schools. But, yeah, I was a Marine Corps infantryman.

Janney: Were you always in 2/4 when you got assigned to a unit?

Smith: Yes. Right after SOI, I was assigned to 2/4. Originally, when I was in boot camp, I was told that I was going to 8th and I in Washington, D.C. to be part of the silent drill team. They had a selection officer that came out during our boot camp phase. At that time, you had to be infantry to do that, so they asked everybody that was a 03 MOS, which is Marine Corps infantry. They separated us and we went through a process of selection. It wasn't a choice for us, really. So, that's how I became an 0311, which is an infantry rifleman. Because of 9/11, shortly thereafter, I was selected along with others because they wanted to beef up all of the infantry regiments and battalions. All of us that had been selected for 8th and I were swapped. I was out of SOI and assigned to 2nd battalion, 4th Marines, there in California. On graduation day, we just got on a bus and took a 10-minute bus ride over to San Mateo and I was assigned to Echo Company, 2nd Battalion, 4th Marines and I spent the rest of my enlistment there.

Janney: Did you actually go through boot camp with any of the guys you deployed with?

Smith: I did. There were several of them. Marcus Waechter is one that comes to mind. He was in my platoon in boot camp. Again, we were in the same Company in SOI. There were some others that I met from different boot camp classes that we all kind of merged up in SOI together that I didn't necessarily go to boot camp with. But, then from SOI, I went to 2/4 and we kept that core group through our first deployment. We went to Okinawa, Japan on the 31st MEU. So, the original OIS, which would have been the actual 2003 push, we were actually in Okinawa, Japan for a year. We were on an original 6-month deployment, but extended for another six months in Okinawa. So, we spent a year there and then came home in the summer of 2003. At that time, we were all junior Marines, and so, we were all Lance Corporals and below on our first deployment. Our senior Marines, who were Corporals and Sergeants that had been on their 2nd deployment, most of them had either got out of the Marine Corps or had moved on to other things, so we became the senior Marines. We got a boot drop before deploying which is when you get new guys to "plus up" those that you've lost. We became the squad leaders and then began our work up, so basically our training for our next deployment which ended up being Ramadi in 2004.

Janney: I know that one of the guys said that Staff Sergeant Allen Walker was actually his DI. Who was your DI when you were going through boot camp?

Smith: I didn't have Staff Sergeant Walker. I knew who he was when he came to us. He came to us during that training phase once we came back from Okinawa. There were a lot of guys that were added to us at that point and that's how we became the Company that we deployed with. But no, I didn't have Staff Sergeant Walker. None of the guys that I served with were my Drill Instructors. I never saw them again. Sgt. Pallencia, Sgt. Lot, and Sgt. Gonzales were my three Drill Instructors at QO Company, but I never saw them after boot camp.

Janney: You guys deployed to Ramadi in mid-February 2004?

Smith: That's correct. I believe it was the day after Valentine's Day when Echo Company, typical Marine Corps, just for operational security, they don't give you an exact leave date ahead of time. They'll usually give you a window, so it'll be a plus or minus three days window that you're going to deploy. We knew it was sometime around Valentine's Day, but I remember on Valentine's Day is when I got the official word that it's going to be tomorrow. We had our bags packed and we were ready to leave any time during that window. We got the official word that we were gonna meet on base, do our final inspections, issue of weapons, and we would be leaving to convoy by bus to March AFB and begin our process to get to Iraq, or actually Kuwait, where we first went.

Janney: Some of the guys flew into Ramadi and some convoyed in. How did you get there?

Smith: Echo Company, being a hero unit, didn't have any vehicles that we needed to take in with us, so all of Echo Company took C130s and flew in to Al Taqqadam Air Base which is only a thirty minute or so convoy from, maybe a little bit longer than that, from Ramadi, where we were at the Combat Outpost. After our two-week acclimatization period in Kuwait, we took C130s and flew into TQ.

Janney: I heard an interesting story from Vergara about the fact that the C130 just barely came to a slow roll before he was told to get off the aircraft. Was that your experience too?

Smith: Laughs. I don't remember it that way, personally. I do remember the landing being a little bit interesting. It was definitely a combat scenario where they go completely dark. You fly at night; low flying aircraft in a war zone is not a good thing - an easy target, so they usually made big troop transports like that late in the evening, in the dark hours. So, we did fly in at night. It was completely dark in the back of the C130 and pretty loud. You start at a high altitude and you kind of do a stair step down. It's basically very drastic changes in elevation in the stair step down until you get to land and try to minimize your low altitude time. What I remember of our flight into TQ is that you would almost levitate, you were dropping so much altitude in such a quick amount of time. We landed, and it's obviously different from a civilian air flight where you taxi to the gate and walk across the tramway and into the terminal. It stopped in the middle of the runway and we ran out the back of it completely not knowing which way to go. They were yelling at us, but you still have the prop blast from the C130 and the noise, and so we're running across the runway towards a large hangar. It was the half dome shaped, Quonset hut style hangar that we were running towards. We spent the rest of the evening there and we didn't convoy out until the next morning.

Janney: Once you got to Ramadi, were you stationed at Combat Outpost?

Smith: Yeah, Echo Company was stationed at Combat Outpost, so we were there. The way the city was broken up was that Echo Company had the eastern half of the city. We were out past the arches which is a geographical landmark leading into the city through the main east-west thoroughfare being Route Michigan which heads on to Fallujah and then Baghdad. The arches were the east gateway into the city. We had that area of operations which was more of a rural type. You would have a cluster of houses in there, maybe some fields, whether they were date palm trees or alfalfa fields, but there was a lot of open area out there in between compounds, almost walled off houses. The Euphrates River was our border to the north. A lot of those fields were irrigated off of the Euphrates River. There were a lot of pump houses along Route Nova which is another landmark - it's the levee road along the Euphrates River on the northern border of our AO. Echo Company had that eastern part of the city with Golf having most of the middle portion and Fox Company having the left or western half of the city. Echo Company and Golf Company were both assigned to Combat Outpost.

Janney: You may remember that John Wroblewski and I stayed at a JSS (Joint Security Station) just inside the arches. We did foot patrols on Nova and Gypsum with 2/8 Marines for several days.

Smith: I remember when you were making your trip plans, I had several phone conversations with Mr. Wroblewski and tried using Google Earth to pinpoint where the actual location was on Gypsum of the incident on 6th April. So, yeah, I remember when y'all were about to head out there.

Janney: Didn't you meet us at the Atlanta airport in 2007 and give Martha Zoller a map?

Smith: No, that wasn't me. Not in that timeframe, it would not have been me.

Janney: Once you guys got settled into Combat Outpost, what were your daily activities?

Smith: We kind of rotated through responsibilities. Between us and Golf Company, we shared responsibilities as base security where we would guard posts. That would last usually a week. Most of our duties kind of cycled through the week. We would have a week where we would be on camp guard, then a week when we would be on patrols, and then a week where we were on raids where we were raiding high value targets. That was always susceptible to change based on operational tempo and what needed to happen. Whether we were doing big sweeps, or we had big fire fights going on, so definitely, during that two-week period in the beginning of April, we threw all that out the window where we were fighting day after day.

Typically, you would have an assignment on whether that day you were going to be doing a day patrol, a night patrol, you may be running QRF, which is a quick reaction force. So, you may be a platoon sized quick reaction force or a squad sized, depending on what was going on that day. So, there's various roles that you could be filling on a day-to-day basis. All of those roles fit into the basic mission of the Marine Corps rifleman, to locate, close with, and destroy the enemy by fire and maneuver, and to repel the enemy's assault by fire and close combat. That sounds like a very memorized saying because it is; that's what we were taught early on in our job as a Marine Corps infantryman. All of the jobs that we were doing there on Combat Outpost fit that mantra. We were doing patrols to locate the enemy and once we found them, we would close the distance between us and them and do our best to destroy the enemy and prevent them from coming back another day. Whether it was us out there doing those patrols or doing raids after high value targets which typically happened in the early morning hours when they weren't expecting it in the 0200 to 0600 time frame. But then, we also had the task of guarding camp.

Janney: During the months of February and March, were you involved in any combat actions, or was it just sporadic pop shots at you guys? Mortar attacks? Tell me what was happening.

Smith: February through March, early on in our deployment, I would say was relatively quiet. We were getting mortared pretty frequently in the evening hours, not so much during the day. We were doing raids and going after HVTs. A lot of patrols were passing out supplies and toys. We had been on SASO, to win the hearts and minds of the local population as a way to beat the insurgency, so we went with plenty of soccer balls and plenty of candy that we would pass out. The February and March months were mainly doing that. Patrolling to win over the local population. Just goodwill. We would patrol out to schools or to a local sheikh's house where we could meet with them about projects to pave a road or just other civic projects. There wasn't a lot of major combat, firefight-wise. Occasional mortaring, some random IEDs were going off. Early on we did lose our XO Lt. Kaler to an IED.

Another of the roles we did have, especially earlier, was doing route patrols where we would patrol the MSRs (main supply routes) through our AO, so down Michigan, up by the power plant, Route Nova, down Route Gypsum, and back around Michigan towards Combat Outpost. It would take the better part of a day to patrol this 10-kilometer circle that we were making daily. This tapered off later because it wasn't effective. But, during those, we were tasked with looking for IEDs and anything that would prevent us from moving freely within our area of operations, whether it be supply convoys or just troop movements. We were finding a bunch of IEDs. Originally, they wouldn't blow IEDs up on us because they didn't want to waste them on individuals, but I think that quickly changed once the enemy realized that if they didn't blow them up on us, they were going to lose them completely to our EOD guys. So, they did start blowing them up on individual guys. XO Lt. Kaler and some of those guys received injuries in that way and were sent home. We didn't have our first fatality in that way for a while; a lot of injuries, but no fatalities early on for Echo Company. For the first bit, February and March, it was mainly patrols and goodwill tactics.

Janney: When you guys were doing that 10-kilometer patrol, were you guys doing that mounted or dismounted or a combination?

Smith: Usually, it was a platoon sized effort, so you had roughly thirty guys out there. A team commander, three squad leaders, and then team leaders. You would be with a radio operator, mainly dismounted, but you would have one guy in a Humvee that was following behind in tow. Just in case we needed to move quickly, or we needed some supplies, whether it be water or supplies out of the Humvee, so we did bring a vehicle with us. Typically, our entire route was done dismounted.

Janney: I forgot to ask you, what was your rank when you deployed to Ramadi?

Smith: I was originally a Corporal in Ramadi when I first got there and then was promoted to Sergeant during the deployment.

Janney: Things were relatively quiet other than sporadic little hit and run type things. When was the first incident of combat prior to 6 April? Or, was there one?

Smith: I can't pinpoint the time frame, but it was sometime in March. We were getting constantly mortared in the evening. I can't remember whether it was 1st or 3rd platoon. I'm sorry, I was a member of 2nd platoon, but 1st or 3rd platoon was out doing a late-night patrol. They were in an area to locate where we were getting mortared from and we had some intel. There was a house that was under construction that we would later take over for an observation post, and immediately to the east of that house was a field. I want to say 3rd platoon was patrolling in that area looking for this and we did start receiving mortars. They found the guys that were mortaring our base from that location and engaged them. I was on a quick reaction force; we took some injuries from that. I think 4 to 5 Iraqi combatants were killed in that contact. Like I said, it was either 3rd or 1st platoon, I don't remember exactly which it was, but I know there were several of our guys that were injured. There was a machine gunner that was on the turret of the 7-ton that received some severe wounds to his legs, but he did survive that. That would probably be the first significant combat. I went out on the QRF, but it was mostly over when we arrived. But that was the first combat that I remember Echo Company having where there was some serious exchange of fires between us and the enemy.

Janney: You have an excellent memory. According to LCpl Aaron Vergara, that was 25th March and it was exactly as you described it. He was in one of the platoons sent out to try to get those guys and he said that they killed five enemy combatants. One got away, and they captured some young guy that begged them not to kill him because he had finals in school a day or two later.

Smith: Yeah, I was actually there for that conversation. I remember that. We were also a little bit freaked out. I still don't know that the guy wasn't a combatant, but I don't think we could prove it. There was a guy that claimed he was an Iraqi policeman that came out of the canal that was right there and he had a pistol on him. Like I said, he claimed to be an Iraqi policeman because it was close to the vicinity of the gas station. It was right between the house under construction and the gas station, there was that field. We were right there on that canal road off of Michigan and I remember it being about 5. They had the mortar hidden under a downed palm tree that was kind of in the middle of the field that was tucked up underneath the tree. So, nightly they were going out that location and digging that mortar tube up, sending a couple of rounds at us and taking off. That night, that patrol just happened to run across them.

Janney: After that, was it just pretty much just the same prior to 6 April? What happened between the 25th of March and the 6th of April?

Smith: To the best of my recollection, it was the same. Just conducting patrols. There were a couple of things, like I said, as far as excitement, whether it be an IED or mortaring or something of that nature. From my standpoint and recollection, it was pretty calm up until the day of the 6th.

Janney: Were you on that patrol on April 5th with Lieutenant Wroblewski? From what I understand, David Swanson was patrolling with Lieutenant Ski that day and they found a claymore strung in some palm trees at the end of the patrol.

Smith: Yes. I remember David being with us. There is a picture of Lieutenant Ski that David Swanson made that you could see him from behind where he was on that patrol. There's a couple of guys in those pictures as well. There's also the picture of Lieutenant Ski kind of knelt down on his knee, and I think it's Calloway the radio operator that is next to him. All of those pictures would have been taken by David Swanson on that route patrol the day before. I would consider that a pretty common patrol when we were talking about common or out of the ordinary. We did find several IEDs, I remember. What kind of struck my mind was when we were right there at the intersection of Gypsum and Nova; we stopped and halted there for a little bit. We were talking with the locals and I can remember them being pretty agitated. I just remember a weird feeling and then that would be the intersection the next day that most of our action took place on 6th April, as far as 2nd platoon, 1st, 2nd and 3rd platoon Echo Company in general, right there at Gypsum and Nova. I kind of look back at that patrol on the 5th as being a little bit prophetic of what was going to be coming up the next day.

Janney: Yeah, that was exactly what I was wondering. I had heard from several guys that the mood of the people changed pretty quickly there over the week or so before the 6th. You didn't see as many kids around either.

Smith: There was a HVT (high value target) raid the day before (author note: night of 4th/5th April. A suspected high ranking insurgent planner and other leaders were captured during a raid this night, although it might not have been the same raid Sgt. Smith mentions. According to OPSO S3 Major Harrill, this HVT leader's capture accelerated insurgent attack plans and caused the insurgents to attack on 6th April instead of their better planned attack to launch in mid-April.) We actually took one of the guys that evening. He was a pretty angry looking fellow. I'd say, early twenties. I can't remember why we brought him in for questioning. Maybe it was his demeanor and the way he was looking at us. He just kind of had that look, when you look at someone and you can tell they don't like you very much.

I think that might have been the main reason that Lieutenant Ski decided to take him. I don't think it changed anything about what was going to happen the next day. I know they were pissed off when we did take him. I think he was the son of somebody important in that area. At that time, he wasn't being detained as a prisoner for more than just questioning. We did have kind of an extended stop there at Gypsum and Nova on 5th April and you could tell that there was something happening. When we went back on the 10th, it was because we thought that those people and that individual in that area were the ones that planned what happened concerning Echo Company on the 6th of April.

Janney: The morning of 6th April, what were you doing? Tell me what was happening.

Smith: Yeah, so 6th April, I was first squad leader, 2nd platoon, Echo Company. My squad was a squad-sized QRF. We were going to be in support of any patrols that needed us that day. Usually if you were a squad-sized QRF, you were going to be in a smaller convoy, maybe two to three Humvees and a 7-ton. A lot of times, we would be taking other platoons out to wherever they needed to be to patrol. Our area of operations was large enough that we often couldn't leave from the Combat Outpost and then reach the far areas of our AO and return on time. It was just a little too much for guys to be effective. So, we would convoy them out there and they would do their patrol and we would convoy out there and pick them up and bring them back. That was kind of what I was doing. I was in charge of that squad-sized quick reaction force and, at that time, we had two Humvees and a 7-ton truck that was assigned to us that morning.

Janney: Tell me, to the best of your recollection, what kind of kicked off that day? What started the action that led the QRF to go up Route Gypsum?

Smith: So, 3rd platoon was out on a patrol near the arches and they had been patrolling for a while. I want to say they were doing the route clearing patrol and were working their way back around. You know what, I need to change that, because I believe 1st platoon was the one doing that patrol because that matters later on in the story. So, 1st platoon is doing the route clearing patrol for that day and then 3rd platoon was just out on a separate patrol in the east half of the city. 2nd platoon is in QRF mode. The way I remember it, most of the action of that day, for the morning time at least, was happening in Golf Company's area of operations. Golf and Fox, I remember there was a lot of distant gunfire, explosions from RPGs, and various things. You could hear it out there. We knew it was happening and you could see the Cobras flying over in air support. You could hear radio traffic coming from Golf Company and Fox Company. They were mainly working together helping each other out. Headquarters Company with Lieutenant Colonel Kennedy and Sergeant Major Booker, and then Weapons Company, they were supporting that effort over there as well. For at least the morning hours, most of that, up until around noon, was all happening in that area of operations, mainly in the heart of the city. To the point that my QRF, my squad level QRF was probably minutes away from leaving Combat Outpost and going into Golf Company's area of operations to help them out.

Golf had various elements that were pinned down and in trouble. They had guys that were wounded, and they couldn't get to. Gentile was one that I remember, he was actually a friend from SOI. He was shot in the back of the head, behind the right ear, and then it came out in his face, the nose area. True warrior right there. They patched him up and he kept fighting for hours. But Golf was waiting for help. So, I remember, my squad was all in the staging area, or the go area, our vehicles were parked right there near our barracks area. I was bouncing back and forth, trying to keep them up to date with what was going on, going back and forth to the command post, speaking with the CO and XO. They were just trying to figure out what needed to be done. They were about to send us out and that's when 3rd platoon on the eastern half of the city, near the arches, got contact from an enemy.

I remember them saying, it was somewhere in the neighborhood of maybe 10 guys, enemy combatants, that they had taken fire from, kind of on the north side of the canal right there at the arches. That's when my mission was changed and Captain Royer made the decision, instead of just sending a squad level QRF, he was going to send a Company level QRF. Basically, everybody that he had left in the base, excluding, I think, one squad from 2nd platoon, who had already been out on a night patrol the night before. It was sporadic groups. The rest of us loaded up in a larger convoy combined of four Humvees and two 7-ton trucks and headed out east towards the arches with Captain Royer. So, at that time, that's when the QRF was my QRF. Lieutenant Ski wasn't assigned to it yet, but then Lt. Ski came out and told me that they were going to beef up the QRF and he was going to go with me at that point. He got on, my squad would be the only squad from 2nd platoon that would go on that QRF and then you had various elements from 1st and 3rd platoon as well.

Janney: So, Staff Sergeant Allan Walker and the other guys in that lead vehicle were the front element of that QRF? Is that what I understand?

Smith: It became that way later on. Originally, that QRF went to the arches and we stopped the QRF right there on the dirt canal road north of Michigan, right there at the arches, kind of where Gypsum begins to go south to north toward the Euphrates. So, right there at that intersection of the arches, we stopped our QRF convoy. We dismounted and then 1st and 3rd platoon elements began searching houses right in there because the enemy that had made contact with 3rd platoon's patrol had broken contact and retreated north. So, we dismounted our vehicles there at that intersection and then started searching the houses, not finding any enemy. It was after a little while, Captain Royer made a decision that we were going to proceed north up Route Gypsum with 1st platoon on the left side of the road on the left lane; 3rd platoon on the right and then 2nd platoon, my squad, with Lieutenant Wroblewski, would stay behind, or in the vehicles, following up on Gypsum, in the vehicles, kind of following in trace. We were going to do what's called a movement to contact, so being spread out online, moving until we made contact with the enemy.

I would say roughly three quarters of the way up that road, and in that general timeframe, is when radio traffic started coming in that we had 4 snipers that were in the tank graveyard up off of Route Nova that had been compromised, were in a pretty bad fight, and had been pinned up against the Euphrates River in a pretty dire situation. At that point is when Captain Royer decided to send 3rd platoon elements with Staff Sergeant Walker over to us in our vehicles. They jumped into the first two vehicles in our convoy, the first 2 Humvees, and then proceeded north on Gypsum in the vehicles. The idea was they were going to go up to Route Nova, hang a left (west) over toward the tank graveyard and go support the snipers that were in contact up there. As we all know, what ended up happening was they were ambushed at the intersection of Gypsum and Nova.

So, in the front two vehicles were two of my guys driving the vehicles. I had Crowley, would have been in the first Humvee, driving the first Humvee. I had Matt Scott that would have been in the second vehicle with Corporal Waechter at that time. Those that survived in the 2nd vehicle, the ones that had gotten out, and entered that small building to seek shelter, were the ones that survived. Most of those, other than LCpl Otey who escaped out of the first vehicle, most of our casualties came out of that first vehicle. I remember a lot of this information is things that I learned after the fact.

I do know, before we started north on Gypsum when we were at the arches, Captain Royer wanted another radio so he could monitor multiple channels. So, he took our radio operator with him. When we got into those vehicles and were following in trace, the only radio we had was vehicle mounted and which would have been in Lieutenant Wroblewski's Humvee. I didn't have a radio, so I wasn't hearing the big picture of what was going on. I didn't originally know about the snipers in contact in the tank graveyard, I didn't know why 3rd platoon was getting into our first two Humvees and leaving. When I stopped, I saw the 3rd platoon come over and get into the vehicles and take off.

I dismounted my vehicle; Lieutenant Ski dismounted his vehicle. I was about to head to him to get the information that his 3rd platoon left. There was, I mean, it was maybe 5 to 10 seconds, that they were on the road before we started hearing small arms and RPG fire. They were taking fire; they were only a couple hundred meters ahead of us, but out of sight due to a curve in the road. I heard gunfire and yelled at my guys to seek cover. Off the eastern side of the road, there was a house kind of behind us 50 to 100 meters away. I wanted to get into that courtyard area of that house. I wanted to take the house and then get on the roof, and then, at that time, that's when all hell broke loose in our area as well.

Janney: Once the lead vehicles made contact, you said you were 100, 200 meters behind them and you sought shelter in this house. Who was in your immediate vicinity? Were you with Lieutenant Ski or near Sergeant Valerio?

Smith: Yeah, Sgt. Valerio was in the area. He had come out as part of Motor T, I guess to help drive vehicles. That was part of the confusion when we were putting the QRF together. I don't know where he came from or why he was in our QRF, to be honest with you, being part of the Motor T. It's funny when you start talking to guys about what happened those days. Each individual's perspective of it, you know.

Janney: The fog of war. I understand where you're coming from.

Smith: Right, right. Just the general fog, the information pieces that they had and what their job was on that day. I probably didn't even know we had Sgt. Valerio with us until after Lt. Ski was hit. I remember seeing him when we had moved Lt. Ski back to the vehicle. In our general vicinity was just 1st squad, 2nd platoon which would have been my squad with Lt. Ski. Our platoon had a Weapons platoon attachment - a machine gun attachment with Corporal Lenz and PFC Calavan. I remember there was another member of the weapons platoon with us, but I can't remember his name and I don't know why he was with us either. Whether that was because he was on a gun or what, so the main body of who was with us was just 2nd platoon individuals right in that general vicinity.

Janney: Who was driving Lt. Ski's vehicle?

Smith: I want to say it was Lance Corporal Brown. Brown might have been driving my vehicle. It was one of those. We had them, because he was not normally in my squad, but you had to have a certain course to drive a Humvee in Iraq, and there were only certain guys that had that. In order to have my QRF that size, we had to take some people from other elements. To the best of my knowledge, because I know Brown was the one who drove the Humvee with Lt. Ski and Calavan, the Corpsman, and Doc Gutierrez back to Combat Outpost after he was injured. So, I want to say Brown was driving his Humvee, but I could be wrong on that to be honest with you. I know Brown drove the vehicle back to Combat Outpost, but I don't know if he was driving his vehicle originally.

Janney: Who was with Lt. Ski when he was hit?

Smith: I want to say that there was somebody in the back of the Humvee that was on a machine gun. I want to say that was a member of the weapons platoon, but I don't quite remember who that was. Like I said, once we started taking contact, I had exited my vehicle and was heading toward Lt. Ski when we took fire and my immediate reaction was to turn to my guys and start getting them out of the vehicles. I had some in my Humvee and I had some in the 7-ton. I started yelling at them, "Get out, get out of the vehicles, and get over to this house." It wasn't until we got to that location, it was in my head that Lt. Ski was going to be coming with us. It wasn't until we got there, and we had made contact with a couple of insurgents that were kind of to our northeast over a little wall. They were firing on us and on the convoy. We weren't there long, less than a minute or two, I want to say, less than a minute, thirty seconds or so, before I heard Lance Corporal Burmaline saying Lt. Ski had been hit. I turned around and looked across that open field that we had crossed to get to this house and could see our two Humvees over there, his Humvee and my Humvee, were the only two remaining Humvees and then the 7-ton was behind it. I looked over and I saw him lying next to that first Humvee. At some point, there was someone on that machine gun, but I can't remember who it was.

Janney: When Lt. Ski was hit, the two Humvees were stopped, and you weren't moving at that point?

Smith: Right. The Humvees had stopped, 3rd platoon had already come over and got in the first two Humvees and left. So, originally, Lt. Ski's Humvee would have been third back and then mine would have been the 4th, and then we had a 7-ton behind us. But, after those two Humvees with Staff Sgt. Walker and Corporal Waechter left and proceeded on to the intersection, then Lt. Ski's Humvee became the first Humvee and mine became the second and we had a 7-ton behind us. We dismounted and, shortly after dismounting, he and I, before we headed off to even speak to each other about what was going on. I assume he was about to tell me why those guys had left when we made contact at exactly the same time. It was a coordinated ambush. The guys that had been ambushed at Gypsum and Nova and that ambush extended pretty much all the way down to us and we had also received fire in our area. Lt. Ski went down pretty early into that ambush, but he had stayed and was on the radio next to his vehicle. So, he stayed with his vehicle. The rest of us, you know, I had ordered the rest of the squad to take cover in the courtyard area of that house to the southeast.

Janney: I'd understood that Lt. Ski was kneeling down next to the Humvee with a map on his knee or looking at the map on the radio when he was hit.

Smith: You may have heard that from somebody else, but that is not my recollection.

Janney: Tell me what happened. Everybody's got a different story. I'm just trying to figure it out.

Smith: No, no, like I'm saying, I never saw that, so I can't verify it for you. It may have happened. But, you know, like I said, I had turned my attention after we started taking contact with my squad, got them out of the vehicles, got them moved up over to that house, began engaging the enemy and then heard Lt. Ski was hit, turned around and looked. So, my personal recollection of that immediate ambush was that I thought Lt. Ski was coming with me. Apparently, he didn't. It was only afterwards when I realized that he was on the radio, due to the round that had actually struck him after going through the handset of the radio. So, he may have been dealing with looking at a map, I know he was on the radio, because of the round and his injury.

Later on, that story gets kind of crazy when Lance Corporal Burmaline tells me that Lt. Ski is hit, I turn and see him lying next to the Humvee. I take off across that field. I didn't really give any orders to my guys. It was kind of a reactionary thing - I took off towards Lt. Ski. Got to him, noticed that he was bleeding from the facial area. He was trying to roll over. I dove down next to him and started trying to assess what wounds he had, and decided that we needed to move him back to the 2nd Humvee. I got up and kind of grabbed him by the flak jacket and started dragging him. About half way back to that other vehicle, I think Corporal Lenz helped me and there was somebody else as well, came over and we picked him up and put him in the back of the Humvee with Doc Gutierrez. Like I said, we noticed that he had an entry wound to the right side of his upper jaw and then an exit wound to the left side of the lower jaw. He was unable to speak to us at that point. He was semi-conscious and was awake, but I don't know if he was really understanding what was going on. He was just trying to breathe. So, we got him in there. At that point, I felt like I had a decision to make. When I look back on it, it was probably one of the toughest decisions I had to make as a leader in the Marine Corps. How to move forward with that, you know. I've got my platoon commander, who I also considered a friend, that's injured.

Janney: And you guys were in the house trying to fight off the insurgents at this point?

Smith: No, at this time, my guys were still at the house. We never got in the house, we didn't. It was padlocked down pretty hard, we didn't have breaching tools, so I don't know if it was occupied or unoccupied. We weren't there long enough to find out. So, the guys were basically in the courtyard of the house; below the walls; firing on the enemy. Me, personally, I had gone to Lt. Ski and then drug him back to the other vehicle. And then, as I said, I had Corporal Lenz help me pick him up and put him in the back of the 2nd Humvee, which would have been my Humvee originally. I had Doc Gutierrez there and he started tending to Lt. Ski.

So, I had to make a choice at that point. I ran back to his vehicle again. By that time, the machine gun that was in his vehicle had gone down. There was a problem with the machine gun - it wasn't loading and ejecting correctly, so our machine gun went down in his Humvee. That vehicle was just kind of sitting out there by itself in the middle of the road. I went back to it and grabbed the radio. I was going to try to make the casualty call off of that radio since it was the only radio we had. As I mentioned, Captain Royer had taken our radio operator with him. It wasn't until then that I noticed that Lt. Ski had been shot through the handset; because once I pulled that handset up, went to push the button on it to speak, the button was messed up. It had basically destroyed the inner workings of that handset to where the button, the spring that you depress to speak, was no longer there.

It wasn't working and there was a perfect sized hole right through the handset. Now, I had no radio, so I went back to the Humvee and Doc Gutierrez was still working on Lt. Ski. Once I went back to that Humvee, that's where I was kind of presented with the decision. I can't call anyone for help, but Lt. Ski needs help now. I've only got one Corpsman. I can't leave this location, you know, partially engaged, so my decision was this: what I ended up doing was we had Lt. Ski in my Humvee, so I got Lance Corporal Brown. I put Brown in the driver's seat. I tell him, "Hey man, you're going to drive back to the Outpost." I knew that south of us was good because we had come from that direction. The risky portion was Michigan, getting in the two to three miles that he would have to drive to get to Combat Outpost. I had to send them back with our only operating Humvee. Gypsum was so narrow that we couldn't turn the 7-ton around, so I didn't want to leave a Humvee and a 7-ton. We couldn't get everybody in the Humvee, so I made the decision to send a driver, a machine gun, the Corpsman and Lt. Ski back to Combat Outpost.

So, Doc Gutierrez was working on Lt. Ski, PFC Calavan was on the machine gun mounted in the back of the Humvee, and Brown was driving. I told Brown not to stop for anything. I told Calavan that anything and anyone that was outside any home in this fight was definitely a combatant. Because why would they be outside? So, shoot at anything and everything that you see outside. I said, "Go!" I had no way to talk to them and they had no way to talk to the Combat Outpost to let them know that they were coming. So, we were kind of in a pickle there, but I knew if we didn't do something, if we didn't make a decision and we didn't start moving in a direction, that Lt. Ski was definitely going to die on the side of that road. So, that's the decision that we made. I sent my only Corpsman away with him and now we're left there in that location to kind of figure out how to proceed forward.

Shortly after they left with Lt. Ski, I heard Calavan shooting on his way out. I don't know if he was shooting to keep people inside or if he was actually shooting at people. He's firing away and they take off. Lt. Cogan from 3rd platoon did come to our location. We basically just hunkered down in our location. I didn't want to move our element because I had no way of coordinating where everybody else was. I didn't want to walk in between fire. I didn't know where the enemy was. I had an idea of the direction the enemy was, but I didn't know where my other units were either. I didn't want to put us in a bad position by moving without the opportunity to coordinate with anyone.

Janney: Absolutely. I can understand there's a lot of stuff going on and you were trying to figure out the best course of action.

Smith: Right. So, we kind of hunkered down there around that house and engaged the enemy as best we could. Lt. Cogan came by at one point with his element pushing north. He didn't have a radio to give me, so he just continued going north towards Gypsum and Nova. It was a brief interaction with him and we just stayed where we were at. Then, there were two Army Bradleys that passed by a little while after. These timeframes really get the best of me as 15 years have gone by. How long a span this was, to the best of my recollection. I mean it seemed like it was instant, but it could have been 30 minutes to an hour. I don't remember. But I know the Bradley came by. I tried to wave him down as he was passing. He didn't stop. He kept going, but once he got to Gypsum and Nova, I could hear some pretty heavy fires going up there and it wasn't long after that they came limping back. They had been hit by RPG and received fire at Gypsum and Nova to the degree that, I remember, one of the crewmen of that Army Bradley had lost an arm, like mid-forearm down and they were trying to evacuate him out of there. But I was successful in stopping them at that point and asked them if they had an extra handset for a radio. They gave me the handset and then I had to make one more trip across that open area. That's why I said I was going to go back. I had noticed Lt. Ski's helmet and weapon were sitting next to the vehicle. I didn't want to leave it there for an Iraqi to make a war trophy. So, I took off across that field again, got to his Humvee, exchanged the handsets, made a call to Combat Outpost to CP and to Captain Royer, was able to get some direction that we needed to continue north on Gypsum to the intersection of Gypsum and Nova, that we had wounded and KIA at that intersection. I had also been told the same thing by the guy in the Bradley. The Army personnel in the Bradley had told me, "Hey, you've got wounded and KIA up there in that intersection. You need to get someone." So, like I said, I made that radio call, grabbed Lt. Ski's stuff and brought it back. We threw all of that into the 7-ton, and I want to say, at that point, that first Humvee was out of commission. We basically took that 7-ton and used it for cover.

We drove north up Gypsum and we dismounted from it on either side, just kind of bounding our way up to the intersection. By the time we got to the intersection at Gypsum and Nova, there was nobody; there wasn't any combat. We weren't being shot at, at that point. Once we passed by Corporal Waechter and his element that were in that building, they came out and met us. Then, we moved further into the intersection and I began coordinating getting the bodies out of the battlefield. That's when we located the bodies of Lance Corporal Crowley, Staff Sgt. Walker, Doc Mendez, and Roberts, I believe, were in that area as well. So, we started coordinating with my guys to remove the bodies and put them in the 7-ton. Once we had placed the bodies, we got in the vehicles and headed south down Gypsum back to Combat Outpost.

We entered Combat Outpost and that's when I met up with Sgt. Coan, who was my platoon Sergeant for 2nd platoon. Coan advised me that Brown and Doc Gutierrez had made it back with Lt. Ski. At that time, we thought the prognosis was good for Lt. Ski. They told us that he was conscious and alert and that they had evacuated him via helo towards Baghdad for further care, but he was stabilized. We got all of our KIA out of the 7-ton into the main building at Combat Outpost and then we kind of plussed up our members.

We got the rest of 2nd platoon and Sgt. Coan came with us at that point and we went back to the intersection of Gypsum and Nova and then locked that area down. Once we got back, we started surveying the battlefield where we had enemy combatants killed. At that time, it's one thing that's haunted me throughout the years, is we found Travis Layfield behind one of the buildings. I guess it haunted me because I felt like we left him the first time that we didn't find him originally. We secured his body and then started working on securing the enemy. We were there for several hours while we were working on that.

We convoyed eventually, in the evening hours, back to Combat Outpost. We were there for just a little bit when we were told that 2nd platoon needed to go to the town hall. Captain Royer wanted to speak with us. We went to the town hall and we were all there for just a minute. Captain Royer came in and spoke to us and told us that Lt. Ski had passed away and that, basically, we couldn't dwell on it at that moment and that we still had a job to do and that he needed us to go back out to that area.

So, later that evening, with Sgt. Coan now acting as platoon commander, and me being promoted to platoon Sergeant, we made a platoon sized patrol from Combat Outpost all the way back to the intersection in the area of Gypsum and Nova. It was a pretty terrible patrol just given what had happened that day and how exhausted we were and the fact that we needed to go back out there. I think that decision was made because the majority of 2nd platoon was not out there for the original firefight other than my squad, and as a platoon we were the freshest. Anyway, we went back. We only had two days the entire deployment that it actually rained, and that evening it rained pretty good, so it was a pretty terrible patrol. But, that's how we ended the day of the 6th into midnight and early hours of the 7th was the patrol back to the Gypsum and Nova area. Captain Royer's idea was we wanted to let the enemy know that they hadn't beat us, that we weren't afraid, and that we would come right back to their doorstep, so we made that patrol that evening.

Janney: After the horrible day of the 6th, you guys went back to Combat Outpost and tried to get some rest after the terrible events of that day. What happened the next couple of days?

Smith: The next couple of days, I know there were firefights throughout the city. Personally, I don't remember being heavily involved in any firefights again for the next couple of days until the 10th. Nothing stands out. I know as 2/4 being the larger unit, there were fights from the 6th past the 10th. Golf Company was in fights. Fox Company was in fights. Weapons Company, the mobile assault platoons (MAP) were in fights. But as far as Echo Company goes, I don't remember us being, especially 2nd platoon, heavily engaged in the days after the 6th. So, the 7th, 8th and 9th until we got to the 10th. Once we got to the 10th, is when I remember our next significant combat.

Janney: Operation Bug Hunt?

Smith: Right. Somebody remembered because that had left my brain.

Janney: One of the guys remembered it, but I sure didn't know until they told me. Another one of your brothers told me that they actually got a translator on the mosque loudspeakers and basically called the insurgents out to fight and said, "Look, we're coming for you" the morning of the 10th. Do you remember that at all or remember hearing about that?

Smith: Right. I don't personally remember hearing about that. Various times throughout the deployment, they would use the loudspeakers at the mosque for different things. I remember kind of an oddity through the deployment at one point. You weren't expecting to hear English to come over those. There's no way to call to prayer. But they started calling us, or speaking to us, to where they were saying if we would turn ourselves in to the mosque, they would pay for our trip home. Basically, trying to entice us to quit. It's kind of comical.

Janney: This was supposedly one of your translators that was calling the insurgents out to fight.

Smith: Right.

Janney: The terp was speaking in Arabic, saying "The Marines are coming for you, let's go out and play" kind of thing. But, I didn't know that they were doing the reverse to you guys, too.

Smith: Yes. That was later in the deployment, but I remember hearing that. Then you would hear about the donkey that was running around with the message spray painted on it.

Janney: I hadn't heard that story.

Smith: I'll see if I can find the Fox Company story, but there was a donkey walking through the marketplace one day that had "FU George Bush" written on it and some other stuff in Arabic. There was this little white donkey that had that stuff spray painted on it. They had found it wandering through the marketplace. So, there were different tauntings that would go on between us and the insurgents.

Janney: Psyops stuff.

Smith: Psychological operations. We'd shoot out their speakers on the mosque towers just to piss them off. It was war. Yeah, those stories on the morning of the 10th, whether that happened, not saying it didn't, but that's not my recollection.

Janney: The day of the 10th you guys were conducting Operation Bug Hunt and you were just going from house to house searching, just trying to pick a fight basically.

Smith: Correct. We had intel that the people that had planned and coordinated the 6th were in the general vicinity of Gypsum and Nova, mainly west of Gypsum, kind of northeast of Fishhook Lake. So, early morning hours, our mortars fired into the Euphrates River, just a show of force type thing, no real target in mind. That's kind of how we kicked it all off, but we were staged and waiting to go off of Nova. Each platoon had a main house that we were going to hit first thing that morning. From there, we were going to spiral out off of that central location and hit every house in our general vicinity until we had searched them all. We were taking people in for questioning.

In 2nd platoon, my role has now transitioned from squad leader to platoon sergeant and Sgt. Coan went from platoon sergeant to platoon commander. We had no officer in 2nd platoon at that point for a given period of time. I want to say it was a month or so before Lt. Martley came in as a combat replacement to Echo Company and then 2nd platoon. So, we hit our house before sunrise, and we started searching several other houses and then we started taking fire from various locations. We moved on to two or three more houses and we weren't far off from where we had started. We were in transition between houses when we first took our first amount of fire. There were some large fields out there intermixed in all these houses, so you would have to walk across these good-sized fields, several football fields maybe in length or width, to get to your next house. We were crossing a field when we started taking fire and so we ran on to this next house and were outside its courtyard wall. Most of the houses there, you saw when you were there, have a six- or eight-foot-tall cinder block wall around them, surrounding that, creating a compound and courtyard. So, we were outside of this wall seeking cover behind trees, engaging enemies where we could see them.

While that's going on, various elements were also moving. In a war, you know your piece of the pie and where you're at, but you don't know where everybody else is and what's going on in their area. But, it wasn't long after we were fighting along that courtyard wall where we were at, Captain Royer and some other guys emerged. They had been stuck in the middle of that field when they started taking fire and had jumped into a ditch which had been a drainage ditch for an outhouse. They had the interpreter, who we called 007, and had David Swanson with them. That's when David Swanson was injured when they were in that field. He had a grazing wound to his arm. But, they had jumped into that ditch which, like I said, was drainage for an outhouse and they were covered head-to-toe in this black nastiness when they came around.

I remember just turning around, not even really knowing who it was, or why they were covered in that nastiness, and figuring out it was Captain Royer. He approached me, grabbed me by the flak jacket and said, "Look at this shit" and he pointed to his helmet. He had taken a round to the helmet and it had ricocheted off the front of his helmet, so we came pretty close to our CO getting shot in the head in that field. We were engaging the enemy all around that area. We got snipers that were up on top of a roof and we got word that we had wounded close to our area. So, I went back to the house we had originally hit with a squad from 2nd platoon, as a platoon sergeant, and set up a casualty collection point. Captain Royer and that command element came with me. That's where you've probably seen the pictures of the interpreter and Swanson sitting on a roof covered in muddy looking stuff. Then, have you seen the short video in David Swanson's documentary "Echoes of War", kind of a slide show?

Janney: Yes, sir.

Smith: So, you see there's a portion where Litke is actually firing. Pictures being played in succession, so you can see him standing up over a wall and shooting. Swanson and the interpreter are sitting there. Those are all at the house where I had set up my casualty collection point. They're all up on the roof. I sent out a fire team from the squad I had with me to go collect a wounded Marine. At that time, I didn't know who it was, but it was Sims. Sims had been supporting a sniper team, a two-man sniper team, and he was acting as their security. Sims was actually shot in the back, in the shoulder blade area and was unconscious, so they called for help. So, I sent a fire team from my Company over to help them out. They brought him back to my casualty collection point and Doc Gutierrez was working on him on the first floor and we had guys up on the roof that were firing. I called in for a casualty evacuation. I called for a helo, which ended up being denied due to the amount of enemy fire in the area we wanted to use as a LZ. I remember being pretty disappointed about that because I felt like we had a chance to save Sims, but we lost him.

Janney: So, you were disappointed that the helo was denied?

Smith: Yeah, I was a little pissed off to say the least. Pretty pissed off. I was pretty close to Sims; a close friend. So, they came around the corner with him on a stretcher. Swanson has pictures of that as well, from the 10th with Sims on a stretcher. That's my team that is carrying him with Doc Gutierrez. They come into our house and so I'm calling in, I'm calling 9 Line which is casualty evacuation report and it gets denied for too many enemies in our area. Doc Gutierrez does his best working on Sims, but I believe Sims was dead before we could ever get him out of there. Finally, an Army tracked vehicle came in to help. I don't know what kind of vehicle this was. It was medical, but it wasn't a Bradley. It was smaller than a Bradley, some tracked vehicle with a Red Cross on it came in. An Army element came and we loaded Sims in it, and they left the area. So, the 10th for me, was, you know, we didn't search a whole lot. There was a lot of fire; there was a lot of fight, but the biggest portion of what I was doing that day, as platoon Sgt. for 2nd platoon was running a casualty collection point.

Janney: Various guys told me that operation was a huge success. One guy said that they saw dump trucks full of the bodies of enemy insurgents rolling on Michigan coming off of Gypsum. So, I know you definitely got some payback there for what happened on the 6th.

Smith: Oh, absolutely. In the little bit of time before we transitioned into more of a casualty collection role, there were several enemies that we had dealt with in our vicinity. So, dump trucks full doesn't surprise me a bit. They were definitely outmatched that day.

Janney: After the 10th, did things kind of quiet down or was it just sporadic fighting?

Smith: After the 10th, from my recollection, it quieted down a little bit, but I would say through our whole deployment, it was just a myriad of highs and lows where you would have a really busy couple of days and then there would be nothing for a while. What may have made it busy was a bad IED explosion like the one that took place on Michigan that killed LCpl Wiscowiche from the combat engineers (author note: occurred 30 Mar 2004) - that was pretty brutal. I was on QRF. It seemed like I happened to be on QRF when the bad stuff was happening. So, I was on QRF that day when that IED went off. That was an IED planted in the middle of the median of Route Michigan, pretty bad one. That IED killed him instantly, so it was a pretty gruesome scene there. Highs and lows like that, when we would have really bad IED attacks.

I was wounded at the end of May on a QRF patrol where we were taking, I want to say it was 3rd platoon, out to the area of the power plant. We dropped them off for a patrol, came back. They called us for pick up, we picked them up and left and were headed back down Route Apple when an IED went off right next to my vehicle. It injured Corporal Lenz and myself. I was out of commission for a couple of days with a fractured jaw and shrapnel to the face. Lenz had basically had his eyes sandblasted, so he was out for several days while his eyes healed.

Various events like that were happening. Various firefights. A lot of our firefights surrounded large operations like Bughunt where we would go out as a Battalion for an operation. It may be an Echo Company AO, it may be Fox Company AO, but basically the entire Battalion would show up and then shut off a large section of the city. We would start in the north sweeping south and each platoon would have a block and we would just start moving. Typically, we'd start early in the morning and then by lunch time, the insurgents would come out and we'd fight until dark. So, during most of those we didn't take a high number of casualties doing those, if any at all. We'd kill insurgents here and there. That was usually where the excitement came from. Long days of firefights were happening then.

There was the incident in which the Governor of Anbar Province, his house was set on fire and his family was murdered. We were sitting out there, and that was after Lt. Marley had come to 2nd platoon, and we were ambushed headed down that canal road next to the mosque near his house. We were ambushed there, but didn't take any injuries on that one. Killed several enemy though. So, yeah, there were just highs and lows. Just kind of a rollercoaster deployment where we would lose a couple of guys.

Another significant one was where we lost the 4 guys from the sniper platoon on that house under construction.

Janney: Did you know Parker, Lopez or Contreras? I know you knew Otey.

Smith: Yeah, Contreras and Otey both came from Echo Company and they were added to the scout snipers. So, I knew both of them. I had met Parker, but I didn't know him that well. Like I said, Lopez, Contreras and Otey, they all came from Echo Company. None of them from my platoon. But I knew them. I had known them all through our Okinawa deployment and then our Iraq deployment, so I knew them all. Pretty tough deal that day when all that took place.

Janney: Let me ask you one other question and then we're going to shift gears a little bit. Operationally speaking, did you know if the insurgents that you were killing were kind of a mix of locals and foreign fighters? I've heard various guys say that there was definitely a mix of both.

Smith: Yeah, I mean, there was definitely a mix. I would say the majority were local. There were definitely some foreign fighters in there. There was a time where Golf Company reported there were black guys with dreadlocks. That's definitely not something you normally see when you're fighting in Iraq. That's not the typical description of an Iraqi, so there definitely were insurgents from other areas. They were being shipped in from Syria and from Iran. It was pretty well known that we weren't just fighting Iraqis.

Janney: Somebody else mentioned that some of the enemy combatants that were killed had Iraqi police ID on them. Did you ever hear anything about that?

Smith: Oh, yeah. We would find it from various times. We would go into a house that we knew we had been shot from. We would go in and find blood trails into the houses where we had shot them and in the closet, you would find Iraqi police uniforms. You would find issued weapons, Glock pistols that were on these guys that were only issued to Iraqi police. So, yeah, we had little to no trust of the Iraqi police or to the IDF that was out there, kind of in a little compound just to the east of us. Those guys, I think it was just whoever was paying them right then.

    I know a lot has changed. I went over to the fire station and we were just kind of reminiscing and telling stories. I said, "You know what? I wonder what it looks like now?" I pulled up Google Earth and so I zoomed in to Ramadi and found Combat Outpost and it looks like a junkyard now. They've got a bunch of junk cars all inside Combat Outpost. It was just kind of crazy seeing it that way. I heard later on, after we left, in the following years when we left, that there was a college across the street from Combat Outpost that was abandoned when we were there that, at some point, they had turned into a kind of outpost. I'm sure there were a lot of changes throughout those deployments.

Janney: Eric, switching gears just a little bit. Tell me some remembrances that you have of the guys that didn't come home. I don't know how well you knew those guys being a senior NCO at that point. You mentioned Kyle Crowley. Did you know him or any other guys very well?

Smith: Yeah, I knew Crowley pretty well. On a personal level, maybe not as much because it was just separation of rank. I say that, but I also say that I felt like I knew him pretty well. Crowley was another one that hurt me pretty bad when I remember the 6th. I remember seeing him being loaded and I didn't know it was him. The whole fog of the war, it's amazing how that happens, you forget decisions you made, even in the moment. I remember getting to the intersection of Gypsum and Nova and we were placing the bodies. I knew that we were picking up our own, but there was so much going on when the 6th kicked off; I'm moving my guys over to a certain area and what we know now, after you've been able to talk to a lot of people, you didn't know at the moment, you know what I mean?

So, I didn't know that the snipers were being ambushed in the tank graveyard. I didn't know why 3rd platoon came and got in the vehicles because Lt. Ski hadn't had a chance to tell me yet. And so, it wasn't until afterwards, days later that I found out, "Oh, okay. That's where they were going." So, when we got to Gypsum and Nova, I knew that they had gotten in the vehicles and gone up there, but for some reason my brain didn't click that Crowley was driving that first Humvee.

When we got up there and we started picking up the bodies, I was focused on setting up security. All the things an NCO has to worry about. Making sure that I've got point guys out. Making sure that we don't get hit again. Are rooftops covered? Are we safe to start this operation? You know what I mean, if we searched it well enough. So, I've kind of got my back to what is going on, but I turn around at one point and I look and I see two Marines. I don't remember who they were, but they were struggling to get a body on a 7-ton. You know, a 7-ton is high up there. It's not like there is this gate that you can let down, right? They were right behind the cab and that's always where we always stored the concertina wire, too. It was in between the cab and the bed of the 7-tons, and we've got the armored plating on the outside. So, I turn around and I look, and they are struggling. They've got this Marine by the arms and ones got him by the legs and they're trying to push him up and he's obviously a deceased Marine. I look and see the nametape on the pants and it says Crowley. It hit me like a ton of bricks at that point because I didn't know it was him and I didn't know he was there. But I don't think I had time to react then. I was pretty numb to it at the moment. It was a moment where I can see it today. I can see it like a still painting and just feel that immediate grief and guilt that you feel. I felt close to Crowley. When we were back in the States before we had left and we were doing all the work up, we didn't have a full complement in my squad. I didn't have twelve guys, I had nine. Instead of having three 4-man fire teams, I had three 3-man fire teams. I didn't have a whole lot of senior guys. Then some of the senior guys, I'm not going to lie to you, some of the senior guys I had, I didn't trust them to be in leadership positions. Not that I didn't trust a Marine; I'm not trying to go against our code, but I needed leaders. Crowley was someone that I thought was a young leader. He definitely needed some help; he needed some work. He was young, he was immature at times, but he had the stuff that I think would have made him a great leader. So, I brought him under my wing, and I made him my 3rd team leader, so he was actually one of my team leaders. He was probably in way over his head for a combat deployment in Iraq as far as knowledge-wise, but he was up to the task and he tried his butt off. There were times when we were on those route patrols where we would find an IED, and we'd have to wait for hours for EOD to get out there to handle it. So, we would just be in security with the road blocked off, waiting, just talking. Talking about home, talking about what we missed about home, talking about everything. So, I learned a lot about Crowley's story and how he got to where he was and troubles he had growing up. I felt pretty responsible for Crowley during that whole thing. So, yeah, that one was close to me. Another one is obviously Lt. Ski. I hold that one really close to my heart and I feel… I've had a lot of guilt over the years, over Lt. Ski. You were here for my ceremony. That's definitely something that I've always struggled with. I mean, how are you awarded something for a failure? That was something that was always in my head for a long time. I've come to terms with it, but for a long time, I considered that day as a failure. And so, why would you be awarded for something like that?

Janney: You did some amazing things that day. I know it's easy for me to say it wasn't your fault, but I hope you have come to terms with that and realize that the survivor's guilt is something that you've got to work through.

Smith: Absolutely. Years and years of counseling and mentoring with other individuals that have been through similar scenarios. I spent a lot of years separating myself from the guys because I did feel, for a long time, like I'd … (crying)

Janney: Take your time, Eric. Take your time, sir.

Smith: So, yeah, I probably didn't go to a lot of the reunions and that stuff because I didn't feel good about it. I felt a lot of guilt about it. I felt a lot of, "I didn't do that right. I could have done that differently" so there's a lot of years there that were pretty tough. Especially the 6th, that's one that haunts me. You know, some of it, I can tell you, I know is irrational, the fact Lt. Ski was out there for a long time, but I looked at it like that was my QRF and he wasn't even supposed to be there. So, the fact that he came was a split-second decision and it changed a lot of things for a lot of people.

Janney: A lot of people's lives were changed that day and you really didn't have direct control over any of that. I think you did an incredible job that day and a lot of people are proud of you for what you did, myself included.

Smith: Right and I appreciate that. I've come to terms with it. It still hurts. I know I wasn't going to change the fact that he wasn't going to be out there with his men. I looked up to Lt. Ski immensely. I've always used him; I continue to talk about him. I've had the honor of being able to go and speak at several Marine Corps balls. Last year I went to Sgt. Major Coan's ball in Spokane, Washington up there for the 8th Marine recruiting district. So, all of those are opportunities that I get to go and speak to Marines about Marines. I get to hear their story and that's my way of dealing with my therapy. It's my therapy to speak about Marines that I had the chance to walk alongside. The guys that I consider giants among men. I know I was lucky to get to serve with some really great guys. Several of them that didn't have the opportunity to come home. It's always great for me to be able to share their stories.

As I said, I think a lot of my guilt, I've come to understand, was irrational. That I wasn't going to change the fact that Lt. Ski was going to be there that day. The way things were going, he was definitely going to make the decision to go. That wasn't something that was in my wheelhouse, but it definitely caused some guilt. And the decision to put Crowley driving the first vehicle. I needed somebody to drive the vehicle, so somebody was going to be in that position. That was a decision that I had to make. It just happened to be Crowley was the one in there. But, as a leader, you look back on those decisions and they're the ones that hurt. You can't change them, you can't stop them, but they definitely hurt. A lot of those guys, I felt like and it seemed like I was always QRF. I felt like I was there when we lost two thirds of the ones we lost. I put them in body bags.

I may not have been close with all of those guys, on a personal level, where we were going out drinking in the States before. I had a different situation then. There's a lot of guys that were single and lived in the barracks and they partied in the barracks and they did a lot of crazy things together. I was married. I got married shortly after I got in the Marine Corps. Actually, I got to 2/4 in October, November timeframe, came home on Christmas leave and I got married then. I've been married to my wife Shelly now since the Fall of 2001, and so I'm coming up on 18 years now. She was with me through all of that time. She's been the rock that's helped me out through all of this. She knew a lot of the people that we've talked about.

Shelly has a story herself about Crowley. The night we left to go over to Iraq, she was carrying something, and Crowley went over and grabbed it from her, just being a good Marine like he was. He said, "I've got that, ma'am." Shelly said, "Dude, why are you calling me ma'am? I'm only two years older than you." She knew him, it hit her pretty hard too having to listen to all of that. When the 4 snipers died on the roof of the house under construction, Parker and Otey and all those guys, they showed that on the news back home and she saw that and it just about killed her. They had video that the insurgents had filmed, and they blurred out faces, but you could definitely see legs and boots and camouflage on, and she could have sworn at that moment that one of those was me. So, she was close. You know, every one of those hurts. It hurts pretty bad.

Janney: Tell me about Lt. Ski personally. What was he like?

Smith: Yeah, I've kept in touch with the family a lot through the years. You know, I called John and Shawn after, but I got away from that. I guess I felt like I didn't want to be a negative reminder for them. But I was trying there, for a long time, to call them on the 6th just to tell them that I was thinking about them and remembering their son. I haven't talked to them in a while. I do talk to Rich pretty often. Several times a year, we contact each other through texts and through Facebook Messenger. In talking to them, Lt. Ski sounded just like them, that New Jersey accent. I remember when he first came to us, when we got our new Lieutenants that day. I mean, I'm pretty bad about that, I judge people pretty quick and I couldn't have been more wrong when I first saw Lt. Ski. Because he came out and for some reason, he was wearing these glasses and I don't remember him wearing them. I guess he wore contacts, but he was wearing the BC, you know, the boot camp brown glasses.

Janney: Yeah, the birth control glasses.

Smith: Laughs. Right, right. He was wearing those glasses and I thought, "Man, who is this nerd? My God, who'd they give us?" I quickly figured out that this was not correct. My first impression was way off. He was definitely a northeast Jersey boy. He had that accent. I remember he spit a lot; I don't know if he used to dip or what. He would roll up these little balls of spit and he would spit them out when he was talking to us. He had that real thick Jersey accent.

Janney: Oh, I know it. I can hear John's voice in my head. We spent so much time together.

Smith: Just like his Dad and his brothers. So, you wouldn't have had to meet him to understand what he sounded like just being around Rich and them. I remember him being firm and fair. He liked to joke around, but he didn't cross that line between officers and enlisted. I considered him a great friend and a mentor. It was one of those lines, the Marine Corps has its lines and that's definitely a line that couldn't be crossed. We did have dinner with Lt. Ski and his wife, Joanna, before we deployed. Shelly remembers that pretty well. He was a good officer, very intelligent.

Janney: Do you have any other particular remembrances of the guys that didn't come back? I mean, I know you shared a little bit about a couple of the guys, but are there any others that you remember particularly well? I don't know if you knew Staff Sgt. Walker very well.

Smith: Yeah, I mean, like I said, I knew them all but it's one of those things that there's a difference between knowing them and being really close.

Janney: And REALLY knowing them.

Smith: Yeah, REALLY know them. Knowing where they grew up, knowing their families and all that stuff is different than truly knowing someone. The platoons have a way of sticking with their platoons, even though you know the guys from 3rd platoon or 1st platoon; you intimately know the guys in your own platoon. So, I definitely knew Crowley well. I definitely knew Wroblewski.

I knew Staff Sgt. Walker and I respected him, too. Earlier in the deployment, we were out on patrol on the Euphrates River and we were checking pump houses. We come up on one that we had suspected had IED building materials. It had a military jacket, helmet, soldering equipment, chemicals that I didn't know what they were, and big sacks. I thought, "This has got to be one" and so I called it in. Captain Royer said, "Set up an ambush on the road leading up to it, and if anybody comes up there and they're armed, take them out." We set up the ambush, and I remember one of the guys, a combat engineer NCO, came up there and it ended up not being an IED making place. I remember thinking, "We are supposed to be doing this the right way." I remember Staff Sgt. Walker approached me after and he said, "Man, you did everything right. It didn't turn out the way you thought it would, but you saw something suspicious and you called it in." He consoled me like a good leader would do, trying to bring about some confidence in a young leader.

Janney: And positive reinforcement. That seems like what a great NCO should be like and what he should do. It sounds like he was a good man and a great Marine NCO.

Smith: Oh, yeah, he was the type that lifted up instead of tore down. You know Marines, we're all over each other, right? We will bash each other. At the end of the day, we love each other and there's nothing that we wouldn't do for each other. But, if there's an opportunity to fuck with each other, we're going to do that too! There would be plenty that would take the opportunity to mess with me about it, but Staff Sgt. Walker was the type that encouraged and, you know, he had the rough, gruff look of a Drill Instructor. That's why you didn't want to mess with him, but he also had another side to him that I got to experience where he was very encouraging to me. And so, that stuck out to me after that incident when he did that. There were a lot of young guys that I knew that had passed. Your Carmens and your Cherrys - those guys. I knew them, I had met them, but I didn't know them real well. It was tough to lose them all. Calavan, that was one that hurt, too. That whole group that was killed and wounded; that was right when I got hurt at the end of May and had gone away to the hospital. I was in there for three days and Corporal Misael Nieto had taken over for me while I was gone as the squad leader, and while I was in the hospital, they had gone out on a QRF thing, where they were picking up a weapons platoon. They were headed back and an IED hit and that's when PFC Calavan died, LCpl Reynosa Suarez died, and Cpl Lee died in that one as well. So, there were several that died in that one. That was a vehicle borne IED attack that hit. That hurt, you know what I mean?

Calavan was attached to my platoon for that whole deployment, so he was a young guy that had a lot of potential. Funny guy that had a lot of stories. We spent a lot of time telling stories and messing around. I know you talked to a bunch of guys. It's different when you're on the NCO leadership side of it - you don't get to do as much messing around with the guys as others do. You know, when you are one of the Lance Corporals or PFCs and not in that leadership role necessarily, so there's a whole different side and element to it when you're part of that high school boy underground. I knew them all; they all hurt, but the ones that were definitely close to me would be Lt. Ski, Crowley and Calavan.

Janney: Eric, can you think of anything else that you want to share with me?

Smith: Not off the top of my head. I'll tell you that it was one hell of a deployment. There were a lot of guys there that put a lot on the line for our country. I appreciate you telling the story; I think it needs to be heard. My opportunity to go and speak at Marine Corp Balls, it's been my therapy, but also, to me, it's essential to tell these guys' story, the ones that didn't make it back, along with the guys that did that are struggling. We have lost guys since we got back. So, the story needs to be told and shared so these guys can live forever through that story. I appreciate you doing it and taking it seriously and taking the time to get the information. I feel honored to have walked amongst giants and most of these guys were young, under the age of 22. Just American heroes. I appreciate your telling their story.

Janney: It's a huge honor for me to do it. The book started out just being about John and my two journeys over there, but after I started to get to know the 2/4 guys, someone suggested I interview them about the Battle of Ramadi. You and your Marine brothers are telling the harrowing story of their struggles, triumphs, and sacrifices.

6 April 2004 Route Gypsum Ambush And QRF

PFC Ryan Miller

Janney: It's 10th August 2019. Ryan Miller, you were a PFC in 2/4 when you guys were deployed in mid-February of 2004. You transitioned through Kuwait like almost everybody did?

Miller: Yes. We spent two weeks in Kuwait. We went to Camp Wolverine and then to Camp Victory, I believe. Once in Ramadi, we went straight to the Combat Outpost.

Janney: I know you guys were doing daily patrols. Were you always attached to Echo, or were you with weapons?

Miller: Weapons platoon is attached to Echo, but it's part of Echo. It's a platoon, but the machine gunners and the assault men attach out to the individual infantry platoons. One platoon is weapons platoon and others stay in the rear, for the most part. I was a machine gunner attached to 2nd platoon. Lt. Wroblewski's platoon.

Janney: So, you knew Lt. Ski pretty well then, I guess?

Miller: Ish. I knew him for a couple of months. I could tell that he was a great person. He wasn't one to use his rank to make you feel like dirt. He was all about camaraderie and he treated you like family, basically.

Janney: Well, you guys were a family. I remember you and I talked about Lt. Ski in Arlington, but I know it wasn't like you guys were hanging out since you were enlisted and he was an officer.

Miller: He had his own quarters with the other Lieutenants. He didn't stay with the platoon.

Janney: Right. Leading up to the 6th of April, give me a general overview of what was going on every day. We you guys patrolling, doing SASO, humanitarian stuff, security stuff, what?

Miller: Yeah, we would patrol every day and then do ops. We were running 22-hour days for the first three months. No days off, you were either on patrol or you were on QRF on post. We would patrol throughout the city, a lot of the whole hearts and minds thing, it was more like a parade than a patrol. So, you would hand out soccer balls and candy and Gatorade and water and all that stuff, which the local parents actually hated.

Janney: Why do you think they hated that? Do you think it made the Marines or them a target?

Miller: It was as if we were indoctrinating their children, type of deal. We didn't really understand their religion, a lot of us didn't know anything about it, so when we would separate the men from the women in the beginning, it was pissing off a lot of people because we didn't know. You know what I mean?

Janney: Sure, I understand exactly what you mean.

Miller: Down the line, we had gotten enough complaints that we had actually taken one of the males of the house and put him with the women so that they'd know that we're not doing anything, which none of us ever did.

Janney: Yeah, of course not. I understand. I had to study about that before I went overseas, the couple of times I was there, because I didn't want to offend anybody, you know.

Miller: Right, yeah, personally, I was there to do a job, so I didn't really look too much into their religion or their beliefs or any of that, but I got a crash course while I was there.

Janney: Yes, I can well imagine. So, as far as the security situation, were you guys getting any direct or indirect fire leading up to 6th April?

Miller: I want to say indirect fire. It was just about every day we'd get mortared. We could usually time, you know, after a while, we knew they were coming in the evening when everyone was getting chow, that type thing. Or, first thing in the morning, and then there would be small, sporadic fire, but nothing real crazy. I was always on point or kicking in a door, so, you know, every time I went into a house, I never had anyone ever pointing a gun at me necessarily.

Janney: So, when did things start breaking loose? Was it not until the actual day of 6th April that things just kind of went to hell in a handbasket?

Miller: Everything was building up. You could tell the community didn't like us there in the first place. We'd go through the city and every single male within military age would just stare you down. Basically, when the kids were around, we weren't too worried about it. We had a pretty good idea that they weren't going to kill their own kids. Most of these people were farmers. That being said, some of them did pick their kids up, use them as shields and whatnot. Most of the stuff we came across in the beginning, it was hard for us to even pinpoint who did it. As far as being mortared or shot at because it was just rural and by the time you would get to the mortars, they had them in trucks, so they would pull off and be gone.

Then our artillery from Junction City came up with this thing that could pinpoint the mortars if they fired three rounds or more, so that was a big one for us because we weren't catching any of the mortar guys. A couple of night ops, we found a mortar site, maybe a base plate here and there or a vehicle with an improvised bed with a big chunk of steel in the bed of a truck so they could fire.

When we were first there, we pissed off a lot of people because we would just kick the door in and search the house. Then word came down the wire that we needed to knock, and in the beginning, we were just confiscating any weapons and any magazines or rounds. Then, after a while, that got squashed and we had to leave at least one AK with at least one magazine and sufficient ammunition to protect themselves.

Janney: How did the day of 6th April start? What is the first thing you remember that day?

Miller: I was on QRF, which is the quick reaction force. We turned out to be the 2nd QRF. We were supposed to be on break, but we got a call that they were splitting up our squad into two teams and sending them out. While our squad was sitting there, the snipers radioed in that they had 20 plus men armed walking online, walking through the field looking for them. So, our QRF went out, which was 1st platoon, machine gunners, most of them. There was Carmen, Martinez; they were the two gunners. Basically, they showed up on site and took immediate fire because we already had snipers that were hit. So, 1st platoon showed up, took immediate fire. Carmen went down relatively fast from what I heard. Martinez had a round ricochet off of his machine gun and hit him in the hand. Then, he jumped off the truck and had a M16 and was shooting that until he ran out of rounds. Then, he threw that and pulled out his 9 mil and was shooting rounds and had someone reload for him because he only had one hand. Then, he was dragging people off. I don't know if anyone has ever really said anything about that, but that's what I got from him.

Janney: Yeah, that's the first I've ever heard of it, but I appreciate you sharing that. The patrols then went up on Route Gypsum?

Miller: 1st platoon responded down Apple there at the tank graveyard. Where we needed to go was on Nova to the tank graveyard. That's where the snipers got hit. I'm still pissed off to this day that we went down Gypsum. I was screaming at the convoy when we left to take a left because that was the quickest route.

Janney: So, you guys started heading up (north) on Gypsum. Were you guys the first element up Gypsum? I know you said that a squad had already been hit at the tank graveyard.

Miller: Yeah, we went all the way down to the arches before we left. We were staging at the gate at the Outpost and my squad leader, who was Corporal Lenz, said, "We have the pine box rule." I asked, "What's that?" Lenz said, "Put them in a pine box before they put you in one. Lock and load, Marine." I racked my bolt back, but we never really ran like that. It was a simple second step from being like that, but that day we knew we were going to get into a firefight. Then, we rolled out of base and we started going past Apple. I knew where the guys were and I'm yelling to my driver, "Go left, go left!" We've got two Humvees in front of us that just keep going. I yelled, "What the fuck are you doing? Go left!" and he wouldn't go left.

We got up to arches and just sat there. I can hear gunfire going on. I yell, "We gotta get over there, what the fuck?" Then word comes down, a bunch of people dismount, and we start pushing up Gypsum. I'm pretty sure there was one Humvee in front of Lt. Ski's and then my 7-ton was right behind his. When we got going down, we got to these two red houses basically opposite from each other and this is where we stopped. We took small arms fire and had everyone dismount because we were just going to walk through it, even though we were supposed to be QRF to our guys getting hit.

A couple of minutes into it, they yelled that Lt. Ski was hit. All the squad leaders run to him – he was right in front of me in the Humvee and this was after him and Royer switched places. Basically, he took over Royer's Humvee for lead truck or some shit. Royer and them decided to walk through a field to meet up with us and when we got to that point, we took small arms fire. No one could really tell where it was from. Lt. Ski got hit, the squad leaders ran to the Humvee and pulled him out. There were like 4 or 5 guys that took him out. He was face down, and they took him back to the other Humvee. It was pretty bad. They got in the Humvee, loaded up, took off, and were shooting all the way down Gypsum there as they left.

We're just sitting there at first. These guys just started stacking up at this house on the right because that's where we thought the gunfire came from. My squad leader, Corporal Lenz, told me to provide cover fire and told me to light up that house. I'm looking at it and we've got Marines going in the bottom; there's already over half a squad in there and I don't know how many people made it into the house, and I don't know if they're upstairs or not. I decided to shoot next to the house rather than light up the whole house because they were going to go through it and clear it anyway. When I first shot, my gun went "Bang!" I said, "Shit" and I racked a round and it went "bang" and I said, "Shit!" I realize my round is hanging where it's not feeding to the gun. I picked it up in one hand and then shoved a 200 round belt across Corporal Lenz's head, pulled it over his head. I stopped right up near the house because I knew guys were in it and then I lit up a path on the back side of the house and dropped my belt. I got yelled at a couple of times to cease fire. But, being a machine gunner, you're trained to keep a fresh belt in. So, I dropped the belt and reloaded.

Corporal Lenz made it to the stack, cleared the house; nobody was in it. While they were clearing it, another Humvee stopped: Staff Sgt. Walker with Crowley driving; Jerabek was the 240 gunner and other Marines in the back and they stopped. We thought, "What the fuck?" I thought the whole convoy was with them and I told them, "We've got to get to our boys in the tank graveyard. We're all good here. We've gotta move." My driver had already bailed out of the truck. I said, "We gotta move!" because they had already started shooting at us and we're the biggest target here. Then, Jerabek was on the gun, which was bugging me, so I told one of the other Marines, who was a team leader of mine, to get on the gun and he refused to do it over and over. So, I told Jerabek to lock and load because he was still in Condition 3. So, he racked the bolt, so I knew he was good to go. Staff Sgt. Walker agreed we had to get to our boys sitting on Nova, and they took off and there was another Hummer behind them with Gaeden and all those boys and none of the other convoy went with them. I had Corporal Smith and Burmaline in my Humvee, so the ranking person for that was the Motor T Sergeant Valerio after Lt. Ski and Staff Sgt. Walker took off. They get up around the corner and a 50 cal starts barking off. I'm yelling at these guys, "Hey, we got friendlies up there? What the fuck? We don't have a 50 cal." Because our unit didn't have a 50 cal. We had a 50 cal on post, but not on the trucks. I said, "We don't have one, that's gotta be enemy." Then, I just said fuck it and started shooting. I didn't shoot directly at the 50 cal because I knew our boys were there; I had a visual on the corner of Gypsum and Nova sitting on top of my 7-ton.

Janney: Which corner? The left corner or the right corner?

Miller: Where it turns at the T.

Janney: I know, but you said you were aiming at the corner at the T. Were you aiming to the right side of Gypsum or to the left side of Gypsum?

Miller: The left side.

Miller: There is a giant field that goes on the left side there and I heard the 50 barking and I heard the 240 kick off. I heard the 50 cal barking and I heard the 240. This went on 5 times, and I know the difference.

Janney: So, that was Jerabek returning fire with the 240 machine gun?

Miller: Yeah, I knew they were in a fight. I just started shooting at people I saw running from the marketplace where they were because there was a big firefight going on and these people were leaving it. I was taking fire from it. I had rounds ricocheting off my 7-ton. I'm telling everybody, "These fuckers are shooting at me" so I'm fucking shooting and I start getting low on ammo. One of the guys runs up, McKinney, and starts linking rounds for me and I just kept shooting. Everybody kept saying, "Cease fire." I just kept telling them to fuck off because I knew our boys were getting chewed up, that they didn't make that corner. Smith asked me if I thought they were dead and I said, "They didn't make the corner, you know what I mean? I don't think they're alive." I just see him drop his head and I just kept shooting.

Then, I was running low on ammo again and McKinney ran out and got rounds from everybody. By that time, Vergara had pulled up to that red house that was on the right in another 7-ton on the Mark 19. He asked me what I was shooting at and I told him. He told me, "If you can hear the 50, walk them on target." I actually kept them left of the 50 cal because I didn't want them hitting our boys. But I knew it was that close. He was telling me, but I said, "I'm left of it, I know I'm left of it." I know I had them strafe out a little farther and walked it down, but that was about it. He tried to walk Mark 19 rounds down on it. I had to put the rounds in the trees on the left side of Gypsum because that's where I was taking fire from and a lot of people were hiding behind the trees. I couldn't get them. I was getting the rest of them that were going down through the fields where they had the weeds, the big, tall ones?

Janney: The papyrus?

Miller: They were trying to run down into that and hide behind it, but I've got a 240 and they can't hide from that. This was roughly about 400 meters out. I only know that from the ACOG that I verified the range with after I ran out of rounds. I had Vergara drop about 600 Mark 19 rounds on target on these people leaving the marketplace. I ran out of rounds. Vergara ran out of rounds. But, they were gone and none of us knew how to drive a 7-ton. I told him there was still people coming through there. One of my other machine gun squad leaders from 3rd platoon gave me his rifle – it was an M16A4 with an ACOG. I took one shot to sight it in on the palm trees and then I waited for the next two to come out. They were looking at the bodies on the ground. The next dude that came out was wearing white and he was looking at the ground and he was in my kill zone, so he was dead. I shot him, at least I thought I hit him – there was a red cloud. Then, the next dude that came out was wearing all black and I waited for him to pass one of the palm trees and I shot him in the melon. That was the last person that I shot that day. I had relayed to Corporal Smith that there was a 50 cal that had just chewed up our boys and it's going to chew up the rest of the convoy and we need armor. We waited for a Bradley – actually two or three of them showed up. Then, we pushed up behind the Bradley into the fucking marketplace and noticed there were two guys hanging out, running around. They ran and got in their Iraqi taxi and took off. I didn't have any rounds to shoot, and I'm screaming at these guys, "Shoot them!" The Army guys, they've got a fucking SAW and a Mk19 and they couldn't hit this car driving away. I'm thinking more along the lines of it's against the rules or whatever. But I already knew that they were not good, and they ended up getting off with a couple of our weapons. They ended up finding them. When we got up on scene, there was a dead guy next to the Humvee, an enemy. Staff Sgt. Walker made it a little way out of the truck; you could tell he put up a fight.

Janney: So, there's an enemy laying there to the right of the truck? And Walker was there?

Miller: There was another one up the road. The reporter came running up right away. I was in the first 7-ton sitting right next to the truck and he starts taking pictures of it and I can hear everyone is just pissed off that he's doing that. I said, "Hey, there's a live grenade at your feet. You better watch your fucking ass." He looked down and about shit his pants because there was a haji grenade with the pin pulled next to him. He took off and we went and collected our own. The Army stayed, but we went back to base and rearmed and came back out and we had found another Marine that had made it behind one of the buildings right there. He was the last one that we found that day. That was about the extent of the 6th there. I mean, I had one Marine come up and he yelled at me because he thought I shot up the Humvee. It really got at me because I knew I didn't shoot it.

Janney: Right. You know, emotions were running high.

Miller: It caused a lot of drama in the Company to be honest with you. We had one guy running around saying I did it. Then he said, "Well, where's the 50 cal?" I said, "It's fucking here. Look around." I said, "Go over in that fucking field right there" and he wouldn't do it. Then I screamed at him about a hatchback taxi behind the buildings, "Search that car. That car's never been here before. I've been through here a hundred fucking times and that car has never been there. There's never been a car back there." One of the Marines went over and eventually popped the trunk on the car and it was full of RPGs and AKs. So, it was where all the people in the marketplace took off, dumped their guns and took off running through the fucking fields where I shot them.

Janney: It's obvious to me they had a DShK because of the size of the holes in the lead vehicle.

Miller: We found it. It still had rounds on the belt. It fucking jammed.

Janney: Wow. So, did you guys find any enemy combatants over there near the DShK?

Miller: No. That was the two fuckers that got away. I'd swear by it. They ended up getting off over toward Fallujah; that's where they took off to. We didn't have any jurisdiction. We couldn't go past the arches is what we were fucking told. On Nova, headed down past the arches.

Janney: I was going to ask you something. Did you see, or could you not see the lead vehicle that Staff Sgt. Walker and those guys were in.

Miller: We never made it far enough up to the corner.

Janney: Not until after the fact?

Miller: Yeah, I didn't see it until we got up on scene. That's why I knew everything was fucked because they didn't make the turn.

Janney: Right. Yeah, they didn't.

Miller: That's the reason why I stayed where I was because I could see if those guys were going to make the fucking corner. When I said they didn't make the corner and we didn't push, man, yeah, it just blew my mind. That's about all I can say about the 6th.

Janney: After the 6th things really ramped up.

Miller: Well, we just decided to shoot more.

Miller: There was a lot more over the intercom calling for war against us out of the mosques and whatnot. And then the IEDs really picked up bigtime. It's hard to shoot at an IED. After that, I was hit with a suicide car bomber once. I've been through many, many, many IEDs. Most of them we caught and they were controlled dets. The 6th really sticks with me a little bit. That's one of the only days that I used my 240, other than in training. Then, after they doubted me, I didn't want to be on the gun any more after that. I can say that I went through over 2500 rounds on target on the 6th.

Janney: I know you did an incredible job. As I said, I think the world of you and appreciate everything you did, and I'm sorry that you guys lost so many brothers.

Miller: Yup. Yeah, that was a bad day. It was a really bad situation for everybody. Nobody was shooting at nothing so…

Janney: The guys that, the brothers that you lost, were you close to any of them? I mean, did you, I know you said you were in Ski's unit, and you got to know him a little bit, but I mean some of the other guys?

Miller: Yeah, I went to boot camp with Jerabek and, I think Crowley. He was just one of the cooler guys. He was just a couple of months ahead of me. Carmen obviously was one of my senior machine gunners. Martinez and I were rack mates all the way up through after boot camp. I knew all of them. Staff Sgt. Walker was a Drill Instructor while I was in boot camp, you know, for one of the other companies and he taught me how to play Spades on the way over on the flight because I had nowhere to sleep because I'm a boot. Fucking everyone else is passed out on the plane so I had to sit there like a dick. He was like, "Hey, get over here, let's play some spades." Laughs. It was me, him, and Rogers and I think there was one other guy.

Janney: So, tell me a little bit about Kyle Crowley. What was he like?

Miller: He was real laid back. But a fucking stellar Marine, you know. He was shit hot, you know, one of the first-class guys that wasn't a dickhead. He was a good dude. He was a good driver. He was one of the guys that I actually preferred as a Humvee driver.

Janney: Tell me a little something about Benjamin Carmen.

Miller: Carmen? He was a really good guy. We didn't always agree on beliefs and whatnot, but he had his way of thinking and I had mine. He was kind hearted; avid hunter; you know, the guy could shoot. That's one thing about just about all machine gunners is they can shoot. That's what I was into, shooting and stuff, so he knew everything in the machine gunner book pretty much that you could read. He was that kind of guy, really good to have as a senior LCpl. He really wasn't that much senior; he was just knowledgeable. Yeah, good guy.

Janney: Tell me about Ryan Jerabek.

Miller: Jerabek – really smart, really quiet, very strong, and just a good person all around. He wasn't one to go out of his way to insult anybody, but he knew his job. Hell, he knew my job too because I taught him. I'd say, "In case I die, you gotta take this machine gun, bro." It was real big to share the machine gun with the rest of the platoon there because we were usually the first to fucking get hit. Yeah, Jerabek was a damn good Marine. He was on the machine gun that day and I had faith he knew the damn thing, otherwise I wouldn't let him go on it. I tell you, he did better than a lot of Marines I know. I could hear everything that he was doing and I was really excited until the noise stopped. It's weird to say that, but I was excited that he was living and fighting hard. I kind of knew what was going on, but didn't exactly, so that's why I shot everything I could. People say I didn't or I just shot at nothing.

Janney: No, I know better because I've interviewed enough of you guys to know about the DShK, about the Soviet 50. I knew that's what you were after when we talked in Arlington.

Miller: I walked rounds down on that fucker as much as I could because I knew those Humvees were going to be coming around the corner. There is a latency when you shoot rounds out. There is a death point that you can't get back. When you put that burst out, you have to know your target. They did this recorded interview afterwards and there were multiple written statements. A few of us had to rewrite our statements because they were just too vulgar in language. Some of us only knew how to say fuck every other word. When you're getting shot at, there's a lot of "fucks" going on.

You know, there are a couple of guys that actually saw some of it. They just haven't said anything. That was that 2nd Humvee. I'm pretty sure Waechter was in that one, Gaeden, Shores, there's a few guys. We had guys getting hit left and right. Yeah, after the 6th, my back really started shitting out on me and nobody really believed me except my platoon Sgt. He let me be a driver for a bit. The two worst places in Iraq were the machine gunner and the driver, so I was happy to go with that.

Janney: I think Matt Scott was in the 2nd Humvee if I remember correctly.

Miller: Yeah, of course, Scott's a good dude. Yeah, yeah, he probably was; he was 2nd platoon.

Janney: Now, did you see Otey run back from the lead Humvee?

Miller: No, no, Otey was with Shores.

Janney: So, in the first or second Humvee?

Miller: They were in the second Humvee. So, Otey and Shores were on one side of the road and then Gaeden and Waechter and all those guys were on the other side of the road.

Janney: Yeah, Matt Scott actually told me that they bailed out of the second vehicle.

Miller: Yeah, because there was an RPG about to hit the truck.

Janney: Yeah, I think, at some point, it did hit the truck, didn't it?

Miller: Yeah, it glanced off of it and hurt part of it, but they had no choice but to bail out of the vehicle when the 50 cal started going. If I remember, Shores and Otey smoked a dude.

Janney: Yeah, but like I said, Matt, you know, told me flat out that's what they did because they were about to get hit by the RPG, so they bailed out and went into a building.

Miller: It was a fucked day. There were a lot of cries from Golf Company that we didn't do that. It was really weird. I don't know how else to explain it. I had plenty of other days shooting my M16, right alongside other people. It's war. This shit gets to everybody at a certain time.

Janney: You may or may not know the answer but, of these guys that you were engaging and killing, was there ever any indication that any of them were foreign fighters, that they weren't local?

Miller: It was both, to be honest. The locals hated us, and foreign fighters were just there to help them. By the time we were towards the end of our deployment, we were disarming these people. They told me I couldn't take their weapon, but they didn't tell me that I couldn't take their bolt. So, I took every single bolt that I could. I didn't care if they had a gun or not, to be honest. We were getting shot at sporadically and we could never hunt them down. It was in that neighborhood, so I did everything that I could in my fucking possible ability to take guns away from them basically without doing it, you know. You can ask multiple people. I would walk in their house, find the guns, go in the spare room and come back out with a fucking gun. We had a pocketful of bolts. We all agreed on it. They raised the price of an AK, allegedly, from $500 to $2,500 in the city of Ramadi.

Janney: It's not like there wasn't another 10,000 AKs for them to pick up or dig up out of the dirt. Sounds like your interdiction efforts were working.

Miller: I mean, it didn't matter how many bolts we took, they had 50s, they had 155s, and had all kinds of shit and wanted to kill us if they could. Then basically, what we were informed of was, "Oh, yeah, the military that was here just basically disbanded and let everyone come in and take everything." So, they dispersed all of the fucking rounds from that National Guard base to the fucking people and from every other base in any city. That's why there's so many fucking IEDs over there. You can't just go out on the street and buy a 155. They were handed out.

When we got hit by a fucking car bomb picking up our platoon, we had just gotten these new Hummers from the Air Force. I was really happy because I was a driver. One of our replacements was a senior Lance to me, Cortez, he was a machine gunner, he was facing rear security, and this car pulls out from a gas station, you know the gas station across from Combat Outpost down the road a little ways? Well, the dude was talking to a fucking Iraqi National Guard, and then pulls out. He's the only car on the road, and then cuts across the median and blows up. My ears are ringing and I look over at Staff Sgt. Wyman and he's fucking screaming at me, so I slam it in first and fucking punch it. We drive through the fucking fire ball; windows all shattered. The turret; I was telling Cortez to turn around because the fucker didn't look right and he ducked down which is a good thing and then that fucking bomb went off and this food truck went all crazy sideways and one of the SAW gunners lit that fucker up all the way across. I was like, "He's already fucking dead." They fucking shot up a truck that basically when the bomb went off, it sent a chunk of shrapnel that cut the dude's head in half. So, the truck just crashed. There were fucking body parts all over the place and they said, "It was an IED; it was an IED" and I'm like, "No, it wasn't." The QRF that came out said, "It was an IED" and I said, "No, it was a fucking suicide bomber" and they said, "No." I said, "What are you talking about? Look around. There's a fucking hood, there's a tire, there's a leg. Give me a break. There's a motor." So, we got lucky on that one, that no one got seriously hurt.

We had two guys that hit us as we were loading up provisions, so we had a whole platoon fucking sitting in back of the 7-ton and one of the fuckers smoked the back of the 7-ton as Sgt. West was about to get up on it. He went to step up. Then, seeing that we were stopped, they pulled back and then slammed into the back of the truck. I don't know if you ever met Sgt. West, but he was a good dude. He was in the Armory for a long time until we got over there and then he started going out with us. Hale was in the back of the truck on the 6th. I don't know, you may or may not have met Hale. He couldn't see what he was shooting at, so I just told him to shoot in that direction. Him and McKinney and Vergara. Hale was in weapons platoon for 2/4. But yeah, he was in the back, fucking shooting. He came over to me and he said, "I don't have nothing shoot at" and I said, "Get the fuck up here and help me out." And he started shooting. There was a lot of people on the 6th telling me to cease fire and, from what I could see, I wasn't about to.

Janney: No, I think you did the right thing. You probably saved a bunch of people's lives by keeping suppressive fire on that position.

Miller: The 10th was like revenge or whatever, if that's the way to put it. We really didn't do shit on the 10th.

We rolled out super early; it was 0300 that we staged; we were moving at 0400 and we were out of the billet at 0500. We had arty firing lume, and mortars firing lume. They were firing lume, so they were just lighting up all of our AO so we could see it. We had a regiment of the Army, I believe, and they had their loudspeakers playing psyops shit and they brought all the Bradleys. Then we went through and we searched almost every single house. That's what we did on the 10th. We went down past the power plant and cut the power. Then, this dude comes out and he's screaming down the street and Staff Sgt. Craig walks up, snatches him up and throws him in my truck. The guy wouldn't shut up; he kept yelling at those guys. He kept praying and yelling and this and that. They parked my Humvee right in the middle of the street. I said, "Anyone coming in the street is fucking dead. Stay out of the street." We heard this 50 bark off. And fucking Cortez yelled, "Do you want to live forever?" He takes off and runs toward the 50. I said, "God damn it!" Now I can't shoot because we have a fucking squad of Marines running up there. This 50 is still barking off. Right before that, Captain Royer was walking down the middle of the road with the radio operator with his fucking bars glaring. I yelled, "Hey sir, no disrespect, but you're going to get fucking shot." Turns out, about 10 minutes later, he got fucking shot in the helmet. He lived. Anyway, we're taking fire, so we radio in that we're taking fire from a 50. This fucking Cobra comes in and runs right over the top of us, fires and is just dumping rounds. This squad of Marines is pushing out and this thing is dumping rounds. They're telling me to shoot, and I said, "I'm not going to shoot my guys." My driver bailed, fucking standing there saying, "Holy shit, lots of shooting going on, and I don't see nothing to shoot." I turn around and this guy is on the floorboard still yelling. I told him to shut the fuck up and I turned around and I see all these bars on the officers around us. I said, "Everybody get the fuck down. There's a 50 cal shooting at us right now." They're looking around. I put my pistol in the holster, and then I just kind of yelled at them and they turned around and were all gone. I want to say, 3rd platoon got into some shit, fucking ended up smoking like 15 dudes in this house. Why there's 15 full grown men in the house, who knows? But they got smoked too and Lt. got a Silver Star for that. Rolled up to the center thing and were sitting at the intersection. I was holding security and I had all these prisoners in my truck. Anyway, we're sitting there, and we've got all these prisoners. I've got 4 fucking prisoners on each side, with fucking three in the middle. We sat there for fucking 13 hours while they're figuring out this whole thing. All the prisoners are fucking enemy combatants, whatever. One of them pisses himself. It was terrible.

Janney: Have you got any other stories that you want to share about any of the guys that didn't make it back that would be great and then we can finish up. How well did you know Deshon Otey?

Miller: Otey? He was a senior Marine from another platoon. So, not very well. He was stern, but he wasn't unfair which is how I would expect every fucking leader to be basically. From what I knew of the guy he wasn't a bad dude.

Janney: Let me ask you a question to just clarify some confusion in my mind about where Smith and Wroblewski were when Lt. Ski got hit. Was it about 500 meters south of Gypsum and Nova or was it closer to the intersection there?

Miller: So, I tried to show you, but they changed it. If you look at the older maps of Gypsum and Nova, it goes down Gypsum and then it turns, there's an "S" turn on the road and then it turns into a "T" which is Gypsum and Nova. We were, about 100 meters, I want to say, south of the fucking S turn. It goes straight; I'm telling you I can see the fucking corner of Gypsum and Nova. It goes from trees to like open, and then a fucking field and then you could see the corner of it. That's right where we were. There is actually a little road, a side road that I was trying to tell our guys about, because I thought we were getting flanked from the marketplace. I thought they were hitting us at the marketplace and then flanking us on that road. But, no one ever made it to that road.

Janney: So anyway, where we did the memorial service for the guys that you guys lost on the 6th, it was right near that S. There's actually a big, huge house and an alleyway on the left side of the road, the west side of the road. The alleyway runs off the end of that house and then curves back south and parallels Gypsum.

Miller: You were right there. If anything, you guys might have been a little farther up Gypsum. There were no walls in that section where he got hit. He literally got hit right before the corner into the S turn. There was a two-story house on the left, a fucking two story house on the right that was red, and then a house right up in front of me on the left was a two-story house that was white.

Janney: So, he was on the left side of the road, or the west side of the road?

Miller: No, he was in the passenger seat on the right side (east side) when he got hit. He was still in the truck. He got on the radio next to the Humvee, got out, kneeled, and they hit the radio and went right through his jaw. None of our Humvees at that point were actually armored. They were fiberglass. They called having armor "an escalation of force." You know when Mattis' fucking LAV got hit, that thing was fucking aluminum. Both of those fucking LAVs were fucking aluminum, and they got blown to shit. They had shit for armor. You know, when our own General is riding around with no fucking armor, there's something logistically fucked about that.

Janney: Yeah, I had no idea that Mattis had even been attacked like that.

Miller: General Mattis was actually supposed to be coming to our base to give a speech or some shit and he ended up sending his number one Lieutenant and then he got hit instead of Mattis. They sent two LAVs and one of them got annihilated basically.

Janney: I did not know that.

Miller: Yeah, they would always hit our guys on those routes. Basically, they took the same route as our chow and all that shit. So, any route that Weapons Company took consistently to bring us supplies or anything would get hit. Weapons Company was the ones that were bringing us supplies from headquarters because they were the truck mounted heavy guns, you know. They would get hit constantly. You know, after the 6th they really made a joke, because they used to have all these guys selling movies and the shitter trucks and the garbage guys, you know, they all stopped coming and all the workers stopped coming. They were like, "Man, we're wondering where they all went." I said, "No, they're all fucking dead, you idiot." They said, "Oh, I never thought of that." I think I saw them in the bushes that day. Pretty fucking stupid. It was a pretty fucked day. There was nothing on our whole deployment there but that was, if you want to know about that day, the 10th, that's my perspective and everyone's perspective is a little different. You know, like Rogers, fucking saying that we didn't have Bradleys there that day. Well, he didn't because he was over at the other part of the 6th. When we rearmed, they wouldn't give me ammunition, just so you know. They told me that there was no more 240 ammo to give me. So, when we showed back up on scene where those dudes took off in the car, I had a 9 mil. And nobody would fucking shoot at them! Except the Army, and the Army missed miserably. I'm pretty sure it was on purpose because it's hard to fucking miss with a machine gun or two. Yeah, it's really hard to miss with a Mark 19.

Janney: Who was it that was telling me that somebody that they had a Mark 19 and the standoff wasn't long enough for the rounds to arm.

Miller: It only takes 13 meters for the rounds to arm. 12 to 15 meters, that's it. I think it's 13 rotations. Yeah, there were a couple dudes taking shots with a 203 and it didn't go off. Most of our heavy machine gun engagements, they fucking went off, at least when I was there. There might have been times when they had the Mark 19 out when I was out on patrol and they did stuff. I remember there was an RPG that went through a Humvee, fucking went through a dude, and it didn't blow up. There's so many IEDs that didn't blow up. We had a missile, like a 6-foot missile shot at us, while we were sitting on post at the graveyard and it hit the hill and it rolled down the hill and we all just went "wow." A big red and white missile. At the time, that was the whole thing with the "Rocket Man" who was a suspect everyone was always looking for. He was basically a dude that knew how to build a rudimentary rail system that you could fire a rocket off of. It would basically just lob it onto you. He was pretty good at it. The Army was searching for him for fucking years. We would always find his leftovers, you know. After he shot that rocket off, we traced it down and we found the spot where he shot it from. It was a frame with fucking rails and shit.

One of the first trips up into the graveyard, outside the outpost, we found a cache. Our first trip up into there, I was sitting on the gun, sitting on this spare tire, and we roll up into the graveyard and hear a "POP!" And I said, "What the fuck? Flat tire?" I looked over and said, "Holy shit, it's a land mine!" I started banging on top of the Humvee, fucking screaming at them. I yelled, "Come on, let's go! Fucking go, go, it's a goddamn land mine!" We ran over an anti-tank mine. It was a big green fucking mine about three inches thick with a big 8-inch pressure plate. Then, we started digging there and found a bunch of weapons caches and whatnot.

We were hit by IEDs and shit after that or small fire fights that we never could really fucking find out who was shooting at us. Then searching houses, I kicked in thousands of doors. They had us, at one point, clearing houses with two-man teams and they gave us about a minute to clear each house. I had a 240, so I was kind of pissed off about it. We got those orders and I said fuck it. I knocked on the door, the dude answered it, I took one step in the house and said, "It's clear." Just shows how ridiculous it was.

Janney: Yeah, that's ridiculous. There's no way you can clear a house that quickly with two men.

Miller: What was happening was a lot of the people said, "Meestah, I can't find my keys." I didn't really stand for any of that. I would just pull out my K Bar and start popping their locks, which worked really well. By the time I was done, they'd usually show up with the keys. I did my best not to sit there and fucking dilly dally when we had 10,000 houses to search.

But, the night of one patrol, I fucking jumped over a creek. I had a M16 with a 203, so I had fucking thirty 203 rounds in my vest and pack. I fucking jumped this creek and my barrel stuck in the fucking ground and everybody just started fucking with me, "Oh, you're fucking useless, blah, blah, blah." I took out a grenade and loaded it and said, "I'm good to go. What do you mean? I've got a 9 mil with a 203. Fuck you guys." Then we got to a house and I cleared my shit and the sniper just let me check out his rifle and whatnot. But that morning, on the way back, they shot the weirdest mortar that I've ever seen; it had fins that kicked out. It was gnarly. It was like a little rocket tip and it had fins that kicked out with a spike behind it. This thing just lobbed over the top of us like fifty feet. They shot it from the college across the street. We were coming in from up towards the Gypsum site on a foot patrol. They didn't launch it; they shot it AT us and missed. That was fucking trippy. I had one fly over the back gate of the Combat Outpost. I was walking back through the barracks and this fucking rocket thing flies over the gate; goes right in front of me and went into the motor pool and fucking starts spinning and then the Docs come out of where their medical office. I'm like, "Get the fuck back in there. It's live!" I'm just screaming, "Call EOD!" Everybody was running around like "fuck," but that was trippy.

We had a mortar blow up about 15 feet off the ground and just a green cloud came out of it. We were walking back from chow. So, I went inside and told the squad leaders about it. We were pretty sketched out about it, me and another Marine. I think it was Cienfuegos. They told us to get our night vision and go inspect it because we were still PFCs, "Hey Private, go look at it." We asked for a gas mask and they said, "We don't have any."

Janney: Speaking of chemical weapons, were you guys going through and finding any caches? Did you find any chemical weapons, mortar shells or anything like that?

Miller: Basically, we had gotten word that Saddam had ordered mustard gas to be loaded into all the artillery and mortar rounds. I mean, we had IEDs go off where there was a fucking green cloud, but we just let it disperse. You know that mortar that went off? A fucking rag came out of the tip of the damn thing. There was the fin sitting there and a rag laying down there in a pit. That's my experience of potentially being gassed by them motherfuckers. It was mostly the IEDs that were leaking green fucking smoke. We're the ones that spawned that fucking mustard gas because we sold it to them. We already knew it.

Then on that first push into Baghdad, they lit all the Sulphur mines on fire. My brother-in-law was in a fucking Army unit and they found chemical weapons. When they called it in, he said these fucking black Hummers pulled up with all these dudes that jumped out with black tactical gear, dudes with big beards and shit. They just took over the scene. That was the last they ever heard of it. Those were full on missiles with fucking warheads.

Janney: It was a hellacious war. I'm just glad you made it home. You have a wife and kids that need you, so I'm thankful you made it home to them.

6 April 2004 Route Gypsum Ambush And QRF

Sgt. Jose Valerio, Motor Transport Chief (Gunnery Sgt., retired)

Janney: One thing I would like to ask you, what made you enlist in the Marine Corps?

Valerio: All right, first name Jose, middle initial S, last name Valerio. Sir, I'm an immigrant. I actually got here in '86 illegally and I did the whole nine yards from the river; got on a train; got on an 18-wheeler, in the trunk of a Grand Marquis and got delivered to my parents. Then, I owe this to my two elementary school teachers; she and Mrs. Lopez said, "Don't be like some Mexicans. You grab a girlfriend; you grab a job; you see the money; you get a car, then you think you've got everything. Education – that's the key. This country gives you that, so take advantage of it." Everybody asks me and I don't know if this should be on record because I don't know if it's true or not, but I'll say it anyway, I'll say it anywhere at any time. My family members have asked me if there was a conflict between the United States and Mexico, which would you back up? I say that's not even a question for you to ask. Most definitely I would fight for the land of the free. The reason I joined the Corps is the following:

I went to the Army; I had a scholarship to college, but they told me I couldn't get it because I was not a citizen, but I'm like whatever. I went to the Army and said, "Look, I got a red card; I'm coming to work here. I got my A number and immigration said they're waiting for my number to catch up and then I'll get it."

They said, "Come back when you have a green card."

I was like all right. The only reason I wanted to join the military was to give thanks and help to continue what the U.S. had done for me, so that nobody would say that I came here for free. It's just to be thankful to the United States. That was my first thing in joining the military. I didn't know if it was Army, Marines or whatever.

When I went to the Marine Corps and said, "Hey gentlemen, I want to join the Marine Corps. This is the card that I have from immigration and they told me they're just waiting for a number."

They said, "All right, we'll pick you up on Monday from work."

I said, "You're going to do the same thing that the Army did, right? You know what? That's fine."

The Marines show up on Monday. Michael, the manager of Black-Eyed Peas said, "Hey V,

Marines are looking for you."

I said, "Mike, I'm joining the Corps. Can I have the day off?"

He said, "Yeah man, take the day off."

I said, "Gentlemen, I'm ready. Let's go."

The Marines said, "What, you're not going to feed us?"

I said, "Yes, of course. Mike, I'm going to feed the Marines."

Mike said, "Marines eat for free."

So, they eat and then we're ready to go.

So, immigration is almost like Black Friday. It's a big line. We walk into that building and I said, "Gentlemen, we have to wait in that line."

Staff Sgt. Tony O'Hendrix, black Marine, tells me, "Marines don't wait in line. Marines go up to the front door. You can go line up if you want, but we're going forward."

So, that's when I definitely wanted to be a Marine. We went in and I show my paperwork. They came back and he said "The Director wants to talk to you." The Director of Immigration asks me if I applied, takes a picture of me, grabs a scissor, cuts the picture, staples it on, and he's typing. He finishes, stapled it, sealed it, and says, "Congratulations, go serve our country."

So, New Year's Day '96 is when I joined the Corps and I never looked back. I would do it all over again. Much later in my career in the Marine Corps, I was with 2/4 assigned to Bravo Company as a Motor Transport Chief to the Combat Outpost. However, Golf Company and Echo were at the outpost, therefore my job was to give them support at Motor Transport. I was a Sergeant at that time. Right now, I am a retired Gunnery Sergeant.

Our job was to protect Marines. Every day I would go out and scrounge metal. My vehicles were armored up more than the ones at HQ. I used sandbags, scrap steel - anything that I could find every day. If I wasn't on a mission, my mission was to go out and get us armor and find anybody that could weld metal. One time, I did both the doors with L shaped steel. I was told by higher ranks, "No, it's too hot. Just get it waist high." We cut them off, but went back to L shapes after Sgt. Conde got hit with an IED. I found a stash of shoulder pads that the Army didn't use for their flaks, so I made what I called combat chaps. I sewed them together and said, "Have gunners wear this to cover the front and back of their legs" because in one of the fights, one of the gunners got hit in the leg.

I do remember April 6th. April 6th started early in the morning when Golf Company was hit in an ambush. I remember going to Golf Company and Echo. Golf Company Gunny was Gunnery Sergeant Jaugan, Staff Sgt. Rodriguez, Staff Sgt. Craig and Staff Sgt. Walker - we all were in Drill Instructor school together, so I kind of knew them a little bit. I went and asked, "Gunnery Sgt., what's going on?" He said, "We're getting attacked and the other platoon just left right now." I said, "Gunnery Sgt., whenever you are ready, our vehicles are ready." He said, "No, we're not leaving because we have to leave a platoon behind just in case the camp gets overrun." I said, "Roger that, sir." From there, I ran quickly to Echo Company COC. Our burden area was just separated by a door. So, I went to my burden area and I went across, opened the door and heard Echo Company's third platoon was getting hit. All my vehicles were already ready. I always kept them ready for QRF. As a matter of fact, I insisted on keeping those vehicles always prepped and ready. But, I hear third platoon was getting ambushed, so I went to the Company Gunnery Sergeant, Gunnery Sgt. Coleman, and at that time I let him know, "Hey, sir, I'm ready, whenever, vehicles are ready, and we can rock and roll. We're here to support." So, he tells me, "Yes, you're taking off with Lt. Ski." I said, "Sounds great."

I've actually driven Lt. Ski (in the past.) We took off on Michigan. Right before we got to the arches, we made a left turn toward the old canal and then we followed the route. Marines dismounted. We didn't see anything, and we didn't find anything. My thought was we got the wrong coordinates or we're not there yet, so Lt. Ski comes back to me and tells me, "Hey, they're at this (certain) spot." He gives me the coordinates. I knew the area because I would always go out every day either with Golf or Echo Company and, if not, I would go out and scrounge items so I could reinforce my vehicles. So, the Marines dismounted, and he comes back to me and I said, "I know where that is, sir. I know exactly where Six is (Six is Echo Company CO, Captain Royer.)" Lt. Ski said, "All right, well take us there." We mounted up, made a U turn at the arches, and then right at Gypsum. I pull over along a canal.

One Humvee almost T-boned me if they hadn't hit the brakes. I said, "What the heck is going on?" We get out, Lt. Ski and myself. Staff Sgt. Walker comes out and he said, "Sir, follow me. I'm going to take you up to Six." Lt. Ski replied. He knows I was known as Sgt "V." He said, "V knows, he's taking us down there." I said then, "Yes, Staff Sgt. I got it." Staff Sgt. Walker, again, we were good friends, we did 3 months in DI school together, so we had a good relationship. Outstanding relationship. I reply, "Yeah, Staff Sgt. I got it." Always giving him the respect because he was the higher-ranking officer. Walker said, "I got it, Sgt. V." I said, "Staff Sgt. I got it, I got it." Then he tells me, "Stand down, Sgt. Stand down. I said I got it and I got it." I reply, "Aye, aye Staff Sgt." Lt. Ski said, "All right, let's roll" and Lt. Ski's going back to the vehicle. Staff Sgt. Walker says, "Hey V, I know you got it, but I gotta take care of all the men." I said, "No, I got you, boss. I got you brother." Now, I talked to him that way because it was one on one. There was no one else present.

So, he becomes the first vehicle, followed by a second vehicle, and then I become the third vehicle. We know Marines are getting ambushed. We know there's a fight, so we're hauling. We are going about 45 – 50 miles in a Humvee and we're going on a snake road. Gypsum turns into a snake road. All I could see was the dust in front of me, but I knew where I was going. You climb in there and go through this, you are able, it becomes nature, you know how to drive with dust. You know how to see with masks. Something happens that you are able to drive at 45 – 50 and go through sand and dust. But, when I saw this, all I hear is, "Bang, bang, bang." Just a spray of weapon noise. I hear it again and I stopped. I said, "Sir, we're getting attacked. We're getting hit." I said, "Get out. Get out, sir, get out."

So, he gets out of the right side, passenger's side. I get out of the driver's side. I take a knee, and I scout the area forward at twelve and I start scanning towards my left. Right around 9:00, there is a laser coming straight towards our vehicle. I see a red dot and I know it's a laser. I just drop on my back and I start hitting that window. I screamed to Corporal Lenz, "Lenz, 9 o'clock, 9 o'clock." Lenz turns his 240 and hits that window. I come back to the rear with Lt. Ski and I say, "Sir, we're getting smacked. We're getting hit" and he said, "I know, but from where?" I just said, "From left, right, front. The only place we're not getting hit from is the back." He said, "All right. Let me get a hold of Six." I said, "That sounds great, sir. I got you."

He goes back up to the passenger side, but the doors are made out of plastic. He opens the door, takes a knee, but he's tall. So, as he takes a knee, I see him looking forward at 12 o'clock. I thought, okay I'm going to cover his back. I turn around and look at 6 o'clock. Now, I'm looking and I'm watching his rear, covering his rear, and I look back to see how he's doing and he's looking at me. So, I'm like, "Oh, hell no." I turn around and now I gotta cover his back. Now I'm facing at 12 and we're facing each other.

He's on the radio and all I hear is, "Six, Six" and the third S, that's it, that's all I hear. In a second, he drops one inch from my foot, grabbing his neck, his throat. I yell, "Sir, what is going on? Sir, what is going on? What is going on, sir?" I see this enemy coming out of this building. "Boom, boom, boom." Snapping, snapping of rounds. I call Medina. Medina is on a 50 cal on my right side on a 7-ton. I yell, "Medina, Medina, 11 o'clock, 11 o'clock." Medina is a Corporal from Puerto Rico. He turns to the left and just destroys the enemy with that 50. "Boom, boom, boom, boom." I said, "Come on, sir. Are you okay?" I take off his hand and blood was pouring. I tell Lt. Ski, "Sir, you're going to be alright. You're going to be alright." I was a combat lifesaver at that time. I worked to stop the bleeding and called for the Corpsman. I tell him, "I got you. You're gonna make it, sir." I saw his eyes and he just talked a lot with his eyes. I knew he had a family. It destroyed me. I saw it in his eyes. I know that he has a wife. Bear with me a minute.

Gets quiet.

I had a little daughter too, at that time. She's now 22. Lt. Ski makes this fist. He makes that fist with his left hand and points it to the sky. Like saying, "I've got faith, Lord. If you're taking me, I know you've got my family." His eyes just said a lot of stuff – they were true and honest. Those seconds, they get imprinted in your life. But, you gotta go. You've gotta roll. I said, "Sir, you're going to be all right. Come on, sir." I start dragging him and he uses his right foot to help me out dragging him. I'm calling and then there's another haji that comes out. I take my M16 and I'm shooting and I'm getting shot at. I put myself on top of him. My M16 runs out of rounds and I pull out my 9 mil. This haji goes running across. I'm looking at him and he's going left. He runs behind a cow. I'm shooting, the cow goes down. I called Medina, "Medina, Medina, Medina." Medina goes back and takes him. I'm pulling and I can hear the bees going by my ears – the buzzing or ricocheting of the bullets. I'm dragging him and I'm pulling him underneath the Humvee and I said, "Sir, you're gonna be alright." I called for a Corpsman again. HMR Doc Rodriguez, I see in the back and I said, "Get your butt up here right now!" He shows up and starts working on Lt. Ski.

That's when Smith shows up and I tell him, "Give me a gunner, give me a driver, and a Corpsman and you take him back. Right now. Take him back to the firm base. Anybody that stops you, if he's not Army, Marine or soldier, freaking kill him. You kill them if they try to stop you and you put my name on it. You tell them Sgt. V said so. Don't stop, and you take him back. Now, on three, pull him back." Ready, set, three, they pull Lt. Ski back. I give them cover. I'm still pinned underneath the Humvee. They go back. I see that they reach the Humvee. I turn around and they're still trying to give him aid. Then I get mad and I said, "What the hell did I just tell you? Take him back or I'm going to start shooting you myself. Load him up and get him the hell out of here right now." They load him up.

PFC Brown was the driver. Just two weeks prior, I had given him a crash course Humvee class and he was one of the great drivers there. So, he didn't get a regular school Humvee class, he only got a one-week crash course. I think PFC Brown is now a Master Sergeant and is a CSO with MARSOC. I met him in 2016, my year of retirement, and we had a great conversation because we had not seen each other since 2004. We met again in 2016 at MARSOC. I was the Company First Sergeant and we met and talked, and it was a great relief, but back to the story.

      They take Lt. Ski back. All I know is my Lt. is down. I don't know where Staff Sgt. Walker is at. I'm still underneath the Humvee. I'm getting shot at and I'm shooting back. The next thing I hear is a bang on top of my Humvee. I pull out my pistol because I know it's got to be short distance and an M16 won't do in a short distance, especially if you are under a Humvee and being pinned down. I yell, "Who the hell is this? You've got three seconds or I'm going to start shooting." "It's me, Sgt. It's me, Corporal Lenz." I tell him, "Put this thing in reverse." He said, "But, you're underneath." I tell him, "I don't give a darn. Put this thing in reverse. Do it now." So, all I hear is the gears shifting, the transfer case shifting. I'm like, "Oh, shoot." I put my boots on the A frame of the Humvee and I grab the shackles from the tailgate, and I pull myself up. Now the Humvee is rolling backwards, and I'm being dragged underneath, holding the shackles.

Janney: Holding on, to the what? You were holding on to the bottom of the vehicle while it was moving in the middle of a firefight?

Valerio: That is correct, sir.

Janney: That's crazy, sir.

Valerio: You're able to see that because on YouTube, "Echoes of War." I'm the one pushing the vehicle back and you can see the back of my flak that there's a rip. It's a small rip. That's right after you look at the picture of Ayon who's in the windshield and I'm pushing back and you'll see my brown hands and you'll see the back of my flak. I never wanted to be in any video, so I would always stay away from those. But they caught me from the back on that one. So, I get dragged back and then I said, "Okay, Corporal Lenz, Corporal Smith, give me numbers." A Sergeant, a Company attachment, he was between the tires of the 7-ton. I don't know if he was hiding or what he was doing. I just call him back, "Get over here. I need numbers. Is your driver up?" He replied, "I'm up, Sergeant V." I said, "All right, check it out. Smith, Lenz, give me two more. We're going to clear this sniper house, the one where the red laser was coming from." He said, "All right, sir." I said, "The rest of you, I need 360 bubble security. Check the rooftops."

      So, by then, the Army's Bradleys are showing up and they're just hauling. They yelled, "Follow us!" We yell, "STOP, STOP, STOP." I'm screaming, "Tell them to stop." We don't have headphones, we don't have comm. They throw me a headset, but it doesn't work. I said, "Lenz, Smith, load up, load up." But the Army was hauling. They were going so fast that by the time we got on the vehicles, they had already made the left turn and they were gone.

      Later, I thought, "You know what, let's roll, man. We're leaving. Don't worry about it. It's on me. I got it. I made the decision, let's go." We're going, but I'm just talking, "Let's go, let's go." As soon as we get into Combat Outpost, I'm blowing the horn, yelling, "Blue Diamond, Blue Diamond, open the gate." We go into the hangar bay. I said, "Get them out of here. Let's go. Off load, off load. Reload." I go straight to Gunnery Sgt. Coleman and said, "Gunnery Sgt. Coleman, I made the decision. We weren't scared, we're not retreating, we came to reload, we're reloading, and we're going back to the fight. I didn't run away." Gunnery Sgt said, "V, shut your freaking mouth. I got you. You're good. Load up and let's go." I said, "Aye, aye, Gunnery Sgt."

      Punched up again and we went back to the fight. By then, everything was already calm. Captain was there. We're just picking up the stuff now, so there's no more fight. Never thought I was going to experience the fog of war. But as I'm walking, Lance Corporal Reynosa, who later gets killed by an IED. He was one of my soccer players because I led the soccer team when we were on deployment. You know, I had told Marines that we would play soccer and he was a great soccer player; midfielder. Reynosa tells me, "Hey, Sgt. V, sorry about your boy." I said, "My boy? What're you talking about?" He said, "Yeah, Lance Corporal Dayniss, he's dead." I said, "WHAT? Oh, hell no. I didn't give him permission to leave the firm base." Some of the Marines took what I said wrong. The reason I said that is because Lance Corporal Dayniss is Motor T and was attached for the last thirty days to Combat Outpost because I would always switch Marines every thirty days. He had come to me and said, "Sgt. V, I'm leaving in thirty days to EAS. I'm done with the Marine Corps in thirty days. I got a baby coming. My wife is expecting me back there." I said, "We're at the Combat Outpost. If we get overran, we need admin and cooks to defend the firm base." He said, "I'll take that chance, but I just don't want to go out." I said, "All right. Every time it's your turn, I'll take your spot." He said, "Thank you, Sgt. V." That's why I said that he didn't have my permission because we had already talked to each other. So, now I got this Marine that I'm in charge of that's dead.

I've got Lt. Ski in my head: How's he doing? Where's he at? This Corpsman, a big white dude (author note: probably Doc Grimes), very friendly, tells me, "Hey, Sgt. V. Lt. Ski made it. He's on his way to Baghdad." I said, "Okay, frigging awesome." Then, I go see our Captain and ask, "Hey sir, how are you doing?" He replied, "It's never a good day when we lose Marines." I said, "I heard Lt. Ski is good to go. He's on his way to the Baghdad hospital (CSH)." Then, I leave him there alone with his own monsters. To me, Captain Royer was an outstanding officer. I always saw him fighting with the higher ups for more vehicles, for more Marines, but was always told no. He's a great man. Never wanted his Marines to die and always had great intentions. I know these things because I would drive Royer and Colonel Kennedy and would hear the battle between them and the higher ups.

Now, I'm wondering where's Dayniss? I go straight into the burden area and look for LCpl Dayniss. I just saw somebody in his rack and pulled him out. I didn't know who it was, but it was him! I say, "How are you freaking alive? Freaking hit me so I know it's you. I heard you were dead." Then, he hits me and I said, "I'm sorry. It's late. Go back to sleep. I'll talk to you tomorrow." I walk out and see the same Corpsman and he said, "Lt. Ski didn't make it." I was shocked, but livid. I thought, "What if I had carried him myself?" What if, what if, what if - this monster has been with me for years.

In the Corps, we learned that things are going to happen, but it's hurtful for the ones that are left behind. People have asked me, "Are you afraid to die?" I'm not afraid of death. I'm just afraid of the ones left behind because they're the ones that are going to suffer. I know what's in the afterlife and I'm not scared of that. I know the Lt. is cool wherever he's at. I've never talked to anybody, but I do blame myself sometimes. But I'm alive, I've got a good job, I've got status, I've got kids. I know I told SgtMaj. Ellis, "I'm not here for awards. I'm here to do my job as a Marine because that's what I get paid for." I'd do it all over again.

Janney: Sergeant, you may not have been born here, but you're a patriot and it's an honor to get to know and to talk to you.

6 April 2004 Route Gypsum Ambush And QRF

Doc Adam Clayton, HMC FMF interview

(Interview conducted via social media messaging)

Janney: When did you join the Navy, and at what point did you become a Corpsman?

Clayton: I joined in 2000 at the age of 17. I had previously done work in medicine, but I wanted to join the Marine Corps Infantry. A close friend told me to join the Navy and I could do both. I have been a Corpsman ever since.

Janney: Very true. So, were you attached to 2/4 here in CA, or were you assigned to 2/4 in Iraq? I'm not sure how that works in the Navy.

Clayton: I was assigned to 1st Marine Division and sent to serve at 2/4. We get orders to the Division then sent to a unit. I was with 2/4 from 2003-2006

Janney: Then you deployed with them in Feb 2004?

Clayton: I did. That was my second of 3 deployments with 2/4. I was assigned as the Senior Line Corpsman for Echo Company in 2004.

Janney: Where was your 1st deployment with 2/4?

Clayton: To Okinawa, Japan in 2003.

Janney: Since you were with 2/4 from 2003 to 2006, how well did you know the men?

Clayton: I knew them very well. As the Senior Line Corpsman, it is my job to know my Marines.

Janney: Yes, sir. Once you deployed to Iraq in Feb 2004, what were your basic responsibilities and daily activities?

Clayton: Sorry, so I don't over explain things. Are you prior military and understand what my role was as a Corpsman? As the Senior Line Corpsman, I was responsible for the health and wellbeing of the Marines. My job was to make sure they were healthy physically and mentally fit to fight, as well as providing medical care for any injuries or illnesses.

Janney: No, sir. I've spent a great deal of time photographing military projects and a 2007 embed with 3rd ID, but 2008 was my 1st embed with a Marine unit. So, explain it as you would to any civilian. As a Senior Line Corpsman, did you patrol with the Marines daily, or have more of a supervisory role?

Clayton: Army and Air Force have medics, Navy has Hospital Corpsman. We are similar in our roles. First responder trauma care and basic medical treatments, as well as ensuring all Marines are medically fit.

Janney: What were you doing the morning of 6 April 2004?

Clayton: We had 9 Corpsman and 12 squads, in the beginning of the deployment squads would go on patrol and I would rotate in to patrol with the Marines. Part of my role is to stay with the Company Commander, who at the time was Captain Kelly Royer. Where he went, I went.

Janney: Were you with Royer and David Swanson on 6 April 2004?

Clayton: The morning of the 6th I was with the CO and we left the base due to a small firefight. We went to the location in the south of Ramadi near MSR Michigan. We got dropped off by vehicle and were walking back when we heard the ambush begin and all hell breaks loose with 3rd Platoon. Yes, I was with them. Radio calls started coming in and we could hear the explosions and gunfire from across the city. At that point, the trucks were sent off with the Marines. We then moved out on foot running north through Ramadi.

Janney: At that point, where did you go? Did Royer and the unit head north up Route Gypsum toward the intersection with Nova? I know the area because I did foot patrols daily with 2/8, but don't know the route you took.

Clayton: We went up through the houses. The trucks took the road up. When the trucks loaded and we sent them off, we got ¾ the way to the intersection, then we started taking fire. As we were running, I saw the trucks pass us. I told the CO the trucks just passed and he tried calling them to stop. That is when the lead vehicle pushed forward to the intersection and the ambush began.

Janney: Were you able to see the intersection of Nova/Gypsum from your position when the ambush began?

Clayton: Not when it began. We took cover when we started receiving fire. We could see the tail end of the convoy when we crossed the main road towards the machine gun that was firing at us.

Janney: The convoy consisted of a Humvee, two 7-ton trucks, and a Humvee at the rear of the column? I'm not sure, so clarify for me.

Clayton: Yes, I believe so.

Janney: Approximately how far was the last vehicle Humvee from the intersection of Nova/Gypsum? Wasn't this Lt. Ski's vehicle and he was with Sgt. Valerio?

Clayton: I don't know how far the tail vehicle was when we passed them and I don't remember what truck they were in. When they loaded up, I was at the first truck. Then, they took off.

Janney: So, the lead vehicle gets hit at Nova/Gypsum and stops the convoy. You, Royer, and the Marines are working your way toward them when you started taking machine gun fire on Route Gypsum? Is that correct?

Clayton: Yes.

Janney: Royer, Marines, you, and David Swanson sought cover due to incoming machine gun fire. Did you seek cover on the left or right side of the road, in a house, or ditch, or where?

Clayton: We began on the left side of the road. The fire was coming from a house on the opposite side. We crossed the road, took cover behind a large rock and then started bounding north across an open field towards the machine gun fire. We pushed through to a row of houses, cleared one and then went to the roof. At that point we could see the intersection.

Janney: Was Royer in contact via radio with 2ndLt Wroblewski at this point (or at any point during the ambush)?

Clayton: At that point, it seemed as if everyone was trying to get on the radio. One of them was 2ndLt Wroblewski.

Janney: I'm sure coms were difficult if not impossible. I understand the lead vehicle's 2 radios were out of action. I'm guessing you and Royer were with Ben Musser during the house clearing process you mentioned above?

Clayton: Yes.

Janney: Were all your Marines okay at this point? The Marines with you and Royer?

Clayton: No one that was with me was injured on the 6th.

Janney: Once you were on the roof and could see the intersection of Nova/Gypsum, what was happening?

Clayton: We could see insurgents scattering as Marines continued searching the area. At that point, the most injured had been taken back to the Battalion Aid Station. Capt. Royer was at that point trying to figure everything out. Soon after, we left the house and went to the intersection.

Janney: Had 2ndLt Wroblewski already been hit at this point?

Clayton: Yes, and he was being taken to the Aid Station.

Janney: Were you involved in treating him or was that Doc Urena?

Clayton: That was Doc Urena. The last time I saw him was the morning before everyone left.

Janney: Did you have a chance to speak to 2ndLt Wroblewski or any of the other Marines or Doc Mendes-Aceves before they headed out? If so, please tell me what was said and who you spoke with.

Clayton: By the time we got to the intersection, Layfield was being carried to the truck. It was honestly just another morning. I couldn't remember what was said.

Janney: I understand. Yes, I can't remember if it was Musser or Sgt. Valerio that found Layfield off to the left of Gypsum about 30 paces, if my memory serves. Does that sound right?

Clayton: The last Marine in that group I spoke to was Jerabek as he took the 240G and was excited and exclaimed, "Sgt., look I'm a machine gunner now!" I don't remember who found him. I know Apple was the one that carried him to the truck.

Janney: Damn. That's tough. I heard from many about what a good Marine Jerabek was, and Musser had trained him on that weapon and trusted his skills with it.

Clayton: Yes, he was, and he stuck it out, firing rounds until his last breath.

Janney: That's what I heard, too. Such bravery. At what point did Swanson get shot during the firefight?

Clayton: I believe he got shot on the tenth when we were pinned down in a sewage runoff trench.

Janney: I'm not sure, but I do remember the sewage trench and Lauersdorf being next to or on top of him afterward.

Clayton: Yes, that is correct.

Janney: Can you tell me any recollections/personal interactions with any of the Marines that didn't come home? It's hard to know these men without these kinds of stories.

Clayton: I wish I had the memory to do them honor with great stories. I do remember 2ndLt Wroblewski being one of the coolest guys I've ever met. Those Marines and Corpsmen grinded day in and day out in miserable conditions, and each day did it again for those next to them. Mendez was the most humble and kind person you'd ever meet. He was an augmentee and from day 1 was amazing.

Janney: I've heard that he (Lt. Ski) was very down to earth and wanted to learn from the grunts, and was very cool with all his Marines. Did he ever mention fishing or anything else personal to you? I'd also like to hear more about Mendez – he's the one I know the least about.

Clayton: Unfortunately, I can recall most of the bad days. The good ones just seemed routine. We didn't talk about fishing as I'm not a fisher. He was a great leader and loved by his Marines and Corpsmen.

Janney: I don't want to sensationalize the bad days or dishonor their memories in any way, but share with me whatever you think is important.

Clayton: Mendez was just a phenomenal human being. One of those people that no matter how bad or hard the job, he would do it with a smile and he would do it well. Never boastful and always kind.

Janney: Did he talk about his family back home?

Clayton: He talked about his mom.

Janney: What do you remember about that?

Clayton: He was a loving son that had a deep love for family.

Janney: Did you know Mendez before the deployment or meet him in Iraq?

Clayton: I met him when he got assigned to the Battalion for the deployment. We didn't have enough Corpsmen, so we got augmented for the deployment. Urena was an augment as well. Of all of us, Urena knew Mendez the best. It was as if Mendez was his big brother.

Janney: I am not sure, but I heard Mendez was killed while rendering aid to a Marine during the ambush? Do you know anything about that, or who would know?

Clayton: Honestly, the truth is we don't and won't ever know. What we have been told and from what it looked like, yes, Mendez was rendering aid. His medical gear was out and there would be no reason for gear to be out if he wasn't using it.

Janney: I understand. Thank you for trying to explain it to me.

Clayton: Afterward when we walked through, the blood splatter on the walls seemed as if he was killed from close range. Which would also speak to him rendering aid and not up with his weapon.

Janney: Sure, as painful as it is, that makes sense. One more subject and we'll finish up. How well did you know Parker, Contreras, Lopez, and Otey? If so, can you share any anecdotes about those men or any of the 34 Marines that didn't make it home?

Clayton: They had all been part of the unit. I spent my days at work with them. Parker was assigned to snipers and was bunked in my open room. He spent most of his time out doing sniper duties. Contreras was our comical relief. I actually just laughed thinking of him. He was a clown in a good way and you could always expect some shenanigans. Last time I spoke to Lopez, he was talking about taking a vacation when we got home. Otey was an old soul. He used to let me borrow his mini discs for old school R&B. Otey was the "blessed one." He lived that day only to die shortly after.

I remember coming back on the 6th and talking to the other Corpsman that received our patients. They had told me 2ndLt Ski had been shot in the face but was okay and was giving a thumbs up as they put him on the helo. When we heard he died, we were shocked. I was with Gonzalez when he died. One morning before a hike, Contreras came out wearing a fireman hat. It didn't go over well with leadership, but the guys all loved it.

Janney: John Wroblewski told me that Lt. Ski passed away at the CSH. John said he spoke to a nurse who sang to him before he died.

Clayton: Cherry had a voice for cadence that was truly unbelievable. We never were told when he (Lt. Ski) passed. We got the news hours later.

Janney: I can picture Contreras in the fireman hat. Musser said he was hilarious.

Clayton: He was hilarious. Once we sent a patient off, we routinely wouldn't get updates for a while. 15 years ago our communication networks weren't the best. We would do all that we could and get them sent off to higher care.

Janney: I've heard a couple stories, but that's the one I choose to think about regarding Lt. Ski. Can you think of anything else you'd like to share about any of the other Marines?

Clayton: Just that they were good men. Young men, sent to do a job we weren't ready for and they did it with a sense of courage that can't be explained.

Janney: Yes, very young and all heroes in their own way. I wish I could have known them, but I will never forget their names or sacrifices.

Clayton: Thank you for what you've done and what you're doing.

Janney: It's an honor to do what I can. I wish I could bring them back, but telling their stories will keep them alive.

Clayton: The best we can do is remember them and honor their memory as men and as warriors. You are doing that for us and for them, so thank you again.

Janney: Tell me your rank now, so I can correctly identify you in the book.

Clayton: Currently, I am a HMC. Hospital Corpsman Chief Petty Officer. Navy makes it hard. Easiest way is Chief or if more official, HMC (FMF).

6 April 2004 Route Gypsum Ambush And QRF

PFC Gregorio Cienfuegos

Janney: It's 17 March 2021. I know you said you joined the Marine Corps, enlisted in the Marine Corps, just out of high school in 2003, but why did you choose the Marine Corps instead of maybe the Army or Air Force?

Cienfuegos: I was actually looking at the Army. My dad was in the Army in the 1970's, so I come from a military family. My great uncle was in World War II. Did World War II, Korea, and then retired. My family has served in almost all branches. I have uncles that were in the Air Force and Navy. One was in Vietnam. My mom's cousin was Recon in Vietnam. It was a decision I made when I was young, about four or five. I was going to do something. I didn't know what branch I was going to pick. In high school, I started looking more at which one was which. Looking at the way the Army's boot camp was, I thought this is looking kind of weak. I said, "Let me check the Marine Corps" and thought that looked harder and a lot longer. I decided I wanted to go ahead and join the Marines and said, "Fuck it. If I want to be a Marine, I might as well be a real Marine and go infantry."

Janney: That is incredibly admirable, and I really appreciate your family for their many, many decades of service. Your email mentioned boot camp in San Diego, and you mentioned several of the guys. Were you in boot camp with them also, or just when you got to your unit?

Cienfuegos: I was in boot camp with Adam Carter. I think Carol might've gone to boot camp with me. Once we graduated from boot camp around August, we had 10 days leave, and then I went home. After that, I met up in San Diego at the airport, took our shuttle to SOI (School of Infantry) and MCT (Marine Combat Training) at Camp Pendleton. When I got to SOI West, I met Marcus Cherry, Christopher Cobb, Adam Carter, Jared Cole, and Chow. I also met Kyle Crowley and Reggie Carr because of our alphabetized last names. I got to know a little bit more about Cobb because Cobb slept on the top bunk and I was on the bottom. Cobb graduated on the East coast. Originally, Cobb and 10 other guys were going to be LAV (light armored vehicle reconnaissance) Marines, but were told at SOI, "You're going to be straight infantry. Whoever told you that screwed you."

Janney: Tell me about Christopher Cobb.

Cienfuegos: He was a pretty cool dude. Really mellow, wasn't really outspoken. He watched a lot, observed a lot of people for the most part. He was a solid dude. I mean, he was pretty cool to talk to. He didn't really hold a grudge. He loved rapping, which was funny. He's gonna be the next rapper. There's this one guy named Little Whitey. He would always be rapping up and down the barracks, or wherever with a beanie on and shit. That was the funny part because he would actually rap.

Janney: Was he pretty good at it?

Cienfuegos: I guess he was okay, but he made me laugh because he kept saying he was gonna stay in the Marine Corps forever. Even though he was enlisted, he would say, "You know what? You're going to start calling me Captain Cobb. I'm going to be Captain Cobb." He was a good Marine. He was a SAW gunner. He loved that SAW, he really did. I knew he grew up in Florida. He was pretty easy to get along with. He said something about his grandfather or his uncles in the Marines or Navy. I met them one time when they came to Camp Pendleton.

Janney: Tell me about Marcus Cherry.

Cienfuegos: We were in vessel line together - me, Cherry, and Cobb. At night, we'd be talking about just the most random shit. He was saying how much he looked like the "Full Metal Jacket" Marine. He said his older brother Andre was a Marine, and had already been to Iraq and had come back. He worked with artillery. So, the weird part with that was we were in Ramadi, the chain of command let Cherry go spend a day or two with his brother. This was just a few days before he died.

I remember the first time I saw Staff Sgt. Walker. When I first saw him, I thought, "Who the fuck is this guy? He's all punked out with his tattoos, his flames, his Misfit shirt, like a punk surfer dude. He had the plaid on with the tube socks with the bands and the big-ass chain wallet. Yeah, he was a pretty cool dude, though. He cared about us.

When I got to 2/4, I met Lieutenant Wroblewski. I thought, "Holy shit, this dude's so fucking tall." I'm maybe 5'9". I guess he was 23, 24; just out of college, OCS. He was in incredible shape and would always smoke us on hikes or runs. He told me, "You're going to be my personal translator." I said, "I want to try, sir." He said, "Well, you better. You did the two-week course. My life's in your hands." So I'm like, fuck. Yeah, I was kind of doubting myself when he said that.

When we got to 2/4, there were seven of us: me, Carol, Cobb, Cherry, Cole, Carter, and Chow. When we got to Echo, we were almost a platoon. I knew Gunny Coleman because he grabbed us at SOI. He met us there and we had a different Captain at the time. It wasn't Royer. Captain Gibson, I think. Sergeant Vasquez was helping us check in. Just like in the movie "Jarhead", we were getting yelled at cause we're in our Alphas, "Fresh meat. Dog food" and shit like that. There were guys just getting drunk by the time we got there and were checking in, because it was already pretty much time for people to be off. So, they were just mind fucking us pretty much. It was cool, you know? I mean, that's just part of the Marine Corps. I thought, "All right, cool. Let's see how bad it's going to really be."

I thought because we all got there together, they were going to room us all together, or in pairs because the rooms were two man rooms. They put me with Lance Corporal Newell. Carol got put with Corporal McGee. Cobb and Cherry actually got their own room together. Carter and Chow got their room together. We were told we were doing a 15-20 mile hike the next morning.

The next morning when they called us for that hike, Lance Corporal Pedro Contreras yelled, "Wake up, get the fuck up" while running up and down, kicking people's doors. I thought, "Is it going to be like every day? Oh shit." It pretty much was, but everybody got used to it. This guy Farmer that I went to SOA with, his brother was part of 3/5, "Dark Horse." He told his brother, "Oh man, you guys are going to 2/4. They're a bunch of fucking drunks, and all they do is fight. Those dudes are crazy. All you hear of them is yelling all the time, no matter what time it is, day or night. They're always yelling." When I heard Contreras yelling, I thought, "It's probably Contreras yelling at people." He was older than most of us. Older than the Lieutenants and the officers. Sergeant Major Booker knew him by name, Sergeant Major Ellis knew him by name. I think even Lieutenant Colonel Kennedy knew him by his name, too. My platoon sergeant was Sergeant Coan and when they were in Okinawa, I guess, words were exchanged between Contreras and Sergeant Coan, and that's when he got busted down from Corporal to PFC. Contreras always said, "You know what PFC stands for? Pedro fucking Contreras. You can't do shit to me. I'll fuck you up." My buddy, Morel Cruz, told me about Coan and Contreras. I guess it was a big deal when he picked up Lance for a second time. Contreras would also say, "What the fuck, man? No one's gonna do shit for you. You're a fucking man now. Figure it out. If you're told to do something, go fucking do it."

When they divided us up, some went to Echo, about 10 went to Weapons Company, and about 10 went to Fox Company, which was a boat Company. Once we got there, we started training like normal. My team leader became a Lance Corporal with Jonathan Brown. The squad leader was Skaggs. When I got there, he was the Lance Corporal and picked up Corporal.

Janney: Were you Lt. Ski's RO (radio operator)?

Cienfuegos: A buddy of mine was with 2/1 and we decided to go to a bookstore and pick up a book on Arabic to get ready. Then, I got selected to go to the terp course for us. The whole Battalion had one man from each platoon go to a two-week crash course in Arabic. The course was taught by a government contractor from Morocco, but he wasn't Iraqi. A lot of his translations and words that he was saying didn't match up with how they spoke in Iraq. They passed the test to get hired, but the dialect was totally different. The words he was teaching us were not matching up. Our Arabic course book sounded a little bit different than what he was saying, but at the same time, it was all pretty much foreign to us. There was one guy, Hadari, I think he was from Fox or Weapons. He spoke Farsi because his family was Middle Eastern. We had another guy Cohen who was part Jewish, but he kept pronouncing the words wrong.

It was me and PFC O'Gwen that went. So, for Echo Company, the guys who got picked were John Huerkamp from 3rd platoon, O'Gwen for 1st platoon, Corporal Lee for Weapons, they had Corporal Lee, Corporal Marks for HQ, Corporal Bowen for Golf, John Hancock and Chadwick for Fox Company. As a young Marine, I got to meet some of these senior Marines that I was in these courses with, so I was trying to learn from these guys. It was more or less memorizing the terms, but the pronunciation was very hard. When we got to Kuwait, we were trying to practice, and it wasn't working whatsoever. The funny part is Staff Sergeant Walker had asked Huerkamp, "How do you say, we're going to kill the baby?" Just like the most random phrases - shit to mess with people's minds. Huerkamp said, "I'll get back to you on that one." When we were in Kuwait, they had us stand at the front of our tent and just fucking go over the phrases: Good morning. Thank you. Open the door. Do you have any weapons? Where is it at? Common courtesies, food, water, stuff like that, plus cultural awareness like don't touch or eat things with your left hand, address the men and don't talk to the women. They also gave the Company smallpox shots.

Around this time is when I met 2ndLt Wroblewski. Sgt. Coan introduced us, explaining that I'd been missing for a couple weeks due to the Arabic classes. I said, "Well, how are you doing, sir?" He said, "So, I guess you might be my radio operator?" His radio operator was actually supposed to be one of my senior ranks, Lance Corporal Kelly. But, during a pre-deployment physical, they had detected an issue with Kelly, so he couldn't go on the deployment with us. So, because that happened, they had this kid named Milczark who was told that he was going to be Lieutenant Wroblewski's radio operator. This kid was from Minnesota, so he had that real strong almost Canadian accent when he would talk. For reasons we won't discuss, he didn't deploy with us to Ramadi. So, the powers that be said, "You know what? If he's going to be the Terp and be with the Lieutenant, we might as well make him be the radio operator and carry the radio, since he's going to be next to him anyways." Calloway and I started learning the radio from the radio operator Corporal Andrade in Kuwait. We were learning the radio again. We learned to figure it out and how to change out the batteries really quick before it loses its setting within 20 seconds. You learn to swap out the battery in about 10 seconds. We learned how to use the different radios, PRs and larger ones, and learn all the nomenclature for that. They really drilled us about having to call for fire and the medevac.

So, Lt. Wroblewski spoke with me and asked, "You're not going to have an accent that can't be understood, are you?" I replied, "No, sir. No accent." He kind of smiled and said, "Well, you'll have to understand my accent. Can you do that?" So, I started laughing. I said, "Okay, sir. I got it." He said, "Hey, are you sure you know what you're doing? All of our lives are in your hands. You can you make sure? Remember, it's all on you." I said, "I'm trying, sir. I'm trying." Like I said, I was second platoon, second squad on our patrols when we were out there. So, sometimes Lt. Ski was with us and sometimes he would be with another squad. If he wasn't with me, he might be with Calloway or Lonis. Even though I went to the course to be a translator, I was carrying the radio on one of the patrols. Lt. Ski called me over because he met with a man in some house. We went into a house and we were checking for weapons and other contraband. I was trying to converse with him, but neither of us understood the other. Lt. Ski said, "What the fuck? You went to a course for two weeks, and you don't know what the fuck you're doing?" I said, "No, sir. They're not understanding what I'm saying." Lt. Ski said, "Were you paying attention in the course? What the fuck's wrong with you?" I said, "I was paying attention. The instructor was Moroccan and his pronunciation and words he taught us aren't the same as the Iraqis." I felt really bad about it, but it wasn't working. After a while, I was able to pick up more Arabic. I talked to some of the other Marines that did the Terp course with me and they also said, "Yeah, man. His shit was all off. So, you gotta use these words, and these phrases instead" and they shared examples with me.

As far as getting to know Lt. Ski, I knew he was from Jersey, knew he was an outdoorsy kind of guy, liked fishing, and liked being outdoors. The first run that he took us on was, holy shit, it was probably 20 miles. All I know is he killed us. He said, "I'm older than you guys. What's wrong with you? You guys are like 18, 19 years old. How come you guys can't keep up?" We're fucking dying running up these goddamn mountains, these hills that they have. People in California, on Pendleton, called them hills. We had guys that came from the East Coast that said, "Those are fucking mountains." California Marines laughed and said, "Nah, dude, that's a hill." Lt. Ski smoked us nonstop.

I don't remember all the bases we stopped in. All I remember is that once we left Pendleton, we went to March Air Force Base, and we stayed there forever until we got on our plane and we flew to Germany. I think it was Rammstein Air Base. I called my cousin and said, "Hey, are you guys good? I'm in Germany." Thirty minutes later, my cousin walked into the airport because her husband Matt Castillo, who was in the Air Force, was stationed out at that base in Germany. So, my cousin shows up and starts talking to me and said, "Hey, what's going on?" You guys are waiting here to leave?" I said, "Yeah." She said, "All right. I threw the kids in the car to come see you. Your mom called me and told me that you're out here." I said, "All right, cool. Thanks for coming out here. You didn't have to bring the kids." She said, "Oh, it's not a big deal. The Autobahn saves so much time." The crazy part is people were saying, "Who the fuck is that girl over there? Why are you talking to her?" I said, "Oh, that's my cousin." They said, "What the fuck do you mean that's your cousin? What is it with you Mexican people having family everywhere?"

I remember when we were in Kuwait, all the officers went out on a big run around the whole base in Kuwait. The enlisted did a separate run with our First Sergeant and Company Gunny. At that time, there was a super crazy dust storm that was just killing us, but they still had us run the whole fucking perimeter of the base. Kuwait was non-stop training. We were training and patrolling.

One night, I got called into the Company headquarters tent, "Hey, the CO wants to see you now." I thought, "What the fuck did I do now?" I was scared shitless. I'm in my greens, but grabbed my rifle and ran over there. When I got there, my uncles were there that were in the National Guard. They're both former Marines. They had deployed in January of '04. So, my uncles ended up talking to Captain Royer. As soon as I walked in, Captain Royer looked at me and said, "You know these guys?" I looked over, and it was my uncles. They were standing there in their Army uniforms. Everybody said it was weird because wherever we stopped, I was running into family.

Janney: When you guys got to Ramadi, what was your first impression of the AO?

Cienfuegos: When we pulled into the city of Ramadi, there were these arches that we crossed over MSR Michigan. We saw the Army unit patrolling with Golf Company doing leftsy-rightsy. As we rolled through, we're like, dude, this place looks like a shithole. There's trash everywhere. These buildings all look like they're abandoned, but then you see people outside of them. You would see these men, our father's age or probably older, just sitting there staring at us. Kids are there just wandering around. We thought it was because there were no jobs for them and they didn't have set school or set electricity. It was just all kinds of craziness. Before we left, they drilled in us the rules of engagement so we didn't get in trouble. Also that we were on a SASO mission. We said, "What the fuck is SASO?" They said, "Stability and security operations." We're supposed to be winning hearts and minds. Our slogan got changed to "no better friend, no worse enemy."

After a while, we knew if the kids were around and they were by us, we were okay. As soon as the kids were gone, some shit was gonna go down. We didn't know what to expect. You know, we weren't shocked. We were ready to do our jobs. But, looking around, we're wondering, "How is this going to work? Are we going to patrol through these cities? What areas do we have to patrol?" Everything was still new to us. We didn't know anything. We finally rolled through the gates of the Combat Outpost. They had some Marines at the gate and we had passed some Iraqis - I don't know if they were Iraqi police who had a little makeshift checkpoint near the arches. Once we rolled in, we didn't have any electricity at the Combat Outpost. We didn't even have beds. We had cots at first. We eventually got some of those kid-sized bunk beds. Once we got into that base, we knew there was no more time to acclimate. We're here, fuckin' get your shit, unloading everything, trying to figure out where we're storing all of our gear, where we're sleeping at. Where's the chow hall at? What food do we have? Where's the water? We didn't know much of anything that's going on. When we were there setting everything up the first time - there weren't many sandbags. That was the main thing. Harden our base and harden our area where we're going to be sleeping. Just a lot of stuff. The Army was telling us about mortar fire and the Iraqis' celebratory fire when they drove around the city doing their weddings. It was like a full-on parade. It was fucking crazy. You would see these Iraqi women done up with all the makeup and a big-ass white dress. There was a guy next to them riding like it was a fucking parade. If there wasn't a convertible, but they'd be sitting on the hoods of the car or where the windshield is, with all these people behind them, and they're fucking just shooting in the air with their AKs and everybody's dancing and singing and running down the streets.

I remember my first patrol, it was at night and lasted about 4 hours. I don't remember exactly the time it was, but we had some of the Army guys with us telling us what to do. We left out the back gate of the Combat Outpost, went through these fields about a mile or two, then we had to go left. It was still a rural area, kind of farmland. We kept going and going. We passed this street and the Army guys said, "You don't want to go down there with your Humvees or vehicles. They call this IED alley. You're always going to get blown up if you go there. Don't go down Nova because Nova's a bunch of IEDs. If you do, you're just pretty much waiting to get blown up." I asked, "How do you guys do your patrols?" They said, "We roll in our Humvees for about 100 yards, get out, walk about 20 yards, jump back in, and keep going." That was totally new to us. We were on foot all the time. The Army had armored Humvees with the sliding bulletproof glass. We had the open-backed Humvees with the soft doors. Because of that, they made us fill sandbags and put sandbags in the back of the Humvees. Sandbags on the bottom of the floors inside the Humvees for the drivers and the passengers. They eventually got this quarter-inch steel for the doors and back of the Humvees. I guess it was supposedly bulletproof, but it wasn't tall enough. The Motor T guys actually welded more steel above it, like another eight to ten inches. For the seven-tons, they ended up designing a long bench, put sandbags all around it and plywood against the sides of the sandbags and used those ratcheting tie downs to hold the plywood with sandbags in between. We felt safe. Later on, we discovered these sandbags ain't doing it. They're stopping a bit, but it wasn't the ultimate protection. It was a false sense of security. That's what all that was.

We were told about IEDs, but we didn't encounter them right away. We didn't know about VBIEDs which are vehicle bombs. As quiet as we tried to be at nighttime using NVGs, the night vision goggles, there were some sporadic lights; street lights would kick on or off. It seemed like everybody knew where we were at, regardless of what time we would leave, where we would leave from. They always knew. The dogs would non-stop fucking give us away, no matter where we were at. I remember we did one of those big Battalion missions, and I stepped on a dog because I didn't see it there. It just started growling and barking at me and almost bit me. I was gonna shoot it, but they said, "Get away from that fucking dog."

They would do a dynamic entry into a house and pull out whoever it was they needed to pull out. About an hour later, we were on our way with whoever they grabbed. They kept a log book of everybody going on patrol and how many patrols you were on. How many people left, who left with what, and how many patrols they were on. From the day you left, the time you left, from the time you got here to the time you left. That was even if you were on a QRF mission, they already knew who was going. Because there was a whole list of what they would submit right before you take off. We had these things called kill cards, and it was your blood type, your last four, and whatever weapons you had on you, and serialized gear, and you had to keep them with you all the time. They had printed out like five of them, and it was pretty much like a piece of paper with a fill in the blank, with a sharpie. You had one in each top of your sleeve, on your blouse. You had one in your cargo pocket or the butt pocket of your trousers. So, the way it worked was they gave you your kill card and your identifying name was the first letter of your last name, your Company name, and last four of your Social Security #. So, that's how they identified everybody. They had a master cheat sheet of who was who. After a while, knowing everybody in your Company, you'd know who was who if somebody's name came over the radio. You're like, "Hey, is that so-and-so?" It'd be good.

The Army showed us how they did things. We thought those guys were a little too ruthless at first, but it turns out they weren't. We didn't know any better. We were told to do stuff the proper way - you can't be getting too hands-on or kicking fucking car doors and telling them to jump out while they're at gunpoint. The Iraqi people knew the difference between the Marines and the Army because of the way the

Army was and the way the Marines were. We look totally different as far as our gear and our uniforms. They were testing us when we first got there. They were like, "Oh, these guys are fucking soft. They don't know about being out here, so let's start flexing with these guys."

The first time there was a firefight, they pulled my squad out. This was 25 March. That's when Lance Corporal Cousins got shot. He was a machine gunner. He got shot, it shattered his leg, and they ended up evacuating him out. We were told they were pursuing some guys and got into a small firefight. I guess what happened was they were on patrol, going out in vehicles and they ran into a mortar team that was getting ready to start launching mortars at the base. They weren't more than 300 yards from the base. By the time I got out there, we heard gunfire going off. By the time we went out there, two guys got medevaced out. They had us patrol the whole area up to this big ass house, almost a mansion about 400 yards from the street. These guys would drive a vehicle, whip out a homemade mortar to start dropping some mortars, throw it back in the vehicle, and just drive away like nothing. They kind of ran into them by mistake. They were patrolling and weren't expecting that to happen so fast. But, that's when that firefight happened and they got to unload on them. I remember going up to these fighters. We thought they were Iraqis. Captain Royer was yelling, "Who the fuck are these guys?" His interpreter said, "Oh, those guys are Iraqis." Then, they had us start searching these dead bodies. We pulled out passports, but they weren't Iraqis. They were from one of the other Middle Eastern countries. I think they were Syrian. I just remember seeing that and I was like, "What the fuck? We got other people fighting us here. That's how it's gonna be from now on." That was the first time being in combat and actually seeing dead bodies of these guys that shot at us. We shot back. They didn't make it, so they're dead now. Once that happened, it changed the tempo. We gotta be more alert. A First Marine Division saying is "Complacency Kills." This incident reminded us to always be alert.

I decided, "Fuck it, whatever. I guess it is what it is. I just gotta do it. If I live another day, then I live another day. If not, then too bad, you know? That's how my life's going to be. So, it's cool." I wanna say pretty much up to April 6th, everybody's like, "Yeah, we're gonna make it home, this isn't gonna be anything." Then, after April 6th happened, everybody accepted the fact, "Yeah, we're probably gonna die. Fuck it, who cares?" You would just go on about your business, bust jokes about embracing the suck. I remember telling Cherry and Cobb and them on 6 April, "I'll talk to you later. See you when I see you, man, or catch you on the other side if some shit goes down." I had no idea that I'd never see them alive again.

I remember there was something very strange with the Nova/Gypsum patrol we did with Lt. Ski on 5th April. It seemed really weird with a lot of the people when we were there. He asked me to go talk to some of these people and I didn't know what the fuck they were saying. They were gesturing with their hands, getting all pissed off and getting crazy. It seemed weird; a little bit different. Some of the people were a bit standoffish. Not many kids running around. When we got done with that long-ass 12 hour patrol the day before everyone was killed on 6 April, I was carrying one of the radios. It was one of those Nova patrols because everybody had to do an IED sweep. Each platoon had to send a squad out for the day to do the sweep, whether it was a Michigan sweep or a Gypsum sweep. We would do it at different times of the day, so they wouldn't keep a set schedule. I remember when we had found some, we were right around the palm groves. We found a fucking bunch of IEDs and we had to sit there for hours waiting for EOD. It was just weird because the river was right there, and then you look across and you're in the palm groves. I would have never described Iraq this way. It looked pretty cool. But, then you look over and you can see its Iraq. It's a shithole. There were crumbled buildings with bullet holes through them with Arabic graffiti on them. We found several IEDs well into the patrol and had to wait a long time on EOD. Near the end of the patrol, a squad discovered several Claymore-type devices hanging in the palm trees, so we had to call

EOD again. Being the radio operator, I had to be careful. I was told, "Hey man, get back, get back! Don't turn on the radio. You can't get too close to the IED because the frequency could make it blow up." Same with the Claymores. At one point, I said, "I want to take this shit off. Can I just dump this radio?" Lt. Ski said, "Nope, that radio is you. It stays with you." I said, "Okay, sir."

I remember the Swanson guy was taking pictures all over the place. I remember that day was hot as fuck, because we were walking and some of the asphalt was coming up as we were patrolling. When Swanson took all his pictures, I think I was only in one or two of them. On that patrol, Swanson took a photo that shows Lieutenant Wroblewski walking with his shadow behind him. I believe I was the Marine on the right in that picture because we were controlling traffic.

Janney: Lt. Ski's dad John loves that picture. It's on the wall in their house.

Cienfuegos: I actually got that tattooed on my arm, too. I got a pretty wicked sleeve. It took years to finish it off. When we came back to base after that patrol was fucking over, we were toast. When we came back that day, something weird was going on. They made the whole Company split up into an Alpha, Bravo, Charlie, as far as geography goes. They also took everybody who were drivers and assigned them different things. So, Kyle Crowley, Matt Scott, Darius Ortiz and Greg Arneson, and my team leader Jonathan Brown were drivers. So, when we got back, they took all those drivers and put them aside. They explained, "You guys are going on patrol. You guys are going to be assigned to a vehicle. If something happens, you're driving this vehicle." "Why the fuck are they breaking everybody up?" He said, "I don't know." When we came back on the 5th, I remember having to go fill sandbags up where they had the mortar tubes. I was sitting there talking with Carol, Roberts, Cobb, and Layfield. We were telling them we were filling sandbags, but we'd bullshit and smoke while we did that. I didn't really smoke before I deployed, but started smoking there. I would have my dad send me Black & Milds or Swisher Sweets. We'd have packs of Newport and smoke those too. So we're sitting there filling sandbags, filling sandbags. We were both sitting with Layfield and he was showing me pictures of his lady. I talked to Cobb and he asked, "What are you going to do when you finally get back? You're going to end up with a lot of cash." I said, "I don't know. Maybe buy a car or something. That would be pretty cool." I said, "What about you?" Cobb said, "I don't fucking know, man. Maybe get drunk. Maybe go look for a good time with some girls or something." That was the last time I saw those guys. We were filling sandbags, and I think that we rolled up a note, put our names on it, and threw it inside one of the bags and left it there.

One of the first things that happened in Iraq at the Combat Outpost was they got a bunch of us to watch these workers, to see if they're doing their job and not fucking around or taking pictures. We called it the Haji Watch. I met one of my seniors, LCpl Reynosa Suarez. We were talking a little bit. That was one of the first days in Ramadi when they were trying to get power working and put roofing on the chow hall. During the deployment, I kept a log of what happened that day, but after a while, I just said fuck it and burned the journal. One of my buddies, Jeffrey Melvin, actually kept a record of everything that happened to us. He's actually still an officer in the Marine Corps now. He was a 203 gunner. He had the 203 on his M16 and they had given him a 1980s grenadier vest to carry all the 40mm grenades. It was ridiculously heavy and the weight made him fall sometimes. We used to laugh so hard because when we'd go out on patrol, if he tripped, he would eat shit so hard. You'd hear him dropping like a sack of potatoes and then you'd hear him, "Fucking country. Fucking hate this place. Fuck you, guys. Why are you guys laughing at me?" That was the comedy part of patrol. We would be betting, "Hey man, how many times is Mel going to eat shit before we actually get out of here?"

On 6th April, around 0200, two in the morning, we went to do a patrol and they had broken up my squad specifically. They combined the first and second squads, almost platoon size. The weird part is we left out the back gate early fucking morning, just after midnight. It was a weird route. We were zigzagging, jumping fences. They had me in the back of the patrol. We were going through our checkpoints. We were by the industrial side at one point. I think we went down "No Name Street" during that. Our final stop was right around the tank graveyard. The moonlight was so bright. It looked almost like daytime. Callaway was the radio operator that day and said, "Be advised, Headhunter is out here." So, we knew that the scout snipers were out here. I thought, "Oh shit, dude, we probably fucked them" because the snipers were doing surveillance. They called in saying, "We see your patrol." Our call sign was Porky, short for Porcupine, but we hated it. So, we walked right by them and went into the middle of the palm grove where we did a security halt and pulled a 360-security perimeter and stayed a bit. Corporal Diggity pulled out his little fire starter, shaved off some magnesium and started a little campfire in the middle of all of us. It was cold as fuck that morning. We saw this house in the distance with goats all around. We saw these sheep herders in the distance. Some of them actually got kind of close and the goats were coming through us. We're like, what the fuck? It left me a little uneasy because that guy was kind of close. LCpl Morel Cruz crashed out for a couple minutes. When he woke up, he was making all these fucking weird noises. We said, "What the fuck's wrong with you? What are you saying?" Cruz said, "I had this weird ass fucking premonition dream. There was all this fog and smoke and these hajis were just running at us and we were trying to put them down. They almost overtook us." I said, "What the fuck are you talking about? Are you having a crazy dream or something?" He said, "Man, this shit felt real." Everybody just ignored him, "Yeah whatever, dude." In hindsight, I think Cruz did have a premonition because Headhunter almost got overrun there. We stayed there for a bit, I think until the sun came up, and then we started moving again. We were on Nova, passed Gypsum, and kept going further out.

So, the weird part is I don't remember exactly what time we came back through the gates. It was probably about 0600 or so. We went to the gates, we got clear, and we went back to our rooms. We tried to take a nap. We slept for maybe 20-30 minutes. Shortly after that is when we started hearing all this gunfire. It was so aggressive that we kind of woke up. I remember them kicking the door open, yelling, "Gear up, gear up, gear up! So-and-so's pinned down!" I remember running out the door and seeing the two Humvees rolling in, hauling ass to the gates. We were right across from the hangar bay and they hooked around the hangar bay so they could pull in facing the gates. So, when they entered the Combat Outpost, they entered one way, went around the hangar bay, and they pulled back in, facing the hangar bay, like you're going towards Michigan again. They're facing Michigan. As I saw that Humvee roll through, I remember looking up and seeing stacks of Marine bodies, just lifeless, just dead in the back of the Humvee. I remember specifically seeing Staff Sergeant Walker's Staff Sergeant of Marines DI shirt. I remember that clear as day. I thought, "Oh, fuck. What the fuck just happened?" Not really in shock, but kind of in shock. I grabbed my gear, and I remember running to the hangar door to see if I could help anybody get up and out. I think that's when I remember seeing Lieutenant Wroblewski - they were trying to carry him out. He was still kind of walking. I remember that bird landing and taking off with Lt. Ski.

They were yelling at us to load up and head out, "We're fucking going out right now!" We headed out. Corporal Smith was the head of the first squad for Echo. He said, "There's fucking all these tangos out there." Or hajis, or whatever word he used at the time, I can't remember. He said, "If they're in the streets right now, fucking shoot them because they're not supposed to be out here. We just got attacked. They're still out here fucking doing this, this, and this." I thought, "What the fuck, man? This is what it is, I guess."

We got to a point and got out. I ran somewhere to the left because they broke up our squad. I didn't know who I was supposed to be going with. I almost got left behind because there was no clear direction. I thought, "Where the fuck do I go now? Who am I going with?" It was very disorganized. It was chaos, pure chaos. I remember seeing my team leader, Jonathan Brown, by the Humvee. He said, "What the fuck are you doing here, man?" I said, "They brought us all out here. I don't know where I'm going." He said, "Go that way" and pointed. I ended up linking up with Bennett and Sergeant Ramos. I don't know how that happened, but I ended up being with those guys. They're like, "How the fuck did you get put with us? You were weapons." I said, "I don't fucking know. It just happened."

I ended up going with these guys and we were in this building. I think we were getting blocked for some reason. We busted the window and tried crawling out one of the windows, but they had barbed wire on it. I had my Gerber and I passed it to them. So, they were able to cut the barbed wire so we could get out of that room or that building. We swept around. We were walking around the area after we secured the area after all the shit fucking went sideways. It had kind of calmed down a bit. We had bagged whatever men we could find; bagged them, tagged them, and put them in the Humvees and we did other stuff as well. If you look at Dave Swanson's videos of that one Iraqi laying there with the vest on with the frag underneath him, I ended up on the opposite side of that guy further down the hill. You know how right there on that part of Nova those houses were street level, but as soon as you went to the other side, they were below ground level and you're almost walking on the rooftops. We ended up scooting out around that guy and came back up on the intersection of Gypsum and Nova looking at that stuff. I remember seeing that taxi cab driver, whoever that dude was, laid out and his car was parked there. He was actually in the car. They made us pull that guy out and they made me move the vehicle. I had to jump in that car, start it up and back it out. I didn't even think it would work. I actually had to stuff myself in there with all my gear on and just pretty much put it in neutral and they pushed it out of the way. I don't know what they did with that vehicle, but they fucking got rid of it. As we were walking around further down Gypsum, I think PFC Eric Israel saw an antenna sticking out of all this trash. We walked over there and that's when he found Layfield. They had covered him in garbage. I was like, "What the fuck, brother? We got someone here. Who's that?" I said, "Oh, fuck, that's Layfield."

My memories are so fragmented. I remember before 6th April, we had our XO, Lieutenant Kaler with us on a Route Nova patrol. That's when he and Carrol got blown up by an IED. At first, I had those fucked up BC glasses. Even then, the goggles they gave us to put over our glasses didn't fit. I had BCs, but I had bought my own pair of glasses once I got to the fleet. I think I went to LensCrafters and paid about $150-$200 to get some glasses. They were supposed to be scratch-resistant, all this other shit because I'm gonna need that. Anti-reflective for the glare and all this other shit. By the end of that deployment, my glasses were all fogged up. They were permanently fogged, which is weird. I actually had to buy some dirt bike goggles and put those over my glasses so I could see through the dust. Anyway, when Lieutenant Kaler and Carrol got blown up, Carrol ended up losing his leg, part of his leg, and some of his fingers. Last time I talked to him was at the 10-year reunion they had in Camp Pendleton. Yeah, all I know is he lives in Colorado now.

I remember Otey from when we were doing training. Yeah, so when I got to 2/4, it was small enough that there were a lot of us, a lot of the senior Marines were actually hands-on with us and the junior Marines that just showed up. The way they broke it down was there's the boots, which were us junior Marines that just came in, and they had the Oki boots, the Okinawan boots. So, my team leader Jonathan Brown, Burmaline, Brooks, Cox, Burai, and Tor. Some of those Marines got sent to Okinawa when they got to the fleet. So, they got to spend part of their first deployment in Okinawa. The fucked up part is all my senior Marines that were there got told by the Marine Corps that they were the bench warmers for the initial invasion. Because 2/4 spent a little more than a year in Okinawa. Yeah, so all the senior Marines that were there, a lot of them got out when they finally came back to Pendleton. When they got into the fleet, that's when a lot of them got out. Some of them stayed in like Lonnie Abalama - he was on his second enlistment there when I was there with him. It was Castillo from 1st Platoon and a couple other guys. But there weren't a whole lot of them. So, they referred to us as boots, not the Oki boots. That's how they broke us down.

But like I said, on the 6th, I remember walking around the corner, and I remember looking over and talking to Donnie Kopatz. I guess he had fallen from a rooftop and landed in a pile of syringes. This dude had some pretty badass ink on his leg from his calf all the way down, and that shit got all kinds of fucked up from those syringes. Kopatz was actually friends with Otey. They're from the same city. Their little brothers were actually friends in high school or middle school and knew each other. They found out they both had brothers in the Marines. They started talking about that. There's a lot of stuff that kind of eats at me because I don't know the exact details of how it went down. See, that's what I'm trying to do. I'm still curious to know. That's my whole thing.

Scott didn't tell me everything that happened. He told me a little bit of it. All I know is I heard them saying that Crowley had no chance in that vehicle. Cobb didn't have any chance. Neither did Roberts. It was just done. That fucking DShK, man. I remember after that, everybody thought these guys don't know shit. But at the same time, after that happened, it was well coordinated. It wasn't just the run of the mill fucking guy who's pissed off because you're invaded by Americans. These guys actually had some sort of training. Well, the weird part is, I don't know how true this is or not, but we got told that Captain Royer got intel from the CIA saying don't let any squad-sized patrols go out. If you're going to send patrols out during this time, send them out in platoon size. And that didn't happen. We still went out in squads. I remember there were some others that were in that vehicle. It was my friend Arnold Chow. He was in one of those vehicles with Zach Shores. I remember also seeing the flyers in big print, saying fucking how much they would pay for Americans, with more for snipers. They were telling people if you fucking surrender, we'll guarantee you get to go home, no harm will come to you and other bullshit.

I remember that day and then going over it and trying to write our AARs (after action reports) and then putting everything away, looking at the vehicles, and it was just fucking horrible. I think the night after that shit happened, we went out for a 12-14 hour patrol. We went all the way out to Fish Hook Lake and back. Some of my buddies from Golf Company were getting into it a lot in the city, and Corporal Bowen held up a rocket launcher, saying, "What the fuck is this? These fuckers shoot these at us all the time. Getting tired of this." I guess it was a Russian LAW-type rocket, like a little tiny bazooka, one that extends and shoots one missile. Bowen was holding that canister. He was one of the guys that went to that Terp course with me as well. All I know is that it was fucked up. We were pissed off.

I get irritated about the fact that people will say, "Well, there's no such thing as weapons of mass destruction in Iraq. You guys went over for something that was fake or false." I'll tell them, "Dude, you can say that all you want, but there were chemical weapons there." At the Combat Outpost where we were stationed, they had blocked off one of the areas and the engineers said, "You can't go in there. That's where we store the UX, the unexploded ordnance." But in actuality, we walked over and looked inside of it. It looked like a little jail cell, but at the same time, it also said that it was a chemical storage facility. The Geiger meter went off the charts there. That's why they didn't want anybody in there.

One of the nights when we were on post, I think it was 7 or 8 April, we were getting attacked on base. One of the NCOs told me I needed a haircut earlier that day. I had my buddy Downing try cutting my hair. Just as he was cutting my hair, we had a mortar land right in the Hesco barrier and it blew up. We got fucking peppered with all that bullshit that was inside the Hesco barrier. He nicked the back of my head because he stopped cutting my hair to dive for cover. When I went back, I had half a haircut. I had a line going across the back of my head where he hadn't started to even trim it at all. One of the other Marines, Lance Corporal Smith, or PFC Smith at the time, said, "Let me in, let me in, let me in." We were trying to run into his door, because his door was directly in front of us. He fucking ran. As soon as he ran to the door, he slammed that door shut. That's when it blew up inside the Hesco barrier, and it hit me. They said, "Oh, you're good. No Purple Heart." I said, "Okay, whatever. Let's just keep going." That was the fucked up part wearing glasses. I had to have them on all the time. So, I was always constantly having to make sure they weren't broken and moving them around. Because all the shrapnel and the rounds impacting close to you, they would hit these fucking glasses. I said, "These motherfuckers are full of shit. They said it's scratch resistant. Anti-reflective, all this other bullshit. These fucking assholes at LensCrafters." So, I was pissed at that. Then, we were all supposed to have inserts. If we had prescription glasses, they were supposed to make us prescription inserts for our gas masks. Before we left Pendleton, I remember we went down to the main site hospital, and they made us, if you had glasses, they made everybody go down there. I can't remember for what. We got our glasses, got our eyes checked again. They said, "You wear these? This is your script?" I said, "Yeah." They were supposed to give us glasses or get them sent to our BAS, our battalion aid station. We were also supposed to get gas mask inserts if you wore glasses, right? They made these newer versions where they had like a spring and they would unfold just a bit, like a prescription lens for your gas mask. I don't ever remember getting those.

One of the times after the 6th and the 10th when we took all those losses, I was with weapons for a bit. It was towards the end of the deployment because I stayed with weapons until we came back stateside. We got hit with an airburst mortar and some liquid came out. So, we had to put on our gas masks. Then they made us walk around the camp looking for mortar remnants. We never found any. But, I remember getting hit with some kind of liquid. I can't remember much more.

I remember my buddy Brandon Lund. I met him in Pendleton because he had just come to us from security forces. He was my roommate for a short time before we left. I remember April 6th was his birthday and he ended up getting shot that day. I have a picture of me and him. The film got fucked up from either the water or the heat or running it through the x-rays because it was a disposable camera. I had like at least 15 of those. I sent them home and I think I've got maybe like maybe two rolls of film out of all that because the rest of them were black.

I remember Cherry. I have a picture of Cherry where he's smiling and is on the phone, too. I gave it to his mom, finally. I even have a picture when we were on the bus ride. Cherry said his goal after he got through this was to re-enlist - he was gonna try being a Drill Instructor because he had a singing voice for cadences. Cobb was always saying he was going to stay in and be Colonel Cobb. We laughed and told him he was bullheaded, but he said, "That's Colonel Cobb, bitch. Get that shit straight!" I also have a photo of me and that kid Milczark sitting next to each other.

Going back to April 6th, I remember Lund getting shot and I was talking to him. I took a picture with him because it was his birthday. I remember when we were preparing to go out on April 10th. I remember Bug Hunts. There was Operation County Fair, Bughunt or whatever the fuck they were. I do remember hitting so many of those goddamn streets and houses. Whenever we hit a house or we went out, it was like fuck it. We're ready to die. No big deal, who fucking cares. Let's do it. Like I got you. You got me. We're good. That's all that matters. Like it doesn't matter, you know. Yeah, and the whole total mindfuck thing was when we got a bunch of the combat replacements. There was this one kid named Jeff Smolenski and when we looked at that kid, we saw Cobb's eyes when we looked at him. If you did a quick glance at that kid, you'd think it was Cobb. That was one of the things that kind of fucked me up. I don't know if it was anyone else, but I know it was like that for me.

Smolenski made it home though. He actually has a crazy story himself. They called it post six. I was on top of the COC at the Combat Outpost. It was the highest post they had and had a 360 view of everything. I think they had a 50 cal facing Nova. They came back off one patrol and me and Ryan Miller had to go up there because Sergeant Thompson told us to go up on the rooftop and move sandbags. Me, Ryan Miller, and Dustin Libby all went to the rooftop and we were moving sandbags. We were laughing because we called across the fucking camp for Jeremy Bennett to look at us. As he looked at us, we all started mooning him while we were on the rooftop. We're all slapping our asses, like, "Fuck you, Bennett. You like that shit?" Then we went downstairs and someone said, "Hey, you guys know there's a sniper out there, right? He's fucking taking pop shots at everybody." We said, "No, but we do now. Fuck it." It was later that day, when we went back on post, Bennet and Smolenski were on one of the posts. Bennett was trying to bulk up. Him and Jones, they were trying to get really big. So, they were counting MREs nonstop. We had jokes that Bennett would tap you out for a MRE shake, the protein shakes they had. Anyway, Bennett and Smolenski were up there. Bennett said he was trying to bulk up and he had fuckin' bowl of hard-boiled eggs and he threw one at that kid Smolenski. When he threw it, it hit Smolenski in the eye, and as he leaned down to pick it up, a sniper round was fired and it fucking hit right where his head had been. There was another round lodged into the 50 cal ammo can where the belt of the 50 cal rounds went in. So Bennett said, "I get drinks for life. I saved Smolenski's life by throwing a fucking egg at him."

After the 6th, there were just non-stop little firefights here and there, almost every day. On April 10th, we went out there like super fucking early morning. Knocked on the door, went through the house, and didn't see anything. There was a separate house outside a little compound. So me, Ortiz, and Madrillejos went in there and cleared it. There was nothing in there. We stayed sitting around there because they pulled the interrogator in there to interrogate that guy. As soon as the sun came up, I was looking in the distance and I saw a little red thing pop up and it was one of those red head wraps, a shemagh covering some guy's face. He popped up, looked at me, popped back down. He was about 150 yards away. I thought, "What the fuck?" I was carrying the little walkie-talkie radio that day. When we looked, he popped up and all of a sudden, I heard a couple of shots. He started shooting at us, from directly in front of me. Then, we started getting more and more fire everywhere. I heard an eruption of fire everywhere. I looked to my left after the first gunshot, and Ian Lopez and Shatzer were lying prone on the ground. I said, "What the fuck are you guys doing? You guys are lying in the middle of nowhere. Get the fuck up. Go fucking run behind a palm tree." Just as I said that, the palm tree to the left of me about 30 yards out just exploded. A RPG had hit it. Nearby was a house with a brick wall around it and on one side of the house was a sewage ditch. In between the fence and the sewage canal were palm trees, and another house on the other side of the canal further down. There was also a small building or pump house at a 45 degree angle from the house – the same little building we had cleared earlier. As we started to return fire, I started taking steps back to the house to lean up against the wall. Just as I did that, Corporal Smith pulled a frag from his vest. I was going to throw it over the wall, because they were saying there was fire coming from inside that yard, but I was sure that there were friendlies over there. So, I didn't toss the frag. Just as that happened, I saw people climbing out of the sewage ditch. I saw the head pop up again and I started shooting at it. I don't know if it was a sure hit, but the firing stopped and I didn't see that head pop up again.

I ran over to where those guys were coming out of the shit canal, and I was trying to help pull them up. I helped pull up Skaggs, Melvin, Lauersdorf, Cruz, and Swanson. The head guy, the interrogator, fucking jumped the wall and broke his leg. So, I was pulling these guys out of that shit, we were getting shot at, and I was standing behind the palm tree. Royer was there, but I don't remember exactly where he was. There was this canal and I was going to jump over it, but I wasn't going to make it, so I just said, "Fuck it" and jumped right in the clear water. I stayed in that canal and was putting down fire from there. Then, we started passing people across and other people started jumping across. We said, "Go that way. Let's put cover fire down." Doc Clayton and I carried the injured guy across.

We all went to this house. Someone called it the Alamo. I don't remember how I got there, but I remember I was running. I got to this palm trees and Calloway was running with Stark, Cullen, and Captain Royer was with them. I saw the house and we came up to it and there was a berm on it next to the pump house. I crept up and I started shooting. Captain Royer popped up because he wanted to see what was going on. That motherfucker took a round square to the center of his Kevlar. His head went back. He looked at me and went back down behind the berm. He took off his Kevlar. He looked at it. He said, "Holy shit. Look at that." Sergeant Collins said, "Congratulations, sir. You didn't die." He shook his hand. I thought, "What the fuck, man? Really, dude." Then they go running back into one of those houses. Calloway was just running. He didn't even shoot. He was just running. His headset was swinging and his gun was just swinging as he was running. From me to that house was probably 30 yards and halfway in between there were some palm trees, I opened up, still putting cover fire down. Then they said, "Okay, it's your turn to run." There was a palm tree halfway between the house and me. I was running to that palm tree, and as I was running to that palm tree, I was shooting. I was running as fast as I could and I ran right into that palm tree, I slammed right into the tree. I was doing a mag load as I was running and just ran into the palm tree. From what they told me, that RPK was on my ass the whole time - it was hitting right behind me as I ran. When they saw me hit the palm tree, they thought the hajis had shot me.

As I loaded up my mag, I spun around the corner, and I had my rifle ready. I rolled around that palm tree and started shooting until I made it to that house. I dove into the doorway. I actually had some hole in my camis, but I don't know if it was from rounds passing through them or not. Then, we were trying to get organized and figure out what we needed to do. I said, "What's going on? Where am I going? You guys want me on the rooftop? Where do you guys want me? Stay here in this room?" Just as that happened, I looked over and see Melvin fucking running and he fucking ate shit at the doorway, too. I thought he got shot by the way he fell because his body went limp when he tripped and he went sliding through the door. He said, "Oh, I'm good!" That's when I saw Degenheart coming in after him. I guess he had got grazed or something. All I remember is hearing, "Get away from the windows and in this house. They're going to fire an AT4. Cover your ears and take cover so you don't get hit with the backblast." That's when Musser shot the AT4.

We dragged that head guy in, the interrogator, who had fucked up his leg, and we tried to cover him, but he's screaming. Doc Clayton said, "Watch his fucking leg." Once the fire stopped, the gunfight completely ended, we started patrolling everywhere, all the houses and stuff. Inside the open brick pump house, we found RPGs, an RPG launcher with about 10 RPGs in it, and there were weapons in there. Someone said, "Pick it up, dude. You know how to use that?" I said, "Not really, but I'll figure it out quickly." Then I was told, "Nah, leave it." I said, "We're gonna leave it? You're fucking kidding me." When we started moving again, I purposely ran back there and that RPG and the rounds were gone. So, I don't know if another platoon snatched them up or somebody fucking took them and ran off with them. So, that pissed me off.

But after the 6th and the 10th and all those other small firefights, we all knew that we weren't going to freeze up during combat. We are going to go all out and not stop. That was one comforting thing that we knew we would keep going until we got shot. During some of those operations, it would be just two of us clearing a whole house even though you're not supposed to do that. If it's just two guys and one gets shot, you're fucked because you'd be in the house by yourself. But, at this point, we didn't have many options and had to do whatever it took until we started getting combat replacements. We were still running two or three patrols a day with different squads and also with my squad as well. We were getting told we were going to get QRF Charlie, which was a relaxed day, kind of like your day off, but it really wasn't because we would still get called out.

I remember Oliver North coming with his camera crew talking to people. As he was interviewing, we started taking mortar fire and we jumped in the Humvees and took off to go see where it was. We couldn't find where those mortars were coming from, but we were jumping out, taking off. The camera was still standing there on the tripod, and it was just getting rocked. The Hummer that was by it had a mortar hit right there, and it just ripped the fuel tank apart, so there was diesel just dripping everywhere next to the camera. We were yelling at the cameraman to take cover, but they were fucking running over here and there because they didn't know which way to go. We said, "Run behind that fucking wall and lay there. Run dickhead, fucking run." That's when we saw all the mortars landing. We ended up getting called out to check it out, but couldn't find the mortar teams though.

We were taking fire on the base one day, so I'm guessing it was somewhere between the 6th and the 10th. I ran upstairs above where 3rd platoon was staying and Cousins was up there on guard. Madrillejos ran up there and I ran up there after him. We were out there on the post and we were just shooting at where the fire was coming from. We were taking fire and we were firing back. There was this guy and he was trying to hide behind his car. He was hiding behind this car. Weapons Company came down and they fucking shot a Javelin or a TOW missile at one of these cars that was parked right in front of our gates because I guess it was a VBIED. They fired one round and it fucking went through the car and it didn't do shit, so they had to do a second shot. I don't know if they used a SMAW or a Javelin to get that taken care of, but it was definitely a VBIED. Once that round hit, it turned into a big-ass fireball, and shit flew everywhere. I don't know if it was new or not, but the EOD guy said, "Yeah, man, they do that shit here. We don't know if it's only Ramadi, but they will put IEDs inside dead animals on the sides of the roads. Like cats and dogs. Dogs especially because dogs are bigger."

One time we were patrolling back from that long wall next to what used to be a college. We would have to jump that wall if we went that way. I had to jump the wall carrying the radio and an AT4. We had been patrolling far back out there in the desert, in the middle of nothing. I was on point with the NVGs and they said, "Where the fuck are you at?" I looked and I thought, "Holy shit. I've drifted away from them or they drifted away from me." We ended up going almost up to where the graveyard was, just shy of that into one big-ass mansion of a house. We busted it open, climbed in it, and we were doing security that night, watching everything. As we were doing that, we got hit with one of those big-ass dust storms. It was almost like twilight with an eerie feeling. It was orangeish pink, and there was just fucking dust everywhere. And as we were coming back from that patrol along Michigan, on the same side of the Outpost, I was walking with my NVGs. We had to turn them off and on constantly because the streetlights sometimes would come on, or the house lights would sometimes come on, so you couldn't see because the lights would drown out the NVGs. As we were going along, there was this hole that was dug on the side of the road and I didn't see it because I was running. Jeez, I don't know how deep it was. All I know is that I was walking, and then I fell down some fucking pit. I don't remember what the fuck happened exactly, but I remember they pulled me up. It was Corporal Gutierrez, the machine gunner from weapons platoon. He said, "Hey, son, you okay?" I said, "I'm down here." He said, "Where the fuck you at?" "Well, I fell down some fucking hole, and I can't get out. I'm too short." He started laughing like it was a funny joke. He said, "It's because you're Mexican. You should be able to dig your way out." I said, "I don't know where the fuck I'm at, man. I can't find my way out either." So, he bent down and helped pull me out because I'm 5'9" and this hole was at least six feet deep. It was pretty deep. My hands couldn't touch the top.

On April 10th, we were guarding a house. Sims was getting carried on a stretcher. Me and Melvin performed CPR on Sims for a long time after everything had happened. Skaggs came in and yelled at us that we needed to get on the radio and call for an air evacuation or close air support. At that point, our radio was pretty much useless. We just had to wait. And as we were waiting, the guys were up on the rooftop still putting down cover fire. Melvin and I were still doing CPR on Sims, because Doc Gutierrez told us we gotta keep it going, "You gotta keep it going. He's not dead." I don't know how long the gap was, but I remember the Army rolled through with their Bradleys. They came in and their Army medics came to grab Sims to take him out on a stretcher. They laid him on it and as they were carrying him out, I remember fucking yelling at the Army dudes because they were just walking like normal. We said, "Dude, what the fuck, man? Hurry!" We almost lost it on those Army guys because they had no sense of urgency. I remember linking up with Lance Corporal Ramirez and Lance Corporal John Huerkamp because we were pushing from house to house. I got left back for a bit and I ended up having to find them. I remember Huerkamp and Ramirez because Ramirez was a SAW gunner and he was putting down cover fire. He was down to his last 20 rounds on the belt for his SAW. So, after that he was just standing there waiting because there was nothing else for him to do. I'm thinking, use your terrain. Whatever you think that would stop a round or would give you some sort of cover, use it. Whether that's a little fucking rock or even a goddamn light pole. Even though it's skinny as shit, it might just save your life.

We were running three or four patrols a day, within 24 hours, until we got the combat replacements, the stop loss guys. At that point, I was getting bounced between platoons, too. They would tell me, "Okay, you're gonna go to the first platoon for a week, and you're gonna be patrolling with them. When you come back, you're gonna be with us for a bit." And then, for whatever reason, I felt like I got blackballed because they sent me to weapons platoons. I said, "Why the fuck did you guys send me to weapons? What the fuck, man?" I was pretty pissed by that point when that happened. But, I ended up with weapons platoon and finished off the deployment with them. At that point, I figured out, if you take an Ambien pill and go to sleep and you get interrupted, you're up for a good 48 hours before you can actually crash back out. Because we were having problems sleeping, I asked about that. The Docs would say, "Hey, if you're cleared for like 20-24 hours of sleep, then I can give you this pill." I popped that Ambien, and I want to say maybe 30 minutes after I had taken it, weapons platoon pulled us out on QRF.

On one of the QRF missions we went out on, I was jumping out of the back of the Humvee because we were getting shot at, and my hand stayed gripping the Humvee tailgate. The AT4 that was in the back of the Humvee fucking landed on it and crushed my right hand. So, the AT4 didn't go off, but whoever had it didn't fucking secure it. When I jumped out, I put my hand on the tailgate and jumped out. And as I did that, the fucking AT4 landed on my hand and it crushed it. Then after that, Melvin stepped on my hand and my hand stayed in the gripping position on the tailgate, so I had to pull it off with my left hand. Because we were on a QRF, I couldn't do shit. Since I was right-handed, I put my right hand down, and I grabbed my rifle with my left hand, and I was just using my right hand as a platform to hold it up, like a post. I had to wait a week or so and we went to Charlie Med. They x-rayed it and said I had a fracture in my hand. I got my hand wrapped. They said, "Just give yourself a week off and then it'll be good." OK, great. You get a few days off, Greg.

Someone gave them the bright idea of setting up in the graveyard to do over watch for Michigan. We went back in there and were looking at some of these tombs and they actually had stored weapons and weapons caches in some of the tombs that were out there. There were also some anti-tank mines that were out there, but they were inert. They clicked, but nothing happened. That shit saved a lot of people by not working. We rolled over them and a couple guys stepped on them, but nothing happened. So, it was one of those, you shit your pants because you thought you're dead, but then you didn't die moments.

We were leaving a place and were going out with weapons platoon. A suicide bomber had packed his vehicle with C4. When he saw the 7-ton with the Marines that were getting ready to climb on, he drove up and detonated himself. But, he had so much C4 in there that it blew the car to pieces. There was next to nothing left. I was behind Dustin Libby and he was getting ready to climb up the 7-ton ladder. The explosion twisted the ladders we had on the back of the truck, and the engine was smack dead. He just stared at it like, "Fuck, what do I do?" We said, "Dude, get the fuck down. Now, let's go over here. Get behind the vehicle." When they went to investigate, that guy had blown to pieces, and shrapnel flew everywhere. The people looked at us like you guys killed this man with a fruit stand nearby, but he got smoked by all the shrapnel. It fucking ripped through this guy and his little stand. Later that night, the feral dogs were all around the blast site. The dogs were actually fighting for the pieces of this guy and the bomber. We could see them with our NVGs. We were watching these dogs fight all night over the scraps, the fresh meat these dogs were eating. So, it was one of those crazy fucking things that happened. We all looked at each other and said, "Dude, that shit just blew up. Are you good?" "Yeah, I'm good." "All right, cool. Let's go." That 7-ton stayed intact and we were able to get it started. We kept that same bent ass ladder on the back of it, too.

I looked up to this one guy, Corporal Misael Nieto. He tried taking me under his wing when I got to the fleet and tried making me work out with him and get me strong like him because I weighed maybe 120 pounds. When he got hit with an IED, it jacked him up pretty good. I remember seeing him all through that deployment until that happened. But, he was pretty messed up from the IED. That same IED killed Calavan. That really hurt all of us when Calavan died.

I think it was April 7th or 8th when my IED went off. Corporal Lindsey from Weapons ended up getting a hole through his head because he was on top of the turret of the 7-ton and shrapnel went through and hit him in the side of the head. I want to say he had a half-inch deep wound on his temple that they had to pack every day with gauze. We kept him there with us while he recovered. I remember when Sergeant Barron got shot in the leg. I think that was April 10th. They had to send him home for that. Flasko was also injured and sent home. I don't know what happened to him. Before April 6th happened, Marcus Cherry was able to go visit his brother in Junction City for a couple days. So did Rutledge. He was from Boston, and his father was a full-time firefighter, but I think he was also a Marine Reservist. He was a Gunny. He was able to see his dad while he was out there as well. It was a weird deal that some of those guys actually had family that they were able to see when we were out there.

6 April 2004 Route Gypsum Ambush And QRF

Cpl Frank Gutierrez and PFC Victor Madrillejos

Victor Madrillejos, PFC (to LCpl) and Frank Gutierrez, Cpl – 5 April 2021

Victor: We went out shopping for the gear making sure that we had everything on there. I remember it being really fast paced. As soon as we got to the unit there was maybe one bullshit training exercise, and we went to March Air Force Base and did that training and then next thing I know we were in Kuwait waiting to go into Iraq.

Janney: Your experience was quite a lot different from a lot of these other guys. A lot of these guys were boot drops that came to 2/4 in September or October before you guys deployed in the spring so I mean you were over there with them for at least a year and Frank it sounds like you were with them longer.

Frank: So, I was in that group that was like SNIF and some of the others, with senior Marines.

Janney: Frank, let me ask you, at what point did you decide to try to get into the Marine Corps and why did you pick the Marine Corps out of other branches of service?

Frank: I joked to a lot of people that it was because I watched a lot of GI Joe as a kid.

Janney: Laughs. I still have all mine.

Frank: Yeah, so do I. So, I had already had an idea that I wanted to be in the military. I originally wanted to be a fighter pilot, but I got glasses in first grade and I really didn't ask anybody or push very hard. To my understanding at that time, you couldn't be a fighter pilot with glasses, so I started looking around and the Marines kept coming up. I had kind of a violent background a little bit when I was growing up, so I wanted to do something that had to do with fighting, and I decided the Marine Corps would probably be the best fit for me. So, I joined in high school. For my 17th birthday present, my Mom signed the waiver and then I was in the Marine Corps at 17.

Janney: Okay, and about what month and year was that Frank?

Frank: I went to boot camp in July of 2001. So, I was in boot camp when September 11th happened.

Janney: Yeah, and I'm sure that changed everybody's perspective. What were you doing that day when you heard the news?

Frank: We were actually at the rifle range getting ready to do the rifle crawl so that was pretty interesting. They pulled us on some bleachers and told us what had happened. It was really surreal because I thought that they were kind of messing with us at first because you know General Striker is just messing with you. They had grabbed the whole Company and they had put us on some bleachers and they kind of gave us a brief in a way I had never heard before because I was still a kid, and I really didn't listen too well when they told us about all the things that were happening. They didn't show us any footage; they just said the fuckers are probably going to war, get ready. Then we had a bunch of people try to go AWOL, which was pretty funny.

Janney: They went AWOL just for one last big hurrah or they didn't want to go?

Frank: I don't know. There are various reasons that people join the Marine Corps, but I think people weren't expecting to actually have to go to war. That was the vibe that was going on at boot camp. The whole tempo changed. The Drill Instructors got stuck on base and stuff like that. The tempo was serious before because it was Marine Corps training, but you could tell it completely changed and so a lot of people got super serious like a lot of us boots. A lot of us changed our perspective and got serious and a few decided that it might not be for them and tried to go AWOL. But, they were all caught and had to go through boot camp anyway.

Janney: Laughs. I'm sorry I'm laughing but that's pretty funny.

Frank: It was pretty funny. And it kind of, you know, so there were guys there that wanted to be there already and shove it home like, "yeah, I really want to be here, fuck those guys!"

Janney: Laughs. And then everybody else is like, "Oh damn, now what? I'm going to get shot at. It's going to suck."

Frank: Yeah. I would say most of the Marines were there to be brave. There was just a select handful that decided that they wanted to find any way they could to get out.

Janney: Now, your experience is a little different than Victor's, but at what point did you get assigned to 2/4 and what was your rank? When was that?

Frank: So, I went to boot camp in July, came out in October, near my birthday. That's why I ended up in SOI until the end of December. I believe I joined 2/4 in early January sometime. I remember because everyone was on leave and one Sergeant sat me down. I believe he was a Sergeant back then, too. He actually met me at the JRC and told me, "Oh, you're going to 2/4; that's my unit." I remember meeting him there and then again later. Me and a few other machine gunners were originally supposed to go to the 1st LAR and then they grabbed me and this other guy and said, "No, you're going to 2/4." So, we went to 2/4 and everyone was on leave and when they came back, I was like the FMG compared to some of the other new Marines like Aaron Vergara. So, I met him really quickly because he was a machine gunner too. He had already been in before 9/11 - he was on the reactive force for September 11. So, he was already in the fleet and I was like the newest of the new Marines, so I got messed with a little bit. I was a Corporal.

Janney: So, Frank was that January of 2002? Is that what I'm understanding?

Frank: Yeah. January 2002 was when I joined 2/4.

Janney: Okay. Victor, let me move back to you. When you got to 2/4, how badly did Frank terrorize you? Was he a mentor or did he make your life hell? That's what I really want to know at this point.

Victor: No, he was really cool. There were some of the senior Marines that were a lot more vicious. He's definitely not one of those guys. There were some guys that were totally intimidated. It's a rite of passage. You've gotta get their respect and so I don't blame them or anything like that. No, Frank wasn't one of those super asshole kind of guys. There's plenty of those guys out there, but he wasn't one of those guys. He was a rifleman, and I was a machine gunner, so what he did was separate most of the time until we went to Iraq. My machine gun group was attached to his platoon. So yeah, that's how we fit together.

Janney: Victor, once you got to Ramadi, what Company were you with?

Victor: Echo Company, 2nd platoon, first squad. Sgt. Eric Smith was my squad leader at the time.

Janney: Frank, what Company and squad were you with?

Frank: I was Echo Company as well. I was a machine gunner for 1st, 2nd and 3rd platoon and then there's a weapons platoon. The weapons platoon breaks up the weapons systems they have, so mortars, machine guns, rockets all get broken up into the other platoons and distributed. I was distributed to Victor's platoon which was 2nd platoon and there were five of us. It was Sean the squad leader, I was a team leader, then we had Ryan Miller as our machine gunner, and we had Calavan as our other machine gunner.

Janney: Okay, you'll probably recognize some of the names on that list. Some of those are Weapons Company guys.

Frank: There's weapons platoon and then there's Weapons Company but yeah, I already recognize. Like I said, Aaron Vergara is a weapons platoon machine gunner. He and I were good friends.

Janney: So, you guys at some point were told that you were going to Iraq. What I understood was initially there was some discussion about whether it was going to be Ramadi or not. Habbaniyah or Ramadi.

Frank: Correct.

Victor: When we got to Kuwait that's when we thought we were going to Habbaniyah because the first word that we got was Habbaniyah and then somewhere along the way, that changed.. In Kuwait, we got pushed and we ended up going into Iraq. So, our orders changed.

Janney: Some of the guys flew in early to get things set up. Maybe these were more senior Marines. Frank, I don't know if you were part of that group or not. Which group did you guys fit into?

Victor: I convoyed in. I wasn't part of the early part.

Frank: I wasn't part of the early part. I convoyed as well.

Janney: Okay, and so when each of you got there, tell me what your impression was of Ramadi when you first got there? I'm guessing that you both were at Combat Outpost?

Victor: Yup.

Frank: Yeah.

Janney: What was your impression of Ramadi in general? What were you seeing? I know you guys had a completely different strategy that the Army did. From what I understood, they just drove around in armored vehicles really fast. Then you guys put feet on the streets and did foot patrols everyday which the Iraqis were pretty confused about.

Victor: Yeah, I think so. You want to go first?

Frank: Yeah, you can go first. That's fine.

Victor: When we first went in, we were there for SASO operations, so we weren't really there to fight and go kick down doors and shit like that when we first went in. The mission was completely different. It was essentially trying to get the government stabilized, trying to help build up the infrastructure, patrol the streets, make sure there are no IEDs in the road and try to get the weapons out of the hands of all the bad guys. I think they were allowed one weapon per family at this time. So, if we found any other weapons, we would take those systems and disable those and bring those back and then, at the end of the day or week, the engineers would make sure each family just had one AK. When we first went in, we thought the fighting was over for the most part. That was kind of the feeling. We got into fights here and there in the beginning. One of our squads happened upon a mortar team and then they waxed those guys.

Frank: Yeah, Eric was part of that team.

Victor: Yeah. I remember that was the first time I'd seen a dead Iraqi, so that made an impression on me. But for the most part, the people were kind of friendly. The city was super alive at that time in 2004. Just imagine downtown Waikiki or a big city like New York. It was busy. Those people were living their lives. They were just going about their day. The children were super nice. They were friendly and they were curious. Everyone was shaking our hands and saying, "To Hell with Saddam." All that bullshit propaganda at that time. In the beginning, we were mostly doing foot patrols. Like you said, the Army seemed a little bit scared to do foot patrols. In hindsight, they were probably correct. I think they just went through with Humvees, did their route, and took the fuck off. They never really stopped and went through and engaged with the population. Our thing was, we were going to mix with the population, engage with them, and show them that we're not the bad guys. We were out there doing patrols every day from the Combat Outpost - we would either push out patrols along Route Nova along the Euphrates River or along MSR Michigan.

Janney: Oh yeah. John Wroblewski and I did foot patrols with 2/8 every day for almost three weeks. So, I know the city much better than I probably want to at this point.

Victor: Yeah, and that's Lt. Wroblewski's dad?

Janney: Yes.

Victor: Okay. So, you know that area. That was our mission in the beginning. We were doing that. We were doing the IED patrols. We were setting up the compound because when we got to the Combat Outpost, it wasn't really secured and it didn't have good defenses.

Janney: Yeah, there was a wall that was blown off of it or something, right? One building just had three walls or something is what I understood.

Victor: Was it the back wall? The back wall had a hole, right? At the LZ, I think.

Frank: At the Combat Outpost?

Victor: Yeah, I think so.

Janney: Somebody said it was damaged. That it was not really what they thought it was going to be when they got there. Everyone said, "What?"

Victor: Yeah, because I think what it was originally used for was Motor T or it was a vehicle repair shop or something like that.

Janney: Yeah, that's what I heard so you guys had to get all that fortified so somebody didn't just drive down MSR Michigan right into the Combat Outpost.

Victor: We would go out on patrol for 12 -14 hours depending on how many IEDs we found and how long the patrol was and how many stops we made along the way. Then we would come back, we would rest for a little while, and then we would fill sandbags all day long.

Frank: Or new Marines would fill sandbags.

Victor: I was a boot at the time, so I was still part of that. I probably filled fucking thousands of sandbags. We pretty much dug an entire Olympic size swimming pool into the bottom of the fucking dirt and fortified the base. So, that's how it was in the beginning.

Janney: Victor, you had mentioned getting the mortar team, and the one that you said Aaron Vergara was a part of occurred on the 25th of March. Does that sound about right?

Frank: Yeah, you know, they would remember better. I don't actually remember the day. I remember getting the call and I remember talking to Aaron afterwards about it.

Janney: I'm sure it's the same one because they got radar coordinates from the Army and they went out and smoked this mortar team and actually captured one guy that begged them not to kill him because he had school exams the next day.

Frank: What I was going to say is that Victor neglected to say that everything was pretty nice, but we did get mortar attacks pretty frequently like Wednesdays or Thursdays. I can't remember what the days were.

Victor: 9:00 p.m.

Frank: Yeah, it was like clockwork. We would get mortared quite frequently. So, they had been sending patrols out just to see if they could find one. Aaron's platoon happened to walk right up on it and that was that.

Janney: I'm glad that they got those guys, but I know that they just harassed you and it was an ongoing problem.

Frank: Yeah, it was pretty continuous, so you got kind of used to it. I know one of our newer Marines, Lloyd, was wounded in that firefight. He had served in the Army previously.

Victor: Yeah.

Janney: Yes, Lloyd got shot in the knee. I think Adam said.

Frank: In his leg, yeah. Aaron Vergara told me quite a bit about that, so I won't elaborate.

Janney: No, that's fine, Frank. Aaron and one other guy that I talked to were involved in that action.

Frank: Your whole first question about what the situation was like in Ramadi when we first got there and how we viewed it is complicated. I was a team leader for machine guns and Victor was a rifleman, so the way I looked at it was a little bit different. I had just been promoted to Corporal during the March training before that, so I was a new Corporal. I remember when we first got there just really looking at the area tactically and thinking about how we would employ machine guns there and how we were supposed to move with the 11s doing their thing. I remember just feeling that initial pressure of having Marines under us. I thought this is going to be very interesting because up until Ramadi, a lot of our warfare training was mostly conventional warfare. We did a few urban training missions too, but mostly it had been like open spaces, in a jungle, up in hills and terrain like that. So, when we first got to Ramadi, it's a very urban city. It's as urban as where I lived in California. You saw people everywhere. I just remember thinking, how am I going to use a machine gun here with civilians and direct my crew to do so? Obviously, patrolling with movement is very difficult with a machine gun when you're going down corridors trying to avoid putting fire too close to Marines.

Janney: Yeah, as they're bounding from house to house. I get that. What were your thoughts when you got there and started thinking about the tactical aspect of deploying your weapon? Get to high ground and provide over watch or bound with the guys?

Frank: I think overwatch. Machine guns, one of the main priorities is suppressive fire, but our main priorities were also main avenues of approach - roads, intersections, and long sightline corridors. Over watch is always the thing if you can get it to provide security, but it's mostly about how much volume and suppression I could put down so that people like Victor could move in, clear houses, get the job done and feel like there's someone watching over them while keeping those enemies pinned down so they can maneuver on them.

Janney: So, you're laying fire down a long stretch of road so that these guys can move.

Frank: Ideally.

Janney: Ideally. So, Victor and these other guys can bound from house to house as they clear them, but that also puts you in the direct line of fire from anybody at the end of the road where you're laying down fire. Tactically, how did you handle that situation? You just seek whatever cover you can find?

Frank: Mostly, we would find corners. This is where I wish Aaron was here because he and I could talk about this because he and I have discussed these issues. I'm sure we'll get into that, but especially on the

6th this was crucial. He might have told you this himself, but machine guns were getting hit a lot when the fire fights actually started. In training, you are told that machine gunners are a priority target, but it doesn't really hit you until it starts happening. A lot of that was because we had newer Marines, and they weren't used to patrolling with the machine guns, like senior machine gunners like Vergara and I were. So, they just stayed on the vehicles when the machine guns were fired because most of the machine guns were initially mounted on the vehicles. They weren't really patrolling with them because they were mounted onto the vehicles. Vergara and I didn't understand why, but part of that is our own fault as senior Marines for not training the younger Marines to take them off the mounts and find better positions to use the machine guns. We had been taught throughout our time in the Marines, but we hadn't really had much time to prep with them. We hadn't had time to teach them these kinds of tactics. A smaller machine gun like a SAW you could dismount and move and find different places to fire from. Whether it be from a house, a corner, or from behind a vehicle. You could move with the gun, but it was uncomfortable because it weighed twenty-five pounds. A 240 Golf is a long, heavy machine gun. But you could still move it around. It was just not the most agile piece of equipment.

Janney: Especially when it's on a pintle on an unarmored Humvee probably for the most part

Frank: As you heard from the other guys, the Hummers weren't really armored at first. I didn't even think anything of it at first. Those first few days, like Victor was saying, we were patrolling through that place all the time and I was that machine gunner on an unarmored Hummer. That's just how we trained.

Janney: So, not even a shield on the front of the gun? It's just on a pintle (swivel)?

Frank: Oh, there was no shield. I thought I was like the Terminator, you know. That's kind of what it felt like.

Janney: Laughs. I'm sorry I'm laughing. It's just that black humor thing – it's just, "Oh, dear God."

Frank: You lose sight of the seriousness sometimes because you're out there for a long time. I literally thought of it like surfing because they weren't slowing down in those Hummers. So, when you'd be holding onto that machine gun, you'd be holding on for your life. It was fun and cool. But later, as machine gunners started getting shot, you realize, "Oh, this probably isn't that smart" without cover, without any type of armor. The whole situation changed because when we first went there, we were all about hearts and minds. We were security, passing out stuff to people, and interacting with them. On occasion, they might take a potshot at you, but you didn't really think much of it. You'd kind of laugh. Later after the 6th, that's when things totally changed, and the mentality changed.

Janney: Leading up to the 6th, what was your recollection of the days leading up to it? I heard that there was some stuff that happened on the 5th, but do you have any earlier memories of anything that made you feel like something bad was going to happen or have any foreboding of what was about to happen? Did the kids' behavior change at all? Tell me what was going on. The Battle of Fallujah is also taking place, so the four Army and Marine Corps Battalions encircling Fallujah is probably what kicked off Ramadi because a lot of those foreign fighters squirted out west to your AO.

Victor: Yeah.

Frank: Yeah, we're familiar with that.

Janney: I think some of them thought, "Hey, let's escape from this encirclement in Fallujah and we'll go to the next town west and kill some Americans." I think that's basically what happened.

Victor: I think that's exactly how it happened. You've got your hard-core guys that stayed in Fallujah. Either they couldn't leave, or they didn't want to leave. They wanted to make their stand there. Then, you've got the Ba'athists and retired Generals in Ramadi that organized the 6th or the 10th essentially. Frank and I were talking about this a couple of days ago. The 5th was really eerie because we were out on patrol. We were doing the Nova sweep and I remember the 5th very clearly because we found some armor piercing rounds and some other contraband at the tank graveyard near Nova. I've got pictures of it, too. It struck me as odd because it was professionally placed there in preparation for what was to follow the next day.

Janney: Let me interrupt you just a second, Victor. You did that patrol with Lt. Ski and the journalist David Swanson that day on the 5th? Was that you?

Victor: Yup, yeah. That was me.

Janney: Ah, so I've probably seen some pictures of you on that patrol.

Victor: You know, it's funny, I probably have pictures that you haven't seen. I gotta find them. I've got to dig them out. I've got a picture of me and Crowley holding those armor piercing rounds. During that patrol, Crowley and I found an old pineapple grenade. We were being idiots. We were throwing it around at each other like a fucking football.

Frank: Like a hard ass Marine.

Janney and Victor laugh.

Victor: Yeah, you've got me and Crowley throwing a pineapple grenade at each other because we found it in the field. We're like, "What the fuck is this?" We didn't know what it was at first, but we figured out what it was. It still had the pin in it, so it was still safe in our minds. It's amazing. We just found a pineapple grenade. It's just like Vietnam, you know. We were just tossing it back and forth to each other. Lt. Ski or one of the seniors said, "What the fuck are you doing?" Laughs.

Janney: I can hear it now. "You guys are going to blow yourselves up. What are you doing? Stop that."

Victor: Yeah. A couple of idiots, you know. I remember that day because we did that. We stopped in the tank graveyard and found contraband. Once we got to Route Nova and Gypsum, that's when things got really weird. The whole area was dead - no people anywhere. That area was mostly businesses like auto repair shops and fish mongers, things like that in that corridor. That intersection was always lively. There were always people out and about, the shops were always open, and people were always walking around. On the 5th, it was just dead. Just completely dead. Everybody was in their houses. Nobody was in the shops. The shops were closed. I remember the hair standing up on the back of my neck like, "What the fuck is this?" I think our entire squad got that feeling. I remember Ski saying something like, "This is fucking weird. Stay frosty." Some shit like that. It was definitely an eerie vibe on the 5th. It didn't feel right. The city didn't feel right. There was just something weird about the city. At that point, we were used to the daily routine of the Iraqis and the way that they conducted themselves throughout the day. It was completely off. The entire city just felt off. People would look at you through the windows, but then they would duck back inside. Kids would run out and their parents would chase them back into the compound

Janney: That's pretty freaky. You were with Lt. Ski. The story I got was Swanson saying there were some claymores or something similar strung in some palm trees that you guys came across.

Victor: Yeah, Frank can elaborate on that.

Frank: It was really weird at that intersection. I don't remember anyone saying anything to me about it, but I remember going there and thinking, "This place normally looks like a flea market and there's nobody here." It was really odd. Even up to that point, I felt like we were being watched the whole time. Up to that point, we were finding more IEDs, too. Just little things you'd see were out of place. It seemed like something was amping up. You could just kind of feel it. Our patrols were taking longer, and we'd get stopped on our patrols because we found something more often. I remember seeing people react differently to us like Victor said. You got the sense that something was going on, like they were trying to test us every now and then to see how far they could get away with stuff.

Janney: The whole idea of claymores strung in the palm trees really freaks me out. That's not a standard pressure plate or triggered IED. That's a pretty hairy trap, and for you guys to find that, I'm sure that really had to put you guys on edge.

Frank: That whole thing was edgy because as the new senior Marine with younger Marines, the thing I was actually most worried about was anti-personnel mines and IEDs. I wasn't as concerned about taking fire as I was about stuff that just passively blows you up. You grew up hearing stories about war; the mental aspect of that unknown really wakes you up. We found those claymores hanging from the palm trees right across the street from where we would have been almost done with the patrol. You could literally see the Combat Outpost and I was thinking, "Oh, we're about to be done and we can get some chow." Sure enough, we found some explosives hanging there and I thought, "Now we're gonna be out here for a while." That's what I remember of that day besides the stop in the field and tank graveyard that Victor mentioned. I remember the marketplace being shut down. It stood out because a lot of those uneventful patrols blend together because it was more of the same thing, especially early on. We did that patrol so many times and a lot of them blend now.

Janney: It's obvious that these folks were watching you guys and tracking your movements and trying to keep track of who was where and when. One thing that I heard is that Gypsum is just the easiest and quickest way to get to Nova, so they tended to go up or down Gypsum every day to get to Nova or come back from Nova. At some point, it doesn't take a genius to figure out, "If I put a L shaped ambush at Nova and Gypsum at some point, I'm going to catch somebody unaware and it's gonna be a bad day for the Marines."

Frank: You said you've been there, correct?

Janney: Yes. I patrolled that intersection.

Frank: You should have also noticed that that area in particular is very open. That was what was interesting about our routes compared to Golf Company. Most of their routes were straight up urban while we had a good combination of both. We had some urban, but also open fields. Gypsum and Nova was a very interesting field to work in because it's this huge open area. They could put some really big explosives there.

Victor: It's like fucking Vietnam.

Frank: Yeah. For me, as a machine gunner too, that's what I want right there, a big area to just go to town.

Janney: Sure, and they took advantage of that by setting up a DShK and RPKs at that intersection. That's why everyone in that lead vehicle except Otey got killed because they had very little chance of escaping that kill box.

Frank: Exactly. Victor and I had this conversation before. I said, "Yeah it's horrible to be on the receiving end, because as a machine gunner, that's ideal for me. That's perfect; that's exactly what I want."
Janney: Yeah, you know, I just can't even imagine what those poor guys experienced. It's just a miracle that Otey got out of there, that anybody survived that. Ironically, he escaped out of the vehicle on 6 April only to be killed on the OP rooftop with Parker on 21 June.

Frank: I knew Otey because I was attached to his platoon on the first deployment. Otey was a good guy. I miss him.

Janney: Yeah, and I've heard so many neat things about him. At some point, I want you to share your memories. Like when Victor mentioned Kyle Crowley and Otey and whoever else you want to share some stories about. That would be great because I've heard some stories, but everybody's got different ones to tell. That would mean a lot to me and the readers to find out more about those men.

Frank: Well, Victor and I were just talking about this the other day, especially about Crowley. In those first few days before the firefights really started kicking off, we had all those patrols and there would be kids and people out all the time and Crowley was quite the showman. He used to dance with them. I remember him dancing on occasions in front of these kids. He was a very lively guy and the kids ate him up. I used to joke to him that he was going to have a picture on a billboard like Michael Jackson there because he would dance like crazy for these kids and they just loved it.

Janney: That's awesome. I've never heard that story.

Frank: It made the patrols a lot more fun because we were out there for hours and hours just walking. You get kind of used to seeing these kids too, but Crowley doing all these fun little dances for the kids is a good memory of that time that I'll always have.

Janney: Break dancing? What kind of dancing was he doing?

Frank: Oh, it was like shuffling and even robot moves and just little stuff. You can only do so much when you have full armor.

Janney: Yeah, that's true. I didn't think about that.

Frank: Yeah, but it was just little stuff for the kids. We would moonwalk and stuff like that.

Janney: And so, would they try to copy you guys and your moves?

Victor: The kids, they loved us after seeing that. They just wanted everything to do with us.

Frank: They were holding them out to us, offering them to us like Simba in that scene from "The Lion King."

Janney: Laughs.

Frank: Yeah, the vibe was totally different when we first got there. What's crazy is our operation tempo was really fast when we first got there. We got no sleep and we were always out patrolling at first. That was just your day and it all blended together.

Janney: Filling sandbags as soon as you get back, Oh yeah, have some water and food, and by the way, you gotta fill some sandbags.
Victor: Yeah, yeah, pretty much. I did find that picture of Crowley on the 5th in the tank graveyard.

Janney: Oh man that would be great. Anything that you own that you can send to me as long as you shot it and you can guarantee that you shot it because of the copyright laws. If I use somebody else's picture that I don't have permission to use, it's a $10,000 per violation fine.

Victor: I'm pretty sure I shot this one.

Janney: So, anything that you have that you can scan or the best quality version of it that you can get to me, the better the chance will be that I can use it in the book. As long as you shot it. That's really cool. That would be neat to have that of Kyle.

Frank: Will there be pictures in your book?

Janney: Absolutely. If you send photos, you just gotta tell me in the picture, left to right, who is who if there's a group. That's easily done if you can remember who all is in the picture. I definitely want to use as many pictures as possible. Somebody, just jokingly, said, "Well, this is a story about Marines. You're going to have a lot of pictures in it right? Because they're not going to be able to read it." I laughed and said, "Ah, that's terrible." Sure guys, as many pictures as you can send. Just let me know who's in the picture and who shot it so I can give them photo credit.

Frank: Yeah, you should have told those guys that you're also going to have a pair of crayons come with it.

Janney: Laughs. Maybe with an 8-crayon box on the back of it? So, the 5th was just a very eerie day. It was quiet on the streets that were normally pretty busy. I don't know who wants to go first, Frank or Victor, but tell me about what your day was like on the 6th.

Frank: Victor and I were talking about this the other day. There's so much that happened that day that there are, especially after this long now, there are only certain things that I really remember pretty distinctly. And who knows, those memories might have blended together too. I remember being called out for the QRF because I know the snipers were pinned down and they were talking about a lot of enemies attacking them. I remember the QRF call. I don't remember gearing up or anything. I just remember I was on the QRF and we had David Swanson with us. We loaded into the vehicles, went down Michigan to the arches which was a big landmark for us. I remember having to dismount there. Victor and I were discussing where each of us was that day. David Swanson was kind of in the back of the fight zone. I was in the very back because I had the machine guns in the very front. For some reason, I was in the back.

David Swanson was sitting there taking photos. I ended up jumping over the side of the truck in the fight zone. The trucks were pretty tall and my rifle swung around and hit me in the face. My broomstick handle, or the handle you can grip the rifle with, actually broke off. I thought, "Shit. That sucks. That shit broke my broomstick." Right after, I felt blood coming down my face where my rifle had hit me in the face. We stopped there and I wondered what was going on. Then, we mounted back up into the vehicle and started heading north on Gypsum towards Nova.

Janney: Yeah, Gypsum goes south/north, MSR Michigan to Nova.

Frank: Yes, we started heading up Gypsum, and all I remember is as we started going down that road and near the very end, there's a curve just before the intersection of Nova and Gypsum. I remember a few pot shots and then all of a sudden very rapid fire. It got deafeningly loud and you could hear the echoes of the gunfire bouncing off the walls, along with the sound of bullets ricocheting off the buildings. Everybody started hunching down and my gunner, Calavan was on a Hummer gun, and Miller was on the 5-ton gun, and I just remember everyone just started to point outwards and started returning fire. It really started getting crazy there.

Janney: If I remember correctly, Miller kind of thought he had an angle on those guys at the intersection. He said he was just pouring fire into what he thought was a good angle on these guys that were near that intersection. This can be on or off the record depending on how you want it, but he said that he got yelled at later because somebody claimed that he had committed some friendly fire at that point. But I know Ryan and I know that he didn't do that.

Frank: Well, there was a lot of fire at that point, so this is that whole fog of war issue. It got very confusing and there were a lot of moving parts all of a sudden. Your brain is snapping too, and everything is slowing down like, "Oh shit, they're firing at me. They're firing at me." So, Miller was on top of the 5-ton and that gun is literally on top of the cabin, so it's really high up. He had a great view. I remember him, not verbatim, but just to the point, saying, "Hey Corporal, I see people with guns. May I open fire?" I said, "If it ain't us, if you think they have weapons, open fire." So, he started putting down a bunch of fire, but I wasn't exclusively watching him because I had two gunners to pay attention to at that time. Just watching where all the 11s were moving. We had another Marine who was a Lance Corporal, but he was kind of new to the unit. He was a 51, so he was an assault man, rockets and demolition man, and he was sort of new to weapons platoon. I kind of knew him and he was right next to Miller, so I asked him, "Can you watch my gunner for me, just help me direct fire a little bit", because I saw that Calavan was on a machine gun by himself. One of the other senior machine gunners was on a machine gun watching the road completely by himself. His name was Carlos Medina, and he had pushed forward. I know you didn't mention him.

Janney: No, I haven't talked to Carlos.

Frank: Yeah, I don't know if anyone stays in touch with him. He was pushed out pretty far ahead and so I remember thinking that he's got no one to assist him, so I'm going to help him as soon as this area is kind of taken care of. Sgt. Smith was giving orders to Marines, getting 11s ready. Somewhere in between that time, I had looked back at the Hummer that Calavan was on, my other gunner, and someone was lying on the ground, and there was a pool of blood. At first, I thought it was a Marine named Berman because this person was very pale. Pinkish light skin and that's when I found out it was Lt. Wroblewski because he was right behind me. I just was looking the other way because I was directing Miller at that time. That's one of the things that I really remember about that stop initially, was all that going on at that time. I remember there were other Marines huddled around him, too. I distinctly remember Eric Smith being right there and I remember Lance Corporal Berman being there. I remember there was a Motor T Staff Sgt. who was under the Hummer returning fire. I don't know his name, but someone said you talked to him.

Janney: Yeah, that was Sgt. Valerio. Eric Smith doesn't really remember Valerio being there, so it's good that you do remember the Motor T Sgt.

Frank: Because I remember him firing from under that Hummer because with all that was going on, I mean, we were taking fire from all sides. So, there was a lot going on. Then, Cpl Smith and Cpl Lenz, who was our machine gun squad leader were there too. He was dealing directly with that Staff Sgt. who was in charge of Motor T. Those two were talking a lot. Smith was talking a lot with them from what I remember, and they were kind of coordinating what was going to happen in that area. So, I was just really concerned with what was going on with the guns because Cpl Lenz was kind of coordinating with Smith on how they wanted to move men and where we should be at. As I said, I then noticed Lt. Wroblewski on the ground. That's why I was asking if you talked to Sergio Gutierrez because he has the same last name as me and we did our first pump together, too. He and I were pretty close. He was a Corpsman then. He was one of the senior Corpsmen there. He was there at that time too, and he knows a lot about what happened there. I think Doc Gutierrez was actually in the vehicle that medevaced Wroblewski out of there. I'm pretty sure my gunner Calavan was on that Hummer as well. But, I know how to get a hold of him and after this interview I will talk to him and see if he wants to even do this. I think this helps the Marines, too. Because, like you said, it's seventeen years ago now and we all have little pieces of the puzzle. If someone is writing it and actually putting those puzzle pieces together, it might make a little more sense. I know the people who were there at that time, so they would each have a little piece of that puzzle. It sounds like you've talked to a few of them but, I know some other people who weren't mentioned. I'll ask them. I'll see. Again, just like it was for us, it's up to them.

Victor: Yeah.

Frank: At least if they agree to it, then yeah, you have more pieces of the puzzle. And what they want to talk about they can talk about.

Janney: And, like I said Frank, you're the only guy besides Sgt. Jose Valerio that has said that he was there that day. Everybody else I've talked to said, "I don't know who the hell you're talking about. I don't remember any Motor T Sgt." So, the fact that you remember him corroborates his story.

Frank: I remember him because, one of the reasons I probably remember him, is, if you talk to him you can tell him I said this, machine guns were mounted on those vehicles a lot, so I had to be aware of the guy because he was the one who was having his guys put more armor on those Hummers. He also came up with, maybe he told you this, this crazy plan to string a bunch of armor together for us gunners to wear so that we were fully covered. I remember looking at this abomination because, if I remember, they were like leg chaps made out of the Kevlar shoulder pads. They were chaps that he had made and fastened together. I mean, it's stupid, but if it works, I'll fucking try it. So, we tried it a couple of times. I remember the guy because, again, as a machine gunner, and him being responsible for the vehicles, you know, we had kind of a working relationship. I didn't really talk to him much, but I remember him.

Janney: He told me that he was lying under the vehicle near Lt. Ski who was also down beside the Humvee. He was there, just like he said, by Lt. Ski's side, in a sense.

Frank: It was the left side of the Hummer, too. I remember distinctly. I was toward the rear of the Hummer and he was lying face down, his head toward the front of the Hummer, and he was on the left side of the Hummer. I can give way more detail, but that is all I'm sharing. There was a Marine down and I remember distinctly noticing the pool of blood that was forming. I went over there and I talked to him. That's why I'm saying I know some other Marines who were right there at that exact point too, because I went to see what was going on.

Janney: Right. And from what I understand from some of those guys, Lt. Ski was at least somewhat responsive and was trying to give people the thumbs up.

Frank: Yeah, and so see this is why I was saying the Corpsman Gutierrez would know even more because, I'm speaking for him and I don't like doing that normally, but until he agrees to talk to you, I remember him telling me afterwards that he had stabilized Lt. Ski really well. He seemed like he was going to be fine until they loaded him onto the medical evac. I believe it was a helicopter at Combat Outpost. Sergio really thought that he was going to pull through and was really surprised that he didn't. He rode with him, he talked with him; he would have that information.

Janney: Tell me Frank, what's your next thought after that happened? Tell me what else happened that you remember that day?

Frank: This is where it gets a little murky because I think I remember us loading Wroblewski onto a vehicle, but what I really remember after that was I went to go help one of the other machine gunners, Corporal Medina, with his machine gun. Because at this point, we were still taking pot shots; we were still under fire. Medina had pushed forward. This is why it's kind of weird and I wish some of the other Marines were here to kind of brainstorm with me because Medina had a full view of the intersection. Because when I went up there with him, we had a vehicle come around from what would be the east direction of that intersection and we lit that vehicle up.

Janney: It was a taxi, right?

Frank: Yeah. We lit one up, but I don't know if it's because Victor reminded me that there was a taxi that ended up going over the side with a body in it. I don't know if that was the vehicle we did because there was so much going on. Sgt. Smith had his men moving and checking houses. Even between that, I went into an alley and there was contact there. So, there were a lot of moving parts going on there and I don't exactly remember the whole order of some of those things, but I believe that was the order.

Janney: Okay Frank, let me ask you this. Did you know that Otey and Null and Matt Scott were holed up in a house or a little shop building somewhere near the Gypsum/Nova intersection? Did you, at that point, know that those guys bailed out of the first and second vehicle in the convoy at the intersection and were pinned down in that building? Did you know that at that point?

Frank: No, I wasn't aware of that. I was talking to Victor about that too because I didn't. I was part of that convoy and I don't remember when the first vehicle headed up. Now, you're telling me there were two that pulled away. I can't remember why. I don't remember seeing that. I remember hearing a lot of gunfire. I was telling Victor that I wasn't sure if that was the gunfire happening at that intersection or if it was directed towards us. There was so much fire going on there. It was really overwhelming.

Janney: So those guys, those two vehicles were part of your QRF that were sent to help out Santiago and

Stayskal and Headhunter II when they were pinned down at Nova?

Frank: I know that was what we were supposed to be doing. I don't remember when they separated because I was in that convoy, so I don't know. That always confuses me and so I was unaware of Otey when they got out. I just remember, again, when I went to help Corporal Medina, we had a full view of that intersection at that point. The Hummer was already full of holes and no one was in it except for the bodies. After that, the only thing that I really remember besides fighting in that little area was a little bit later. I don't know how much time had gone by because we pulled up to that intersection and started collecting Marines that had fallen. I was one of the Marines that actually had to pick them up. I don't know if you talked to Sgt. Smith about that, but when we first initially started loading some of those Marines onto that 5-ton, one of the people he found was Crowley and he did not want to carry him. He said, "That's my team leader. Can you do it for me?" and I said, "Sure." So, I put his body onto that 5-ton along with several other bodies. I mean it was a really big wakeup call and a huge reality check because I have a lot of bad memories tied to a particular moment. That's why I don't remember too much in between because that was terrible.

Janney: I'm so sorry. I can't even imagine.

Frank: I don't know where I got the wherewithal, so…

Janney: At that point, you're just exhausted from fighting. What time of day was this? Like 10:00 or 11:00 in the morning? What time was the QRF activated?

Victor: I remember it being like 8:00 – 10:00 in the morning.

Janney: Okay, Victor, so tell me what you remember about that day. You told me that you were initially assigned as Lt. Ski's RO (radio operator.)

Victor: Yeah. The first thing I remember was we got the QRF around 8:00 to 10:00; somewhere around there in the morning. I remember we got in the vehicles. We got to the arches and we hooked up with Captain Royer. Captain Royer's radio in his Humvee got shot and it wasn't working, so he needed a radio operator and he grabbed my ass.

Janney: Let me stop you right there for a second. Didn't he normally have an RO? You would think a Captain would have an RO.

Victor: I don't know who in the hell was his RO, but I think his radio got shot. It was in the Humvee and it got shot or something.

Janney: So, his RO's radio was down because it got shot, so he just said, "I need another RO. Victor, you're it."

Victor: Yeah, yeah. I pretty much got told to go with him. Ski told me to go with him essentially and that's when I detached from our squad and our platoon, and I went off with Captain Royer, David Swanson and Doc Clayton.

Frank: That looks like weapons platoon.

Victor: I don't know who all was there, but I got some pictures of it. We had the interpreter "007" with us. So, what I remember was when we got to the arches, the Captain and other guys were already there kind of panicking. His radio got shot, so I got attached to him and then that's when I separated from Frank and the rest of the platoon, and I went off with Captain Royer and them as his RO.

Janney: Let me stop you again. I don't mean to interrupt you, but there's some things I need to clarify. From what I understand, Royer and you guys were getting some fire from the east side of the roadway and Royer took off after those guys? Is that what I understand? He shot off east or to the right of Gypsum and he said, "Let's pursue these guys." Is that true?

Victor: I don't know exactly that portion, but I know if you're looking at the arches coming from Combat Outpost, the buildings on the left hand side going towards Route Nova, we started heading in that direction through the houses. We might have been chasing somebody. I don't know. I wasn't a squad leader or anything like that. I was fresh out of SOI. I was just some dumbass boot and I was just trying to do exactly what I was told to be honest with you.

Janney: Sure. I'm not asking you to question what he was doing. It's just that I heard that he took off after somebody took some potshots at him or something which was really a distraction from what you guys were trying to accomplish anyway which was to get to Nova in the first place as the snipers' QRF. Again, I'm not here to question anybody's command decisions or put anybody in a bad light. I'm just trying to figure out what exactly transpired. So, I'm sorry, go ahead.

Victor: I don't know exactly what the reason was. I thought that we were probably heading there to save somebody. He had the radio for the most part. I was just his walking RO essentially. The situation that I understood, we were running to Route Nova to save a squad. I think it was the snipers. I'm not entirely positive on that, but we started bounding through houses and then we got a call and then we started just running like at full blast trying to get, as fast as we could, to Route Nova to save somebody.

Janney: That was Santiago, Stayskal, (redacted) and Ferguson.

Victor: I think it might have been. It might have been that we were trying to get to them. A lot of people were trying to get to them also. But, before that happened, we were going through houses. We took one house where we took POWs – not POWs, but whatever. I don't even know if we kept those fucking guys.

Janney: Unfortunately, some of them were in the first and second vehicles of your QRF. Unfortunately for them, let me put it that way.

Victor: We started running towards Route Nova and then I remember just running and running and running. Just trying to keep up with this fucking guy because he was running full speed. You know, I've got this fucking radio on my back and a 10' fucking antenna.

Janney: Right. How much does that thing weigh, like thirty or forty pounds? Plus, you've got extra batteries.

Victor: I don't know. It was the big one and two rechargeable batteries which are 7lbs. each. The heavy batteries were 7 lbs. each. I'm not a fucking big guy.

Janney: So, you've got about 60 lbs. of radio, plus your IBA, your Kevlar, whatever else you've got to carry and a weapon.

Victor: I'm just trying to keep up with this guy. That was my main fucking mission and end goal. That was my focus. Just keep up with this motherfucker - that was all I was trying to do. Trying to keep up with him and we were getting shot at a lot of the time. Then on the way there, we stopped and turned around. This must have been like an hour or so, I don't know, the timeline was kind of crazy. We had been running for a long time, but we stopped, and we turned around and we ended up running across this open field. I remember we were getting shot at from the other side of the field. We bounded across the field and tried to get across. I remember seeing one of our guys shoot one of the Iraqis. In my head, I thought, "Fucking awesome." Then we ran across this field and I remember a big wall that we had to get over. Again, I had the radio and everything and I thought, "I'm not going to be able to get over this fucking 10' wall" or whatever the fuck it was. I just looked at it and thought, "There's no way I'm going to get over that fucking thing with this radio." For some reason, we just ran that fucking day. I think me and Frank tried to figure it out. We probably ran like two-three miles from where we started at the arches to Route Nova. It was at least three miles from what I remember, just forever.

Somehow, we ended up on Gypsum and Nova. I don't remember if we came up Route Gypsum. We definitely didn't come from the Nova side. We didn't come from the bush side. We came around that building that we ended up being stuck at. We came around and we made it into the building somehow. Along the way we were running; we were running and then I was trying to keep up with this motherfucker and I was throwing up and I was like, "Fuck this. Just go." I was looking back, and I was checking to see if I was the one that was lagging, but I remember there were other guys behind me, so I thought, "Oh, I'm good. I'm fucking running and there's other guys behind me." "Okay, I'm not the slowest one." So, I kept trying to keep up with Captain Royer and the whole time he was just saying, "Keep up with me. Keep up with me." I remember my breaking point was when he turned around and said something like, "Keep up with me. Marines are dying. Just fucking find it and let's go. Just keep up with me." At that point, along the way, there was another Marine named Shaver. Brad tried to stay next to me because he saw that I was struggling, and he kept asking me to give him the radio and I told him to fuck off five times. We were running like that, and it just clicked in my mind that I was like, "Fuck dude, I'm not going to make it. I'm not going to make it." I'm not going to keep up with this guy, you know. Then, I just told Brad to go ahead and take it. Shaver took it off and as soon as that radio was off, I just felt so much freedom. Up until that point I wasn't really able to shoot back. We were running and being shot at from everywhere, but I wasn't able to shoot back because I'm carrying the radio. My whole thing was to keep up with this guy. That was my only goal was to keep up and keep up and keep up. As soon as I took the radio off, I felt instant relief and I just started firing back whenever I could. I just felt so much more powerful, and I was shooting. Even before that day, I hated the radio. I have no idea why they chose me to be the RO. They chose the smallest guy. I just always felt like it was a joke or something.

As soon as I got it off, we edged up to that house on Gypsum and Nova. Along the way, we ran into Sgt. Ramos from machine gunners and Bennett. I don't remember if we hooked up along the way, but somehow we ended up with them. I remember one of the comedic reliefs of the day was Sgt. Ramos telling every single fucking Marine that he came across the story of how he was taking fire and hid behind this palm tree to make himself a small target and then the rounds were just bouncing off the tree right next to him. The way that he told it was just so fucking funny.

Frank: With his accent.

Victor: Yeah, with his accent and the way he told everybody that he saw that day the same story. That was the only funny moment of that day. We ended up on that rooftop for I don't know how fucking long, but it felt like forever. I think we had a couple of guys with us that were wounded. I could be wrong about that, but we didn't move from the rooftop for one reason or another. From where I was on the rooftop, we could see the Humvee that was hit on Gypsum and Nova. We could see the Iraqis. We were right on that bend on Gypsum and Nova and right in front of us on the right-hand side of the road, there were two Iraqis laid out, the Humvee, a taxi right on Route Nova and Gypsum on the left hand side on the embankment that came down. Then, there was another taxi on the right hand side and there was a dead Iraqi all the way under that. There were bodies everywhere. Dead Iraqi bodies. We didn't see any of our Marines because I think they had already come and gotten them at that point. I think Frank was there before us. We were trying to piece this together the other day, too. So, Frank was there earlier and got the bodies and then they took off. Then we got there afterwards, and ended up at that house. I remember I was up on top of the rooftop. I was there with Doc Clayton, Captain Royer and some other guys. There were hajis running from the far side of Nova. If you remember Nova in that area, on the other side of the road it kind of went down. There was an embankment there, like a dike near the river.

At least three guys came and ran out of there and over by the buildings on the far side of Nova on the right hand side. The Captain said, "Fuck it. Light 'em up." I asked him something stupid like, "Which one?" He must have said, "All of them" and I just let off. I don't remember anything after that. I remember the sun was going down. I remember snippets of that day. One of the really clear things that I remember was the sun was going down. The sky was bright orange. There was a mosque on our left-hand side and then we heard them start blasting evening prayers. I remember thinking to myself, "This is so fucking surreal. I gotta be dreaming right now. This has got to be a nightmare. This hasn't happened. It's just so beyond fucking weird."

Janney: Yeah, you've been fighting all day and then all of a sudden, the sun is going down, a beautiful sunset and then you hear the call to prayer.

Victor: Yeah, the call to prayer. It was quiet before that for a while. It was fucking weird. I checked and I only had one magazine left. We stayed there at that house for what seemed like forever. Some Marines came and eventually rescued us. I think that was Frank and those guys that came to the house. That's when they found Jerabek and then we took off after night fell. We went to the chow hall when we got back and I remember eating like a fucking zombie. I kept thinking, "What the fuck is going on?" I didn't sleep at all that night and the next day we went out to the Fishhook Lake. We chased them down in the Fishhook. That's what I remember about the 6th. I wasn't near Lt. Ski or anybody like that, so I didn't actually see a lot of the bodies up close. I just saw a lot of dead Iraqis.

Janney: Sure. I'm glad that you didn't, and I wish Frank had not had to. I know it was a terrible day. Let me ask you guys together, there was an Echo Company squad that was made up of Cherry, and Roy Thomas, who lost an eye in that engagement. There was a 2/4 Echo squad that pushed west on Nova that actually broke the back of that engagement where those thirty guys were trying to kill Stayskal, Santiago, (redacted), and Ferguson. If Carmen and Cherry hadn't come along with their 240 blazing atop that Humvee, Santiago said he and his team would probably have been killed. Stayskal was already down with a serious chest wound, and (redacted) had an arm wound. Do you remember? I know it happened because Santiago and those guys told me, but where did those guys come from? They obviously weren't part of the QRF that went up Gypsum to Nova, but they came down Nova from the direction of Gypsum. (Author note: This interview with Victor and Frank took place before I interviewed Brandon Lund and Ramon Barron who were members of the Echo squad on a Nova IED sweep that formed a makeshift QRF and rescued Headhunter II. Quetglas, Lund, Barron, Martinez, Pedro Contreras, Carmen, and Cherry were some of the members of the team that drove their Humvee into the midst of the attacking insurgents that were enveloping the sniper team and saved Headhunter II. Tragically, Carmen and Cherry were KIA during that.)

Frank: I can answer that question, partially. So, from my understanding, there was actually a QRF sent before us. I could be getting it wrong, but I think either first platoon or third platoon were part of that effort. The other machine gunners that I talked to were already halfway there. They were already past all that and took fire, too. I'm going to guess it was them because Roy Thomas was there. He was part of another platoon that must have already gone up. From what I remember, unit wise, our QRF was not the first QRF to go out.

Janney: Hey Frank, let me run something by you. It's something that just popped into my head. At some point an earlier patrol walked by Santiago and those guys found an IED. Thomas and those guys had called EOD, but then they also heard the radio call for a QRF from Santiago, so maybe they were sent to deal with the IED, but they became a type of QRF?

Frank: It could have been possible now that I'm thinking about it. As part of that QRF call, I was told that some machine gunners had already been hit, too. I think it was Martinez and Carmen. They must have gone to help Santiago before we could get there because they heard the QRF call while they were already out there running an IED sweep.

Victor: Yeah, if it was like 8:00 – 10:00 in the morning, it would make sense that there was an IED patrol out.

Janney: Okay, so they were IED patrol and became a QRF.

Frank: Yeah, they were probably already close enough since they were already out there. It's probably what happened. I know there were already Marines, especially machine gunners, under fire before we got out there. I'm pretty sure that happened first. I know the machine gunner who was there, and he was part of Vergara's platoon. I don't think Vergara was with them – he was on Gypsum. Maybe another machine gunner, Ryan Miller, was there? Machine gunners were getting hit like crazy.

Janney: That's what I understand. I said Cherry, but it might have been Benjamin Carmen. I know Vergara and Ryan Miller were on Route Gypsum near you guys during the Gypsum ambush.

Frank: It might have been Cherry. I'm pretty sure Carmen and Martinez were the machine gunners. So, if they had been hit, someone else had to get on the gun.

Janney: So, at that point Cherry got on the gun and he also got hit?

Frank: He might have, yes, because if there is a downed machine gun, somebody's got to get on that thing. Lays down so much volume of fire that someone's got to get on it. The machine gunner was probably down at that point. I am remembering that now that I think about it. Before we even got there, I had heard that machine gunners were going down like flies.

Frank: That's what it sounds like. I believe it.

Janney: Victor, I'm sorry I interrupted you again. So, we think we solved the mystery of the squad that saved Santiago and those guys - Marcus Cherry and Benjamin Carmen were involved in that effort. Tragically, both of those Marines were killed in the process. Tell me, you guys were on top of that building with Royer and you were just kind of keeping the area secure after the gunfire had died down. The call to prayer at sunset was surreal and created an eerie feeling.

I know an Army Bradley rolled through that intersection at some point. Frank, maybe you remember that. Earlier in the day, it came along, but retreated because one of its soldiers had his arm blown off by a RPG. They didn't have time to rescue anyone or have much effect on the gunfight. Otey, Evan Null, Matt Scott, and Shore were holed up in a little building next to the intersection.

Victor: I don't remember.

Janney: Somehow, they got a ride back to Combat Outpost. I'm not sure, maybe I'm mixing that up with some other incident, but do either of you guys recall finding those three or four Marines in that building and giving them a lift back to the Combat Outpost?

Victor: I don't remember that.

Frank: I don't remember. Yeah, I forgot to mention that earlier, but after we loaded the downed Marines, we went back to Combat Outpost. We grabbed more Marines. I was talking to Victor, I remember Musser joining us. I remember Musser getting on the vehicle because he said a Nordic prayer. I remember that was the sort of place that actually does make some sense. We went back out, but it's really blurry to me. I don't know if that's maybe when we would have picked up more people. My brain was full at that point.

Janney: The one thing that I do want to do is see if we can push through and get at least through the 10th, Operation Bug Hunt. What do you guys remember of the rest of that period?

Victor: Bug Hunt revenge.

Janney: The 10th was the day that was basically just a citywide push. The whole Battalion went through kicking in doors and looking for caches and insurgents.

Frank: I just have one part I forgot to mention about going back to the Combat Outpost to get more Marines and drop off all the Marines. That's all I really remember about the rest of that day.

Janney: Sure. That makes sense. Victor's kind of wrapped his day up with the call to prayer. That's a pretty eerie end to the daylight hours of the day. That's chilling. I can hear that in my mind and it gives me chills just thinking about it. The next day obviously you guys were taking stock of what happened and there's got to be some kind of action to show the insurgents that, "Hey, Marines are not afraid to fight. We're not soft" like some of the Iraqis thought previously. What do you guys remember from that day?

Victor: Again, I really don't remember the 7th. I remember we went out right away. I didn't sleep at all on the 6th. I remember just looking at Crowley's rack and just zoning out like, "What the fuck?" We went out on the 7th to the Fishhook, moving to contact essentially. Finding these guys and hunting them down was the mission. I don't remember anything particular about that day though.

Frank: One thing I remember about the days in between the 6th and 10th is that we stepped up. We didn't get much sleep. We decided that we were going to keep going to that area to see if we could find more enemies to kill. I can't remember if it was the 9th or the 7th, but that's when Ayon passed away. I was on the QRF for that and was really sad about that because I actually got to know Ayon quite a bit in our patrols we had before since I was on the machine gun all the time. When I went to that QRF and found out about Ayon, it was pretty soul crushing as well. (Author note: LCpl Eric Ayon was KIA near the Gypsum/Nova intersection on 9 April 2004.)

Janney: Yeah, it's so ironic that Swanson took a picture of him trying to move that vehicle at Nova and Gypsum, too. Swanson shot that photo through the windshield as if to say, "I'm going to make you famous" and then, tragically, a couple of days later, he's dead. That photo is haunting anyway because of the juxtaposition of Ayon within a vehicle that 8 of his friends were just killed in.

Frank: Yeah, he was expecting a child at that point too. He had told me about much he was looking forward to all that. He told me about how much he loved his wife.

Janney: I had no idea his wife was expecting. That's even more terrible.

Frank: That's all I really remember about that. I remember going out to Gypsum a lot in those days. Like I told Victor, I did a QRF and they didn't have a machine gunner, so I ended up staying with another platoon I don't normally stay with just to be a machine gunner. I don't really remember too much about the days in between other than almost getting hit by a piece of concrete from an IED that they exploded over there. I thought, "Whoa, those fly really far. I almost had a meteor hit my head."

Janney: To your knowledge, were they using command detonated IEDs or were they pressure plate?
Frank: No. I was with the machine gun, so I didn't get that close to it. They detonated another IED they found because there was the one that killed Ayon. We were still in that area and they had found another one, so they detonated it. I just remember all the chunks that came up because it was a particularly big one.

Janney: Wow. Any idea if you were dealing with any foreign fighters or were they locals?

Frank: I got a good look at some of them, but it's not like I know what the enemy looks like other than wearing gear and having a weapon on them. I couldn't tell you where they were from. They weren't us and they were trying to shoot us, so we shot back.

Janney: I just didn't know if whoever was in charge of that Company or squad might have searched the guys for papers and said, "Yeah, these are Syrians." I've heard varying things about that. It's usually the officers that have information. I didn't know if either of you knew.

Victor: When we took detainees, if they had identification, we would write all the information on the back of their shirt: where we found them, what they were arrested for, what their names were, if we found IEDs, but as far as what nationality they were, I don't think we really knew a whole lot, especially during that time. We couldn't really tell them apart. Like you said, a lot of them were probably fighters from Fallujah. I think a lot of them were people that lived in Ramadi and thought, "We get to kill Americans. Let's go ahead and take our shot."

Frank: Yeah, I think it was a combination of both. I know for sure some of them were local. Again, I could speculate that they just took the opportunity since people were already shooting at Marines to get in on the fun. Because there were some people that I definitely knew for sure were local. It was an opportunity for them to get their shots in at Marines. I'm sure that some of them took advantage.

Victor: On the 10th, I might be mixing up the beginning of the operation with the end. During the operation, we ended up in a house where we were pretty much pinned down next to an open field. David Swanson was somewhere around us. I remember we were taking fire from three sides of the building and then there were other Marines in a bigger building across this little canal. The building was like two or three stories and then the one that we were in was one story.

Janney: Was Royer there along with Ryan Miller?

Victor: No, this was before they got there. We were fighting from the building and heard over the radio that Royer got hit in the head.

Victor: I don't remember much about the beginning of the day. Did we start that mission with "Hells Bells" or not?

Frank: I'm pretty sure the psyops were going, so yeah.

Victor: I remember we kicked off super early in the morning before the sun was up. I was in a position where I was watching over this house and the family was sleeping outside. Some Iraqis do because it's hot inside, so they were sleeping outside on their lanai or porch. I was in this bush just watching this house and then boom, the mission started. Artillery and mortars began firing illumination rounds and then "Hells Bells" started playing. Then "boom" - you started hearing motherfuckers kicking in doors. As the lumes were going over and the music was playing, this family started waking up and started sitting up looking around. I was looking through my ACOG. As the illumination would go over every couple of seconds, it would light up my position and then I would just drift into the shadows again.

Janney: Yeah, that's very cool. Just seeing them being lit up as the lume was set off.

Victor: I remember that mission then. We were just pushing through houses and then we ended up in that house after Royer and those guys came out of the shit trench.

Frank: No, maybe not because that was Swanson. Swanson was with us; he joined us after.

Victor: We ended up at a single-story house and were taking fire. We got the call that Royer got shot in the head and then they came back over the radio and said, 'Oh, he's fine." What the hell, that was crazy. They ended up coming into that building we were in. They got hit and then were low crawling through what they call a canal, but really it was a sewage canal. I don't know if you remember those in Iraq? Irrigation ditches full of shit. Yeah, it wasn't a canal. Trust me. They came through and then one of my friends, Geoffrey Melvin, came in and they were just covered in shit because they were just low crawling through one. They got to our position and came into our house. This was the first safe position for them, so they were just exhausted and thirsty. I remember Melvin coming up to me and his face was just covered in shit and he tried to grab my Camelbak to drink out of it. I said, "Get the fuck away!" I think the tip came off and I said, "Okay, I'll waterfall." I opened it up and waterfalled it for him so his shit covered face wouldn't touch my drinking water.

    We ended up leaving that house and hooked up into another three-story building. We stayed in that position for a long time. I remember the canal was pretty wide and a lot of guys were trying to jump over it, and they were getting stuck in the canal. When it was my turn to run across, I was running and when I got to the other side, somebody was watching and they said that when I was running, they could see rounds hitting right behind me. Someone's rounds were chasing me down. I didn't have any idea about it. I said, "Oh, that's fucking cool, I guess." We ended up in that house for a long time. The main thing I remember about that house was that Litke was there on the rooftop. Lomas was there and one of them shot a cow with a 203. This house was in front of this big field, and we were trying to get Army helicopters in to evacuate one of our guys, Sims, later on.

We were in that OP and we found out about the other guys and Sims in it. We got them out of that situation, and we had them stabilized on the street because we were still taking fire from across the field. It was kind of like the same thing with Route Nova. There was a little bit of a ditch, so we were kind of huddled behind this ditch and we were all shooting from there. I remember Doc was working on Sims.

Frank: Everybody was taking turns doing CPR trying to keep him alive and I think we eventually got him in that house, right?

Frank: Yeah, we got him in the house.

Victor: I remember during that time when we were still outside is when Swanson got shot and he was pissed. He got shot and it grazed him. We were non-stop from the 6th and I don't think any of us had slept at all. We were stuck in the house for a long time and on the bottom floor, there was this room full of pillows and this TV that was on. We were watching MTV or some shit like that and taking rounds through the windows, but we had guys securing the house. We had guys at the front door, the back door, guys up on the rooftop, and then we had so many guys that we were trying to get a little bit of down time during the firefight because you're constantly on the gun sometimes.
There were at least five of us down there and we just passed out. We were just so exhausted. Rounds were coming through the windows, we were still in the middle of a firefight, but we were just so exhausted that a few of us nodded off.

I remember Gunny came through and looked at us. Usually, he would chew us the fuck out, seeing guys just racked out, especially during a firefight. He looked at us and I think he was as tired as we were at that point, and he didn't say anything. He just went to the rooftop and made sure that we had security. Later that day, we ended up pushing off the rooftop after the Army eventually came and got Sims. They came in with an AAD, I think it was?

Frank: Yeah. Some sort of armored vehicle.

Victor: They came and got him, and then we pushed out across the field. Then, almost all of the Marines had either a cigarette or a cigar in their mouth during that push across that field. I remember everybody was like, "Fuck it. If we are going to die today, then we're going to smoke." I remember everybody, even the people that didn't smoke, were smoking. I think even Lomas smoked that day. We ended up pushing across the field to where we were taking fire, a move into contact kind of thing. We ran into this one house and cleared it. That's pretty much all I remember about the 10th. I don't even remember coming back to base or how we got back.

Janney: Frank, what do you remember about the 10th?

Frank: As Victor was saying, I do remember the beginning of it being surreal because that music came on, the flares were being fired and were drifting down. It was very moviesque. "Hells Bells" playing and I just remember thinking, "That can either be a really good song or a bad song to be playing right now" because the Marines are really pissed off. It was morning and we're amped up and people are going to be breaking some doors down. I remember looking at that and thinking about it for a second, but after that morning part, I don't remember too much. I do remember that it wasn't until around daybreak that the shooting actually started. I started hearing lots of gunfire after the sun came up.

Victor: Yeah, it was later on.

Frank: Yeah, and I was next to a vehicle that had detainees in it. So, they had already detained people by early morning. Swanson and Royer must have already been hit earlier because I remember Swanson joining us where we were at. He was covered in mud, looked pretty dejected and was smoking. I think that is shown in one of the pictures of him. We had makeshift security around this vehicle where we had these detainees. As the firefight was going on, you could hear some of them talking. But, as the distance closed, you could hear the gunfire getting closer, kind of like a thunderstorm approaching. I remember doing Marine Corps logistical checks in my head because I didn't want to run out of ammo. On the 6th, I didn't even have a grenade. I didn't have a normal grenade – I only had smoke and I almost ran out of ammo. A lot of us were running low on ammo on the 6th, so when this firefight started, I remember looking at everybody and checking myself. I'm like, "Dude I've got grenades. I've got smoke. I'm full of ammo. I'm ready to go. We're ready to go this time."

    I remember going around and checking other Marines to make sure they had ammo because the other machine gunner who was in charge, Corporal Lenz, was on the Mark 19. That's the automatic grenade launcher and he was doing his best at suppressing whoever was firing rounds at us. I remember Maxwell, but he was going crazy with his SAW, squad automatic weapon – he was shooting, he was suppressing. I remember sitting there for a little bit and taking stock, not really seeing anybody, but keeping alert and making sure everybody was good. That all changed because we went into a house.

    At this point, it was a weird break that I had in the middle of the gunfight, because I still had warming layers on because it was really cold that morning. We got into this house and I thought, "I've got to take a position." I started coordinating fire and thought, "I'm really fucking hot right now. I do not want to pass out from heat exhaustion in the middle of this. Do I really want to take off my gear right now?" I had a full fleece on under my gear and long johns, so I went up the flight of stairs that had a landing, but saw a window. I went further up and then took all my gear off, took the warming layers off, and then put my Kevlar back on. I said, "Okay, good to go" and I went up to the roof.

    I don't remember how we all got there, but Sgt. Leaky was there, my gunner Calavan was there, and I remember Litke being there. Then it became like a fucking video game because there were heads popping up and gunfire going off, all in different directions. I had an ACOG on my rifle. I was one of the few people that were lucky enough to have an ACOG 4x scope, so it was like a video game up there. We were taking a lot of fire and also had the machine gun up there. David Swanson took some really good photos on that roof. At one point, we were worried about getting overrun, so all hands were on deck on top of that roof. I remember telling David Swanson, "I don't know what's about to happen here, but it's looking pretty bad right now." I'll never forget the look on his face because I didn't know if we were going to get overrun or not. I've never seen anyone's eyes that big. My gunner kept lighting people up with the machine gun. I had a lot of ammo because I didn't want to make the same mistake as last time, so we had plenty of machine gun ammo and as long as that thing didn't go down, I wasn't too worried about it.

    At one point, we got the call that they were going to be bringing in a medevac to get Sims.

Corvedo and I got all the Marines on top of the building and we laid down suppressive fire. We coordinated and fired all together. It felt like a movie. We laid down a shit ton of suppressive fire so that the Corpsmen and Marines could cross this really big open field where those cows were at and got blown up. We had to clear the field because we thought we were going to get a helicopter for Sims. There were a lot of us on that building. I was on the top the whole time. That's why when Victor was talking about chilling at the bottom, I thought, "That fucker!" We were all out there shooting and keeping security on the top and this guy got to take a break. Lucky bastard.

We had security on top of that building and then they called in to give cover for Sims. We got Sims in the building and at that point there was an 11, his name was Litke, so he was decently friendly with Calavan who was my machine gunner. I told Litke to take his 203 so he could cover. Tactically, that is a good combination to have a machine gunner and a guy with a 203, so he could cover the dead space with the grenade launcher that the machine gun can't hit. So, he stayed up there with Calavan. I went downstairs to go see Sims because I had done my first deployment with him. Doc Gutierrez and a bunch of us were all taking turns doing CPR on Sims.

I remember going back up to the roof. I just remember most of my day being on that roof and it was like a video game other than that horrible moment. There's a bunch of pictures of it that David Swanson took of all of us up there. Of course, the one photo I'm in is not flattering, but who cares. I look stupid as fuck. Also, I took a picture of David Swanson with Corporal Brown up there, too. All I really remember about that day is really being on that building, on that rooftop and just lighting people up for quite a long time until the Army showed up to get Sims out of there. It looked pretty bad at one point because that building was right in the middle of a huge firefight. From what I could tell was happening, there were enemies everywhere. I mean we were doing fine up there. No one was really getting hit. By then, I think, since everyone was tired, and we got that first one out of the way, I feel like the Marines really became Marines because they knew what was up. Everyone was moving a lot better; it was a totally different fight that day to me.

Victor: Because we were hunting prey that day.

Frank: Yeah, we were hunting and I was prepared for it. I had already had in my mind what was going on by then. Yeah, I think most of the Marines got some good time there because I just remember after we got back everyone had a story. Every single person had some sort of story about that firefight.

Victor: Yeah.

Frank: It was quite the day.

Victor: Not the way they did the 6th. They started taking potshots and using IEDs. They would fight us for a little while, but then they would take off. They never went full-fledged at our ass anymore.

Frank: Yeah, after that it was more hit and run stuff, exactly.

Janney: Talk about your brothers that didn't come home with you. You can talk about Crowley. I know Frank may have some great stories about Sims, and I've heard some really amazing things about that young man. Frank, I think you mentioned that you were with Otey at some point.

Victor: I knew Ski pretty well.

Janney: Since you were his RO, you obviously spent a lot of time with him. Maybe more than anybody else I've talked to. I know he's a Lt. and you are a PFC or Corporal, but yet you still have to work as a team, so I know he shared some personal stuff with you at some point. Tell me what you remember about Ski.

Victor: I remember him being a no-nonsense kind of a guy. Really responsible. Really on his shit. I looked at him like I thought he was the best Lt. in the entire Company. What you think an officer should be. He embodied that to me. To be honest, I was a boot, so I was afraid of the fucking guy. He was always business most of the time. He was on his shit most of the time. That's mainly what I remember about him. He was pretty fair, but he was stern. He would get on my ass if he thought I was fucking around, like throwing pineapple grenades. Overall, I liked the guy. I think everybody else liked him. I don't think he had any enemies. Crowley, I knew a little bit more. We were friends for a little while. I didn't know him extremely well because he was senior to me. He was one of the boot drops I think, so he was a couple of months before me so he thought he was a little bit saltier.

Janney: He was a Lance, right?

Victor: Yeah, he was a Lance Corporal. I was a PFC at the time. I don't remember if I got promoted in country. I got promoted to LCpl in country. He was actually my team leader. I remember that's a big deal for a boot to be a fucking team leader, you know. That's huge. It was me, him and Matt Scott. That was our team.

Janney: Oh yeah. Matt and I talked a number of times. I like him a lot. He's a good guy.

Victor: Yeah, yeah. Scottie's cool. He's got sources.

Frank: Yeah, he's got sources.

Janney: Matt was on Gypsum. He was one of those guys in that second vehicle that was in the building with Otey now that you mention Matt. Yeah, Matt Scott, Null, Shores, and Otey were in that building together.

Frank: Scott was stuck with Otey and Null.

Janney: Exactly. Matt had a bad day, but survived. Everybody did, but those four guys in particular were stuck there for hours. Nobody had a good day that day.

Frank: Those are two good guys to get stuck with. I did my first deployment with both of them, so I know Otey and Scott.

Victor: Yeah, Scottie is funny.

Janney: Yeah, he's a good guy. I think the world of him. He just had a baby not too long ago.

Victor: Yeah, he called and told me. Crowley and I were friends. He was from Oakland and I was from Sacramento, so we kind of had that in common.

Frank: Most infantry Marines have some sort of thing that they bring into the Corps with them.

Victor: Yeah. We got along pretty well. We were always on patrol together. Again, he was my team leader, so me, him and Scott, we spent a lot of time together. Him being a boot team leader, he kind of looked at us and said, "Don't fuck me up." So, it was a team effort with all of us. It wasn't like he was telling us what to do all the time. He was a really impressive Marine for his age. I'm pretty sure if he was alive today, but he definitely would have been a squad leader at some point in time. He was that kind of guy. He was on his toes. He was good.

Janney: I heard that he really liked music. You told me about him dancing for the kids.

Victor: Yeah, we really clicked on that. We listened to a lot of music. He grew up in Oakland, so he didn't really grow up around country white dudes. So, we would make fun of the country white dudes together. People listen to country music and shit like that and we would hassle them. The funniest story that I have with him is probably with the pineapple grenade that time on the 5th. For sure, that was one of the highlights.

Janney: I can't even imagine. If I had been Swanson and saw you guys doing that I'd have said, "What in the hell do you think you're doing? You're going to get blown up."

Victor: Yeah, we were always doing stupid shit like that. Like Guti said, he was always joking around with the kids. He was an overall good guy. Not a whole lot of people knew him because, again, he was a boot, but me and Scott knew him well.

Janney: Yeah, and that's what I found. A lot of the boot drop guys came in late and nobody had a chance to get to know them before April.

Victor: That's kind of the thing about it. Even within the Company, there's a lot of guys that I don't know because you get into the unit and then you go into war, your entire world revolves around your squad and platoon.

Victor: Yeah, you don't really interact with a whole lot of other people up until that point. I knew guys in Golf from boot camp and SOI. A lot of guys that are mentioned in Joker One and those documentaries, I know those guys for the most part, but you only really know the guys in your squad and your platoon.

Frank: The other thing is you get to know people on your first deployment. Unfortunately, our first deployment was in a combat climate, so that's usually when you learn who you're working with the most when you first actually start doing training ops and then you deploy.

Victor: Yeah, you kind of know who's good and who's not good.

Frank: That's what quality is about. They perform.

Janney: That's a tough way to get to know somebody, too. You find out quickly who's good and who's not. Victor, who else can you think of that you want to tell me about?

Victor: Again, those are probably the two that I knew the best from that deployment. We had other guys that we lost in another deployment, but for that particular deployment, those two and then Calavan. I knew Calavan pretty well. He was really a kind of goofy guy. When we got back one time, I just remember him sitting in his rack watching Friends and just giggling to himself like a schoolgirl.

Victor: I knew Simon really well. I've got a lot of stories about that guy. You could write a whole book on Simon Litke.

Janney: I heard that about Pedro Contreras, too. Did either of you know Pedro?

Frank: Yeah, I knew Contreras. I have a funny story about meeting him.

Janney: I'd love to hear it. I have a story that I heard about him wearing a fireman's hat at PT.

Frank: Sounds like some shit Contreras would do. When I first met him, we were having a field event that's boxing and pugil sticks. I'm brand new and I don't know anybody, and my last name is Gutierrez. He sees my name tag and says, "Hey, Gutierrez, come over here." I said, "Hey, what's up?" He introduces me to all of these Mexicans. He asked, "Do you speak Spanish?" I said, "Nah, I'm a poser Mexican." He said, "Get the fuck out of here!" Laughs.

Janney: That's hilarious.

Frank: Yeah. It's pretty funny. I'd see him around after that. When I was in the Marine Corps, I didn't really hang out with too many people. I kind of stuck to myself a lot of the time, but I did get a lot of pictures of those guys. Contreras was another one of those funny motherfuckers.

Victor: He was always fucking with me.

Frank: He was like the Latin version of McIntosh. He goofed around a lot.

Janney: I heard that he was the class clown basically.

Frank: Yeah, he was a funny guy. But, I'll never forget that he rejected me. I had a pretty good laugh about that.

Janney: Laughs. Just because you didn't speak Spanish.

Frank: Pretty much the same thing that Victor said about Crowley. I remember him being very, for lack of a better word, spirited. When we were on these patrols, especially since all of these guys were new, it was kind of a drudge for a lot of them, but Crowley always had energy. He was always into what he was doing, and he brought good vibes to people around him. He just had good, full energy about him. He always made patrols fun. I didn't get to know him that well because he came in, and these guys came in really briefly before we went to combat. Those few patrols you're out there with everybody, you get to know people pretty quickly. You're walking out there and you're just talking to people or you have to sit because we found something, and then we end up sitting there for a while. Just like Victor was saying, he was a team leader for a reason. He had good awareness, good energy and solid presence for a young Marine. Understanding what was going on around him. He probably would have inspired a lot of Marines if he had more time.

Janney: I've heard so many great things about Crowley. You said you knew Otey a little bit.

Frank: He was part of third platoon and my first deployment, like I told you, weapons platoon is tasked out to particular platoons based off of their weapons systems. So, on my first deployment, I was with the third platoon and Otey was in the third platoon along with a bunch of others. Again, I didn't hang out with a lot of people. I used to stick with myself, but you talk to these guys and you spend a lot of time with them doing these training ops. I remember him, McCall and another Marine named Henry and I have a picture of them all. But you know, it's just like any place, there's cliques. You've got like the little Mexican clique, the country white guy clique, the black guy clique. We have our squad base on these ships. On these ships there's no room. They call them racks for a reason. Have you ever seen a military ship berthing area for Marines?

Janney: I've seen pictures, but I've never been in one.

Frank: Like literally, the next bed is above you by three or four feet. You sit up and you're hitting the other person's rack. We're all crammed into this little area and you open the door and where the stairs lead to the rest of the ship, there's these four black guys just hanging out, Otey among them. Having a good time, being them. Him and this guy named Henry who got out, but was a senior Marine to me. They would just be talking and having a good old time. It was fun training with them.

Janney: Who else can you remember?

Frank: Well, especially with the new Marines, I met Chris Cobb.

Janney: Yeah, I know of Christopher Cobb and I interviewed his mom as well.

Frank: Yeah, I got to meet him at the machine gun range just before we left. One of the few quick trainings we did is we took all the 11s who were SAW gunners on a course to learn how to use 240 in case any of the 240 gunners went down. Also, to show them that their weapons systems can be adapted with ours because we carry the tripod with a traversal elevation mechanism. It's just a fancy name for this part that connects the tripod to the machine gun and has some notches on it that allows you to control going up and down. The SAW is able to connect to this same system as the bigger machine gun, the 240. We got to do a training op with them and Cobb was one of those guys. He was kind of a goofy dude. He was a little lost like a lot of new Marines are because I was trying to direct him on to fire. He couldn't hit the target worth a crap when I first got a hold of him, but by the end of the day he could lay a good burst on some pretty far targets which was pretty cool to see. Especially with a new Marine when you know you're going to go to combat soon. That's the quick memory that I had with him. I got to meet Layfield for the first time in the barracks.

Janney: Okay. Yeah, tell me about Layfield. I know his mom pretty well.

Frank: Yeah, so see this is almost kind of a sour thing for me a little bit, but I met him in the barracks because I was listening to some heavy metal music and he came by, walked in and said, "Hey, Corporal. I didn't know you knew this band." It's called Loco. It's not a very popular metal band, but it's pretty good. I said, "You know who they are?" "Oh, yeah." So, we started talking and I found out that he was actually from the same city as I was in California.

Janney: Oh, that was weird.

Frank: So, Crowley was from Oakland, Victor is from Sacramento and Travis is from Fremont, which is where I was from, too. So, we started talking a bit to each other because he went to Washington High School and I went to American High School. These two schools are rivals. Same district and the same schools see each other all the time and we used to give each other shit because the American's mascot was the eagle; his was the Huskie. I said, "Man, my eagle would fucking rip your Huskie apart." I got to hang out with him a little bit and just talk with him. I told him when we are on leave when we get back from this deployment, we should hang out. Corporals are not supposed to hang out with lower Lance Corporals or below. They do, but you're not really supposed to – it's an etiquette thing with leadership. He said, "Hey, when we get back to Fremont, we'll hang out." Unfortunately, he never got back home. My mother worked at the same plant as his mother because in Fremont before Tesla took over, there was a huge car plant there in Fremont, the General Motors Company, so a lot of people from Fremont, that's their job like Detroit and other areas. There's a car plant that supplies a lot of jobs for people. So, Layfield's family, some of them, worked there, too. My mom had talked to them, but I never had the courage to talk to Layfield's mom afterwards.

Janney: Yeah, I understand. His mom is really sweet. You would love her though.

Frank: Yeah. I've seen her car go by because she has that memorial on the back of the truck. I just couldn't work up the courage. There was another Marine who was there, Corporal Bum Lee. I don't know if you heard about him. I did my whole first deployment with him and he was one of the people that I actually hung around a lot. I have a load of pictures and I'm actually really good friends with his sister. His sister is actually my child's godparent. I couldn't have talked to Layfield's mom because I went to bring Lee's stuff back to his family and that experience was just too hard. Actually, meeting the family was one of the hardest things I did, even harder to me than combat. So, the thought of even talking to Layfield's family, I just couldn't bear to do it. So, I never did it. I've just recently become a father. When I went to bring Lee's stuff back to his family, the first thing his mom did was smell his clothes. As a parent now, I can understand why we do something like that.

I didn't really have much courage to talk to some of the families because I also made a poor attempt at meeting Calavan's family. He was my gunner and I was there with him when he passed away, and like Victor was saying, he was a goofball. He was pretty funny. You mentioned that you had talked to Sgt. Damien Coan. If you ever talk to him again, he'll verify this story. The very first time Calavan came to our unit, he said, "I'm looking for Sgt. Coon. I'm supposed to find Sgt. Coon." Sgt. Coan was right there and said, "You mean Sgt. Coan?" Calavan said, "Oh yeah, Sgt. Coan." Yeah, Calavan had a really deep voice too, which really made it even funnier. He was kind of a weird guy. Another funny thing is, if you ever talk to Bennett, he had a funny "interaction" with Calavan. Bennett was a 51, an assault man. He was pretty new to the unit, too. Calavan had come in just before and found out that Bennett did jujitsu. Calavan said, "Oh, I can take him." I said, "Calavan, you're dumb." A person who knows how to fight and a person who thinks they can are two different people. He still tried to take on Bennett and Bennett beat the shit out of him. So, that was pretty funny, too. He had a lot of bravado. Calavan and I talked a lot. He used to tell me about these people he cared for back at home and about his family life.

Victor: I went to boot camp and SOI with Jerabek. I didn't really hang out with him too much.

Frank: I do have one great story about Jerabek that everyone will like. When we got to Kuwait, we were doing some training with weapons platoon. We were doing some hand-to-hand training. We knew we were going to be doing those sustainment operations, basically police work. We needed to refresh on hand-to-hand stuff because you're probably going to be touching people. One of the drills we were working on was just getting peoples' hands off the muzzle of our rifle by doing a circular swiping motion which is a really basic maneuver. Jerabek had a grip of steel. No one could break his grip. Everybody was talking about it, "We can't break this guy's grip." So, Captain Royer came over and said, "I'll break his grip." We all thought, "I don't know about that."

Royer lets Jerabek grab the muzzle and he starts full on yanking his whole body, throwing the rifle around with Jerabek attached, flinging him around everywhere and he would not let go. He could not break his grip. It was really funny because he was pulling Jerabek back and forth, up and down. Royer really went for it, but Jerabek could have been a fucking rodeo bull rider. He did not let go at all. He didn't give up. His arm was bending backwards and in ways he couldn't even probably bend, but he wouldn't let go.

Janney: That's amazing and hilarious. So, what did Royer do when he couldn't break his grip?

Frank: Royer looked really dejected and just walked away.

Janney: Laughs.

Frank: It was great. We all did a high five. Jerabek said he wouldn't let go and he did not. Good job.

Janney: What a great story! I bet Ryan had a big old grin on his face from ear-to-ear at that point.

Frank: He had that silly smile wearing those military issue (BC) black glasses. So, that's what I remember about Jerabek.

Janney: There's another story that I'll tell you real quick. It's not funny, but on 6 April when he went out on that QRF, somebody gave him the 240. Jerabek was so excited and said, "Guess what! I get to be a machine gunner today." He said that to Doc Clayton and Ryan Miller before he headed out. He was just so proud of himself. You know he went down fighting. I've heard multiple people say that Jerabek was just hammering down with the 240 that day and didn't quit until his final breath. So, he really was a machine gunner that day.

Frank: I saw the fight in him. I know how much fight he gave, so yes.

Janney: Especially with him not letting go of Royer's rifle.

Frank: That made us proud.

Janney: Yeah, that's amazing. Can you think of any other guys that you have stories about?

Frank: I need a list. My brain is hurting right now. Quite honestly, I would not do this again. It's nothing against you. I don't really talk about this ever.

Janney: No, I understand.

Victor: Frank and I were trying to figure out how the hell Crowley ended up in the vehicle on Gypsum and Nova. Did you pick up on that in any of your interviews, or no?

Janney: No, but wasn't he one of the few people that had the certification to drive that thing?

Frank: Yeah, he was a driver normally.

Victor: He was a driver, yeah.

Janney: I think he was just one of the drivers, so he got tapped. Now, why was that the lead vehicle? From what I understand, it didn't start off being the lead vehicle. It'll become clear at some point.

(Author note: Crowley was a certified driver, so was one of the few drivers available that day. According to Sgt. Valerio, Staff Sgt. Walker ordered Crowley to drive around the other vehicles to become the lead vehicle. There is some question about whether it was 2ndLt Wroblewski or Staff Sgt. Walker, who wanted the 240 machine gun equipped Humvee up front rather than the SAW equipped vehicle that had some armor shielding around the gun as the lead vehicle. Tactically, someone wanted the heavier machine gun up front in the lead vehicle. In any case, the lead vehicle was driven by PFC Kyle Crowley, the 240 machine gun was manned by PFC Ryan Jerabek, the vehicle commander was Staff Sgt. Allan Walker, and the radio operator was LCpl Travis Layfield. Everyone in the lead Humvee was killed except for LCpl Deshon Otey.)

6 April 2004 Route Gypsum Ambush And QRF

Cpl Shawn Skaggs

Janney: It's 1st September 2019. When and why did you decide to enlist in the Marine Corps?

Skaggs: I had my first kid when I was 16. Graduated high school at 18, nowhere to go, and wanted to be a better father than my original father was.

Janney: So, you joined at 18? Once you got out of boot camp, were you always in 2/4 or were you with another unit and transferred to them?

Skaggs: Correct. I was actually an SOI guard in charge of the guard system on 9/11.

Janney: What was your actual job in the Marine Corps? Were you Infantry?

Skaggs: Well, I was an infantry rifleman to start, and as soon as I hit 2/4, I became an infantry Sawman, so a SAW gunner for my first two years.

Janney: When you guys deployed to Ramadi in February of 2004, what was your rank?

Skaggs: When we deployed in February, I was a Lance Corporal. Before we stepped foot into Iraq in early March, I got promoted to Corporal, and I was second platoon, second squad leader.

Janney: Who was in your platoon?

Skaggs: Eric Smith was the first squad leader, and at the time John McGoody was the third squad leader. Jonathan Brown was my first team leader. A guy named Maxwell was in my squad. I chose him as a SAW gunner. I had Tim Freggos who was a rifleman. I had Stiff as a rifleman. I can't remember all their names. There's 13 of them.

Janney: Once you guys got to Ramadi, tell me what was happening?

Skaggs: We had a rotation set up where we'd do a couple days on guard and a day where we'd go out and do an MSR sweep, walking 24, 27 clicks in a day, sweeping for mines and IEDs on the roads, all on foot. If the schedule permitted, you'd get a day off to do laundry and rest. You're always on QRF, so you'd either be QRF 1, 2, 3, or 4, depending on what we were doing. The early time was just patrols. We were just getting out, trying to figure out the area, trying to establish a presence in the AO. Let them know we're here and are the new guys that took over from the Army.

Janney: You guys were doing that in squad and platoon size elements?

Skaggs: Yeah. I did a number of squad patrols where myself and my other 12 Marines went out and we patrolled around for three or four hours at a time outside the wire. Depending on the actionable intel, we'd do a MSR sweep, look for IEDs on the road, and some intel advised that we may be attacked this day or need to be cautious of this area and we may need more manpower to go out with a platoon. Most of the time, it was a squad level event with yourself and your guys, maybe the Lieutenant or the platoon Sergeant would go with you to get eyes on to see what was happening.

Janney: So, basic security, SASO, hearts and minds type stuff and then looking for IEDs?

Skaggs: Yeah, the general gist. About March 28th was our first encounter with the enemy and it was ours as a Company. Third platoon, second or third squad was out doing a remote patrol and they actually ran into a mortar position March 28th and got in contact. One of the Marine machine gunners was shot in the leg and wounded and was evac'd. We ended up killing nine, captured one, and one got away. We seized the mortar position which the Army told us about previously, saying that they got bordered every Thursday evening at about 9. The Army just would hide inside their rooms and shelter and wait for the mortar attack to be over. Our CO wanted to be more proactive and we went out and started looking for them as soon as we took over. We found that position near our base and they had a bunch of machine guns set up to defend it, but our guys were better

Janney: What was happening in the 10 days leading up to the 6th of April? Pop shots? IEDs?

Skaggs: A little bit here and there, but you saw a lot more suspicious activity. It just kind of died off. I was on patrol with Lieutenant Wroblewski and David Swanson doing a MSR sweep on April 5th. We walked down Nova and detected some IEDs in the road. We backed up and set up perimeter security around it, waiting for EOD to come. Finally, EOD blew all the IEDs in place. We moved to the corner of Gypsum and Nova which was the big corner for April 6th. The Lieutenant was talking to David Swanson, and we're all keeping our eyes open because it's always beneficial for us and there was nobody out. Normally, there's hundreds of people in front of shops, mechanics have things going on, kids running the streets. We hit that corner on April 5th, and there was nothing. Nobody out. All the doors were closed. Everything was shut down. It was the oddest thing to us that there was nobody in the streets. So, April 5th was really an odd kind of indicator of something's wrong, something's going on at that AO, because people that are normally out are not here. Then, obviously, April 6th was the very next day.

Janney: Besides the IEDs that you detected, somebody found claymores strung in the palm trees?

Skaggs: My guys were on the flanks that day, but I don't know if we ever determined it was claymores, but there were devices in the trees.

Janney: I'm curious about what you thought about David Swanson.

Skaggs: It was mixed emotions at the time. I'm like, who the hell is going to come out here and try to do a story on us in the middle of nowhere? At the time, nothing was going on back then. Fallujah kicked off and the insurgents wanted out of Fallujah. It was weird, but he was a great dude. I mean, honestly, I think he was a down-to-earth, good-hearted guy. I got to know him pretty well the next three or four days after that. We were sitting side by side on April 10th. Captain Royer got shot in the head, in his Kevlar, and Swanson and I were two feet away. We talked about it a bit. We were on a rooftop together, bullets were flying, and he got shot. We were side by side.

Janney: That's interesting to know that you were with him that day. Charles Lauersdorf was there.

Skaggs: Yeah, he was our Intel guy and always ran with the CO. I was actually on a knee when that RPK opened up on us and started spraying the wall behind us. Luckily, they're not that great of a shot. They hit the wall and ended up actually hitting the RO's antenna. We all jumped into what I call a shit trench because they had this irrigation shit trench, the sewer trench next to the house. We jumped into a two-foot shit trench and laid on our back for 3 to 5 minutes as bullets rang out above us. We tried to figure out where to go and what's our best place to get around, but the head guy broke his leg jumping off the wall when the bullets started flying. We ended up having to stand up, turn around, and crawl face first, do a transfer about 100 yards or so to get back to the rest of my platoon, which was on the other side of the wall. Then we had to get up and go around again on the other side of the wall, and move to safety.

Janney: That's an ominous photo David Swanson shot that day. You can see all the muzzle flashes in the tree line across that field from you. What were you doing on the morning of the 6th of April?

Skaggs: Lt. Ski was that kind of role model with his Marines. He was the dad. I think after we did that patrol on April 5th, we got back and he was doing a debrief. I remember this clear as day, because we'd come walking in after spending about 14 hours outside the wire. We found the IED and set up a perimeter so we didn't get hit. It took forever to get that done, and then finish our route and get back. It was already past dark, and I was walking in the gate. Everybody was tired as crap. Captain Royer walked up to him and said one of the posts had seen something in thermal imaging outside the base. I think it was actually a herd of goats or some shit, but he wanted the Lt. to take us back out and investigate. Lieutenant Ski looked at me and said, "We ain't doing that. My guys just got done doing a 14-hour patrol, 27 clicks, and are dead on their feet. We're going to do a debrief. The guys can get something to eat and get some water. You can find somebody else to do it." So, Lt. Ski loved his men and stood up for them.

 We had different views during the debrief that night. He got a little irritated with me because my guys were on the flanks. My squad broke up into halves on each flank that day and I would walk up front next to the engineers. The engineers were the guys who were in the very front of the patrol searching for bombs. I'd walk up front and just talk to them and kind of talk to my guys on the flanks to make sure that they were okay. He kept getting irritated at me because I was so close to the front and broke it down to me logically that if a bomb went off, it would have killed an engineer and a squad leader. He kept wanting me to hang back, so we butted heads a little bit.

 The evening of the 5th, our team got tasked with a one squad patrol, 6 AM to 9 AM. I don't know if me and him butting heads was why he assigned me to that patrol. I sat up most of the night writing a five-paragraph order, getting everything ready, to get my guys ready. Eric Smith came to me and asked for Jonathan Brown to be one of his Humvee drivers for the next day for QRF. He needed an extra driver, and Brown was a decent Humvee driver and smart. I agreed to let him have Brown, who was my best teammate.

 So, the next morning at 6 AM, we got up. My other 12 guys, we go out on patrol. We go patrol around the river basin, checking pump houses for weapon caches. We're walking around Nova. We do a huge loop and we come back to base around 9 AM. There was nothing happening by 9 AM. We debrief 9:30 to 10, grab some chow, and go to sleep. I remember the morning of April 6 being dead asleep in my rack, and Jonathan Brown kicked my door in, screaming and yelling, "Get up, get your clothes on, and get ready to go. It's war. This is it." So, I jumped up, and I heard everybody outside. I see two seven-tons, and a couple Humvees come rolling in. It turned a corner and there was a stack of covered bodies in the back of it. Blood shot out all across the concrete and gravel as he was pulling around into our med bay. Right then and there, as a 22-year-old man, my heart sank to my stomach. I jumped back inside my room really quick, put my clothes on, and went from being in my underwear to being fully in battle gear in probably 30 seconds, less than a minute.

I ran outside, and Brown was already in the next room getting my squad ready, getting it worked up. I ran out and we immediately loaded the trucks. That's when Brown told me that Lt. Ski was shot in the face, had been medevaced, and they don't know if he's going to make it. So, you imagine looking at a guy that's like your father, you respect him like your dad, and you hear that he was just shot in the face. So the anger, the emotions, everything you feel, it's breaking and pulsing through your body. For lack of better words, you want some kind of revenge. You want to take the person that did that to this man that you cared so much about, and you want to put him in the ground.

We load up on the trucks. We loaded anything and everything. Gunny Coleman came out and opened up the armory. We were all getting extra ammo, extra grenades. We had no clue if fighting was still going on. We just went and it was over. It was mixed communication. I remember dropping off the trucks, and we rolled out there with everything and everybody we had. From that day on, we could carry every AT4 and everything else we could just to lay waste to the enemy.

So, we got there, and the intersection was clear. Nobody was there. We spent the next hour or two securing the area, searching the area, looking for any of the bad guys, picking up all the dead enemies, and putting them in a truck. That's when Swanson got that picture of Nathan Appel picking up one of our brothers. He put the body bag over his shoulder and he went to the truck. They all got murdered in that intersection. Marines, police and the enemy, dead. It's just a feeling that washes over your body that you don't ever shake. It's just ominous.

Well, after the 6th, I think it was the night of the 6th or the night of the 7th, Captain Royer and I met with Sergeant Coan who was our platoon Sergeant. He sent us out on patrol to re-gear our mind towards combat and not think so much about losing men. We also lost Crowley who was one of our Marines that day. They put us out on patrol to ease our mind and focus on the surroundings, so that we're not wallowing in sorrow. We did a patrol, maybe on the 7th, a couple hour patrol, just walking around at night. Again, no action, no nothing.

We supposedly had actionable intel that stated that certain people in certain houses were involved in the April 6th ambushes on our Company. Of course, we went ahead and geared up and got ready to act on the intel. We were rotating as a Company, and each Company, each platoon had a certain sector of houses to clear. We all rotated squads, so one squad would be the entry squad, one squad would be the inner security, and one squad would be the outer security. We rotated two houses, 10 of the 10. The very first house on one of the 10 was my squad. We got that because he said my squad was the roughest, toughest, and didn't give a shit. So, we hit the first house, which was our house, and quote, unquote, was involved in the April 6th ambush of our guys. We hit that house and didn't find a single weapon, single explosive device or anything there.

He made me continue house to house to house. By the time we got to the third house, it was my squad's time to be out on security. I was known for being out in front of my dudes. I was leading my squad across the field. When I got down to the corner around this brick wall, a block wall behind us, I was walking and I ran into Captain Royer, his whole entourage, and the Doc. I knew that our house was more secure, so they're gonna be talking to him about what was going on, and all of a sudden, firing broke out. We started getting shot at from every damn direction. That's when they shot off the radio antenna, hit Royer in his Kevlar, and the only thing we could do was dive into the shit trench. We're trying to figure out what the hell's going on. I remember Doc Clayton crawling across there, exposing himself to enemy fire to make sure Royer wasn't actually shot in the head. Doc Clayton went back and got under cover.

The CO told me to stand up and turn around until he crawled back to my position. I told him to lay down cover for me so I didn't stand up for nothing. I stood up, turned around, and started going back the other way. I was on the little PRRs, the headset radio, talking to my squad. Before we stood up, the CO had his rifle laid out on the ground. I remember standing up and hearing bullets explode over our heads, snapping, cracking, popping and hitting the bricks around you with chunks of brick flying back in your face.

As you dove down forward as quickly as possible, getting back out of the line of fire, again, now your face first in somebody's shit, and then we crawl back to the corner. One of my SAW gunners posted up on a tree, shooting. We got up, turned the corner, dropped on the road, and we were going to jump over. We tried to jump, but we couldn't. I'm wearing full battle rattle, covered in shit, and I go to jump this fence. I caught the tip of my foot on the fence and ended up going face first into the ground. Before I could move, I saw my buddy, Harry Smith, and he grabbed me by the "oh shit" handle on my back and started dragging me out of the way. I started yelling at him, "What the fuck, you gonna let go of me?" He grabbed me and said, "Man, you okay?" I said, "Yeah, I'm good." He said, "I thought you were shot, man. The way you went down like a sack of potatoes." I said, "No, I just tripped. I'm good."

I wind up getting everybody behind me. I was trying to secure a house which was probably about 100 yards away. The CO, right before he saw me, was kind of screaming, yelling, kind of disoriented. They were telling me to get back to this house. It was clear to me we needed to get back to this house. We started running back across this 100-150 foot field. About halfway across, I look back, and Doc is back there by himself, dragging the head guy that broke his ankle jumping off the wall. I run back to Doc and grab the other side of this guy and start helping him drag this guy across the field.

There's a chest-high water trench that they have dug for irrigation. We had the shortest guy down in the water trench, jump in the water trench, and I won't forget this day in my life. I'm standing on top of the other side of the water trench and he opened up. Anything he sees moving, muzzle flashes, he didn't care, he just had to shoot. Then, he helped drag me, Doc, and this head guy out of this trench. It was so high, we couldn't even get enough leverage to be able to get up and over on our own. We were sitting there for a bit trying to get a game plan going. I almost got killed that day.

The CO that day was running around and he grabbed my squad and was inside the house. I used to grab my squad and tell them what to do. I was naturally just talking to these R.O.s. We had two R.O.s. One was on Battalion, and one was on the Company frequency. I was telling them, "These guys hate close air support. They hate helicopters. We can get birds in here to get them off the rooftops. They're going to flee a lot faster than we're going to fight house to house." The CO came downstairs, flipping out, saying something. I looked him in the face and said, "You need to get on the fucking radio, use your fucking rank, and get some fucking close air support." My platoon Sergeant tried to grab me and jerk me to the side, "You ain't fucking talking to CO that way." I was pissed and said, "This motherfucker's not doing his fucking job." Sgt. said, "He's doing his job."

I went off into the house. He ended up getting on the radio, going back and forth, for a long time. We were getting heavy fire from that house. Everybody else went right down to one of the rooms. We ended up shooting a rocket from one house to the next house. Ben Musser decided to jump out the back door and shoot this rocket. I'm in the adjacent room shooting out the window. We decided that we were going to line up 3rd squad and 2nd squad, kick in the doors and attack the same house. We went to hit these fuckin' same houses and took off running across to them. It's probably 100 yards, 150 yards between the houses. We got up running and the machine gun opened up and started laying fucking rounds everywhere. It was me, Joe McGee, and 3rd squad's Doc who was with us. We left the house by ourselves.

We three made it across the whole fuckin' field and I was the first guy into this corridor area. I had to turn left into that area of the house. As soon as I turned left, there was a guy standing with an AK. He started fucking blasting around at us. All we could do was hunker down behind the brick and wait. We pulled pins on grenades, threw them in the door. Grenades went off. Smoke everywhere. He fired another mag. We did it again. He fired another mag. But we couldn't see him because of the smoke. All we could hear was the racket of the AK. We knew something was coming. He went back behind the walls and started shooting at us again. At that point, our Doc actually got shot in the leg. He was behind me trying to come to me and ended up getting shot in the leg. He rushed out to kind of the courtyard, up next to this brick wall, and then he was shot. I threw a couple more grenades in the house.

By the time that we did that, the squad had enough time to catch up, to focus on us and not on the other building. We threw two more grenades and then Joe McGee ended up killing the guy. Joe McGee started calling for me. The guy was dead already, but had a RPK that he was manning in the courtyard. They were hiding behind that brush fence shooting at us. We captured another guy and took the firing mechanism out of the gun so they couldn't take it and use it again. At that time, we got a radio call that John Sims was dead in the back house which Eric Smith was occupying.

Later, on 21 June, we hadn't had a radio check from our sniper team in three hours. We were kind of concerned and they sent a QRF. We heard across the radio that Tommy Parker, Otey, Lopez and Contreras were all found dead on the roof of a house they were using for an OP. I had a younger daughter. We were just about our kids. We don't have a way to get home to see our kids, hold our kids again, hold our wives. I broke down. I broke down in the middle of the evening mostly, right there in the field. I'd had it up to here with this shit. Losing all my buddies, losing my friends. It was a hell of a day for me.

Janney: It sounds like you knew Tommy Parker pretty well. Tell me a little bit about him.

Skaggs: If I'm not mistaken, I think Tommy was from Kentucky. He was a country kid. I'm a country kid from Missouri. We talked about home a lot. He was just a down-to-earth guy. He was a great guy. I'm sure a great husband, a great father.

Janney: What about Lopez, Contreras, and Otey? Did you know those guys very well?

Skaggs: Yeah, I did. I knew Contreras the best out of all of them. Contreras was one of the best guys I ever served with. He's a Hispanic kid from Mexico or wherever else and I'm a white kid from MO. I'd always have him try to teach me Spanish and told him he'd have to do something to make money one day. In fact, I used to tell him to teach people some Spanish. We had a hell of a time together. It didn't matter what kind of mood you were in, he'd cheer you up.

Janney: I heard he was quite the joker. Somebody told me a story about him showing up in a fireman's hat for PT or hike, and he got in a little bit of trouble for that, but it cracked everybody up.

Skaggs: Oh, yeah. That was Contreras. That guy didn't care. I mean, seriously, that guy, I miss that dude. He made me laugh all the time. No matter what he did, I mean, that guy would make me laugh. He would make me chuckle even on the worst days. He had that broken English. He'd say something, you know, it'd be half English, half Spanish. He called it Spanglish. He would just make you smile, make you laugh. You'd give him all you had if he needed it. It was nice to be able to be here with a man like that.

I knew Otey. He was in third platoon and almost always stuck with them and hung out there. I knew he had great plans outside of the Marine Corps. I think he wanted to be a rap producer. He had dreams. Bigger than most people's dreams. He had a smile that you could sell for a million bucks. I mean, a great man. Not quite as funny as Contreras, but a kind-hearted, easy-going guy. I mean, he made it out on April 6th. He ran out and hid in the building with bullets flying all around him, and he made it. Then just a couple months later, he gets killed. He was going back home in like four or five days. Yeah, that was a terrible thing. He survived, but he didn't survive. I mean, it's a mixed question that everybody asks themselves. Like, how can a man be so lucky one day to survive such shit? Then you got something so routine, and wouldn't be easy to perish.

I mean, don't quote me wrong, there ain't nothing about being in combat, in a war zone, that's easy. Nothing. It all sucks. Every day sucks. You do the best you can every day to hold yourself together and inspire your men. Make it through the day, ensuring that your choices don't kill somebody.

Janney: Did you know Crowley or Ryan Jerabek?

Skaggs: Crowley was in my team up until his first one. He was one of those kids that were just kind hearted. He was young and very cool, which most of them were. They did exactly what they were told without questioning. Such bravery, courage; he had it all. He was a guy that you could obviously see grow up and knew he had great things in store for him. He was a fantastic man.

Janney: What about Ryan Jerabek?

Skaggs: I knew him, I didn't know him. He was a machine gunner within our left wing. When we got to Iraq, they decided to make our weapons platoon attached to different platoons there, so we could facilitate more missions and be more flexible with the missions that came about. We didn't have a primary weapon platoon. I think we were attached to the third platoon at the time that he was killed. He's a great kid. I'm judging by his character. He's a guy that was brought up into an ambush with two Humvees, and he's manning a machine gun without a slip of armor. This steadfast guy stood there under heavy fire from all sides and manned his machine gun trying to protect his Marines that were around him until his last breath. What else could you want from a warrior?

I mean, so did many, many men on April 6th. They sacrificed themselves for other people, like Lt. Ski. I mean, Lieutenant Ski probably thought, next time I'm going to have a RO in an open ambush. Because Captain Royer wanted an extra radio operator, he took our radio operator that day to have three operators with him. So, Ski was listening to the radio, and again, no one knew what was going to happen. But Ski was stuck with manning the radio to try to get help for our guys in an ambush. That's how he got shot. He had his radio operator, and Ski could do these intellectual things, to communicate it from safety. But, he had to take a knee next to the damn Humvee to man the radio, to communicate, to try to save his guys, and he got shot in the face.

Janney: What about any of the other guys like Layfield, Carmen, Staff Sergeant Walker?

Skaggs: All of those guys are the guys you know. I mean, you know, but you don't know. We all ran in passing, we all talked, we all joked. But, as far as details about their personal life or how they grew up, no. Because, again, in a platoon, you typically hang out with the guys that are around you all the time. On the 6th, our big loss was Crowley. Most of the guys that were lost on the 6th were weapons platoon or third platoon. I mean, you know the guys and know them well. They didn't go out easily. They went out fighting. They all lived and sacrificed their lives to try to save somebody else. I'm sure when the shit hit the fan, he was there and that helped me. He was doing everything he could as fast as he could think of what to do to take care of his guys. I have no doubt about that.

Well, everybody thought Ramadi was going to be nothing, right? We rolled in there in 2004. We all laughed. We were kind-hearted, free-loving spirits. It's not going to be a big deal when we come here. The push really wasn't that big of a deal. Other units had a whole lot of casualties, but this is going to be a joke. We're going to have one night where we're going to see a whole lot. Then, it turned out to be like a horror movie. Every day I left the base, I'd say a prayer and hope it's not my day, walk out the gate and you do what gotta do. I would say that I was blessed and I served with heroes. I did. I served guys that were unrelenting heroes that were willing to sacrifice themselves for other people. Whether they made it home or didn't make it home, they were the type of guys I served with that would put their lives before anybody else.

6 April 2004 Route Gypsum Ambush And QRF

Doc Keith Grimes, HM3

Janney: Today is 10th May 2020. Keith, tell me why you decided to join the Navy?

Grimes: Oh wow. That's a little bit of a convoluted question. So, as I was growing up, I was rebuilding motors and transmissions and air conditioning and things like that. I was driving down the road on a very late evening, wet, and raining. I grew up in San Diego. I came across a vehicle that, in front of me, slid off the road and flipped down a 100 ft. embankment. And so, after I pulled that guy out and got him all sent off to the EMTs when they arrived, I realized that I could pretty much fix anything or everything, but people I didn't know how to fix. So, my father was an Army Major, he was an artillery officer as a matter of fact. I always had the intent of going in the Navy or going in the military per se but medical became the forefront after I realized I could pretty much rebuild or fix anything, but I didn't know how to fix people and I felt that was really something that I wanted to go into considering especially that my mother was a registered nurse doing trauma.

So, I went to actually join the Marine Corps as a matter of fact. I was going to join the Marine Corps Reserves after I planned on going to college and I realized at that point because I hadn't done enough research, I found out real quick that the Marines did not provide their own medical and they used Navy Corpsmen. My father told me "Whatever you do, don't go in the Army because you can get paid the same by going into the Air Force or being more comfortable." So, what do I do? I joined the Navy and volunteered to go with the fleet Marine grunt units.

Janney: Yeah, so you just went in the back way instead of just directly signing up.

Grimes: Pretty much. I wanted to know what I was made of but by the time I was twenty-two I was working at the casino making really, really good money and they offered me a job with a $50 grand salary; full benefits if I were to stay onboard and not join the Navy. I declined because I wanted to travel the world. I was bored with college, and I wanted to learn how to fix people. I wanted to see personally what I was made out of. This was before 9/11 and all that. As things were ramping up, I was hoping to get in soon enough, maybe to go to Kosovo to really kind of see what I'm made of. How do I react under extreme circumstances and can I do what I need to do to be able to fix people. That's kind of what got me in the Navy.

Janney: Do you remember the month and year?

Grimes: I joined in February of 1999. That's when I went to boot camp. I think I signed up, God, I think I spent a couple of months there. So, when I went into MEPS, they were fighting me because they were trying to, they gave me the nuke test, they were trying to make me a nuke, they were trying to make me a sonar tech. I was fighting with them because I didn't want to be any of those things. I came in specifically to be a Corpsman. I went to the Army as well and they guaranteed me that, "Hey, we can give you a shot at Airborne, Rangers or whatever. You can go to Ranger school as long as you meet the physical core requirement and you'll be medical." I went to the Navy, the same thing, and so when I went to MEPS, I said, "Hey, I want Corpsman." And they're like, "No, let's take the nuke test. You only missed it by one point so you can retake it. You can be a sonar tech, you can be an electrician. You can do whatever you want, but Corpsman is not available." So, I told them that basically, "Look I'm going to go out (I don't know where this came from and this might backfire on me) but I'm going to go out into the hallway and I'm going to buy a Pepsi and I'm going to come back and if you don't have Corpsman available at any time or any way, then I'm going to go to the Army and I'm going to join them." Because I was rolling in with a 91 ASVAB or whatever the hell I had, this is what I want. So, I came back, and he said, "There's your fucking Corpsman." I said, "Okay, I got what I want. How soon can I ship out?"

Janney: I'm not as familiar with the process of the Navy. I know you go to boot camp and then, you know, the Marines go to SOI, so after you got out of boot camp at which point do you start training as a Corpsman?

Grimes: Okay, and so, now my story gets interesting. I went a very unusual route. I joined the Navy as a Corpsman with guarantee. I went to boot camp in February of '99 and while I was in boot camp they said, "Hey, the Presidential Ceremonial Guard is doing a presentation, if you want to go you can go." I don't know what the hell that is and I talked to my RGC's and I'm like, "What's that?" and they go, "I don't know." And I go, "Well, I'm going to check it out." So, I go to this thing and the Presidential Ceremonial Guard did their presentation and I'm like, "You know what? I'm 6'2", I'm in good shape, I'll volunteer, whatever." And I got chosen. So, my Hospital Corpsman A school was postponed for two years. I went to the Presidential Honor Guard out in Washington, DC for two years, undesignated under the caveat that if I failed, then I go to the fleet undesignated. If I succeed, I can pick whatever rate I want and go to that school after two years.

The kicker is that I can't take any advancement rates, so I was kind of stuck because I had college credit at this point. I was stuck because they screwed up my transcripts. I came in as an E1, but then I was stuck as an E3 for a year because I couldn't advance without taking a test.

Because I spent that extra year as an E3, I was basically an E3 hanging out on Presidential duty traveling the country and especially DC for two years doing Presidential Ceremonial Honor Guard. Then at the end of that, they said it's because you're an honors man, a master ceremonial whatever they called it at that time; it's changed since then.

They said, "Hey, what do you want to do?" "I want to be a Corpsman." So, then I shipped off to Corpsman A school. Now, the track for Corpsman is normally you go to "A" school and once you get your designation as a Corpsman at the end of that school, then you choose to go specialties. Surgical technician, x-ray, P & T, all of the different advanced schools if you want to. But I wanted to go FMF, so I asked to go FMF, but they didn't give it to me because there was enough going around at that time. So, I got sent to work at Balboa. I begged, pleaded, borrowed, and humped legs to try to get into the emergency department. Luckily enough, I spent three years in their emergency department doing ER and trauma over at Naval Medical Center, San Diego before I had a chance to even go with the Marines.

Janney: I bet that was some good real-life experience in what you were going to do later.

Grimes: It helped. I was pissed. I am not going to lie. I was so mad that they sent me to a hospital even though I was asking for something that nobody else really wanted and I wanted FMF. I wanted to be a grunt. I wanted 1st Marine Division out of Camp Pendleton and they said, "Nope, you're going to Naval Medical Center, San Diego. Have fun." So, I made it a point to get every single credential I could between EMT. Basic Life Support, Advanced Life Support, Basic Trauma, pre-hospital trauma, advanced trauma, I did everything I could think of. Neonatal, you name it, I got that certification, so I want to know everything including advanced cardiac life support. Things that I was not allowed to go to, but they allowed me to audit it with doctors and surgeons for advanced trauma just because I was asking, and I would not leave them alone. But it worked out well because at the end of that tour the detailer herself came in and said "Hey, what do you want to do for orders?" "1st Marine Division." I'm like "You know what, okay." That was easy, ready to go. I got 1st Marine Division, so I shipped off to field medical service, no, FMTB, I'm sorry. Back then, it was field medical service school. Now it's become field medical training battalion out of Camp Pendleton. After that, I was sent to the Division Surgeon's office and I was chosen for 2nd Battalion, 4th Marines.

Janney: So, at what point did you get assigned to 2/4?

Grimes: I had a buddy of mine that worked, he worked in the hospital as well. He put in a good word, "Hey, we want this guy to come to 2/4 with us." Because he got there about six months prior to me and he said, "Hey, we want this guy." So, he kind of hooked me up and that way I ended up assigned to 2/4. I want to say that it was August of '03, I got assigned to 2nd Battalion, 4th Marines and immediately they said, "Hey, you're going to go over and jump into Golf Company" and within a few months I ended up taking over Golf Company as a senior line Corpsman in 2003.

Janney: You didn't go on that Okinawa deployment then?

Grimes: I went on the Okinawa deployment in '05. I think they were on deployment when I first arrived there, and I suspect that was the 31st MEU at the time.

Janney: Right, they were.

Grimes: Okay, so I'm sure I remember. So, I know they were in Bridgeport when I arrived there. Then they ended up, right after they got back from Okinawa, if I remember correctly, and so I did the Iraq deployment and then the follow-on Okinawa deployment after that. Okinawa, Philippines.

Janney: You obviously went with them in mid-February of 2004 when they deployed.

Grimes: I convoyed in. I was an ADVON. Because I was ADVON on the way in, so I convoyed in. So, we flew to Kuwait, hung out and so I rode out from Kuwait into Iraq in the back of a Humvee with a gun mount. We convoyed all the way in. It felt like forever because we were sleeping on the side of the road in the gravel and hanging out in the back of the truck for what felt like days.

Janney: Now, from what I understand, there was not a lot of action going on in the convoy. There were some IEDs and stuff like that. Do you remember anybody getting injured on that movement?

Grimes: Not on mine. I believe Fox Company; I think they took contact a little bit. We were not all together at that time. I know Fox Company, they had some helos running in. I'd have to talk to Ray Parra. He was in on that one. Have you spoken to Raymond Parra?

Janney: No, sir. I have not.

Grimes: Okay. He was the senior line Corpsman of Fox Company. He was in the convoy when, I don't know why or how it happened, but Fox Company was a little bit separate from us from the main convoy. We did not receive any contact or anything. We had a couple of issues like Marines throwing water bottles out or MREs out while we were on the fly and kind of accidentally picking people off with the water bottles because people were trying to raid our trucks for whatever they could grab when we stopped. It was kind of a different world at that time. We didn't take any convoy casualties or really any IEDs that I recall all the way in. It was pretty mellow the whole way in. We had a couple of tense moments, but nothing that I remember of any contact that I can recall at the time. But, I had a very small purview at the time because I was really managing my particular convoy at the time and my Marines at that time.

Janney: When you got to Ramadi, were you assigned to Echo or you said Golf Company initially?

Grimes: I was in Golf Company. I was a senior line in Golf all the way in. We were hanging out in Junction City for a very short while and then we pushed out to Combat Outpost. I was one of the first units to arrive at Combat Outpost because I was a senior line Corpsman with the CO. It was even before Echo Company rolled in there. We were just setting things up and digging in fighting positions.

Janney: Yeah, filling sandbags is never any fun.

Grimes: Oh yeah, and I was young and dumb. I know a lot of Corpsmen shirked out of that particular one, but I was in there digging and doing it and whatever is necessary because I needed to finish making my name with those Marines like, "Hey, whatever you do, I do. I'm here. Teach me your tactics, teach me what you're doing. Why are you doing fields of fire, here and here? What is your escape plan? Where can I evacuate people?" I was still in the learning phases of the whole thing. Because you've got to remember that I got there in late August of '03. We pushed out for Iraq in February of '04, so I was still pretty green when it came to Marine stuff, shall we say. I wanted to understand tactics in movement and fighting positions, field of fire and designated left and right laterals. All those things that I was trying to learn and so I learned it while I was digging fighting positions, "All right. What's our immediate action; our remedial action; what do we fall back to in case we get overrun?" Things like that.

Janney: It sounds like you did your homework.

Grimes: I was trying to learn as quickly as I could because I knew we would get thrown in the fire. That was the first deployment of my whole career.

Janney: Once the rest of Echo and the rest of 2/4 got there, I know there were some familiarization patrols with the Army unit that was there.

Grimes: Laughs. That was interesting. Yeah, we did a bit of leftsy-rightsy (familiarization patrols.) I didn't really have a chance to really liaise with their leading medic on the Army side of the house. I want to say that they were a reserve unit or something.

Janney: There was a National Guard unit and then there was 82nd, I think.

Grimes: I think it was 82nd. I know it was an Airborne unit and they were sloppy. It was part of the National Guard unit that I remember the most. They would be walking around, and we'd be on patrol or whatever and these guys, they were the machine gunners. They would set their machine guns down while they were hanging out and smoke and joke and I thought, "What the fuck are you doing?" It really threw me for a loop - either this place is pretty cold or you're not taking it seriously. I don't understand why you're not on point ready to kill anything that comes at you. You know what I mean? It really threw me for a loop. We're walking around and we're driving around waving at people, "Okay, we're trying to win hearts and minds." Got it, but like these guys were setting their weapons down and moving five feet away from them as they were talking and whatever.

Janney: I think it was a combination of both. That they hadn't seen a lot of contact, as well as the fact that they were just lazy from what I have heard from some of the other Marines.

Grimes: That's kind of what I got as well. I got zero contact with their medical people. I mean zero. I mean we had to build the medical department on that Combat Outpost. They didn't have a designated landing area for a helo. They didn't have a designated medical treatment area. They had like a small room, but it wasn't really set up. They didn't have fall back plans for when we get a mass casualty – where are we going to put them? They didn't have any body bags. "Well, why don't you have body bags? Do you have body bags that double as carriers?" No. "Why don't you? I don't understand." There was really a kind of confusing time of us trying to understand a very different dynamic. It was like a different paradigm of onboarding procedure versus what I had come to understand about how the Marines operate. If that makes sense.

Janney: It does make sense. I was embedded with 3rd ID and there's a world of difference between the Army and the Marine Corps. There's no comparison.

Grimes: Yeah, I agree. I got pinned down probably three times by the Army. I can tell you stories about that one. The way they operate is very, very different and so the turnover was very lackadaisical. It was, "Okay, we're going to have to rewrite this script because this doesn't work for us." They barely went outside the wire. I remember once talking to the Army guys and they said, "We only go out once in a while. We stay close to the Combat Outpost, and we don't go deep. We don't go overnight." They had almost strict non-contact type rules of not going into the city and doing anything because they didn't want to piss off the population.

Janney: Exactly. Yes, that's what I've heard too. They did a lot of mounted patrols whereas the Marines usually did dismounts and so they were very different.

Grimes: They did not do any foot mobile anything. I do not remember one time doing leftsy-rightsy with the Army. They never did foot patrol. It was always full and fast. We go out, we come back, and we're done.

Janney: Yeah, that's pretty much what I heard as well. I mean even when they were doing the familiarization patrols, it was usually mounted and they didn't really want to get out of the vehicles.

Grimes: Like ever. I did nothing foot mobile until we were like, "Hey, we have it. This is what we are going to do."

Janney: As far as supplies that the Marine Corps had available for you, did you have everything you needed in your new triage area at Combat Outpost?

Grimes: Oh, Jesus. Not even close. Because I came from the ER at Balboa, I knew a lot of people, not only at that hospital but multiple other outlying clinics and I moonlighted a lot at different trauma centers around San Diego and so, I knew a lot of different people. So, I'm glad I did this now. I don't know if you've ever seen those, it's like a six-foot sea bag with a zipper on the side. And it's a massive seabag. When I rolled in country, I had one of those and a standard seabag and they were full of stuff: two cases of IVs, antibiotics - everything from Augmentin to Doxycycline, everything. You name it, I had as much in there as I could because I was concerned about not having what I needed. When we were in Kuwait at Junction City, I was trading bags of thousands of pills of Doxy (Doxycycline), Keflex. I was trading for different meds, so that I had a huge variety of different meds, whether it be Motrin, Tylenol, all the basics. Benadryl, Sudafed, you name it. I wanted to make sure I had it. Also, we were trading out for antibiotics because we didn't have enough. We had some, but none of us were comfortable with what we had. So, we were trading the outgoing Army units and they were giving us scopes, drugs. Not controlled - we didn't trade for any narcotics or anything. Everything else, we were trading like crazy trying to offload from them as they were leaving the country so that we had it when we were going in. It just wasn't enough.

So, when we arrived at Combat Outpost, because it was so anemic, I was glad I got yelled at for that big ass bag. We were very glad that we had so many extra supplies because that had to sustain us until we started getting logistical convoys a couple of weeks later. We weren't sure what we had. I had one or two, I think they were called cat one tourniquets back then, because we've gone through so many iterations of it, I had one or two tourniquets just on me. Most of my Marines didn't have very much of anything except for their IFAK (Individual First Aid Kit.) So, I started grabbing rods, or heavy-duty metal dowels and things like that, and we started making tourniquets out of whatever we had around so we could stock them in specific pockets on all of my Marines including myself. I usually carried ten of them on me. For that reason, we didn't have enough. We did a whole lot of make do with what you have. Also, we had no armor. We had wood and sandbags. That's it. We didn't have shit.

Janney: The crazy thing is the 82nd, some of their Humvees were armored and then they had Bradleys too, which you guys did not have access to.

Grimes: We didn't get their armored Humvees either. We just had the standard ones until they got them from the Air Force. I don't know where they got those armor upgrade kits from. So, after we acquired a couple of armored plates randomly, we used tie down straps, jumped on top of the 997 ambulance and we strapped them onto the sides, so that, if anything, we could sit on the bench, lift our feet and duck our head and that would give us a modicum of armor for when we were transporting patients.

Janney: That was smart it worked out like that, but that's a lot of foresight.

Grimes: Yeah. Well, my doctor was pissed because the armor covered the big red crosses on the side of the vehicle. It covered them completely, so he was livid, "Oh well, according to the Geneva Conventions, we have to have the crosses visible." Well, yeah, but you're not the one that is riding in this thing and every time I've been in this thing, it's getting holes punched into it because they are shooting at us specifically. So, the armor really saved us when we were bouncing around. I can go on and on with stories on that one.

Janney: That required a lot of effort. Late February, early March, tell me what's going on. Are you going out with your Marines everyday or every other day? What are you doing?

Grimes: Yeah, for the most part. We didn't have enough Corpsmen for all the missions that were going out. So, what would happen is, although I was assigned to headquarters platoon in Golf Company, I had two Corpsmen in each of my squads, but I had four platoons. So, needless to say, I was short on Corpsmen, so I was kind of running through and these guys were going out all the damn time. We had squads from each platoon going out without Corpsmen completely. Because of that, and I'm not really one to hang out, I was bored out of my mind. I would rotate out with my Corpsmen, so if they were doing three days of patrolling, I would jump in and relieve them one of those days to give them a break. If a squad was going out or a platoon was going out, depending on what they were doing, we only had two Corpsmen per platoon and we had three squads, so I had to jump on one of the squads to give medical coverage for that extra squad as we were going out operating. Once we started patrolling, I only remember two days of being in on the actual Combat Outpost itself and not out patrolling or doing whatever. It got really hairy because I was going out on patrol and came back, and another squad would get ready to go out. I'd say, "Hey, do you have a Corpsman?" "Nope." "All right, give me a second to reload on food and water and I'll roll out with you." So, I'd do back-to-back to back outside the wire just to make sure my guys had medical coverage.

Janney: I heard the tempo of Echo Company was just unbelievable. I heard they worked harder than just about any other Company.

Grimes: Yeah, Echo was pretty rough. Golf, we were out there a ton and there were different AOs of what they were covering, like sections of the AO. Echo had a more rural area, whereas Golf was covering some of the urban type stuff, but we had to bounce out to some of the suburban stuff and cover the train station and some of the different things like that. We were all tag teaming, but I'm sure I operated with Echo at one point or another. We were kind of trading back and forth just trying to keep each other afloat because the Corpsmen were getting rolled. They were getting worn out because the Marines, they were on a 3 and 1, I think, so they were either on security, they were on a down day, they were on a patrol day and night ops. But the Corpsmen, there's only two per three squads, so the Corpsmen were constantly going. So, I was constantly relieving my Corpsmen to give them a break even though I was getting burned out a little bit. The guys needed a break because they were stuck on security or doing whatever. It was interesting as we were trying to establish our foothold and figure out how this is going to go. It was kind of questionable on the deck plate when we were out bouncing around. The leadership may have known. I'd get little snippets from the CO, the Company Gunny and the First Sergeant, but most of the time, I would try to be out there as much as I could so that I could be at that point of friction where my skill set was most applicable. I didn't want to be the guy doing paperwork and evals the whole fucking time. That's not how I'm built.

Janney: What was the first day that you remember that you or your Corpsmen had any contact?

Grimes: Let's see, um, I had been on back-to-back ambush patrols or whatever for just days on end and I was tired as all hell. I came back on the early day of April 5th and I happened to be on the Combat Outpost on the day of the 6th. I was pissed as hell because I was trying to roll out with the QRF when they got pushed out, but one of the other Corpsmen jumped on and they pushed. I was basically stuck on the Combat Outpost with nowhere to go on the day of the 6th when everything started rolling out. It started really ramping up and there's gunfights going on in the whole city. So, until we started receiving casualties, I was hanging out by the COC listening in on everything that was going on. I thought, "What do we have? What do we need? Do we need to call in helos to have them standing by?" All hell broke loose. That was a crazy fucking day since most of Echo took the brunt of the 6th. Golf took a lot of it too, but when Echo got hit, we were getting casualty calls all the way in. When we were finally able to fight our way in and get to those casualties, they rolled in and myself and one other of my guys, we were there jumping into the back of the Humvee with all the bodies, checking pulses and hoping that we could find somebody that was still alive in the back of the Humvee. There was just a pile of Marines in the back of this truck.

I had to rapidly set up the morgue. So, the relaxation/entertainment room that we had, right next to the TAS, we had TVs and watched movies and whatever. Well, I turned that into the morgue and fired up the air conditioners on full bore and started bagging and moving bodies into there as I was checking them. So, I'm up there on the truck checking pulses and working my way through all my buddies, for Christ's sake, and once we checked them, we were moving them off and putting them in the makeshift morgue right there on Combat Outpost until we could move them out.

Janney: I know that you mentioned that you treated Lt. Wroblewski.

Grimes: Yeah. God, I'll never forget that. I remember when he came in. Gunshot to the face and it was kind of chaotic. I was working with a lot of people at that time, but I popped over and we had one guy doing CPR over around the corner. If you have Marines coming in and they are basically dead and you don't know if you can work on them or you expect it, you kind of put them out of view of the Marines. So, we had one guy over there doing CPR and attending them. Then, we were dealing with random Army guys coming in with hands blown off or whatever. Then Lt. Wroblewski came in and I think I had Lt. Phan working on that one. I fucking hated that guy; he was completely incompetent. We took away his gun and locked him in a closet at one point. I don't know what day it was on because as a physician, he was completely useless for what we needed. He was trying to do acupuncture to stop traumatic bleeding. Yeah. It was a ROUGH couple of days.

I couldn't tell you what day or what order Wroblewski came in. I just remember him because we had to rip open the cric kit and throw a cric in there and the Lt. was kind of fucking it up, to be honest. We finally got that thing in, got his airway secured. He's breathing and we said, "Okay, you're good to go, you're stable, birds on the way" and then we loaded him up on the litter, ran him outside waiting for the 46 to land. It finally landed and I had my guys help me carry him out and I did the turnover with the Wing Corpsman who was the air Corpsman, the Corpsman on board the helo when we put Wroblewski in. I took him in, and I strapped him in and then I gave the turnover to the guy and made sure, "Hey, watch his airway. It's secured, but it may become unsecured. You need to watch his airway." Because it kept tripping, he was kind of fighting it a little bit. He was pushing it out and that was my concern. We strapped him down, again, got him set and I looked at him, "Hey, you good? You can breathe?" He gave me a thumbs up. I talked to the Corpsman onboard, and he said, "Yup, I got him." I said, "Watch that fucking airway. It keeps moving because he's fighting it a little bit." Just natural body movements. Then I disembarked the helo and went back to treating casualties. Unfortunately, that was the last time that I saw Lt. Wroblewski. I was pretty pissed when I found out that he had expired on the flight. I have personal theories on that one, but don't know for certain.

Janney: I've heard a couple different stories. One, that they didn't pronounce him until he got to the CSH (Combat Support Hospital in Baghdad.)

Grimes: Yeah. You have to have a physician pronounce, so even in flight they are not authorized to pronounce official death.

Janney: That's probably why they said that he passed at the CSH instead of on the helo.

Grimes: Exactly. My suspicion is not only did he have a bit of damage and bleeding and whatever. My concern, my fear, the ones that I feel that we stabilized him. I had him. I looked him straight in the eyes and said, "Are you good? Can you breathe? Are you good?" He didn't say anything, but he looked at me, gave me the thumbs up and relaxed. Then I found out he expired on the flight. My personal theory, I have no evidence, I have nothing I can prove on that one. I think the Corpsman onboard that aircraft sat down, strapped in, which is still protocol, but I don't think they were watching that airway and I think the airway was compromised and in one way, shape or form, it may have come out or something. That's what caused the Lt. to not be able to make it. Because he was stable when I put him on that fucking bird. He was clear, I was talking to him, he was responding well. He gave me a thumbs up; basically, I'm good.

Janney: Yeah, I think everybody was really shocked when he did expire.

Grimes: Yeah. I'm telling you he was fucking stable when I put him on that bird. I know it because I was the last one of 2/4 to see him go. I know he was relatively stable. Could there have been more damage? Sure. I can't blame the Corpsman or the flight crew. I don't know what happened on that flight. But he was clear, he was conscious. He was stable. He was well-managed. Everything was locked in. He had his IV; everything was good when I put him on that bird and walked off that aircraft; he gave me a thumbs up. Yeah, I know he was in pain, but he was gutting it. Like, "Yeah, I'm in pain, but I'm good to go." He responded to everything I gave him. Good to go and then, the next thing I know, they pronounced him. What the fuck do you mean they pronounced him? He was stable when I put him on that goddamned bird. So, he's one of the ones that really sticks out among all the other Marines that I treated because most every Marine in Echo and Golf Company that died or was injured on that deployment, I was either there when they got hit or I was working on them in the Combat Outpost. Almost 75% of them. Because I had a ton of trauma experience compared to everybody else. I don't know. I wish he would have made it, but I just don't know. I can't tell you for sure. I wish I had more evidence either way to nail that Corpsman, but then again, what are you going to do. It's combat.

Janney: That's right. I know more about the guys that didn't make it that day than I do anybody that was wounded. Who else did you treat that day? Do you remember?

Grimes: Ah, Jesus. You know what, it was like a flurry. It was back-to-back to back-to-back people, including the Army coming in. I had a Bradley come in that they had so much damage to them, too. I remember the Bradley came in and the gunner's hand was blown off. He had put a tourniquet on his own hand. I want to say he was on a 40 mic up there. So, with the Mark 19, he tourniqueted one hand and kept shooting with the other. So, I jumped up on top and I'm the one that lowered him down to get him down into the treatment area. As far as I know, he lived. Lost his hand, of course, because it was gone. Yeah, that day was crazy. The next four or five days were just nuts. The 6th, I was in Combat Outpost treating Army and Marine Corps. Then, every day after that, I was on every mission, convoy, or whatever I could get out on. QRF, you name it, I was out there. I didn't come back for a couple of days. Like, I literally did not come back to the Combat Outpost for a couple of days. Because I'm not going to be fucking stuck on this Combat Outpost again when the skillset that I have could be better used out there and doing what I do. Shit, there were times. I remember Langhorst coming in, I remember Aldridge coming in. I remember certain people. Sometimes it was just so hard to tell who I was working on because it was chaos. Not so much the situation, but we're still busy cutting, stripping gear off and working on wounds; I didn't have a chance to look at their face because we were up to our elbows trying to save their life. I remember a good couple of them. It's hard. I remember grabbing a couple of Marines, and these are hard core guys. I think it was Abbott and a few of my guys, and I said, "Hey, I need some help holding this guy down so that I can work on him." These are Sergeants and senior Corporals, squad leaders, some of the hardest Marines that I know. After we were finished treating those patients, it was like, "Doc, don't ever ask me to do that shit again. I can't fucking do it. I'm not going back in there." I said, "Damn, man, Okay. I gotta figure this out." It was pretty nasty.

I had a few guys go combat ineffective because their feet were so jacked up that I had to lay them up for a few good days with flip flops, just so that their feet could dry out, so they could actually walk on them again. So, the foot powder thing is no joke. I want to say we each had one or two pair of boots. We usually had one that was desert intemperate, but we were operating so damn much. If we weren't on patrol, we were digging. If we weren't digging something, we were on security. So, they really didn't have a chance to dry their feet out. There were only one or two people that really didn't physically manage their feet very well. So, I had to lay them up for about 5 – 7 days just to dry their feet out because their skin was almost sloughing off from moisture and athlete's foot.

Janney: Almost like a version of trench foot, I guess.

Grimes: Yeah, pretty much. Yeah, damn near. That's kind of what it looked like.

Janney: I know you said that you were involved in some of the patrols after the 6th. Tell me what you remember about those days.

Grimes: Busy. I only came back for a couple of days on that one. I want to say on the 7th, I rolled out with 2nd platoon, I believe. We went out and I think that was a standard patrol. So, we did a foot patrol, came back, 3rd platoon went out. God, we were busy, man. I kind of jumped from squad to squad in those days because every time there was a gunfight or whatever, and I believe on the 7th, was when Rossman got hit. Rossman got hit, so I jumped from 1st platoon to 2nd platoon or somebody, I lost track, I'm not sure. I lost track to be honest because I carried not only a squad radio, but a PERK 119 standard radio on me and. I bounced from squad to squad intermittently, so I remember working on Rossman. I worked on him, fought our way back. We only took two vehicles back to Combat Outpost, dropped him off, grabbed more ammunition and whatever and fought our way back in to link up with, I think, 4th platoon to continue going. It was chaos because I was jumping between so many squads that day. I think I operated with three different people that day. Then, we went out on an ambush that night. I was up for like four days straight. It was kind of blurry once Rossman got hit; that was a nasty little engagement. I was with the squad at that point, I think it was 2nd squad of Golf Company. And 2nd squad of 2nd platoon. I don't fucking remember. It's so bleary right now.

Janney: Did that happen on Easy Street or Nova or where were you guys at, do you remember?

Grimes: I got in so many gunfights that day, I don't remember where we were. I don't think that was Easy Street. We were rolling down a road and we had a translator there. We were looking around and one of the Iraqis came out and said, "Hey, don't go down this direction." That was like a perpendicular left direction from where we were traveling. I remember that we were all kind of looking at each other. We should not go down that road right there. He said, "Yeah." Okay. And so, of course, we hung a hard left and went down that road because somebody was trying to, hopefully, engage us down that road, so we were going to attack them. We were about three quarters of the way down that road and that's when we got hit. Rossman took a round to the chest. All hell broke loose at that point. God, we took over somebody's house. We were shooting our way through. You know how there they have the brick walls, and they have the steel doors or whatever?

Janney: Oh, yeah. I did foot patrols there every day with 2/8 for two weeks.

Grimes: Yeah. So you know how the houses are set up. Reynolds, God, that guy cracked me up. That guy was like an ox. He would pick up everything and throw it. He was just a corn-fed ox-type. I remember dragging Rossman back and we took cover behind a gas tanker. It wasn't exactly a planned idea; it was available. So, we hid behind the vehicle as I was working on him and that's when the guy came around the corner and I popped him with my 9 mil because I was working on Rossman at the time. So, I got Rossman; we ran back to link up with the squad because we were a little bit disconnected. There's Reynolds running and bashing his body up against one of those doors trying to get into this house. I'll never forget this with Reynolds, the fucking ox. Probably popped about 9 rounds into this door lock just to kick the fucking thing open, so that we could drag Rossman in there and finish stabilizing him. Then, get a casevac in there, so I could get him loaded up onto a vehicle and get back to Combat Outpost. I'll never forget Reynolds bouncing off this stupid door multiple times, but he didn't give up; he kept trying. This guy's like 260 and, I don't know what the hell, he was 6'2" or 6'3". He's just bashing his body against this door trying to get it open and I just ran up and shot the fucking thing because I didn't have time for that shit. He just looks at me and starts laughing after we kicked the door. We're going into the courtyard and then take over the house so I could finish treating Rossman. Ran him back, loaded more ammunition, and talked to Sgt. Major Booker there. I'm covered in blood at this point and Booker still shook my hand. I said, "Hey, I'm covered. I've got shit to do." He said, "Okay." Grabbed a couple of cans of ammunition and threw it into the back of the truck and then fought our way back. Then, I think I jumped in with 3rd platoon, 2nd squad. I jumped three squads that day, so it was really kind of weird. At least on the 7th and 8th. The 8th and 9th were kind of interesting because we were traveling around the city and I want to say we were doing a cordon; what the hell do they call it? A cordon search?

The vehicles would come out of nowhere, like randomly. At that point, it was a free for all. If they're not in their house, if they were coming at us in any way, shape or form, we killed everything within 400 meters. That's kind of what we all discussed because of the amount of chaos and the amount of firepower and antiaircraft weapons and things that were getting thrown at us. If we get shot at, we're going to kill anything we come across. We would have people who would do pot shots, so we would chase them for multiple klicks, kill them and then move on. Sending the message, "Hey, if you come at us, we're going to find you and do whatever we have to do to kill you and then we're going to go on our happy way. We're not going to stop until you're dead." We did that multiple times those following days of people taking pot shots. They continued to shoot at us and we continued to shoot at them until we finally took them down. Then, we'd go back our happy way. There were a few times, I think it was on the 9th or 10th, one of my machine gunners was a wee bit overzealous. A taxi came around the corner and he starts lighting this fucking thing up with a 240 he had. He's unloading rounds into this goddamned car and I'm tapping him going, "No, no, no. We're not doing that today. That was yesterday. Today, we're not trying to shoot at anybody until they engage us directly." The Marine, I forgot who it was right now, he said, "Oh, my bad. Good to go." I don't get to shoot at everything that moves today. We moved on our happy way.

Janney: Yeah, the 10th was Operation Bug Hunt or one of the Bug Hunt ops.

Grimes: Yeah, something like that.

Janney: I've heard from various reports that there were dump truck loads of enemy combatants.

Grimes: We had enemy combatants inside the ambulance. The local ambulance would come in and drop off fighters and then roll out. Those were the days that the local police turned on us, so we had to engage them as well. I was on the Combat Outpost, and I got a phone call, "Hey, casualty inbound. Shot to the back of the neck." Or the back of the head, whatever. I got pushed out and we arrived on scene, and this is the luckiest individual I've ever met in my life. Of all the trauma that I've ever seen and done, this is truly the luckiest individual that I've ever met in my life. He took a 7.62 round to the base of the skull, three inches below on the neck. That round, it's a matter of distance, trajectory, angle, but everything lined up perfectly. The bullet itself impacted dead center, just left of his spine at the base of his skull, traveled around and left a groove all the way around, and popped out just about two inches to the left of his nose, out of his cheek. The angle and trajectory and distance and velocity, everything worked out perfectly. The skin kept that bullet under until it popped out right before his nose and this guy only had a very fleeting flesh wound.

So, I'm looking at him expecting a dead guy or at least catastrophic damage to his cheekbones, orbits blown, whatever. No, he's like talking to me in Arabic. I think I've gotta send you back to the rear. He had an RPK. So, I took his RPK and four drums of ammunition and I kept it. Because usually, if I'm working on somebody, I sling my rifle behind me, and my pistol usually goes on their chest. I grab their rifle or whatever they're carrying and I put that next to my pistol, so that I don't have to deal with reloading if I'm working on somebody and somebody pops out. I can use different weapons until I have reached that point of having to reload when I'm working on them. So, I end up grabbing his RPK and slinging it over my shoulder. I have a great picture of it somewhere I should probably send it to you. I kept that RPK for about two to three weeks before the Shawanis, or whatever they called them, demanded through my chain of command, the return of their weapon, so I had to finally give it up. Now that was a good time. I patrolled with that thing, my M16, and my M9. I'm not going to give up a fully automatic, nice suppressive fire weapon unless I have to. They were kind of mad and I had to give it up at some point. It's against Geneva Conventions, but this is a different world. I used to own an AK before I came into the Navy, so I knew exactly how to operate and clean the action on that thing. Tear it down, clean it, I know all that shit. So, I made use of that. My CO said, "Hey Doc, you know how to use that?" "Yep" "All right, good to go." He didn't say a word after that. He was cool with it. It was a good time.

Janney: Going back to the 10th, did you treat John Sims?

Grimes: Ah shit, I'm not sure. Do you remember what injuries he sustained? I don't remember.

Janney: He was shot, I want to say, under the armpit, under his arm so it was a chest wound through that way, and it missed his body armor. They were trying to do a helo evac but the LZ was too hot and they couldn't get him out. I don't know if he expired on the battlefield.

Grimes: I want to say I treated him. I remember an entry and exit wound on an individual. Here's the thing about the hot and cold LZ. A Marine Corps pilot was a casevac of opportunity, "Well, is it hot or cold?" "Well, it's hot." "Oh, okay." And they would still land. They might land a lot faster and take off a lot faster, but they would land. The Army, unless it was cold, would not land at all.

Janney: Maybe it was an Army bird then.

Grimes: I don't know because half the time, we just told the Army bird, "Yup, it's cold. Just drop in fast before it becomes hot." Even though we're taking shots as they're fucking landing, we were trying to get them in and get them out because we had that brick wall that would protect them and then we would have a circle of guys around them and cordon to give them suppressive fire to help keep them from getting hit.

But the Army guys, if it was remotely warm, they would not land. Which was frustrating as all shit for us. I remember a kid that came in, God, you're saying under the armpit and that sounds really familiar. God, there were some guys that had so many bullet holes. I'm thinking it may have been a different guy, because I know that somebody ended up, there was a hole below the right armpit that had partially severed the aorta and we worked on him for a while. We found out later that the aorta was severed and there was nothing we could do. But, we were pumping fluids and trying to patch holes as quickly as we possibly could. It may not have been him, I'm not sure. As I said, it was like casualty after casualty after casualty and after ten to fifteen, it blurs together and I'm literally putting fingers in holes and trying to stem bleeding and just trying to do what you can with what you've got.

Janney: Yeah, I can't even imagine, Doc. I'm okay with blood as long as it's not mine.

Grimes: Yeah, I know. I've been up to my elbow in people's bodies. I've actually massaged the heart directly with my hands because there was a hole in the stomach and dropping intestines in. You name it, I've done it pretty much. Sometimes you don't have time to look at their face. You don't really make it a point because they are all anatomical objects to be fixed, so you have to compartmentalize that emotional response of, "Oh shit. I borrowed a DVD from him yesterday." I need to get him fixed and then I'll figure it out. That's kind of how I operated. No matter what, I'm just working on people. I worked on Iraqis. I had to work on a guy that I fucking shot. I specifically shot him twice. We rolled up on him and I had to fucking fix him. Now, I know a lot of Corpsmen that were like, "Finish him." I might not like the job that I have to do sometimes, but even if I know I shot him, I still could do what I could do to fix him, but managed my supplies carefully. Because I knew I had to continue operating. You don't always have to like the job you do and that's kind of the caveat that I didn't like of having to treat my own people that I wounded. I shot people and then I had to treat them. It sucks, but that's the job that I picked.

Janney: Yes, absolutely. You're a good Corpsman and that "do no harm" thing.

Grimes: Yeah, that's a really fine line sometimes and it's a tough call sometimes. I know this guy was just standing there shooting at me and now you want me to patch him up. So, I learned real quickly that one good bullet deserves another. So, if I shot somebody, they were finished before we rolled up on them so that I didn't have to waste my supplies on them. You put yourself in that position; I'm going to finish it. That was an enlightening moment of having to treat a head wound that I personally inflicted; to save the guy - maybe we can get some intel out of him. That was an enlightening experience. I realized one good bullet deserves another and two bullets wasn't enough, so if you had to do five, you do five, but you make sure they don't get up. I don't have to waste my supplies on them, so that I can use them in a gunfight later today on my own Marines.

Janney: After those 4 or 5 days in early April, more of the same? Tell me what you remember.

Grimes: We took a lot of rocket attacks. We had a lot of ambushes. There'd be days, especially in the beginning after April, I would count between 9 and 12 IEDs going off every single day. It was just constant. We would get mortared. One time we got mortared and a mortar popped through the roof of one of our berthing areas and it was sitting there spinning on the deck. Because of the way it had hit the concrete roof on the berthing area, it had broken off the fuse. So, we had to call in EOD to come get this damn thing. I remember running into the room because we heard the pop, and there's this mortar round. It had to be a 60-mic spinning in the middle of the room. I'm like, "Oh, shit! Go ahead and get out of here." We called EOD and they just grabbed it and got rid of it. EOD does their thing eventually when they finally show up.

Yeah, a lot of random gunfights. We got translators to antagonize them a little bit. I don't remember what we got the translator to say, but he would not want to translate what we were telling him to say. He had the bullhorn to just bring them out. After all of this, we were looking for a gunfight. We were full of rage. We were getting vindication on this. If you're going to fight us, come do it now. So, all the translators that we went through did not want to translate what we were telling them to say over the bullhorn because we were trying to start a fight. Like, "Hey, if you're going to come out, come out now." We lost a good number of translators unfortunately. We usually found them in the river later. It was the nature of the beast considering that we were a free fire zone and units were not allowed to travel through the Ramadi AO without permission from different commands, so we knew they were coming. We almost shot up an Army convoy because they randomly decided to drive through the AOR. That one almost got ugly because we were setting up an ambush on that road. Here comes the Army. Lights on, hauling ass. So, we almost lit them up because all we could see was headlights and nobody was supposed to be here.

Janney: Yeah, that would make for a really crazy day.

Grimes: It was right after the four Humvees were stolen from the Army camp somewhere. I know one of them had a 50 cal, one had a Mark 19. I don't know what else the other ones had, but there were four Humvees that were stolen from an Army camp and, no shit, not two days later, we're patrolling along MSR Michigan, and suddenly we have multiple headlights hauling ass coming down the road and you can kind of tell, all right, that kind of looks like a Humvee. We thought it might be the stolen ones, so we were getting ready to light them up because we weren't sure. Nobody was supposed to be there. If anybody comes through that city, whoever was operating in that area would be made aware of it. They'd give us a call from Bastard or Joker, "Hey, by the way, be advised units coming through at this time and this time window." "Okay, copy, good to go." But, we didn't know about these guys. We were about to light up this whole Army convoy because they were fucking stupid. Thank God we didn't because we thought, "Ah, this doesn't quite look right." But, that was right before we got pinned down for like six hours as I'm lying on this roof smoking a cigarette and having 50 cal SLAP rounds popping through the roof because the Army was stupid and shot at everything just because they happened to see us at an over watch point on an OP.

We were on an OP, I forgot where we were, but we were on the roof. We were doing over watch, checking everything out. It was the middle of the day, and an Army unit came through, and they happened to see my Marines watching them. I'm sure they saw the helmets or something, but they go into a rapid fire minute and they start shooting everything in a 360 degree direction. They are literally shooting through the building and 50 cal slappers are popping through the roof where we're standing. So, we're doing the star clusters and everything and we're trying to call the radio. Every time we popped a star cluster, the cyclic rate of the fire reaching our building would go up. I have the PERK 119 on my belt and the radio operator is saying, "We're going to kill them. We're going to engage this unit because they are going to kill us." We're saying, "No, no, no. Don't engage. Don't engage." So, we're all hanging out, laying on the roof smoking cigarettes because there was nothing else we could do. I have SLAP rounds popping through the roof all around me. We got Battalion telling us, "Don't kill the Army unit." The Army unit is continuing to shoot at us even though we are doing the pre-set star cluster signal of friendly fire. They kept shooting more and more. Then finally, they connected the dots, and we got a hold of that unit and told them, "Hey, knock it off and move on." Between the SAW gunners and the SMAW guys - I had two SMAW guys with me, and they were getting ready to light up each end of this convoy. We were so angry we were going to decimate this Army unit. We were going to pop the front and the back, and then just kill everybody in there because we were so mad. We got pinned down for hours because they're fucking stupid. Thank God we didn't look back, but at that time, we were just full of rage, piss, and vinegar.

It happens. I got shot up by a couple of Army guys and they're like, "Oh, my bad." "What do you mean, my bad? We're operating on parallel streets moving in the same direction and you're popping rounds at me? Are you out of your fucking mind?" "Oh, sorry!" "Bitch, that's the third time that you've shot at me. I'm going to kill you just for spite." Then, we'd move in other directions. I did not have very good interaction with the Army half the time, except for a couple of good armor guys and Charlie Med. They were awesome.

Janney: You got any funny stories?

Grimes: God, there are so many things. As a Corpsman, I usually bounced around. I bounced around with security. I would bring them soup and bread and whatever, just to keep them warm because it was kind of cold. Well, I walked into one of the buildings where we were berthing at Combat Outpost. I walk in and there's a ring of Marines around the perimeter. I'm not sure you want to put this in the book, but you can keep recording. There's like 4 or 5 guys standing in the middle of the room. Their pants are around their feet and all I hear is chanting and hooting and hollering. I'm like, "What the fuck did I walk into now?" As a Corpsman, my life revolves around: "Don't touch this. Don't have spider fights and snake races. Don't play with the bullets and do dumb shit. Don't have knife fights just because you're bored." I'm constantly giving advice, "Don't do this and if you do, I'm going to have to stand by because you're going to hurt yourselves." Kindergarteners with guns. I fucking love them, but goddamn they kept me busy sometimes. So, I walk into this room, and I see all this going on. I had about 10 seconds of, "Yeah, maybe I don't want to be in here for this." I turn around and roll out of the room. So, I give it a second and I ask one of the guys, "What the fuck are you guys doing in there?" The Marines that I saw with their drawers around their ankles were having a competition or a betting match for time and distance for who can shoot farther or faster. I'm like, "Are you kidding me. Really?"

Another time I walk in and a Marine comes in, "Hey Doc, theoretically speaking, if I have a live round and I put it somewhere I shouldn't, is there a way to remove it?" I said, "All right. Come here and sit down, talk to me, tell me the story, don't play hypotheticals. What did you do?" Well, there was apparently a bet involving a 50 cal and a rectum and if you push an object far enough into the rectum the body's natural suction will pull it in. So we had to retrieve this live 50 cal round out of this kid because he was trying to outdo the other people with the bet. I said, "We're not documenting this. Just come here. Let me get this fucking thing out. Don't touch it. Don't say anything and I won't say anything. Is it a deal? We'll say this never happened." So, I had to fish this 50 cal round out of this kid because he wanted to win the bet. Apparently, he did. So, we had to get this live 50 cal round out of this kid without the doctor knowing because the doctor would throw a goddamn fit. "Are you fucking kidding me? Really?" Dumb shit like that.

Another funnier example was third platoon. I went into their berthing area, and somehow, some way, they had set up in a little building with a little courtyard. All these assholes are sitting there in their silkies soaking in a kiddie tub, enjoying the sun, and having a great old time. Like it was a fucking beach party. All right. I'll play along. Where'd you get the kiddie tub from? It was an inflatable kiddie pool! They were having a good old time, just lounging in the sun, tanning and hanging out in this damn kiddie pool. Are you fucking kidding me? All right. I'll be back later once I'm not on duty. Okay, fair enough. Shit like that. You never know man, you never know.

We were dealing with all of the bugs and having the PMCs (private military company) come in the flight bay which stunk like high hell. God, we couldn't even shower for a good couple of months because our water guy was supposed to be delivering fresh water so we could set up our showers. They were solar showers, right. Well, I popped the top on it and this water didn't smell right. I opened the lid and there's a fucking fish swimming around our water tank. What they did was they took the water truck over and got water directly from the Euphrates River. They poured that directly in there to save themselves money. I'm gonna dig this fish out of the water tank, so we can actually take showers. So, we couldn't take showers for a couple of days which made me super popular.

Janney: I heard it was a catfish that was swimming around in the tank.

Grimes: It was a catfish! Yeah, that was part of my crew. I was in there going, "Uh?" We had to climb up and check all the tanks and then we had a water truck show up. We super chlorinated the shit out of it and so this driver was pissed at me because I was making him drive circles around Combat Outpost just to shake up and oxygenate the water, so the chlorine sterilizes it. Because I had to super chlorinate it so I could clean the damn tank out. Oh man, they were mad as shit at me. Then, the water truck driver ended up in the Euphrates, too. He was a shady dude.

So, the HET teams - Human Exportation Teams. There was a team of four to six Marines and I'm hanging out in the berthing area and I'm cleaning my weapon and I get a call, "Hey Doc, you're needed over at the showers over by this building." I go over there and he's a Staff Sergeant. He said, "Hey, I'm so and so with the HET teams and we have a couple of enemy combatants that we captured. We need you to go in there and take a look at them, do a physical and declare them as healthy, fit, and undamaged and then just sign this form." All right, I'll play along. What the hell is the HET team, so he explained it to me, and I pop in. It was in one of the random buildings with tiled showers and they had them locked up in there. So, I give these guys a quick physical, "Yup, he's fit, he's strong, he's not damaged, not shot, tortured, or anything like that. Good to go." All right. Thanks. Do you need anything else? Nope. We're good. Okay, so I roll out. A couple of hours later, Staff Sgt. comes over, "Hey Doc, I need you to do an after-action physical on this guy." I said, "Okay, which guy?" He says, "The guy you looked at before." I said, "Well, which one? I looked at three." He said, "I don't know. This guy. Here's the form. He's physically fit." I ask, "Where is he?" He said, "Oh, he's in the room." I said, "All right. Let me go see him." HET guy says, "No, no that won't be necessary. I just need for you to let me know that he is unharmed and not bleeding and stuff." So, I say, "Okay, I need to go see him." HET Sgt. says, "Yeah, no you don't." Ah fuck. I said, "Okay. You guys are part of the spooky people. Where's the form? Yup, sure. Where's the guy?" HET Sgt., "Oh, he's gone now." I said, "You've got to be fucking kidding me. You know what? I'm an E4. I don't get paid enough for this. I don't know what to fucking do. Give me the goddamned form and I'll sign off on it. He's fine." And they rolled out, doing whatever the OGA and the ODA and all the other people do. They take those guys, and they fucking roll out. I never saw those people again.

Between that and the kids. I had to pop a kid one time, so the CO was pissed off and he brought the sheiks and all them in. They're laying out the body of this kid that a couple of us shot up because the kid had an AK47 shooting at us at an OP. I said, "Hey, look, if you're going to come out, you're going to get shot. No matter who you are or how old you are, if you shoot at us, we're going to shoot back." Now, you are causing the death of these young kids. This kid had to be fucking 13, maybe. Of course, I had to fucking shoot him or else he would have shot me.

I have so many stories, Greg. I get called out. I was on a secondary patrol and an OP reported some kid shooting at them. So, we came around the corner, the kid is shooting at us, and we're all shooting at the kid and he dropped. This was a trash dumping area and I was literally fifteen feet from the kid. I could not see him, because after we shot him, he dropped like a sack and blended in with all the trash and debris in the area. The First Sergeant happened to be on patrol with us and said, "Hey, Doc go check him out." I said, "You know I'm the Corpsman right? This guy could be booby trapped or whatever. Are you fucking kidding me?" First Sgt. said, "No, no, no, Doc, you'll be fine. He's dead." So, I finally found the kid. Do you know what a dead man thump is?

Janney: No.

Grimes: Okay. So, if you're not sure somebody is dead, especially an enemy combatant, then you flick them in the eyeball as hard as you can. Nobody can tolerate that level of pain without flinching. So, what I did was I rolled up on this kid and I put my 9 mil in his ear, because even if he moves either way, the ear canal kind of holds the barrel enough to where I can blow his head off and not worry about it. Put my pistol in his ear and dead man thumped him just to make sure he was dead. That way, I can roll him over to see if he had a grenade or anything under him. Now, you're not supposed to send your Corpsman in to do that shit, but I just had that kind of luck. I put the pistol in his ear, dead man thumped him, he didn't move, searched under him, no booby traps. Rolled him over, he was very dead.

We continued our patrol. That was the one that I almost got fucking popped because, later in that patrol, I had to treat the guy I shot. I shot him, we rolled up on him. I throw him in the back of my gun truck, and as I'm working on him, bullets are coming in and bouncing between the armored plates on my truck as I'm trying to work on this guy. On that one, we heard a loud pop and myself and my aid gunner, who was on the gun mount up there, jumped out of the truck as a RPG was flying a few feet over the truck because they were trying to take out the truck. "You've got to be fucking kidding me." That was another hairy day. Weird stuff.

Did you hear about the reporters that got embedded with us? We had a couple of reporters that embedded with us. One of them was a female that was completely stupid. I cannot remember her name. Female reporter wouldn't listen to anything; she runs off and does whatever she wanted to do. I want to say a few months after we left, she ended up getting captured, and finally released. She was from NPR or somebody, I forget who. She was arrogant and fucking stupid. It was painful because every time we got stuck with a reporter, half of the time, they would attach them to me or another Corpsman, so that way if they get hit, medical's right there, fix them. Work on them, whatever, get them out there, and keep them alive. Thank God, I only had to deal with her a little bit. Then, she went out with another unit and said, "Hey, stay in the truck while we investigate something." She grabs her cameraman and she just runs out, running amok, away from the unit, no protection, and damn near gets her ass shot because the Marines are trying to figure out how to get her back in the truck. She was belligerent and absurd.

Well, one day, I got a former Marine, an independent reporter of some sort. I don't know if he was with a paper or something, but he was independent, I remember that. This guy must have been no older than thirty. I was told, "Hey, Doc, he's with you." Fuck, here we go again. We ended up going to the Agricultural building and we were setting up and we came under so much goddamn fire. They were trying to overrun this building. I was on the backside popping claymores. It was that bad. I asked, "You got a weapon?" He said, "Nope." I said, "You know how to use one?" He said, "Yep, I'm a former Marine." And I'm like, "Fuck yeah, here." So, I grabbed an M16A4 that one of my SAW gunners happened to have and gave it to him. I'm like "All right. You're going to shoot out this hole and here's your right and left lateral limits. If anyone comes in that area, you're going to shoot it outside the building." He said, "Okay." He dropped his camera, and he was shooting with us as we were trying not to get overrun at the Agricultural building. Well I left him there, he was on the second story, I think, and I dropped onto the mezzanine that looked over everything. As I'm shooting over the balcony type thing, it's like concrete, I'm shooting over the concrete trying to pop somebody that keeps shooting at me and next thing you know I'm lying on my back, like ten feet back because they had fired an RPG at that particular wall. Blew the wall to shit, our bells were rung as we're trying to get everything back together and get back shooting, but the reporter is damn near losing his goddamn mind, but he's still shooting, God bless him. We ended up getting out of that one, but we popped every claymore we had just to keep that building intact because they were getting really close to over-running us because we were shooting as quickly as we could. Another crazy day and I think that was May something. Wasn't too long after April. Crazy days. What else have you got?

Janney: Do you have any remembrances of any of the guys that didn't come home?

Grimes: Yes, quite a few of them.

Janney: Tell me what they were like.

Grimes: There was a couple that were really bad. The worst was Deryk Hallal and probably Nick Aldridge. Those are probably the ones that hit me the hardest because we would hang out and bullshit with them a lot. I know Hallal; he and I used to always fuck with each other.

Janney: He was a tall guy like you, right?

Grimes: Yeah, he was a tall dude. I'm pretty sure he had to be 6'2" like I am. He and I would always fuck with each other about who's taller. Well, he went out with one of my very junior Corpsmen and during the firefight, he was with 3rd platoon, that was Damon Rodriguez and all them. Hallal got hit. When this first happened, I wasn't there for this one, I was on a different side of the house. It was the 6th when they took fire, and Hallal took a round. I'm not exactly sure which round he took, but I know the one that killed him. I know that he had a round in his hip. They were moving Hallal from one building to another. Well, that one hit me hard man because I liked Hallal. That hit really hard because Hallal was a good dude. Just a really good guy, but a great Marine. I'll never forget Hallal, man.

    Aldridge was another one. Nick Aldridge. He died. There are so many of them. I'm kind of going through the list here that I remember and how it went. He was foot mobile, and I was vehicle mobile. Before we went on patrol, I was fucking with him and said, "I know I owe you a DVD, but I'm going to borrow a couple more movies because I'm bored." He said, "Okay, fine, Doc." So, I ended up borrowing a couple of movies from him and then we both went out on patrol. He was foot mobile, and I was in a vehicle. In the convoy that I was in, he was in the center median, foot mobile, and through lack of being able to see anything, because it was a super dark night and a communication issue, one of our Humvees hit Aldridge. So, we ended up medevac'ing him back. Basically, it crushed his skull for the most part. And so, I'm sitting there working on a guy that I had borrowed a damn movie from trying to get him back up on line and we lost him.

Janney: What was Hallal like though? That's really more of what I meant.

Grimes: Yeah, Deryk was a good dude. Most of the time, I was bouncing around with everybody else. We'd go on a field op or we'd practice squad rushes, I'm messing with different Marines and talking shit because they were like, "Oh, you're just a squid." I had to hold my own. Specifically, it was hard because it was so soon after I got to the unit that we deployed. I didn't have a ton of time to really get to know a lot of the Marines in that Company because there were 850, I think we had. In that whole battalion, and because I operated with so many different people. Then, obviously a couple hundred in Golf Company. I operated with so many different people before we went on deployment. Then during, I knew them well enough, but not where I'd say, "Hey, we're close friends. Come over to my house and drink beer." I would go to the barracks and drink with them and goof off. I had to sew some of them up because they were doing something stupid and not tell the chain of command. We're all hammered and I'm doing sutures in the barracks because I'd get a random phone call of, "Hey, we cut ourselves" and fuck, yeah, here we go. Being young and dumb, it was a very loose line of what I should be doing and what I should not be doing as a new Corpsman. So, I'd go in there and patch them up, fix them up, throw in an IV, and get them squared away. Then, I'd disappear and not document anything, so they didn't get in trouble.

Looking back as a Navy Chief now, I think, "Okay, I shouldn't have been doing that." But, a lot of it was just I want to take care of my guys and make sure they don't ruin careers over stupid adolescent pranks and throwing knives at each other's feet because it's funny. I knew everybody enough to where I knew their mannerisms. I can tell you how each one walked. If we were on a night patrol, I knew who was who based upon how they walked and how they moved. Because you can't see shit at night, so therefore, I had to identify them by their mannerisms in order to identify who I needed to talk to or where I needed to go if something were to come up. I can't tell you that I knew them all really well personally. I knew them personally, but I had to keep that bit of a distance there because I knew them medically, I knew them physically, I knew their mannerisms, and how they walked and what to expect out of them. "Okay, this guy is a common idiot, so he might do something dumb, but this guy won't, so he'll take care of him. This guy took well to training." But personally, there's only so much I can personally get engaged with because I had to, by proxy of the job, remain somewhat of a third party so that in the middle of the night they're not asking me to do illegal or immoral shit because they did something dumb. It's a really weird line to walk. That's for sure.

I knew them more operationally than I did personally, just because of the fact that I just got to the unit, and we deployed out. I knew some personally and I had personal relationships with a lot of them, even the ones that got shot or died or didn't come back. God, there were so many of them. I got my first rifle because they refused to give us rifles at first, but McPherson got hit by a bicycle IED. It took out his jaw and he got medevaced out. I ended up cleaning all of his tissue off of this rifle and said, "I'm keeping this so that I can have a rifle." Because I was operating with just a pistol up to that point when we would go on patrol. Little shit like that, I don't know it's kind of hard to explain on some fronts you know.

Janney: Yeah. I heard the story about that bicycle IED.

Grimes: Yeah, that was Mack. That's how I got his M16A4. That's how I got my rifle. Finally, the CO said, "Okay, issue all the Corpsmen rifles and NVGs." That's what convinced him. I said, "I cleaned all the tissue off this and I'm fucking keeping it." He said, "You know what, Doc? Okay. I get it." I kept it. I just didn't turn it back into the armory after I checked it out. The armorers did not want to clean all the tissue and skin and blood off of the damned thing. So, I took it and cleaned it up and made it serviceable and that's what I used for the whole deployment. You do what you've got to do. There's not enough rifles to go around and they weren't willing to give them to Corpsmen at first, so I finally got all my Corpsmen issued rifles after that.

Janney: And NVG's.

Grimes: Some of us got NVGs. I got the monocle. Some of us got the ones that go on both your eyes. I forgot the name of it. I always had the monocle, and I used the shit out of it, that's for sure. That's why I always did the night missions because I had the monocle. You could shoot with it, so it made it handy. You could work on people with it.

Janney: Didn't you find that it affected depth perception?

Grimes: Laughs. It fucked up your depth perception, that's for damn sure. You know how many holes and random things I fell into just because it looked like a dark spot of mud? But, you learn real quickly. You go, "Okay, is that a hole or is that just a dark spot." What I would do because I always had the small LED lights that were set to blue, with the blue filter on it. So, that way if I had my NVG on and there was enough moonlight, I could work on somebody. If there wasn't, then I would cover up, slip the NVG out of the way and I could flip on my blue light. Then, I could tell the difference between blood or mud or water with the blue light. That's why we kept red or blue lights in the back of the ambulances. Because blood turns purple under a blue light, but mud stays black. Water obviously looks clear. If you have red, you might not be able to tell, it's either mud or blood because of the red lights. If you flip it to blue, and that's why most 997s have blue light inside the box automatically because it turns blood purple, but the only problem is you can't see veins very well. So, you flip to your red flashlight and that way the blue and red light turn purple, so you can see veins if you have to. If not, you do it by feel.

Janney: Wow. Okay. That's very interesting. My wife is a NP.

Grimes: Oh yeah. So, I'm sure you get all kinds of information. Some of the Marines do not understand why we have a blue light in the back of our ambulance, "Oh, it's just Navy." "No, it's medically proficient. Here's why. You just can't see veins very well." Because cities that have drug problems will put blue lights in their public bathrooms because with a blue light you can't see a vein because obviously the vein is blue. But with a red light you can, so if you have your red filter flashlight, you can use that to help you find the veins, but 9 times out of 10, you're doing it by feel anyway because you know where the grooves are. So, you just leave the blue light on.

Janney: Okay. I got you. I was walking around JSS Karama where John Wroblewski and I stayed with Captain Martin, the guy that was in charge of that JSS. I had a green head lamp on and it was pitch black, there was no moonlight and I'm tripping over things until I flipped this light on. I said, "How come you don't have a headlamp, Captain Martin?" He said, "Because the snipers shoot at the guys with green headlamps." Laughs.

Grimes: Laughs. If you're out walking around, you don't want to turn on a light at all. You just learn to operate in complete darkness. It's interesting.

Janney: Yeah. So, I turned that headlamp off and didn't use it for the rest of the time I was there.

Grimes: Laughs. That's funny as hell. It's like cigarettes. You can see cigarettes from I don't know how far away. That bright flash. Oh, yeah.

Janney: And smell them even further.

Grimes: Oh God, yeah. You can smell them for hundreds and hundreds of yards. I don't know what the norm is, but yeah, you can smell them coming.

Janney: I've heard you can smell cigarette smoke for a mile if the wind is just right.

Grimes: Yeah, that's why when I was in Iraq, I used to smoke, but I flipped to dipping because lying in ambush you can't smoke, and I don't want to smell like it. You can smell it from so far away. You can smell them coming if you are downwind. It's plain as day. Especially if you can smell the body odor, you can smell the cigarettes. You think, "Oh, okay, there's somebody out bouncing around, give them a minute." And then no shit, they pop around a corner and you're like, "Oh, yeah. There you are. I knew you were there somewhere." You got me thinking about a good story. I was out on nighttime patrol one day. I had a squad radio on and I had my main radio. Well, we're walking along and as we were moving through the city, we'd pick up other people's squad frequencies periodically. Over my head set I hear, "Nah, nah, man. I can blow his dick off. What are you talking about?" These two snipers were going back and forth because there was a guy trying to bury an IED and one sniper was saying, because he was behind him, "No, no, I can shoot him through the face, and I can blow his dick off. I promise I can do that." "No, no, I bet you can't do that. I bet you a hundred bucks you cannot blow his dick off. I swear to God, you can't do it." And so, these two are going back and forth and I'm like, "What the fuck am I picking up?" and I'm cracking up. Within a minute or so, I hear the gunshot and, "See, see, I told you. You owe me a hundred bucks. I told you I could do it." Taking this guy out who was burying an IED and they just shot him through the lower back and blew his pecker off just on a bet and then finished him. This is our world right now.

Janney: Yeah, I'm sure you would appreciate some of the stories that Jesse Longoria and Jonathan Woods and Santiago have to tell..

Grimes: Longoria. Goddamn. He and I talk periodically.

Janney: Jesse is a hoot. I love that guy.

Grimes: There were a few guys. Robbie Wagner, I used to always operate with him. It would be the middle of the night, "Hey, Doc, Doc, if I get stuck with a needle and then I stick myself somewhere else with the needle because you're trying to give me an IV or something, will I get AIDS?" "Oh, God, everybody wake up. Let's do the STD brief and let's talk about this because this is fucking absurd."
You're keeping me up all night long asking me random questions. Fucking Robbie Wagner, goddamn. He cost me a lot of sleep over random questions. The gist of the stories is that there are so many of them and they blur together. I don't know what you're looking for in your book and I have to tread that fine line of what do you need to know to fill in blanks and what are the oddball things that people should probably know.

Janney: I get it. You've done an excellent job. I appreciate your time.

Grimes: Yeah, absolutely. Anytime you have a question or it's something that I may have been involved in feel free to reach out to me. I have no problem answering questions and at least filling in blanks, not only from the medical side. I jumped so many different squads and operated on so many different people. I made my mistakes here and there, but at least I can fill in some blanks that maybe people might not know. Like Wroblewski and things like that. Just because I was the guy on hand and happened to be there at the time.

Janney: Did you know if Doc Urena was with Lt. Ski on Route Gypsum?

Grimes: Yeah, yeah, I know Urena. I want to say Urena was with him because Ski was…

Janney: Everybody got jumbled up that day because of all of the shit that was going on.

Grimes: I just don't know if he was with him on that day. I forget who exactly I had assigned Urena to.

Janney: I know Doc Urena pretty well. We haven't gotten down to the interview part. If I remember correctly from our first conversation, he tried to render aid to Lt. Ski when he first got hit. I'm just vaguely remembering that. I just didn't know if you knew or not.

Grimes: I'm familiar with him. He was one of the eleven Corpsmen that worked for me at the time. Urena got there; he was really young. We had Cahill and all them. We had a variety of junior guys at the time and because I was doing a lot of the command element and stuff and making sure that was dealt with, "Hey, we need to have our P & T's here. We need to have people on patrols." I didn't see Urena a whole lot because he was kind of thrown into the fire. I'd tell the Corpsmen, "All right; go with them, get to know your Marines. Learn your craft. Let me know if you need anything. I'll be back with you shortly. I'm busy."

Janney: Did you know Mendes-Aceves very well?

Grimes: I knew him well enough. He and I got along. We were just different personalities.

Janney: I just didn't know what you remembered about him because not many people that I've talked to knew him very well. Tell me a little bit about him.

Grimes: Yeah, well, so he was stationed at the hospital and got assigned to us at the last minute before we rolled out on deployment. He knew his craft. He was a gun fighter. He really jumped in there. He was hardcore about his Marines. He knew his medical craft well. I never had a question about what his capabilities were with his unit. Never had a question. Never had to worry about him. And I know he was Echo. We all kind of interacted a little bit. Periodically, Doc Adam Clayton and I would do a pass down with what's going on with each other's units and how we can support each other. I remember finding Mendes' body in the back of that truck when I was checking pulses and everything. At least I know that the guys that stripped his medical gear off, we ended up getting them later because I ended up recovering his trauma shears. I didn't know him super well because he was attached and then we deployed. It was really, really rapid because he was stationed at Balboa at the time. Good dude. Operationally, very, very proficient. From all the Marines that I talked to, he went down with Staff Sgt. Walker, as far as I know. They fought to the very, very end. He died working on Walker, if I understand correctly.

Janney: Okay. I had heard that. I just didn't know if you knew that end of the story, too. I heard that he was rendering aid when he went down.

Grimes: My understanding is that they made it out of the vehicle and Otey was trying to get out of the ambush zone for that matter. A lot of the guys were hit in that one. I know Walker and Mendes made it out of that vehicle. I know Walker was hit and Mendes was working on him on a side road or an alley or something like that. My understanding is that, from the stories, at least how they were found, Mendes was still in the process of working on Staff Sgt. Walker when they came up and they got them. That's my understanding of it. I wasn't there, so I'm not sure.

Janney: That's more information than I had. I just heard that he was rendering aid. I've never heard to whom.

Grimes: As far as I know he was working on Staff Sgt. Walker and he was shot from behind while tending the wounds on Walker. That's what I understand. I've heard that from a few people. Especially how the bodies were found because we had to go back and fight our way back in and pick up bodies that we retrieved.

Janney: Yeah. That was a terrible day.

Grimes: That was a rough one. I remember when I found him within the stack of bodies, his face was intact – it was not like he was shot through the back of the head. I'm not sure how he took rounds because I was going through people so fast trying to find anybody that might be alive before I bagged them and dealing with all that.

Janney: I appreciate you sharing that about him. That's a part of the story that I didn't know.

Grimes: So, now we need to confirm that. That's just what I heard from a few people about how the bodies were found.

Janney: That makes sense. Nothing is definitive, you know what I mean. The fog of war. If that's what you heard, then I'm assuming that's what happened.

6 April 2004 Golf Company

Sgt. Glenn Ford

Janney: Today is 20th April 2020. Tell me why you decided to enlist in the Marine Corps.

Ford: So, I initially went in 1992. I was in the delayed entry program for about a year and my initial thought of going in was, I wanted to race bicycles competitively in college, but I went to go and try out and basically found out that I was good enough on the bike to race at the collegiate level, but I didn't do enough homework because I was too busy out riding my bike. I come from a military family. I grew up overseas living in Japan and was exposed to Naval and Marine officers. My Dad was a civilian working for the Navy Exchange system and I thought back to those guys that I would see at the O club in their dress blues. I remember my Dad always telling me when I asked him, "Hey, what do those guys do?" He would say, "Oh, those guys are crazy. All they do is shoot stuff and blow things up." Well, they were actually Marine pilots and I realized, who doesn't want to shoot stuff and blow things up. That was post-Desert Storm; Desert Shield time and seeing that stuff on TV, sitting there thinking that would be something that I would want to do. So, I enlisted in 1992.

Janney: Okay, so you had been in continuously from then?

Ford: No, I had broken time. So, I was in '92 through 2000, and that's when Walker and I had first become friends. I re-enlisted in late '95 and checked into a new unit and Walker was there. We were both Lance Corporals together in 3/4. I decided to get out in 2000 and that was when Walker went down to the drill field. That last four years we were great buds. He would go to all the bike races with me because I started racing again. He always said that he was my soccer mom, he was the guy pinning my numbers on my jerseys. He would go to a feed zone and hand out my water bottles to me and do the feed stuff. Pretty much, at that time, we were in Twentynine Palms, so we basically had a free place to stay with my other friend from high school in Huntington Beach. So, that's where we were on the weekends when we weren't in Twentynine Palms.

Janney: In 2000, you were thinking about getting out. What was your rank at that time?

Ford: I was a Sergeant at that time.

Janney: So, you had gone from Lance Corporal to being promoted to Sergeant. Now, had Walker been promoted to Sergeant as well?

Ford: Yeah. At that time, when I got out, he was a Sergeant as well. He had accepted orders to go to the drill field. That's when he went down to San Diego to become a Drill Instructor. So, in the two and a half years that I was out, we were still hanging out. He would be working in cycles and whenever he had some down time between cycles, we would go hang out. Either at my apartment or at the bar that another good buddy of mine was running down there close to the Depot.

Janney: At which point did you decide to re-enlist?

Ford: So, in '01 when everything went down with 9/11, that was the kind of the process, the thought process of going back in. But, it was because Walker had said to me, "Hey, you either spend your time with other people that race bikes or other Marines that are still in." Like himself. I said, "Well, what are you getting at?" He said, "Well, it has never left you. You still live this disciplined lifestyle. You're going out and you're doing your training, racing every weekend, racing four times a weekend. This is the Super Bowl. This is what you did for the first eight years to prepare for war. Why don't you look into coming back?" So, it took a while because I had the whole re-enlistment process of bringing back former guys or what they call "retreads". They were putting that process back together in a more efficient way. So, then it was springtime 2003, and at this point I had called every recruiting office starting in San Diego and worked my way up all the way to Northern California where I grew up. I said, "Hey, I was initially enlisted, and then I was a NCO. I want to come back in. I was a Sergeant. I was a 0351 and a 0311. What can you do?" The boss in the office basically said, "Hey, we'll work on your package. How soon can you get up here?" So, when I got up there, it was about 48 hours later, and he said you need to retake your ASVAB; you need to score a 50 or higher to be able to be accepted back in. The only downside of this is that you are losing all of your time in grade as a Sergeant." At that point, I believe, it was two or three years that I had in grade, so if I were to go back in, I was basically starting at day one as a Sergeant, but the records showed that I had the experience for it. So, I went and got the physical and took the test. Then in August was when my approval came back in.

At that point was when Walker was coming toward the end of his time at the depot - that Fall, while I was on the recruiter's list waiting for orders to my unit. I had already got them and I called him up because he said, "Hey, as soon as you find out what Battalion you're going to, give me a call because I can set it up with the monitor when I can go there too." That's when I called him up and I said, "Hey, I'm going to 2/4. They're leaving in February or March." He said, "All right." He called me up about a week later and said, "Hey man, I'm standing in front of the monitor. Where is it you are going again?" I said, "2/4" and he looked at the monitor and said, "Hey Gunny, can I go to 2/4?" Gunny said, "Heck yeah, you can go to 2/4." So, that was when we linked up there about a month later in December.

Janney: Okay, so December of 2003?

Ford: 2003. It was around Christmas because I remember they were looking for people to stand duty on Christmas and I volunteered for Assistant Officer of the day because I was one of the few guys that was there. Everybody else had gone out on Christmas leave knowing that the deployment was coming up. The crazy thing was the night I pulled duty, I ended up pulling duty with Walker as platoon commander. He had just checked into the Battalion himself.

Janney: Was Walker on duty then, too?

Ford: No. He had checked in the following week, I think. We checked in really close to the same time.

Janney: Tell me a little bit about Allan Walker. As I said, a lot of the guys that I've interviewed served under him, and maybe not even directly under him. One thing that I'd heard that he said was his main goal was to make sure that everybody got home alive. That was one of his goals.

Ford: Yeah. So, the funny thing is, that statement right there was the last thing he had said to me. My squad got ambushed in March when we got there; it was the first decisive engagement that went to kinetic at nighttime. It was a night ambush right there on Route Michigan, right down the street from OP Hotel. There was a green bridge and we had turned back just past the circle by the Saddam Mosque and we basically got lit up. The ambush was kicked off with RPGs that overflew us. It was a contact rear and we got into a fight. It lasted about 10 – 15 minutes and we patrolled back in. I was all ramped up and the guys had done really good. I had one guy that got hit, Yansky, in my squad. He got a through and through in one of his arms.

So, I was up, making sure he was good while the guys were standing down. I went to the chow hall and Walker was sitting there. I said, "What's up, bro?" and he was just kind of in his thoughts doing his thing. I knew he was preparing for something. I sat down, and we shot the shit a little bit and I got up to leave and I said, "Hey, I gotta go check on my guy Yansky" and he said, "Hey, give him my best." Well, Walker knew who Yansky was because he was Yansky's Drill Instructor at the depot. I patted him on the back, and I said, "Hey, I'll check up with you later." The last thing he had said to me there was, "Hey man, no matter what you do, make sure you take care of your boys. Make sure you're looking out for them. Make sure that if they are doing something, you actually do the work and write them up for an award." That was pretty much the last serious conversation that we had when we were kneecap to kneecap at a table. You know, like a couple of buds. Even though, at the time, he was a Staff Sgt., and I was a Sergeant, it was one of those things where it was two old buddies sharing some time together. There wasn't a lot of time for the guys in this platoon to get to know him because, you know, he had just checked in in December.

Janney: Exactly. Walker was the DI for a couple of the guys I interviewed

Ford: Yeah, so it was Yansky, Irwin, Zimmerman; there were a bunch of guys in Golf Company that he was their Senior Drill Instructor for. A lot of the guys in Golf Company were freaking out when they saw him. Because he was walking through the area and I was in my room and I yelled down, "Hey, what's up fuck face!" He looked up and he just started laughing. He said, "What's up, dude?" I looked down the catwalk and all these guys there in Golf Company were just frozen like, "OH MY GOD!" I said, "What are you guys doing?" One said, "Sergeant, that was our Senior Drill Instructor." I said, "Well, good. Don't disappoint him. Go back to what you were doing."

Once, we were either on the convoy or on the truck ride into Ramadi, I said, "Hey, why were you guys all freaking out when you saw Staff Sgt. Walker walked by, and I said hello to him?" They said, "Well, he told us that he was a cook. We never knew that he was an infantryman." That was typical of Walker because he knew that those guys, with what was going on, would constantly be asking him, "Hey, what's it like to be in the grub?" "What did you do for your job?" That was just like his character to tell them a complete lie that he was a grub and he wasn't a grunt. He did that to get his point across that, "Everybody's got a job in the Marine Corps." It doesn't matter what you do because not everybody gets to pick what they do. So, to make it a level playing field, he told them something that wasn't near the truth at all. But, he still kept the training relevant to everybody that he wasn't just an infantry guy making infantry guys - he was making Marines. That's what made perfect sense to me. Those thoughts didn't dawn on me until after that deployment when I went out as a recruiter, and I basically had to sell the Marine Corps and not sell the job.

Janney: Yeah, every Marine is a rifleman, so that means a lot.

Ford. Yeah.

Janney: You were in Golf Company or Weapons Company? What Company were you in?

Ford: I was in Golf. There were only a handful of us that knew Walker as well as I did. Me and Walker were in the same platoon together for four years at 3/4. Reagan Hodges was in Weapons Company in 2/4. He was in our same Company in 3/4. He was another guy that knew him well. He was roommates with Walker at 3/4.

Janney: Well, that's good to know. I'm scheduled to interview Reagan soon.

Ford: Oh, yeah. Great guy. Then, there's Winston Jaugan. He was our Company Gunny at 2/4 Golf Company. He was a squad leader in Kilo Company in 3/4. So, there was a small clique of us from 3/4 that basically got put back together in 2/4.

Janney: Did you guys convoy into Ramadi or did you fly in?

Ford: We convoyed in from Camp Victory. I believe it was a two- or three-day movement that we did on the back of 7-tons. I believe Echo Company did the same movement, but their movement was a little bit staggered from ours. I think we got there a day or two ahead of Echo.

Janney: Were you at Hurricane Point or at Combat Outpost?

Ford: We were co-located with Echo Company at the Outpost.

Janney: You told me a little bit about that action in March. I don't guess that had anything to do with the guys that were sent out to try to catch the mortar men on 25 March.

Ford: No, that was a different situation. My ambush happened on the night of 23rd of March. I heard later on from the Intel guys that the ambush was kind of one of those probing ambushes to see what the capabilities were for us to be able to fight at night. We never really got it confirmed, but just from the debrief that I did and what the guys were telling me as part of the debriefs, I think we tagged like two or three guys that night. But, by the time we went to go check for the bodies and stuff like that, they were gone. They had even policed up all the casings because my squad went the next time we were on day ops through a couple of the buildings to see if we could find any of the casings and they were all gone as well.

Janney: Yeah, they were pretty good about dragging the guys off, I heard.

Ford: Yeah, they definitely were. We definitely saw blood stains and stuff like that, but what they basically figured out was these guys are not going to be afraid to fire back. Because from what we had heard from the guys at the Outpost, it sounded like a whole Company had fallen down and it was just your one squad. We were definitely giving them hell because we were throwing 40 mic grenades down and all the SAWs were going and we definitely tried to maneuver, but we didn't want to give up the original objective because my squad was basically running a recon patrol along the route that the Company was gearing up for going out. Because as soon as we got into the wire, there was another platoon that was loaded up in vehicles. They were going to do a hit on a house, so my squad's job was to go recon the route and go by the target house - we got hit on the way back in. We were about 600 meters to a klick down the road from the Outpost when we got hit.

Janney: And that was in toward the city? You were coming from the mosque, you said.

Ford: Yeah, we were down by Saddam Mosque. We had come out right in front of it. We took a right on Michigan and we were patrolling back. It was typical - we were on the home stretch and the last leg of our route and it was going to take us right into the gate at the Outpost and then all hell broke loose. Because one of the things that happened was, as we were walking, the lights went out and as soon as the lights went out, I said to myself, "Oh, shit." That's when we saw the RPGs fly over us and then impact in front of the squad. Everybody turned around and I was walking with the last fire team talking with my radio operator to let them know that we had just hit our last checkpoint and we ended up turning around and started firing. Then my platoon Sergeant, Staff Sgt. Calloway had gone to the right and I went to the left because Yansky got hit in that first volley of fire. Then, it was me and Doc Vagnorise. We ended up pulling Yansky off to the side into cover. The Doc started working on him and that's when I started trying to pull the other two fire teams back to get online and start to fire. I was with the last fire team that was there and we were trying to get everybody online to attempt to maneuver. Once we gained fire superiority, I pushed one team down and back out. They couldn't find anything, so they ended up strong pointing a house that had a second floor. They were looking down an alleyway and we were doing the sweeps, trying to find anything. We weren't able to find anything. Then the Company did send out a vehicle to pick up Yansky and they medevaced him and we continued the foot patrol back in after the contact was over.

Janney: Did they actually medevac Yansky to CSH in Baghdad?

Ford: No, they stabilized him at the Outpost and then they flew him out. Actually, you're right, they did fly him out and I think he went to TQ. He might have gone to Baghdad, I'm not solid. All I know is that a couple of weeks later, he was back because it was a through and through. After he had gotten back, he still had some nerve issues, but they did a couple of surgeries. He ended up going on the deployment after we had gotten back. He was in Echo Company at that point.

Janney: It sounds like that 23rd March action was the first real combat that 2/4 experienced.

Ford: From talking with everybody after the fact, once we started getting a little more down time, I think it might have been Colonel Bronzi or somebody said they started calling me the Night Rider. I said, "Well, what the heck is that?" They said, "Well, we call you the Night Rider because you did the last night engagement that the insurgents would have and you guys have done so good that nobody else got ambushed at night." I thought that's a badass nickname. I'll hang on to that one.

Janney: Tell me what happened next after the 23rd.

Ford: Yeah, I would say if they fought on 25 March and got the mortar team, it was probably another proving type of event because the next big thing that happened after that was April 6th. Because after the ambush that I was in, it was kind of treated like, "Yeah, that's going to happen." Let's record it and it was just something that happened. We were doing our job. What's the next mission, you know? What are we doing after that? That's when April 6th happened.

Janney: What I understand about the 25th was they got some radar information from the 82nd Airborne and that's how they got a grid to go out and try to get those guys on the 25th of March.

Ford: Yeah, that sounds accurate because I know that there were a couple of situations, that one in particular, where the counter battery was pretty effective, so anytime that we got indirect fire that was when we started realizing that we would hear these artillery guns going off, and we said, "What the heck? Where's that coming from?" I would run over to our COC and say, "Who's shooting artillery?" They said, "Counter battery. They've got a lock on where we're getting some indirect fire from." I'm sure they might have gotten a grid to get action on it if they already had a platoon out or if they had artillery up already.

Janney: Tell me, were you guys patrolling every day? A lot of the guys told me they were basically doing a combination of mounted/dismounted, kicking trash looking for IEDs. What were you doing in the days leading up to the 6th?

Ford: For the most part, during that period, we were on foot almost all the time because we were still in, I think what Colonel Bronzi called it, our acclimation to the city. So, it was going out handing out candy, shaking hands. One of my objectives was trying to find people that we could communicate with to say, "Hey, if you've got any information for us, we want to know." You know that kind of a thing. At that point, we were still given the freedom as squad leaders in Golf Company to pick our routes and then submit them for approval by a watch officer. Our patrols were going maybe 4 – 5 hours at a time. So, you can probably imagine we were patrolling maybe 7 – 10 klicks in that period.

Janney: And you were operating in the Sofia district or in the east, the Gypsum side?

Ford: We were on the Combat Outpost side of the city. We were in the urban area. We were in the neighborhoods. We were in the industrial park which was really shady because it was basically like a big ass parking lot that was like a storage facility that had random garages that were locked. Some of them were open during the day. They had auto shops working on stuff and just dumping oil out on the sand. Some shady characters were in there. My only thought was, "Okay, what the fuck are they building in here? I'm sure they're putting together IEDs." So, on one of our Battalion operations, our Company was tasked with going through there. We just started popping locks and opened up the garage doors and we were finding money and weapons, IED making materials, you name it.

  The areas that we were patrolling, the hairiest part was going up and down the neighborhood streets, which you do, patrol after patrol. You've got to throw something in the mix to kind of break up the pattern. So, sometimes we would skirt around the neighborhoods, going through the fields and stuff along the railroad track and then cut in on the back side and then patrol back through the urban part. Or go through the urban part and take the rural area back. Echo's side of the area seemed to be more out in the country. The couple Battalion ops that we did in their area had a lot of wide-open fields, canals that were dug, a lot of stuff like that. So, they definitely had a different look in their area compared to ours. So, basically our side was the urban area that went from Combat Outpost all the way to the government center on both sides of Michigan.

Janney: Tell me that you were doing on the 6th of April? What was going on?

Ford: We were the QRF for the Company that day. I think we had just come on. We were doing shifts in the ready room. We would spend 4 or 6 hour shifts there. Basically, we were just hanging out. Then, next thing you know, all hell breaks loose. They came running over to us, snatched me up and said, "Hey, get your boys ready to go. Go back to your house and let the other squad know they need to be ready to go." We went over and got them all up. The platoon met up right there in front of the ready room and then we got pushed out. We loaded up the vehicles and then went out. They dropped us off and said, "Hey squads, you've got this way, this way and this way as far as the streets. Start pushing towards the government center and clear everything."

We just started clearing houses. At that point, all I had heard was that we had taken a couple of KIAs from 3rd platoon, and we were maneuvering to link up with 3rd because they were strong pointing a house at that point. We started clearing. Normally, I would send a pair to do over watch covering us from the street. About halfway through, Weapons Company rolled up as Battalion QRF. I saw Hodges in a vehicle going down the street. I said, "Hey, what the fuck are you guys doing?" He said, "We're going to cover your movement down the street because we've got the heavy gun on top. We're just going to be giving you some extra firepower as you maneuver down the street." Weapons moved with us as we swept towards 3rd platoon's position.

We were searching through houses. At that point, the guidance was, "Take whatever you find as far as weapons or IED making materials. If you do come across something, call it up into the Company and we'll push whatever outfit your way, whether it's EOD or interpreters." We didn't end up finding a lot. We confiscated a couple of AKs and a pistol. But, the overall mission was to link up with 3rd and help them get back.

Janney: What time did you start hearing the ambush on Gypsum? 1000? 1100?

Ford: I would say it was late morning. At that point, there wasn't a lot of sharing between the Companies of what was going on. We had known that Echo was getting hit kind of at the same time that our guys were getting hit. All I know is what was coming out of our COC was all about what 3rd was going through that day and that we did have KIAs on scene. Echo was doing their thing. So, my focus at that point was, "Okay, everybody needs to be doing what they are told." We were told to get up on the trucks and make sure that everybody is ready to get into the fight. And we were.

Then we went and did our movements down our lanes, down our streets. The whole Company was online for however many streets there were towards 3rd platoon's position. At one point, I think it was the 6th, there was a Sergeant from Echo Company that was rolling around. He was bringing wounded guys from Golf back to the Outpost. We said, "Where are your other vehicles?" He said, "Well, we knew you guys were getting hit, so I just jumped in a vehicle to help out." I said, "Okay. Where's your driver?" It was just him driving this truck from Echo Company, helping us out. I said, "Okay." As we were going through, we were shooting quite a bit. We were returning fire. I had a couple of guys that smoked a couple of guys on some rooftops with 40 mic grenades that they launched out of their rifles. They would get an ID on something and would start shooting that way. You can probably imagine. I'm running between fire teams going, "What the hell are you shooting at?" Or it's coming over the headset, "Hey, we've got two guys here. I see weapons." My standard reply was, "What the hell are you waiting on?" and you'd hear them opening up. We got to the consolidation point and did a Company movement back to the Outpost.

By that time, it was starting to get dark and I got all my guys back to the ready room. Between myself and the other squad leaders, we made the call, "Hey, we release you guys." We got called out for the remainder of our shift in the ready room." Everybody was cool with that. The guys started laying down to rest. The adrenaline would leave them as soon as we got back.

So, at that point was when I started walking around. I started asking guys from Echo Company, "Have you seen Staff Sgt. Walker?" The guys there said, "No." I walked into the COC, and I got stopped by Colonel Bronzi who said, "Hey, what are you doing? You're supposed to be in the ready room." I said, "I'm looking for Staff Sgt. Walker." At the time, I didn't realize what was going on, but Bronzi said, "Sgt. Ford, get your ass back to your ready room and take care of your Marines." That was when I basically found out. Somebody came and told me, "Hey, Staff Sgt. Walker is dead." I said, "What the hell happened?" It came out later on that he had gotten killed in that ambush going toward that house. When he dismounted, he was telling those guys to move, and they were basically calling him into the fight, to go strong point that house. They drew him into that house and they had a DShK 50 cal machine gun mounted in the window and basically mowed down his platoon as he was moving towards that house.

Janney: Yeah, that's pretty much what I heard too. They had a well-planned, L shaped ambush at the intersection of Nova and Gypsum, too. I don't know if Staff Sgt. Walker and the other guys that were in that first vehicle were part of the QRF that was sent to Headhunter II sniper team: Santiago, Stayskal, (redacted), and Ferguson were in contact. They were doing route watch on Nova to look for IEDs and those guys got hit pretty hard. Thirty guys surrounded the snipers and were trying to flank and kill them essentially, so Headhunter II called for a QRF. I don't know if Staff Sgt. Walker's team was part of the QRF or if that was just a patrol that he was leading.

Ford: No, it was definitely I think a QRF since they were rolling platoon deep in vehicles.

Janney: There was also an IED on Nova further west where Nova curves down into the city, so there was a lot going on that day. But, I think they were trying to draw Marines up Gypsum because, as I said, they had that DShK set up there at Nova and Gypsum. So, it was just a terrible situation. I don't think there was any way that they could have avoided that. I'm really sorry about your friend. I have heard vicariously a lot of wonderful things about Staff Sgt. Walker, but I have not had the opportunity to talk to anyone that knew him like you did.

Ford: Oh yeah. I appreciate it. They named a building after him at the Depot. It's the facility where they take care of injured recruits. They named that new building after him. So, there's two versions. The guys that knew him from the Depot called him "The Beast" because he was just a big, hardworking guy, a big dude. All the guys from 3/4 that knew him when we were all Lance Corporals together, we called him "Wookie." Short for Walker.

Janney: Laughs.

Ford: You'll get two versions: "Oh, are you talking about Beast?" "No, Wookie was first." So, you'll get those conflicting names there. He was always that guy, even when we were Lance Corporals, he was always that guy that was taking care of and making sure that everybody was good. Always number one when it came to acknowledging whatever job we were doing. The guy definitely cared about being a Marine and loved what he did.

Even up to the point of talking me into coming back in, putting it in a way that I would understand. "Hey, your life has been about racing your bicycle. That takes discipline. That is a person that wants to surround themselves with certain people." He said, "This is the Super Bowl. If you want to be a part of it, you've got to come back in. It's definitely something that we can still do together." I thank him every day because I ended up riding it out. I retired back in 2014. To find guys that want to be Marines, they always tell you in recruiter's school that you can recruit in your own image, but I find that 50% true. I think for a lot of the guys that were fresh off combat tours, I think you recruit in the image of your friends, your brothers. Being able to share his story like that and to know that his legacy will live on. I knew by talking to you when you had said that there weren't a lot of guys that knew him, I knew that was going to be the point. Because he had only been there about thirty days before anybody was starting to work up and preparing to deploy. So, even his Marines under him didn't have the time to get to know him because it was so early in the deployment.

Janney: Exactly. As I said, I've talked to a couple of guys that Walker was actually their DI, but a lot of the other guys didn't have that experience with him. Didn't know him as well. I heard he was an incredible mentor though. He was very hands-on about trying to instruct his Marines and try to do what he could to make sure they were doing the right thing and staying as safe as possible, but yet still going after the enemy.

Ford: Yeah. That's something that we had learned coming up together. We were kind of in that mid-'90s, you know, we joked about "the crucible Marines." We were that kind of early generation where it was okay to question authority as long as we did it tactfully. We were the "wanting to know why" generation. He told me some scenarios when he was a Drill Instructor and there were ways that he had handled some things. He was always the guy that was always one step behind in rank. When I checked in, we were both Lance Corporals, but then a couple of months later, I picked up Corporal. So, he was one step behind then. He picked up Corporal about a year later and I picked up Sergeant really quickly. He observed how I did things. He told me one time while we were drinking some beers, "You know, I got pulled aside by one of the Company officers when I was a Drill Instructor. It was a Lt. or a Captain, and he said, "Why did you handle it that way? Dealing with that recruit?" He said, "Well, that's how a good buddy of mine would have done it. He would've asked the questions to understand about the man before he just gave him the stock response that we would get." He would do that. Walker would want to know you as a person instead of your rank or where you were from. The guy loved to read. He would read a lot of philosophy, poetry. Things that you would never expect a Marine to be reading. It was because of his parents. I think his parents were educators and it was something that they had passed on to him. What it did for him was that he had this desire to understand the person versus just the Marine. He would take different things from different people, like we're taught, from their leadership styles and he would come up with his own way. I think that's where we paralleled each other. Because on one of our deployments, we were roommates. We would come back and we would be joking, "Hey, did you see what so and so did today? Can you believe that?" "Yeah, good thing we made the call to fix it before it got up to the boss because he would just mess that guy up." So, we kind of shared our philosophies.

Janney: In my humble opinion, that's what makes a great NCO - being able to understand your Marines and not just dispensing some random punishment, understanding the mentality, where that young Marine was coming from. I think Staff Sgt. Allan Walker exemplified that.

Ford: Yeah, Walker was one of those guys where everybody was equal to him. A lot of senior enlisted and officers judged the book by its cover, but we were different. If you saw him in civilian attire, he looked like a skinhead. He shaved his head. He had tattoos. He wore that type of clothing. He wasn't the racist type of skinhead. There's a group of skinheads called SHARP. It stood for Skin Heads Against Racial Prejudice. He was the most non-racist person you could ever meet. It irritated him to no end if anybody were to talk down or negatively about anybody, because to him it didn't matter. He was in the right organization for that because everybody was equal. Everybody was judged on who they are as a man, who they are as a person, and I think that's one of those things that we looked at in every single person - that this guy is on our team. Doesn't matter where he is from; what his color is, what his religion is. What matters is what can that person contribute to the team? I think the knowledge base of the things that he had studied and learned about and his upbringing in Lancaster or in Palmdale, contributed to his success as a Marine.

Janney: Yes. He sounds like an incredible man and exemplified what a great NCO should be.

Ford: Oh, yeah.

Janney: So, after the 6th, were you involved in Operation Bug Hunt on 10 April?

Ford: Yeah. After that, there were a handful of battalion operations. Bug Hunt was kind of the bigger named one where we were given our sectors. Things kicked off early in the morning, as they all did. It would kick off right at the first morning prayer at 0430. We would basically have to clear our sectors. There was another scenario where you were given a neighborhood street and our little strip map, we would draw where our sectors were, and we had to hit every house. As we hit every house, we would mark those houses off. It was searching for weapons caches, IED making materials, or large amounts of money. I would say on that first one, April 10th, there was a lot of anger involved.

So, we were not the nicest guys. We were doing everything from shooting through locks to get in the compounds, to kicking doors down. Physically snatching guys out of their houses, throwing them out on the front yard, and then flipping the houses upside down. I think at that point, it didn't really register that we were making the scenario worse for us. I think as the deployment went on, we had to switch back into the mode of, "Okay, now you can't be shooting through locks. Now, you have to knock on the doors and they have to let you in. You can't just be throwing guys out onto their yards. You have to actually be nice to them, walk them out, and tell them to wait outside."

I think there was a lot of growing up and a lot of understanding the big picture at the squad leader level. The Company Commanders did a really good job in entrusting us and the last thing that we wanted to do in leading the squads was let the Company Commander down. So, whenever he was giving us our intent or the intent came from our platoon commanders, we did everything that we could to make sure that we were doing it right, because we didn't want something to bring any kind of discredit to the Company. That's what I figured out about Golf Company that I was with, and I'm sure it was the same with Echo. I think even more so on the second Ramadi deployment was the bond that everybody had. For the most part, the majority of the guys in the Company still keep in contact with each other. I'm out here in PA now, working for one of the guys that was a squad leader in 2nd platoon - working for a company that he and another Marine officer started. I think that the lessons that were learned there, going through that kind of hardship, speaks volumes to everything that I preached as a Marine recruiter. That, "Hey, there's a bond that will be with you forever. You're going to make relationships for the rest of your life." Anybody that you served with or been a part of will give you the shirt off of their back and take care of you.

Earlier, we were talking about the routine. We had day and night ops which were basically patrols around the city or through our Company sectors, QRF, and then you took a day off. We had security and then you had a day off, which was basically your time to do laundry and rest. Rolling into June-July into the hot time of the year, we started establishing these observation posts or OPs. We had OP Ag Center, which was the Agricultural Center right down the street from the Outpost. So, a platoon would be in charge of manning that.

We established the OP at the Agricultural Center which was basically a higher learning facility.

They had a big library in there, but they had shut it down. So, we would go up into the elevated floors to look down onto Easy Street and Route Michigan to make sure that those areas were free of movement and people weren't setting up IEDs. It was also across the street from what we called the soccer stadium which was a big parade ground for the government center they had there, and then we also looked down Route Michigan towards the government center.

One of the other things that happened was we established that Golf Company would send platoons to do rotations going and guarding the government center with the Triple C guys, who were government contractors. They would do a lot of the diplomatic movements with the leadership of the Anbar Province government officials. Then Golf Company, along with some Iraqi police, were doing the physical security at the government center. So, we had eyes on from the Ag Center, and then down at the government center. We did that for a couple of months, but then they kept targeting the Agricultural Center, taking pot shots from the neighborhood behind it.

Because of that, we established another observation post called OP Hotel which was across the street and down a little bit. It looked down onto the market on that side of Michigan. That was the hotel that looked down over the same area where my squad got ambushed back in March. So, we did quite a few IEDs, watching them getting placed and doing kinetic detonations, which were shooting at the IED to blow them up. We got rocket attacked there a couple of times. There were a couple of VBIEDs (vehicle borne IED) placed in front of OP Hotel. Anytime we'd set an observation post, it was targeted in some way. If you've seen any of the 2/4 videos, the Battalion video, there's a couple where you can see a couple of those VBIEDs getting blown up right in front of the window.

Janney: I have seen those, yes. Did you know Lt. Donavan Campbell? So, you've read his book, "Joker One." Is that pretty accurate?

Ford: Yes. I think in chapter 36, there was a situation where one of his Humvees hit one of my guys, Aldridge. It was the first book that was pretty accurate. You know, coming from his point of view. I think he did his guys well as far as telling everybody's story from his platoon.

Janney: Yeah. I think he wrote that his biggest regret was not taking the time to write more guys up for the heroic stuff that they did. He wrote that he regrets that more than anything.

Ford: Yeah. 1st platoon, those guys were a bunch of jokers.

Janney: Literally, huh?

Ford: Yeah.

Ford: I think that as a whole, from the top down, it was probably the fastest learning experience that anybody can go through in combat. I'm sure that happened throughout history, all the way back to any war being conducted by the Marine Corps. Everybody gets really good at their jobs exceptionally fast and the lessons learned are pretty amplified. I went from spending eight and a half years in a pre-war environment during the Clinton era and then going into a situation where you're being tasked with going and leading missions - doing all the mission prep and order writing and coordination to take your squad out are quite different jobs. Your twelve guys out on your own, and then occasionally getting thrown in with an interpreter, civil affairs guy, or an Intel guy. I think the biggest take away that I got was, "Hey, this is pretty cool to get this type of responsibility." Having a level of knowledge to utilize every skill set available to you whether it was going to be calling in air or indirect fire or calling in medevac; having the ability to do that and while taking care of your guys. It took the joke of, "I'm just a dumb grunt" to a whole new level. I used to love hearing that on recruiter duty from these non-infantry types, "Oh, you're just a dumb grunt." I'd say, "Okay, that dumb grunt has gotta be able to do all of these different things while being responsible for all these guys." I think there was more pride than that being truly a joke. As a kid joining after Desert Storm and learning all these things during that time, by the time you become an NCO, you take those things for granted, these things that I learned. When you take all those things that you learned to all of your resident classes and schools, you find out that you actually have to use them. It just gave me so much more value in the whole education process that the Marine Corps provides.

Janney: It's pretty commendable keeping all those balls in the air at the same time.

Ford: I imagine you're finding that out. I don't know if you've done other units, but I do know one thing that you will find is that the relationship of that generation of guys from that specific 2004 deployment rivals anything that is out there. It doesn't matter what Company you were with or where you were in command. There were guys that I had served with in 2/4 that when I came back in, I was in with 3/5. I had Rod Schlosser there who was my boss when I was working in the operations section. He was Operations Chief, a Battalion XO. Mark Halton was a Battalion XO in 3/5. Even at the time, Colonel Kennedy, retired General Kennedy, we crossed paths in Afghanistan, and he said, "Hey, this is a 2/4 reunion here." There was an automatic trust because we fought together in Ramadi and although we're in a different battle space in Afghanistan, there's a higher level of trust.

Janney: You know, the Marine Corps is a small world in that sense that you served with guys in multiple deployments like Kennedy and Schlosser.

Ford: Even now, in the civilian world, when I am retired. I'm employed by one of the guys that was one of my peers in Golf Company. We're doing similar stuff where we're still supporting the war and building logistical stuff for them, but I think, "Wow." It's like working in the family business, where your boss is your brother kind of scenario.

Janney: What was your rank when you retired?

Ford: I retired as a Staff Sgt.

Janney: I know a lot of the guys in Fallujah were encountering foreign fighters. Did you have much experience with that in Ramadi or was it more locals?

Ford: I would say yes. We had eyes on a couple of scenarios when we were at the Ag Center. Right across the street, there was a taxicab station or a bus stop. We called in a few times when we saw some guys that definitely looked like they were from out of the area getting off the bus. I would do the debrief when we got back from that rotation, out of the watch, and submit it to the Intel section. I think it was 1st platoon or 2nd platoon that encountered a bank, there were some check stands that were there, there were these white dudes with red hair and were defending that bank. We went in and cleared it and found a bunch of weapons in that bank. So, there were quite a few reports that I'd read about. Briefs going out on missions or preparing for missions by the Intel guys that yes, there are foreign fighters in our area. These are the things you're going to want to take a look at. They were also generic things like, "Look for guys wearing Adidas sneakers and track pants." Shit dude, Iraq plays soccer, of course they're going to be wearing Adidas and track pants. It wasn't exactly the type of intel that you wanted to hear at the time, but when you look at the big picture it was, "Okay, there is some financier somewhere that's buying track pants and Adidas sneakers for these guys to wear."

Janney: Now, Lt. Campbell wrote about a couple of Africans they were trying to capture, that they did, in fact, capture. I was just curious as to what your experiences were.

Ford: As far as 4th platoon, we didn't really snatch up any kind of foreign fighters. If anything, we had probably engaged quite a few. It got to a point where if we knew that if we did hit somebody, the Battalion had shifted to, "We're not going to waste energy to go down and confirm." It was going to put us at risk.

Now we definitely had seen people that were there that were not of Iraqi descent; they were Syrians or Jordanians, but they were definitely not from the area. There were a couple of scenarios where we had gone into houses in the latter part of the deployment where we were closer to speaking terms with people because they started recognizing us. We were able to build certain rapport with certain households and we would ask what was going on through an interpreter, "Where are all the kids? Where are the women?" They would say, "There are people from out of the area here." They would show us on a map that this house, this house and this house, they're out of town, but there are people there now. So, we would go look, report back, and there would be a package built to go do a hit to go check on them. Sometimes they were there when we did the raid, or they would be gone.

I know my squad identified a couple of houses where there was one guy in particular that I had visited a few times. I met him at the government center. The guy spoke perfect English and told me, "I want you to get rid of these bastards in this city." I said, "What are you talking about?" He said, "There are a lot of foreign fighters here and I want you to clear them out so I can have my city back." I asked, "Well, what do you do here?" He said, "I'm a Catholic in a majority Muslim city." I said, "Oh, really?" He was educated in England and I asked him for his phone number and turned it into the Intel section. A couple of weeks later, we were doing one of those Battalion ops and there I was in his living room in his house. So, there were a couple of scenarios where we were able to build rapport with certain people and they were able to be turned into assets.

Janney: Let me ask you one final question. The 23rd of March, is the last time you spoke with Staff Sgt. Walker?

Ford: Let me see here. Yeah, it was that night.

Janney: Can you think of anything else that you need to tell me or want to tell me?

Ford: No, at this point I think I've done a pretty good mind dump. I just want to thank you for being another guy that's telling our story. Like you said, the guys at Fallujah did some phenomenal things. The guys in Ramadi kind of just fell into the mix and got lost. I think there are a handful of people like yourself that have heard that story and it's compelling enough of a story to be told. It just means the world to a lot of us. It seemed like, for our generation that were there in '04 that bad things happened in Fallujah and that was a massive effort for the Marine Corps, but it's nice to see that this story, the Ramadi story, is being put on that same level. Because the fighting was just as intense.

Janney: You lost more guys in 2/4 than any other unit in Iraq in a 5 day battle. Those guys need to be remembered and honored. It's an honor to even attempt to tell the story and I hope I can do it justice.

Ford: Yeah. We pound our chests about being the first Marine Battalion there. But, several cycles later, during the time that Chris Kyle was there, there were two infantry Battalions there.

Janney: Exactly. When 2/4 was there, you had a lot of ground to cover with not a lot of Marines. You said that at one point you were with 3/5?

Ford: Yeah, so I was with 3/5 for the Sangin deployment in 2010. That should be your next project - the 3/5 deployment to Sangin in 2010.

Janney: The neat thing about 3/5 is that those guys recovered Tommy Parker's rifle.

Ford: Well, it's an honor to share the story so that it can be told and give more insight to Big Al and his legacy and getting his story out there. One thing I will add, is that in one of those weird, crazy and sick conversations that he and I had, he went out the way we had envisioned each other's way of going out. He just got into the fight. That's a reality that everybody had come to expect and question that had been asked hundreds of times after we got back. The question to me as a recruiter was, "How did you deal with it?" and "How did you accept death?" Basically, "Hey, it's a reality." You understand when you sign that contract that the inherent risk is coming home in a box and if you do it, there's a right and a wrong way to justify it. That's one thing we both said was, "Hey, we just want to be doing the right thing leading our guys into the fight." Not running away from it. All we can do is continue to tell our story to keep them around. It's not just the Walkers out there, it's those young 18-19 year old Marines. Those are the guys that we really feel for, especially when they were in the Marine Corps a year or so ago.

Janney: I hope I can do them all justice. That's my intent.

Ford: I know you will. Jim Booker, you know, he's come out here and seen us a couple of times. You know, that's the crazy thing, all the guys in the Company, we still keep in contact with him. He carries around a lot. Every time I see him, he always reminds me that when he was prepping Walker to go out, he remembers distinctively that he was wearing his Drill Instructor school class shirt underneath this cami's and that's how they sent his body home. That just tells you the pride that he took, not just in being a Marine, but being that Drill Instructor. Molding young men.

6 APRIL 2004 GOLF COMPANY

SGT. DAMIEN COAN

Janney: When and why did you decide to join the Marine Corps?

Coan: I joined the Marine Corps in my junior year of high school. I chose the Marine Corps for two reasons: 1) My older brother joined a year before me. 2) To be honest with you, the recruiter for the Marine Corps just did a better job than everybody else. The Army recruiter was starting to talk to me about discipline and what the Army can do and he just looked like a slob in his uniform, so I stopped paying attention to him about ten minutes in. The Marine Corps just appealed to me more than anything else. I contracted in June 1995. I went to boot camp in June 1996.

Janney: Were you always with 2/4 or did you get assigned to 2/4 later?

Coan: I went to 2/4 in 2001. So, I had a couple of duty stations before that.

Janney: So, you had the opportunity to know, maybe better than anybody, the guys that you deployed with in February 2004.

Coan: Yes, I did.

Janney: So, once you got to Ramadi, what you and your men were doing. And what was your rank at this point?

Coan: I was a Sergeant and when we got to Ramadi, it was securing main supply routes from IEDs. There were a lot of foot patrols, some vehicle mounted patrols, but most of our stuff was on foot and a lot of standing post, alternating posts on the Combat Outpost.

Janney: Where were you and what was happening the morning of 6th April?

Coan: So, the morning of 6th April, I was on the Combat Outpost. I was out with one of the squads with Corporal McGee and his squad on an ambush overnight on the 5th. We had gotten back to the Combat Outpost early that morning before everything kicked off. Lt. Ski and the other two squads from the platoon were on the QRF that morning, so when everything went off, I was actually asleep at the Combat Outpost, myself and the second squad. Then, we all got woken up and jumped in the second run of the QRF that went out. By the time we got out there, everything else was pretty much done. I was on the Combat Outpost when Brown came in with Lt. Ski and we got him helicoptered, medevaced out from the Combat Outpost. Then, I scooped up the rest of the Marines and went back out. Eric Smith, Corporal Eric Smith, at the time, was serving in the capacity of the platoon Sgt. with Lt. Ski on that QRF when all that went down, so he was out there with them to control the platoon when Lt. Ski got hit.

Janney: So, Corporal Smith and Sergeant Valerio were in Ski's vehicle when they were part of that QRF, when they got hit on Gypsum?

Coan: Yes. I believe so. I can't remember where Sgt. Valerio was, but Ski definitely was, yes.

Janney: What actually started the action that day? I've heard different stories. There was an IED near the tank graveyard and the snipers got attacked. What do you remember about what kicked it off, what Ski and Staff Sgt. Walker and those guys were sent out as a QRF. What started the events of that day?

Coan: I honestly can't remember what the exact event was that started it. I do know that the QRF was called out there in response to some elements of the sniper platoon that were under attack and were kind of pushed up against the Euphrates River. They were going out there and trying to reinforce and pull those guys out of that mess. I know that was part of it, but what exactly triggered the whole thing I honestly can't remember. There was so much stuff that happened that day I can't remember what pinpointed everything. You know, what was the number one trigger that started it all?

Janney: After the QRF was ambushed at Nova and Gypsum, you said you gathered up your Marines and went out as a second QRF going north on Gypsum?

Coan: Yeah, by the time we got there, it was a wrap. The fighting was over. The Marines were just cleaning up our guys, doing the whole intel thing. Taking pictures of the insurgents and weapons.

Janney: So, not necessarily particular to this day, but just in general, were the insurgents that you guys encountered and killed in combat, do you have any intel about whether those were foreign fighters or were they locals? I heard stories that some had police ID on them. What do you know?

Coan: There was, my understanding is there was a good mix of local and foreign fighters. You know, the exact intel on it, I don't know what's been declassified at this point.

Janney: I'm not asking for specifics, just general knowledge.

Coan: Generally, I think there was a mix of local and foreign fighters. I know that some of those had come over from Fallujah. You know, Fallujah was a main focal point during that period in Iraq. With all the military presence in Fallujah, by comparison, Ramadi just did not have the same US forces present so they were kind of pushing into Ramadi and trying to operate out of there.

Janney: So, those guys were just kind of squirting out to the next place on MSR Michigan? They said, "Oh, here's a good place to make a stand."

Coan: I think so, yeah.

Janney: On 10th of April, you guys conducted an operation. Tell me whatever you can share.

Coan: So, the 10th of April started off very early in the morning. I remember how coordinated it was with everybody in the Company kind of hitting and breaching doors at the exact same time. Very well organized from our standpoint. The fighting kicked off pretty early. We got two or three houses cleared before the fighting really kicked off. My recollection of the day's events was I don't think the enemy, the insurgency, had planned to fight that day. I think that once they saw how en masse we were and how much ground we were covering, I think they were in react mode when the fighting started. They hunkered down pretty good. There were a couple of good machine guns and fortified positions that we had to clear out. It took all day. We were out there all day long. Clearing, fighting, and capturing some folks. There were a bunch of weapons. Captain Royer took a shot to the helmet that day. I remember that was the day that John Sims got killed. I was there trying to medevac him out. One of my squads had scooped him up and the Corpsman was doing life saving steps and we had a Bradley fighting vehicle come get him. That is what I remember most about that.

Janney: Yes, sir. Was that the day that the journalist David Swanson got wounded?

Coan: Yeah, I believe it was.

Janney: After the events of that week, was it pretty consistent fighting or did things calm down?

Coan: From what I can remember after that week, I wouldn't say it was consistent fighting, but it didn't really get calm either. They went more toward guerilla tactics and didn't want to fight. That was the last time that they really tried to go toe-to-toe with us and really put up a big fight. After that, there were sporadic ambushes, lots of IEDs, but not a whole lot of firefights. After that day, we didn't have a bunch of Marines massed up out there.

Janney: Yeah, I understand they took a beating after the 6th. They probably didn't want to mess with you guys after that.

Coan: Yeah, they took quite a beating. I can't remember exactly what the numbers were, but I know that my platoon took out at least a platoon's worth of insurgents over that four-day period. I think the Company as a whole, Echo Company, and then the Battalion as a whole, the whole Battalion, 2/4 as a whole, took out a lot of the insurgent fighting force in that four-day period, so I don't think they would have had the strength to keep operating in that manner.

Janney: What do you remember about Parker, Contreras, Lopez and Otey when that happened?

Coan: We were up on the graveyard on the observation post watching MSR Michigan for IED prevention when we got a call that they hadn't checked in on the radio in a while. They had missed their radio check-in window or two. I grabbed part of the platoon, got a squad down there and went and checked it out and that's when we found them. Corporal Dagoody was the first one up there and he was unfortunate enough to be the first one to find them. The whole thing was not good.

    I knew Otey. Otey was in my platoon on the deployment before that. Otey was one of the Marines in my platoon then. Good Marine, great guy. I knew Contreras. Contreras was one of those guys that would help you break the monotony, the boredom, and the misery of being an infantryman, whether you were in Okinawa or anywhere in between doing training. He was one of those guys, great sense of humor; give you the shirt off his back kind of dude. Those two I knew the best. Parker, I didn't know very well coming from the 2nd platoon. I didn't know the other two well, but I knew Otey and Contreras pretty well.

Janney: I heard DeShawn Otey was a pretty quiet guy until you got to know him.
Coan: He was a pretty quiet guy. He didn't say a whole lot unless you really got to know him. The thing that I remember about him was his smile. His smile could brighten up a room. He was kind hearted. A good man that could just walk into a room and smile and help brighten your day. That's just the kind of guy he was. It was pretty crazy that he made it through the 6th only to have that happen to him a couple of months later.

Janney: Yeah, that was heartbreaking. I read his after-action report from that day and it's just almost the ultimate of tragedies to escape that and then that happen. What about Lt. Ski? You being a senior NCO, you probably had a good bit of interaction with him. What do you remember about Ski?

Coan: I was his platoon Sergeant. We didn't have a Staff Sgt. So, I knew Lt. Ski and his wife Joanna, probably better than most of the Marines outside of the squad leaders that knew him equally as well. Lt. Ski was the lead from the front Marine, a great platoon commander. If that hadn't happened, he would have gone far in the Marine Corps. He had a good reputation just coming out of the infantry officers' course. He was a great, great man. He liked listening to Pantera. Pantera was one of his favorite bands. He liked to turn up Pantera when driving a Chevy truck around. He wanted nothing else than for his Marines to be successful and safe. He cared about the Marines a lot. He did everything he could to shield the most junior Marine from the bullshit that you can sometimes get from the higher headquarters and genuinely cared about their safety, mission accomplishment, and being successful Marines. The young Marines that were married, had a vested interest in them being successful husbands and fathers as well. Overall, I don't think you could ask for anything more from a second lieutenant coming right out of the IOC that got put into the fight.

Janney: I heard he was really eager to learn from his NCO's, yourself included.

Coan: He was. He asked a lot of questions and was always looking for opinions, but he also knew where the line was, too. He knew when it was just time for him, as the platoon commander, to make a decision and not necessarily seek advice or ask questions. There is a time and a place for all of it and he did a really good job very early on understanding where those lines were, and when he didn't have time to ask for opinions or advice, he would say, "This is what we're doing." He had a good balance of it. He was very eager to learn from everybody, not just the squad leaders and myself within the second platoon, but even some of the other platoon sergeants and the Company Gunny Brendan Coleman, now Sgt. Major retired. He'd seek advice from him, too.

Janney: I heard he liked to fish. Did he ever talk about that with you?

Coan: We didn't talk about fishing too much. I do know that he liked to fish, but it wasn't something that we talked about much. When he and I talked, we talked mostly about the Marines. If we weren't talking about the Marines, we were talking about his wife Joanna or my wife and kids, or kid at the time.

Janney: You guys didn't have very good comms being in the situation you were in. Did anybody get to call home very regularly or was that kind of a rarity?

Coan: Well, it was every couple or three weeks. The Company had one or two satellite phones that would rotate through the platoon, so you'd get a phone call home every two weeks to a month. I can't remember how regularly it was, but it wasn't too often. Any time you got, it wasn't a lot; maybe twenty or thirty minutes to call a family member. That was about it. We didn't have internet cafés or anything like that that I saw later on in my career.

Janney: So, no email at all. I guess it's kind of a stretch that you had email at the Combat Outpost.

Coan: No. We started to get some email capability towards the end of the deployment. But, it wasn't something that every Marine was able to do. There weren't computers set up for every Marine to get on email. We did start to get a little bit of that capability towards the last couple of months or so.

Janney: Tell me a little bit more about some of the other guys that you have recollections of. Did you know Staff Sgt. Walker well?

Coan: Knew him a little bit. In fact, Staff Sgt. Walker was a Drill Instructor in First Recruit Training Battalion, Bravo Company, before he came to 2/4. When I left 2/4 and went to the drill field, the MCRD San Diego, I intentionally tried to get and was successful at getting to Bravo Company. Staff Sgt. Walker was a beast. He was very helpful to me as a Sergeant, being a platoon Sergeant. He was one of those guys that I would talk to if I ran into a situation with a Marine. He was one of the Marines that I would talk to get some advice from about how to handle it. Big, big heart. I didn't know him real well. He got there maybe in September, October. We deployed in February, so I didn't have a whole lot of time to get to know him real well.

  Let's see, Crowley. Crowley was a funny kid. He used to make me laugh all the time just with his jokes. He had this thing where if you'd say his name, he would do this funny thing. You'd say, "Mr. Crowley" and he'd go off and do some Ozzy Osbourne riff.

  I'll tell you the Marine that I remember the most and that I still talk to is Eric Smith. He was the first squad leader, filling in as the platoon Sergeant on occasion. He was my platoon Sergeant for the twenty-five days or so that we didn't have a platoon commander. I had to be the platoon commander until Lieutenant Marley got there. With him as my platoon Sergeant for about three weeks, he kept me sane through that whole deployment. Very calm, cool, and collected. Very good leader; great Marine. He has done great things out there in Dallas – Ft. Worth area fire department since he got out of the Marine Corps after that deployment. Phenomenal guy. Great Marine that really helped me out. I appreciate all the Marines that I serve with, but I appreciate him quite a bit. He really helped out.

6 April 2004 Golf Company

Battalion Gunnery Sergeant Winston Jaugan

Janney: When and why did you enlist in the Marine Corps?

Jaugan: I joined the Marine Corps in 1987. I'm originally from the Philippines. Around 1986, that's when we, as a family, went to the United States and we ended up in Louisiana. Then, after almost a year working some bullshit job in Louisiana, I told my father, "You know what? I want to join the military. Give something back to this country. This is my last opportunity." I was supposed to join the Air Force, but the Air Force took all my paperwork and the recruiter made it a hard job, so I ended up joining Marine Corps. I signed up to be 0311 infantry. From there, I went to boot camp, SOI, and other training. Then, I switched from a 0311 and became a machine gunner. Then, from 1st Marines, I had orders to go to sea duty, which was the USS Carl Vinson in Alameda. Stayed there for two and a half years. Then, I went to 3/4. Became a squad leader, machine gun section leader, then a platoon Sergeant, and then a platoon commander. Then from 3/4, I was shipped to MPC (Military Police Company.) By that, I mean I only stayed at MPC for a year and fought it. Then, I went to the drill field for three years where I had a successful duty on the drill field. I did a green belt (author note: the following definitions courtesy of SgtMaj James Baum - Green belt is the Kill Hat Instructor whose job it is to make the recruits miserable and get them strong), a J (author note via Baum: J is the platoon Sergeant that is an experienced Drill Instructor teaching drill and sets the standards for the platoon), Senior Drill Instructor (author note via Baum: platoon commander), and became a Chief Drill Instructor. My last couple of months, they made me the Battalion S4 Chief.

After the drill field, I went to 2/4 in February. I checked in at 2/4. I was about to take my leave, but instead of taking my one-month PCS leave, I only took one week off. Then, I did one field op with Golf Company, and then from there, I went to Advanced Party in Kuwait, and then from Kuwait, I went from Advanced Party to Hawaii. That's how I got left. I got there in February 2004. I remember talking to Colonel Bronzi, well, Captain Bronzi then. The first time I met Bronzi, I can't really forget my first meeting with him. I reported to him, introduced myself, and this is what our conversation was: I told him, "Sir, just give me a chance to show you what I got. My job is to keep you and the First Sergeant alive." He told me, "Carry on." That was it. That's all the conversation.

From that day until I left Golf Company, not once did Captain Bronzi give me an order or anything. You know, I was working hand-in-hand with the Company XO, Lt. Scott. We worked together to improve training, especially during the time in Ramadi. I tried to do my job to the best of my ability. In my opinion, being a Company Gunny is a dream come true, especially going to war as a Company Gunny.

When I was a Company Gunny, when I got there, there were only two of us that had combat experience. It's me and the First Sergeant. Everybody was freshmen coming from SOI. We have Second Lieutenants, and then we have Campbell, who's actually an Intel officer. He's not an infantry officer. So, when I got there, it's just me and 1st Sgt. that had combat experience. During our deployment, especially during the earlier part of our deployment, I would usually go out with squads on a patrol. Just to give those guys confidence, seeing senior leadership with them. Just tell me a rank that doesn't have a fucking privilege, you know?

So, we started off doing that. I'm gonna fast forward a little bit here. One time, I went out with one of the squads of Joker 3 on a night ambush patrol. I got hit that night about 2300 by an IED around April 2nd, a couple days before April 6th. There were three of us that got hit by that IED: LCpl Tambunga, LCpl Wagner, and me. So, I got hit that night, and then the same night, back in the Combat Outpost, the doctor said, "I can't fix him here." So, they gave me a couple shots of morphine, and they sent me to an Army base. I've forgotten the name of that base. Then, I was supposed to get shipped to Germany. But, I saw an Echo Company convoy at the Army base midday or late on 3rd April. So, I just hopped on the convoy and went back to my unit later that day. I went UA (Unauthorized Absence) because I had to get back.

This is where I'm coming from. Every Marine and Golf Company Marine is a testament to this. I love every fucking single one of them. I do. There were days that I didn't sleep at all, like three nights in a row with no sleep. Just to make sure the men were okay. Because the rotation of the patrols is 24 hours. The rotation of the teams, two rotate, and I make sure when they come back, they see my face. When they leave, they see my face. The way I did the QRF, I wanted to make sure that I'm going to be the first out on the QRF with the CO. I got the medics with me. I got the senior enlisted assigned to the Company with me at all times. So, as soon as we hear the COC, I'm gone. I'm with the CO. Usually, I'm right beside the CO or I got Taylor and Charlie. I just love those Marines. I don't know how else to express that. I love those Marines. Throughout that deployment, I kind of put my family to the side. About once every two weeks, you get a SAT phone call. We only had two SAT phones. You can call your family, but usually, if I got my chance, I would give it to a Marine, "Hey, you take my call." I put 10 minutes on the phone for whoever kicked ass that day. He should be rewarded, "You got 10 minutes of my phone time." Usually when I call, I talk to my wife for maybe five minutes, "How's the dog? How's the grass? How's my truck? Fine." That's it. You know, I had no time for emotions, because I'm more focused on my Marines and Joker 6.

I tried to take care of every one of them. Every single one of them. There were days when we had mass casualties or we had deaths. I mourned with them, but just enough not to be complacent. I always snapped out of it. Gotta go out, got a mission. Gotta make sure they are all okay. If some of the platoon Sergeants are actually going to go out because they just got hit, I go out with them. Then, I came back. You know, and that's pretty much it. I love those men, you know what I'm saying? Joker 6 and the XO, they love those Marines. I guess that's why we were so successful. Well, I know we had some KIAs and a bunch of wounded, but we were still successful because of the Four Horsemen - we called ourselves the Four Horsemen: CO, XO, First Sergeant, and myself. We all four worked together real good. You know, we were all on the same sheet of music and we all watched each other's backs. It's all about commanders with them (the Marines.) You know, there are times I go off to an outpost, you know, with a squad and sit there for six hours just to stay with the squad. I just hang out with them, just letting them know that, "Hey, I'm here. I'm just one of the boys." If there's a fight, there's a fight. So, that's pretty much it as far as for myself.

As far as the Marines from Golf Company, I can tell you this, I haven't seen a bunch of Marines that when a firefight comes, they all yell like Indians. I mean, they're so happy. It's like Christmas or something. It's like New Year's Day. They're going out on a fight. I haven't seen a single Marine who hesitated or became a coward or anything like that – none of them. They all wanted to fight. They all loved to fight. That's why we were successful as a Company during those times, during that deployment. We were spread out thin, though. SgtMaj Booker did the best he could with the number of men we had. I mean, there's only 2,000 of us. Providing security, winning the hearts and minds. We built a cohort, but it's only 2,000 Marines unseated with a population of more than 500,000 so we were spread really thin, but we were able to fight, we were able to kick ass and that's why I'm really proud of those guys, you know, really proud of them.

As I said earlier, I got hit on April 2nd at night, and then I hopped on the convoy and got back late the next day. It was midday or late the next afternoon of April 3rd when I saw the convoy. There were three of us that got hit by that IED: LCpl Tambunga, LCpl Wagner, and myself. Tambunga shattered both his eardrums, Wagner got hit in the hand, and I got hit in the eye. I still have shrapnel in my eye. As a matter of fact, I remember when I hopped on that Echo convoy, because Echo and Golf, we all stayed in the same Combat Outpost. I think it was Golf Company guarding the gates at the time, and when we came in, I was still in PT gear without full combat gear. I was that way, green on green, because my camis just got ripped; the Doc cut them off when they were treating me. So when I came back, the Marines were just smiling at me. I guess I looked funny with PT gear and no combat gear. At first, they said, "No, no, no!" Sergeant Major Booker was in Blue Diamond, but when Colonel Bronzi first saw me, he gave me a really good handshake. He shook my hand really, really well. Colonel Bronzi just pulled me in tight, and said, "I had no idea how I was gonna replace you. So glad you're back." I said, "Yes, sir. I'm back. I'm not going anywhere."

Later in the deployment, around the first of August, and 2/4 is coming back in September. My wife's about ready to give birth around September. So, around August, they were sending the Company Gunnys back to the States for an advance party. I told Colonel Bronzi, "I'm not going back. Send one of the platoon Sergeants back. I'm staying with the Company." So, my wife found out about it about four years later. I slipped. Actually, it kind of slipped. That's how she found out I volunteered to stay over for the RIP with 2/5.

I stayed with Colonel Bronzi the whole time. When 2/5 came in for the RIP (Relief In Place), I wanted to make sure that I was one of the last ones to leave Combat Outpost. I met the rest of the guys at the staging area. I kind of helped them transition in. I shadowed a Company Gunny in 2/5, and then I went out with some of the patrols with our squads and the squads of 2/5. I went out with Colonel Bronzi and there was a time when they got hit. They were doing a patrol. They got hit. We still had a vehicle. So, Company Gunny was there and I took one of the 2/5 Marines back to do the casevac. I did the medevac and then went back to Combat Outpost. Pretty much showed them how the Marines of Golf Company work. You've got to be aggressive.

As a matter of fact, I got a surprise visit from a couple of Marines a while back. One came from Wisconsin, and one from Reno - just showed up in my garage and we had a few drinks. It was a surprise visit for me. It was a nice visit from them, so that kind of made my day. But, we are very tight. I live in the states, but we still talk to each other. I still talk to Colonel Bronzi on a regular basis. Every now and then I go to his house. I live in Sacramento. When I go to his city, I always visit him. So, pretty much, we are all brothers. Up to this point, we're all brothers. We still stay in touch with each other.

I have a lot of memories. Here's what I can say though. Maybe in a couple of days I can write down the stuff and we can talk again. I want to talk to you about those guys and honor their sacrifices, about how brave those Marines are. I try not to forget every one of them, especially all those KIAs and the wounded. Those are amazing Marines. I'll never forget them, especially Hallal. I remember Hallal the day before he got hit. There's two of them, Hallal and Flaw, and both of them were over six feet tall. Every morning, I would always talk to them and kind of make jokes with them. Almost on a constant basis, you know. It broke my heart when I had to open up the body bag and it was Hallal. I guess I wasn't really ready to do this interview today. I want to gather my thoughts together and I want to be more specific about every Marine, especially the squad leaders of that Company. They are my heroes. They made me who I am today, those Marines. The Battalion Commander made a comment to the XO or the CO, I don't know who he talked to, but the fact that the Company looked up to me like not only their Company Gunny, but more of like a father figure. I want to make sure that I speak about every one of them. There's quite a few of them who really stand out. There were a lot of Bronze Stars given to the squad. I want to tell their stories. The CO's story, when he got hit by a grenade during April 6th, how he led and led squads, fire teams, to get to the point to rescue those Marines who were pinned down in a house. To get to that spot next, he rammed that vehicle into a house just to clear that house so he could establish a base of fire, fire support for the attack to move up. Those stories, I want to write them down to tell you that story in detail. How brave those men were, from the lowest to the highest. Their bravery – that's why I love those guys. I really want to talk about those Marines. I'm going to talk about those squad leaders, fire team leaders, what I've seen. It's just unbelievable what I've seen from those Marines. I couldn't believe that there are such human beings that brave, and at the same time, loving.

Wroblewski, Zoller, Quinn, Janney

J. Wroblewski

J. Wroblewski

Canon Andrew White

5 April 2004 - R to L - LCpl Derek Calloway, 2ndLt Wroblewski and 2 unidentified Marines
(Photo courtesy of David Swanson)

J. Wroblewski with RCT1 Marines - Fallujah

RCT1 Marines & J. Wroblewski - Al Faris

MajGen Kelly, J. Wroblewski, BGen Mills

J. Wroblewski, MajGen Kelly
Memorial Service

RCT1 Marines
Front Row L to R - Michaels, Wroblewski, Bond, McMullin
Back Row L to R - Gordon, Sloan, Crosby-Carlsen, Martinez

MajGen Kelly, J. Wroblewski, BGen Mills

MajGen Kelly, J. Wroblewski, BGen Mills, Col Welsh

MajGen Kelly and John Wroblewski

J. Wroblewski, MajGen Kelly, PSD Marines memorial service - 6 March 2008

J. Wroblewski and MajGen Kelly

John Wroblewski and MajGen Kelly - PSD Marines memorial service - 6 March 2008

2ndLt Wroblewski memorial shirt

MajGen Kelly, John Wrobleski

Cpl Jonathan Yale – Navy Cross
6 Seconds to Live by Gen Kelly

Wroblewski with 2/8 Marines - Rt Gypsum

John Wroblewski and Madrassas students

Greg Janney and John Wroblewski - Camp Victory

J. Wroblewski at LZW

J. Wroblewski

J. Wroblewski and MajGen Kelly - PSD Marines memorial service - 6 March 2008

John Wroblewski and MajGen John Kelly

The Magnificent Bastards in Ramadi and A Father's Journey There

John Wroblewski

6 April 2004 Golf Company QRF

Cpl Nick Kelly – Weapons Company

Janney: It's 21st of May, 2020. I know your family has a rich history of service in the Marine Corps, but why did you decide to enlist in the Marine Corps, and when was that?

Kelly: I was in high school from 99 to 2003, but I grew up, from birth till middle school at Camp Lejeune, North Carolina, because my father was a retired Lieutenant Colonel in the Marine Corps. So the Marine Corps was natural to me because I grew up around it. I grew up seeing Marines PT and going to see the different vehicles that Marines drove. I was just kind of accustomed to base culture, so it wasn't really a culture shock to me as far as growing up. I got into high school, about 11th grade around 2002. I talked to my dad and I told him I wanted to enlist in the Marine Corps. He asked what MOS I wanted. I told him that I wanted to be an MP because I always wanted to be in law enforcement.

Spring break rolls around in 2003. That's my senior year. I went out to New Mexico to go snowboarding with some buddies and while we were on spring break is when the invasion of Iraq happened. I remember watching CNN at the time and it was showing the NVG footage, all the tracers going off. I kind of reflected while I was snowboarding and I just decided that I'd rather be in the infantry than a MP. I made up my mind that when I got home, I was gonna talk to my dad about it. I was in the garage lifting weights and I was really nervous to ask him that I wanted his consent for me to join the Marine Corps. I asked him and to my surprise, it was the easiest conversation I've ever had with him. I asked him for advice, and he said just do what you're told to do, don't get in trouble, and you'll be fine. We went down and talked to the recruiter and he warned me that once I gave up that MP slot, that it was done for, but 03 could be guaranteed. I signed the paperwork and graduated in 2003, and 10 days later, off to boot camp I went. I have a son now who was born about nine months ago, and I'd like for him to continue our family tradition.

After boot camp, when I went to SOI, I signed up for 0331, machine gunner, so I went through machine gunner school. We graduated from the School of Infantry and I reported to the 2nd Battalion, 4th Marines, Weapons Company in September-October 2003. So, my dad was getting married and I told the NCOs, "My dad's getting married and I want to be in his wedding." I had just gotten there and it pissed everybody off because they were thinking you're pulling the whole, "My dad's a Lieutenant Colonel" thing, you want special treatment. Back then, you could actually go in the hole on leave. So even though you hadn't earned leave, you could actually go negative days in, and you just paid it back as the year progressed.

When I got back, we were gearing up to go to Iraq. They had this new concept of a MAP platoon, which is a mobile assault platoon. So they're basically getting rid of two or three tow missile mounts and then equipping them with heavy machine guns. That's how I got stuck with the third platoon of MAP 3. A lot was different with Intel nodes in every Company and other things. When we deployed, we flew to Kuwait, acclimatized for a few weeks, and then convoyed into Iraq. The only thing remarkable about the convoy was the front of the convoy hit an IED near Babylon.

Once we got to Ramadi, we were stationed at Hurricane Point. We did leftsy-rightsy with the Army unit we were replacing. I was assigned to the platoon commander's vehicle, which was up in the very front vehicle. I remember doing it primarily at nighttime. I do remember seeing soldiers in their Army uniforms, putting Iraqis up against the wall, kind of like stop and frisk type stuff. It was the first time in my life that I'd ever been outside the States, but it was really dirty and chaotic, especially downtown. As far as the locals' reaction, a lot of them seemed kind of confused by our presence. We were wearing the new digital camouflage. There was nothing, no hostilities or anything, but everyone looked at us like a circus coming through town. At that time, the first half of the deployment, it just may come as a shock to a lot of people, but we didn't have any armor plating around the turret. It was literally an open top, V plate armor like the 1980s style.

Throughout the whole deployment, they would send these line companies in to, you know, section off a part of the city or neighborhood and conduct searches. A lot of time, Weapons Company, we'd go in the wee hours of the night, we'd help cordon off that part of the town or the neighborhoods within Ramadi. The dismounts would dismount and they would set up fighting positions around the vehicles. I just sat up in the Humvee and waited for any kind of attack. I had that Mark 19 automatic grenade launcher. The Mark 19 is probably one of the most awesome weapon systems in urban fighting. We actually had 81mm mortarmen that didn't want their Mark 19s because of the myth that the Mk19 wasn't going to be practical in an urban environment so, they gave us a bunch of SAWs and 240s. After our first engagements, they saw how awesome the Mk 19 weapon system was, so they complained and got Mk 19's back.

Hurricane Point certainly got a lot of indirect fire. I guess they were using little islands off of the Euphrates River to mortar us. We used to rotate in and out of guard duty at Hurricane Point, then a rotation up on top of the dams, and we had machine gun positions. One early morning, me and LCpl Gordon were up there and heard rockets. I didn't know if they were using mortar tubes or makeshift iron launchers, and I don't know what kind of rockets they were using, but we did hear them ignite from the Euphrates. About 20 seconds later, there's several impacts that were hitting around the hooches at Hurricane Point.

On the morning of 6 April, we were resting in our racks. The platoon commander came in and told us we all needed to mount up. In hindsight, it's funny looking at the information that was shared with a 19-year-old PFC. When we would leave the base, I had no idea where we were going, what we were doing. We launched out of Hurricane Point. We went to the government center. We were basically the QRF because it was our week to be QRF. I understand there was a push on the government center by some of the insurgents. They had RPG teams that were trying to make inroads there at the government center. I think this was about 10am.

Janney: So did you hear the firing on the eastern part of the city at that point? Were you cognizant of that?

Kelly: Oh, that's funny you ask that. So that'll bring you to one of the more terrifying moments of my life. So, we got to a government center. We didn't have any engagements up until that point, but Lance

Corporal Pepper had a 50-caliber machine gun, and they called him up to an alleyway. He just lit it up with the 50-caliber. I remember being kind of jealous, "I wish they would have called me and my Mk19 to engage." Then, we get the order to push to go over to Easy Street, where the 3rd platoon of Golf Company was pinned down. I didn't know at the time, but they had taken a lot of casualties and KIAs.

We leave the government center, and we're approaching Easy Street, which is by Saddam's Mosque. That's when I heard all the fighting going on over there. So, at first it was distant automatic gunfire. You could hear explosions coming from the rockets. But, as we got closer and closer, it got louder and louder. That thought and those sounds, they'll always stay with me - we're getting closer and closer to this danger. Actually, I'm getting chills right now. It's just kind of terrifying that you're moving closer and closer to this. I had no idea what we were about to experience.

We came in on the opposite side on Easy Street, so not by Route Michigan, but on the other end which goes by the soccer stadium. Our convoy is about five Humvees, all with heavy guns. We were at this corner to get on Easy Street, and I saw these two insurgents. They were walking down the street about 100 yards from us. They weren't super close to the fighting, because they both had their weapons slung. It took them by surprise to see Marines coming their way. As soon as they saw us, they hauled ass inside of a courtyard. By that time, I had the Mk19 rotated on them, and I engaged the Mk19 and the other four vehicles in the platoon engaged the building with their heavy guns. Mk19s, two 50 Cals, and a 240 Golf all lit up this building.

Our job was to push and make contact with 3rd Platoon, Golf Company. They were all pinned down within a house in a courtyard. They had several that were seriously injured and they already had, I believe, 2 KIAs. Our heavy guns were to suppress the insurgents and also try to get those wounded guys out. We pushed up Easy Street and eventually made contact with the third platoon of Golf Company. I remember coming up Easy Street and there were rockets skipping down the street at us. It just felt like gunfire coming from everywhere. I have distinct memories of seeing the line, the 0311s up and on the street taking fighting positions behind carts and different stuff in the shops. It was basically one big gunfight going on.

I can't remember the Lieutenant's name, but the platoon Sergeant who was in the thick of it was Staff Sergeant Damian Rodriguez. He ended up staying in, and he got out as a Sergeant Major. They called our Humvees up to where those Marines were pinned down. I'd already been told that the insurgents were using taxi cabs to bring wounded out and to bring new fighters in. We rounded the corner and one of our Marines was dismounted on the ground. This is kind of when realization hit me. There was a Kevlar on the ground and a few M16s on the ground, not abandoned, but were left there on the ground with no one holding them.

A guy named Cox was throwing the stuff at us so we had it accounted for. Cox basically took the helmet and threw it up to me. When he threw the helmet up to me to catch it, it was halfway filled up with blood. I remember there being a bullet impact strike through the helmet. We got the gear, and where we were at in that corner is where those Marines were pinned down. A taxicab comes around the corner about 50 yards in front of me with about four Iraqis in it. They hit the brakes and I engaged the taxicab with my Mk19. I put about ten grenade impacts on the windshield and all over the taxicab.

There was a guy named Gentile, pronounced Gentilly, but spelled Gentile. He's Italian. He was shot in the initial part of the ambush, and then had a bullet strike the back of his head with a severe facial injury. They put him inside of my Humvee, and he's got bandages all around his face. He was sitting down below me on the left and was kind of hunched over. There was blood just coming out of the bandages. A Hispanic Marine whose name I can't remember had a serious hand injury. So, we got those two wounded guys and we threw that vehicle in reverse and punched back up Easy Street. We met a high back Humvee headed to Blue Diamond, so we put the Marines in it. There was also a tall Marine there that was KIA. Another machine gunner named Pepper and I loaded him into the high back and they headed to Blue Diamond. Those are the last of my memories of 6 April. I think the battle just kind of died down from there. It was pretty late by this point, about 4pm or later. The strangest thing I remember is that we were out there over 4 hours, but it seemed like 20 minutes.

The next day, I remember we got up and we went back out for more patrolling. We were down over there off Route Nova, which is basically an MSR that parallels the Euphrates River. We were way down Route Nova. Early in the morning, we were in a neighborhood, and the Marines were just basically searching houses. We had several engagements down there. On the way back, we got ambushed. We're cruising down Route Nova, and the opening of the ambush was an RPG that was shot at my Humvee, which was the front Humvee. I imagine they wanted to disable the front vehicle to stop the convoy. The RPG missed and detonated on the berm right next to our Humvee. We stopped in place, rotated onto the village there with a bunch of palm tree groves. We just lit the village up with the heavy guns where they were firing from. The dismounts got out from MAP 3 and they started to maneuver onto the village and assaulted it. That lasted about ten minutes.

There were some guys or a guy up in a palm tree. The platoon commander told me, "Engage the guy in the palm tree." I engaged the top of the palm tree with the Mk19 and the guy must have fallen 30-something feet to the ground where he lay mortally wounded. All the palm fronds from the tree were falling down, too. When my fellow platoon mates were assaulting, one of our Marines, Lance Corporal Cummings was shot through the chest. It missed his SAPI plate, so it was a through and through his chest. The Marines grabbed him, brought him back to our line of Humvees and loaded him up, and then we made a push back to Blue Diamond to the field hospital. There was a lot of stuff that happened on Route Nova. It was not a good place to be. We always used to joke that IEDs would just grow naturally on Nova like potatoes. That was the end of the day of the 7th.

From what I remember, we didn't get another engagement until the 10th, for that big Operation Bug Hunt. That was a city-wide push. The line platoon got in another engagement somewhere over there off MSR Michigan, near the soccer stadium, if I remember right. Again, we were QRF, so we had to go out there, and give them some heavy gun support. I think our mission was to over watch them as they patrolled back over to Combat Outpost. We got into several firefights on the way back to the Combat Outpost. The Mk19 saw a lot of action that day, too. It was definitely a day of payback after the events of the 6th and the 7th. We gave it to them that day.

After the battle of Ramadi, the rest of the deployment is memories of different engagements. Just kind of hit and run type stuff. Another day that stands out, if I'm not mistaken, is July 21st. I remember we got in a lot of engagements that day, right around the Saddam mosque. We were flying down Route Michigan. I'm not even sure of our purpose, but we were headed toward Combat Outpost. We were over there by that fighting position, I think it was an old hotel, three or four stories high. There was a car bomb that went off that was actually captured on some reporter's camera. They were in Sgt. LeCharge's truck. Our lead Humvee was approaching the car, and I just had a gut feeling, that's a car bomb. So, we pass it, and nothing happens. The second truck passes it, nothing happens. The third truck passes it, and the car explodes. We didn't have any casualties, so we just did a self-assessment. Then we kept pushing. We got up to Saddam's Mosque on Route Michigan and we got ambushed from a real tall building off to our right as we were going towards Combat Outpost. We got near the traffic circle. There were insurgents coming out, and they were shooting rockets at our Humvees, but they weren't hitting us. There were IEDs all over the road. For some reason, they were not being detonated. We just engaged.

The rest of the deployment was more IEDs, hit and run tactics, and a few ambushes. Being with the Weapons Company, we brought so much heavy weaponry to a fight. They wouldn't stand and fight any more after those five days in April. I can only imagine that if you started a fight with us and you're getting pounded by this heavy weaponry, you're probably going to make a strategic retreat. We had ambushes at nighttime. We had more ambushes during the day, but no sustained fighting for hours on end or anything like that.

You wanted to know about some of my brothers, so we'll start with Marcus Cherry. We went to SOI together. He was a 0311 (infantry) and I was a 0331 (machine gunner), so we were basically split after SOI. I went to machine gunner school while he was doing the 0311 route. He had some mutual friends that I had that all went to 6th Battalion, 1st Marines. You know, they were all in Phantom Fury in Fallujah. So, we all had mutual friends because we were all basically in the same boot camp group that came in together. I just remember going to parties out in town, him being at the same parties and us talking.

Let me tell you about Sergeant Kenneth Conde. He was actually in my platoon, so I was out there when he was shot on 6 April. That day, we were out on Easy Street. We kind of pulled back onto Easy Street during the fighting, right before we linked up with the Golf Company guys. He came around the corner up to our Humvee, and he had his sleeve cut off of his arm. I think it was his left arm, but basically a Corpsman had cut the sleeve off. He had a battle wound dressing on his shoulder. It wasn't in the exact same alley that he was in when he was injured, but talking to all my friends after the fact, they were assaulting down the street. He took a round to the shoulder, and from what I was told, he and his squad just pushed and basically wiped out those insurgents in that alleyway. I heard it knocked him down or he staggered back when he was hit, but he got right back up and said, "We got stuff to do." I remember he refused a medevac out of there. That unfortunately caused him some problems later, but that's a brave guy there. Take a round and then just kept fighting. I knew him pretty well. Conde was one of the Sergeants in our platoon when we got out of SOI. He was a mentor-type teacher, basically a salty Sergeant that knew everything that he was doing. Our relationship wasn't a buddy relationship. It was more like a student-teacher relationship. He's the one that actually taught all the MOUT warfare training to our platoon and the Company.

Another one of my good buddies, Cpl Gordon, another machine gunner, was on the 50 Cal in the very back of the platoon. When we got back from Ramadi, they had us all in formation and wanted to know if there were any volunteers that wanted to go to 2/5 as combat replacements. I'm not saying that I'm the bravest or anything, but I was the only one that raised my hand. I wanted to get back there and keep going. I looked over at Gordon and he raised his hand after I raised my hand. They ended up not taking anybody as replacements, but afterwards I asked him, "Why'd you raise your hand? He said, "Well, if you're gonna go, I'm gonna go. I don't want to, but I will."

We had another Marine named Corporal Ryan in my platoon. He's from New Jersey. He was a combat replacement after the Battle of Ramadi. Sergeant Conde had actually already been killed. Corporal Ryan came in. We hit it off pretty well, even though Corporals aren't always nice to Lance Corporals. I remember he liked to read books a lot. I love to read books. We would talk about books we'd read. I'd be on firewatch cleaning weapons during the day while everyone else was asleep and he would come out and talk to me. He told me about fighting in Afghanistan and I told him about the Battle of Ramadi. It was leave time for him because he was staying over with 2/5. In the middle of the night, a helicopter came to Hurricane Point and he gathered all his shit. I never got to say goodbye to him or anything. I did consider him a friend, but you know, it was last minute. They woke him up and said, "Bird's here, you gotta go." I woke up in the morning and he was already gone. I never got to see him again. He was killed in Ramadi in front of Hurricane Point when 2/5 was attacked.

  I guess one last reflection. I'm not trying to be a Marine Corps fanboy, but after going through that part of my life, I've really come to realize how awesome a fighting force the Marine Corps really is. It's actually amazing looking back on it, talking to some of the guys you served with about holding off insurgents and saving the day. I'm convinced a very small number of Marines could really put the hurt on a large number of insurgents or other trained military. That's something that I took away from the whole experience. As I get older and I think about it, it's that brotherhood. You know the deal. You're brothers with those guys. If you ran into one of them right now, it'd be like no time had passed at all, even though it's been 16 years since then.

6 April 2004 Golf Company QRF

Sgt Marvin Endito – Scout Sniper

Janney: Today is 8 Dec 2021. Tell me why and when you decided to enlist in the Marine Corps?

Endito: I enlisted in the Marine Corps because I just wanted to get out of here. I live here in Thoreau, a little small town in New Mexico and there's not much here. Not much opportunity. A lot of people here end up sticking around here. It's said to say, but a lot of people, nothing becomes of them. They don't leave the state, they don't leave the town. I wanted to do something different. I grew up in a family of five, so it's not like we had everything, but at the same time my parents provided for us. They worked hard. They instilled a lot of good in me. I have family that served in the military, in the Marine Corps specifically. My uncle served in Vietnam. He passed away probably right around the time that I was born and I never met him, but I knew about him and I know him from pictures and stories from my mom. I think, really, from a young age, I was probably fifth or sixth grade, we went to go visit him at his gravesite. We cleaned it up. It was probably Memorial Day. I think that one incident planted a seed in my mind. When I got into high school, my interest got deeper and deeper and I chose to enlist in the Marine Corps. So, when he passed away, his family had been living near us so his kids, his wife, they grew up away from us. Although we had some contact with them, we weren't close with them. It's something that I had been wanting to know.

Janney: You went on to serve honorably and in his memory, so that's a wonderful thing. I'm sure he's looking down somewhere saying, "Yeah, that's a chip off the old block."

Endito: Yeah. When I was young, my mom, my aunts and I went to visit him. They were talking about him, and even at a young age, I could see that they were proud of him, that he served. He did something good for the family and for his country.

Janney: Absolutely. When did you enlist in the Marine Corps?

Endito: It was my senior year in high school. This would be 1996. I started seeing my local recruiter probably at the beginning of the school year and kept talking. Meeting with my parents and by New Year's Eve, December 31, 1996, I signed my papers and everything at MEPS here locally and I was ready to go.

Janney: Did you graduate early and report or do early enlistment?

Endito: No, I signed my papers and everything, but I had to wait until I graduated.

Janney: After boot camp and SOI, what unit did you report to?

Endito: I reported to 2/4. 2nd Battalion, 4th Marines in December of 1997. I was always in 2/4. When I got to 2/4, I got there with Longoria and Santiago. We were together from the beginning all the way. We had gone to a sniper platoon together. We went to school together. We went to combat together and we all got out about the same time.

Janney: Wow. That's a dangerous crew right there. Before you guys got ready to go overseas to Iraq, what were your thoughts on the boot drop? You had a lot of green guys coming in and I know you had to work really hard or maybe you didn't being a scout sniper. You didn't really have as much to do with getting those guys ready to go, but tell me what your thoughts were about that.

Endito: My situation was a little bit different. After we got back from Okinawa from our year long deployment before Ramadi. A couple of months after we got back from Okinawa, I got fapped out to go to the military police unit there on Camp Pendleton. I spent five months with the military police up until February. During this time, we were supposed to be working just getting ready. Working with our new guys and prepping and instead I was stuck with the military police Company. That was just not a good situation and I hated it. I begged to get back to my platoon because I knew I needed to work with my guys. I needed to work with the people that we were going to Iraq with. With our platoon, we did have new guys that were coming to our platoon, but they were not young and new, like the boots that came in - the guys that came with the boot drop, the young guys. They had been in a while, but they were new to our platoon. They're new to us and it was important for me to work with them; training is everything and I really didn't get that opportunity because I was stuck with the military police Company.

Janney: When you deployed to Iraq, what was your rank?

Endito: I was a Sergeant.

Janney: Some of the guys flew to Ramadi to get things set up, and then some convoyed in. Which group were you in?

Endito: I left early. I was what is referred to as an advance party. So, there's a group of us that left in mid-February then the rest of the unit they got there late February I think it was. So, I left early. Me, another Marine from our platoon, and our platoon Sgt. left early. So, we actually flew in. We flew into Kuwait and from there, we flew into Ramadi.

Janney: Who was your platoon Sgt. at that time?

Endito: Gunnery Sgt. Lindsey.

Janney: When you first got to Ramadi, were you at Combat Outpost or where were you stationed?

Endito: We were at Hurricane Point.

Janney: What was your impression when you first got there and were taking over from 82nd?
Were you doing your familiarization patrols with those guys?

Endito: Well, from the get-go, it was quiet. Nothing was really going on. Iraq was not the hotbed that it was moving forward. A year and a half before, George Bush with his banner had declared the war was over; mission accomplished. Prior to leaving, we knew that we weren't going to do much. Just some patrolling, a lot about stability, and helping the area to rebuild and to try to get the Iraqis on their feet. Showing up, I don't think a lot of us expected too much. It was going to be a lot of patrolling, handing out basketballs, school supplies and just kind of showing a presence. SASO – I think everybody expected that. When we got there, we met with the Army unit's, our counterparts, and they told us it had been fairly quiet. They hadn't been seeing a lot of action, if any. It was fairly boring. Not a lot of action, a lot of patrolling. Just kind of going around and meeting with the people.

Janney: I heard they did a lot of mounted patrols. The Marines patrolled on foot.

Endito: Yeah, they had some light armored vehicles. They weren't patrolling on foot as much as we did once we showed up. I think they had Bradleys and some type of armored vehicles. Yeah, it was mainly using the vehicles to maneuver throughout the towns.

Janney: In sharp comparison to your unarmored Humvees and 7-tons.

Endito: Yeah. Absolutely.

Janney: Yeah. That's bad. When you guys transitioned into command of the AO, when did you notice things start to change? When did you notice that things were starting to be different? Less people in the marketplace, maybe the people were less friendly, or did it even go down like that?

Endito: So, once everybody arrived, this was probably late February or early March, it was a good month or month and a half where we were patrolling and going out and about. Got a good feel of the city as a whole being on our feet. It was quiet; nothing really going on. We did lose a Marine early on. He was with the engineers. We lost him early on, but there wasn't much fighting. There wasn't any type of heavy fighting, or we weren't really being informed of bad guys in the area that we needed to address. It went from being quiet, being boring, same old patrols, to just all of a sudden, the city turned completely upside down on April 6th. That really is my recollection where things started to turn is April 6th.

Janney: What were you doing the morning of April 6th?

Endito: The night of April 5th and many nights before that, weeks before that, we would go out at nighttime. We would go out and patrol. We occupied OP sites, and houses along roads just to watch and see if there's any enemy activity. Is there anybody placing IEDs? Watch the roads, watch houses and we'd do that a couple hours after midnight. We'd get out there at midnight and come in at 0400 or 0500 and that's what happened the night of April 5th. We went out probably about midnight; maybe 2300 and we came in early in the morning on 6th April. We got back to Hurricane Point about 0400, maybe 0500. Clean up, account for gear, make sure everybody is good, get a little something to eat and go to bed. We would go to bed about 0500 and sleep in until 1000 or 1100, and do the same thing over and over again. But, the morning of 6th April, we all went to bed at about 0500, 0600 or whatever it is. I've thought about this over and over. I'm thinking it's about 0900, somewhere around that time when our platoon Sgt. Gunny Lindsey, he about kicked down the door and came running in, just yelling, "Get up, get up, get up, get your gear on, we gotta get out there. Shit's hit the fan. We have a platoon from Golf Company that's been engaged. They have two KIAs already. We gotta get out in the city. Get your gear on. Team Leaders come meet with Lt. and me here in our little area. Let's go, let's go, let's go." We're waking up wondering what's going on. Honestly, within a matter of about ten minutes, we were in a Humvee flying down Michigan toward the sound of gunfire.

Janney: So, did you guys link up with Joker? You weren't attached to Echo at that point?

Endito: At that time we were working with Battalion command because when we left Hurricane Point, we left with the Battalion Commander. We left with Sgt. Major Booker, so we left with the command element from Hurricane Point. We left and headed down towards the government center. That was our first stop. We dismounted and I know we got on the rooftop and after being there for about two or three minutes, they told us, "We gotta go." That's when we started moving towards where the Golf Company Marines were engaged, maybe a couple of klicks away.

We took vehicles up to the government center. Then, after the government center, it was just a movement to contact. We were on foot and we were just moving. Not necessarily running, but just about. We were just moving, moving, moving because they were still getting engaged. From the government center, we could hear the gunfire. We could hear the explosions; we could see the smoke that's going up into the sky here and there. We were moving towards them and as we were getting closer and closer, it was getting louder and louder as we were moving. Just moving, moving, and once we got to the vicinity where all the fighting was taking place, one of the most memorable things for me was the grenades. Bad guys were just running around on the rooftops, and they were just throwing grenades. Just chucking them and running away. So, there's really no point in stopping and trying to fight or anything because they were just throwing them and running. They were disappearing, so by the time the grenade would fall to the ground, they were already gone. We just kept moving and moving and moving. We got to the area where they were, but at this time, they had moved on. The fighting had moved and spread and whatnot, so once we got to the area that we needed to be, it was pretty chaotic. We got there and I remember that an Army unit was there. I think there was a Bradley in the area, a couple of Bradleys in the area. The Marines from Golf Company that were getting engaged there in the area that we were supposed to be looking for and trying to reach, were gone at that point. But, we were in the area that they were in because the streets were just covered in brass, so it was very apparent that the fighting was there. At this point, we weren't necessarily getting engaged. The fighting had moved on. I remember we stopped for quite a while and at that point it was just trying to orient ourselves and see who was around us. Try to link up with folks, see who is over there two or three streets over. Just try to get an idea of what it is that we need to do, where we need to go.

Janney: I can't even imagine. Like you said, I know it was chaos. The fog of war. Comms were problematic that day, and you didn't have any air and it was just a bad day.

Endito: Yeah. Chaos.

Janney: Was the rest of the 6th just like that? Everybody in the QRF on Route Gypsum got ambushed. That's when Lt. Ski got hit and the QRF that was sent for Santiago and his guys never even made it. It was another Echo Company squad that actually came to their rescue.

Endito: Yeah, so being there on the street, we're still trying to wake up. We were in bed thirty minutes ago and now we're out here on the streets still trying to figure out what's going on. Is this for real? I wasn't too aware of what was going on throughout the whole city. It was just bit by bit we were kind of getting information that there were attacks going on throughout the city. It was everywhere; it was widespread, and it took a while for information to come in and for us to get a good picture of what was going on throughout the whole city. Throughout the rest of the day, we were just kind of moving around. Just moving from place to place trying to link up with people. Trying to gather people. Sgt. Allen Holt lost two Marines that initially got contact and had WIA.

So, that's the area that we were in. They got contact, they had two Marines KIA and a couple of WIA right in the area. It was kind of early on when we showed up there and were just kind of waiting and trying to gather ourselves. I had known that Holt's squad had lost two Marines already and we were just kind of waiting. I can't tell you what we were waiting on or what the situation was, but we were just kind of sitting there waiting to move. Just down the street, we were facing one direction and coming towards us all alone, walking towards us was Sgt. Holt. I'll never forget this because this is one of the most vivid memories that I have of Ramadi, and this memory that I have, it's the Battle of Ramadi for me. When I saw Sgt. Holt is walking towards us, he's covered in blood, his camis are just soaked in sweat and blood, he's just holding his rifle in his right hand and he's just kind of slowly walking towards us. I could see that he's just exhausted. His eyes are bloodshot. His face was flushed. He was just walking towards us all alone. He got close to me, and I said, "Hey Holt, you good, man? You good?" and he just kind of gave me a nod like, "Yeah I'm good." He just kept walking and made his way onto the back of one of the 7-tons or Humvees and he's just kind of sitting there waiting to go back to Combat Outpost or wherever. But, I saw that and I thought, "Holy shit." He looked like he had been in a fight.

The rest of the day was a lot of movement. A lot of chaos. A lot of moving parts. I think 1300, about noontime is kind of when things started to die down. Things died down and after that we were able to move back over to Combat Outpost. We went over to Combat Outpost, so it was probably about 1400 – 1500 at this time. We got over there and we got a chance to refill our water bottles. We got a chance to get something to eat and it was at that point that, as a platoon, we got info that Santiago's team got hit, and Stayskal had been medevaced. We had another Marine from Sgt. Longoria's team that hurt his hand. He fell or something, so was out of the fight. So, it was at that point, because I was a team leader of Headhunter 5, because we had lost a couple of guys and we needed to kind of move and shift bodies around, right there alongside the Humvees at Combat Outpost, my team was disbanded. I went to Sgt. Longoria's team Headhunter 3, and then one of my other Marines, Corporal Parker, who eventually got killed a couple of months later, went over to Santiago's team. The other two Marines that I had left stayed behind and they eventually made up another team with replacements that we got later on.

Janney: So, your teams got rearranged, cobbled together to make partial strength Headhunter teams. At that point, did you guys go out again?

Endito: Yeah, because at that point we got word that a couple of Marines were unaccounted for. I think there were two or three Marines unaccounted for. So, at that point, it was regroup, everybody ready, we're going to head back out there and search for these Marines. I don't know if that's something that we did right away, but we went back out searching for these Marines that were unaccounted for. So, at that point the fighting was done, and it was just a matter of let's go look for these Marines that were unaccounted for.

Janney: Was that on Gypsum or another area?

Endito: That was in the area where Golf Company was at. It was more Golf Company AO on the south side of the city. They were found later on. I don't remember exactly when that was, but they were recovered.

Janney: At some point the day ended and you guys were able to try to resume the little bit of rest that you didn't get to start with. What was the day of the 7th like? I know they started the citywide raids that night. What were you doing on the 7th?

Endito: This is weird. I've always talked to people that I keep in contact with and they tell me that their memories are shot. They're missing tidbits here and there. That's the case with me. I'm thinking long and hard about the 7th - where I was, what I did, and I know that I was with Sgt. Longoria's team at that point. I think we went out. We did patrolling and stuff like that or it was that day that I had to go back to Hurricane Point. Because I know at some point, I had to go back to Hurricane Point to get all my stuff because I was at Combat Outpost where Sgt. Longoria's team was. I don't know if that occurred on the 8th or if that occurred on the 7th. For the life of me, I just can't recall that.

Janney: That's completely understandable. I know you guys were exhausted and there was so much stuff going on. I'm not asking for exact specifics. Just whatever you feel like sharing.

Endito: Whatever day it was, either the 7th or the morning of the 8th, me and Parker went back to Hurricane Point to gather our belongings because we were moving over to Combat Outpost. We took the long train and, on the way, we had to stop at Blue Diamond. I know we made a stop at Blue Diamond because they had to take care of something. We had to wait there for about an hour, so I remember I got out and I was just kind of sitting around and I'm filthy. I had been wearing the same clothes for however long it's been and I'm just exhausted. I feel like complete shit and I'm just kind of sitting there and a Gunny or a 1st Sgt. that I knew from 2/4 years before came up to me. He recognized me and we were just kind of talking and he told me, "Man, you look bad." I'm just sitting there and I kind of looked at him, "Yeah, I guess I do." From Blue Diamond, we finally got back to Hurricane Point. We gathered all of our belongings and we moved on back to Combat Outpost that same evening. There's a day there that I might be missing and I'm sorry I can't cover that.

So, after that, our mindset was completely different. Now, we're in the fight. After the 6th, after the 7th, and then on the 10th I believe, we lost another couple of Marines. Now I'm with Sgt. Longoria's team, Headhunter 3, and I think it was on the 10th that we pushed out from Combat Outpost towards Echo Company's AO. It was early morning, just door to door knocking just trying to flush out whatever bad guys were left.

Janney: Operation Bug Hunt is what I was told.

Endito: Yeah, it's either Bug Hunt or 1, 2 or whatever it was. Yeah, it was Bug Hunt. It was early morning, 0200 or 0300. We stepped off and just cleared the city in Echo Company's AO. All morning into the afternoon, fighting erupted here and there. I believe it was the 10th that we lost Sims. Lance Corporal Sims from Echo Company. We had been out there all day. You're asking me about the next thing and my next memory after what happened on the 6th, but this is probably the most prominent thing. We came back, I think it was on the 10th, so after we went out and did what we did, patrolling and fighting throughout the day. We got back to Combat Outpost, I'm thinking about 1400 or 1500. Our team, everybody's good, we're all good, and got back to the COC. We're waiting there in front of the COC, and at Combat Outpost, one of the Marines from Echo Company came up to me and said, "Hey, Sgt. Endito, did you hear about Sims?" I said, "No, what's going on? What's up?" He told me, "We lost Sims today." My heart sank to the soles of my feet. I was absolutely saddened to hear that we lost Sims because before we left for deployment, I was with the MPs. Like I mentioned, I was with the military police Company, and the whole time that we were down there for five months, Sims was in my squad, so he was also FAPed out to military police. Yeah, so Sims was in my squad, and he was my Marine for those five months that we were down there. I got to know him. I think it was right around Christmas time that both of his parents were in a bad car accident, so he had to go home for that. Then in February, we reported back to 2/4 and left for Ramadi, so he wasn't officially my Marine anymore. But, when I heard that we lost Sims, that's honestly the saddest day throughout the whole time that I was in Ramadi. Everybody else was around, so I held it together, but I just wanted to crawl into a hole somewhere and cry. I was so sad. I was absolutely devastated to hear about Sims passing away.

Janney: I'm so sorry about your friend. I interviewed his mom, Margaret Kellum. It was a really tough interview to get through. I'm so sorry for your loss. As I said, he sounded like a great guy. Tell me some stories about Sims. I've heard a few funny stories, but I'm sure you can enlighten me as to a few others. I heard he had a great sense of humor and was just a fun guy to be around.

Endito: Yeah. His Southern accent. One of the things that I remember most about him was one night at the Christian roller skate rink right up at the 62 area at Camp Pendleton. This is when we were with the MP Company. We were out there that night because they were manning the gate. I think I was on patrol with somebody, so we had a vehicle, and we were just kind of going to these different posts and checking on them. Making sure everything was okay. We got there and it was probably 0200-0300 in the morning, probably a Tuesday or Wednesday and it was quiet. It's boring and we were just kind of standing around outside just talking, drinking coffee, smoking, whatever. Sims is over there in the bushes and he's doing something. He's just over there. He's kind of talking and he's trying to woo someone or something out of the bushes. It happened to be a raccoon. There was a raccoon in the bushes not too far from where he was at and he was trying to get the raccoon to come to him. Whatever he was doing, it was working. The raccoon was getting closer and closer. Whoever was on guard there with him said, "Sims, what the "F" are you doing over there? Get that raccoon out of here." They chased the raccoon away. I said, "Yeah, Sims is from the South. Probably happy to see a raccoon out here."

Janney: Yeah, you may have had a pet raccoon if he'd kept at it. Did you hear about the incident when Sims was involved in renting a U-Haul truck and they all rode around in it drinking?

Endito: Laughs. No. Oh my goodness. I lived off base, so I didn't get to be there and hang out with them in the barracks or anything else like that. The one other prominent thing that I really remember is when, like I mentioned, his parents were in a vehicle accident, and this happened right around the holidays. It was either around Thanksgiving or Christmas time. It was bad because it was both of his parents. I had to make sure that all of his leave papers were good to go and approved and everything was set to get him on leave to go home and be with his family. At the time, he was just really worried and it was sad because he thought he was going to lose his parents, or that he could have lost both of his parents. I always think about this. His parents ended up losing him.

Yeah, so aside from work, the only time I spent with Sims is when we were actually getting ready to report to work or we were on duty. Most of that time, I was in a different area and he was on post in a different area. But, he was my Marine. I never had a bad experience with him. He was a good Marine. He did what he was told. He did his job well. I did not have any issues with him. He was a good kid. He was younger than me, so I refer to him as being a kid. But, he was just a good guy. When we all got sent back to 2/4, I wished him well. Once we got to Ramadi, I would see him every now and then, just in passing, "Hey, Sims" - just a quick hello. He was always very respectful. Just a good guy.

Janney: Yeah, somebody described him as a Southern gentleman as I mentioned before.

Endito: Yeah, exactly. I never labeled him that way, but absolutely.

Janney: Tell me about Tommy Parker. You're one of the few people that I've interviewed that actually served with him. So, what do you remember about Tommy?

Endito: Parker was on my team, Headhunter 5. Once they got to our platoon, I only spent a good two months actually being around my guys and training with them. Then, I got moved over to military police and I spent five months there. Once I got back, it was probably just a couple of weeks that I spent with them before we left for Ramadi. So, I didn't get a chance to work with him much prior to deployment. I knew he had a family. He was married and had a daughter. He lived off base. He was a good guy. Christian guy. He didn't hang around the barracks much. There's not many crazy stories that I know of him doing anything crazy around the barracks because he had a family and he was living off base. He was a good Marine. Yeah, Corporal Parker. He was our Humvee driver, so he had to go attend a course to get certified or licensed or whatever to be a Humvee driver. It was me, him, Helton, and Corpsman Leija. That was my team. Parker was very close with Helton. They got along really well and they knew each other prior to getting into the sniper platoon, so whenever we were doing training or not, those two were always together. I actually paired those two as my Bravo team, so if we ever got stood up, it was me and Corpsman Bach as team Alpha, and team Bravo was Parker and Helton. He never gave me any issues. He was a good Marine. He did what he was told. He did his job. Unfortunately, there were a couple of situations when we were in Ramadi when we had to go out, like the morning of April 6th, Parker got stuck being in a Hummer because he was a Hummer driver. The rest of us were on foot and we were moving. We were popping house doors, and climbing rooftops. One night, he was stuck driving the Humvee. I know he didn't like that, but he had to do it.

Janney: Do you have any other recollections of any of the guys that didn't come home?

Endito: Not anybody that I was really close with. Yeah, that's kind of thing. Because a lot of these kids were really young and I really didn't get a chance to know a lot of them personally. Unless we were in the same platoon, everybody else is where they were at. These kids were just new. They were very new, so a lot of the younger ones I didn't know. The DI, Staff Sgt. Walker was new. He came from the drill field and had just gotten to 2/4 and Echo Company, so I didn't really know him too well. There was an incident I remember. We were at Combat Outpost and this was probably March. We were getting some food, breakfast or maybe dinner. I don't know exactly what Marine it was and I don't know what he did that got Staff Sgt. Walker going, but Staff Sgt. Walker turned into a Drill Instructor. It was a scene. In the middle of the chow hall, the Marine was standing there at the position of attention and Staff Sgt. Walker was just going off on this kid. We were just kind of standing afar with a smirk on our faces. It was shortly after that day that Staff Sgt. Walker was killed. I don't know what that Marine did to bring the Drill Instructor out of him, but that was a sight to see.

    I'm getting to know a lot of these Marines that we lost, little by little, over the years through all the Gold Star mothers. I've gotten to know Mrs. Layfield pretty well and Mrs. Jerabek. It's important for me to try to maintain contact with these Gold Star families. I want to know them and want to know how their sons grew up. It's important that we do that. The same thing with Parker. Parker was my Marine. Officially at that time, he was Santiago's Marine when what happened. But, I still very much consider him to have been my Marine.

    When those Marines were killed, it was really hard to comprehend and understand exactly what happened. We still don't know. What circumstances occurred that they lost their lives? Nobody made it out of there, so nobody knows. I think that makes it really hard because we just don't know what happened. We just don't know how it went down. I did meet with Parker's family a couple of years ago, I think it was Memorial Day, about five years ago. I went to visit them and it was absolutely hard because I think they were hoping that I would bring something new to them. Something unanswered, something that could clarify for them what happened. But, I had nothing. I know as much as they do. His wife showed up. His daughter showed up. It was wonderful to see them. His daughter is all grown up. The last time I saw her before we left for Ramadi, she was still a baby. She was grown up and sat there at the kitchen counter and listened and took in whatever I had to say about her dad.

Janney: After the 10th, what do you want to share with me about the rest of your time there?

Endito: Yeah, some ups and downs. Successful stuff that we did around the city. More fighting, of course. There were times throughout the next couple of months where it did get quiet. We were just kind of patrolling, just doing the same things over and over. Then, all of a sudden, fighting would erupt somewhere. I guess another highlight for me would be in August. Hancock was there.

This is when Corporal Powers was killed on the 7-story building. I think the day that Powers got killed was August 17th. They were on the 7-story building. Prior to that, they had been receiving sniper fire here and there leading up to that event. The evening of August 17th, Powers was killed instantly with a headshot. That's a heck of a shot for whoever it was with him being on top of a 7-story building. After he was killed, me, my partner, and Hancock's squad were sent up to the rooftop. Hancock's Fox Company squad's job was to resume watching Route Michigan for any enemy activity or IEDs. Me and my partner were tasked to find out where this person or persons were shooting from who were taking sniper shots at our guys. We were looking east from the 7-story building and it's a needle in a haystack. These shots could be coming from anywhere.

We went up there and did what we were told to do. Powers had been cas evac'd already. They took his body, but they didn't clean up anything. So, when we got up there it was still a mess – that's one of the things that's stuck with me the whole time we were in Ramadi. Almost 24 hours later, the insurgents tried to take us and the building down by placing explosives right under us on the 7th floor. They tried to bring the building down on us and the other occupants, but that didn't happen. It destroyed that building pretty badly and it's amazing that the building didn't collapse. It was a 7-story office complex, so there were people just coming and going, working there. Doing whatever they were doing all the way up to the 7th floor. We were situated on the rooftop on top of the 7th floor. So, it could have been anybody walking in there with a briefcase, putting explosives here, explosives there, and just walking on out. It's amazing the building didn't collapse from the explosion. We didn't find the sniper or the bomber as far as I know.

Janney: I'm glad that 3/5 Marines recovered Parker's rifle.

Endito: Yeah. That all kind of ties together what happened with Parker's team, what happened to Powers on that rooftop, and probably a couple of other Marines from our unit with that sniper rifle and the recovery of the M40 rifle by 3/5. Just kind of amazing how it all came back around.

Janney: I'm glad the 3/5 sniper team saw that long gun in the car and killed the guys that had it.

Endito: I went to school with the team leader that was on that recovery mission, so it's absolutely wonderful that it was them that recovered it. The whole situation sucks and it's shitty.

Janney: I heard the rifle is at the Marine Corps Museum at Arlington now.

Endito: At the moment, yes. Back in 2016 or 2018, we had a get together back at Pendleton and at that time, the rifle was there. The armorer was able to bring the rifle out to us.

Janney: That was a solemn moment, I'm sure.

Endito: Yeah, absolutely. I think really looking back now, it took a while to come to this realization. Longoria told me this. A couple of years ago, we were sitting around. I hadn't seen him in a while. We linked up and were sitting around drinking beers and we were talking about our experience in Ramadi. He mentioned our time there and particularly all the loss, the 34 that we lost, everybody that was wounded in action, people who lost limbs and whatnot. Everything was absolutely terrible, but everybody that came behind us benefited from what we experienced, and what we went through. All the losses that we sustained. We showed up with Humvees that were "armored" with plywood and sandbags in between, and that quickly changed. There were a lot of other learning points throughout the whole time that we were there, and everybody else that came after us benefited from our experience, from our losses, from the things that we did right, and from the things that we did wrong.

We lost way too many Marines, but because of what we went through, because of what we experienced, a lot of Marines, and even the Army that were in the area, benefited from what we went through. I know we lost guys when we were in country, but we also lost guys after our return. There's a couple of Marines that succumbed to their injuries since and also to suicide. We also lost Marines who returned to Iraq, like Sgt. Major Ellis, and deployed again.

When I found out about Sgt. Major Ellis, it was incredibly sad. My heart absolutely sank, and I cried. Sgt. Major Ellis was an absolutely wonderful Marine to serve under. He was our First Sgt. That day when we were on the rooftop of that 7-story building and it blew up, it took a while for them to get us down. We finally got down to the bottom. After QRF arrived and they secured the area around the building, we got down and one of the first Marines that I ran into was Sgt. Major Ellis. He was standing there with his mouth wide open, staring and looking up at what was in front of him – this blown up 7-story building. He had a disgusting crazy amount of dip in his mouth, and he stopped me and said, "Hey, Sgt. Are your boys okay?" Whenever he said something like that, it was very genuine. He absolutely meant it. When he stopped me that day, he wanted to know if my guys were okay and he was absolutely sincere about it. I told him, "Yeah, First Sgt., we're good. We're all good. I think everybody is good." Then, we kept moving to another area. But, when I found out about Sgt. Major Ellis, I cried like a baby. Wonderful Marine on the verge of retirement and then, just an unfortunate casualty of war.

Janney: Have you got any other stories about Sgt. Major Ellis that you can share with me?

Endito: Sgt. Major Ellis came from the Recon community. He was a Recon Marine. So, being in the sniper platoon, he had a soft spot for us. Throughout training and our time in Ramadi, he always took the time to make sure that we were good, that we had everything that we needed, just checking in on us. Again, every time he did that in passing, or if he took the time to stand around and shoot the shit with you, it was always very genuine. Always. Absolutely genuine.

Janney: Everybody loved him. I've never spoken to anybody that didn't love him.

Endito: Absolutely. He was that Marine. He was that Marine that, no shit, gave a shit about his Marines.

He was probably like this with everybody. I'm saying that he had a soft spot for us in the sniper platoon, but I'm sure he was like this with everybody. Always took the time to say a quick hello, ask how we were doing. Ensuring that your Marines were well equipped, and well prepared. "Did you eat something? Are you good? How's your family?"

Janney: He was definitely a Marine's Marine.

Endito: Absolutely. The look alone, with his flat top. He belongs on a poster for the US Marines.

Janney: Can you think of anything else that you want to share with me before we sign off?

Endito: I think the one final memory that I have of that whole experience was coming home. Because with everybody that we lost, with all the people that were wounded, while we were out there, I guess I was fortunate to make it through the whole deployment and come home unscathed. I mean, literally not a scratch on me. I came home and I'm good. It was the plane ride back that hit home. We left Kuwait and I think we flew to Germany. From Germany, we flew to Ireland, and from Ireland, we went to Baltimore. Flying into Baltimore, the Captain came on the loudspeaker and said, "We are approaching Baltimore. Temperature is whatever it is. Welcome home to America." There's photos: people are clapping, and everybody is feeling good, and we're landing back on US soil. Everything is good to go. We left Baltimore after spending an hour or two there and we flew straight back to Riverside to March Air Base in Riverside.

    As we were approaching March Air Base, we were starting our descent. Again, the Captain came on the loudspeaker and said, "We're descending. We'll be landing pretty soon. The weather is whatever it is. Beautiful California. Weather is good. Welcome home, Marines." When he said that, it was just quiet. The plane was just completely quiet. No cheering. No clapping. No nothing. It was just so eerie. It was so quiet. Personally, for me, I was sitting there and I didn't care to cheer. I didn't care to clap because all I could think about was the Marines that weren't coming home with us. The Marines that didn't make it home. The Marines that we lost. The Marines that got medevaced. Just how ugly it was. I'm just looking out the window and I'm thinking, "I fucking made it home." I looked around at everybody else that was sitting there and I'm getting that same feeling from everyone around me. I'm just looking around and everybody is kind of looking down or just not showing that much emotion, but that's what I felt. There was really no need to celebrate because we were coming home without 34 Marines. People had lost limbs. People that got their jaws blown off. Reflecting back on just all the ugliness from the past 7 months, there was no need for me to celebrate. I think that was the same with everybody else.

Janney: Well, I'm glad that you did make it home and I'm glad that the rest of your brothers did make it home. I'm sorry for your losses. Truly. I'm trying to get you guys to write the book. I'm not writing it, I'm just recording your stories. I hope it will help the families.

Endito: That's wonderful. I appreciate you taking the time to do this.

Janney: I appreciate YOU because, like I said, without you and 40 of your brothers telling me their stories, there wouldn't be a book other than just John and I going over there a couple of times.

Endito: That was wonderful to see. So, 2008 was when you went back out with Mr. Wroblewski?

Janney: Yes, sir. It was March 6, 2008. Almost four years to the day that the ambush took place.

Endito: Yeah, that was wonderful to see around that time. 2008 is right around the time where I was starting to admit that, yes, I do have PTSD and yes, I need to get some fucking help. Starting to reconnect with people that I haven't talked to in years. Right around that time is when I kind of started to pick myself up off of the floor. Just seeing something like that, seeing that that took place with you. You heading back out there was a piece of it - one of the things that got me to think, I really need to get my shit together and get some fucking help. Because Lt. Wroblewski didn't make it home. I'm fortunate, I'm here. I had a son at that time and I really need to get my shit together.

I guess really nowadays a lot of the focus is on the aftermath. What everybody has experienced since coming home. What happened, happened. People we lost. That's not going to change. It's a matter of those of us who are still here to move forward and to make the best of whatever it is that we've got going on. That's just kind of how I see it with my combat experiences. Having been through years of therapy. It's going to be there, all the experience that we had, the images, all these bad thoughts. Bad memories, they're always going to be there, but it's now a matter of how we choose to deal with it.

I mean now, moving forward is a matter of us looking out for each other still. We're not in combat anymore; we're not in uniform any more. We're not the young studs that we were. But, a lot of us have families. A lot of us are doing good things. Business owners. Some of us are not doing so well. I think it's really important to keep that mindset that we still need to rely on each other and we still need to be there for each other and support each other. I try to do that the best I can. I'm not the best at it. When I'm down, I'm down and I'm pretty much useless. But, if I can support anybody else or be there for anybody else, I try to do that the best I can.

Yeah, so personally for me, where I try to keep my mindset is, "Let's move forward. What's next? What can we do tomorrow? What can we do to better our situations, especially for those who are still struggling?" I'm still struggling a lot and it's very hard for me to ask for help. I don't ask for help. I'll admit to that. I need to, but I'm just trying to focus on moving forward.

Janney: It's definitely not a sign of weakness to ask for help. You know that, right?

Endito: Yeah, but it's hard to do. It's absolutely hard to do. I try my best if there is a reunion going on to be a part of it, or if it is just a small get together. I think the person I visit a lot is Longoria. Me and him, we are very close now and when we were in the Marine Corps. He's one of my go-to's if I'm really in trouble. Do you know Joe Hayes?

Janney: I don't know Hayes.

Endito: So, Hayes was with Golf Company. Lately, he's really been my spiritual go-to because my relationship spiritually with God has really been an issue. So, he's been helping me out there. I'm always down for seeing somebody because that's up there with one of the best forms of therapy. Just being around the guys that we were in the shit with. It's always good to see somebody and I've made a strong effort to try to visit our 34. They're all over the country, but I've been to Arlington. I've been to see Layfield in San Francisco. Cox, I think, is there. I've gone to see Parker. I've gone to see a couple of other Marines, but I still have a lot on my list. And I'm always down to visit if anybody is in the area. Let's get together because it's always wonderful to see them. Just like in combat, when shit was getting heavy, it was always good to see a familiar face. You know, just seeing them to say, "Hey man, you're still alive. Shit, I'm still alive." That's the same feeling nowadays when I see somebody I haven't seen in a long time. It's, "Shit, you're still alive. You're doing good. Well, I'm doing good, too."

Janney: Yeah, just the stories that I've heard about the guys. I don't know how well you knew Pedro Contreras. But, I laugh just thinking about some of the stories that I've heard about him.

Endito: Yeah, he was loud. He was a loud dude. I remember him from way back in our Okinawa deployment. We were out in the jungle, and he was just always loud. Like, "Dude, shut up!" But, he would be doing goofy stuff. He would be doing funny stuff, so it's hard to tell him to shut up and be quiet. We're out in the jungle: it's wet, it's cold. We've been out there for days. Some days we're hungry. It's miserable, and there's Contreras over there being loud and doing something goofy.

Janney: Were you there the day he showed up for the morning run with a fireman's hat on?

Endito: No, no. Laughs.

Janney: He got in a world of shit about that, but it cracked everybody else up.

Endito: Yeah, that guy was a ball of sunshine. I didn't know him personally, but from afar I did.

Janney: Yeah, I'm sure he got on the NCOs nerves, but he cheered up the rest of the men.

Endito: Laughs.

Janney: The rest of the guys loved him because he would make them laugh in spite of their misery.

Endito: Absolutely.

6 April 2004 Golf Company QRF

LCpl Kristopher Privitar – Weapons Company (First Sgt. on date of interview)

Janney: Today is 23rd April 2020. Tell me why you decided to enlist in the Marine Corps.

Privitar: I was actually going to go into the Navy to begin with. I was going to go be a part of their thermonuclear program. Was all ready to go. I wasn't contracted yet, but with that job lined up essentially. Then, the Navy guy said I had to go do MEPS real quick to get weighed in and do final stuff. I said, "All right. I'll go." The next thing I know, this guy is trying to have me shipped off. So, I said, "Whoa, whoa. I'm not ready to go yet. There's supposed to be some time." So, it pissed me off. That day, the Marine recruiter called me and told me what he had to offer. I was pretty pissed off at the Navy guy, so I said, "Yeah, all right. We're going to do this thing." He said, "We need some of your documents to run to make sure you're a citizen." I said, "Well, the Navy guy's still got it." So, he said, "All right, so give him a couple of days to get it back to you. Just let me know." A couple of days goes by, nothing happens, so the Marine recruiter calls me and asked if he gave it back to me yet, and he hadn't. So, what he did was he actually went over there and confronted the Navy guy for my documents, and he got them. That is when I first met him when he had my documents in his hand. If he was willing to do that for me, for somebody he really doesn't know, then this is something that I want to be part of. He was a Marine infantryman and he tried to talk me out of going into the infantry, but he couldn't do it because I wanted to be in the infantry because that's what the Marines are known for. They're not known for playing the flute or something like that. They're known for being infantrymen. Especially at the time that I enlisted in August 2002. Everything was going on, so that's what I wanted to be a part of.

Janney: So, it was after 9/11. Did you have any family members that were in the service or you just felt inclined to do that?

Privitar: My step grandfather was in the Army. He didn't go anywhere, but was in. I had an uncle who actually still works with different organizations like the Marine Corps. He was in the Marines, but I didn't know that until after the fact, so that didn't really compel me to do anything.

Janney: That's still admirable; wanting to serve. That's an amazing thing and I wish we had more of that in today's youth. After you got through boot camp and SOI, were you assigned to 2/4?

Privitar: No, I went straight to 2/4. They were currently on a MEU doing stuff in Korea and where I met them was in Okinawa in transit to Korea. So, I had gone straight to 2/4.

Janney: You did that Okinawa deployment before they went to Iraq?

Privitar: Yes, I did.

Janney: Okay. At this point, what was your rank?

Privitar: PFC at that point. I think I got promoted to Lance on that deployment.

Janney: When you come back from Christmas leave, they're telling you to start getting ready. Some of the guys have told me that initially they thought they were going to Afghanistan. Did they do a lot of urban combat training before you went, in case you were sent to Iraq?

Privitar: We did a lot of urban training. We were going out to March Airfield and doing a lot out there. I don't remember rumors about Afghanistan. I think we always knew that we were going to Iraq. We did a lot of convoy operations because we were the Weapons Company, so we had vehicles. That's when that whole Janet Jackson thing went down. We were in the field, and they had the Super Bowl while we were at March Airfield and that's when that happened. It was funny.

Janney: When she had the wardrobe malfunction?

Privitar: Yeah. I'm pretty confident that's when that happened. We were at March Airfield for a while for that. We had some guys come from 5th Marines that were in F1. I know one of my platoon was Sgt. Lechard; he came from 5th Marines and jumped to an appointment with us. So, they may have passed that to them over there, but I remember them coming to us beforehand.

Janney: You get to Kuwait mid-February. Did you convoy to Ramadi, or did you fly in?

Privitar: We were with Weapons Company, so we were charged with providing security and escorting the entire Battalion from Kuwait into Ramadi. It was a three-day movement and pretty rough.

Janney: Yeah. Did you guys get hit with IEDs and small arms fire during that movement?

Privitar: I think a couple of IEDs had gone off. Some small arms. Nothing as significant as what was to come. I remember taking contact a couple of times. One of the vehicles up front.

Janney: You got to Ramadi taking over for a National Guard unit and elements of 82nd I think. Did they kind of take you around and give you the lay of the land to learn the routes?

Privitar: It was an interesting time during the turnover, because once it came time for us to take over, they were kind of watching us do it. We were patrolling down roads that they wouldn't dare go down in Bradleys and we had these non-armored Humvees with no real armor. Just going down these roads that they were afraid to go down in Bradleys because they would get heavily IEDed and attacked and stuff. We found that rather comical. You're afraid to go in a big ass armored vehicle and here we are in unarmored Humvees.

Janney: I heard that a lot. That you guys didn't have armor and that the Army guys were mainly doing mounted patrols and didn't want to go to certain areas of town because they were afraid.

Privitar: They didn't want to go down Route Nova. They didn't want to go down Route Michigan. They didn't want to go down any of those. Here we were just going down them because they had to be patrolled, so we could control the area. They didn't seem to get that concept.

Janney: Yeah, I heard that they pretty much sat on their base and just rode around occasionally. I'm sure that was a big change for the populace when they started seeing Marines on foot patrolling areas that had never been patrolled.

Privitar: It was an interesting change from the populace interacting with us when we first got there.

Janney: How were the people in general? Were they cautious or were they receptive to you?

Privitar: It was really hard to tell. Knowing that we weren't fighting a uniformed enemy; we were fighting locals essentially. They looked like locals. They didn't wear uniforms, so we were cautious with them. This could be somebody that is trying to see what tactics we were using. How we were interacting with them and try to plan around it. We were still trying to give that positive aspect of, "Hey, we're here to help you." So, it was a really intricate kind of thing to do during that time.

Janney: A very delicate balance. It's a different strategy, yet you don't want to start an insurgency.

Privitar: Right, especially them seeing us walking around on foot was a big change for them, too.

Janney: Exactly. You guys were at Hurricane Point? What was happening then?

Privitar: Yeah, we were at Hurricane Point. We didn't get too much contact when we first got there. I think they were still trying to feel us out, especially with us doing something different than the Army. Towards the end of March, it started kicking up. We started getting mortared more frequently and more accurately. Before, they would land wherever and didn't come close. Now, they were starting to hit on the outsides of the camp, some inside the camp, and sometimes it hit our Conex boxes. The attacks also started picking up towards the end of March, varying from small arms to IEDs. We also killed a high value target towards the end of March and that's when everything broke loose.

Janney: I heard Col. Kennedy and Sgt. Major Booker were in a couple of vehicles and someone launched some RPGs at them. Booker jumped out of the vehicle and fired a couple shots with his M14 and killed the guy. Some higher up was upset because it was his cousin that Booker killed.

Privitar: Yeah, that happened a lot. "That was my cousin." "Well, he shouldn't have been a dirtbag."

Janney: What were you doing from 25 March to 6 April? Were you patrolling every day?

Privitar: Absolutely. We had a rotation where a platoon would be on day patrol, one platoon on night patrols, and one on camp guard. Our platoon was on patrols a lot even during our guard time frame. I think we actually picked them up a little bit because that's when they started utilizing the 3rd platoon for additional patrols on missions because things started kicking up a little bit. We realized that, "Oh shit. We just poked the hornets' nest. Shit's about to go down." So, we did that a lot and also in support of Echo, Fox and Golf to kind of give them vehicular support. Give them some crew served weapons on the ground, some heavy weapon support.

Janney: When you say crew served, were you guys fielding 240 Golfs or SAWS or 50's?

Privitar: We had 50s, Mark 19's, 240's, and also had TOW missiles. Because we were Weapons Company, we had the crew served weapons. We had all of them out there.

Janney: There was quite a bit of activity on the 5th and my theory is that a lot of that was generated by the fact that the Marines in Fallujah were attacking, so the insurgents pushed west to Ramadi.

Privitar: Yeah, that's kind of what we gathered, too. Basically, they were flushing them out of Fallujah, so then they started coming over to Ramadi. We already had our groups of individuals to deal with and then we had the ones coming from Fallujah, too. There were just so many of them right there. I believe that's what led to what happened the next week because they had so many individuals there to fight us. That's why they picked that time frame.

Janney: What were you doing on the morning of the 6th of April?

Privitar: That was a very interesting day. I was on the north bridge with Lance Corporal Cox, first name Aaron. All of a sudden, we started getting mortared. At that point, we were kind of used to getting mortared and we're like, "Oh shit, here we go!" What was really interesting is they were walking them on to us on the north bridge. They started off at a distance and then they got closer and closer. I looked at Cox and told him, "I think we're gonna die, man. I think we're fucked." Because there was nowhere that we could go, we were on top of the bridge. We couldn't just jump off because then we'd be screwed that way, and there was nowhere else to go. So, it was just, "Oh well, fuck it." We were facing outboard, because at that point, they started trying to rush into Snake Pit which is where Fox Company was. So, Fox Company was having to deal with them trying to come into the compound. I was talking with a friend of mine, "Hutch", David Hutchinson. He was on one side of the base, and insurgents were sprinting toward the base trying to get into the compound. Myself and Cox were on the other side, taking them out on the other side of the walls with 240s. I was DM (Designated Marksman) at the time, so we were just trying to maintain that side of the wall while the platoons were on the other side that day. All hell broke loose for a couple of hours and then we had some other problems across the river. Over on the south bridge, they were trying to attack Snake Pit as well. So, our people on the south bridge had a 240 engaging insurgents on that side. That was an interesting day. I can't remember all the details. I'm undergoing treatment for chronic PTSD because I'm having memory specific issues. Essentially, my brain has erased certain things

Janney: Sure. I don't want to push you back into bad memories either.

Privitar: I don't mind because I think it's appropriate for people to know and understand. If you don't know about it, then it happens again. History repeats itself also. It's an important part of Marine Corps history, 2/4's history. I don't mind talking about it, but I can't remember some things.

Janney: I completely understand that. I know it's a difficult thing to talk about and I appreciate you spending the time with me to try to flesh it out tactically and operationally. There were firefights all over the city on the 6th and you guys were getting hit pretty bad, too.

Privitar: Yeah, one of them was particularly bad. I remember the day, but I don't remember the date if that makes sense. Echo Company had a patrol go out with a full platoon. They had gotten ambushed, and they had gotten knocked down, too. I believe it was three Marines and a Corpsman. Everybody else was KIA. My platoon got called in to go reinforce them, to try to take them out. I remember hauling ass down there, just trying to get there. We were getting engaged on the way there. There were IEDs on the way there, but we just kept pushing. We needed to get them out of there. I remember showing up and going through a portion of the road and two technicals came behind us and closed us in. I was thinking, "What's going on? Oh, fuck. Is this it? This is where everything is going to happen." We showed up and were trying to figure out what was going on. Once we figured out where everybody was at, it was just a constant battle just trying to get to the Marines. Trying to get our fallen Marines to vehicles. We were fighting for a couple of hours, and we were able to retrieve everybody. We loaded them into the vehicles. Then, all of a sudden, our trucks took off. It was just us on the ground. That was it.

Janney: WHAT? That sounds like the 6th. They left you behind? They just took off without you and you're sitting there?

Privitar: Yeah, it was just us on the ground because they had taken everybody else to Charlie Med. This was the same day that Sgt. Conde got shot in the shoulder (author note: Sgt. Conde was shot on 6 April.)

Janney: I remember hearing about that story. He was hit pretty badly, but he kept getting up, saying that he was ready to fight, and refused medevac.

Privitar: Yeah. I can't remember if it was before or after, but that day Sgt. Conde got hit in the shoulder. So, we were there, and we were just looking around at each other like, "Oh fuck. Well, what now?" I can't remember how long we were there. It was just a constant fight. I remember running out of rounds and having to get a magazine from another Marine that was there, because I started going down an alleyway and we had guys coming through. We had to move to different positions. I remember Captain Bronzi was there, too. I remember linking up with him and he said he needed to get to a certain point. So, me and my guys essentially escorted him there, coming down different alleyways and taking him there. Then shortly after that, when our vehicles came back, the Army came with them with their Bradleys. I remember seeing the first one come around the corner. I was so excited to see them. I said, "All right. We're going to be fine. We're going to make it." Then, the first gunner dropped. From my understanding, a sniper took him out. As soon as that first gunner dropped, all the rest of the gunners buttoned up. So, now all we had was vehicles with no guns out coming to us. We're like, "Fuck, man. We have no guns right now. We just have ours here."

Janney: Plus, you were already condition black on ammo, so that's terrifying.

Privitar: Some of us got over to the Bradleys, trying to squat down on top of the 50's to be able to engage them. Because they were buttoned up, we couldn't get inside. But, we needed the guns and a couple of Marines jumped up there, manned those. I remember running up to the back of the Bradley to try to go get ammo to resupply my guys. The whole time, I could hear rounds hitting between my legs, but I had to get to the Bradley. When I got there, this Army Colonel came out of the back of the Bradley. I don't remember his name, but I'll never forget the character. He came out of the back of the Bradley. He was wearing yellow dishwasher gloves and carrying a shotgun. He came out and said, "What do you boys need?"

Janney: Laughs. That is crazy. What was the deal with the dishwasher gloves?

Privitar: He was a full bird Colonel, too. That was the funniest thing in the world. I guess he just wanted to have something on his hands to prevent shrapnel burns or something. At the time, we were pushing men to wear gloves to keep minor shrapnel burns from injuring their hands essentially. So, I'm seeing it and I'm hearing it and I'm grabbing ammo and running. Then what happened to someone wound up calling some Cobras (attack helicopters) in. The Cobras did a couple of gun runs and cleared out the streets. That was the only way we got out that day because they were able to get us some air support. We were fighting for many hours. It seemed like forever.

Janney: That's pretty hairy. 2/4 didn't have a lot of air support because it was all devoted to Fallujah. It was tough to even get any. In my opinion, you guys needed a lot more air support than you had.
Privitar: Yeah, it was crazy. I think that was the only reason we made it out that day because they had us surrounded. I think they had us surrounded for about three or four blocks of the area.

Janney: So, you guys were fighting from a house at this point?

Privitar: No, we were in the streets. We didn't want to go in the houses because at the time, they were starting to booby trap the inside of the houses, too. We didn't want to go into the houses just because we didn't have any way of detecting it. We didn't have Warlocks (electronic IED countermeasures); we had nothing. We were on foot in the street because they took everybody to Charlie Med.

Janney: Man, that's crazy. No cover and just getting hammered. What happened next?

Privitar: Yeah, we went back that day, but they didn't push us back out that day because of everything we had gone through. The following day (7 April) we were back out around Nova, and we were just constantly getting hit. Nova was a little bit different from the rest of the roads. The rest of the roads were in urban areas and Nova was up on a berm in a more rural area. We were going down and got hit with an IED and some small arms fire. We all dismounted. I had to yell at the CO because he was shooting over our truck. Our truck was named Old Yeller because it was the only tan truck in the Company. We used to say that Old Yeller ran on blood because the truck would be running like shit for a few days and it would start running just fine as soon as we went into a firefight. Anyway, the CO was shooting his M16 and shot the hood of our truck. He shot through the hood of our truck and so one of the guys had to yell, "Sir, you're shooting our truck. Fucking stop!"

We bounded forward toward some buildings. Psyops was with us too and they got their terp out. He had a shotgun with him, and he started shooting up in the trees. We said, "What the fuck is he doing? They're in the building. Why are you shooting the shotgun into the trees?" The next thing you know, a guy fell out of the tree. This is when the insurgents started attacking us from the trees. Before this time, we didn't know about it. He either saw them or he knew something that we didn't, but he hit a dude hidden in the trees. We started to engage guys in the building. We made our way down there. Crew served weapons were hitting the buildings and we eventually made our way there. There were 8 – 10 individuals there. As we were going up, Marshall Cummings got shot in the side. So, we had to get a grip and once we cleared the area, we got him over to Charlie Med. After we took Cummings to Charlie Med, we got tasked with another mission. Go do something else. As soon as Cummings got patched up, we dropped him off at Hurricane Point and went back out again.

Janney: Sounds like it was a hell of a day.

Privitar: Yeah, it was an interesting time because we were on QRF at that point. It just seemed like there was always something going on. We'd get called out to help and it was weird.

Janney: Yeah, if you're QRF you know you're going to be in contact when you get called out just because of the nature of that mission. I know the week of the 6th through the 10th was intense. Like you said, all hell had broken loose basically. There was another big operation, Bug Hunt.

Privitar: Yeah, it was Bug Hunt. We had a few of those. We had gotten called out to go supply the outer cordon with our vehicles. Our dismounts were assisting, I think it was Echo or Golf Company, who were the primary for it. Yeah, our Company was tasked with that.

Janney: They were doing house sweeps and you guys were in support of that movement?

Privitar: It was a scary time. Going into houses, being part of a two-man team and going into a house. We weren't seeing a lot of booby traps in houses yet, but that's when they started doing it. Just one of those things in the back of your mind. It wasn't a really prevalent thing yet. They were doing the carpets, starting to do things like that.

Janney: By using a pressure plate type device or something, I guess?

Privitar: Yeah. We were smoking them that day. At that point, if we knew it was a hostile compound, you started clearing the room with a grenade. That's how it was. Then, as you made your way through the rooms, if you ran out of hand grenades, you would go to the trucks and get more. If we were taking fire from a house or compound, we obviously knew it was hostile. We didn't bother sending the guys in first. We'd throw a grenade in first, and then go in and clean up afterwards.

Janney: That sounds like good tactics to me. After the 10th, did things calm down a little bit?

Privitar: In comparison to that timeframe, it calmed down. However, throughout the rest of the deployment, it was kinetic. It just wasn't to the point of how it was during those days in early April.

Janney: So, Kris, tell me some stories about some of the guys that you served with. Funny stories are always a great thing to share if you've got any memories like that that you want to pass on.

Privitar: I remember Cox did something stupid. I remember Cox being angry as shit because they hit us with an IED one time that was stuffed inside of a dead donkey. He was pissed off because his fucking vehicle smelled like dead donkey after that.

One time, the Company was playing a dodgeball tournament, then maybe an hour or two after our dodgeball tournament was done, we had mortar impacts right inside where we were just playing dodgeball outside of our hooch. I remember thinking, "Okay, well, we're not playing dodgeball anymore."

We had a Xbox. We used to play Halo a lot. Our way to decompress was to play Halo. There were several times, because the weather there in Iraq, the sand and everything, that the TV or the Xbox wouldn't work. So, we would have to play short straws of who took the TV and Xbox to the local repair guy. We had a local who had a shop there that would sell us movies and cigarettes and things like that. So, we would take it to him to have it repaired and then so once it was repaired, we would all exit the hooch. The one guy that got the short straw had to go inside the hooch and turn on the TV and turn on the Xbox in case it went off.

Janney: Man, that's horrifying just thinking about it. Thinking that it might explode.

Privitar: Yeah, luckily, we never ran into that. We had gone through a couple of shop owners because we had shitter cleaners come on base, too. We'd escort them on and a couple of times they wound up being bad guys. We ended up killing a couple sets of shop owners and shitter guys because they were fucking fighting us. So, it was always the luck of the draw if they were going to plant something inside of the TV or inside the Xbox.

Janney: I know you've got some other stories. Did you know any stories about your fallen brothers that you can share with the readers to help them know about those guys?

Privitar: Benjamin Carmen. Carmen was one of my best friends at the time. He was an assault man with me. I want to say he was in the turret when he was killed on 6 April.

Janney: I've not talked to anybody else that knew Carmen well, so tell me about Benjamin.

Privitar: He was an odd individual, but fun. He had a girlfriend back home. He would talk about her a lot. We would hang out in our room and go do our thing. We weren't those types of people that liked big crowds of people, so mainly just he and I would go out and explore California. Whenever we would come out to Combat Outpost, I'd try to go see him and make sure he was doing okay. Things like that. Whenever I would get care packages in, because I know he got care packages from his family, we made little trades from our care packages and stuff. He had an odd sense of humor. We would play video games together all the time. Things like that. As I said, he didn't have a lot of friends. He was pretty much my good friend during that time frame. He was pretty private.

As I mentioned before, Sgt. Conde was my squad leader. I still go by a lot of things that he did and said, even to this day. He was always taking care of us. One time, I got screwed over on leave. I went on recruit assistance, but they charged me leave instead. So, when we did that leave block during Christmas, I filled out that leave shit and he kept it to himself. That way I didn't get charged leave to make up for it. He was always making sure we had what we needed to take care of ourselves. Make sure that we are mentally healthy, spiritually healthy. Always checking on us because he actually cared about us. One thing Sgt. Conde did what I've always done, and I've expressed to my Marines, don't ever ask your Marines to do something that you are not willing to do yourself. That's something that I do even to this day as a First Sergeant. I'm out there with my guys during working parties and things like that because if I expect them to do it, then I should be able to do it.

Janney: That's a sign of a great NCO and I'm glad that you took that lesson from Conde and have carried that on to your Marines. That's really cool.

Privitar: When July 1st happened, it just was a blur. The same IED that killed Sgt. Conde was the same one that injured me. I had some shrapnel wounds from that. I was in disbelief when they said he had been killed. I was combat aid at the time and patched him up a little bit before the Corpsman got there. Then I went and got security. I didn't think it was possible because he was just a machine. Like he had been shot and kept going in April. I was in disbelief when it happened. I think the whole platoon was shocked. We just couldn't believe that it happened.

Janney: You know, when a man like that goes down, I know that it makes the rest of you realize that, "Hey, you know, it can happen to anybody."

Privitar: Yeah, it hit our platoon really hard. We had to take a couple of days after that. They didn't send us out which was really rough for the Company because, again, we were still pretty kinetic during that time frame. It was all pretty hard, too. Sgt. Conde was our front runner. He is what represented MAP3 really well. Takes off at your heart.

Janney: It sounded like he was a beast, no doubt.

Privitar: Yeah, he was.

Janney: How about some of the other guys that didn't come home?

Privitar: I didn't have a lot of interactions with the other guys in the Company, but Jeremiah Savage was a mortarman. He was one of those guys that if you needed something, he would go out of his way to help you. Whether he fixed your car, went over some kind of training aspect, or just to hang out and bullshit. He was always there for everybody. He was a go-to guy. I remember him going around and making sure that we were good. The day he was killed, he was on the 50 cal and we hadn't put up the heightened armor yet. We hadn't welded the tops of the Humvees yet. His being killed led to that because it was easy for a piece of shrapnel to hit the exposed men. It got to his upper body; that's what killed him. But, that's what led to us getting the improved armor.

It was a really interesting time because it seemed like we would come up with a concept or an action to counter what they were doing to us. Then, they would come up with something else to counter what we did. We got to where we were able to spot the IEDs regularly, so then they started putting them in the road as they were building the road, as part of it. They started putting them inside the blocks as they were building the roads. That's when they started using VBIEDs because they didn't do that at the beginning. Then they started making VBIEDs and there's no real way to guard against those because you don't know where they are.

Janney: Sure, and it's a human piloted bomb. It's hard to get inside the mind of somebody that's driving a suicide vehicle or the mind of the guys that were designing the IEDs. I can't even imagine.

Privitar: Yeah. It's really hard to continuously counter what they were doing because then they would counter us. It was like, "Fuck man. What are they going to do now?" Then, they started using AP (armor piercing) rounds. A friend of mine, Billy Webster, was sitting down in the turret, and they shot the back of the turret. It went through the turret into his Kevlar, went around the back of the Kevlar and went out the front of his Kevlar helmet. They saw that we had armor on our trucks and they thought, "Well, now we'll use the rounds." Those could penetrate our armor. We thought, "Well, fuck, what are we going to do now?"

Janney: But Webster didn't actually get hit? It just kind of skipped around inside his Kevlar?

Privitar: It didn't go through his head. It went into his Kevlar and followed the outline of the helmet - the path of least resistance. He was injured, but my good friend Billy Webster survived.

Savage getting killed hit the 81s platoon pretty hard. Morris getting killed hit them pretty hard, too. Because, like Weapons Company, the platoons were pretty tight. So, whenever we lost somebody, and I don't mean to have more significance than other Marines that were killed, but it seemed to hit the platoons a lot. Really hard during that time frame, so we coped in other ways.

Another way we relaxed was to watch the Sopranos. I just remember us being excited at the time because we were stuck on watching "The Sopranos" series. Every Tuesday night, we would try to watch another part of "The Sopranos." After "The Sopranos," we went to "Smallville." We started watching "Smallville" because that was what we had.

We kind of knew the Army's PX schedule at one point. So, we would always try to make sure that we had to do a resupply from JC (Junction City base) and knew the schedule so we would be the first ones in there. We would know what was where, so we could buy everything first before the Army guys got to it. Then where they put their chow and shit, for the chow line stuff, so we would go over there and try to bargain with them to try to get different hamburgers or whatever it is that the Army had there. Same thing with Taqqadam. We would go to Taqqadam just to boost morale. Go out there for half a day, go eat at their chow hall and things like that.

I remember, one time, going to Taqqadam and we got in a firefight for a couple of hours. It was a pretty bad one. I don't remember the day, but we were all dirty and some of us were bloody. We got to TQ and all we wanted to do was eat. That was it. We were going to eat and then fucking go back. The chow worker kind of had to let us in because somebody in the Marine Corps was talking and was doing like a PNU or whatever, and we said, "Look, we are from Ramadi." They said, "Oh, okay, go on in and eat. Just stay quiet and stay in the corner." We just wanted to eat, so we went in there and sat down. I remember the Sgt. Major Estrada pointed back at us and something to the effect of, "You guys don't be like these fucking dirt bags back here." Pointed at us. We thought, "Whatever, we don't care. You guys fucking live here. You can kiss our ass." We just didn't fucking care. Normally, we would argue back, but at that point, we just didn't care. Our chow at Battalion was actually pretty good in comparison to what I first had, compared to other places like Combat Outpost. We had a pretty good chow service. I couldn't complain about it.

6 April 2004 Golf Company QRF

PFC Justin Weaver - Weapons Company

Janney: Today is 8th May 2022. Tell me when and why did you decide to enlist in the Marine Corps?

Weaver: Going through high school, I listened to all the Marine Corps commercials on TV and everything. Everybody in my family is prior military. My uncles are both in the Army, Air Force veterans. I kind of just wanted to follow in their footsteps, and the Marine Corps caught my eye. You know, the flashiness of the uniforms, the pride of the tribe, we call it, you know, being part of the Marines. That caught my eye, and I didn't have a real great family growing up, so I sought that out.

Janney: When did you enlist? What month and year was that?

Weaver: I was in 11th grade when I joined the delayed entry program. It would have been 2002. I left for bootcamp on July 15th of 2003 and I graduated boot camp October 2003. I joined the infantry. I went to Camp Geiger for infantry school. After that, I was sent to Hawaii, Kaneohe Bay, before going to 2/4. I was at Kaneohe Bay with 3/3 in Hawaii. We weren't even there long before 2/4 needed bodies for the workup to go to Iraq, so we got shipped to Camp Pendleton, California. We arrived to 2/4. I arrived to 2/4 about six months before we left for Ramadi.

Janney: What was your rank at this point, Justin?

Weaver: I was a PFC at this point.

Janney: When you deployed overseas, did you convoy in, or were you part of the advance team?

Weaver: No, I was part of the convoy. We went to Camp Victory in Kuwait, and we were there for a few weeks, basically just doing a little training stuff, getting ready. Then, I was part of the convoy team into Iraq, not part of the advance team.

Janney: What was your first impression when you got to Ramadi? Did you guys hit the ground running, start doing patrols?

Weaver: We were told that Ramadi wasn't that bad. We were told that we were going to be going over on a hearts and minds humanitarian mission, putting generators in schools, handing out medications to hospitals, stuff like that. We were told we were there to win the hearts and minds of the locals. When we first got there, it definitely seemed like that was going to be the case. Everybody was smiling and waving at us and was friendly towards us. That quickly changed within a few weeks' time frame.

Janney: Do you think that part of the change was because the 82nd Airborne that you guys were replacing was doing a lot of mounted patrols and you guys were doing foot patrols?

Weaver: Yeah, I definitely think we had a lot more of a presence than the Army did there. You rarely saw the Army patrolling through the city or anything. They mostly did convoys, especially once we got there. It was very rare to see any of the Army guys in the city.

Janney: So what was your first patrol like?

Weaver: Well, I'm not going to lie. I was terrified. You're in a combat zone; you're scared. But I trusted the guys around me. I believed that they were all capable individuals. I felt comfortable with them. We didn't get a lot of training beforehand, but we all meshed really well together. So I felt pretty comfortable. I was nervous, but at the same time, like I said, I felt comfortable around the guys I was with. I felt comfortable in the leadership I had.

Janney: So what Company were you attached to at that point?

Weaver: I was with Weapons Company. I originally came over to be a mortar man and we weren't allowed to use indirect fire weapons or anything there, so anybody that was within the mortar platoon got changed to be a 0311 (infantry) or 0331 (machine gunner.) I originally started as just a regular rifleman there. I eventually became a 0331 (machine gunner) once my good friend Jonathan Hurley got hurt. He was our first casualty, him and Dan Holm. I took over being a machine gunner for Hurley.

Janney: So were you on a 240 or a SAW (5.56 NATO caliber Squad Automatic Weapon?)

Weaver: I originally started out on a 240 Golf (7.62 NATO caliber) machine gun, but once we got bigger weapons, it changed. Because we were all just changing over, the mortar platoon didn't have any heavy machine guns yet. We didn't have any Mark 19s or 50 cals yet. So I was on a 240 Golf for probably three weeks, and then finally got changed over to a Mark 19 (40mm automatic belt-fed grenade launcher).

Janney: Where were you guys stationed? Were you at Hurricane Point?

Weaver: Yeah, I was with Weapons Company. We were at Hurricane Point. Hurricane Point was originally Uday and Qusay Hussein's vacation palace there in Ramadi. We were originally there. We stayed on the palace grounds. We didn't really operate inside the palace. Y

Janney: When did you notice the mood of the people start to change?

Weaver: Well, like I said, the first couple weeks, everything was calm and peaceful. The locals seemed to like us at first. We almost got into that lax attitude of things being calm and it wasn't going to be bad. I started believing the whole Hearts and Minds thing that we were going to, you know, we were just going to be there to win Hearts and Minds. Then, one day we were leaving Snake Pit, where Fox Company was, that's a base right outside of Hurricane Point. We heard something that sounded like a tire pop. You know, none of us had ever been hit by anything before. We thought it was a tire exploding. There was a small IED that exploded next to one of the trucks, and that's when Holm and Hurley got hit. That was the end of

March, right around there. Because a week later would have been when April 6th happened. That's around the time frame that I started to worry about things. When Holm and Hurley got hit, that hit us that this is real life, this is really happening. Then of course, like I said, the first week of April turned out to be a real shit show for everybody.

Janney: Yeah, that's when Fallujah kicked off and a lot of the fighters were kind of forced out of Fallujah or squirted out as they were being encircled.

Weaver: Yeah, the first push through Fallujah was going at the same time and a lot of the fighters that were coming through, Ramadi was the main central point for fighters being pushed through the country and whatnot. So, we got a lot of that runoff.

Janney: What are your recollections about April?

Weaver: My first recollections of April were, we came out of Hurricane Point one day and the streets were just deserted. There was nobody. That's when we kind of started to get a little antsy and we were worried. I remember on April 6th, we were coming down by the Farouk Mosque there. We came around a turn, I'm not sure what the streets were, but we came around a turn and we were ambushed. There were some people laying in this little fenced off area. I don't know if you'd call it a park or what you'd call it, but it was a little fenced off area that had like, some grass overgrowing in it and a big cement wall behind us. They had a guy with a machine gun down in that grassy area. Then, right next to that grassy area was a little alleyway where they had more insurgents at the other end with RPGs, grenades, that kind of thing. That was the first time we ever took actual contact. It was scary definitely, but in that situation you kind of react on instincts and your training.

Janney: What time of day was that?

Weaver: So, that would have been early in the day. I'd say probably around 11 o'clock.

Janney: Had you heard any fighting going on anywhere else in the city?

Weaver: Yeah, we had heard that Golf Company had been ambushed there and we knew that they were pinned down, that they were having a pretty rough go of it.

Janney: Had you heard anything about Echo and their QRF?

Weaver: As far as we knew, Echo was en route to help Golf Company. But the radio chatter for me wasn't exactly clear. I was in the turret the whole time, so I didn't get a lot of the radio chatter that came through.

Janney: Sure, I get that. Alright, so what happened during the ambush? What can you remember?

Weaver: Well, I can remember we came around the corner and the guy in the grassy area lit us up with an RPK machine gun there. He lit us up pretty good. Savage was the first one to react, Jeremiah Savage. Before any of us knew we were even being hit, he had already sprung up and shot the guy in the grassy area. We all dismounted. Like I said, we reacted a lot better than I would have expected people being shot at would react. We all dismounted and moved straight to the attack.

Janney: So tell me what happened next, Justin?

Weaver: We dismounted and like I said, there were guys in an alley there. They were right beside the grassy area. We dismounted and I was with Sergeant Garcia and 1st Squad. We proceeded to move through the grassy area. He sent myself and Dan Ackles up over a wall where they were fighting us. I had climbed up over the wall and the insurgents were behind a white taxi cab there. They were preparing more ammunition and more grenades to use against us. I shot one of the guys that was there and that's when they noticed us on the wall and started to fire back. We moved back to the Humvees and started to fight the guys in the alley there. We fought them for probably 10-15 minutes before we finally managed to overcome them.

Janney: Alright, so what happened next?

Weaver: After that, I believe we made contact with Golf Company. We moved to make contact with Golf Company.

Janney: So, you were moving north, basically in the general direction of the Euphrates at that point?

Weaver: Yeah, we were traveling, I guess it would have been north. We would have been headed towards the north bridge, or away from the north bridge, but on that side of the Euphrates, yes.

Janney: All right, so did you guys link up with Golf, or what happened next?

Weaver: Yeah, we actually ended up making it up to Golf and helping them get things under control a little bit. Then, we ended up linking up with Echo Company, who was in a fight on the other side of town.

(Author note: At this point in the interview, poor cell phone connection ended the call and the interview. I reached out to Justin Weaver again after that, but we were not able to do another interview.)

6 April 2004 Golf Company QRF

Cpl Reagan Hodges – Weapons Company

Janney: It's 7 Feb 2024. Why and when did you decide to enlist in the Marine Corps?

Hodges: I joined the Infantry Delay Program my junior year in high school. I was confident when I was eight years old that all I was going to do was be a Marine or a cowboy. I wasn't much of a cowboy, so the Marine Corps was my next option. I didn't want to go to college. I wanted to go to the Marine Corps. I enlisted in 1993.

Janney: What rank were you when you guys deployed to Ramadi?

Hodges: I had just got busted down. I was a Corporal. Then, I got demoted for getting in a fight with that guy out in town that SgtMaj Booker told you about. So, I was a Lance Corporal when I went over there. But, as soon as we got over there, even though they had just busted me down, they meritoriously promoted me back to Corporal.

Janney: You said earlier that you were kind of known for getting in fights.

Hodges: Yeah. I never backed down from one anyway, that's for sure. It was my second time in the Marine Corps, and I was a little older. I had fun with it. I like to mess with people. It was a good time most of the time, but I ended up getting in a lot of fights for it. It's funny looking back on it now. It makes for great stories, I guess. If you can take a punch, it makes for a great story. I hope these youngsters don't try me anymore. A lot of times in the Battalion when we'd have something that needed to be handled, when somebody needed to go talk to a guy, they would send me to kind of influence them - just explain things to them to straighten them out.

Janney: What Company, platoon and squad were you in?

Hodges: I started off at Golf Company and then I moved to Weapons Company 81's platoon. I went to a 0341 Mortarman. We were stationed at Hurricane Point.

Janney: What do you remember leading up to April the 6th? Was there anything really important besides the birth of your son that took place there in February and March?

Hodges: At the time we went over there, we didn't really understand how serious anything was. I was prepared for battle. I loved being in the Marine Corps. I only wanted to be in the infantry, and I wanted to experience that. I never was the kind that would pray for war, but if there was a war, I would pray to be in it. I didn't like being in the garrison. That's where I got in trouble. I like being in the field. I like getting ready for war. I didn't like any of the other stuff. One of my Gunnery Sergeants, when I got NJPed, said, "This is the kind of guy that we just need to put in a glass case and just break in case of war." I was suited for the field. I was suited for combat. When I was in the garrison, I liked to mess with people. I didn't end up getting in fights all the time. They said, "That's just Hodges. That's just the way he is." So leading up to the war, I always had that mindset that it's going to be on. I was really hard on my guys. I was a lot older. I was 27, 28. They were 19, 20. So, I was on their ass a lot, even though we weren't that far apart in rank. We were far apart in experience and age, and they respected and listened to me. I didn't take advantage of that situation. Well, maybe I did at times.

So, I'm going into it mentally prepared, but there are two things that happened that really just woke me, not just woke me up, but just slapped me in the face. One was when my buddy McPherson got his jaw blown off. So, when McPherson got his jaw blown off, that just flipped a switch, a switch that you didn't even think you had. Like sometimes you think you can fight until you get punched and then you realize either you can or you can't. Well, you think you want to fight until something really happens and then you realize usually you don't have control over it. Either your body's telling you to go fight or your body's telling you no. It's one or the other. I always wanted to go forward. That is what I want to do. So, it really put me in the right mindset. The situation here is no joke. Then, the second incident that happened was when we were on QRF, quick reaction force.

We got called out into town. We didn't know exactly what happened. We knew there had been an attack. Somebody's Humvee was down. People were hurt. People might have been killed. We didn't really know. So, we get the call, and we get into town to a disabled Humvee. I'm not sure what Company. I believe it was Echo Company. They had been attacked with a RPG and small arms fire, and they had somebody killed. We get up on the scene. It's kind of eerie. The town is really quiet. It's late at night or early in the morning just after midnight. On the mosque loudspeaker, all of a sudden, you hear they're doing one of their battle prayers or something like that. I thought, "This ain't gonna be good." Suddenly, all the electricity, all the power in the town just goes off. I got my team and I'm loading a couple of them in the back of this Humvee that's been destroyed. It's been turned over, but we flipped it back over. It's been rolled. I didn't know the severity of anything in it. We didn't know if anybody had been killed. We just knew something wasn't right. It's a QRF - we don't have time to sort it out. Back at base, I had just got yelled at for slapping one of my Marines. I almost got in trouble. I did a gear inspection before that mission, and one of my Marines didn't have a flashlight on him. I slapped him upside his head. I said, "What are you doing?" I ended up giving him my flashlight. So, that comes back into play when we get to the scene. I'm at that Humvee. I know that the top is kind of crushed. The windshield was smashed. Now, I'm inside of the Humvee. You have to remember, it's really dark. All the lights are off, and I can't see out of this windshield. So, I take my hand and I'm trying to rub the windshield. It just feels really weird. I don't know what it is. I don't have my flashlight now because I gave it to the other Marine. I pulled out my camera, and I was just going to take a picture because the flash would light up the inside. I just needed a quick shot to see where I'm at, so I could get adjusted and figure out why I couldn't see out of this windshield. At this point, we had no KIA. None. When I took that picture and the flash lit up the inside of the Humvee, literally seven inches in front of my face, it was covered with blood and human tissue. That was why the windshield felt weird. That was one moment in my life that affected me tremendously, partly because it was our first KIA, and just the realization that at this point, I knew that no one except Echo Company knew that this Marine was dead. It's a feeling that I will never forget. It was just a shock to me.

I'm the only one in this truck now besides the dead Marine. We had Marines in the back, but since the back was torn up, they took them out. Now, they're towing me in this vehicle. Like I said, I couldn't even see. My heart sank at that moment. It hurt my heart for the family. I remember thinking to myself, somewhere thousands of miles away back home, a family has no idea that their worst nightmare just happened, and I'm sitting right here in the seat that it happened in. I just kept thinking, they have no idea about the news that's coming their way. Then, I started thinking, I wonder what they're doing right now. Because they're probably just carrying on with their day, hoping for the best. I just felt so bad, because they had no idea of the hell they were going to experience soon. That just put me in a whole different mindset, maybe for the rest of my life.

In the background, you got this guy speaking over a loudspeaker talking about how they're going to kill us. When those two things happened, it was a full-fledged realization - it's 100% go time. It didn't matter if we're going to the PX or walking to chow. I was hard on my guys. But, guess what? They might have not liked it then, but they love it now. When they see me now, they thank me for it. We were fucking good. My fire team got picked to do all kinds of shit that nobody else would even think about doing. They didn't have to ask us - if somebody's in a house needing help and we have to go in there, my team was just going. We were ready for whatever needed to be done. I had a bunch of young guys that didn't know whether they respected me or were afraid of me, but they did what I asked. I cared about them and loved them even though I was hard on them. We reacted really well together as a team and we got to do a lot of stuff based on that performance.

All I cared about was being a productive infantry Marine and I wanted a deadly fire team. I wanted us to all be as safe as possible, but safety is sometimes an afterthought. I wanted to be a good NCO and leader. I hope I was. I wasn't trying to be anything other than who I was. There were situations where I would go into houses instead of sending someone else. My number one guy was a guy named Steven Lelong, a dude from L.A. I just remember him telling me stories about how his dad had passed away, he had a mom and a sister at home, and a special needs brother that he really loved. I thought, I can't send that guy into that house because he might get killed – he's gotta take care of his brother. They respected me for doing that. So, when I told them to do something, they did it without hesitation because they knew I would do it.

Back in garrison, when we were in the field training and I would do it like that, they would think I was just trying to be an asshole, be a ball hog, so to speak. But, they understood it a little bit more as time went on in that deployment. Maybe they realized it then, but they didn't express it to me until our 10-year reunion. A lot of my guys came up and thanked me for that style of leadership, which doesn't always work for everybody. But, it worked for the people that I had. It worked for me. They all thank me for it now. I stay in contact with every one of them and I love them to death. A lot of times, they'll send me messages, just randomly thanking me for the way I was. And I thank them for being who they were, because honestly, it looked like I was a brave guy out there. I was 27 and I didn't have much to lose like they did. I had just gone through a divorce at the time and I didn't give a shit. The kids that I have now, I didn't know they were being born because they're from a prior relationship. I've raised them since they were three, though. So, I didn't go there with the mentality of "I have kids at home. I'm going to be a father." I might have not been as brave if I knew I had more to lose. I tell those guys, "I just went there by myself, cowboying it, and you guys had families at home and were 19, 20. I don't know if I would have been that brave if I would have been 19 or 20 with a wife and kids." Maybe, but the fact is, I don't know. I just know what I did. I know what I did at 27, 28. But, the point of that is, I never got to tell them how much I really respected and looked up to them and admired them for all they did. I mean, they didn't hesitate. I know part of it was because they were afraid of me, but maybe that works. But, they reacted so well, and they did everything. They never second-guessed or hesitated. In a lot of those situations, if they would have hesitated, it would have gotten somebody killed. Thank God that they didn't. That's what I admire about them. That's all I wanted to ask of those young kids. Young kids that've been told everything to do. They've been told when to eat, and when to wipe their ass since they've been in the Marine Corps.

On 6 April, we were Golf Company's QRF. We saved their ass. They love us. My call sign was Tex. When we would roll up, they would love it. We came rolling in with vehicles, with the big guns wiping everything out. Those were my boys. I loved going to QRF when it was Golf Company. Not because the fighting was any easier, just because I had a relationship with those guys. We got in firefights to go QRF somebody else that needed help in a gunfight. On the way there, we would get attacked or ambushed. So, it was nothing to get in three fights on the way to a fight. You're involved in one little fight and you're going to another one and IEDs are going off all around you. And then when you get to where you're going, shit really hits the fan. We lived like that for a week like that. I was in Weapons Company QRFs since that was our week to be on. We didn't miss anything. We hit every area of operation, and we're talking sometimes 18, 20 hours a day.

My fire team and squad were down there in the city to help Golf Company. We started dancing down some road after some fighting. It started off as probably half a dozen or more insurgents. We were basically following them down to town, trying to advance on them. They were stopping and shooting, and we were after their ass. We ended up killing them off here and there until we got down to two of them left, we think. I ended up going to the top of a building. We've got them in a neighborhood. What they were doing is they keep going to these houses that civilians are in, kicking in the door, just taking that house and using it to fight us. What we're not doing at that time is we're not just unloading into the houses because we don't want to kill innocent people, even though that would have been the easiest thing to do. We always knew for the most part whether we were in a good neighborhood or a bad neighborhood. If they lock you out, they don't want you to come in to help. Sometimes, they're just generally afraid. But, other times you have people that want you to come in - they are there to support you. So, I just didn't want to take a chance on, first of all, it's illegal. You can't do that. Just the moral aspect of it mattered to me. So that's the situation I was telling you about earlier. It was hard for some of us. But sometimes we'd have to go in those houses. And that's when I had a, I would a lot of times set my men up on the outside and say, "Hey, me and Lelong are going to go in. Your guys stay on the outside if anybody pops up. That's your responsibility." Because I just felt better with us going in, a little bit older guys.

So, I went to the top of the three-story building. We had two more terrorists out there about 150, 200 yards away. I'm on the third floor, so I can see kind of where they're maneuvering. There's a berm about two feet high on the ground there. So, if you're level to the ground, you could hide behind it and could low crawl nobody without anybody seeing you. But, if you're elevated like I was, you have the advantage. I could see that they're low crawling and they had no idea I'm up there. I'm about 20 feet high. I end up shooting these guys about three or four times. I know I hit one guy in the ass and one in the ankle, and I'm not sure where else they're hit. We're kind of paused in action right there. Everything's kind of quiet. We give Captain Weiler a call and he comes and links up with us. As he's linking up with us, they get behind me. We're advancing. So, I know that I have two people down ahead of us. We're lined up. We're going to patrol up there, and we're going to retrieve the weapons and anything we can from them. As we're patrolling up, I take the point. I'm way out in front of everybody else. Captain Weiler is behind me with everybody else. As I'm walking, I see one of the men that I shot and he's got an AK still in his hands laying there. I was thinking, he's going to wait until I get close. He's either going to throw a grenade at me or he's going to just start unloading on me again. But, he sees a bunch of people coming at him right now. So, he's just lying there. I look back, and I tell Captain Weiler, "I'm going to shoot this guy." He told me not to. We're walking up, and now this guy's body is moving like its flinching. I don't know if it's just the body doing that or if he's just moving. A lot of people, after you kill them, their bodies still make those weird movements. So, I didn't know. I'm going up to this man, and what I did wrong was I focused on him too hard because I didn't know if he was alive or dead yet.

I get pretty close to him, and to the right side of me, I hear a gunshot. I turn and look and I'm no farther than 10 feet away from a dude that has a RPK on bipods buried in the ground in a fighting position, and he's unloading on me. I turned toward him and charged him full speed without any hesitation. I don't know why I didn't fall to the ground. I just took off running at him while shooting, and I ended up killing him. I shot him eight times - we counted. He started shooting at me when I was 10 feet away, and I ended up killing him when I could see the whites of his eyes. He shot at me 29 times and missed me every single time from that close. I got a Navy Commendation medal with a V for valor for that. SgtMaj Ellis apologized, saying it should have been a Bronze Star. I don't care. I have that moment, that memory with my brothers, plus I got to go home.

This was when Captain Weiler was wounded. He got his Purple Heart from that incident. Because when he hit the ground to take cover, one of the 7.62 rounds fired from this RPK hit right next to Captain Weiler's face on a rock, and the shrapnel from that bullet hit him in the face. I ended up just going after the guy and killing him. I was so close, I could have stabbed him. Then, my other buddy Lelong, came up from the side and ended up shooting him a couple times, too. Like I said, he was my number one guy. I was the only one on my feet at that time until Lelong stood up. I truly believe that the reason I didn't get shot was God saved me. I choose to think that's a God thing, but that's just me. It's gotta be a God thing. There's no other reason that the guy didn't wipe us all out with the RPK. That's just lucky, because there's nothing I did well. That's just pure luck. That's God. That's whatever you want to call it. That dude could have wiped us all out, because we would have all been laying on the ground. My grandma said, "Reagan, 10 feet away with a RPK? God kept you safe." Yeah, it was just crazy. I'm walking through it in my brain right now, and I still get chills. I have chill bumps now, I swear to you. I remember taking a video after that, and somebody's asking me, "Man, how do you feel?" On the video, my skin has just turned solid white. I just remember the adrenaline. It was a different feeling than I've ever had. The only thing I could say was, "Wow! My mom almost collected my death benefits." And I started laughing. Everybody was laughing, including me. We were just done for that minute, but we weren't through fighting for the rest of the day. It was insane. I don't know how I could be so close to somebody shooting at me with a bipod and miss. I've relived it a million times in my head. I've come to this conclusion. I put myself in that man's shoes. He's looking up and sees a load of Marines walking towards him, straight in line. He knows something's about to happen. He's dying right there. He knew he was going to die because even if he killed me, there's a squad right behind me. I think that he just got nervous and couldn't hit his mark. This happened on 10th April, because I remembered that it was my sister's birthday.

On one of those days, I think it was 6 April. We got hit. We got hit and it was pretty good. We were in a good firefight and that's when we got the call. I was telling you a while ago we were getting in fights on the way to fight. We got a call to go across town. We were coming to Golf Company. They had a couple of KIAs and a couple of guys wounded in action. We were on our way to them driving down Route Michigan, the main road. I just felt like we're about to get hit, we're about to get hit. We're cruising probably 30, 40, 50. There's a group of people on the side of the road, and the people in the buildings next to us were really up high. There were a lot of people out that day and most of them were adult males. I said, "This is not good. We are going to get ambushed. I can feel it in my bones." We're hauling butt through there and I see a little girl about 10 or 11 standing on the side of the road. That's when a massive IED went off. They timed it wrong - it went off maybe a fraction of a second too early. When it went off, I was looking right at that girl. The bomb had to have been right behind her. That girl just disintegrated. She was nowhere ever to be seen or found again.

I didn't have an armored Humvee the whole time. I did not. I have a funny story with Sergeant Major Booker, he grabbed me and asked, "Reagan, are you still sending that girl money?" I said, "Yes, Sergeant Major. You're fucking making me." At this point, I think SgtMaj Booker had started to see things differently. I think he just didn't know I was going to be good in combat, but now he sees me like I'm fucking out there doing my job. He started to look at me in a different way, not like this guy that gets in trouble all the time. Booker said, "You don't send her another fucking penny. If she says anything, you let me handle it." He said something like, she's not getting any more of this blood money - you're out here sweating and bleeding for this.

Like we discussed about SgtMaj Booker's comments about me in his interview, when he saw me at that reunion years ago, I was just happy to see him. I think maybe in his mind, he was still worried that I had a problem with him when he asked me, "Well, Hodges. What's it gonna be?" I think he thought I was gonna fight him right then. That wasn't even on my mind. Maybe he didn't know where we stood at that point. I sure didn't hold a grudge because of the things that had happened before we deployed. SgtMajor had every right to be upset about all the stupid shit I was doing. I just have so much respect for SgtMaj Booker. I love that man. He's a war fighter, too.

Like I said, I used to get in a lot of fights, even with guys I really liked. Maybe I'm bipolar, because afterwards, I'd pick them up and go buy them a beer. They might not be as cool with me as I am with them, but that's just the way I am. I'll never forget Staff Sergeant Walker. When Walker was a boot, when he was a PFC back in 1995, he was my roommate. I got out after four years, but Walker never got out. I told you, I liked to get in a lot of fights. Walker was huge. When we were roommates, he was probably 19 and I was about 21. I was a big brother to him. I would say, "Hey, we're going to go to the club tonight. Some shit is probably gonna happen. We're gonna go, and you're going to have my fucking back." I do say that just to see if people are going to be about it. Walker would say, "Let's go." I mean, he was the guy. If you needed a little backup, that was your guy. So, we were really tight. And he'd throw down. He was one of the tough guys.

Back at Pendleton, when Sergeant Major Booker looked at me, stood up, and said, "I'm about to NJP you." He's in my face and said, "Maybe when I court-martial you, you'll have the balls to step outside the gate and fight me." That's when I stood up, looked him in the face and said, "We don't have to wait. We can go right fucking now." My Lieutenant and my Staff Sergeant just looked at me like, "Oh, shit" because they know how I am. Sergeant Major said, "Sit your ass down." So, I sat down. He starts going on his rant again, and I just ignore him. Booker said, "Get your ass outside the office." So, I'm standing outside the office at parade rest. I look down the hall, and I see somebody checking in. I see Staff Sergeant Walker walking down the fucking hall, checking in. So, Walker comes walking down the hall, and me, not having any filter, I'm just happy to see him. I said, "That's my boot! Walker, what the fuck are you doing?" Walker said, "What are you doing?" I said, "Dude, you're not gonna believe it. I just got fucking NJPed." He said, "You're shitting me. I think that's what you were doing the last time I saw you."

So I got NJPed. I was on a 30 day restriction. My last check-in is at 2200 every night. You've got to be in uniform. You got to check in and do extra duty and you can't drink. You can't go anywhere. But Walker, he was the staff NCO coming into my barracks. So, while I'm on restriction, Walker and I would drink beer together, for two weeks. Yeah, he was my boy. When he got killed, it just tore my heart. Me and Sgt. Ford. We were all Weapons Company back in 1995, 1996, so we all knew each other. When

Walker died, I kept in touch with his aunt. We got linked up on Facebook. Walker is one of the guys I love the most. He was a really good friend. I have known him since he was 19. I could tell you stories about him having my back. If I was getting in a fight with four or five people, he would just be there. Walker would say, "I don't want to do this." I'd say, "I'm not asking." And then, he would just be there. That's the kind of guy Walker was. Like, fuck it, let's do it. He was loyal to the bone. He was a drill instructor. He's just one of those guys that I've known since he was a teenager and he died on the battlefield where we fought together. It's very sad, but kind of special at the same time. I just hate that they rode into that kill box and there was no walking out of it.

Speaking of kill boxes, I was in the QRF sent to the OP where Tommy Parker and the other Marines were killed on 21 June. We pulled up. I wanted to know what was going on. Everybody else stayed in the truck. I got out of the truck. I went to the top of the building. I had seen the bodies that just got removed. Just seeing the human tissue and blood on the wall where the radio man was sitting, and looking at all the brass. They were just straight up executed. I remember walking up there and thinking, God, this is the final resting place of so many people at one fucking time. It pissed me off because I'm thinking, how did this happen? You could tell whoever was on the radio probably got shot first because there was blood all over the wall where they were sitting on the stairs.

Janney: In my opinion, they should never have been on a fixed OP 24/7 without setting security downstairs, especially since the Iraqi brothers had people working on the house all the time. I understand the "projection of force" order, but that's not really how you should deploy a sniper team.

Hodges: I felt like they had some kind of a false security with these people because the Iraqis would come up on the roof with them sometimes, correct? They falsely trusted them. They were up there having a conversation with them, and shooting the shit. You could tell from the last link of the stairs to get to the top, to the roof. That was where they were sitting with the radio, which, if you're going to have security, you need to have it lower, because your security can't be at the bridge. It can't be right there when you're coming, because the next step was to the rooftop. I think when they shot him, the other two just started shooting the people that were on top of the roof.

I was with Adams and Martinez when we went to get some chow afterwards and the Army Sergeant Major made that comment to us about the sniper team being killed. The Sergeant Major wasn't wrong about our guys getting killed, but it wasn't the time to say that. We had just got through putting them in body bags. We were covered in blood and just wanted a bite to eat. So, we went over to get something chow and that guy chewed us a new ass. I threatened to kick that Sergeant Major's ass when he said that. Sergeant Major Booker really had our backs when he called that guy and yelled at him. It was funny in a way, "Well, my Sergeant Major called the Army Sergeant Major and threatened to kick his fucking ass." That's literally what happened. And you're talking about a Sergeant Major that didn't like me at the time. Booker didn't like me at that time. If you do anything in my interview, I hope you can let people know the level of respect that I have for SgtMaj Booker. He gave me one of his 2/4 challenge coins when we met at that Texas gathering. He said, "I had one of these left, Reagan. I've had it for years and I want you to have it. I saved it specifically for you." It was an emotional moment for me. He's a hard motherfucker, but I swear he had a tear in his eyes. He introduced me to his wife that day and he said, "This is Reagan Hodges. He wanted to beat my ass at one time." I laughed and said, "Don't get it twisted. He wanted to kick my ass, too." She said, "So, you're Hodges?" I thought, "Oh, God. I don't want to hear it."

At the very end of our deployment, they started to get some armored vehicles in there. In order for you to get an armored vehicle, you had to basically give up your fire team. I went through a couple of fire teams, but I always had the fire team that I had, I always loved and I always was happy with. If I had to move somebody around, I could move them around. I always felt really confident with my crew. We had a high back Humvee, just a high back, and no armor. Even a 5.56 shoots through that. It was my decision to keep the unarmored Humvee, because I didn't want to split up my fire team. If I went to armor, I'd have to drop one or two people, because you can't fit them all in. You can fit them in a high back, but in an armored vehicle, you can only fit basically three people, the driver, and one in the turret. I wanted the extra guy. So, we're driving around town in the only unarmored vehicle. I got a cow skull wired to the front of my Humvee. And we're going through town. We're ripping it up. We did notice that we seemed to get targeted for IEDs more than anybody else, maybe because we didn't have armor. Sergeant Major came to me one day and said, "Hodges, don't you think you want to take that skull off?" I said, "For what? Because it stinks? He said, "No. You're making yourself a target out there with that thing." I said, "Sergeant Major, we're the only fucking vehicle that's not armored." Booker said, "Yeah, you're probably already a target. Keep the fucking thing on there. I don't care." I'm looking forward to meeting you in Cali. I wish my other kid would come, but he's in the Naval Academy. He has a track meet. He throws a Javelin. He has a track meet against Army Eagles. He said, "That's the only track meet I can't miss, Dad." He's going to follow in Dad's footsteps. We're going to set it up where I'm the first one that gets to salute him.

6 April 2004 Combat Outpost

Cpl Patrick LeBlanc – Echo Company

Janney: So first off, tell me when and why did you decide to join the Marine Corps?

LeBlanc: Right before September 11th, 2001.

Janney: So, why the Marines? They definitely have the best uniforms, right?

LeBlanc: Well, yeah, there's that. I don't know. I guess I've been over the top with other things in life. I figured if I was going to do the military, I might as well go all out. Enjoy the best.

Janney: Once you got through boot camp, what was your MOS or your job for the civilians?

LeBlanc: When I got out, I went to infantry school in North Carolina. But, that's right when September 11th happened. They said, "It's for real this time. We're going to war." So, when I was at infantry school, I said, "I've got to test on everything." It was basically a test on one weapon system, and then if you do well, you go on to another one, and then you go on to another one. So, it was like, screw this, give me everything. I did all of the systems, did really well, figuring I'd need it. Then when I graduated, they took the five top guys and sent us to Washington, DC. We didn't get to go to the fleet. So I was pretty bummed about that. I was in Washington, D.C. for two years before I went out to California with 2/4.

Janney: So, it wasn't until you went out to California that you were transferred into 2/4?

LeBlanc: I got there in October 2003. That's when I met Roy and everybody, all the guys you met.

Janney: Right. All right, and so then you guys deployed to Iraq in February of 2004.

LeBlanc: Yeah.

Janney: Once you got to Ramadi, were you at the Combat Outpost or where were you?

LeBlanc: Yeah, we were at the Outpost.

Janney: What was happening there in Ramadi in February or early March?

LeBlanc: Well, when we first got there, not a whole lot was happening. The army unit was real lax about everything, even when it came to patrolling. I forgot who came to see us in Kuwait. I think it was either their First Sergeant or somebody like some senior enlisted in the CO. They were saying, "Oh, you guys will be fine. You don't get attacked. Just smile and wave at everybody. You know, the heart and minds thing. When we trucked into Ramadi, we were the first convoy that wasn't attacked. So everybody was pretty positive about stuff. The first couple weeks of foot patrols and convoys and everybody was pretty excited. We're grunts wanting action and all that, but you know we're excited that nobody was getting their asses shot, so it's kind of mixed feelings. We were giving out soccer balls and Gatorade and trying to find Intel on stuff that was moving through Ramadi from Syria going east into Baghdad and Fallujah. Because Ramadi was like the first big city before you had to go into Fallujah and Baghdad and points that way.

Janney: At that point was MSR Michigan open or did you guys have checkpoints set up?

LeBlanc: Yeah, we had checkpoints from day one, but we weren't really finding anything. Some of the truck drivers and taxi cab drivers are kind of pissy, but other than that, I guess, well, you know, the initial push of the unit before us, everybody was kind of used to it. You know, the locals.

Janney: I'm assuming that all the bad guys were either locals or were already in country as far as foreign fighters at that point, if you guys weren't seeing much coming in from Syria.

LeBlanc: At that point, yeah, that's what I figure is, you know, and the only time, and even for the rest of the deployment, the only really time we tangled with the locals is if we, you know, accidentally screwed their house up or accidentally shot the shit out of their cow. Most of our engagements were foreign fighters coming in.

Janney: When you say foreign fighters, is it because the combatants killed had papers indicating they were foreign?

LeBlanc: I don't remember a whole lot of local entanglements, so when you say foreign fighters, and I've heard this time and time again, but yeah, there's a lot from Syria and everywhere else. It wasn't a whole lot of people from the city. You know, you get rocks thrown at you from people, but you can't really blame them. As I was saying, you get pop shots taken if you walk by a house that just a couple of days before, one of our mortars landed on their house. Yeah, but it's not too many locals that fought with us.

Janney: That's understandable. So I heard, too, that the Army told you guys that you can pretty much count on being mortared on a certain day around chow time. Was that your experience, too?

LeBlanc: Sure. A good pattern was the chow hall getting hit, the mail trucks and the supply trucks getting hit. It's like they weren't interested in hitting transport carriers until later on. They wanted to hurt morale more than anything. Our water truck used to get hit all the time. I wonder if they thought it was fuel or they knew it was water. Either or. It pissed us off. But yeah, I don't remember the roof of the chow hall staying intact more than a couple weeks at a time.

Janney: So, did you guys suffer any casualties during that build-up to April?

LeBlanc: The only one I could really remember was Wiscowiche, the engineer, when he found the IED on MSR Michigan on 4 April 2004. I think Whiskey was our first casualty of war.

Janney: At this point, what was your rank?

LeBlanc: I got over there as a Lance Corporal, E3. I was kind of like a senior Lance Corporal because I already had two years in. You know, Senior Lance Corporal's almost as good as the Sergeant. I was in a funny position because I had two years in, but I was new to the unit. They did their little Okinawa tour. Everybody knew everybody. And, you know, I already didn't, I came in right before the deployment. I had a top secret clearance and everybody hated me. But Sgt. Coan and Gunny Coleman told me, "Oh, you gotta do this. You gotta do this." Because they expected a whole bunch of me. They wanted me to be a team leader. They wanted me to do all this crap. When I came, there was a boot drop that had Litke and Arneson. I was in charge of all the new guys because nobody else wanted to do it. They were all my little chickens, even if they weren't in my platoon or in the Company, they still all followed me around. When we got over there, I was a SAW gunner. So, if you had a SAW, you were like the assistant team leader. I don't remember exactly when, but at some point I became a team leader. It was right between April 6th and the 10th because I ended up with Sims' rifle when he was killed on the 10th. But, I hardly ever used it because I just preferred carrying a SAW.

Janney: What were you doing the morning of the 6th?

LeBlanc: The second platoon, most of us were at the Outpost for the 6th. There were a couple of drivers and a couple of squad leaders out to do the MSR sweep with the other platoons. Then, some of us were on guard. We were either on guard or in the hooches when that shit started. I'm still fuzzy on what kicked it off because I was listening to it from far off and not out there. I've heard 50 different stories, but I think they funneled into the marketplace there and something kicked it off, either that heavy machine gun that was up on the hill, basically an anti-tank device. When we finally got up there, we found a whole bunch of ordnance. Then, it quickly spread out towards the river and the tank graveyard. So, it was just a shit storm.

They started pulling people like the cooks to get up on guard and then we loaded in trucks to go out. But this was sometime after, as we already heard there were casualties and the trucks were already coming back with people when we were getting orders to go out. By the time I got out there, it was pretty quiet. But, they were still loading our guys up into the trucks. There's sporadic engagements. It was a fucking blur for the next couple hours. It's pretty foggy because you're processing it, at least for me. What the hell? This is for real.

We started walking, doing patrols because we had intel that the fighters that broke and ran were around Fishhook Lake. We spent probably the next 10 hours going around that whole entire lake looking for people. We were just shouting, angry, belligerent, kicking in doors. I remember getting back from the foot patrol, we all went to the chow hall, because we were mounting up to go. It was our turn to go out. Our weekly schedule was one platoon had foot patrols and MSR sweeps, one would do OPs, and then the other would do guard duty. So, that was the end of guard going on patrol. We're all sitting in the chow hall after going around the Fishhook Lake. That's when Coan told us that Lt. Ski didn't make it. Walker didn't make it. That's actually when I realized that it was my birthday. It was after midnight, so 7th April was my 21st birthday.

Then we started the week of foot patrols. That next couple of days, we were just hateful. Just, you know, stopping cars, searching houses. It almost felt like that would lead up to something like the L.A. riots. Basically, I really don't care what you have to say. You look like a bad guy. We were finding a lot of weapons caches at this time; taxi cabs full and the trunks would have RPG rounds in the back and the guys had five different sets of papers. So, we knew something was coming.

On the 10th, we did a big operation. We started real early in the morning, going door to door. But yeah, we were going house to house. Someone said, "Let's do a real hard knock on this house." Right when I started running, the damn sole of my boot gave out and I fell. I thought, "Oh, this sucks." I got up and we kicked this one door and the people were all pissed off. Or, we'd go through the house, start flipping mattresses, looking for guns and shit, just being dicks to the people. And then, I don't remember how far off it was, but then all of a sudden, just gunfire everywhere. It did not stop for the whole entire day.

Janney: So, were you finding a lot of IEDs before the 6th?

LeBlanc: No, there wasn't. We heard about it. Our quick little build-up in California - people were talking about them. But, it was still almost like a foreign concept. It was like, yeah, okay, whatever. That sounds a little too high speed for these guys.

Janney: Did the IEDs ramp up after the 10th?

LeBlanc: Yeah, you wouldn't find them on every patrol, but at least once or twice a week, twice a week. The engineers got good at finding them. We got good at spotting the signs and symptoms – what it would look like if there was one on the side of the road, how the ground would look disturbed. The kind of places. Piles of trash, dead animals. If nobody's standing anywhere around an area that's a good indicator that something's terribly wrong.

Janney: How well did you get to know the guys? I know you were new to 2/4.

LeBlanc: I got to know them pretty quick. Unfortunately, it seems like a lot of the guys that didn't make it home were the ones I made friends with.

Janney: Do you feel like sharing any personal stories or recollections of those guys?

LeBlanc: One thing I would like to say about Ski, because he was new to the unit too, we kind of got there around the same time. Coming from D.C., enlisted and officers didn't talk at all. Washington, D.C. is very black and white with enlisted on one side and officers on the other. But it's like he went out of his way to introduce himself to me. It was just weird, it kind of caught me off guard how pleasant he was. Then, he found out that I was from Connecticut, and he was from up north too, and it's like he wanted to connect with people. It was just weird getting that from an officer. If we'd be out in the field, he'd come over to us. We're jaw jacking or smoking and he'd ask, "What did you guys think about that exercise? I want to get your point of view." We would look at each other like that's some shit officers aren't supposed to say. You respect a guy for that big time.

Janney: Tell me about some of the other guys. I know you mentioned Walker. Did you know him?

LeBlanc: He wasn't in my platoon, but we still got along because we talked a lot about music and about the California lifestyle. He was into that kind of stuff. He was a funny dude after hours. Completely different from the Drill Instructor. Yeah, because he was a D.I. before he deployed to a line unit and I heard that if a 13 year old skateboarder could be a giant Marine that would have been him. He was just like a cool big kid. It was sometime in April and we were in chow hall. We got into this heated argument about the ingredients in a shepherd's pie and people thought that we were in a legit argument about something. It was just me fucking off. But, we both had our differences of opinion. He didn't think you put carrots in it. But, you gotta have carrots in a shepherd's pie, right? But, he didn't think so. He was ready to fight about it. It was just funny.

Janney: Did you know Ryan Jerabek or Kyle Crowley?

LeBlanc: Those were good guys. Crowley actually became a team leader right around the same time I did. But he was a junior Marine. He was on the fast track to go places. He'd ask me for advice on some things, like if he was doing something right, and I'd just tell him, "Just do what you think is right. If you think it's right, it probably is, because they picked you to be in charge." He was quiet. He tried to be funny, but I think a lot of the other guys were pissy at him because they got there at the same time as him, but he was already a team leader. So he kind of buttoned up a little bit. But, he was good to go. Ryan Jerabek was really quiet. So, I didn't get to know him a whole lot. He was on the other end of the compound. So, I would just see him in passing, really.

Janney: What about Cherry or Benjamin Carman, did you know those guys at all?

LeBlanc: Cherry was another one who was really locked on. I think he got to be a team leader too, or he was trying to be. But he was real motivated, real physically fit, and knew how to get the job done. A lot of the guys that we got in that boot drop, unfortunately that was a lot of our casualties. That was a good group of kids. It's kind of sad that I call them kids because they're like a year or two younger than me. The Marine Corps has a sad habit of doing that. That was a good group of kids. There weren't a whole lot of slack asses. The one that got me the most was when the snipers got killed, that's the one that's affected me the most. When Otey and Contreras and Lopez and Parker got killed, I think that's what bothered me the most.

Janney: Do you want to talk about that? So, I understand that Musser, and I can't remember the other guys on his team, or the team that he was with, they pretty much rotated out of that fixed position. They would do 24 on and 24 off, and were rotating out with Parker, Contreras, Lopez, and Otey.
LeBlanc: Yeah. Every goddamn day. That's just something snipers don't do. I never have understood that, but I mean, for whatever reason, Royer wanted them in that fixed position. Royer - that fucking guy. That's a whole 'nother book. But, Royer wanted us to be seen, he wanted us to make a presence. I just have nothing good to say about him.

Janney: How well did you know Otey?

LeBlanc: He was my buddy. He was one of the few that actually became my friend. Otey, Roy Thomas, and Huerkamp. Those are the three guys that actually welcomed me when I first got there. Deshon was a good dude, classy guy. Probably the most genuine guy out of the unit. If he said something, he meant it. There wasn't anything vague about him. I think Roy's the one that told him about my wife being black. Otey never believed me about that for the longest time. But, then he finally saw her when she picked me up one day. After that, it was like you're cool. You're in. We can be friends. At first, Otey would sit on the third deck and just wait for my little red Dodge Neon, with Kim in it, to make sure it was her. Him and McCall would sit up there like, who is this motherfucker? He was a really good dude. I don't know anybody that didn't like him.

Janney: All right, so what about the other guys on that team? Parker and Contreras and Lopez?

LeBlanc: Parker was real quiet, but we respected him because he was a sniper. You could jaw jack with him for as long as you could, but then he was out doing something. But, he would take the time to talk to you. Most snipers were dicks. They'd look at regular grunts like, whatever. Parker was a good dude. Yeah, I talked to him a bunch. Just about little shit, you know, where you from. That means a lot when you're over there. It breaks up the time. So, you value those conversations and passings. We were all taking a like a public relations picture for a 7-ton one time. I think Coca-Cola sent us a bunch of shit and you know they wanted a Company picture. We just happened to be smoking by where the picture was so I said, "Come here, motherfucker. Stand over here with me." That was a couple weeks before they were killed. It was just one of those times we were out smoking in our little designated area, just talking about shit. I still got that picture somewhere. Lopez, he was real quiet. But when he did talk to you, you're like, what the fuck did he just say? Because he'd come up with some Shakespearean retort to something real stupid you'd say. I thought he didn't even know how to talk at first. He was real quiet.

In contrast, Contreras was the complete opposite. He would never shut up. You'd want him to stop talking. I lived out in town. But, I'd still stay and help them on field day and whatnot. Sometimes, we'd be out there really late. It'd be 11 o'clock at night and everybody's dog ass tired. He's out there just singing his head off, cleaning his room, blasting crazy ass mariachi music. He's just going like an Energizer bunny.

Janney: So did you know Carmen or Layfield, any of those guys?

LeBlanc: Unfortunately, I didn't get to hang out with a whole lot of them, because I wasn't at the barracks after hours much. I just remember that group being really locked on. They were ready. Every last one of them wanted to be better. Every day they would ask five different senior guys, "Hey, what can I do to be better?" I just was not used to seeing that. It was just really good. It's a damn shame we went through what we went through.

Janney: Tell me about the homecoming when you guys came back to the States.

LeBlanc: Well, we went to Camp Victory again for a night. Then we flew to Kuwait and then we flew back. We were in a 747 or something, some big-ass nice plane. But it was just fucked up because they loaded it, and the whole back of the plane was empty. I just remember how spooky that felt. Those seats should be full. Yeah, we got back and we had a homecoming thing. It was good to be back. We got back in September and then I was getting out in July. So, they put me on guard. A camp guard with a couple of the other guys. I went for two days on, three days off.

Janney: Did they offer you guys any counseling to kind of decompress?

LeBlanc: Hell no. Everybody was getting drunk and going UA, fucking off buying fat ass motorcycles and getting divorces. There wasn't any type of counseling and that's truly regrettable. You gotta talk about it. That's what helped me a few years ago. Well, it took me a few years, but I got hooked up with the VA. Did some group shit which didn't work, medications which were bullshit, and then I found a real good therapist and started prolonged exposure therapy, which is fucking horrendous in the beginning, but it has a good outcome. As long as you can get through it, it's worth it. It's you and your therapist and you just talk about the worst experiences over there. They record it. You just close your eyes, put your head down and you just talk about it, kind of wander off. The therapist kind of reels you back in. You talk about it for an hour and then you go home and you listen to it. Then, next week you come in and you do it again. I did it for about a year and it helped.

6 April 2004 Combat Outpost

LCpl Ryan Savage

Janney: It's 19 August 2019. Tell me why and when you decided to enlist in the Marines.

Savage: Why and when? That's a good question. I guess I wanted to be in the service just like my grandfather and my uncles and then I kind of wanted to prove other people wrong, because when you're in high school and you don't have a backbone and everybody was saying, "You're not going to be worth anything." This Marine recruiter went to the same high school and he talked me into it. I said, "You know what? I want to try it" so I did it and here I am. A Marine veteran, go figure. I joined on June 4th, 2001.

Janney: Once you got through boot camp, what was your job?

Savage: I actually signed up for security operations, but that obviously fell through because I got pushed into the infantry. Which was fine, I wasn't traded, so that was going to be my specialty. But, other than that, basic infantry was my go-to.

Janney: Were you always with 2/4 or did you get assigned to them at some point?

Savage: Yeah, I was. I was with them the whole time. After boot camp, I went to SOI, school of infantry, and after I graduated, I went straight into 2/4. Got put in Echo Company. I was doing well there and then, in June of 2003, I was asked if I wanted to go to Headquarters Company to help out with securing the platoon. Usually in 2/4, if you're asked if you want to go, you're doing a good job, but if you're told, then you're doing a bad job. I was asked and not told, so I thought that was a pretty good thing. So, that's who I was with in Iraq - Security platoon in 2004 in Ramadi.

Janney: Once you guys got to Ramadi, what was the general situation?

Savage: Day to day was providing security and escorting diplomats if we had any attack commanders around different bases, doing supply routes and trial runs for the different outposts. That's what we pretty much did on a day to day basis. I was the machine gunner 100% of the time I was out there. They put me through a special machine gunner school before we went out. I was on the 240, the Mark 19, sometimes the SAW, but mostly it was a Mark 19 or the 240. I was on a convoy most of the time.

Janney: Leading up to the events of the 6th of April, from what I understand, you guys were, besides your responsibilities, the unit was doing a hearts and minds mission, handing out soccer balls and Gatorade.

You were transitioning from 82nd Airborne?

Savage: We were doing a lot of that stuff. We visited a lot of the sheiks, I think maybe two of them I've met. I don't remember exactly how many of them there were. Went to some apartments and we were doing something there, and then we had to leave right away. There was some intel that some of the insurgents were on their way there and we had to get the platoon commander out of there.

Janney: What were you doing the morning of 6th April?

Savage: 6th of April, I know I was on a convoy. It was real early. We got back, I can't remember exactly from where; we did like a night ops or something. We were just running supplies somewhere if I remember correctly. Then we got back and they switched us out, so I went on to the front gate, posted at the front gate because they wanted to start rotating us from convoy to manning the posts. I guess someone complained that they didn't want to be on post anymore, so we just started doing that. So, they had us do that and that's where I was on the morning of the 6th was at the front post, the front gate. Then we started hearing a bunch of news about what was happening. On that day, that morning, the Army actually set up a post outside the gate. I think there was a roundabout there, a small one, not a huge one, they had a tent out there. The Army was just posted out there and I didn't hear too much Intel about what was really going on until after the fact. I had just gotten off the convoy and put on post. I wasn't happy, but it's a job.

Janney: Did you hear any of the gunfire that was coming from the ambush of the QRF?

Savage: I heard some explosions in the distance. I do know that I was on Hurricane Point and there was a convoy being set up, staged to go, help and assist if needed. They didn't send us out, but we were on standby. I was pretty much geared up and ready to go, but I guess they didn't want us to go out there in the whole mess. Because we only heard bits and pieces of what was going on and we were getting anxious and wanted to go out there and help our buddies and stuff, but we didn't quite get to go do that, so that's unfortunate. I can't remember exactly how long after, but I know that there was a cordon search conducted. We went helping everybody else with cordon and knocks, going house to house and searching for weapons and stuff. There was a whole bunch of stuff found. My team didn't actually find anything which surprised me. We just didn't see anything; we just didn't find anything.

Janney: So, that was probably on the 10th of April when they did Operation Bug Hunt.

Savage: Yeah, okay. That sounds about right. If I remember correctly, yeah.

Janney: After that, you guys were pretty much in combat off and on through the rest of the deployment?

Savage: Yeah, that sounds fair. I know my experience was a lot of mortar attacks and IEDs. Those are the two major things that I was exposed to. I may have been involved in maybe two firefights, but most of the time it was just insurgents trying to blow up a convoy when they can to kill several people. I remember my first one was on Route Nova - coming around the corner and just BOOM, an IED just goes off. One of the guys almost got paralyzed because he lost consciousness – it almost snapped his neck. Good thing he was still able to walk. Other than that, there were a few others here and there. I guess that's what you get when you are on a convoy. You never know what's going to happen around a corner.

Janney: Since you were with Echo most of the time, tell me some remembrances of any of the guys that didn't come back. I mean, how well did you know Staff Sgt. Walker, Ryan Jerabek, or Kyle Crowley?

Savage: Well, Staff Sgt. Walker, I didn't know him very much well. I don't think he came to us until late. I do remember Lance Corporal Sims. I was his roommate in Okinawa, and we got along pretty good. Crazy as all get out. He was really good at ground fighting and wrestling and crazy stuff like that. Spoke his mind. He was a pretty cool dude. He was quiet, but when you get him in a very comfortable situation, he was out there. He'd talk your ear off, but a lot of times he was quiet, but he was pretty funny. He did a lot of crazy shit.

Jerabek, I didn't really get to know him. I talked to him a few times. I think we may have gotten into an argument, but kind of cooled off a little bit. We both weren't the type to just duke it out. It was just one of those in the heat of the moment kind of deals where we said, "Fuck you." "Okay, that's it, all right, I'm done." Other than that he was a pretty cool guy whenever I was around him.

Janney: I heard Ben Musser had worked with him on the 240 Gulf quite a bit, and he was super excited that he was going to be a machine gunner on the QRF that day.

Savage: That's what I heard.

Janney: Yeah, Doc Clayton said the last conversation that he had with Jerabek was that Ryan said, "Look Doc, I'm going to be a machine gunner today" and was all excited about that.

Savage: Yeah, I heard that story too. This past April, I finally got to make it out to my buddy Chris McIntosh's place.

Janney: I met Chris in Hampton, Georgia in April a year ago. He's a nut. He had a kilt on and kept flipping his kilt up showing everybody that Scots don't wear anything under their kilts. He's a pretty funny guy.

Savage: Yeah, he's pretty funny. I finally got to go up there and I saw a few of my other friends and we talked. They had stories and we shared some of our experiences and stuff. It was a pretty cool little event going on. I knew Otey. We were roommates for a little bit.

Janney: What was Otey like?

Savage: He was kind of quiet, but if he got along with you, he would talk to you a little bit. I wouldn't say he was a loner, he just kind of knew what he wanted and would just go hang out with friends in other parts of the barracks. He would go hang out with friends somewhere else out of town. He was always going out. He was a little talkative if he was comfortable around you. That's how I perceived him. He didn't live messy or nothing, so just like John - they were pretty clean and liked things clean, which was good. He was a pretty cool guy. We got along well. I think he thought I was kind of weird and I kind of am, I have to admit that.

Janney: What about some of the other guys that you knew? What about Benjamin Carmen?

Savage: Carmen? Not as well as I'd hoped. He was a pretty cool guy too. I knew Contreras.

Janney: Tell me something about Contreras.

Savage: He was the life of the party. Real funny and he'd talk your ear off. Wasn't afraid to speak up.

Looked after the little guy, that kind of deal. He'd give you the shirt off his back. Two or three years ago, my nephew had a thing at his junior high school for veterans on Veterans Day. My nephew, of course, did me. What was funny is right next to mine, one of the other students had written something about Contreras. I thought, "Well, that's pretty cool." I didn't expect that this kid would know who this guy is. I don't know if he was a relative, but it seems he had to be a relative, otherwise I don't know how he would have known him. The majority of the stories were about Chris Kyle and other ones that are named most of the time. I just thought that it was neat that right next to mine was a story about Contreras. I thought, "Holy shit, that's pretty cool."

I remember I was going on a convoy and we were coming around and I saw Contreras and the other guys, Otey and Parker. They were in Motor T. I don't know what they were doing, but they were in the Motor T lot and they saw me and said, "Hey, how's it going?" You know, waving at each other and giving signs and just kind of cutting it up a little bit from a distance. The next day, I heard the news that they all were pretty much assassinated and I'm like, "Holy shit." Just the last thing that I remember us saying was, "Hey, what's up? Be safe!" That hit me pretty hard. I didn't get the whole story until I read about it a few years later. Then, we lost the sniper rifle at the same time, too.

Janney: Right. Parker's rifle. I know 3/5 recovered Parker's rifle a couple of years later.

Savage: Yeah, it was a couple of years later. Of course, that started to really tick me off reading about it. I was working in a jail and saw a magazine that had two medics that were with us on the cover.

Janney: Which medics are you talking about? Mendes, Doc Urena, Clayton?

Savage: Ah, damn it. I can't remember. He was holding a rifle.

Janney: So, what did the story say about them?

Savage: What did it say? I really don't know what it said.

Janney: You said it upset you. Was it just the fact that the picture was in the magazine?

Savage: The thing that upset me is that usually when someone leaves something in the jail, they forget about it, they just kind of leave it there. I was reading it, and someone walks in and says, "Put my shit down. Don't be touching my shit." I said, "Dude, this is about my guys. It has nothing to do with you. This is about one of my guys finding something that they have been searching for a long ass time. You're not going to let me have that memory? That's pretty fucked up." So, I ended up stealing the magazine with him not looking because he pissed me off.

Janney: You deserve to have that a lot more than he did.

Savage: Yeah, that's what I was thinking at the time too. Laughs. That's what ticks me off. I think someone was trying to say that they were sleeping on the job and not paying attention. I heard that and I thought, "No, that doesn't sound like them." Someone told me that Otey was in a fighting stance when he was laying down when they found him. He was quick to react.

Parker, I knew a little bit. To me, he was a little quiet. I was quiet too. I've been in some sessions to try to get things out and talk about stuff. That's kind of why I reached out to you. This is a good chance to test it out and kind of get back in the deep water. Maybe I need to talk about it with somebody and see where it goes.

There was another guy, Jeremiah Savage. I know that he was a little crazy. We interacted a little bit. Our mail got mixed up. He'd get my mail and I would get his mail and we'd go and switch them and make fun of each other about what's in the mail and things like that. He was a cool guy. I liked him. It was unfortunate, what happened. I hate to say it out loud, but it probably should have been me or my convoy that that happened to. It was just because it was so weird that we left that area, went back, and we made it back to Hurricane Point. Shortly after, we hear the convoy he was on got attacked in the same area that we were just in. You know, what the hell just happened?

Then, SgtMaj Joseph Ellis. He became my First Sergeant towards the end of 2003. Liked him a lot. He was like a father figure. At first, I didn't like him. I thought, this guy is an asshole. Then, eventually I started realizing that this guy is kind of like me. You know, he was a good leader. Looked out for his guys. Sometimes, I can still hear his voice in the hallway because he'd always call out my last name. You know, "SAVAGE!" "Yes, Sergeant." "Get your ass in here!" "Roger that!" I would do whatever it is that he had me tasked to do. He was a good person. Gave me a lot of good advice and helped me out on a few personal things. He was a really good guy. It was crazy because I hadn't spoken to him for a while. He died in February 2007, right? So, in January when I got my Sergeant rank, I thought he would probably be proud of me if I let him know. "Hey, I just got promoted. What do you think?" You know, to just kind of cheer him up. I was trying to look him up in February. That's when I first saw that he had passed away in Iraq. I'm like, "Holy shit!" Yeah, so, I wasn't having a good day. Neither were the inmates.

Janney: I'm sure he was proud of you or would be proud of you. I'm sure he knows somehow.

Savage: I was in my supervisor's office a lot that day because some inmates pissed me off. I went off and let it all out, not crying, but just anger. "He's gone." I was in the supervisor's office and they said, "Hey, what the hell is going on with you?" I told them what happened, and said, "Oh, we understand now." It was a good thing that the supervisor was also a Marine Corps vet because otherwise he wouldn't have understood. But he understood, so they let me go home early.

Janney: What was your rank when you first deployed over there and were you the same rank at the end of the deployment?

Savage: I was a Lance Corporal when we deployed to Ramadi and when we left I was a Corporal.

PIGS (Professionally Instructed Gunmen) and HOGS (Hunters of Gunmen)

Sgt. Jesse Longoria (HOG)

Janney: So first off, Jesse, tell me why you decided to enlist in the Marine Corps.

Longoria: I'm from Bastrop, Texas. I graduated from Bastrop High School in 1997. My entire four years in high school, I was in Naval Junior ROTC. From my time in middle school, I knew I was going to join the military. I just didn't know at that time there were different branches. I just thought everything was the Army. So my intention was always, I'm going to the Army. Then, when I joined ROTC in high school, my commander was a Navy commander, a captain, and his enlisted personnel was a MEPS Sergeant in the Marines. What struck me was his demeanor, his uniform, and just the way that he approached things, and he was in shape. That's when I realized I want to go to the Marines. So, from my freshman year till graduation, my mindset was on joining the Marines. That's where that came from.

In the 90s, my brother was also in the Marines. He graduated two years before me. He was Intelligent and graduated in 95, and I graduated in 97. We had the same recruiter, Gunnery Sergeant Wynn - he was Staff Sergeant Wynn at the time. He ended up being in a show on HBO. I think it was "Recon." But, he was one of the characters in that series. When I decided to join the Marines, the recruiter asked me, "What do you want to do?" I said, "I want to shoot guns and run around in the woods." He said, "Oh, you want to be infantry?" I said, "Well, sign me up." He said, "No problem." Sure enough, I signed up 0311.

In August of 1997, I shipped out to boot camp and MCRD California. That's basically how I decided I wanted to be a Marine. Growing up, I worked at Sonic Drive-In in Texas and I've always felt like I grew up in a small town. I was seeing people and everyone just always seemed to be doing the same thing. I thought, "I don't want to be like that." When I found out I was going to be going to bootcamp in California, coming from Texas, I thought, "Ooh, I'm already doing big things at the age of 18." Graduating boot camp, you get this ego of yourself. I graduated in Deltas, which is our blue trousers and our Charlie, which is the tan top. Leaving boot camp on the airplane and going back to Texas for my 10 days of leave, I felt like I was a badass, and very proud of being a Marine. Definitely was looking forward to SOI, which was the School of Infantry.

When I got back after my 10 days of leave, I was very motivated and still very proud and happy that I had joined the Marines. Boot camp was not very challenging for me at all. I just learned real quick, keep your mouth shut and do what you're told and you'll be just fine. I started on Halloween of 97, October 31st and we graduated around December 12th or 13th. We got assigned to the 2nd Battalion, 4th Marines. We're in our Alphas graduating SOI. They say, "Well, you're not going far from here. You're just going over the hill." SOI was in the Cape Horno area. We were going to San Mateo which is just one huge hill. Once you get over the top of that hill, you start seeing the lights of San Mateo which is the home of the 5th Marines, and which is also known as Area 62. So, these 5-ton trucks pull-up, and the Marines that were being stationed with 2/4 jumped on the back of 5-tons with their sea bags and we were going up that hill. Started coming down, and we reached the Company office where we checked in. The Marines started off loading the five-tons, started checking in, and during that time, it was Christmas and everyone was on Christmas leave, so they told us to go somewhere for 10 days and then come back and check in again. We were just like, "What?" So, that's what we did. We left. Fortunately, I met a couple of buddies in SOI that were from the Los Angeles area. One of them was Sal Lopez. He's still a Marine Corps Master Sergeant right now, as we speak. We became good friends in SOI because his last name was Lopez and mine was Longoria. Our bunks were just right on top of each other. So, we end up being pretty good friends. He said, "Hey, I'm from Long Beach. You can come stay with me for the next 10 days." I think by then I had already blown through my boot camp money, so that's what I did.

When we came back in 10 days, we checked in and got assigned to Bravo Company, first team. From that point on, I went through a bunch of schools and course. Then in 1998, we got deployed out to Japan as part of the 31st MEU. And from there, we jumped on a ship. During this time, there was an operation called Operation Desert Fox. Kuwait was on the defense from Iraq with Iraq conducting airstrikes into Kuwait. So, we were sent there to defend Kuwait and we were on Mutla Ridge for about three months. Then, we went back to Japan and later went back to the States. From that point on, we became senior Marines. I became third squad leader for Golf Company, first platoon. We deployed to Japan about a year later. While in Japan, I decided to reenlist because my first four years was already coming up.

There was a flyer to take the sniper course. I signed up for that. I passed the ENDOC and then joined the sniper team around 2001. So, I was on the 2/4 Sniper team. From that point forward, from 2001 until 2005, when I was medically retired, I was with surveillance target acquisition, which is basically the sniper platoon. Snipers were attached to H&S Company, the Headquarters and Service Company. So, from that point on, I was basically training as a sniper to do those types of missions. I spent the last five years of my Marine Corps career in a sniper platoon.

When we deployed to Kuwait with 2/4 in February 2004, we were supposed to do a two-week acclimatization, but we only had a couple days to do that because they wanted us out there earlier than the Battalion. We flew out to Iraq in CH53s and jumped on 5 tons to Blue Diamond in Ramadi where we stayed a week or two. Then, we got a call down from the headquarters saying that they wanted a sniper team in Combat Outpost, which is where eventually Golf Company and Echo Company were going to be. I had a four man team: myself, my assistant team leader Sgt. Felix Corona, my radio operator Cpl Jason Finch, and Cpl Andrew Stotts. I always felt that I had a well-rounded team. Corona was, I called him my Mean Marine. I wasn't much of an aggressive person when it came to yelling or cussing at people or chewing them out. That really wasn't much of my nature. I was more of a nurturer. I had a different type of leadership style than some other Marines that you will probably come across. I was really more about trying to be fair and listen to my teammates, to come up with the possible best answer to whatever it is that we're facing. Corona was my muscle man, the yeller. He could PT, he'd chew people's ass. He was my go-getter. Whenever I needed something done, I'd just tell him and it would be done. Stotts was my intelligent person. He was super smart when it came to the SATCOM, which was a big deal back in the day. In 2004 now, it's probably just normal. He was smart on the radio. He was just a very intelligent person, and whenever I had something technical or something involving radios, Stotts was the man and I'd always ask him. So, we were a real rounded team. Finch and Corona bumped heads a lot and that was kind of hilarious to watch them. Finch was a very short piece and Corona knew how to push his buttons. It was pretty funny at times. Sometimes I'd say, "Hey guys, come on, stop bullshitting around. But, that was kind of our humor in Iraq, watching those two guys.

Two things I'm very proud of: 1) All my guys came home. 2) We still have a pretty good bond, even now. Myself and Finch were hurt, but everyone else was good to go. We still shit talk and bullshit as if we're back in Iraq. But, we managed to hold it together as a team. When we got to Combat Outpost, we teamed up with an Army sniper team that was out there. The first thing that came out of their mouth was, "This is the worst city we've been in." The sniper mentality is how many kills you can get, right? They said, "This is the only city that we haven't got a kill in." For us, since we hadn't been in the initial invasion, we thought, "Of course, they're going to put us in the city where nothing is happening" because we were so hungry and so eager to get out and do big shit, until it happened. Then, we realized, it's the real shit.

Initially, we did some patrols with the Army snipers and they were good to go guys. I think they were Airborne, I believe 101st Airborne. They showed us the routes and the places with possible high positions. After two or three days, they ended up just saying, "All right, guys, it's yours now. Take it." So, we did. We started out just doing little patrols until the rest of the Company and escort Company arrived. So, my team was the first team to arrive at Combat Outpost. We were kind of getting the feel of it and there was really nothing much going on. Hearing from the Army snipers that this was the only city with no action gave us, I wouldn't say complacent mindset, but we weren't too worried about anything.

About a week later, Golf and Echo Company started arriving at Combat Outpost. I tell everyone that even though you think that you're watching someone in their yard, they're actually watching you. I mean, that was their home. They knew that the Marines were arriving. I want to say the first night that the majority of Golf and Echo arrived at Combat Outpost, a firefight broke out. I remember jumping on top of the main building in Combat Outpost. I used a weapon called a M40A3 and I had a PVS-10, which is a day and night scope. It's a little bit heavier than your normal scope, but the benefit for this scope was day and night use, so I can use that same scope during the day, then flip the switch and I can see at night. I thought that was pretty cool at that time. As Marines, we tend to worry about how much weight we're carrying because you're carrying a weapon, you're carrying ammo, you're carrying water, and then other gear, plus your body armor and Kevlar. For some folks, the PVS-10 scope wasn't their choice because of the weight. But, I chose to keep that scope because I figured the majority of our operations were going to be at nighttime but I could still use it during the day.

When you're done with a mission, there's nothing more rewarding than just coming home with your weapon, cleaning it, putting it away, knowing that the next day you're good to go. You don't have to worry about taking off the scope and putting on another scope. For me, it was easier. The extra weight was worth carrying, plus you didn't have to worry about re-zeroing either. So, our first firefight we experienced was a couple days after Golf and Echo arrived at the Outpost. On the rooftop, looking through our night vision scope, we can see silhouettes and muzzle flashes. I was so new to this and folks were trying to get their combat action rhythm. Marines want that combat action rhythm. I think a lot of folks probably found it within probably 48 hours of arriving at Combat Outpost. My first shooting at a silhouette is what I consider my first firefight. But, I don't know whether I hit him. The next day, the Marines went out and said there was blood here and there, but nobody was found. So, that was my first combat experience. It was dark and I don't know what, if anything, I accomplished. It could have been someone else that was shooting. I don't know. So, I would never claim that was a confirmed anything.

So, the first couple of weeks, it was random stuff like that. McPherson was our first casualty, that I remember to a T. I think he was attached to a weapons platoon in Golf Company. I still talk to McPherson to this day. He was the first one that made me realize that this shit was serious. He was leaving Combat Outpost to do a patrol via Humvees, just kind of a show of force. I remember just being outside my room there. Once the Humvees left the Outpost, they got down the line, and I remember hearing a blast. I thought, "Damn, whoever got that shit got fucked up." Then, maybe five minutes later, four o'clock in the morning, my buddy comes zooming back in. I said, "What's going on?" He said, "Matt got hit. He's seriously injured." I said, "Are you fucking kidding me? I just called the dude." Then, I remember them sending some of the folks out to go find his lower jaw. Thankfully, they found it. Obviously, he got medevaced home and they were able to re-attach it during surgery. That was the first time that it really sunk in for me. An IED or bullet has no eyes and can get anyone. Mac was the first one to get injured and wasn't part of the whole April 6th and 10th events, and he felt like he didn't do anything really just because he got injured so early. The one thing that I reiterate to Mac, and for me it's the honest truth, I tell Mac, "Because of you, it made a lot of Marines take this shit seriously before it happens to one of us. That's something that you should be very proud of. That made us realize the shit's real and anyone can get hit. It was just a matter of minutes from when you left and then you were a changed man for life." He's a very good guy. He's doing very well now. He's married. And, we talk very often.

So the 6th of April popped off and we were just coming back from a mission. We were already back in our room and I kept hearing gunfire. Then I hear Marines yelling, inside the Combat Outpost, about to leave the wire. Humvees kind of zoomed out of the base. I thought, "I've never heard this shit before." The guys were sleeping. So I get out, and I'm just in my trousers and a shirt, no shoes on. I'm walking and say, "Man, what's going on?" They say, "Some of the Marines are pinned down. Some Marines have been killed." I said, "Are you fucking kidding me?" I went back in and woke the guys up. I said, "Hey, let's go. Let's get on top of the posts." We had these posts, like little towers. The posts on the towers were filled two feet high with sandbags. I told the guys, "Get up! We have guys being hit, guys being killed. Let's jump on the towers!" Because we have a problem that our sniper rifles can take advantage of. We weren't far from the city and you could see some of downtown. You can kind of hear what's going on in that direction. We jumped on the towers and we started doing our observations, hearing over the radio about the KIAs and support here, support there. It was very chaotic. At this time, they told us to be looking out for people giving signals on the rooftops, on the ground to the insurgents. Our rules of engagement were if you see that happen, feel free to engage. So, at that point, our team observed maybe seven or eight folks that we saw that were doing that, guiding the other folks. We did our best from our view to take out as many as we could. I think we took out seven or eight. The 6th is when Sgt. Holt's Golf Company squad got pinned down. After the firefight was all over, everyone came up to the Combat Outpost. We started to see the Marines come in, and their camis are bloody.

One Marine that stood out to me in particular was Sergeant Holt. Holt was actually in my platoon when I was in Golf Company as regular infantry. So, I knew him. But, when he came back and had lost a couple of his guys during this time, his eyes were just bloodshot red and he had that thousand yard stare. I mean, he had it. All I can do is just say, "I'm sorry about whatever happened out there." You need to do your best to try to comfort someone. But, when you just see them like that, it's very difficult to try to help out or to feel that you're helping someone. It's very different. But, that sticks out to me in particular. After the 6th, it was just anger. Just so much anger with what happened to Golf, to Echo Company. Revenge, it is revenge. You know, you took out some of my Marines, and that's it. It's fucking on. That seems like the mentality for the majority of the deployment, from when I was there to July. The anger and revenge we wanted for our brothers.

As far as my team, we were missioned to manage the posts for Combat Outpost. During this time, I think Echo Company got ambushed. The word was that they were going to try to overtake Combat Outpost. So, anybody that we saw that was an enemy threat, we would take them out. We were on our post, and there was a white car that pulled up, and they popped the trunk. These three guys came out. Actually, there's probably five of them. Popped the trunk, and they started handing out weapons. And, you know, we already had the green light, you know. I popped the first one, he dropped. Then, they started scattering. The funny thing was we had no idea where it came from because they were in an alley and there were some trees and bushes. I think they thought no one could see them. After the first one got popped, it was funny, whatever you want to call it, but I'm going to say funny because they kept running closer to me. So, their silhouettes kept getting bigger and bigger and bigger. Then, I popped another one. Then, a third one. Sure enough, they kept running closer and closer to the point that by the time I got him, his head was about the size of a damn bowling ball through my scope. So, we ended up taking out three, and the other two got away. Then the next day, one of the squads went back and the bodies were piled up there. So, we got them.

Then, I want to say, the eighth, ninth, and tenth, we were just kind of on and off again. We were just on so many different missions. Everything didn't calm down, but I think the more impactful things were the grunts that were kicking in doors, looking for weapons and insurgents. We were just more on post and basically just observing folks that we felt that were a threat or were signaling to help out the insurgents. Our job from that point on was just taking out those folks, doing over watch for the whole freaking unit. During Operation Bug Hunt, Sgt. Corona, my senior assistant, took out one of the top guys. You remember the playing cards for the top 50 insurgents? Corona took out one of the top ten guys during Bug Hunt. Basically, we just did a lot of over watch to try and keep the Marines safe. The grunts kept security and the other 11 guys were kicking in doors and doing their thing. I can't take credit for any of that because that was them.

I knew Tommy Parker and Otey and Lopez and Contreras. Parker was in our platoon. Back in San Mateo, he was a basketball player. I love basketball, so we played basketball a lot. Good kid. We would talk to Tommy pretty often. He would come by our room because we were lucky enough to have a room for the four of us. But, at that point, the location that we were at was a big space. So, folks sometimes came to us just to get away from their area because they were in bunks, kind of like being on ship. Tommy would come by often, ask questions, and watch a movie with us. He was a good operator. One thing that we always tried to instill in him was, "Hey man, just never forget that things can happen. Don't get complacent. Just keep doing the right thing and everything will be alright." Tommy had a smile that would light up a room. Funny guy, but very respectful. His dad was military, and he was just a very respectful guy. Very athletic. He just seemed to always do the right thing. It got to the point where, we felt as leaders, we'll give you your own little team. I think it was Headhunter II Bravo. We were very confident in him. He was a good Marine, a damn good Marine. He loved cars. He had this Miata that he was trying to fix up. I stopped by his dad's house in Arkansas about 10 years ago. His dad has a little storage and he kept his Miata just the way Tommy had it. He loved sports, loved cars, was a respectful guy, and a damn good Marine.

On 21st June 21st, we were in one of the OP hotels in downtown Ramadi. It was one of the shitty ass hotels that the Marines go to observe stuff. We had the radio on. At that time, we were an eight-man team in the OP Hotel. When those words came across, you know, Tango, Oscar, Mike, Mike, Yankee, Parker, KIA, you know, all of them, we were just like, "What? KIA?" It was just so freaking hard to hear over the radio, and they weren't that far from us either, which was the part that sucked worse, because knowing we couldn't do anything about it, but yet we were so close to them. By then, it was just too late.

Just hearing it over the radio was just alarming and sad; anger, revenge, just a combination of emotions.

From June onward, it was just kind of steady operation stuff. We were going out observing. After Parker's team got ambushed, they made us operate in eight man teams. After that happened, I brought along Lopez, who I mentioned earlier, who was my best friend from boot camp all the way till right now. We're still good friends. I can't say we're best friends now because we live so far apart. He's still in the Marines and I think has already been on 10 or 11 deployments. We still talk probably two or three times a week. So, he joined my team and on 24th July, we were on one of the houses not far from where Parker's team was killed, and we were doing our observation. Our radio kept acting up throughout the day. Somebody came over the radio and said, "Hey guys, come back and let's change out the batteries in your radio. Then, we'll send you back out in the morning." So, we've been watching the MSR all day. We may have been there a couple of days already. We're going to be heading back to Combat Outpost and my job as a team leader for both of those teams was to designate on which Humvee each individual is going to go in. The standard was four Humvees will come out. As a team leader you know they're always watching us. We think we're watching them, but they're watching us. With our sniper rifles, we're an easy target. They think, "That's the one that I want. I want the one with that rifle because I know he's the one that does stuff." I always carried my rifle and so did my partner, Lopez. On this mission, I don't know if someone else carried Lopez's sniper rifle, but I always carried mine. On this particular day, I said, "One sniper in the first Humvee, three in the second Humvee including myself, Lopez, and Finch. Then, two in the third, and two in the fourth." It's hard to mix up when you've got an eight-man team in four Humvees. Eventually, some of them get doubled up.

So, in the days before this, my buddy, assistant team leader Corona, had watched the movie, "Man on Fire." He kept telling me I had to watch it, so my mind was on watching the movie when I got back to Combat Outpost. The radio kept acting up, and Finch needed to call in to enter the wire, so he asked if he could use my flashlight, which I always kept in my right cargo pocket. I bent down to get the flashlight so he could use it to get on the radio. My right arm is holding my rifle and it's exposed above the Humvee side. Actually, I don't even think we even had any type of armor at that time on the Humvees. As you know, with a sniper rifle having a longer barrel, my arm is more exposed. I always kept the bottom of the sniper rifle on top of my foot because it gets really bumpy, so I cushion the rifle so my dope (rifle adjustment) doesn't get thrown off. This whole time, I was trying to figure out the radio shit and the flashlight, and literally, all I see is a flash. The flash, then I smell burning flesh, which was mine. Then, I just felt wet from my blood. I look across and I see Finch and Lopez. Their faces are just bloody. I already knew my arm was fucked up. I just felt it. I knew the adrenaline was high. But, I still feel like it ain't right. But, I'm looking across and I see their faces just red with blood. So, I'm thinking, my arm is what's going on with their faces. Immediately, I'm just like, "Fuck, I'm all right." Then, Logan tells me he's hit. Finch tells me he's hit. I already knew I was hit. This is where it gets foggy. I just remember going in and out of consciousness. I've heard from different folks that I wasn't okay. I thought in my mind that I was still trying to do my shit, but I jumped out of the Humvee and apparently, I was going in circles. I couldn't hold my rifle anymore because my arm was basically split in half.

The doctors were asking me if I could still feel my fingertips, which I could, which is the reason why they initially saved my arm, but my fingertips were touching my elbow. What had happened was an IED blew up literally right behind me. Supposedly, it was a 155mm mortar round. It blew up on the same MSR, the main supply route on Michigan that we observed the whole fuckin' day. That's how sneaky these motherfuckers are. You have fuckin' snipers that are watching this fuckin' road all fuckin' day long. That's how we knew it's their fuckin' country, man. They could've buried that thing days before, but it's hard to say. But, someone close by had to detonate it. When we got back to Combat Outpost, the adrenaline was still there. Corporal Griffin was holding my arm together the whole time after I got injured, all the way back to the Outpost. Lopez knew that I was fucked up because he could see the bone in my arm. He fucking just jumped in the other side of the Humvee, threw my ass back in that Humvee, and drove my ass back to Combat Outpost. This happened about 2200 or 2300.

I kept my cool until I got to Combat Outpost. But when I got to the Outpost, the gymnasium lights inside were bright. I looked down and then I saw my arm. That's when I lost it mentally and I started yelling, "I'm in pain! I'm in pain!" I remember them stripping me down, several of them, I believe. There were four cots in the little hospital bay for the injured folks. Whoever's the most injured gets those cots. Whoever has a chance of surviving gets those cots. Whoever doesn't have a chance, unfortunately, they just have to stand by. So, they all three of us - me, Logan and Finch on the cots. I remember hearing a lady from CNN, I think, coming in asking, "What happened? What happened?" This is where Corona, the fucking muscle man, the fucking asshole, tells that lady, "Bitch, get the fuck out of here." Then, from that point on, the morphine kicked in. I just remember the helo coming in, it was nighttime, and I remember the dust particles hitting my face from the chopper blade wind. I knew my time was up as far as serving in Ramadi.

Got to Baghdad the next morning, woke up, and there was a guy missing a leg and someone was missing an arm. I was in the middle between both of them. I just knew I wasn't in good company as far as that goes. I remember trying to get up, but I had so much morphine in me that I fucking threw up all over the fucking floor. I saw this guy come in sitting in a wheelchair, bandaged the fuck up, looking like a fucking elephant head. He said, "Pretty fucked up what happened last night, huh?" I said, "Who are you? I don't even know who the hell you are." It was Finch. His face was swollen, he had two black eyes and bandages across his head. I went to Germany for a week, then went to Bethesda for three or four days. Then, I ended up in Balboa. I got the majority of my surgeries done in Balboa. In February of 2005, I was medically retired.

The outcome was I ended up losing my right arm. I held on to it for a couple of years and then eventually, it got to the point where I had so much metal in it and severe osteoporosis. My hand kept shrinking because my fingers couldn't move. I basically just kept my hand for cosmetic reasons for a while, because I was 25 when I got injured and I was worried about my looks to be honest. So, I kept my hand for a while even though it looked fucked up if you stared at it, but I always work. Even in the Texas hot summers, I was wearing fucking long johns. I was totally insecure about my scars, but it got to the point where the metal was breaking through my skin. Then, a big hole developed in the palm of my hand where you could literally see through my hand. The doctors said, "The longer you keep the damn thing, it's gonna keep eating away more and more." Eventually, in 2011, is when I got it amputated. But, I went on for as long as I could. Finch and Lopez ended up getting shrapnel to the face. The majority of the shrapnel was bone from my arm. Finch got it worse than Lopez, and he ended up getting shrapnel to the brain, and he was eventually medically discharged for short-term memory loss. They jokingly say now that I put fat in their head from my arm.

Pigs and Hogs

Cpl Jonathan Wood (PIG)

Janney: It's 22nd April 2020. When and why did you decide to enlist in the Marine Corps?

Wood: Honestly, it was back early in 2000. I was just sitting there, seeing what was going on with my life, and I didn't like where I was going and what I was doing. I always thought about the military, I always have, and then just one day, I said, "You know, enough's enough, quit thinking about it and do something about it." I went and talked to the Marine recruiter in late 2000, and joined the Marines, April 10th of 2001. I want to do something with my life. I grew up in the small city of Lindon, Utah. It's just north of Provo about 10 miles. I was a West Coast Marine at Camp Pendleton. I know I wanted to be infantry. When you think of a Marine, what do you think? You don't think of a guy that's sitting behind a computer desk or in an air-conditioned office. You're thinking about somebody that's out there on the front lines doing some stuff, taking care of and defending America. That's what I wanted to be. So I chose infantry.

After basic and SOI, what I ended up doing is I went through weapons because my MOS was 0341. I was a mortarman. At the two-year mark, I was in the 2nd Battalion, 4th Marines, and we were deployed to Okinawa, Japan. While we were there, I found out that the snipers were going to be doing an NDOC. I said, "You know what? Let's do this." It's just one of those things I put in front of myself to accomplish. I said, I can do this. And I did it. So I went through the sniper NDOC in Okinawa, Japan in late 2002. The NDOC itself was the tryouts to try to even get into the platoon itself. It was just barely over two weeks. The first week was knowledge. And sleep deprivation like a mother - all they were doing was just cramming knowledge down your throat and not allowing you to get much sleep. The second week was filled with training, filled with work. After that second week was over, the initially established platoon sat down with the people who were trying to get into the platoon, going through the ENDOC and decided who they wanted to take and who they didn't want. I was one of the lucky ones that got picked. At that point, once you are in the platoon, you are one of two Marines - either you are a HOG (hunter of gunmen) or you are a PIG (professionally instructed gunman.) I was a PIG. In order to become a HOG from a PIG, you actually have to go to sniper school. After you graduate sniper school, then you become a HOG, then you can call yourself a Marine sniper. I was never afforded the opportunity to go to sniper school. So, I can only say that I was with the sniper platoon. I cannot and will not call myself a Marine sniper.

Then stop-loss, stop-move came through. So, 2/4 was deployed in Okinawa for an additional six months. I stayed with Fox Company in the weapons platoon until we returned home, at which time I was moved from Fox Company to H&S to become part of the sniper platoon. July of 2002. We returned home after the stop loss in June of 2002. There was a two week post-deployment leave. Once I came back, I immediately reported to H&S Company. And that's when I became part of the Sniper Platoon, roughly late July, early August of 2003. We knew we would probably be deployed, but we didn't know where. There are many times in this interview, I'm not going to be very happy and it has nothing to do with what you are asking me or anything. It's a lot to do with our direct chain of command, the people that were in charge of the snipers is specifically who I'm talking about, our platoon sergeant and our platoon commander. I think that's a lot of the reason we didn't know things, why things went south in a lot of ways. It wasn't until 60 or 90 days out that we knew it would probably be Iraq. As far as up-training for Iraq, we had more mountainous open environment training than we did urban.

We flew into Iraq a little earlier than the Battalion. I was with Team 1, with the chief scout, Sal Lopez. He was the chief of the whole platoon and he wanted me as his Bravo. He wanted me as his Bravo is equal to where he had Newland as his spotter. I had Andrew Lord as my spotter. So, I was with Alpha, but we were not attached directly to a Company. We would be here with Fox for a couple of missions, a week or so. Then, we'd be transferred over to Echo or Golf. Sometimes we would be at Hurricane Point with H&S for X amount of time. But for most of my deployment to Iraq, 90% of it or more was at Combat Outpost.

When we first got there, I'm just speaking for myself, but at home, we'd be used to the American way of life – pizza, movies, you know. All of a sudden, you're over there and there's no movie theaters, no pizza to order, no swimming pools. It just takes a little bit to get adjusted to Iraq. When we arrived, it was the mid-eighties. So, it wasn't too much of a difference from back home. I still remember this vividly. I was outside having a smoke with a good friend of mine Kevin Olick who was in the platoon. Suddenly, dirt started to land on us from a mortar. We're sitting there going, "Oh shit, dude, this isn't training anymore." This is when it's like it really starts to hit you. This is the real shit, dude. This isn't a fucking video game. This isn't training where you're going to be done at five o'clock. You can't hit restart.

I know that was a pretty common occurrence for indirect fire, mortar fire, to hit Combat Outposts and other areas around chow time. We could almost set our clocks by when they would mortar. It was a fucking ghost town during the day. Nobody went outside. We always stayed indoors or stayed behind the sandbags, you know, a fortified position. You didn't see anybody out there walking around playing catch or any of that kind of shit during the day, no. We were pretty much doing anything and everything that the battalion commander asked us to do, or the COs of the Company, or XOs of the Company asked us to do. But 90% of what we did when it came to that, we would be out and about ahead of the movement of the Company. We would provide an over watch for them, and we'd be in radio contact with them to let them know if it was clear, if there was anything or anyone suspicious that we'd seen. We would radio back down to them to let them know what was going on.

By mid-March, shit hasn't hit the fan quite yet. But at the same time, we were never able to leave the Combat Outpost or any post of that nature without being known to the public out there that we were leaving. We could never do anything secretively without them knowing that we were moving or we were doing something. You thought you were watching them, but they were watching you. We were watching each other pretty well. But, when I would leave with me and my team, and we always left with a minimum of a four-man team, we were being watched. Usually, we had two teams that would move with each other with space between us, and what we would generally see is just people out there doing what people do in Ramadi. There were a few times that younger children or adults would approach us. We weren't standoffish and weren't like get the fuck away from us, but through our body language, we would let them know don't get any closer to us and don't come talk to us. The one thing that I have seen, and I also found out later, especially with the adult men or teen males, they would not pay attention to our sniper rifles or our M16s that we had on us. They always would watch where we kept our handguns. They would always watch the handgun. They were more afraid of handguns and knives than they were of any other weapon. I can't say this for a fact, because I didn't do the research, but I was told by some Iraqi native peoples that the reason for them being afraid of handguns was that Saddam's regime always used handguns when they killed people. Most families had AKs, so that's just what dad does when he goes and shoots up in the air when there's a celebration or whatever.

So a lot of those days, like I was saying, 60%, 70% was just pure, utter boredom, interspersed with the 30% of chaos. If you were doing over watch for a patrol, you moved around. We had set OPs like OP Hotel and the Ag Center. I didn't go to the Ag Center all that much, but I've got some stories about that fucking place. Went there a handful of times, I'd say three to five times. The general OP that my team went to was where we were overlooking on the east side of Michigan. There by the gas station was where our OP was located. It was a house that was under construction, about a click and a half further east from the OP house that Parker and his team used.

The 6th of April and after, those few days were utter and just pure fucking chaos. Marines dying, a Navy Corpsman dying, all this other kind of shit's going down. What I specifically remember is at the time I was out there at Camp Victory with H&S and we were told real quick, "Hey, we're going out. We got to go out there. We're under fire. We're taking casualties. We got to get them suppressed. We got to do some shit." This was very early in the morning. I just remember slapping on my gear and we got pushed to a certain point where we got off. We started moving in formation.

I just remember looking up and seeing the Sergeant Major of our Battalion, Sergeant Major Booker. He's a really good dude. I know him personally. He's a good man. A very, very damn good man. One hell of a Marine. So, I just remember looking up and Sergeant Major Booker was there, 30, 40 yards away from us, myself and Lord, my spotter. We were moving and all of a sudden, we heard small arms fire. We didn't know if the bullets were coming at us or were being fired away from us. You only knew where the shots were originating from. So we ducked down, we took a knee. I do remember that Sergeant Major looked at myself and at Lord knowing who we were. We were two snipers. He just looked at me and said, "You two are coming with me." We started doing little movements down. I cannot remember what road we were on, but we were walking to the east. When we attached with Sergeant Major, we started going up little two lane roads to the north. I know we were south of Michigan, but I couldn't tell you what road it was. I know there's a major road we were walking to the east on, but it wasn't around Michigan. We were south of MSR Michigan down in the industrial district.

I just remember myself and Lord moving with Sergeant Major Booker. At that time, we left the rest of the Marines we were with behind us and so automatically, I knew that we were secured to the south of us, to the back of us. So, our focus was always to go down side streets and always forward and always watching each other's movements. You said security, and this guy bounds. I know a grenade came at us one time, and small arms fire. Thank God, none of us were hit by either one. I remember looking down the roads and seeing people with man scarves. You know, the turbaned things - the red and white or black and white. I remember seeing a couple of individuals with their faces wrapped in them. I did not have the opportunity to return, or not to return fire. I only saw them for a split second and they got out of there, which is a smart thing for them to do. We ended up getting to an intersection. I know it must have been a pretty big intersection because I remember Sergeant Major telling myself and Lord to freeze, sit down, and set security, which we did. At that time, other Marines caught up with us. I just remember holding security until we were told to pull back.

We didn't see anything. You could hear gunshots in the distance. There was nothing right there on top of us. You could hear gunshots. We sat there, myself and Lord, holding that intersection. Again, I don't remember the exact times or anything like that, but it seemed like it was a good half hour to 45 minutes before we were released because of who we were, what platoon we were with. They had other 0311s who were more prone to the urban combat environments than say what we were. That's why they brought them up and also pulled us back for who we were. They don't want to risk those assets. It was daylight, but not toward dusk when we were pulled back, maybe mid-afternoon. So, they pulled us back and out. We were out from early morning to mid-afternoon. By the time it all wrapped up and we were done, it was always in daylight hours, though.

On the following day, we were mainly doing over watch. That's what our job entailed, and that's what they wanted us to do. It was just me and Lord and there were times we would be separated. Intentionally, we had to be separated from everybody else to bring less attention to ourselves and our movements, to try to stay concealed as much as possible. But in order to get on top of a vantage point, 99% of the time that was somebody's home. I came home with a clean conscience. I never forcibly entered a person's home. I would always announce my presence. It's not like you're sitting ringing the doorbell and knocking politely on the door. We made our presence known very clearly and they knew that we were there at the door. You weren't shooting locks off and kicking in the door. Every time before we set off on these missions, they're telling you what the SOPs are and the ROEs are. We were always told to never sit there, talk, and engage with a female.

I remember this older lady - she had to be in her late 50s, early 60s. Of course, she doesn't understand a fucking thing I'm trying to say to her. And at the same time, I'm not allowed to physically touch her or manage her in any way. Situation's dictating, some shit's hitting the fan, I have to get on her roof. She needs to get out of my fucking way; I'm getting up there. But, if shit's not hitting the fan and I need to get to an elevated position, it wouldn't be as urgent. But, that home was where I needed to get, but she refused to open the door. I encountered a lot of that.

I remember myself and Lord, again, it was just me and him. We jumped over a wall, moving through someone's backyard to the next house. When we jumped over the wall, we interrupted a family having dinner in the backyard. At the same time, how in the fuck are we going to know when we jump over this wall, we're going to be interrupting a family having a dinner? Thank God nothing happened. Nobody panicked. Everybody kept a cool head. And ice. You learn on the fly real fucking fast in an environment like that. When Lord and I jumped over the fence, we just saw the fright in their eyes. There were multiple males there. They just spoke in Arabic and told their wives and children to go inside the home, and through body language and very broken English, we just understood each other really quickly. We're not here to harm you. We're just moving through. We left the property soon afterwards.

One time on one of the missions, we were doing over watch and it was just me on this roof. Lord was on a different part of the roof. This young boy about 9 or 10 came up, looked at me and said, "Mister, mister. Would you like some fruit? Would you like some water?" It was pretty cool. I said, "Yes, yes, I would." He came back up and gave me some fruit and water. I didn't feel like he was trying to poison or harm us in any way. It was a good moment. It was a good thing. I remember him coming and sitting by me. It wasn't like he was trying to distract me, or stop me from doing my job because he wasn't sitting there always asking me questions or wanting to touch the gun or any of that other kind of shit. It was nothing like that. It was just pure childhood curiosity, innocence. I didn't want to be an asshole and give this young man a bad memory – to think negative about any foreigner. I remember taking the time and talking with him for a little bit, but then I did let him know after a couple of minutes that, "Hey, I got to get back to my job and I do appreciate you bringing me some fruit and water." What's wrong with taking 30 seconds out of your day just to make somebody's day a little bit easier and let the kid know that not all of us are assholes, you know?

We're still doing over watch. We're back at our OP and our platoon sergeant came in and said, "I need X number of bodies to go over here and do a little working party." I thought, "Okay, dude, let's go do this." It was myself and maybe six or seven other guys. We all ended up getting in some Humvees and 7 tons, and we were moved over to where they had the bodies of our Marine brothers. Our job was to move the bodies. Sergeant Major was there. It's very sobering. You can't put into words when you know what's in the bag that you're picking up. That's someone's father, that's someone's brother, that's someone's son. Knowing that and then having a few moments of communication with your family back home is precious, because someone else's family doesn't know that their son, their brother, their husband has been killed, but you do.

We were out there at Hurricane Point which was one of Saddam's son's palaces. I didn't know if it was Uday's or Qusay's. This is where we were staying. There were toilets there, but no running water. We have these port-a-potties outside. I had to go relieve myself and we got mortared. I said, "Fuck, no. I'm not dying taking a shit." I kicked open the door, wiping my ass and grabbing my M16 all in one motion to get the hell out of there.

It's getting hotter. The early months of the year during Monsoon are not too bad, but it's getting hotter and hotter. By now, just the air temperature was 136 degrees F. The Battle of Ramadi was over at this point. At this time, we would have been moved to Combat Outpost. Our leadership got word from higher-ups that they wanted us to get out there and find a different OP to watch to the east of Combat Outpost, out there towards the gas station. Myself, Lord, Newlin, and Chief Lopez pushed to the east and find one house that wasn't as far to the east as we wanted, but it still had a good elevation. We could see the house where Tommy was killed.

Then, further to the east on the north side of Michigan, there were two houses right next to one another that met the criteria of what we were looking for. When we first entered the house, the house furthest to the west of the two, we encountered two gentlemen. The house was under construction and one of those guys was the owner of the home. The second was the contractor building the home. But the contractor did not speak any fucking English. The homeowner did. He wasn't pissed off or rude to us, but he let us know, please do not pick this home. Because if you take this home, you're putting me and my family in jeopardy. I told him, "Don't worry about it. I'll go back to my chain of command and let him know that this house is not what we are looking for. We will not be back." We punched back later that day and I told my chain of command, "Let's not worry about that house." I told them what me and the gentleman had discussed, and that there were better options. Long story short, they chose that house. That one's burning my fucking mind, because guess what happened? Yeah. He and his family were killed. Again, I'm not blaming this directly on my chain of command, but at the same time, I almost have to. Because they are the ones I reported to. I voiced my concern, but my main concern wasn't just the safety of the family. It was because there were other buildings that offered better vantage points and better views with no construction going on. I guess why they chose that home was because it was closer to the gas station.

Speaking of Tommy Parker, he was a cool dude. He was a smart man. Mechanically and everything else. We were sitting outside about seven o'clock. I always had to have coffee. I'm just one of those people. I don't give a shit if it's 140 degrees outside, give me my cup of coffee or I'm an ornery son of a bitch to be around, and he knew that. This was one of the rare occasions that we were around each other in the morning at the Outpost, just shooting the shit. He told me about his wife Carla and their daughter, Laura. I wasn't married, but I had my daughter and I'd tell him about her. We'd always talk about what we were gonna do once we got home. As far as working on cars, I had this 1967 Chevy Van 90 - the "Mystery Machine" is the best way to explain it. Tommy is a lot more mechanically sound than what I am. We were talking about how we're going to work on it when we got back to the States. Tommy talked about his Miata, that little fuckin' rice rocket. He had a motor he wanted to put in it. He loved that little car. Neither one of us would sit there and harp on the nasty, bad things going on around us. We'd always look forward, concentrate, and stay positive.

We all saw Tommy's parents when we got back. That's kind of a bittersweet story. They were there to greet us. They took us all out to dinner for pizza, the whole platoon. It was awesome to see the love that his family had for us, even knowing that their son was not amongst us. They loved us all. They understood that this was Tommy's other family. They've already had the funeral of their son, but they wanted to come out and welcome us home, because we were their extended family. It was just hard to... it was sobering. You get home like that and to have one less than you left with.

When it came to me and Tommy, we always had a good mutual understanding for each other. I was issued an M40A3 which had a rail system just in front of the scope, which allowed you to be able to put on a night optic. Tommy's rifle was an A1. He did not have the ability to mount that. So, what myself and Tommy would do is when I would come and relieve him, I kept the A3. He and his team walked back to Combat Outpost with the A1. When he came back and relieved me, I left my M40A3 with him, because he stayed there for 24 hours and I stayed for 24 hours. Well, like everybody knows, half the day is only light and the other half is darkness. I need to have abilities and capabilities to fire and engage at night. What fairness would it be for me to take my A3 with me and leave Tommy with nothing? So, I gave him every fighting chance that I would have had. So, technically, it was my A3 that was taken from Tommy when he was killed. The 3/5 sniper team recovered my A3 in Fallujah when they killed the dudes that had it. Now, the rifle is either at Quantico or the Marine Corps Museum and will be displayed at some point. Our platoon Sergeant, Gunnery Sergeant Shane Lindsay jumped my ass when we got back after Tommy's team was killed. Jumped my ass because I left my A3 with him. Why the fuck didn't I have it and why did I have Tommy's A1?

But, I tell you what, if I could go back, knowing what would happen, would I have changed anything? No, I would not. I wouldn't have. I still would have gone off and given him my A3 or anything else I thought would have helped him and his team. We had to take care of ourselves. That's what we did. That's what we tried to do was take care of ourselves. Here's the deal. Parker should have never been on that roof to begin with. I should have never been on that roof to begin with. We shouldn't have used a house under construction as an OP 24/7 for weeks on end. We should have kicked the Iraqis out or posted security on the first floor if we were going to use it as a fixed OP.

There's kind of a funny, weird story about how I broke my ankle, even though I didn't know I broke it at the time. We were out there at Hurricane Point and all of a sudden, Gunny's running in there saying, "Hey, you guys got to get out on the wall. We got somebody that's approaching it trying to push something over the wall." So we all grabbed our shit and we started running out of Saddam's son's palace down the road to the east to where that gentleman was. Do you know how, in the middle of intersections where cars turn, they naturally push little gravel piles together? I was running, it was dark, and I didn't see it. My left foot hit the gravel and slipped. I crashed like a fucking champion. I went to the ground and slid towards Sgt. Endito. Endito, one of the other snipers, saw me and was laughing as he ran by me. I couldn't get back up at first. We ended up running into each other a little bit later that day and he said, "You know what you reminded me of? Do you remember that Kevin Costner movie where he shoots the charging buffalo and it skids and stops at his feet? You ate shit and skidded across the ground just like that buffalo." I said, "Fuck you, dude!" Given the situation, I didn't know it was broken. We didn't have a full-on hospital right there where we could do x-rays, so all the Doc did was, "It looks like you got a bad sprain. Put this cast on it, here's a couple of ibuprofen. Drink some water." He slapped me on the ass and I'm out the door. That was some of the worst pain I've ever felt. The only reason I found out that it was broken is when I got back stateside, I was doing some shit and I rolled my left ankle. It didn't pop, but it hurt pretty bad. I got an x-ray and they asked me, "When did you break your ankle?" I said, "I'm not sure I ever did." He said, "You see these lines that are a little bit gray and a lighter color and fuzzy? Those are fractures that have healed." I said, "Whoa, the only time that I've ever hurt my ankle with that kind of severity is when I was in Iraq."

Another funny story is one time we went to the top of the hospital that was to the north of Michigan, right there by Nova, where Nova makes its bank. We were up there in our little area and we started to get hungry. There's 8-10 of us. We saw a little market to the east of us and said, "Let's send some people down there to get us some chow. So, whoever draws the short straw gets to go get chicken and Cokes for everyone." Given the environment, that's a pretty risky little endeavor. We provided over watch for those fucking poor bastards to go down there and get everybody some chow. To this day, that is the best chicken that I ever had over there, with their pickled vegetables.

PIGS AND HOGS

Sgt. Shawn Spitzer (HOG)

Janney: It's 28 April 2020. Why did you decide to enlist in the Marine Corps?

Spitzer: Honestly, I was into a lot of drugs and my dad caught me with a lot of drugs and I wanted his respect, and I just knew I needed to make some changes, so I went into the Marines in September 96 and then got out of the Marines in July of 2000. And I went back into the Marines after September 11th. I felt the Lord gave me a few dreams. So, I went back. I was a Sergeant when I got out the first time, and I lost rank to a Corporal when I came back here, because I'd been out more than a year. I was out about a year and a half. So, I lost a little rank. My first MOS was 0351 which is explosives and wire-guided missiles. But, I went to sniper school in April of 98 and passed. I think Gunny Booker was the instructor, but he had just stopped instructing the class right before mine. I was a sniper for a good two years or more. When I came back. I was a sniper with 2/5. Then, I got switched to 2/4. Thank God, I scored really high and passed sniper school the first time, so I was a HOG, a sniper, MOS 8541. I was a team leader at 2/5 for those two years. I was the only one of the group. I had the highest rank. Everyone else had been there for two, three, four years.

I got to 2/4 around February 2002. I went on the Okinawa deployment and was roommates and teammates with Milo Afong. When we deployed to Iraq, some of us were split up a little bit at times. We flew to a base outside Fallujah. I think we drove from there to Ramadi on the back roads. We got there a little ahead of the Battalion. We met up with the 82nd, and got with their snipers. I thought they were really good. Marines are kind of trained to arm themselves, but I thought they were locked on and they knew what they were doing. They understood things and gave us a good grasp. But I thought, for me and my team, I was boxed. I was near Hurricane Point. There was a good turnover and I thought they did a good job. So the one thing I remember going into Ramadi was when we drove in there, in the center of town about four or five in the afternoon, they had all the sheep. They were butchering the sheep in the market, and I'm thinking, "What are they doing butchering all these animals?" There were so many animals in the center of the city. This is just bizarre. But, it was because a lot of them don't have refrigerators. So, as they're coming home from work, they get a cut of the meat and they take them home to cook. I was also thinking the city would be much smaller, but Ramadi's 400,000 or more.

The snipers used whatever info was best for each of us. We only had two laser range finders in the platoon. The most senior people got them, and I wasn't one of those. So, I had to use a map. Everywhere we went, I would print off an 8.5x11 map and get my ranges off of my map. I had a M40A3 with a regular scope. Olech was another sniper with me and had an A3 with a big night scope. I was a team leader, he was my ATL, assistant team leader. I didn't want the night scope, so I gave it to him. So what's interesting is that we had already been there for a week or two. We were kind of itching to get out. The very first night we went out just to do reconnaissance, just check out the city at night. We were kind of off the main street, and we were looking at the main street, and as we were approaching it, there were two guys with AKs just sitting there in the middle of the street, and we were about to shoot them. But, as we were getting closer, it just didn't feel right, and we walked up on them, and they were actually Iraqi police. You couldn't tell their uniforms unless you were like 10 or 20 yards away, but they weren't even paying attention. I took the little Arabic course, the mini-Arabic course. I wasn't fluent by any means, but we kind of talked to him a little bit, and then we left. But it was interesting, even the first night.

And then after, Fox's area was kind of like the playground compared to Echo and Gulf. They were on the battlefield. There were more Christians in our area, and it wasn't extremism. The people's view of that town was kind of what local mosque they went to, and it was usually geographic. So, if you live on this block, you go to this Muslim Mosque. I don't know if that's the full case there, but you would walk down the street, one side of the street on your right-hand side would hate you, the left-hand side would love you. So, it was just interesting. I don't know if there's any Christian churches there, but I kind of spoke the language, and I'm a pastor, too. I ran the Bible studies in Okinawa when the chaplain left. So, I kind of kept up on religious stuff. I also taught the Islam classes for our platoon and stuff. So, long story short, I was an atheist, and I was finally healed by the Lord of Power to become Christian. The whole war was really religious stuff in the view of the mosque and how everything was going, and a lot of people weren't connecting the dots. I tried to keep up with those things because I thought they were important. I did meet a couple of Christian, but they were pretty poor or lived in not nice areas. Yeah, the Middle East is terrible. This is off topic a little bit, but when Milo wrote his books, we were best friends. We started a nonprofit ministry for injured troops. It kind of evolved and some of our events have been on Christian TV, like TBN and the JTTV, which is kind of like the Christian MTV. We get professional motorcycle riders to do jump shows for Veterans Day and some of those things. We also met some people who were leading thousands of churches overseas in different groups and we helped with survival training in case they ever needed it. For example, if pastors are getting persecuted and they have to disappear.

In Fox Company's area, there weren't a lot of roadside bombs, unlike the other ones. So, we worked with the Fox Company CO and XO. Anything they had for us, we would do it and help them out. We were spying on certain blocks or certain areas. We'd watch roadways. We were doing a little bit more reconnaissance than the other teams. The other teams were more just trying to get to the roadside bombs because there were a lot of them. There were some in our area, but not like the other ones. They had shifted to almost all just watching for roadside bombs 100%, where we weren't doing it that way. I don't know how much I should say, but I was asked to take a picture of a guy at his house through my scope for intimidation. I don't know if the Marine Corps wants that out or not, but that's just how we did it. There were certain tactics that we could use to get people to play along. At several points, a lot of the leaders in the community would be brought into the meeting to meet the committee and they would go over stuff. Some of them were playing along, but most of them were against us completely. Most of them were Saddam Hussein loyalists and didn't want us there. But, let's say, strategically, somebody up high really wanted them to play along, that they needed this leader for certain reasons. I was used to intimidate the guy, to take some pictures of him, "Dude, we're not going to play around with you." I was used to take some pictures of some high-ranking people there that they really needed to try to get on board to make something happen. We didn't do much of that, but a little of that was done.

I didn't speak more than a few words of Arabic. My Arabic was so bad that I'd joke with the people we were talking to. I didn't have an interpreter, but could use one of ours if we needed one. There were a few people that spoke a good bit of Arabic. Ammar Hekmati was one of them. There's another guy or two that really picked it up. I could only use it to break the ice, but not enough to get any information from them.

There was one incident pretty close to where Fox Company was. Fox was right across the street from Hurricane Point. There was an engineer attached to Fox and an RPG was shot at him, actually flew through the window and killed him instantly. But Fox, for the most part of March, wasn't too busy. A little bit here and there, but it just wasn't that busy. I know Combat Outpost was getting mortared pretty regularly and this was because they were so close to the farmland on the outskirts of the city. We were more in the center of town, so folks couldn't fire a mortar at us from just anywhere. We'd only travel between bases maybe once every two or three weeks, and then I'd be able to catch up with Longoria and the others, or if we had to fix the scope or something. So, we weren't always in communication with each other to know what was happening in their AO. They had their missions, we had ours.

Honestly, our AO never really changed. We kind of stayed just flat. The morning of the first week of April, we had a little skirmish here and there, and a roadside bomb here and there. But I guess the easiest way to think of it is, our area just geographically wasn't a hot spot. The Outpost's AO was a complete hotspot, because they both had the government center and Echo had the farmland in that area. Route Nova connected our area to Echo. It was a little busy, but we just weren't that busy. The first time it got really busy was that first week of April. On 6th April, we were out that morning and heard over the radio about Santiago's team being under attack. I was trying to stay up to date. We tried to get involved, but they weren't going to let us cross Company boundaries. We heard an Army tank was going out there to help them out. That got to us. You could hear the gunfire in the city. Another good friend of mine in Golf really got into it. He was a team leader and two guys with Golf got severely wounded and I think one died.

We were expecting about half the guys to die. They're doing house-to-house and Vietnam- type searches. We were given a mission and the corridors got set early the next morning, around 0300. We were tasked to go out in front of everyone. I remember we lined up and had to go out in front of them. It was early, still dark out. It was literally the longest hundred yards of my life because we got put in the biggest firefights we had been in with Golf, near the government center. Then, they started clearing behind us. What happened was the insurgents had pulled out that night and had kind of taken their stuff. So, it was kind of a blessing in disguise. When we encircled the area, they took out that knife, because they thought it would probably get really bloody for everyone.

It's just kind of interesting. One of the things that really struck me as the most interesting thing about being over there was the chess game being played back and forth. For example, the Army took Humvees everywhere. The Army did not like to walk at all. They were kind of lazy, but Marines walked everywhere. So, what the insurgents did was they would plant smaller roadside bombs, like just a pound of TNT equivalent in the ground and try to blow them up instead of using big bombs, because then they wouldn't need as much explosives, kind of wasting the materials. There was always just a cat and mouse game. You could kind of play it out in your head - well, this is going to happen and this is going to happen. You could kind of see four or five chest moves ahead in the game before it happened. For me, war was just a constant chess game. We would change the way we did something, and they were going to do something else to counteract your new strategy. Like the IED that injured Longoria was obviously a smaller IED as compared to one designed to take out the whole Humvee.

I was also hit with a bomb and got evacuated. I think mine was 4th May or 5th May. I was in Fox and I was the only sniper that didn't have any kills. I kind of needed to get a kill. We kind of needed to get a kill to get that monkey off our back. Just because it was kind of making us look bad, because we weren't busy like everybody else. So, I set up on the corner, just south and west of the hospital, to watch the corner of Echo's area and the hospital. There's a big intersection there. We set up there to watch the area that night because we just needed to get a couple kills. When we were done, I actually requested permission to stay there for a few days because I knew if we stayed there for three or four days, we'd get to kill. They rejected it. They ordered me to come back in. We'd actually taken four extra guys from Fox Company with us so we could watch a lot more sections of road. We could watch about a mile of road that way. So, my other team was more even with the hospital. We were kind of south and west. We were really close to Echo's area, right on the border of it. When we pulled out in the morning, they pulled out onto Route Nova, which is right by the hospital. One of my team guys, Miller, got hit with a small roadside bomb. It was really close to the big hospital and the college there, kind of right together. He got hit right by the hospital and we heard it since it was about 400 yards from us. So, we came around the hospital to try to help him. We couldn't see the location because there's lots of stuff in the way. We came around Nova, really hurrying and I got hit with a roadside bomb, too. It really incapacitated me. It was a small roadside bomb, about a pound. I had to rush off the side of the road where there's a 30-foot embankment. To my right was the field, and the Euphrates River was down there. To the left was the hospital and the university there. They were using either the hospital or the university to set the bombs off. So, they were watching the roads with a kind of a phone detonator. Those are the only places they could have done it from was the hospital or university because it's the only elevated position. They lit it up right by our feet. It went off about a yard away. I had to run to the fence on the left and just kind of hunker down because I couldn't do anything. I thought I lost my eye, but the trauma of the blast just shut it. I got a lot of dirt and shrapnel on my right side.

When I got hit and medevaced, SgtMaj Booker and Colonel Kennedy met me at the Army Medical Center near Hurricane Point. It really struck me how much they cared. And, what it was like as a Commander to see all these young kids getting hurt. I've seen it, too. I was probably 26, a lot older, and I hated seeing those 18-year-old kids that didn't know much getting hurt. But, I thought from Kennedy's perspective, when he was there for me, that his job kind of sucked. Watching all that happen kind of sucks. So, they sent me off. I got airlifted to the CSH in Baghdad where they took a lot of dirt out of my eyes. Then, I got sent to Germany and they did the same thing there. Then, I got sent back to the States. I saw 30 to 40 wounded guys there. I actually got put in charge of them, and I hated it. At the base there, they were still making them do some work, moving furniture and other tasks. Some of these guys were broken, just broken. I was one of the least injured. I was in the best condition of almost all of them. But, me and another guy Hayes, who was the team leader for the two Golf Company Marines injured that I spoke about earlier, were at the hospital. Hayes was injured by a grenade during that firefight. I have a Marine Corps Times article from that month that described our base in Ramadi as the worst base of the Iraq war and the only thing they were comparing it to was Vietnam. On the flight from Baghdad to Germany, the medics were saying that was the most people they've ever had on a flight. It was completely full, when before it had just been a trickle.

Then, another guy and I went back to Ramadi. During the time that I was injured and recovering in the States, Longoria got hit. When I got back, I went to where Longoria had been, to Golf Company, and kind of took Jesse's spot. I stayed there for about 45 days before we ended the deployment and came back home. One thing that's really amazing about Jesse's story is what Lopez, the chief scout sniper, did when their Humvee got hit by the IED. They were hit and everyone was hurt. Finks took some shrapnel to the brain. Longoria's hand was hanging off. It was a pretty big bomb. Lopez, the chief sniper, something happened and he had to take over being the driver. He got hit with shrapnel and was really bleeding from the head a lot. He took over from the driver and drove them back to base. I think Lopez ran the Hummer into the building or a concrete barrier when he got to base, and passed out from blood loss right as he got there. He was really the hero of the story by getting everyone back. He saved the day by getting them medical attention quickly rather than waiting for a casevac. Santiago was at that base and told me about it. But, Lopez really saved the day.

People asked me about how I felt about going back after being injured. I was neutral. I kind of didn't care. There's part of me that really wanted to get back to help my guys, but part of it, too, because of that article describing Ramadi as such a hell zone, I didn't want to go back, to be honest. Honestly, the Lord gave me a dream to go back, and so I did. Two things about going back: 1) I got to Battalion HQ at Hurricane Point. At F-2, the Intelligence Staff Sergeant was telling me about some of the mustard gas that the insurgents were using. The battalion had been hit with about four to six different mustard gas explosions over the deployment. It was older mustard gas, so it didn't really have much effect because it was too old. The next morning, I went down to where Echo Company was and they had just been hit by a big mortar attack. Right as we pulled in, the mortar attack was ending, and I'll never forget the smell. There was just dust in the air, like a miniature tornado, and for the life of me, the smell was just so different. You never smell anything like that. I've blown up a lot of explosives, but the smell of those mustard gas mortars was just so distinct from explosives. I'm thinking, "What have we gotten ourselves into?" 2) The other thing that I really noticed there was a huge difference between Fox, Echo, and Golf. When I went back, because there's been so much death there, it's almost like death was running through their veins and it wasn't like blood anymore. The Marines were more trigger happy and it was because we have to save lives, no matter what. It was different from Fox because we'd had a few guys injured and a couple guys died. But Golf and Echo were surrounded by death which is a lot different than a guy or two dying. So that's the one big thing I noticed, the difference between Fox and Echo and Golf.

I knew the guys that died in the Euphrates River since I supported Fox. Fox Company is a boat Company. Well, Fox Company leadership wanted me to swim out to this island on the Euphrates because they thought the insurgents were storing weapons there. Honestly, I felt the Lord told me not to do the mission, but I didn't know why. So, I told them, "Two of my guys didn't have enough time in the build-up to get them the swim training they needed." So, ten of the Fox Company guys did it, and two of them drowned trying to get to the island. These guys were the best swimmers in the Battalion, easily the top one percent. When they drowned, the whole Euphrates River got shut down for 3-5 days. They ended up recovering the Marines though.

I'll tell you one thing that some of us have an issue with. There should have been more awards presented to the men. The guys that deserved the awards are the front guys, the people that are fighting every day. Even Lt. Donavan Campbell of Golf Company mentioned this issue in his book, saying that it was his biggest regret. What's interesting is that Fox Company gave me a meritorious with a combat V, basically like a thank you. My award is a "nobody's ever heard of this" award, way below a Bronze Star, but these other guys didn't get anything. None of the other snipers down at Golf or Echo got anything, which surprised me. They saw so much more than we did. I think they really deserved it, especially the Marines that were out there fighting in the streets. They really deserved some Bronze Stars, Santiago maybe a Silver Star. There's some people that really deserved some awards there. Because I remember when I got there and I told them, "Fox Company gave me this little award." I mean, you've never heard of it. I've never even heard of it. Fox was just saying, "Thank you for the great job you did for us." Those guys were bummed because I got something small, and I felt their pain. Those guys deserved some freaking medals, especially Santiago. You know, so it's a shame that they didn't get the recognition that they deserved.

7 – 10 April 2004 including "Operation Bug Hunt"

Sgt. Marc Coiner

Janney: This is 31 May 2019. Just tell me a little bit about yourself.

Coiner: I was originally born and raised in South Africa. Then, at the age of 25, I got out of the South African Navy and I came to the States, took about a year off, and then I joined the Marine Corps in 2001.

Janney: What led you to want to join the Marines after being in the South African Navy for a while?

Coiner: Well, the South African Navy used to have a Marine Corps, but they did away with it, so I initially wanted to be in the Marines. Coming to the States, luckily I was still an American citizen, a dual citizen, so I could fulfill that dream by coming to the States.

Janney: Right. And where did you go to boot camp?

Coiner: Boot camp, I went to MCRD San Diego.

Janney: When you went through boot camp what was your MOS? What did you train for?

Coiner: When you get out of boot camp you get assigned if you are going to be in the infantry or if you're going to do supply job, admin job and that pretty much dictates where you go to after that in other words all the infantry guys, they go to SOI which stands for the School of Infantry. If you're non-infantry, then you go to MCT, Marine Combat Training school. After you graduate from that school, either MCT or SOI, you're issued orders to your new unit that you will join. After SOI, I graduated, was a 0351 and was sent to San Mateo to 2/4.

Janney: All right, for all the civilians out there, what is a 0351?

Coiner: 0351 is a designated anti-armor assault team. Basically, what the anti-armor assault team does within the infantry platoon is he takes care of any light armor, explosive devices, can also construct barriers to blockade the enemy, set up ambush sites; things like that. So, the 0351 will have a bunch of explosives or rocket launchers with them, the small rocket launcher.

Janney: Like a Javelin?

Coiner: Yes, in an extreme case, the 0351s were trained in using a Javelin rocket.

Janney: At what point did you get assigned to 2/4?

Coiner: Directly after School of Infantry, I was given orders to go over to 2/4 in Aug 2001. I think my first two years were spent on a Pacific, well basically, the 31st Marine Expeditionary Unit. The 31st was tasked to patrol and be in the area of the Pacific Ocean, so could be anywhere there like the Philippines, Thailand, Korea, Japan, Australia, places like that.

Janney: You deployed with 2/4?

Coiner: Yes, we deployed to Okinawa with 2/4, in the beginning of 2002. We were only initially supposed to spend six months there as a standard deployment length, but that was right after 9/11, so they told us we need to stay behind for another six months in Okinawa and that's when the initial first invasion of Baghdad was going on at that time. Oh, as a matter of fact, I just remembered too, when we were in our second six months stay in Okinawa, we reacted to a hostage situation in the Philippines. It was a missionary and his wife and the Mujahidin had taken them hostage. We were reacted to sort that situation out. Just before we got to the area, we were told to back off because the Filipino government wanted the opportunity to take care of that situation. Unfortunately, we know how that played out. He was killed, but his wife made it out alive.

Janney: So, you deployed with 2/4 to Ramadi. Was that in December or January of 2004?

Coiner: January of 2004. We landed in Kuwait and I think we spent probably about two weeks in Kuwait getting everything together and whatnot to drive north up to Ramadi. So, I think right at the beginning of March, we ended up in Ramadi because my birthday is on the first of March and I remember filling sandbags at two outposts in Ramadi.

Janney: That's a helluva birthday present. The only good thing was maybe it wasn't 120° F yet.

Coiner: Laughs. You're telling me. Yeah, it was still pretty temperate then.

Janney: Were you always assigned to Echo or were you assigned to a different Company?

Coiner: No, I was in a different Company, but originally, I was in Golf Company for about three years, and then I was transferred over to H&S Company.

Janney: What were you basically tasked with in H&S?

Coiner: In H&S, I was a platoon Sergeant for what they call security platoon. Basically, the security platoon was tasked with the safety and security of the Battalion Commander or any VIPs within the battalion. The XO or sometimes we had media come through. We also did convoys, convoy security for mail runs, when guys got mail and on two occasions for a hot meal, we pushed security for that. And, of course, back and forth convoys from one camp to the other because when we were in Ramadi, 2/4 was split up into three different camps. You had Hurricane Point, that's where headquarters was, Fox Company went to a compound called Snake Pit, and then Golf Company and Echo Company were tasked out there to Combat Outpost.

Janney: Now, was Combat Outpost just inside the east arches of Ramadi on Route Michigan?

Coiner: Yeah, that's correct. Headquarters, Hurricane Point was that big, I don't want to call it a palace, I don't know what it was before, right on the corner when you face, where the Euphrates River splits, right on that point. Yeah, that's where H & S Company was stationed. Just across the road at that circle, just in front, was Snake Pit. That's where Fox Company was based at.

Janney: When you guys got there in March, what was the general situation there in the city?

Coiner: It was pretty unknown. When we got the briefing back at home, we were told it's going to be a stability mission where they had to win the hearts and minds of the people and we were warned about too much collateral damage. Check your fire; we were given all kinds of classes on rules of engagement. It was supposed to be just a kind of peacekeeping; "let's calm things down here" kind of mission. So, nothing too high speed. We were never given the impression that we were just going to go in there and kill some bodies. So, we just started setting up camp. We really didn't fortify all that much. Just the basics. We started running little convoys, just little operations just to get used to the city. Get used to the lay of the land; the different alleyways and things like that just kind of get ourselves acquainted. We took over from an Army group. I think it was because they were part of the Big Red One Army Brigade that had these compounds and they were holding them for us. When we got there, they moved out and went back to Junction City, I believe it was called. They kind of gave us a little bit of a rundown of what was going on and the lay of the land. They took us on the 50-cent tour, but when they left, we were kind of on our own. I guess the Battalion Commander started setting up meetings with the local sheiks and leaders.

Janney: Based on what I heard, it was relatively quiet until everything broke loose in Fallujah.

Coiner: Right. It was absolutely quiet. People may speculate, but there had been word going around that they would talk to the Army, when the Army was at the Outpost, they would shoot at them and have fun with them, but when they saw other guys move in with different color uniforms and Humvees, they maybe just stood back and watched. Who are these new guys and what do they do? So yeah, they were just sussing the situation out to see what was going to happen.

Janney: You know, another thing that I heard was that at some point the patrols got kind of on a regular schedule. I don't know if that's your experience or not, and that some people warned against that, but yet that still took place. Is that your experience also?

Coiner: Absolutely. I mean, it was morning and nighttime patrols or convoys and things like that just were on the regular. I'm guessing because I'm not a higher up, I was just a Corporal at the time, I didn't know any better, but I'm guessing that the higher ups weren't given the Intel that it was a hostile area. It was just pretty much low key. I'm guessing that's as a result of Fallujah being the key city as it was at that particular time. Fallujah was more seen as a threat and Ramadi was just some little city that we needed to hold and keep control of.

Janney: So, basically when they started their actions to push the insurgents out of Fallujah, my understanding is that a lot of them just moved west and took up positions in Ramadi. Is that what you're guessing?

Coiner: Uh, no. Ramadi was actually a stopping point for them because, I mean, they were literally bussed in, you know, like Greyhound tour buses. They were bussed in from Jordan and Syria and all surrounding places. They would use Ramadi as a pit stop because once we started on our bug hunts, Operations Bug Hunt, which are basically sweeps. We were just doing cordon knocks, searching houses for weapons and things like that. We found a lot of houses that had huge stockpiles of blankets and food, and a couple of compounds where we actually found weapons buried in their yards. So, we weren't too sure why this was going on, but when Fallujah kicked off, it all came to light. There's your "ah hah" moment.

Janney: In your experience, were a lot of the insurgents foreign fighters?

Coiner: Oh, yeah. A lot of them.

Janney: I've interviewed some people that have said that, who knows, they're all dressed alike, we didn't stop to check their papers after we shot them because the papers were in Arabic anyway.

Coiner: Right. A lot of that information will be kept from them because it's really on a need-to-know basis. Call it lucky or unlucky, but I was in H&S Company, where Company headquarters was and that kind of news travels fast because when you kill a guy like that, he's bound to have some sort of documentation papers on him. In his wallet, on his person. We found a bunch of Intel on them and there were a lot of guys from Jordan and Syria.

Janney: Yeah. That's what I heard as well.

Coiner: I can't say whether they were paid extra to come and fight. I don't know, no clue.

Janney: What about any Iranians? Did you hear of or encounter any Iranian Quds Forces guys?

Coiner: Yeah, we shot a bodybuilder. I don't know why. I guess he had an argument with some locals there. I think he was competing for Mr. Iran or something like that, a bodybuilding competition. He went to his car, opened up the trunk, pulled out a weapon, and decided to shoot up this little store. So, we took it upon ourselves to protect the store and take him out. Once we got to digging through his wallet and his vehicle, we came to find out he was actually Iranian and almost celebrity status. Good thing we didn't kill a celebrity.

Janney: When did you guys start having more issues? Before 6 April or was there a build up to that?

Coiner: There was a buildup. I think that the first major conflict was when a Lance Corporal, a driver, took an RPG to the neck and obviously it instantly killed him. They shot up all the other Humvees. After that, very sporadically, we would hit an IED, and as a matter of fact, that's the timeframe McPherson hit an IED which blew his jaw off. McPherson was another 0351; he was one of my better friends.

Janney: Forgive me for not knowing, but did he survive that?

Coiner: Yeah, he survived that by some miracle. We found his jaw and put it in an MRE bag and taped it to his chest and then sent him off to Germany. He's still going through surgery after surgery. At least he can somewhat chew, drink and talk. He still talks pretty well.

Janney: Well, thank God for that. I didn't know that you guys were able to recover that.

Coiner: It took some time, but we found it by God.

Janney: What were you doing the morning of 6 April?

Coiner: We were resupplying Fox Company with ammo and whatnot and it came over the net when Echo Company was hit and the shenanigans that were going on at that particular time. That's when we quickly organized a QRF on standby to see where we could help with what was going on. It took some time for the CO's and everybody to coordinate and figure out what was going on and who was getting shot from where. Once we got the call, we headed on out to the graveyard. I think at the time, there were two squads trapped in buildings because everything was shot from one side to the other.

Janney: Did you personally know Lt. "Ski" Wroblewski?

Coiner: No, I knew him in passing.

Janney: So, basically, you were not part of that QRF, you were part of another QRF sent out to try to quell the situation?

Coiner: That's right. I guess anybody that was not out on the roads was pretty much organized into a QRF. They pretty much hit the hornet's nest with a baseball bat and everybody jumped into action.

Janney: Being in H&S, you probably didn't know the Echo guys that were killed that day?

Coiner: No, no.

Janney: Following 6th April, was that when Operation Bug Hunt took place or was that prior to?

Coiner: That was prior to.

Janney: And then after the 6th?

Coiner: After the 6th, they launched a bunch more cordon knocks. On the regular, we were just busting doors down, looking for these guys, looking for weapons, or anything. Any kind of clues like those blankets for example. We tried to find stuff like that and we found a bunch, holy cow. They hung out, I mean the guys that survived April 6th, 7th and the 8th, the guys that survived got pulled away to the hospital and got treated and I guess they decided to hang around and keep taunting us I guess. I don't know. No telling.

Janney: Are you getting a lot of indirect fire at that point? Or was it hit and run stuff, ambush type situations?

Coiner: Yeah, sporadic ambushes here and there, some more IEDs and yeah, indirect fire. That was just regular and getting mortared. Every now and again rockets would come through the roof. That happened to Gulf Company when they were out on a QRF and they were sleeping in a storage area and one of the rockets came through the roof and landed in the barracks area. Luckily it didn't detonate.

Janney: Oh, thank God. At what point did it calm down or were you at those kinds of levels of action throughout the rest of your deployment?

Coiner: Not really. It wasn't as intense as the 6th, 7th and 8th of April, but it was constant. Maybe in a week, you would get a day off. But, for the rest of the week, you were fighting. It was pretty much every day that we were fighting some sort of conflict.

When they sent in Second Battalion, 2/5, to take over from us in November, when they were taking over our spot so we could go home, I don't know how or why, but the mujahideen decided to step it up a bit. We were busy taking the 2/5 guys around and giving them the 50 cent tour to let them get acquainted and then there was a big attack going on. It was crazy. Unfortunately, I was involved in that because, at that time, 2/5 went to Combat Outpost, Hurricane Point and Snake Pit, they were holding it and they sent all our guys from Fox Company, Echo Company and Golf Company, they sent them back to Junction City for decompression before they get deployed back home. But, the security platoon and the other guys from the other companies stayed behind to drive the new guys around and show them, "Hey, watch out for this alley. Check this guy over here." Just give them the layout, and we hit a big old firefight again. Fortunately, at that time, we actually had access to the Army's armor, you know, Bradleys coming out to help us. Helos came out to help us out. I guess that's the one aggravation that I hold, that the whole time that we were in Ramadi, if we asked for a tank, if we asked for artillery, if we asked for a helicopter, it was like you needed an Act of Congress. I guess maybe all those assets were tasked out to Fallujah at that time. I don't know. I was not privy to that. But, it was like sucking blood out of a stone to get any kind of asset out there to help us out. For the most part, it was just 2/4; us, us and us. The three of us.

Janney: Yeah, I've heard much the same. Getting any air assets was like pulling teeth.

Coiner: When it got too hot there and got attacked, my God, there were like four or five Bradleys out there. There was even an Abrams tank out there. Blew my mind. I wanted to say, "Hey, where were you guys about two months ago?"

Janney: Yeah, no doubt. I wonder why that was. You guys didn't have any kind of armor basically, as far as up armored Humvees. You guys were riding around in 7-ton trucks with sand bags.

Coiner: Well, we started off with canvas doors when we were in Kuwait. But, while we were waiting and staging to get ready to drive into Ramadi, they gave us these quarter-inch panels to screw onto the doors. Yeah, quarter-inch mild steel. It wasn't even ballistic steel. It was an L shape. I guess your head got a little bit of protection. But obviously, McPherson was proof that it didn't work. I've actually got photos of one of the Humvees that got ambushed, and you can literally see all the bullet holes in the unarmored door of the Humvee, you could put cigarettes in the holes because the AK rounds went right through there like butter.

Janney: That's ridiculous. I don't know why they didn't think about that before they sent you there.

Coiner: I guess that validates the remark that we honestly had no idea what we were getting ourselves into. We honestly just thought, "Hey, this is no big deal." Win the hearts and minds, kiss the babies, wave at the adults, give them chocolate, balls and candy, and move on. It wasn't like that.

Janney: Is there anything else that you want to share about any of the guys that didn't make it back? I didn't know if you were close to any of them besides the one fellow that got wounded.

Coiner: Yeah, Todd Bolding, that was another one. I mean, I always feel a sense of guilt a few times because when Matt got hit, he was my best friend. Then, Todd Bolding got hit with a RPG. And Nicholas Aldridge, he got hit by a Humvee in the middle of the night. You know, in those situations, I was never present. You want to always think to yourself, if I was there, could I have done something to prevent that? And that question can never be answered.

Janney: That's true. From my perspective, I understand survivor's guilt. My son-in-law was in the 173rd Infantry at Restrepo in Afghanistan. They lost a lot of guys. He has terrible survivor's guilt and thinks he should have been able to do something. I pray that you guys can work through that.

Coiner: Well, I managed to. I had a hard time in the beginning. It was a bad time for me. But one of my Marines, Hamby, and his wife, Andrea Hamby, posted something on Facebook about the thing that they call MRT, which stands for Magnetic Resonance Treatment. This post is going on to say that they'll treat veterans for free. At that point, I had given up on the VA and all this medicine that the VA was shoving down my neck. And group therapy, kumbaya sessions, and CBT therapy, it was not working at all. So I figured, I'll give this a shot. If this doesn't work, then nothing will. I tell you what, I am so grateful that I did that. That was back in 2014. It's made a significant change in my life. MRT is using a magnet, kind of like an MRI magnet. They do an EEG to see where the brain activity is, what brain activity is going on. The dormant parts of your brain, which is normally with depression and PTSD, is up on your frontal cortex. Those synapses don't die, they just go dormant, they go to sleep. You know, during a traumatic event, like a plane crash, combat, car wreck, whatever. They place this magnet on your forehead, it pulses, and basically sends a signal to these synapses so that they can wake up and start firing again. It hyper drives those synapses, so they basically receive that message about 50 to 75,000 times in a second. What happens is that the frontal cortex wakes back up again, glucose starts working again, starts getting fed into the brain again. All those coping skills and the ability to feel good about yourself and good about any situation and see the positive side of things, you start to remember how to do that for yourself.

Janney: Yes, that's fantastic. My wife is a psychiatric mental health nurse practitioner, and I'm not sure she's familiar with that, so I'll have to let her know about that. She's a big proponent of EMDR, but if it didn't work for you, then it probably doesn't work for other people as well.

Coiner: I guess it's just like medication. Medication works for some people. It doesn't work for other people. Right now, the Department of Navy has actually accepted MRT as a treatment down in San Diego in Camp Pendleton. They're actually using it, some of the guys down there. I think back in the day, in the 90s, yeah back in the 90s it was called TMS, transcranial magnetic stimulation. And there's all kinds of videos on YouTube about it.

Janney: I really appreciate your time doing the interview. It's an honor to be a part of putting everyone's stories together to explain what you all went through during the Battle of Ramadi.

Coiner: I really appreciate it from my side. I really appreciate it because I guess it's a bit of a thorn in all of our sides where Fallujah and Ramadi were going off at the same time. Fallujah got all the press, all the glory. Ramadi was not even heard about. I don't think anybody really knows about Ramadi. I've mentioned it before to some of the guys I work with and they never even heard of it. To think that Fallujah was just maybe five miles down the road west of us, and there were three or four Marine battalions taking Fallujah. We were alone in Ramadi, the same size city and we had no assets. So yeah, it kind of blows our mind how we can be forgotten. When people bring it up in the media or in books and things like that, it kind of gives us a little sense of what we did mattered.

Janney: I cannot thank you enough for everything that you went through. I certainly will never forget the names of your fallen brothers.

7 – 10 April 2004 including "Operation Bug Hunt"

LCpl Ben Musser

Janney: Tell me a little bit about why you wanted to join the Marine Corps.

Musser: My family had a history of joining the military. My grandad was Army enlisted, then became an officer, and served in Vietnam as a commander in the 101st Airborne. We just have that service in our blood. Back in junior high, I decided that I wanted to be a Marine because they were the baddest, the toughest, weren't just generic guys; they were special. I started going into my recruiter's office in junior high and they told me to come back in five years. Laughs.

Janney: Well, that's outstanding. When you finally were able to enlist, did your parents sign early for your or did you enlist when you were 18?

Musser: Yeah, my parents signed early. I still went up to the recruiter's office. I worked in the recruiter's office for a couple of years before I graduated. I helped get them into my high school and did a lot of stuff with them and I was kind of one of them and we all just waited on the day when my parents could sign. I had to technically be one of the seniors, so I had to wait. On the first day possible, we went and signed contracts. Three days after graduation, I shipped out to boot camp.

Janney: So, they kind of adopted you and took you under their wing. Did they give you any knowledge about what you needed to do to be a success at boot camp or as a Marine?

Musser: Yeah, they gave me a lot of good advice and I had researched on my own. They were not typical recruiters with me, they didn't blow smoke up my ass because they knew that I was hardcore into it and was smart enough to know what I was getting into. I had some of the best scores on my ASVAB that ever went through there and wanted to be a Marine guard. I knew exactly what I wanted and it wasn't a spur of the moment thing.

Janney: Yeah, it sounds like they were more like big brothers or uncles rather than recruiters.

Musser: Yeah, they were more like big brothers and friends. We were close like that. We would hang out and party with each other in the evenings. Doing stuff that I should have done when I was in high school. I grew up in Independence, Missouri, a suburb of Kansas City.

Janney: So, once you enlisted did you go to California or to Parris Island?

Musser: California. I was a Hollywood Marine.

Janney: Did the knowledge that the recruiters give you, did they help you through that process?

Musser: Oh, very much so. There was one key thing they told me that I tried to pass on to everybody else going in. That boot camp is all fun and games and DIs are always messing with you and doing crazy stuff. They told me that the basic part of it is that every time they do something to mess with you, that there is a purpose behind it. If you understand the reasons, then their games and stress that they give you isn't nearly as bad. There is a game where they take all your stuff and dump it in the middle and mix it all up and then they give you like 4 seconds to get your stuff. You run and pick up handfuls of stuff, which is impossible. Then they make you get dressed and you might have two left boots, one size eight and the other a size thirteen, pants that don't fit. Stuff like that. It's designed to mess with people and stress them out, but all of that is tied to combat. If you're in the middle of the night getting mortared and your stuff is everywhere, you just grab what you can, get it on and deal with that later. A good example of one they loved using is we would have bathrooms in there and would have three shitter stalls, so they would make the entire thirty-man platoon fit into these three shitter stalls. There is like 3 inches of breathable space and 4 people have something stuck down in the toilet, it was that kind of thing, where you're packed in there like sardines where you should not physically be able to fit that many people. We would do stuff like that and there was still a reason behind that. I love using the example of, you remember at the end of Black Hawk Down where there wasn't room in the vehicles and they all had to run out of there. That's what happens with Marines.

Janney: Laughs. You guys would make it work, right? "Adapt, improvise and overcome."

Musser: My bunkmate in boot camp, we were the two that it didn't matter what they were doing to us. It didn't stress us out. We would find humor in absolutely everything instead of getting all stressed out and freaking out like the vast majority of the platoon. That was something that I carried over to Iraq into combat and it helped. I had a bunch of guys tell my mother this once we got back that I was the person that would get people to kind of chill out and breathe and not freak the "f" out every time something happened while we were in Iraq. I would crack jokes at really odd times because it would help everybody else cope and deal with stuff that was happening.

Janney: When you got into the Marines, what was your job?

Musser: Basic infantry, 0311. I crawl through the bushes and shoot at people - that's the main job.

Janney: What month and year did you enlist in the Marines?

Musser: Well, I joined in May of 2001 and then we deployed at the beginning of 2004. I worked my way up. I was one of the senior Marines when we began that deployment. I was a Lance Corporal when we were first deployed.

Janney: So, a lot of the PFCs really looked up to you for direction and, of course, the NCOs kind of funneled down information for you to pass down to them.

Musser: Very much so. I have a stupid high IQ, so I knew my stuff very, very well. I was one of the best SAW gunners and taught ridiculous extra information about that weapon. I was widely looked by everybody as the best and most knowledgeable SAW gunner to pass on information to the newer, younger guys, especially with that weapon system. I knew pretty much the entire rest of the job. I was very well known as a field Marine more than anything.

Janney: When you guys deployed to Iraq, did you go to Kuwait for 2 weeks and then to Ramadi?

Musser: Yes.

Janney: Initially, you guys were doing security patrols, and hearts and minds type stuff, handing out soccer balls and meeting and greeting, basically trying to get the lay of the land after you transitioned into this city from the army unit that was there before. Is that correct?

Musser: Right, and one thing that might be a funny story that you probably haven't heard. When we first got there, we were taking over from the Army unit that was there at the Combat Outpost. I got selected as one of a very small handful of Marines as added security to go along with the Captain Royer and with Lt. Ski as an added ring of security to go with them. We went on a patrol with the army in their Bradleys just to see things and to show the officers around. We drove down Route Michigan and they stopped their Bradleys. That was my first combat op with Lieutenant Ski.

Janney: So you were basically part of their PSD or in charge of their PSD?

Musser: There were one or two other Lieutenants that were there, but I was there specifically assigned to Lieutenant Wroblewski's security.

Janney: So, you were in his immediate circle. You probably got to know Lt. Ski fairly well.

Musser: Yeah, in fact, there are a couple wonderful stories about Kuwait when I fell "in love" with Lt. Ski and thought he was an awesome dude. The first one was before we deployed. We were doing a lot of training and had seen a couple of brand new, butter bar 2nd Lieutenants and most of the senior enlisted people. By senior enlisted, I mean non-blue branch Corporals and up, who made fun of the brand new Lieutenants because they would come in thinking they knew all sorts of crap, and you've got Lance Corporals that knew what the hell they were doing and were a whole lot better. Lt. Ski was humble from the beginning. He came in and said, "Hey, they taught us to do this and this in officer school, but how do you guys do it?" He said "Show me how you guys do it" instead of just diving in there all hardcore and saying this is how we're gonna do it, even though most of us knew that was a horrible way to do things. That was the first one. Then, when we were in Kuwait, it was 2300 and somebody came over to our tent and said, "LCpl Musser, Lieutenant Ski wants to see you." I'm like, "Oh, shit, what the fuck did I do?" I'm freaking out because the Lieutenant wants to see Lance Corporal nobody over here. I go over to his hooch, go in, and say, "Yes, sir. What do you need?" Lt. Ski said, "Hey, you listen to Pantera, right?" I said, "Yeah." He said, "Do you have any Pantera CD's? I really want to rock out to some good music." I kind of started laughing. Just look at the Lieutenant asking to borrow a couple Pantera CDs from me. We got him some CDs. That was my transition to, "Hey, he's not just the smart guy, he's fucking cool too." Laughs.

Janney: Those are some great stories. I appreciate you sharing those.

Musser: Before he passed away, he was far beyond, in my opinion, the best 2nd Lieutenant that I had ever met. The guy was just incredibly, incredibly smart at his job and still humble with it. A cool fucking guy. Other guys would say, "Damn, dude, you're in love with the Lieutenant" and I would say, "This guy is just fucking cool and you guys should be happy to have a Lieutenant like this.

Janney: You guys were getting indirect fire daily and people popping off shots at you occasionally?

Musser: When we first go there, for the first couple of weeks, we would get mortared every night. You could set your watches by it. It was honestly a little bit of a joke, and we would get a little pop shots here and there. Nothing real major at all. I think the first major thing we got into was a month or so in, we had one of the platoons drive into an ambush in the middle of the night just right down the road on Route Michigan and my platoon was QRF for it, so it was basically completely and totally done. Everything was totally done at that point, but we showed up to clean up. This was the first gunfight that I'd had for Echo Company, well, the first real gun fight, since we got there. It was a really well laid ambush for our guys that didn't go well for the bad guys. Our guys just lit them up and obliterated all of them. But, we went out there and were on clean up. I was a combat aide then. I was kind of a first response, back-up medic. In addition to the rest of my duties, I would be able to help the Corpsman. They taught me a little first aid to help assist, especially if something happened to the Corpsman. So with that, I had an extra little bag on my gear with medical stuff, and I was one of the only people out there that had any sort of rubber gloves for dealing with bodily fluids. So, I got asked to float bodies up into the body bags. Our Corpsman was actually working on one enemy combatant that basically had no chance to live. He was badly injured, but our Corpsman said, "Not only am I required to, but I can use the practice." He was doing some work on him and I was loading up the bodies right next to him and everybody was in this, "Holy crap, this is something brand new that we are dealing with dead bodies right here" and it's very somber and people were freaking out. A couple of people were trying to take pictures and the Gunny would yell at them because there were random flashes in the night out there in this field. I'm sitting there loading up bodies and everybody is freaking out and my twisted, fucked up sense of humor, I load this one guy into a body bag, looked around, and said, "Who wants to go on a working party?" I took the dead guy's arm, raised it and waved it around, saying, "I do, I do" and everybody just started laughing. That was the crack in the ice for everyone to just snap back into reality by cracking a joke with this dead Iraqi body right here. That's where everybody would tell stories of, "Yeah, dude, you're the guy that got us all to snap back into reality and deal with the situation" and not get overly stressed by stuff like that. I volunteered a dead body for a working party.

Janney: Yeah, I grinned when you said that because I can see the humor in that. You've got to laugh, otherwise your response might not be appropriate. At this point, we are in mid-March, late-March on the timeline. Basically, because you had had an ambush and things are kind of ramping up, did your position change any as far as your day-to-day patrols? Were you guys doing day to day patrols, or doing patrols at the same time of day, what was going on daily?

Musser: We were still doing squad size patrols which, personally, I absolutely loved. The Army guys said that we should go out with big forces and patrol the main streets. Instead of doing that, we did a lot of small size patrols that went through people's yards. It brought us a whole lot closer to places that they were not expecting us to go. We still did a lot of Route Michigan patrols, but we were walking down the damn road sweeping for IEDs and we had conflicting opinions on methods of war fighting. I sat there and bitched about it non-stop too to the couple of people that would listen. They were walking down the road kicking stuff hoping that it wouldn't blow up on us with mine sweepers and shit like that and then when it goes off, everybody shoots up the whole hillside and hopes to hit a trigger man. That is a horrible way to fight a war. It's taking on guerrilla warfare the totally wrong way. That's what the Army was doing. They would go out, but that's the advice they told us. If an IED goes off, light up the damn whole hillside. That's fucking stupid.

Janney: Yeah, that's a lot of exposure for you guys in the process and it's just not necessary.

Musser: It's horrible. Basic math skills here, you won't want to go for the most kills to win a war. The more kills, the better kill: death ratio that you have, the more likely you are to win the war. If you sweep for IEDs, you're gonna lose one, two, three, five guys at a time and maybe hit one trigger man. It's not an effective way to fight a damn war.

Janney: Yes, and that's a big maybe on hitting somebody because that's the whole point, is to not get hit after they pop an IED on you guys.

Musser: That's where I want to say that I really think I am the only person that served in Echo Company that actually agreed with some of Captain Royer's tactics and methods because he was doing more stuff like that. He was taking squad sized patrols and not going down the damn main roads. He did some other dumb shit, don't get me wrong, but his tactics like that, I think, I still argue with people about it, everybody else absolutely hated the guy. I got along great with him the couple of times that I interacted directly with him. But, it was because of stuff like that, we were getting face to face with our enemy instead of spraying up hillsides. Which, you know, does turn into losing more people, but we're actually doing the damn job that we're supposed to be.

Janney: Right. As far as April 6th, what were you doing that morning?

Musser: I was sleeping. I passed out. We had been working the night before and so I woke up to all hell breaking loose. They were trying to load up a giant QRF and everything had more or less already happened that day. They took a portion of the platoon; I think it was one squad. My squad was still in bed sleeping at that point because we had different sleep schedules.

Janney: Miller said yesterday that you guys were basically working 22 hours a day.

Musser: Oh, yes, very much so. We were non-stop so whenever stuff needed to happen, it would happen. So, you sleep when you can. That was very important. I woke up and I could still hear some shooting and stuff going on. When I got out of the rack and they were loading everybody up that they could in a couple 7-tons, so they could go out there. It was absolutely a cluster fuck at that point. There was a whole lot less organization and they were trying to get everybody out there to help secure the situation. Then, we ended up sitting there for far too long instead of going out and doing that because they had to get some sort of organization with that, so they could keep up some numbers and make sure that they didn't lose anybody. So, when we finally got out there, I got out there to the corner of Gypsum and Nova and was probably the first responder on the scene after everything was done. There was no more gun fighting. I never shot a single round or anything like that. But, I got up there and we started to secure the area and we were right at the intersection where the lead Humvee got obliterated. I'm trying to think of a good way to describe the location.

Janney: I did foot patrols with 2/8 every day for a couple of weeks so I know that area really well.

Musser: Ok, so that T intersection where that lead Humvee got shot, you're looking at the T, straight off to the left there was a mechanic shop right there on the left. I pushed out on security and right behind that, like right up next to the building, there was a giant mound of trash. At that field right there, there was nothing but trash. I looked over and saw the body of a Marine. I got a little bit of extra security to help me out and I went and checked on that guy and it ended up being Travis Layfield. I found his body still with all of his gear on. So, I was the one who found his body. We finished securing the area and I loaded Travis Layfield's body up into a body bag there. We had a bunch of cameramen, David Swanson, some other guy was out there at that point, taking a bunch of pictures and I loaded this guy up into this body bag. I really didn't know Layfield at all. He was one of the junior Marines in one of the other platoons. I, honestly, found out who that was much later. One of the senior Marines picked him up and was loading him into the Humvee and the camera man took a picture of him carrying the body bag and it was a big center foldout in Time magazine in the big article that they did about us. My shadow was right there, that was when I loaded him up into that body bag.

Janney: I've seen that shot that Swanson took. That's a hard thing to look at. I can't even imagine what it was like being there, but I've seen the photo.

Musser: It was 10 times harder telling his mother, "Hey, that picture happens to be your son." Because I don't think anywhere in the caption did they say who that was.

Janney: No, and probably wouldn't for many reasons. Wasn't Layfield in the lead Humvee?

Musser: He was one of the one's in the lead Humvee. Some of this is stories from the other Marines, so this is what I've put together, when the lead Humvee got shot up, a couple of people that didn't die in the Humvee tried to jump out and get to cover. He was probably thirty paces from that vehicle. I think he ran around the first building that he could, the closest spot. He would have been on the driver's side, jumping out that way trying to take cover around the building.

Janney: On the left side of Gypsum

Musser: Yeah. A fun story that you're not going to get anywhere else, um, it was two days, I think it was one or two days after April 6th, I was on the camp guard for the Combat Outpost, on the very front camp where I was watching Route Michigan and I saw three eighteen wheeler dump trucks. Have you ever seen an eighteen-wheeler dump truck or rock hauler? These dump trucks – I saw three of them drive by stacked to the top with bodies that we killed that day. The horrible, inhumane part of me, whatever you want to say, as shitty as it was that we lost that many people that day, we had a 20:1 kill/death ratio. We killed so many more of their people than they did of ours. They set up this perfectly executed ambush. I'm trying to think of a decent way to say this, but we walked right into the mouth of the lion and then rammed right through the back of his head. Something that I saw as a relatively young Lance Corporal and Caption Royer, he kept doing stuff like that. He would lead us into stuff knowing that was an ambush because that was the end result That got us into more gunfights where we could actively engage the enemy and kill them instead of wandering around trying to get blown up kicking trash looking for IEDs and spraying the damn hillside to kill a triggerman that just blew us up. That is basic warfare. You have to find a way to actually be able to fight your enemy. That was my argument with most of my brothers that were there. That's what Captain Royer, and frankly, I don't look bad at him for the vast majority of decisions. He got us involved in the fight instead of spraying hillsides. That was my mentality, also. I was like, we have to be in the way of danger to be able to actively engage with these people and to be able to fight them. Otherwise, you're looking for roadside bombs and you're shooting at people after the IED detonates.

Janney: Was Eric Smith the vehicle commander in his Humvee?

Musser: Yeah, he was the first squad leader. I actually went to boot camp with that guy. He lives somewhere within about thirty minutes from here in Texas. We just never really stayed in touch. Yeah, that guy, that guy right there, he was a poster boy Marine. There was a whole group of us that went to boot camp and he was the most senior out of our group in his billet. When I got meritoriously promoted, right after this, because he was a very brand-new young squad leader that ended up being in charge. He did a damn good job of it. I give that guy a shitload of respect. Yeah, he was in that vehicle and ended up taking over command of that vehicle when Lieutenant Ski went down. I don't know a whole lot of details about what he did out there, but he kept all the rest of the guys engaged and doing what they needed to do.

Janney: From what I understand, he was trying to get Ski evacuated and Doc Urena was giving him aid as he could and they thought that he might be okay, but it just didn't work out that way.

Musser: Yeah, I think basically, that it was just a whole lot of that. I don't really have a better way of explaining that. They were trying to do what they could and trying to get them out of there.

Janney: So, after the 6th, how did things change at the Combat Outpost and in the city?

Musser: It was an absolute light switch. Up to that point, we were all hearts and minds. Every one of us carried a bag of chocolate with us when we would go on a patrol. The little kids would run up to us and say, "Meestah, Meestah, chocolate." We would give them chocolate, and we were all friends, shaking hands and giving out soccer balls. On April 6th we saw firsthand, just how much, not just the enemy combatants, but the entire city was against us. During that, there's plenty of people, you know, there's plenty of gun fights that followed. They had an absolutely honed system. Seeing that textbook ambush that was masterfully planned, you could see on April 6th, after I got out there and was first on the scene trying to secure the area after everything was done, it was still a hot area. We saw three bodies, and that was right in the middle of the street. Most already had their weapon gone, the weapon that they had. There were no shell casings, almost anywhere. They had this one really efficient method, where you would shoot somebody and that person would fall down and die, someone would come out and pick up his gun and carry it off to pass on to the next person. Kids would pick up shell casings, other people would come and pick up the body and get it out of there. They had a method like this where, not everybody was pulling the trigger and shooting at us, but everybody in the city was involved. On April 6th, we had people in that same market that we were talking to the day before, they all knew what was going on and they all closed up their shops and got the hell out of there before it happened. So, that was the lightbulb moment if you want to put it that way. That was the lightbulb moment when we really understood that this whole fucking city is against us. These people are blowing smoke up our asses, they're not our friends. Everybody in the city knew that shit was coming except for us. We saw that clear as day, so, there were no more hearts and minds on April 7th. Zero. We went out there and you guys fucked with the bull, and now you're gonna get the horns. No better friend, no worse enemy. We're done being your friend. When we were going out like that, we knew then that all of you wanted us dead. We're here, bring it on. That was how we felt immediately after April 6th. The mentality of our entire Company, battalion, completely changed from hearts and minds bullshit to now you're getting the worst of us. Now, you're going to see the devil dogs. You're going to see us pissed off now and that's pretty much how we carried out the entire rest of the deployment. April 10th was when we got to emphasize that a bit. On April 10th was when we got to actually illustrate that and we went out just absolutely trying to start a fight. We went out intentionally trying to get into a big citywide gunfight on our terms. And we did. I got a medal for combat valor that day. I was the first person in the battalion to get a medal for combat valor.

Janney: That's outstanding, man. What was the circumstance in which you got the medal?

Musser: Yeah, I love telling that story. This is after Lieutenant Ski. Sgt. Coan had become acting platoon commander and Cpl Eric Smith had taken over the role of platoon Sergeant. We went out on this patrol and we started off very, very early in the morning and we kicked in some house door and that was part of our plan and searched the house and didn't really find anything and we were basically going to go around and kick in a bunch of doors and search until they decided to fight us. After that first door, we were moving across this big field. I was half-way back in this patrol going through this big field and we were the first ones that took fire. Somebody started lighting up our whole platoon and the entire platoon immediately hit the deck and half of them were in this trench full of sewage shit, so half my platoon got covered in shit. Me, being of the crazy mentality that I am, I didn't hit the deck. I could tell that they were more toward the front of the area, still very close to where I was, but they weren't shooting at me. So, I was the only single person in the entire platoon that did not immediately hit the deck. Instead, I kind of took off running to my two o'clock position from where I was to this house that was probably fifty feet away. I wanted to get in there because that was going to be a hell of a lot better cover and somewhere where we could establish return fire. Burmaline, I think he was a Lance Corporal, at the time, Jason Burmaline, we used to call each other by first names, so I can't remember any of them. Laughs. He was a fire team leader from one of the other squads, he saw me running over there and he grabbed his team and followed behind me and he got to that house right behind me with his team. We went in there and cleared that house and got up on the roof and started establishing return fire at the people who were firing at the rest of the platoon.

Well, that kind of pissed them off, so they turned and started shooting at us. We were up on top of this roof top and it has about a waist high concrete wall going around the edge of it. Burmaline and the rest of his team took cover behind this wall. This is the story that I tell everybody - my dumb ass, I sat there and could hear bullets whizzing right by my head. I stand there like a statue and I get this big shit eating grin on my face and I thought, "Wow, this feels just like a paint ball game." I used to play paintball a whole bunch when I was in high school. It was a true passion of mine and I was very, very good at paintball. I always kicked everyone's ass in paintball. I basically said that to myself, so leaned forward and started shooting again. I shot one of the people that was shooting at us and threw a grenade and blew up another guy from that rooftop. Burmaline stood up and we continued shooting and they led us to a secure spot where they got cover and the shooting started to die off. We got so much more secure than the shit trench. I was actually on a M16 at that point and named one of the designated marksmen for the platoon and one of the first people that got the ACOG site on their M16. I was also one of the top shooters in the Company but, as far as rifle quals, I had a better rifle qual than most of the snipers.

I had gone to one of the first infantry optics courses. They created this course right before we went to Iraq and taught us all about the ACOG, thermal vision, and all the different optics that we use in the infantry. The brand-new ones. They taught us how to hide from thermal and all that fun stuff. They spent a good amount of time teaching us all about the ACOG, so I knew a lot about it and how it was actually operating instead of just putting one on my gun and figuring out how to use it.

Janney: Yeah, that's good that they gave you that training.

Musser: I think there were only two people from the Company that went.

Janney: Was that just because of size limitations on the course, or because…

Musser: Yeah, because it was brand-new, they wanted to see how it worked out and stuff like that. There were only a couple of people, this was the first class of that course that they had offered. So, yeah, I got a lot of cool training on that. I learned how to hide from thermal - that was the coolest thing I took in that class. I totally loved the ACOG, it had this little chevron that was illuminated, but you could still see enough through there to be able to see your target and effectively range your target based on that. The little lines that go underneath that are supposed to be even with the man's shoulder so that to light somebody up and know how much you need to elevate for the range.

Janney: You being the designated marksman, did you guys have 300-yard zeroes or 200-yard zeroes or what? So, you were sighting in at 25 meters for a 300 yard zero?

Musser: I believe so. The way the ACOG worked then, I can't remember this off the top of my head, the little chevron that was illuminated, that was set up to be, I want to say, either 200 or 300 yards. I think it was 200 yards from where that one was supposed to be. Then, with the little lines under, that would be like 300 and 400 or whatever, and so on further down. It wasn't like on the iron sights like where you had to zero to a certain range. The ACOG is set up for a certain range.

Janney: One other thing I wanted to ask you, when you guys had gone out, and the enemy combatants that you did find, did you notice that there were any freedom fighters? I had heard from another Marine, that some of the guys that you guys were killing were actually cops and had police ID.

Musser: Iraqi police? We got a lot of that stuff. There wasn't any unit between the Iraqi police and the Iraqi National Guard or civilians. There was corruption and we saw that everywhere we went that there wasn't any group that was totally safe and on our side. Everybody had somebody that was trying to kill us. We had a lot of intel. Personally, if there were two dead bodies there, I couldn't tell you if one was Iraqi and one was Syrian. But we got a lot of intel saying, hey, these people that we just killed were Syrian or, the people that we were fighting, there was an influx of foreign fighters coming down to help out, stuff like that. I couldn't tell you the difference between any of them if my life depended on it. I never ran into anyone, "Hi, my name is Ahmed from Syria." I personally cannot think of any dead bodies that we saw that were Iraqi police or National Guard. I know plenty of times they were shooting at us but I can't think of any specific time where I personally saw it whereas other people were engaged with them.

Janney: So, after the 10th, did it continue at that same level of conflict?

Musser: Yeah, definitely. On the 10th, that's when I got to know Captain Royer really well. And the house that I had secured, we had set up kind of a base security there. They sent my whole squad back to the original house to help weapons platoon people that were still there. That's when I got in real good with David Swanson. That guy was freaking out. He was sitting there; he had been shot in the shoulder or something and was asking if he got a Purple Heart. We kind of laughed at him. We weren't allowed to smoke when we went out on patrol, but I always kept my cigs with me. David Swanson chain smoked, like 10 cigarettes, and he kept asking me for cigarette after cigarette and we still joke about that to this day. He sat there and smoked half my pack of cigarettes and I tried giving him one of the extra rifles that was on the back of the Humvee and he almost took it. In the back of that Humvee, I grabbed an AT4 tank rocket and I was like, "Hey, this might come in handy." We had gotten to the point that we were like, "Hey, we're not fucking around with the rest of all that dumb shit. We will do what we want to and go fight." There was a little bit less accountability on who was supposed to be keeping track of that AT4, so I took it.

We ended up going back to that house that I had secured and there were a whole bunch more people in there as soon as we got back to that house. There were basically two houses that were actively engaged in shooting at each other, probably 75 yards apart from each other. A whole house of Marines shooting at Iraqi combatants and I got into the house that I had originally secured and from the far end I see Captain Royer leaning out the back door shooting all by himself and the first two rooms were stacked full of Marines shooting out the windows. I said, "Okay, I'm going to cover this door" and I ran over there and said, "Here sir, let me cover this." I don't remember exactly what I said, but it was basically you can do your job and I can be the gunfighter. The Captain's job is a whole lot more important than pulling a trigger. He can be on the radio or be doing a whole lot of other shit. So, he told me that we were fighting these other guys and there's a heavy weapon on this far window to the left from how we were looking and it's got a bunch of our guys pinned down. So, I said, "Hey, can I use this?" and I pointed to the AT4 and he said, "Yeah, sure, that's a good idea." And so, I started running out that door and he grabbed me by the little handle on the back of the flak jacket and yanked my ass back inside. "Hang on, let me get some support fire for you." He said, "Pass off your weapon and prep the AT4." I handed off my rifle to somebody and prepped the AT4. Pulled all the safety pins and the sights and all that fun stuff and he talked some people upstairs and to the rest of the house and coordinated some supporting fire. He said, "Okay, we're ready." Then, he yelled, "Okay, supporting fire!" I ran out that little door right between the two buildings that were actively engaged in shooting at each other. I thought, "Hey, I'm the biggest target here." I aimed that AT4 through this 2x2 window 75 yards away and shot that anti-tank rocket through this tiny, little window, and blew their asses to smithereens. I jumped up, and I remember throwing that AT4 tube against the corner of this brick building and, in celebration, I ran inside and high-fived the CO. Not proper, high-fiving the CO, but so did five other people. I got my rifle back and said, "All right. Now, let's go secure it" and I was totally leading the fucking day. It was pretty cool.

Janney: That is pretty cool, man.

Musser: They said, "Okay, let's do that" and we spent five minutes getting some shit set up so we could go over there. We went over to that house and my NCO from my squad, I believe he was the squad leader. It might have been my team leader. we stacked up to go into that house and I prepped a grenade and threw it into that house and it blew up. Then, we both went around the corner to go inside and secure it. Then this dude jumped out with an AK, right where the damn grenade had just gone off and started shooting at us and my M16 got hit, right on the gas tube along the top of it, at the little laser pointer thing we had on there.

Janney: I know what you're talking about.

Musser: The bullet went through that and hit the battery gas tube on my M16, effectively making it a single shot M16. And part of that, I don't know if it was shrapnel or the actual bullet ricocheted off, hit Degenheart - he was my team leader or squad leader. It hit him in the arm, so we both pulled back out of the way and a couple of other people went up and prepped the grenades and threw them in there. I stepped back and bandaged up Degenhart. He took off his blouse to do that, so the whole rest of the gunfight we're joking that he was going all Vietnam style. He was just wearing a flak jacket and no undershirt or anything like that. Totally showing off his arms. We had a bit of a big, long, good laugh about that. Then we went into that house and finished securing it. The other guys had already blown it up. That's when I learned that you always need to cook your grenade. That's how I learned, the fun way. Cook your grenades and then throw them in. You wait about 1 ½ to 2 seconds and then throw it. That way they don't have a chance to react before it blows up. We got in there and secured that house and they had the belt-fed AK74, and somebody was trying to take it apart, because, at that point, we threw out all of our standard operations. We used to confiscate any more than one AK in a home. Every house was allowed to have one AK. If they had any more than that, we would confiscate them. That got way too much work, so after, I believe it was after April 6th, we would just take the bolt out of the weapon.

Janney: I heard that.

Musser: We would just take the bolt.

Janney: I heard that from Miller, too. Miller said the same exact thing.

Musser: We would subtly, very, very, subtly, take them out of every weapon that we could, even if they had one operating weapon for the house, we would try to take the damn bolt out of that. Because the whole thing was that they were just using them against us.

Janney: Exactly.

Musser: They had this AK74 and were trying to take the bolt out. Nobody knew how to disassemble that weapon. I ran over and started disassembling and the spring popped out, smacked me in the face and busted my lip open. I said, "Oh, son of a bitch." Royer came over and slapped me on the back and said, "Don't worry about it. I'll give you a Purple Heart for that!" Laughs.

Janney: Laughs.

Musser: I took the bolt and stuck it in my pocket, and it was pretty close to the end of that day. Most of the gunfighting went on elsewhere at that point. But, we did lose John Sims that day. He was probably the person that I was closest to. We got over back to the original house and the Corpsmen were in there working on him. He was a good friend of mine that I hung out with a lot. He was laying on the ground and they had all this gear off. He was bare-chested and I didn't see a spot of blood. He looked like he was pretending to take a nap by then, and they continued to work on him. I was like, "Hey, is he gonna be done for this fight?" They said, "He's not doing good." His gear and stuff was there and I asked, "Then, can I take this? He's not going to need it?" They said, "Yeah, he's done with this fight." So, I took his rifle and I left mine there so that I had a fully operational weapon. Then the Army kept saying that we had too hot of an LZ even though we totally secured the area. They finally brought us some tanks and shit and we got out of there. That was a huge thing, for me, in hindsight, months and years after, I think about John Sims, out of everybody that died. The person that I was closest to. I saw him lying there, possibly dying, and I didn't give it a moment's thought before I thought, "He's got a gun that he's not using and I could use it." I swapped that out and used it for the rest of the day.

Janney: Well, that's what you had to do then.

Musser: It was very inhuman. That was the mentality of a very uncaring person.

Janney: You can't go into action with a single shot M16, so you did the right thing. I wouldn't second guess yourself about that. I'm sure you don't.

Musser: It's better now, but for a long time, I was like, "Damn, what a cold-hearted prick I was." I knew I had to do it, but it was a whole matter of, "Hey, at least acknowledge that" kind of thing. It took me a few years to get over that.

Janney: So, I interviewed Sims' Mom. Tell me a little bit about John Sims.

Musser: Yeah, that guy was awesome. He was huge into wrestling. One of my other guys, Jason Rogers, was technically married, and had on-base housing. But, then his wife left, and they didn't do any paperwork, so he still had this base housing. We basically used it as a bachelor pad for about a year or so before we all went to Iraq. Sims was working. He was working with the MPs as a camp guard during this whole period. So, he had a little MP badge and he was roommates with Jason Rogers in this little house. The rest of us would go there on the weekends, sometimes during the week. We would totally have this bachelor pad where we would party and party when we were off work, 24/7, as much as we could there. That's where we really hung out the most. One time, we rented this big ass U-Haul and put a bunch of shit in the back. We put in a bunch of chairs from the barracks, and had somebody drive us out. We were all getting shit-faced in the back. You know how the back of the U-Haul has the Mom's attic thing? The extra shelf up there. He climbed up there and set two little chairs down on the bottom and put this snow board between them. He did this wrestling impression where he launched himself off of the attic and totally belly flopped off this snowboard, thinking that it would break. He just bounced off. I don't know how the dude didn't break three of his ribs. That shit was the funniest thing that I ever saw. He was pretty short, shorter that the rest of us Marines, short, stocky dude. Usually pretty quiet, but he had this genuine, absolute love for WWF and he had been into stuff like that. We talked about stuff like that from time to time. I was into that kind of stuff, but not super hardcore. That was always my favorite memory of John, him jumping off this little ledge, belly flopping across this snowboard and being catapulted to the deck.

Janney: I heard he had a Southern accent.

Musser: Oh yeah, he had, not like a true Southern, but like a Georgia southern. I don't know if you know the difference.

Janney: But anyway, I heard he was a great guy and I appreciate you sharing that story with me.

Musser: Absolutely, he was an awesome dude.

Janney: Yeah, Kim Leblanc said that she got to know him a little bit and she described him as a true Southern gentleman. Do you have any more stories?

Musser: Oh, yeah. He was a really decent, good guy like that. There's one story of the snipers. That's one that always hits home harder. You know, I had a list of every one of the guys that passed away tattooed on my arm. All of them were very important to me. But I always felt more strongly about the snipers, the 4 snipers that died just because of the circumstances. Actually, only one of them was an actual sniper. He had taken a bunch of the 0311s and attached them to the sniper platoon to be added security to them. I was that also. I was one of the additional security or whatever, that got attached to the snipers. I had also done the sniper NDOC when we were in Japan, then failed it. So, I had a very healthy, huge respect for the snipers. But I also knew a lot of the snipers, the actual snipers and so I volunteered, and they were happy. Sergeant Coan was happy to get rid of me. We butted heads a lot. So, I was attached to the snipers and we had these 4-man teams. My team was two snipers and two added security, and the other half of our team was the 4 snipers that got killed. There was Tommy Parker, Pedro Contreras, Deshon Otey and Juan Lopez. I knew all four of those guys. I knew them all very, very well. They all went to boot camp with me. I had known them basically my entire military career, including Tommy Parker - he was from, I think, Fox Company. Anyway, he was from a different Company, but I still knew him because I went to SOI with him. The other three were from Echo Company and I knew them all incredibly well. Half of them on the side. But we were all part of the same team, I think part of the same team. It was a really weird operation. We had two 4-man teams that were all part of the same team. We were doing this 24 hour rotation on this building that was a couple hundred yards from the Combat Outpost. We were doing this 24 hour OP, where 4 of us would sit there for a day and then the other 4 would come and replace us, just back and forth, the same damn kind of rough schedule where, in the middle of the night, those 4 would come and replace us. This is the one part that I actually do talk shit about Captain Royer because, apparently, he has no idea what snipers are supposed to be doing. Part of our orders were, when we would go out and transition on the top of this building, we were not supposed to be covert at all, we were supposed to all walk around when all eight of us were there and be totally visible. Giving at least the impression that we had at least a squad sized element securing that building. Even though it was a 4-man sniper team that, you know, snipers should be fucking covert.

Janney: And not always in the same fixed position.

Musser: Yeah, you understand about snipers. We were supposed to be very covert. Then, we were supposed to, the full 4 of us that were leaving, were supposed to sneak out of there and pretend like we couldn't be seen walking down the damn road back to the Combat Outpost. That last day when I was up in that position, being up there for 24 hours, we would go 50:50. It wasn't like sleep time or anything like that, but two people would be up watching the road and the other two would get something to eat and relax for a minute, because you really had to stand on this tiny, little ledge to be able to see. The concrete wall around the edge of that rooftop was like five feet tall. There was this little one-foot ledge right by that edge, so you had to stand on this little one-foot ledge and lean over, or stand up on your tippy toes. Three of us scaled this and actually watched the road. It wasn't where we could all sit in some kind of culvert and watch the whole road kind of thing. So, we would watch 50:50. The day before we had been briefed on these two brothers that owned the home. They would allow us to use this home that was under construction, but they were still doing some little shit to it. Most of the home was already built. But, these people were letting us use their home for this purpose. So, the day before all the guys got killed, I was sitting up there. I was one of the main ones actively watching at that point, and these two brothers showed up. I saw them signal that they wanted to come in and they were outwardly showing that they were not aggressive or anything like that. So, they came up on the rooftop and seemed to check on us basically and they offered us ice. Which was so fucking weird because it was the middle of summer in Iraq and I didn't think anybody knew what the hell ice was. But, they offered us ice for our drinks and stuff like that and said that they were checking on us and then left.

Janney: Did that seem kind of weird to you?

Musser: It was weird, but not to a threat level.

Janney: Weird, but not alarming, is what you are saying.

Musser: Yeah. We might have seen a couple of weird things like that where this is not a typical, random Iraqi person kind of thing. We went into this house once that had a whole bunch of computers and it looked like an office building. I'm like this is totally not typical and we did a whole shit load kind of extra Intel during the search. It was that somebody ran a business and had a whole lot more money than most of the other people. It was on that scale. It caught my attention because it was odd. I was looking for it and I saw no overt, open threat. I felt absolutely no threat from it. They were just coming up there being cool.

So that night, Parker's team came out to replace my team, and they were having a whole bunch of radio trouble and we were doing our thing where we were standing around in the middle of the night up on this rooftop and they were having all sorts of radio trouble. They couldn't get their radio to work quite right. We offered to swap them out and leave our radio because it was working. And they said, "No, I'll get it working, it keeps doing this stuff and that stuff." They had two different radios. They had a little handheld back-up radio. But, we were all sitting there talking and I remember Parker specifically talking about it. We were all discussing that we were all going to end up dead. This is not how you use snipers; we're pretending that we are decent Marines and if anybody tries to fight that order, and it's only half of that and easier killing. All of us said, "This is a suicide mission." Parker was explaining that in a tactical mindset about all that exact same stuff. We all said, "This is fucking stupid - a sniper team with eight people standing around so everybody can see us." If they want us to watch the road, let us watch the damn road. Go find a covert spot and move and go to a different place, whatever and shit like that. The way it should be done. Nobody would listen to the snipers on how to do their job. Royer was just kind of doing it his way. So, that was one of the last things that we talked about with Parker and his whole team. Like I said, I had known Parker for a long time. You remember in the movie "Saving Private Ryan" when that sniper was up in the bell tower? At the end of it, he yelled, "Parker, get out!"

I used to always do that to Parker every time I saw him. I would say, "Parker, get out!" It was our little inside joke. That was the last thing that I said to him as I was walking off the rooftop that evening. The next day, someone woke me up and people were rushing around like on April 6th. They said our snipers just got lit up. That, of course, got my attention because that was the other half of my team. So, I go running out. I didn't go back to that house, but the rest of my platoon, or at least my squad, was on the QRF and they went out there and secured that house. Most of them told me what they saw when they got up on that rooftop. How their bodies were laid out and stuff like that. Everybody was saying that they all fell asleep and the same two brothers went up there and basically did a quick draw, pulled out pistols and executed all four of them. They use that as a teaching tool in sniper school. They continually use that as an example of what not to do: these guys did this and they're all dead because of it. That was something that always pissed me off, because we all knew that it was a suicide mission no matter what happened. I knew those 4 guys. Those were the four of the best Marines that I knew. They don't fall asleep on watch. They were doing their 50:50 and people were making assumptions based on how they found the bodies after they had been ransacked and pillaged through and all their stuff taken. You're making a judgment based on all of this stuff, and how it should have been done different. They were talking about Parker's boots were off, like the other fuckers couldn't have taken his boots off after they shot him. After I got out, I actually did a FOIA request (Freedom of Information Act) on the case because they did actually do an investigation through NCIS on the deaths of those four guys. I still have it somewhere in the house. I have the investigation done by them. They go over where each one of those guys was shot, where their bodies were laid out, the whole 9 yards. Knowing what I knew, being the last person to see all 4 of those guys alive, I can piece the situation together a whole hell of a lot better than pretty much everybody else. Come to some better conclusions than half the shit that they were guessing at. They really pissed me off because at no point did anybody really ever ask me anything about it. Even though I was the last person to see them alive. In the NCIS report, it says the last person to see all of them alive was the fucking gate guard at the camp. I said, "Hey, dumbasses, they came and transitioned with 4 other Marines, so there are four people that saw them alive."

That always really bugged me and then I found out that they used that example of those four guys for years and years at sniper school, talking about them in different training. I always knew full well that Parker wasn't asleep, and Parker was the first one that died. You can tell that just by the report. He was shot point blank, straight in the head where he had no time to react, so I was able to piece this together, those same two guys came up there and were greeted. They were up there visiting, not fighting their way up onto the rooftop. We all knew who they were at that point. They were greeted and they pulled out their weapons. Parker was the first one that was shot before he could react or do anything. Then Otey, you could tell, was trying to react, but didn't have time to; he was the second one shot. Then Lopez, he had made a few steps over and was starting to engage. Contreras actually got a couple of shots off as he was trying to engage, and got halfway across the rooftop and was trailing blood through that. So, he took a shot or two and was actively engaging them. Then all their stuff was ransacked, and the radios and Parker's sniper rifle was taken. We actually found out later on, I think it was two different Marines who were killed by the rifle that Parker had. I know Caleb Powers of Fox Company was shot while guarding his camp. His was a real sniper shot. They matched up some ballistics and the bullet had come from Parker's rifle. The enemy snipers took that rifle and used it on us for a time after that. I know they actually recovered that rifle a couple of years later. 3/5 Marine snipers in Fallujah killed somebody and found that rifle. They matched up the serial numbers and it's there in San Mateo, I believe, in the battalion headquarters. They actually got that rifle back.

Janney: Yes, I heard that. I'm so glad that 3/5 recovered it and killed the guys that had it.

Musser: That was a really cool thing. You know how a lot of the military wear that rest in peace memorial bracelets, KIA bracelets? I had a whole list of people that I could have worn that bracelet for, like Sims, but I always wore mine for Parker because he was the leader of the team. They dumped all that responsibility on him. They said that he and his whole team fell asleep and he was in charge and they all took away the hero status that everybody else got. We lost 34 people over there and they were all remembered fondly the way that they should be. Like Lieutenant Ski, that guy was a fucking hero and was an awesome dude. He's got his gravestone in Arlington and remembered the way he should be. Going out the right way. They didn't do that for these other 4 Marines. Mostly, Parker, because he was the leader and very much dishonored their memory when they shouldn't have because they came to a lot of bullshit assumptions, so yeah, that, far and beyond, was one of the biggest problems I had, or impacts, I guess, with the people that we lost, was when we lost Parker and everybody just kind of blamed him and said he did wrong. They didn't have a fucking clue because they didn't talk to the right people and know and didn't know the situation. They blamed Captain Royer and his tactics for everybody else's death except more for this one because they all blamed Parker and said that they all fucked up and went to sleep and shit like that, they all jumped to a bunch of conclusions, because like, they found his body with his boots off. And half wrapped up in a poncho liner, like, it was in the middle of the Iraq fucking heat. You're not gonna wrap up in a fucking poncho liner to stay warm in the middle of the damn day. There's a whole of shit that went on with the radio logs that they collected. Lopez was on radio detail and trying to make the hourly radio checks. His logs don't match up don't match up to the ones that were at the command post. I knew damn well, because that was half of what we were dealing with, that they had shit radio communication that wasn't going through. So, those guys sat out there for way longer, being killed, than they should have. I'm not saying that radio checks would have gotten somebody out there to check on them still alive, but they would have found their bodies before they sat there a couple of hours. Because they sat there dead for probably at least three hours before they launched the QRF and found them which gave the little shitheads all the time in the world to dig through their shit and collect what they wanted to and do whatever they wanted to. The craziest part is nobody heard any of the gunshots. We should have been more than close enough - our guards at the Combat Outpost should have heard those gunshots. But, at the same time, this is Iraq and we heard gunshots all the time and we don't go launching the damn QRF. Those shots came from where one of our OPs were, and they should have managed to get in touch with that OP. Maybe we should put two and two together.

Janney: Just my opinion, but they should have had security posted downstairs. First off, they shouldn't have had you guys in a fixed position that you guys went to everyday. But, secondly, they should have had security downstairs because that's just a disaster waiting to happen as far as leaving the downstairs unsecure.

Musser: Yeah, that's a normal thing that's not possible with that. Snipers, honestly, shouldn't have security at all. Their job is totally different, and it shouldn't have been a position where they needed security. They could've still sent us out in a 4-man position and had, like, two people as security and two snipers doing their job. We're out somewhere covert. For security, we needed more than 4 people to be able to have peace and security on the house.

Janney: Now, it's an OP instead of a sniper hide.

Musser: Exactly. Which is not what you use when you have a whole Company of Marines to use for an OP; use the snipers for something useful. None of the tactics ever made the slightest bit of sense or had any sort of backing and, yeah, all this was just stupid as shit to me. I saw brilliance in a lot of the rest of the tactics from Captain Royer that nobody else saw and could not, for the life of me, find any sort of half-way comprehensible tactics involved with the way he was using our snipers. I had been attached to the sniper team for at least a couple of weeks before that and most of the ops that we were doing were a whole hell of a lot more sniper-like. Where we were taking different positions and stuff like that. We would meet up and then bounce to a different position and then it was a whole hell of a lot more effective and Intelligent other than, "Okay, now we're just going to use this place" and do all sorts of stupid shit. I mean those deaths were the biggest ones that impacted me because I knew all four of them very closely. Just the way that they treated those deaths and tried to use it as an example. Shortly after that, they tried to put half the blame on the other Marines that were security, "Okay, none of you grunts are going to be attached to the snipers anymore" which disappointed me. I just really liked working with the snipers.

Janney: Hey, what about Otey? Did you know him very well?

Musser: Yeah, I knew him. I would say that Lopez is the one I knew the least out of all of them. I knew Otey and Contreras really, really well and you couldn't tell them apart. I know this sounds half-way racist, but I always thought that Otey was the classiest black dude that I knew. He very much liked stuff like hardcore ghetto rap and stuff like that. But that guy would like, talk to you in proper English and stuff like that. And treat you with respect and so I always had this opinion of Otey. He was one of the calm, classiest black dudes I have ever met. I was proud to know him and stuff. I'm not racist at all, anything like that. But, there are some people that you get along with better than others. He was the easiest person to get along with, too. You could walk up to that guy and he didn't see color.

Janney: Yeah, it's funny how Kim LeBlanc talks about how Otey didn't believe that she was married to Patrick LeBlanc. He would watch when she picked up Patrick every day at the base because he said, "No, there is no way you are married to him." It's just funny how she told that story and he would stand there and watch to make sure it was her car that was picking up Patrick every day. Anyway, that's just kind of a funny story that I've heard about him.

Musser: Yeah, I'm real close with those two also. I knew Kim pretty well. I went over to their house a couple of times and stuff and I'm still in contact with both of them on a regular basis. Kim blows up my fucking Facebook feed with crazy shit. She's a whole different person on Facebook. Laughs.

Janney: It's weird how people do that. So, now, somebody, I can't remember who it is, somebody's got a copy of the after-action notes, the journal, that Deshawn Otey wrote down after 6 April. Have you ever read that?

Musser: I have not and I would love to. He was the only person that lived from the 1st Humvee.

Janney: I know, I know. That's the ironic thing about it. The fact that he wrote all that down and it would be an incredibly unique insight into what he experienced. I don't know who's got it, but I bet Bob Gibson can remember who's got it. Somebody had it at the Hampton, Georgia reunion and I wish I had known that they had it. Those guys didn't really know me very well back then, so they hadn't really opened up much. It wasn't until this last reunion that they realized that I had honorable intentions and, you know, wasn't gonna besmirch their reputations or that of the Corps. I just wanted to tell their stories, but somebody's got a copy of that - he hand wrote it after that engagement on the 6th. The fact that he escaped all of that, jumped out of a truck and didn't get killed like everybody else did. I will definitely get a copy of that and send it to you.

Musser: I would love that. There's a couple of things like that that I absolutely love having. I would think that if you ever talk to Ian McCall or Ramon Barron, those two were really close with Otey also. They would be, if they had the AAR or don't have it, they would be, Mac and I would very much like to read it.

Janney: What was Barron's first name?

Musser: Ramon Barron.

Musser: Contreras would get on your nerves, and you would yell, "Shut the fuck up Contreras." He had no filter whatsoever. We would bring ladies to the barracks, and he was on the first floor. As soon as we would walk into the room, he would come out and just yell, "Ram it in their ass!" Laughs. We would say, "Shut up Contreras!" He was a funny guy. He was cool like that. Absolutely selfless. When we were in Japan, we did a bunch of general warfare training and we did one of the SERE courses (Survival, Escape, Resistance, Evasion), which is generally reserved for pilots and higher ups and Navy SEALS and shit like that. But, it's the search and rescue POW kind of training. At one point, all of our groups were captured and they were doing their little fake walk to the prison camp, whatever, and we staged a whole little escape plan, but we needed somebody to take one for the team. Contreras was going to go jump on this lead guy that was holding us all captive. As we were marching, he was going to jump up on him and tackle him and try to take him down while the rest of us escaped. Contreras jumped on his fucking back like a monkey. We all managed to run off and my squad was the only one that made it away clean. Everybody else had somebody else that got captured. If one person on the squad got captured, then the whole squad must stay together. We were the only one that made it away thanks to Contreras taking one for the team.

Janney: Jumping on his back like a monkey. That's a great story! Laughs.

Musser: Thanks to Contreras. Have you talked to Anthony Allegre?

Janney: Anthony and I have gotten to know one another pretty well and I have plans on doing that. We have not done an interview yet, but I feel like he will do it. He said that he would in Arlington when I was spending some time with them there.

Musser: Remember me telling you about all the people that we killed that day? Most of them were from Anthony. He was up on top of one of the 7-tons with a Mark 19 grenade launcher and was just lighting those motherfuckers up. That guy kicked some ass that day. He got revenge for every one of our guys that died, and then some. I don't know if it was revenge because it kind of was.

Janney: No. I didn't know that that's what he did during that firefight. I know him but we haven't talked about his story yet. That's good to know, at least I have that insight.

Janney: Anybody that helped with the book that I interviewed; I want to give a copy of the book at no cost, so you've got one since you helped me trying to get it all put together.

Musser: Yeah, I would definitely love that once you get it all done. Then, I can call you up and nitpick on details. Laughs.

Janney: Laughs. That sounds good. Feel free to do so. Maybe I'll let you look at the first draft and you can say, "No, no, this is totally incorrect. You need to go back and check your notes." Laughs.

Musser: You want to know something that is incredibly crazy and funny? The story that I was telling you about the snipers? For at least over ten years, the other three Marines, the two snipers and the one infantry guy, that were on my team, I had no fucking idea who it was. For the life of me, I spent days and weeks with all of them, one of them I knew very, very well. The other Marine, because he came from my Company and for the life of me, there was a mental block or whatever you want to call it, but I had no clue who I spent weeks with and who the other people on my team were who saw Parker and his team last. Until that ten-year reunion, I was standing there over by the basketball court, they had some stuff inside, and I had just met Parker's wife and daughter, and I was telling the story about it. The guy next to me said, "Yeah, dude, I was there." He had that same thing that he could not remember that he was on that same team with me. This is a dude that I knew better than Sims or Parker or anybody else. And he said, "Yeah, I was on that team." Neither of us could remember that we were on that team together and who the other two snipers were. We finally found who the other two snipers were, the two snipers remember working with each other, but the four of us could not remember who the fuck was on that other team. We all remember Parker and his whole team, but none of us could remember it, even though we all sat next to each other for a couple of weeks. It's fucking weird.

Janney: That is weird. But you know the mind is a crazy thing. How your perceptions of things maybe change, or you lose stuff like that, I don't know.

Musser: That was the weirdest one out of it because that was the dude that I went and hung out with all the time, through SOI, and I got to know him pretty well to know that. We would hang out every weekend and stuff, and he was one of my two close friends that I always hung out with all through SOI and the first bit of being in the fleet and could not remember that that was the guy sitting there next to me, attached to the sniper team. And he had the same thing that he could not remember: that was me that was with him.

7 – 10 April 2004 including "Operation Bug Hunt"

Cpl Logan Degenheart

Degenheart: I believe it was October of 2000 when I enlisted. I always just assumed I was going to go to college. I always did really good in school and took every math class I had in high school. I always loved math and every science class. But you're young. Then, somebody asks you that question, "How are you going to pay for it?" My family was well off enough that they could have paid for me to go to school. Dad kind of brought it up to me. I grew up with him fishing, hunting, and out in the woods all the time. He said, "How about this? There's ways to make money working in the wilderness doing what you've grown up doing. Or, you can go into the military and get them to pay for school." I said, "I don't know. I was going to go to college." Dad said, "Well, you can go to college after the military on the GI Bill." I said, "Yeah, okay. I suppose I can give it a shot." That's how I ended up in the Marines. I was still in high school, 17 years old. My parents had to sign a waiver in October when I signed the delayed entry program. I left for bootcamp on June 18th of 2001. I graduated high school at the end of May, and I had about three weeks before I left for boot camp. I turned 18 on Black Friday when we met our drill instructors in boot camp.

Janney: How did 9/11 affect you and your training?

Degenheart: We graduated from boot camp on September 13th. The morning of 9/11, we were getting our haircuts for graduation. I was standing there in line, and you're not supposed to be looking around, but the barbers had a TV on in there, and we're all eyeballing it out of the corner of our eyes, trying not to get yelled at for catching glimpses of it. But I mean, shit, we hadn't seen TV or known what was going on in the real world for the three months we were in bootcamp. I don't think it really hit us what we were watching on TV - it's just news. You don't realize where it is or what was taking place. We marched back from getting haircuts and they put us up on a quarter deck, and then they came out and eloquently explained it to us. They didn't say, "We're going to war." That's not quite the way they laid it out, but we all knew what it meant or what it could mean. I had some good conversations with my parents prior to leaving and enlisting. I've had family in the military that I didn't know about. When I enlisted, I didn't realize how deep the Marine Corps was on my dad's side of the family. I knew my grandpa was in the Army Air Corps and that was about it as far as what I knew of my family being in the military. I knew one of his brothers was, too. When I went to enlist, there was a second cousin of mine whose husband was a Gunnery Sergeant when he retired. I found out there were 12 to 14 guys on that side of my family who've been in the Marine Corps. I didn't enlist because I knew it was a legacy thing, but it's funny what you learn after the fact. It goes back to being young and not paying attention to what's going on around you. Obviously, you've got to go in there prepared for the fact that you may have to go to war. It's something you have to prepare yourself for. I think if you were going to join the military and think, "I'm just doing it to skate through and I have no intentions of ever going anywhere," war could rock your world. The Marine Corps is not the branch of service that you go in thinking, "I'll never get deployed." Especially since I enlisted as 0311 infantry.

    Once I got out of SOI, I was assigned to 2/4. In bootcamp, I went through a screening or two for MARSOC to have a chance to be selected. They lost a bunch of guys out of 2/4 right before we graduated, so they needed a bunch of bodies. The government puts you where they need you. Interestingly enough, the guy I signed up with for the delayed entry program were in the same bootcamp, SOI together and we both went to 2/4. We ended up in the same platoon in 2/4, too. Yeah, so we deployed to Okinawa in June 2002. I was promoted to Corporal in December 2003 and was a Corporal in Iraq.

    We came in on C-130s from Kuwait to Ramadi, and ended up convoying from the airport. I don't remember which airfield we flew into. I know it was a hell of a ride coming down on the C-130. It was pretty fun. It wasn't a long convoy ride to Ramadi in the trucks and was uneventful. It was hard to say how long the ride was when you're all keyed up.

Janney: What was your initial impression of the AO? Did you do leftsy-rightsy with 82nd?

Degenheart: Yeah, we did that for a week or so, maybe ten days. So, on first impression, I was kind of shocked. It would get cold at night, and I expected there to be camels all over and be warm all the time. But there were nights on guard that our canteens would get ice in them. You get an idea that you built up in your head of Iraq and what it's like, and then you actually get there and there's quite a bit of greenery and water around Ramadi. I think the transition from a daytime temperature of 80 to 90 degrees F to the low 30's we experience at night made it feel much worse. That's a huge swing. I was in Echo Company in 2nd Platoon at the Combat Outpost.

    Back in the states, I was assigned to the Company armory. I was in that until we got into Kuwait. Then they told me that I needed to go to a line platoon. It was rough mentally because I wasn't with my team during the workup. In Iraq, I'm a fire team leader and I'm trying to gain rapport. I knew most of the guys, but gaining rapport with them before you go into a combat zone is important. It was kind of a messed up situation and I think maybe they could have done it a little sooner. I went into the platoon and was out there with the squads, on guard around the Combat Outpost and doing the patrols like everybody.

    The first time we got any significant contact was during a response to a mortar attack (author note: 25 March based on Vergara's interview.) We were QRF (quick reaction force.) We had been getting mortared regularly at chow time about 2100 every Wednesday or Thursday. The 82nd soldiers said, "Oh, yeah. Happens all the time." I'm sure they thought they'd test us to see what these new guys in different uniforms were like. So they ended up sending third platoon on vehicles towards dusk, the day they thought it would happen. They ended up over towards the soccer field, east of the Combat Outpost. The patrol ended up coming up on them right after they got done shooting some mortars. Actually, it happened right next to where they started building the house that the snipers had got killed in. But, the patrol came up on the mortar team and got into a gunfight with them. We got sent out as QRF. By the time we got there, most of the fun stuff was over. I remember my first experience of getting out, and 3rd Platoon's Doc had some wounded Iraqis behind this berm. I came up over the berm and Doc said, "I need your help. Hold this bandage on him." I'm thinking, "I don't know about this." It's tough to wrap your head around the fact that we've got to give them aid, even though they were just shooting mortars at us and shot a couple Marines in the gunfight.

It was definitely a quandary. A few minutes ago, they were trying to kill us. The reason they're wounded is because they were mortaring us, but got caught. So, I tell the Doc, "Sure, I'll hold it on here really tight." It was a real mental game. I know we're supposed to render aid, but it's hard when you just heard mortars flying overhead, plus Lloyd just got shot in the knee. As far as the wounded insurgent, it was a futile effort at that point. His guts were hanging down off his hip. The Corpsman was obliged to do something for him from a moral and Geneva Convention perspective.

After that day, we didn't have many engagements until 6 April, other than some IEDs that we were finding as we were sweeping the routes. It always seemed like second platoon was getting hit constantly. It seemed like anytime they went out, they were getting shot at or ended up in bad situations a majority of the time. I'm sure it was just a coincidence, but honestly they did. Second platoon ended up in some of the worst situations. I guess third platoon was just lucky. We got hit by an IED one morning. They were dropping us off at Route Nova to sweep Nova over to Gypsum. We had nine people up and no one got hurt. The insurgents hit it too early and it didn't do anything.

On the morning of 6 April, we weren't the ones out doing the patrol and weren't the QRF. We were the tertiary force. We were probably out the day or night before, so the 6th started off as our rest day. The standard routine was a day of patrol, a night of guard, QRF, then a rest day, but you were QRF standby on rest days. So, I don't know if we were in between that and guard, but I know when third platoon got hit and they started just drawing everybody out. They ran out of vehicles and there's five of us that didn't fit in any available vehicle. So, I ended up staying back with a couple other guys. I was there when they brought back the five-ton and we were unloading the guys that got hit out of the back of the truck. I remember carrying, I assume it was Staff Sergeant Walker, because he was the biggest Marine in our unit. I helped take him out of the truck and carry him.

So, we were still back at the Combat Outpost. They brought everybody back and unloaded them. I saw the helo fly in for 2ndLt Wroblewski and watched them carry him out to the helo. They brought a few more vehicles back and then we all loaded up and we ended up with the whole Company out there at Gypsum and Nova sweeping, and cleaning the shit up. I was there with Apple when we found Layfield and carried him back to the Humvee. That's David Swanson's photo of Apple with Layfield over his shoulder. We found a taxi with a bunch of explosives and weapons in it. Then, we went back to the Outpost.

Janney: I talked to Ryan Miller, and he said that one of the taxis there had a trunk full of weapons.

Degenheart: Yeah, that's the one I'm talking about. There was one there that had numerous AKs, RPGs and propellants for them, RPKs, you name it. It looked like it was driving on Nova and then went over into a culvert. It had its front wheel down into the culvert, so I don't know if it was brought in during the situation or if it wrecked. It almost looked like they had wrecked it as they were getting there. We ended up getting back to Combat Outpost and it was about dusk when we left to go on a night patrol. We patrolled around the north side of Fish Hook Lake, came down around the east side, around the south bottom, and then went back to the Combat Outpost with no action or any contact. They usually wouldn't try to take on an unknown size force in the dark. It was a mental blow to all of us after losing Lt. Ski and all the other guys. There were a lot of guys that said, "Man, it's fucked up." It had been a long day and we were wiped out even before the night patrol, but we had to do it. We couldn't let the insurgents think that they had won.

At that point in the deployment, we had been sweeping Nova, Gypsum, and Michigan every day -

about a 10 mile loop every day. We were getting three hours of sleep, so it really wasn't out of the norm for us to just get three hours of sleep at night. You're almost conditioned to it at that point. I don't remember what the next day brought, but I'm sure we got to sleep for a little bit in the morning when we got back from the night patrol. Then it was right back into the routine.

    I don't remember anything much about the next day or two. The 10th is the next significant day I remember. I know Golf had some stuff going on the 7th, 8th, 9th. It's hard to recall almost 20 years later, but I'm pretty sure we ended up in Golf's AO one day. On 10 April, we started cordons and searches at around 0200. We were going to our third house and I remember seeing a guy run from a building in front of our squad - from one building to an outbuilding. I yelled up to one of the other fire team leaders, "Did you see that guy?" He said, "Oh yeah. I did too." Then, all of a sudden, that's when you started hearing shots going off. Me and Dagoody (another fire team leader) went into another house and cleared it, but one room was padlocked and we didn't have time to pop the lock. We got up onto the roof of that house, started engaging and giving cover for our guys moving forward. Everybody else was to find cover. In no time at all, they told us to bound back to our first house because they had that set up as our OP (operation post.) They wanted us to secure around it. So, we moved back there. Then they told us to push out more again. It was one of those lack of communication issues that happen sometimes. We started moving back out towards that building. I think LCpl John Sims had already gotten hit at that point. One of the guys who was there with Sims on the roof told me he thought the bullet that hit Sims came in from low and came up underneath the edge of his body armor. The shot came up at an angle. That's why we came back to that building where everyone was working on Sims because we were going to secure an LZ for him. But, the LZ was really hot. They called Sims' air medevac off because they said the helicopter couldn't land due to enemy fire in the area.

    Then we started moving back out. I ran by some little bushes we had stopped near. They almost looked like lilac bushes. I ran to the house that I and Dagoody had originally first cleared. I pop back out the door to cover from a house off to our east. As our guys ran from that same bush, I could see bullets hitting the porch pillars and the front porch. I kind of leaned out and started cover firing through the window at this other house. Then, a bullet hit me in the shoulder. I got a little less exposed, but still provided cover fire for the last two guys that had to come over. Doc Clayton was in the house there with me, so I said, "Doc, I think I got hit. It hit me in the left deltoid area." Doc said, "Let's get your gear off to make sure we can see an exit wound and see where you're hit." Doc checked me out and said, "I don't see anything other than the entry right at your deltoid. We'll bandage it up and you can keep going."

    This is when LCpl Musser shot an AT4 at that house where we were taking fire from. He fired it from the south side of the building we were in. After he did that, we all ran over there and went into the entryway of that house. There was a big room that was unfinished and then the rest of the house. One of our guys, Thomas, threw a grenade around the corner. Not Roy Thomas. We had another Thomas in our platoon. He threw the grenade in and everybody's standing there. I said, "Well, follow the fucking thing. We only practiced this a hundred fucking times." I go barging through the door and then Doc Moore comes in with me into this empty room. I'm popping into the corner, but you can't see anything. It's just an absolute cloud of dust and smoke. From the next room, this guy sticks his AK around the corner and just starts unloading. Doc goes out one of the windows and ends up getting shot in the thigh. I pushed LeBlanc back out the door.

    There was one of those doors like a screened door outside the entry door, but it's all metal. When I pushed LeBlanc through, his canteen caught it and closed it behind him on me. This guy is just unloading this fucking AK. All I see is a little archway there. I just got up behind this archway. I'm holding my gun against my chest, muzzle down on the left-hand side with my right hand up here by my chest. I remember watching this brick wall just getting tattered. It's just disintegrating from the impact of the AK bullets. I'm thinking, "Fuck, one of these bullets is gonna catch me in the femur or something and I'm just going to go down." Suddenly, he stopped shooting. I think, "Holy fuck." I throw the metal door open and bound outside. I'm thinking that I'm even alive is a miracle. I started looking around and I got hit across the top of the right hand towards the bottom meaty portion of it. It didn't hit any of my long bones in my hand, but just tore the skin open. Musser wrapped it up for me. At that point, two other guys bounded in, and they ended up taking the guys. One guy was standing on the stairwell with an AK or RPK, and they shot him. They detained the other guy.

Royer had just gotten in there with us. He had originally been in the house that Musser fired the AT4 from like I mentioned. Then somebody called the First Sergeant and they were on their way back with vehicles for casualty collection and extra guns. We all had to bound back over there for security and we did that. The First Sergeant told me to get in the vehicle. I said, "First Sergeant, I'm fine. I've got a couple of wounds." He said, "Well, if you're hit, you need to go get everything cleaned out, get it taken care of." So I ended up getting in the 5-ton and we headed towards the Gypsum-Nova intersection. The truck got hit by an IED on the way there. It blew the windshield out of the five-ton, but nobody got hit with anything.

We got to the T intersection at Nova and Gypsum and the Bradleys had it secured. They came in and picked us up and took us to another place for medical treatment. I don't remember where they took us, but it wasn't the Combat Outpost where all my shit was and all my brothers were. It might have been to Junction City where the Army had a hospital. I know Texidor was on the flight with me, but he was pretty critical with a gunshot wound to the thigh, so I never saw him after we landed. I had a chance the next morning to use the Sat phone while I was in between appointments to just let my folks know before anybody called them. The staff kept saying, "You should call." Sometimes words can get mixed up and sound worse when somebody hears something secondhand back in the States, so I called my parents and a friend of mine just to let everyone know I was okay. I believe I called them Easter morning. Anyway, I was there for a day and a half, and then they flew me back into Blue Diamond. Then, I convoyed back to the Combat Outpost the next morning.

More of the same off and on for the next five months. Action would happen in spurts. You'd have weeks where nothing was happening and then they would hit you with something, like an IED followed up with small arms fire. Honestly, after the tenth of April, they didn't really try too much with us. They definitely never stood and fought us like that again. It felt like the majority of the rest of the deployment was sitting in OPs over Michigan. That was about it. We still did patrols out through there, but we never tended to have much conflict. They kind of quit doing the sweeps on those huge loops. We had received a vehicle from the Army that sent out frequencies that were supposed to kind of electronically jam signals to IEDs and possibly detonate them. So, we kind of quit putting patrols out to that extent. It was much safer than kicking trash piles looking for IEDs.

Janney: You told me that you finally got that bullet out of your arm, but it was some years later.

Degenheart: I ended up getting it out in December 2009. It was the steel armor penetrating core of a .30 caliber Russian round. I made a necklace with the bullet on a sinew rope and wore it for quite a few years, but the sinew started to wear out and I was afraid I would lose it. So, I put it away in a watch box. I'm not even sure where it's at, but I need to find it.

7 – 10 April 2004 including "Operation Bug Hunt"

LCpl Omar Enrique Morel

Janney: When and why did you decide to become a Marine?

Morel: I decided to become a Marine during my childhood. I'm from the Dominican Republic and was born there. Came to the U.S. when I was 12 with my stepdad, my mom, and my sister. My stepdad was an Army veteran. He spent 28 years in the Army. Later, he passed away and I saw how his funeral went, the honor guards and all that. I thought that when I graduated high school, I wanted to go ahead and join the military. I decided to join the Marine Corps. I went in after I graduated high school and went to Parris Island, SC on June 10th, 2001, and graduated boot camp on September 7, 2001 which was a couple days before 9/11. When I think about 9/11, the main thing that comes to my mind is I was living in the D.C. area, so you could see the smoke from the Pentagon that day. Before it all happened, I was talking with my grandma, and she asked, "Why are you joining the service? Aren't you worried about going to war?" I said, "No. We haven't been to combat in many years. I'm just going to go in, do my time, then go to college." Then, 9/11 happened while I was on boot leave and that changed everything. But, the main reason why I wanted to join the Marine Corps was my stepfather. The way that he was treated during and after his service. That was my first encounter with the U.S. military during the funeral. He's buried now at the Arlington National Cemetery in Area 50.

Janney: What did you think when 9/11 happened? What were your thoughts and the thoughts of your fellow boot camp graduates?

Morel: Well, when it happened, everybody was confused at first. I went to my high school that day, because we didn't really know how big this thing was. So, my recruiter and I went to my high school to speak to some of the students there about the Marines during the break. Then I went home normally, and that's when I found out that it was more like a terrorist attack than anything. I thought the Marine Corps is always the first to fight, so I was already getting in my mindset that I might actually get to go there. So, on Sunday after 9/11, I went to North Carolina to SOI. While we were there, they were telling us the whole time that you guys are probably going to go to Afghanistan from here. What ended up happening was 90% of the class ended up going to 2/4 in California. The rest of the class stayed on the East Coast and went to Afghanistan.

I got to 2/4 in November of 2001. I spent my entire time in the Marine Corps with the second platoon. As soon as we got there, we had to get our gear, our weapons, and get ammo. At that time, there was a threat that there was going to be an attack on the West Coast, that they might attack the bridges on the West Coast. We were the QRF, quick reaction force, and say, if something happens, we were going to be the ones that were going to take over. So, we spent a couple months on that, and then after that happened, we started getting prepared to go to Japan to start training because we're going to be part of the 31st Marine Expeditionary Unit. We got to Japan in July 2002. We were supposed to rotate out by the winter, by December. What ended up happening was that since the U.S. was getting ready to start combat in Iraq, they froze everybody in place. So, we ended up staying in Japan for almost one year, around 11 months. We did a whole lot of training in different countries and all of them were in the Pacific. We got back in June of 2003. When we got back, there was only one unit that was back from Iraq. The rest were still over there. So we felt that we didn't and wouldn't get to go to combat. The other battalions, 1/5, 2/5, 3/5, joked that we were the Battalion that doesn't go to combat. That bothered us. We felt like we hadn't done our part.

By that time, all of our senior Lance Corporals and Corporals had left the Marine Corps. So, we were taking the senior positions when it comes to junior enlisted personnel. We became the team leaders and squad leaders for the new group of Marines that were dropped to us. Right after we got back, we got to meet General Mattis in a large meeting in the gym. He knew that we were a little upset that we didn't get to go and be part of the invasion of Iraq. He told us, "We need some Marines to go back to Iraq, and you guys are going to be the first ones to go." I thought, "Yeah, right" because we didn't know that there was going to be a troop buildup to go to Iraq. After that, we were training mostly between ourselves because we probably had enough Marines to total one platoon. We didn't have enough Marines to do much. So, we would do small training like how to be a team leader or a squad leader. We only had one platoon commander which was Lieutenant Bartlett at that time. We didn't have any other Lieutenants at that time. There were only 20-something of us for the whole platoon of Marines in Echo Company. After that, we started getting more Marines. Around the fall, we got like a batch of Marines dropped to us. The officers came in before, which included Lieutenant Wroblewski and the other officers just before our boot drops.

We started training because we knew we were going to deploy somewhere. That was the word on the street. We started training. We trained hard. We started training with vehicles. That was the first time we trained with vehicles because Echo Company used helicopters. We were what they call a helo company. Whenever we needed to go somewhere, it was going to be on helicopters. In Iraq, we're not going to use helicopters. We were going to ride around in vehicles. We started training on 7-ton trucks. None of them had armor, but we thought they would give us something better once we got to Iraq. So, we train and train through February. In mid-February we shipped out to Kuwait to go to Iraq. We spent a couple of weeks in Kuwait training, getting used to the environment. Pretty much all we did was train, PT, and get information about the area we were going. We didn't exactly know where we were going until right before we deployed when they said, "Hey, you guys are going to a place called Ramadi."

The Battalion got divided on the way to Iraq. Almost everybody had to go by vehicle from Kuwait to Ramadi. Echo Company was airlifted from Kuwait to Anbar province where Ramadi is. We were put on C-130s a couple days after the main body of the Battalion left and flew in. We got to where we were going to be staying, which was the Combat Outpost in Ramadi. There were still Army personnel there and they had armored personnel carriers. They had everything. When we got there, the only thing we had was unarmored Humvees and some unarmored seven-ton trucks. The first thing we did was grab plywood and sandbags to use for makeshift armor. We would put two layers of plywood on the sides of the truck bed and fill the gap in-between with sandbags, plus we put sandbags on the truck floor in case of land mines or IEDs. There were no ballistic windshields and there was no armor plate. When we got there, all the heavy-caliber weapons that we got, like 50-caliber weapons, were left by the Army. The Army left those to us so we can keep them in the watch posts that we had inside the base.

We were one of the first platoons to go out. The 82nd took the second platoon out on patrol the second night or the third night we were there around mid-March. We went out in a loop by the soccer stadium and came back in the back side of the Combat Outpost. That was our first experience. During that, we found an Army helmet in a park-like area near the road. We thought it might have been booby-trapped. It was just odd that there was an Army Kevlar helmet sitting there. They picked it up because it wasn't a booby trap. Before the Army left, they told us, "Every Thursday evening, you guys are going to get mortared. We're like, "Whoa! If this happens every Thursday in the evening, how come you guys haven't caught this guy yet?"

After the army left, our very first firefight was during a patrol. They were driving in the vehicles and made a turn near where the four snipers would be killed on June 21st. They made a turn there next to an open field and found the guys that were getting ready to launch the mortar. There were a couple of people that got hurt from the first platoon. There was one driver, Anderson, from the second platoon that was with them. During that firefight, a couple guys from 1st platoon were hurt and some of the enemy combatants were killed and one enemy was captured. I remember seeing the bodies of the people that were killed inside the facility room we had in the Combat Outpost where we would take the enemy combatants when they were killed until we handed them over to the police. Second platoon was not involved. After that, there were no mortars coming in every Thursday night because the mortar team got eliminated that night.

From there, things started evolving slowly. The next big thing that happened was there would be an IED here, an IED there, and people started getting hurt. Then, we had our first casualty when Wiscowiche, one of our engineers, got blown up and killed by an IED. It got very serious after that. Not too long after that, the reporter David Swanson joined us. We were a little worried because a lot of stuff was happening in Fallujah and in Baghdad, but nothing was happening near us. So we felt like we were lucky that nothing was happening to us. Then, April 5th, my squad leader came to me to discuss a patrol. I was a team leader for the second squad of the second platoon. He said, "Hey, we got to go and do a night patrol. We're stepping off around midnight on April 5th/morning of April 6th. I said, "Okay, so what do you need me to do?" He said, "I need you to plan the route we're going to take." While I was getting my team ready, I was also in charge of creating the route. We left the Combat Outpost around midnight and we headed towards the left side of Michigan towards the east. It was an all-night patrol and didn't come back till daybreak. We went east along Route Nova and went through the tank graveyard. We went to every single spot where there was major combat on April 6th.

Right before sunrise, we always stopped to do a security hold. We had this belief in the Marine Corps that the two times you're going to get attacked the most often is during sunrise and sunset. Sunrise was about to happen. So, we did a security hold in the tank graveyard. There's a lot of palm trees around. We felt like that was the best place to do it because we were about a mile from the base and we were just going to stay there, rest a little bit before we went into the base. The sun came up and we were going through some alleys before we hit a road and other alleys. Usually when you do that route, you always see people on the street. You always see the butchers that are beginning to kill the goats and sheep for that day's sales. That morning, we didn't see anything or anybody. None of that. Everything was closed. There was nobody outside, nobody waving. Usually there'll be kids coming in, getting candy from us and asking us for food and stuff. There was none of that that day.

We get back to base between 7 and 8:30 in the morning. I ran into Lieutenant Wroblewski and he asked me, "Hey, did you see anything?" I looked at him and said, "Sir, I didn't see anybody out on the street." He looked at me and said, "Wow, really? There's always movement outside. That's strange." That was the last time I got to talk to him because I didn't know what was about to happen after that. But, he was obviously concerned there was no movement, there was nobody outside. I went ahead and went to bed. It was already daylight by the time I went to bed. I woke up because I heard heavy gunfire from the snipers that were on the watch towers on top of and around the Outpost. The way our little Combat Outpost was the biggest building was in the middle of the COP. That's where all the staff is, all the headquarters people and all the higher-ups. We would stay in the buildings that were next to the wall. I hear the snipers firing and the heavy machine guns on the towers going off. Then, I see the snipers bringing out the Barrett 50 caliber sniper rifles. I thought, "Whoa. Why are you guys bringing the heavy guns?" At that time, we couldn't really hear gunfire from Golf's AO and elsewhere. We were so far away from where the action was that we couldn't really hear everything that was going on with everybody else. We still didn't even know what was happening. All we saw was that the snipers and the people on the roof were engaging some figures out in the woods outside of the COP. It didn't really hit us until the vehicles started showing up with our casualties.

By the time we woke up, everybody else had left us because we had just spent all night patrolling and had just gotten back. We woke up and said, "Where's everybody at?" "Oh, the rest of the platoon went out to help the first platoon." I told my guys, "Get ready because we're probably going to go out next." Someone told me, "There's no vehicles for you guys to go out in." The only vehicle left was an ambulance with a big Red Cross on the side of it. The fighting was way too far away for us to try to make it to them on foot. I remember the first vehicle that I saw that I knew something bad had happened was a seven-ton vehicle that came in. He couldn't go through directly to the Battalion aid station, but went around and across to where we were at. It was surprising because they were just speeding inside the little COP, but then when they passed, I saw a Marine hanging out from the bed of the 7-ton. Half his body was hanging out. I remember that he had blonde hair, but his head was all red and he didn't have any armor on. The only thing he had on his upper body was the skivvy shirt we wore under our blouse and body armor. He didn't have his blouse, his bulletproof vest, or his helmet. All I could see was that he had blonde hair and his head was all red, and he was just covered in red. The back of his skivvy shirt was also dark red. I knew right then that something bad was happening, that people were dying. Up until that moment, we did not know.

Whenever we had a casualty previously from an IED, which is mostly what we would get casualties from, we would call the Army for a helicopter casualty evacuation. We used to get mad at the Army because they seemed to have a hard time landing their Blackhawk helicopters in the little LZ that we had set up inside the Combat Outpost. One time, they claimed it was too small for them to land. So, after the vehicles came in, which included an Army Bradley vehicle, two CH-46s came in and both of them landed. They started rushing our casualties into the 46s. At this time, we don't know who's hurt, who's killed, or what has happened. By this time, I think it might have already been the afternoon. When the 7-tons started coming back in, the first thing people were telling us was, "Hey! Grab grenades, grab grenades." Gunnery Sergeant Coleman had a room right next to where we were sleeping and that's where most of the ammo was. We went in there, grabbed grenades, grabbed ammo, and just grabbed everything. We got on the 7-tons and they took us to Gypsum and Nova which was where the ambush had happened and where most of Echo Company's fighting was happening.

When we got to the Gypsum and Nova intersection, most of our guys were on the rooftops, including Captain Royer. In the distance, about 500 yards away, was a large crowd of people, probably the whole neighborhood, just looking at us and what had happened. The first thing I noticed when we arrived was 2 Humvees that had been ambushed and all of the blood in the Humvees. You also see some of the insurgents that were killed, laying on the road, and on the side of the road. There were between 3 and 5 guys that were left. The guys were telling me that the Iraqis had been picking up their bodies and taking them away, so we didn't know how many insurgents had been killed. At this point, there's only about 4 bodies left behind. I remember that one of them was wearing a 1980s-era US camouflage flak jacket. Another thing we saw there were pineapple hand grenades that hadn't exploded. There were RPGs that were just sitting there that hadn't blown up. There were the dead insurgents and unexploded ordnance scattered all over the area. Our job at that point was to go ahead and secure the area.

    We got people from EOD to gather and dispose of the explosive devices when they showed up after nightfall. Before they arrived, we found another Marine off to the side of the road. David Swanson took a photo of one of the Marines carrying the fallen Marine to the truck. That was the last Marine making it off the field that day. So, EOD came in, put all the ordinance in one spot and blew it up. We went back to the COP. At this time, we still had no idea who was hurt or killed. We just knew a lot of people were hurt, but we weren't the only ones that got hit. Golf Company was also hit. That was another reason why we couldn't get help from the other Companies, because they were also being engaged, especially Golf Company and Weapons Company. Weapons Company was Golf Company's QRF that day, and Golf was fighting in the city, so we were on our own on April 6th.

    We grabbed something real quick to eat, and then Sergeant Coan came in to talk with us. Coan was the platoon commander for the second platoon, so it was Lieutenant Wroblewski, followed by Sergeant Coan, and then the squad leaders for chain of command. Sergeant Coan came in and told us that Lieutenant Wroblewski was hit and unfortunately he didn't make it. This news really upset everybody because Lieutenant Wroblewski was probably the best platoon commander that we had had because of the way that he cared about the platoon. We respected him greatly so we were always trying to do our best so he would be proud of us. So, that really hit us hard that night. Because the second platoon had the fewest losses, we were tasked to go out and patrol that night. Third platoon and First platoon took big hits with the number of people they lost. So, we were told to go ahead to get up, gear up, and go back out.

    So, we went on another whole night patrol. We went out around midnight, and didn't come back till daybreak. The only thing that we noticed that night was we ran into our four-man sniper team who were also patrolling. We didn't even know that they were out there like that. So the snipers passed us, and we had a young Marine named Coleman that stepped onto some live wires. He yelled out when he got shocked and we almost thought we were getting attacked again. After our patrol, we came back to the base, and then we got briefed on what had happened and everything, and how it happened including tactics that the enemy was using: using vehicles - when an area was cleared, they would drop off new people with vehicles and hit us again. They were also using ambulances to transport fighters and carry weapons and ammo. At this time, we thought we had really lost this thing because the whole time we were there, we thought about winning the hearts and minds of the people. After April 6th, we decided that we were taking the gloves off. We're not going to be that nice anymore. I'm not going to go out and wave at every single body that I can. One thing I remember about April 6th is that when everything happened, all the imams started yelling over the mosque loudspeakers, ""Jihad, jihad, jihad!" So, in my opinion, what they were doing is calling all the jihadists to come out and fight. This happened after the firefights started. So, they probably thought, "Okay, we got them pinned down. All the jihadists come out and fight these guys because we have the upper hand."

So, the next couple of days, Echo Company, we didn't see much except for an IED that happened at the intersection of Nova and Gypsum on April 9th. That's when the Marine driver, Ayon was killed. Ayon is the Marine that's pictured in David Swanson's April 6th photo behind the bullet riddled Humvee windshield trying to start the ambushed vehicle. What happened with Ayon was he came out to check our disabled truck and he got hit with an IED and was killed. . But, during April 7-9, Gulf Company was still getting hit in their area. So, we decided to do a Battalion-sized mission in the Golf Company area. We pretty much went through every single street in Golf Company's AO, doing a city-wide cordon and search. We got some intel that the people that planned April 6th were in a certain neighborhood about a mile or so away from Nova where the ambush took place.

We went in there around four in the morning. Nobody was awake. We were just going in there and started going into houses. The person we saw was this young man, and I'll never forget him. When we drove into that neighborhood, a young man saw us. As soon as he saw us, he ran. He must have been around 18 or 19. He ran into some alleyway and kind of disappeared on us. So, we started hitting houses. The very first house that we hit was probably the largest house we had in that neighborhood. We went in there and searched everything. By the time we were finished searching, it was already daylight. We were leaving that house to go around the back side of the house to go to another house. That's where we got hit on April 10th. Myself and my whole squad were attacked. The 10th was pretty much my very first firefight because during April 6th, by the time we got out there, everything had already happened. The only thing that saved us on April 10th from having a whole group of 20 something Marines get killed was that when we were walking through, we met up with the CO, Captain Royer. As we were walking on the back side of the house and when they opened up on us, there was an irrigation ditch right next to an eight foot tall cinder block wall. We all went into that irrigation ditch for cover because this all was farmland and there was no other cover anywhere.

We jumped into the ditch and started fighting back from there. This is when Captain Royer got shot in the helmet, but he was okay – it was either a spent round or a ricochet, otherwise he might have been killed. One of those guys that was not part of our platoon or unit, decided to jump from the side of the wall that was perfectly fine onto our side of the wall and he broke his leg, like straight up shattered his leg. After we fired back for a while, we were able to get up and run across to another house. We were able to get up and then run across to another house that we decided to use for cover and a strong point. Myself and another Marine had to carry this guy who had a broken leg. That's when one of our 2nd platoon guys, Cpl Logan Degenheart got shot in the arm.

We got to this little square home that only had one storage room. So, we had all of the 2nd Platoon, the Company Commander and his people there, too. We were thinking, "Well, okay, what do we do next?" We started looking out of the windows to see what was going on. The whole Company was fighting. We saw where these people were shooting at us from. We tried to gain fire superiority using two SAW gunners and had some success with their 249s. LCpl Ben Musser found an AT4 rocket, so he went up on the roof of the house to fire it at the insurgent house. We provided a little cover fire and then LCpl Musser fired the AT4 across a field several hundred yards wide right into the window of the insurgent house. It was an incredible shot as the AT4 isn't a very accurate weapon. The rocket flew right into the window and exploded inside the house, killing most of the insurgents inside.

We left the house we were in first and assaulted over towards the house where the combatants were firing at us from. As we were running toward them, some combatants came out and we saw face to face from a distance of about 20-30 feet. We just opened fire on them and overwhelmed them. One guy probably took about a hundred rounds just from the squad automatic weapon. It took him a while to fall down because the impact of our bullets kept him standing. We killed the eight guys that exited the house to fight us. By the time we got inside the house, we had to kill all of the insurgents that hadn't been killed by the rocket because they continued to shoot at us. When we finished clearing this house, we decided to start a sweep and take the fight to the enemy. We started finding RPGs and many other weapons during our sweep. We kind of knew that the people that we were killing in this neighborhood were the people that were killing us a couple of days earlier. During this time, we had one casualty there in the Company who was Lance Corporal Sims. I went to boot camp with him. Sims was shot in the chest somehow. Melvin and I started giving him CPR until we could get him a casualty evacuation. The Army brought in an armored vehicle to evacuate him since the Army wouldn't land a helicopter on the field nearby because they said the LZ was too hot. We loaded all our wounded and Sims on the armored personnel carrier. Unfortunately, I think Sims passed away en route to the Junction City hospital.

By the time that the wounded were evacuated, there was no more fighting. We even had helicopters flying over making gun runs that were giving us a hand. We were a little upset that the Army helos wouldn't take our wounded out though. If we'd had a helo for Sims, he might have survived, but it's hard to know for sure.

Then the next day, we had a memorial service for all the guys we lost during April 6-10th. Until this point, I don't think any of the enlisted had an idea just how many men we had lost until we saw this single rifle there with all their dog tags on that rifle. It was terrible and we were devastated at the loss of our friends. Then, the realization kicked in that we were just at the beginning of our deployment and we still have almost 6 months left. But, from there on, they knew that they can't go toe-to-toe against us in a firefight. At this point, they started using more and more IEDs than anything else. Then, IEDs weren't working anymore, so they started using VBIEDs (car bombs.) On the first car bomb they used, I believe they killed four men. That's when Anthony Allegre was injured, on the very first car bomb in either late May or June. One of my best friends was hit too in that same attack. He was the shotgun driver. The driver was Anderson and that's when he got his Purple Heart.

What happened was they parked their VBIED on the side of the road. The Marines were doing a lot of the same things every day. The enemy would watch us and plan attacks based on our patterns. So, every day we would go and relieve a platoon of Marines that was on the observation post in the cemetery. Every day we would drive there, relieve them, and come back a couple hours later and relieve them for a few hours. We were switching on and off doing that. And when the car bomb happened, it was near a gas station and the 2nd platoon was the one driving Weapons Company. Allegre was part of Weapons Company. When they got hit and those four Marines were killed, there was nothing left of this car except for a piece of the engine block.

Now, we changed our tactics. When we saw a car parked in the middle of the road like that, we would avoid it as much as possible and call the Iraqi police. If the Iraqi police didn't show up, we would go ahead and fire a rocket into the car. We always felt like the Iraqi police were working on the side of the insurgents anyway because there would never be anywhere near a firefight. So, the Iraqi police wouldn't show up. We would go ahead and fire an AT4 at that car. At this point, the insurgents changed tactics again and started using suicide car bombs. They would drive the VBIED right into us and detonate it. Right in front of where the first car bomb hit when Allegre got hurt, a second guy drove into a patrol and blew himself up. But, when he drove into us, instead of all the explosion coming to us, the blast went the opposite direction. By some miracle, the force of the explosion went the opposite direction. The only people that got hurt were all these Iraqi guys that were standing next to this little gas station. There were pieces of the suicide bomber everywhere and only pieces of the engine block left.

The next big thing that happened was the four man team with sniper Tommy Parker getting executed on the rooftop OP on June 21. A group of second platoon was asked to go check on the snipers after they missed a second radio check. For security purposes, they wouldn't call out KIA or injured names on the radio, but would use the first letter of their last name and the last four of their social security number. After being with these guys for months, you began to learn who was who from that letter and number. When they announced that the whole sniper team was executed, that was really hard because all those guys were really liked by all the men. You know, it was one of the controversies from the deployment. We all knew that they shouldn't be putting snipers, especially a four-man team, in the same place every single day because something like that might happen. Especially in a building that was under construction where you have people coming in and out all the time. That was one of the things that we always criticized. After the snipers were executed, we would start getting hit every Wednesday mostly. We would get fired upon, but mostly around Golf Company's AO. By this time of the year, the temperature in the middle of summer was like crazy hot, over 120 degrees F. We're just doing the best we can in the heat with the guerilla attacks and IEDs. Something would happen and Golf Company would ask for reinforcements, so we would go in there and help them out. That happened all the way to the end of the deployment.

Some of the people killed after that were probably killed by enemy snipers using Tommy Parker's M40A3 sniper rifle. We knew that because the men were shot from a long distance away and were killed by a single rifle bullet. Ballistics on the bullets proved that to be the case for some of them. We also had two guys that drowned during a mission that were swimming across the Euphrates River.

After that, we came back home and that's when reality checked us. While you are out there, your whole mentality is they're not dead. They're just back home, you know, just hanging out and stuff. Because you don't want to accept that your buddies have died. You didn't want to take that feeling out on the streets. Because the day you do that, that's when you were going to get killed for not paying attention to the danger around you. I said, "Oh my God, the day you don't have your mind where you're supposed to have it is the day that either you or one of us is going to get killed because you walked past an IED, or didn't notice an IED because your mind wasn't in the place where it was supposed to be." But, we finally got back and that's when the reality of what had happened truly hit us. Before we got back, there were a couple of people talking to us that said, "Hey, you guys. Come to us and let's talk about what happened during the deployment." We said, "Nah. We don't need to talk. We're good." But then, you really start thinking about it, the violence, all these guys are gone, and you realize these people are not coming back. They are really gone. Then, you start having survivor's guilt, like what could I have done better to have kept my friends from being killed. You try to live in a way that would honor the folks that never came back and, and just try the best you can every day.

After I came back, it was just about time for me to get out of the Marine Corps. I had less than a year remaining in my contract. So, I didn't even help the new Marines that were coming in. I just pretty much handed over my title to the guy that was coming up and said, "Okay, here's your new responsibilities. They're your Marines now." I thanked everybody that was getting ready to get out. We were sent to different areas to do different things. We didn't even get a chance to say goodbye to all the guys in 2/4 that we deployed with because they deployed to Japan. But, the one thing that I can say, everything that we went through made a difference in my next job when I got out.

I went back to the DC area. Right away, I got a job offer at the headquarters of Veterans Affairs. So, everybody that I was working with was a veteran. Everybody that I've worked with were mostly veterans. But there's one thing that you can see between my guys from 2/4 and any other unit that I see is that we try to stick together. If somebody's hurting, we're going to go in there and say, "Hey, are you okay? You want to talk?" We keep track of each other and care for one another and are there for our brothers. We don't want to see anybody else do something to themselves because they either have survivor's guilt or they have an issue, a problem. So, we look out for each other and always want to make sure that everybody's cool. That's why we have these reunions every year, that's why we have this Facebook group where we can chat in case somebody's having problems. We're always there for each other. That's one thing about 2/4 that seems to be different from the people I meet that served in other units. I'll ask other veterans that have retired, "Would you still talk to the guys that you served with?" Most will say, "No. Why would I? I just want to leave and that's it."

I forgot to mention, right after we came back from Iraq, we got to meet the family of Lieutenant Wroblewski. I remember me and LeBlanc kind of stood out because we were the only two dressed in civilian clothes. LeBlanc and I were already getting ready to get out. We were short timers by then. We're the only people outside of uniform because we're doing this work with veterans. We didn't really get to speak with Lieutenant Ski's family at that time because I'm out of uniform and I didn't feel like it was appropriate. I decided to let the guys in uniform tell those stories.

Janney: No, man, it'd always be appropriate. Like I said, John would love to hear from you at whatever point you want to share your memories of Lieutenant Ski. He's a very down-to-earth guy. Like I said, I spent 10 days with him in 2007 and almost a month with him in 2008 when we were in Ramadi and he's the most laid-back guy you'd ever imagine and he loves you guys. You're his son's brothers. Do you have any stories about the guys that didn't come back?

Morel: Let's start with Contreras. He was the clown in the Company. He was the one who was always in trouble. He would just make everybody laugh. He got busted down like a couple times. So, every time he got busted down, he changed his nickname. He got busted down to PFC, so he would say, "I'm PFC Contreras, Perfect for Cleaning." I remember on April 10th when, in the middle of the firefight, he was running through us and he did a combat roll. He just did a roll. We said, "Hey, you okay, you okay?" He said, "Yeah, man. I'm fine. I just wanted to do a combat roll in combat." I said, "Dude, you scared the heck out of us. We thought you got shot and that's why you fell." Contreras said, "Nah, I just wanted to do a combat roll in combat." I spoke with Contreras the night before he was killed and he was telling me, "I'm so motivated. I know I want to be a sniper now." Hearing that he was killed right afterwards was just so sad. You see Contreras, and you feel like he's just a joker and just wants to do his time and get out, but this time, he found something that motivated him to stay in the Marine Corps. It's a tragedy.

Pedro Lopez. Fun fact - me, him, and about eight other guys got the same tattoo on the right arm. Lopez and about nine of us decided to get a tattoo in Japan all together at the same time. We all got the same thing, which was a tribal arm tattoo that has USMC in it. So, that was one crazy thing we all did in Japan during our deployment there. Lopez was very quiet. He was a little short Mexican guy, very quiet. He got married right before we deployed. I believe his wife is still in Mexico. He was from a small town in Mexico. We were pretty much a group of friends in the same unit at the same time. He was just quiet, a good Marine, and never messed with anybody. Always stayed out of trouble.

From the second platoon, PFC Crowley was one of the junior Marines. He was very small. He probably weighed like 100 pounds soaking wet. When we met him, we said, "Wow, dude. Can you carry the gear?" He was one of the Humvee drivers. So, that's one of the things he was doing. I heard he had recently been promoted to team leader, so he was definitely motivated.

Janney: Did Kyle Crowley ever talk about his family much or did he share anything with you?

Morel: Nah, nah. You know, being such a young Marine in a different squad than mine, he probably felt like, "Okay, if I talk, I might get in trouble. I'll just keep my mouth shut."

DeShon Otey wanted to be a rapper, I believe. This other guy named McCall and Otey were always together writing rap lyrics and stuff. I met one of his best friends last year. You know, I got to talk a little bit about him with his friend. I didn't really get to know Tommy Parker that much because he was a sniper and was doing what he was tasked to do. Calavan was a great machine gunner. He was a good Marine. He was hard-charging, a strong Marine. Calavan got killed when Allegre got hurt with that car bomb. He was, I believe, the machine gunner on the vehicle that day. Carmen got killed on April 6th near the tank graveyard. He was another quiet guy. He was killed when his squad was rescuing the sniper team by the tank graveyard. I didn't get to know him that well because he was an attachment to us. So, we only got to see them whenever we got to go out, because they were the only ones allowed to drive the 7-ton trucks. Lance Corporal Cherry, he was one of those kids that was going to go places. So, the first thing we noticed was, by the time he got to us, he was already a Lance Corporal. We were like, "Oh, he's going to go places." Because usually, they're Privates or PFCs when they come to us. By the time he got to 2nd platoon, he was already a Lance Corporal. He was sent to 1st platoon because they needed more bodies.

One thing I forgot to mention is that on April 6th, most of the guys that got killed had a radio on. Cherry had a radio, I believe. Roberts had a radio. Lieutenant Ski got shot through the radio handset he was talking on for the Humvee radio. The enemy was really going for the people with the radios because they knew that the most lethal person in the Marine Corps is whoever has the radio. People with radios get all the reinforcements to the fight and coordinate the fighting. Unfortunately, I didn't know Ryan Jerabek very well. He was a respectful young Marine and wanted to learn. We knew that every Marine under SSgt. Walker would be fine because Walker was a former DI. I believe that some of the guys that were in our unit looked up to him because he was their training instructor. They were literally intimidated by him. SSgt. Walker was a great leader. Doc Mendez-Aceves was also killed on April 6th. I believe he was helping another Marine when he got killed (author note: Interviews and evidence at the scene indicate that Doc Mendez was rendering aid to SSgt. Allan Walker when he was killed.)

Corporal Lee was killed by a car bomb. Lee, myself, Lopez, Ramon, Garcia, and a group of 12 of us weren't citizens yet. So, we all went in putting our papers in to become U.S. citizens while we served the USA. So, when we were in Iraq, we still didn't have citizenship. So, that's one important thing. We served with people that died for this country that weren't even citizens. That's how much they love this country. There was a group of about a dozen of us that weren't citizens yet. I got my citizenship after I got back. After I left the Marine Corps, they gave me my citizenship. Contreras and Lopez were not born here, but I'm sure that they acquired their citizenship through the enlistment. I think all of us went in together, turned in our paperwork to become citizens through our service, but we were shipped to Iraq. That's how it was. When we came back from Iraq, we started inquiring about getting our citizenship and that's when we got it, most of us.

HQ, H&S, INTEL MARINES

CPL CHARLES "DEUCE" LAUERSDORF, HQ-INTEL

(LATER PROMOTED TO SERGEANT, NOW 1ST SGT., USMC RESERVES)

Janney: The first thing I wanted to ask you is about your recollections of the guys that were killed in the ambush on 6 April, 2004. Did you know any of those guys?

Lauersdorf: I didn't know them well, unfortunately. I was still trying to prove myself. I was a POG, person other than a grunt. I was showing up to infantry and trying to kind of prove myself a bit, but April was the first chance I had to do that.

Janney: What about Lieutenant Wroblewski? Did you have any interactions with him at all?

Lauersdorf: Actually, I had more interaction with the officers than I did with the infantry guys, because I would talk to the platoon commanders quite often, briefing them, and then they go back and brief their guys. Not on a personal level, with me being a corporal. I did talk to him more than I did a lot of the other guys. Part of my job was to brief the patrol leaders. If I wasn't going out on that patrol, my job was to brief them on the most current enemy situation, things to look out for, indications and warnings, that type of thing. I remember every time I spoke with Lt. Ski there, he was always on point as far as paying attention, and asking questions if necessary. He actually gave me enough respect to let me know, "I trust what you're saying and I'm going to take it as the gospel and that's what we're going to do." Not many officers were like that. I've spoken to thousands and thousands of officers since then, and a lot of them are just like, "Okay, I'm bored. Let's just get on with this, because I know more than you do." Lt. Wroblewski wasn't like that.

Janney: In your capacity as an intelligence analyst, reading the stuff online that I've seen, you've made some comments about the fact that the maps were not up to date. Also, I'm not sure if you were the individual that had done this or not, but I understand there were some individuals that had cautioned them to change the patrol routes, and in some cases, that wasn't done.

Lauersdorf: Absolutely. That was in the interviews by the New York Times. Again, being a Corporal, although I believe I was a Sergeant at the time of the interview, I didn't have a career on the line. So, I could say whatever the hell I felt. I was mad because as soon as I left the unit and got back in the States, I took orders to Defense Intelligence Agency. Immediately, I'm thrown in there and I'm doing basically imagery analysis in Ramadi. I'm like, "What the crap is this? I can get really awesome imagery that's days old." In Ramadi, it's not days old, but it's months or years old. Yet, I was forced to use imagery that at that point was three years old while we were in Ramadi. It just really pissed me off. I remember us having to do the same IED clearing routes, the same one, every single day, and yet there'd be casualties every single day. It doesn't take an Intel guy to say, "Hey, we need to change things up a bit." But the orders coming down from the Battalion Commander, LtCol Kennedy, didn't address my concerns. I think there was a lack of leadership at the highest levels and an ignorance to tactics. You don't have to go to Scout Sniper Employment Course to know you don't put snipers in the same hide, every day, at the same time. You move them around, and that's not what happened at all, but our Company Commander failed in getting Battalion leadership to understand that. We had to keep going the same routes. After that NY Times article was published that publicized my disdain and the intelligence failure, I got a call from my boss, who's a Marine E7. He said, "Hey, get ready to testify before Congress because tomorrow you're going to go down to the Pentagon and talk to DIA and Congressional Affairs. They're going to basically brief you on what's going on. What it comes down to is your Company commander got fired, and they want to talk to you about the intelligence shortfalls that you experienced. They want you to testify." I said, "Okay, great. Let's do this. I'd love to get the word out. Let's hope they change something." Well, I was all set to go. Then, a couple days before the actual testimony, I got a call and was told to stand down. The only one testifying is now Colonel Kennedy. Again, if you're a Colonel, you obviously have a career that you're looking out for. So, you're going to have to go talk to them, and you're going to say exactly what the Marine Corps wants you to say. Whereas, if you get a Sergeant up there who doesn't have a career, he's going to tell you exactly how it is. That might piss off quite a few people. But that never really ended well.

Janney: No, I'm glad you did. I appreciate you sharing that with me. When you talk about the Company commander that got fired, that was Royer?

Lauersdorf: Correct.

Janney: Is there anything else you want to share about that?

Lauersdorf: The lack of armor is why so many of those guys got killed that day. It wasn't just the fact that the armor wasn't there - it just wasn't making its way down to us, for one reason or another. But, as far as Captain Royer getting fired, I always speak real highly of Captain Royer. Actually, I just talked to him the other day, just briefly, in an email.

But if you talk to a lot of other Marines, especially a lot of the junior Marines with Echo, they hated him. They called him Captain Casualty, but I don't think they realized he wasn't in it just to get his combat action. He truly cared for the welfare of his Marines. He's not going to say, "Okay, guys, this sucks, but this is what your leadership wants us to do." You're just not allowed to do that. It's not a leader's job to like the policy, just enforce it. Another reason the younger Marines didn't like him, from what I could tell, was because we were collocated with another company, Golf Company. They had much more lax rules and defense postures. For example, when they stood duty, they didn't have to wear heavy and uncomfortable armor, despite the very real dangers. Capt. Royer, however, would enforce the full armor rule. When things are uncomfortable, it's hard to say, "You know what? He's just trying to keep me safe, and seeing as how often we are attacked, it makes sense." He knows what it comes down to. If you get complacent, you die. So, let's not do anything needlessly to cause casualties. But, because he was just following orders from higher ups, like walking the same patrol routes every day, you think it's your Company commander and place blame on him. You think it's his idea. So, it honestly generated a lot of ill feelings towards him. I did spend a lot of time with him, one-on-one, just sitting in his office talking, just man-to-man. Whether it be about family or whatever, about intel, what's going on, my thoughts, or my recommendations. I thought real highly of the guy. I know that wasn't really the question, but just kind of thought I'd throw that out there.

Janney: I don't know the man, and I don't have any opinion about that. I just thought I would get your input about that. I know that was a little bit of the controversy swirling around the whole situation. That's not the focus of this interview. I do appreciate you sharing that with me. I did want to ask you about that. I was curious if nothing else. And you know, not changing the routes is a story and I think that Dave Swanson reported on it or somebody else did that. You know, they went through the marketplace on 5 April 2004 and basically told the merchants to close their shops - they were going to kill Americans the next day. So, they knew. They had placed two IEDs on Route Nova, too. The attack on Headhunter II at the tank graveyard prompted the sniper team to call for a QRF. The shortest distance from Combat Outpost to Route Nova is up Route Gypsum. The huge QRF left Combat Outpost and turned left (north) on Route Gypsum. The insurgents had already set up a textbook L shaped ambush at the Nova and Gypsum intersection with a DShK Com-bloc 12.7mm machine gun, RPKs, RPG teams, and gunmen on the rooftops at that intersection. Based on my interviews and evidence found after the fact, the insurgents sat there most of the morning in those positions waiting for whomever came up Gypsum. Marines found the DShK jammed, but with a belt of ammo in it next to a huge pile of cigarette butts and fresh human feces. Whoever drove into that kill zone was doomed unless they were in armored vehicles, and the lead vehicle was an unarmored Humvee. The insurgents were there waiting for them. As you know, everyone in that Humvee was killed in the 6 April 2004 ambush except LCpl Deshon Otey, who somehow managed to escape out of the vehicle unharmed.

Lauersdorf: The enemy will adapt, and they will learn. Whenever you have a pretty small area of operation to begin with, and there's not a whole lot of roads, you can probably guess. If you only have three roads, you can probably guess which one they're going to take. It's Michigan, Gypsum, or Nova. That's pretty much the extent of it, right? You can't say, "Oh, wow. They got lucky on that one." You take the same route all the time. That's not going to help anything. But, you try to maintain that presence. I don't think it was just bad judgment. I don't necessarily know if that's at Company level, but I'm pretty sure that was Kennedy's. Because I know a couple of times that behind closed doors, while Calvary never sat, that talk, Colonel Kennedy, it kind of gives an indication there that there's obviously some frustration. Obviously, I came to Ed and never said, "Well, that's good. Where's our armor?" I asked the Commandant, and of course, it all rolled down the hill and we got fired. But, you know, nonetheless, it's just, yeah, it was really cordial. But we were dealing with, as far as the entire situation, we weren't dealing with a dumb enemy. A lot of people just think that they should have been able to pull off what they pulled off. Well, now that we have a lot of guys from outside coming in using a lot of training ground, before they move out to Persia. We want to do it with just some low-level surgeons and there's definitely some bad guys there that knew what they were doing - they were well trained. So hey, you know these guys took me even if we tried to be as random as we could. They would have set up, you know, and hey point here, here, here. There's no other option. So, I think it would have happened one way or another. Does that make sense?

Janney: a lot of the interviews that I was conducting in 2007 when I was there, the guys were saying that a lot of the people that they were fighting and eliminating were foreign fighters. Did you see that in Ramadi or were those locals?

Lauersdorf: I think it was a nice mix. I'd say that much. You can tell, your little pop shots, that's probably just some local guy who's trying to make a name for himself, establishing "street cred," if you will. But, in our very first firefight we had as a Company, they would show up and actually had some pretty good fighting positions, and they stuck to it until death. If you find somebody in a sustained firefight to a point to where they know they're going to die, then chances are you're not dealing with somebody from some small, cut-off town in Iraq. You're dealing with somebody that came from somewhere else that had training, or they were from that area and have some military training. That's a level of commitment that a "hanger on" doesn't possess. There's definitely a foreign fighter influence there, a foreign fighter network. You can see it just in the way a lot of them would buy into it. Just look at the ambush on 6th April, the impressions and effects they had. How they had the higher ground and that one major Michigan position, plus the ambush set up at Gypsum and Nova. That's not something some untrained locals would put together themselves.

Janney: Exactly. Just from my untrained perspective, that's what I thought.

Lauersdorf: The SIG (significant acts) board that I had to create in the operations center showed the things that were happening and what they needed. There's an IED that went off here. There's an IED that went off there. There's a firefight here, another there. In that way, people could see the trends. Of course, if you look at the routing, they were all along those routes. That day, I took a picture of it, and it just says, "Location: Ramadi. Event: All Hell Breaks Loose." That's essentially what happened. Captain Royer and I and the rest of the base and the headquarters team were out there running and we're getting into firefights. We're trying to make our way to that ambush site. Obviously, we weren't able to make it there on time. But, it was just one chaotic day. Unless your leg was broken, you were out by the gate and loading up. The only people we left back were either broken or sick.

Janney: At some point, you and David Swanson got paired up. There's a picture that I remember seeing where David is lying in a ditch next to you and you can see enemy muzzle flashes out in the palm tree line. That's a pretty intense picture.

Lauersdorf: Yeah, I have that picture. That photo is one of my favorite, prized, cherished photos that David took. That's the face of war right there, you know? And the fact that he was able to get that shot off along with a bunch of other shots, in that kind of chaos, obviously shows the kind of guy he was. Rather than trying to stay down, burying his nose in the raw sewage, he was just going to town. It was amazing that he could take shots like that. That was absolutely nuts right there. Plus, later on during that firefight, Swanson was grazed in the arm by a bullet.

  That photo was taken at the very beginning of it and I always make a joke. Because if you look at my rifle right there, the ejection port cover was still closed. Whenever I show my friends or maybe when me and people are talking about that, I say, "That looks cool, but who's the dumb POG there who hadn't even started firing his weapon yet? Yep. That's right. That's this guy." I hadn't even started firing yet. But at that point, that's when the ambush had just started. We weren't sure where everything was coming from. We knew we had guys in front of us and guys behind us, and we weren't sure where the friendlies were at that point. Fire was coming from every direction.

Janney: It's all good. I probably would have been the guy with my nose in the sewage. I don't know. It's hard to say until you're in that position.

Lauersdorf: I think you would have been in the same position as David. I think you would have been out there taking action shots, too. It just takes a certain type of person to go overseas to a combat zone in the first place. But, you know, just like you said, you never know how you're going to react when you're in that position. I remember we were running to get to Lt. Ski and where they were at, Capt. Royer noticed his radio operator falling behind. He was this little small guy. Capt. Royer turned around to him and said, "Do I have to take that radio from you?" He's trying to motivate him. And this kid said, "Sir, I think so." It was pretty funny. He stayed with it and no one ended up having to carry his radio. But, I don't think any of us could have stayed with Royer anyway while trying to carry that big radio. You never know how you're going to react until you're in that position.

Janney: I know it's hard to say, but I would have liked to think that I would have done as good a job as David. Who knows? So, we had GPS coordinates from another man in Echo Company. I can't remember the guy that John Wroblewski got the coordinates from. I don't know if you've looked at the pictures on my blog, but where we did the actual memorial service was in an alleyway in the marketplace that veered off north and then westerly behind some houses in a cinderblock lined alleyway. To your recollection, is that where Ski was at?

Lauersdorf: I don't remember the alley. I just remember the main road, the shops to the left and the right, and then you kind of went up the hill to where the 1st Humvee was hit in the ambush by the 50 caliber Soviet-style machine gun at the intersection of Gypsum and Nova. But I don't remember the alley, though.

Janney: It was like a pretty wide spot in the marketplace. Probably three or four hundred meters before you get to the intersection of Nova and Gypsum.

Lauersdorf: So, it was right before where the ambush was?

Janney: Yeah, it was right before (just south of) the intersection of Nova and Gypsum.

Lauersdorf: So, you're talking like right where the ambush was then, right before the intersection?

Janney: Yes, it was not too far from that. It was maybe 300 meters or less south of that.

Lauersdorf: Yeah, I think I know what you're talking about.

Janney: Anyway, there was a little concrete cinder block lined alleyway with a bunch of houses of different heights all along the ends of the alley and at the entrance to the alley. Then there was a big palm grove to the, I guess that would have been the north side of the alley, away from the marketplace.

Lauersdorf: Right, right. Were you not able to go to the marketplace?

Janney: We drove through the marketplace to get there. But, according to the GPS coordinates that were provided to John, Lieutenant Ski was shot on Route Gypsum very close to that alleyway, or at least that's what we were told. But then again, we were with Major General Kelly too, so I don't know from a security standpoint if they thought it was safer to do the memorial service in the alleyway instead of in the middle of the marketplace.

Lauersdorf: Oh, right. I thought that Lt. Ski was actually on the street there next to the vehicle.

Janney: I was always concerned about that. I wondered if just for security reasons, they chose to go in the alleyway and do the memorial service. So, maybe the GPS coordinates were off or something. I've got an aerial map that was provided that shows where every man fell that day. That alleyway closely corresponded to where Lieutenant Ski was shot.

Lauersdorf: Is that what you said on your blog?

Janney: No, the map is not posted because I was afraid I was going to get somebody in trouble for posting that. Because it was a military document that somebody had gotten and shared with John in spite of the fact I'm sure that they weren't supposed to do that.

Lauersdorf: Okay, got you. Are you concerned that it may be classified?

Janney: Well, that was my major concern. My biggest goal is just to tell the story of these guys that fell in combat that day and to honor their memories. I didn't want to get anybody in trouble or dishonor the Corps in any way.

Lauersdorf: Right. Well, I mean, just on that note, if you ever come across anything that you think you'd like to use, but you're not sure, if you want to just send it to me, I can let you know if it would be a problem for anyone. I'd have no problem with sharing that info with you and helping you out in that regard.

Janney: I would definitely like to scan that map and send it to you.

Lauersdorf: Yeah, that's fine. I thought from an intelligence standpoint, I could make sure. I was going to give you my two cents and you could do whatever you want with it.

Janney: Yeah, I would appreciate that. Like I said, I never thought about sharing the map because I was concerned that it would get somebody in trouble. I wasn't supposed to take a Gold Star parent to a combat zone where their son was killed. The D.O.D. frowns on taking parents of deceased Marines to a combat area. So, I basically had to sneak John in on that second trip.

Lauersdorf: Oh wow. Were you ever afraid at any point that they would shut it down?

Janney: Of course. DOD doesn't allow that type of trip. However, I thought that it's better to ask for forgiveness than for permission. I told John Wroblewski on the flight back to the States in 2007, "I don't know how I'll do it, but if I ever get a chance to go back to Iraq, I'll take you with me." As soon as I got back, I started planning the Ramadi embed and submitted our paperwork, listing John as my photography assistant. So, in March of 2008, when I got to Baghdad and the Marine PAO told me that we couldn't go to Ramadi, the jig was up, so to speak. I had to tell him why it was so important that we get to this specific spot in Ramadi. As I said, MSgt Ellerbrock made it happen. We had no media support. We had no security other than what 3rd ID provided on the way into Baghdad and what RCT1 and 2/8 provided to us when we were in Fallujah and Ramadi. I felt very blessed to be a part of that and to get John Wroblewski to the site where his son got killed and to honor the memories of the other fallen Heroes. That was the whole purpose of the trip.

Lauersdorf: Right. Well, the purpose of all of us here is to do one great thing or many great things. Not the only thing. One of yours is right there. Without you having the credentials you have, without you being the person you are, and the background you have, you know, it's a 99.9% chance that trip would have never happened unless it was 50 years from now and everything was both stable and safe, or Ramadi is a tourist destination, which I don't see happening anytime soon. It's amazing that you were able to do that, and you had the testicular fortitude, if you will, to pull it off, and the cunning and the willingness to get it done. Basically, you tried the conventional route and it failed. Let's try something a little unconventional just to make it happen. So, I applaud you for that. I have nothing but respect for you for achieving your goal.

Janney: I don't ever want people to forget those men and the sacrifices that you and your fellow Marines and their families made. That was the whole point of getting it done and that "adapt, improvise, and overcome" ethos. I never let anybody tell me no or stop me from fulfilling my promise to John.

Lauersdorf: I'll tell you what, that's one hell of a way of doing it. I mean that's no easy feat. I can understand from the DOD standpoint. If your mission goal had gotten out there and that basically this is a sanctioned trip, then what's going to stop all the other parents from trying to do it? That's not something that a lot of other parents should try doing. I know that if my sons were killed in combat, I'm going to go to that site come hell or high water. I will make it there somehow, some way, at some point. Now, would I want my father or those son's mothers to go? No, absolutely not.

Lauersdorf: But, you were able to pull that off in the way you did. I think that's genius.

Janney: Well, I don't know about all that. It was a promise that I made to John on the way back from Iraq in 2007. I felt I had to honor that commitment, even though John told me later that he thought I was just being nice and never expected me to follow through.

Lauersdorf: I feel so proud to have served with all those guys. But when you're 18, 19 years old, taking a round in the leg, it's like, "Fine." I remember being in the command operation center (COC) one time during one of the first ambushes, Ray called in his own 9 line (combat casualty medevac.) He was a radio operator. He was wounded. And yet, he called it his own medevac 9 line. A guy manning a 50-caliber or 240 Golf machine gun getting shot and still laying down rounds to protect his brothers and make the scene safe to evacuate himself and any other wounded. They never walked away with Bronze Stars or Silver Stars because everything happened so fast and nobody was thinking, "Hey, that was really heroic. Way to go. Good on you." It was just, "Let's keep each other alive and let's get the hell out of here." People just did amazing things. It's just absolutely phenomenal. So, for you to be able to go on record and basically write about a group of young men, I don't think you could possibly pick a better group than those guys.

I created a website while I was there in country to basically connect the spouses and the mothers because they had no way of staying in touch; they didn't have the Wives' Network. A lot of these guys were 18 years old. They didn't have wives. They have mothers that care about them. So, I created this website, kind of like a blog, while I was out there. Colonel Kennedy ended up making me shut it down because he was concerned about operational security, and I understood where he was coming from regarding OPSEC (operational security.) But, at the same time, this website was the moms' and families' lifeline to their kids and what was going on. I was the Intel guy, so I knew what I could and couldn't share. I'm honored to have created it, and I would create it again even if I was told, "We're going to bust you down to Private." I would still do it. During that time, I created a memorial picture and posted it. By that time, all the families were notified. That picture that I made is actually still on many of their profiles. Picture of them together on that deployment is their main profile picture. A lot of my good friends that I still talk to from the Marine Corps were from that deployment. I spent three and a half years with Marine Special Operations Command, so I made a lot of great friends there. But, I've never experienced the level of camaraderie that I had with Echo 2/4 in that nine month period compared to any other unit I've ever been with.

Janney: The hardest thing I've had to do with this book is interviewing some of the Gold Star parents. I just feel an incredible responsibility to do as much justice to their memories as I can possibly do, and the mission and the book are a big part of that.

John and I stayed at JSS (Joint Security Station) Karama for part of our time in Ramadi with about 50 Marines and about as many Iraqi Army soldiers. Other than the unknown loyalties of the Iraqi soldiers, I never felt safer. When we were there in 2007, we were with the US Army's Third ID (Infantry Division) for the most part. I've spent a lot of time with the Army Rangers doing stuff with them back in the States. But, I never felt safer than when I was in the company of Marines. There's a difference in the way they handle themselves. The Marines are like big kids in the Humvee, but when the Humvee doors open and they step out, they get deadly serious.

Lauersdorf: Oh, absolutely. After I left 2/4, I went to DIA (Defense Intelligence Agency), where I'm around civilians and joint forces. It's nothing like being with a bunch of Marines. Now, my Reserve unit, PACOM, Pacific Command unit, is me and one other Marine. That's it. There's not another branch I'd rather be in when I walk down those halls because no one messes with us. No one tries to get in our way. You hear the 30 guys say, "I wanted to be in the Marine Corps, but…" But the way they look at you and the way they talk to you, it's just so different because of their level of respect for the Marine Corps. But then on the flip side, there's no camaraderie in that unit. All they care about is awards and what they're going to have lunch that day. If you could make me an O4 in the Army right now, or an E6 in the Marine Corps, I'll keep my E6 all day long.

Janney: Exactly. I understand at least a part of what you're talking about. There were times in 2007 that I was definitely shaking in my boots, but when I was there in 2008 the fact that we were embedded with Marines made me feel that if anything happened, I was a lot safer with the Marines than I would be with an Army National Guard unit. I know the security situation was a lot different in 2008 than it was in 2007, but that's not what I'm talking about.

Lauersdorf: I can tell you right now, the insurgents and the fighters and even local criminals felt the same way about the Army while they were there in Ramadi. Because whenever we were taking over from the Army, when we first showed up, the soldiers would say, "We never go in these areas right here because every time we do, we get into a firefight. So, we just stay away. We don't go to certain areas because it's bad. There might be an IED on that road." So, you guessed it, those "bad" areas are where we started patrolling the next day. The day after that, another area the Army wouldn't patrol.

Janney: It was almost funny in a way when we were there in 2008, because a lot of the 2/8 Marines were saying, "We're actually kind of bored. Things have calmed down. Nobody likes to be in combat, but we miss that in a sense."

Lauersdorf: Well, it's adrenaline. That adrenaline rush you feel in combat is the biggest adrenaline rush, and the whole reason I went on that deployment. During my first Iraq deployment, I never saw any action. A lot of my friends did, but I didn't. So, I thought, crazily, "I need to see action." Then I left Iraq the first time and we went to Japan. While I was in Japan, I begged to go back to Iraq with a ground unit. Well, as fate would have it, they said, "Okay, we're gonna let you go. We don't want you to, but you can go. We're gonna send a T-80, temporarily, to the 1st Marine Division." So, I get to 1st Marine Division, and I'm happy as can be. All right, I'm on the ground side. It's all I wanted. Then they said, "Okay, we're gonna send down a regiment." Then the regiment went down to the 2nd Battalion, 4th Marines. I'm thinking, "Hey, this is great."

Then they took me down to the lower echelons, which is what I wanted, and I ended up with 2/4.

Interestingly, my dad was with Echo Company, 2/4 in Vietnam and was actually wounded in action. It was grenade shrapnel, and we might as well just call it a scratch compared to everyone else. My Ramadi deployment quickly became one of those, "Be careful what you wish for" situations. Because you talk to a lot of young Marines and they say, "I want to kill people." Well, what do they not realize? If you're killing people, if you're shooting, that means that bullets and harm are coming your way too, which means there are good chances if you don't get it, one of your friends will. So, is that really what you want? Is that really what's important to you? You're part of your people. It's still gonna feel that way until they experience it, and then that's not so much fun. There's no replacing that feeling, that major adrenaline rush. But, there are two sides to that coin.

I knew when I did the New York Times interview that it was going to cause fire, but I didn't care. I was pissed off. I was pissed off about all the guys that I was with that were killed. I was pissed off that there was a lack of training, a lack of armor, a lack of current digital imagery for analysis to save lives and time. There was obviously a big disconnect when we weren't getting to imagery in Ramadi. There was no criticality for me to get to imagery that I was getting in D.C. But yet, down at our level, we couldn't get it. I remember planning a raid one time and saying, "Let's post up at this house." Well, when we get there, the house isn't even there. I lost a lot of credibility.

I remember some German reporters came down and wanted to do an interview. I was talking to them and I could tell right away the way they were asking the questions was they were trying to slant the story that the U.S. didn't need to be in Iraq in the first place. It's just a big waste of time, money, and lives. I could just tell by their line of questioning that was the conclusion they were trying to reach. That's never been my reason for anything I did in the Corps. My job is not to define the mission, but to execute it. I think a lot of the fallback is, because you look at what happened to General McChrystal and what he said, even though it's a book, people will automatically try to twist something wrong that you say, and I'm going to get in trouble. I'm not concerned about that. I think what you're doing is totally different. It's not some article about Iraq and the church. I mean, it's obviously not the journey of John. Take it however you want. I think it's great. And so I think a lot of the other guys will too. This isn't just another article or some other book about Iraq. It's about something that they can all relate to.

HQ, H&S, Intel Marines

Sgt. Pete Rosado, H&S Co. (Gunnery Sgt., retired)

Janney: Today is 23rd April 2020. When and why did you decide to enlist in the Marine Corps?

Rosado: I enlisted in the Marine Corps in October 1987. I was in college and I felt that I wasn't going anywhere. And I had great grades in college. But one day I did realize what I was doing, I thought that I was doing circles and I was. So, one day I woke up in the morning and instead of going to class, I went to the recruiting office. The recruiter started telling me things. I said, I don't want to hear anything. I just want to leave. I had something going on at the time, so I asked the recruiter if I could leave in July of the following year, in 88, which was nine months from that day. And July came around, and I got on the plane and left. I went to Parris Island, did the recruit training, and I wanted to travel, because like I said, I felt that I wasn't going anywhere. So, I went to boot camp and I made a mistake that I entered the Marine Corps as an open contract guy and that really backfired on me and I was given food service and I really, really regretted the fact that I went to join the Marine Corps as an open contract. The MOS was extremely easy and I never thought it was a challenge for me. And I tried to move out of the MOS several times and it never happened. So I just kept doing the best I could. And I was always looking for, I don't know, maybe places that people will say you don't want to go there. You know, so one time in 2001, I decided that I wanted to go overseas, and our monitor told me I don't need gunnies and food service overseas. I said, "Okay, so I'll tell you what I'll do. If you don't send me overseas, I'm going to go to a unit that sends me overseas." That's how I ended up in 2/4.

Janney: Okay, so what unit were you in before you transferred to 2/4?

Rosado: After boot camp, I went to the food service school in Camp Johnson, North Carolina. From there, I went to Iwakuni, Japan. It was a good duty station. I was busy there. I went to Australia and I went to Korea. I went to the Philippines. It was pretty cool. From there, I went to guard embassies in Quantico and I went to MSG school in Quantico, Virginia. From Quantico, I was assigned to the embassy in Cairo, Egypt. While I was in Cairo, this storm erupted. It was August 2nd, 1990, I'll never forget. Everything changed. My life changed. Everybody's life in the military changed that day. From Egypt, I went to reopen the embassy in Kuwait in March of '91. We went there kind of heavy. Once things got settled, half of us were reassigned. I went to, I can't say this name on the phone or we're going to get cut off. So, I'm going to spell it for you. I went to the embassy and S-A-N-A-A. Okay. I hated that place with a passion, Jesus Christ. I really disliked the place. It was awful. And from there, in 1992, I went to the embassy in Copenhagen, Denmark. It was fun. I mean, don't get me wrong. I loved it. That's what I wanted to do. You know, being back home, I was born in Puerto Rico, I felt that I was going nowhere, to be honest with you. You know, I felt like no matter what I did, I felt like I was in the same place over and over, non-stop. So, it was great for me to go around.

Janney: Pete, let me ask you this, though. Why did you choose the Marine Corps versus the Navy if travel was really what you were looking for?

Rosado: It was a calling. I'm the first-generation Marine in my family. My father was in the Army, but my father was totally against me joining the Army. So, when I joined the Marine Corps, they didn't know. I did it behind my family's back. I never told them. I just told two friends of mine, that was it. And I could trust them, so I know they would never rat me out or anything like that, and they didn't. But my mom couldn't figure out why I was doing certain things. Because one day I started packing all the little things that I had; I didn't want nobody to mess with them. So, I started packing them, and she saw the boxes. So, she started asking me, where are you going? I just told her, I'm going to the Marine Corps. I'm leaving next month. My father came along, and my goodness, talking about being a father, nothing came out of that. He bashed me, "You're not going to do anything with your life. You're not going anywhere." I totally ignored him.

Two months went by and then I left for Parris Island, Platoon 2075, Delta Company, 2nd Battalion. Then from there, I went to food service school, and then I went to Iwakuma. I went to embassy duty, and then after embassy duty in Denmark, I went to El Toro in California, here in California. When I got to California, I don't know, I just really loved it. When I got here, I told myself, whoa, this place is awesome. I'll never forget. It got so awesome that I stayed living here. It's a beautiful state. I've been here since 93. It is a beautiful state. Yes. It's a shame they're full of communists here, but that's a different story. The politics here. They're too left-wing for me. I'm not a left-wing guy. So, after I got to El Toro, I became a Drill Instructor in 1996. After 1996, I went there. They sent me, I didn't want to go there, but they sent me to Marine Corps Air Station Miramar, which is technically El Toro, but you know, with a different name. But they moved from Irvine, California, they moved down to San Diego. And it was a good air station, I'm not gonna lie to you. I enjoyed it. That's when I told you that the Marine Corps was not sending me overseas at all. Then I chose to go with the infantry. I had friends in the infantry and you know, I was not totally going into a dark room when I went there. I didn't know anything, but I knew a few things, how they executed their business and how they deployed, et cetera. Because what happened was I didn't go overseas. I just went from El Toro and I was reassigned to 2nd Battalion, 4th Marines.

Well, 2nd Battalion, 4th Marines, at that time, they deploy every 18 months. Every 18 months they went on a pump. In the 5th Marine's case, they flew to Okinawa, and they became part of the 31st MEU which is a deployment, because once you get there, you get on the ship. I did a field dock about two months after I got there. We went to Operation Seahorse Wind in Monterey, California. It was an operation between the regiment and the aircraft wing, a combined operation. I learned a lot there. It was free. It was something new to me, extremely new. I have never done it before, nothing like that. I really gotta like it. The Army Manchus, that's their unit, their base. They have helicopters and I think I saw tracks there, and I think they were Bradleys. They had a motor pool, but they had aircraft there, helicopters. We were there for about three weeks.

I got back on a Saturday and on Monday morning, that's when 9/11 happened. So, I tell you right there that the day I left was on September 9th of 2001. It was that fast. And again, everything changed after that. The tempo changed, the attitudes changed, the workload changed. Everything changed. Everything. Everything that you knew how to do in the military changed that day. We deployed to Okinawa in 2002. We were supposed to get back to the States at the end of 2002. But the first wave was going to Iraq. So, we got stuck in Okinawa for another six months. I didn't have a problem with that. I liked Okinawa, we got on the ship, and I didn't mind the ship. During all days, because of my MOS, you get to see every Marine that is within the unit. Every time we went to the field, you got to see the Marines and you talked to them. Of course, you know all the staff NCOs and I knew all the officers. I knew most of the NCOs of each section with Echo Company, Fox Company, Golf Company, and Weapons Company. I think I knew about 95% of the old H&S people. We got back from there around May 2003 and shortly after that we found out we were deploying to Iraq within the year. I talked to everyone and what I learned was how they did it and what they did.

In my MOS, I went to Iraq on the first wave. Well, when we went to Iraq, everything was different because we did OIS-2. OIS-1 was a totally different operation compared to what I did and what they did in the beginning. It was very hectic for my section. We were not in control of the food production. We had to go and hold the food from an Army facility. From there, we distributed the food into three little FOBS twice a day. Rarely, we did it three times a day. It was two times a day, nonstop. I also had Marines in the Combat Outpost, in the Snake Pit, in Hurricane Point, and I had one Marine in Junction City because that's where the facilities were. He was my liaison between me and the Army. Again, nobody has done this before, so nobody could give me a one-second point of instruction on what to do. In the beginning, every day was a different way to do it.

We used to breed horses when I was a kid and had a system of how to move a horse from point A to B to C and back to A. One day, I got on my computer and did a little Photoshop of how we could accelerate the process of the boxcar in the morning, the logistics train. Instead of doing stops in every FOB, we took a longer convoy and as we drove by, a vehicle entered the FOB, but the main train kept moving. For example, the minimum convoy was four vehicles, so we ended up going out with six vehicles. Vehicle six will exit on the Snake Pit, vehicle five on Hurricane Point, and the rest of the vehicles will go to the Combat Outpost. Every time we went to the Combat Outpost, it was a bigger challenge. Not for me, because the Combat Outpost was an extension of the battalion because of the demand for the ground guys there. So not only us, all the other Marines and sailors, the Corpsmen, they used to deliver supplies or ammo. Going to the Combat Outpost was really a logistic run, not only a chow run.

Janney: I guess you were taking the food in those big plastic containers that kept the food hot?

Rosado: Yes, every day. I got lucky because in seven months there, nobody got sick from foodborne illness. That was one of my biggest concerns. If you don't clean those containers perfectly, when the Marines eat it, they get sick. We never had a case. Because when that happens, it's not only one guy gets sick. It's the whole section. You have no idea of the amount of work the Corpsmen and these infantry guys did every day. You have no idea the respect that I have for these guys. The battalion commander, the sergeant major, all the coordination and all this stuff they have to keep in the back of the mind because, you know, they have the other bases and the air of operation. The CO, the XO, Major, and the S3, they have to coordinate everything. Every convoy that goes out, you gotta go through them. Everything that you do that involves anything, you gotta go through the chain of command. It doesn't matter what it was.

Before we deployed, we had a meeting and asked, "Gunny, can we do chow for three different places?" I said, "Sir, we can do it. We'll have to do it." But there is a question there because I didn't have enough food service gear to support the operation. We overcame this because the Army gave me a big hand. Captain Tipton loaned me his equipment. I signed for a vast number of food containers. That was the only way that I was going to do it because we didn't have enough equipment to allow me to do that.

Captain Tipton died, died right there. A mortar killed him. When I found out, it bothered me because he lent me a hand. You don't forget people like that. So, food service three times a day to all the facilities, it seemed impossible. Once you deliver breakfast, those containers stay there until dinner. On the dinner run, you drop the food and pick up the containers from breakfast. So you had to have two sets of everything and then it's two sets times three for all the different locations.

Janney: Speaking of mortars, I heard that the Iraqis like to mortar you at chow time.

Rosado: One time, we had a guy in the battalion that was artillery, and I called him one day. Because I noticed the first mortar landed short from the metal roof and the second mortar landed on the same line towards the roof. I said, "You know what? These guys are aiming for the metal." I think the Iraqis learned that when the convoy came around, they were going to eat, so they obviously had eyes on the facilities and operations. Toward the end of deployment, I was in the Combat Outpost for about a month, and it was raining mortars nonstop. It was bad. Many Marines got a Purple Heart that day. At that time, Gunny Coleman got hit by a mortar and got shot on the hip. I felt the explosions, the sound blasts in my chest and I was inside the building. I ran and started putting tables in the vertical position and told the Marines to get behind the tables. I put the tables on the corner of the building, that way we knew we were not going to get hit from the rear. It damaged the building, the port-a-johns, and damaged vehicles. It punched a hole through one of my refrigerators and destroyed it.

Janney: Tell me more about your time in Ramadi. You had your hands full but stayed on top of everything.

Rosado: We did the best we could. You get used to the tempo. You're working seven days a week. I used to get up at 0545 and we didn't go to sleep until midnight. It was busy. It was hot. I think the hottest day that we had was 156 degrees F. The next day I put a thermometer out and took a photo of it at 130. I went on patrol with a weapon a few times with Echo Company, a couple of times with Golf Company. I went on patrol with Weapons Company about three times. You get in a Humvee with the windows up and 150 degrees outside, you've got the combat gear on, but you get used to it. Sometimes the Marines would lower those windows and I'd say, "Put the windows back up." "It's hot." "I don't give a fuck. Put the windows up."

Janney: Tell me about some of your recollections of some of the officers and NCOs that you knew.

Rosado: I only have great stories about those guys. You realize the officers were working long hours every day. I went on patrol with a Lieutenant in Echo Company. The officers in Echo and Golf were busy. They were working 24 hours a day. I'm not lying to you. They go on patrol, go to the government center, and inspect cemeteries, just everywhere. The number of tasks these guys were executing on a day to day basis was just amazing. I saw this.

Janney: Of the times that you went out on patrol, did you ever have contact or IEDs?

Rosado: Yes, we had all of that. The most impressive thing that I had was one time we got ambushed on the left by mortars and we had people on the right. I don't know where they came from, but the Bradleys from the Army showed up. That was totally impressive. It was like a freaking war movie, but in real life. It was 30 August 2004. I saw that Bradley came out of nowhere, and there was a tiny building on the right. This guy just shot about three or four rounds of the 25-millimeter cannon and that was the end of it.

Janney: Did anybody get hurt on any of those patrols that you were on?

Rosado: Yes, we got a guy that got shot in the leg and the round broke his femur. Yeah, he was medevaced that day. Some guys got shot in the pads. We had a mortar that hurt a guy on the side, but he was okay. I never went on patrol when a Marine got killed. I remember I was going to go out with Weapons Company, and somebody told somebody, and I was ordered by the Sergeant Major not to go on patrol, and I said, "Okay."

On 6 April, I was inside the wire in Hurricane Point. We couldn't leave. Everybody that was involved left, from the CO down, the grunts. The rest of us were there just standing by. All my Marines were standing by if they needed to send more people out. I knew that for those four or five days, it was just constant combat. But, on the sixth, the first day, we lost the guys, too. "All hell broke loose" is what they wrote on the board at Echo on the 6th. We had another bad day on the 7th when we lost another. On the 6th – 12 men. The 7th of April, one guy died. On the 9th, one. On the 10th, one. It was 34 guys, but it was 35. People may not say that, but I always say 35, because my First Sergeant Ellis went back as a Sergeant Major and got killed in the second deployment. I remember when he died, I really felt that in my heart. I just couldn't believe it. You're talking about a guy that you think is invincible and he got killed.

Janney: So, tell me what SgtMaj Ellis was like. What do you remember about him?

Rosado: He was the First Sergeant, came to us from SOI East, and we deployed with him, and he was a great guy. He was one of the guys that you knew his presence was felt. He was a good person. Every time you talk to him, he'd talk to you. After I left the battalion, he was a Sergeant Major. I told him, "I really don't want to leave. I wish I could stay." We talked and he said, "I understand." He was blunt. He'll talk to you, "Hey, Master Sergeant, how the fuck are you?" He was that kind of guy. It was great. It was a guy that I could talk about anything for a long time. He would shoot the shit. He used to love hot sauce. So, every time I got a hold of a bottle, I would give it to him: Extreme, Texas Beach. He was physically fit. This guy was all that, tall guy, healthy, a PT monster. He was a Recon Marine, all that super stuff that less than 1% of the Marines get. Well, he's done it.

I got promoted at the end of my time, so the battalion didn't rate me anymore and I was forced to move on. He deployed with us, but he also got promoted to Gunny. So, the position fell automatically to Gunnery Sergeant Sullivan. Later, he became a Warrant Officer, and I lost touch with him. I haven't talked to him in years. I went to Fort Lee, Virginia for a leadership course in my MOS. So, I missed him. When I got back, they were gone and weren't on deployment. One day, I was looking at the plaque that we have from 2/4 and we had a staff sergeant named Stephen Parkinson. So, I gave him a call and he said, "You know, that Sergeant Major got killed yesterday." I felt like something swallowed me into the earth. I said, "What?" I closed the door of my office, and I got mad. I got angry. It really bothered me when I heard about his death. I talked to Sergeant Coleman at the time. He became a Sergeant Major later. He told me what happened. It was a bomb, a VBIED and then he bled to death. Let me tell you something that First Sergeant Elliot used to say, and I'm going to verbatim quote him, "Fucking Marines, they do some stupid shit, but we got to love them."

I met Chris Kyle once. He was introduced as a Delta Force guy, not as a Navy SEAL. When they told me he was Delta Force, I thought he was an Army guy. It was not until the movie came out that I started searching and said, "Isn't this the guy I met that night?" Yeah, he was cool, but I was told that they were Delta Force. It was just a handshake. "Hello, thank you. How are you doing?" It was very informal. It was fast. Remember, these guys came to do a job, not to say hello. During the movie, the actor is clean, but Kyle looked like he's been out there, so you could tell he was a working guy.

You know, a lot of people only do 20 years, just set their mind for it. I wanted to do 30 years. I was promoted to Master Sergeant in '05. Then in '06, I went to the doctor because I was feeling like I was dying. They found two tumors in my thyroid, so they removed it. I wanted to stay, but I started gaining weight. Four years later, I say, I gotta get out of here. This is not going to fly. It just destroyed me and took everything away from me. I was never able to PT the way I used to PT. It was career ending. I pushed for retirement and that was the end of it. I retired in 2011. I mean, the difference between now and then, I think that I'm not even 20% of the man that I used to be.

Master Gunnery Sergeant Mirarchi was another great guy, along with Staff Sergeant Layton. He was in Weapons Company. Those guys were the QRF, the quick reaction force. Every time there was a gunfight, those are the guys that go out there with the heavy guns and help. Sergeant Garcia was another one. Sergeant Garcia actually got to shoot a LAW or Javelin. I used to see Sergeant Conde every day until the day he died. It was sad when he passed. He got hit in the forehead. He'd been wounded in the shoulder prior to that. So, he was still going out there on patrol, and was on patrol the day he was killed. Everybody talked highly about that guy. Everybody. There were a couple guys - Cpl Munez, Cpl Fernandez. There was a guy named Crink. I knew all those guys from the ship. What happens when you're in the ship? You get Marines to help you in the facilities. You get to talk to them. Every day, because you got to task him or you got to find out what's going on, "Hey, I want you to stop what you're doing. I want you to do this." So, you get to interact with them. You get to know them. When we leave the ship and you see them back in the bay, "Hey, come here. What are you doing?" Then you see them in Iraq, and next thing you know, they've been killed. I remember two guys we lost. Savage got killed in Iraq. He actually worked for me in our section for a couple of weeks. Then, Otey who got killed on a roof with three other guys. Savage was married, and he had four kids, if I remember correctly. I believe it was an IED that killed Savage. The Marines knew him, and they were sad, "What the hell? What's going on?" "Savage got killed." I said, "What? No!" You think that you know these young guys will go back home, but not all of them did. Jeremiah was a kid, a young Marine, a good guy. I have nothing negative to say about him at all. Savage was the guy that did everything he was told, and that he needed to do. Destiny, I guess.

Like I said, one day that we got hit, about 25 mortars landed in the Motor Pool. They were aiming to get the largest concentration of Marines. I want to tell you something that I was wrong about, and I'm glad that I was wrong, but I was 100% sure that we were going to get attacked on the 4th of July. I was able to acquire barbecue, so we did a barbecue for the 4th of July. I told my Marines, we're going to get hit. It's Independence Day, we're going to get hit. Nothing. Not a round. Maybe they didn't realize the significance of that day to us. I figured they knew what the Fourth of July means to us. We were making barbecue, so that smell had to be all over town. Thank God that I was wrong and nothing happened that day.

HQ, H&S, Intel Marines

PFC Carlos Segovia, H&S Co. (promoted to LCpl)

Janney: Today's 22nd April, 2020. Why did you decide to enlist in the Marine Corps?

Segovia: I enlisted in the Marine Corps in February 2003 for various reasons, including patriotism, family, and the benefits the military offered. An older cousin served in the Marine Corps and I admired the brotherhood and connection within the Corps. I finished boot camp in September of 2003 and attended M1 school in Camp Lejeune. NC. I graduated from the school on November 27th, 2003, and was a supply admin, 3043 MOS. I was selected for a unit and checked into 2/4 Marines, CAT 7. I remember November 22nd, it was close to midnight. Got picked up by my Corporal. One of the first things my Corporal told me was, "Don't even bother. Don't get comfortable. We're going to get deployed soon. We're going to Iraq." So it was a nice welcome to the unit, kind of a pat on the back. Grabbed my suit bag, took me to my barracks, and that was it. It was just kind of getting ready for deployment a few weeks down the road, or two months, really, because we checked in November, and we didn't really leave until February.

From the very moment that I checked in, everything was all Iraq. My Lieutenant, my Staff Sergeants were all pretty communicative in regard to what we needed to do to get ready and what we were going to do and where we were going. Didn't really have any details in regard to when. I think we knew it was soon. To be honest, it was a blur. Crazy looking back at it now. 17, I think years now, it's unbelievable. There was a certain sense of urgency with everyone. Everyone was moving a little bit faster. I got there right during the holidays. I got there for Thanksgiving and was there for Christmas. There was a lot of, "Spend as much time as you can with your loved ones." In between we're training and getting ready. There was a lot of field training. There was no downtime. All our time was spent in preparation for this. Being a PFC, my head was spinning, trying to wrap my head around and just do what I'm told and do it as fast as possible. So, I didn't opt out, because usually after school, they give you the option to opt out and take leave, go visit your family. I didn't opt out for that. I just said, "Let me go straight to the unit, and let me save my days for something else." One or two Marines checked in with me. I remember when the rest of the Marines arrived. It was just busloads of PFCs. Being a part of H&S, the line Companies were trying to continue that training from boot camp and SOI because they had very few guys.

Very few guys that I remember and very few guys that were in the line Companies had any combat experience. So, there are a lot of first timers. Even if they were E-4, E-5 guys and they really had the experience of being in combat, they were trying to get the Lance Corporals and PFCs ready. We didn't know what the hell we were going to face. All day and night, they would train. I would hear the weapons inspections, drill, and room inspections. It was just day and night. Hey, you're a boot, you gotta understand how we do things. Not only in the fleet, and then obviously the preparation for deployment. I remember it was a little different for me because I was being mentored by the Corporal that was in charge of the admin side of supply. So, I would be pulled out of training. You have to understand how to do these systems. But, 2/4 overall was training. They're always in the field. I remember weapons constantly being in the field. Echo was always in the field. I remember doing the Battalion hikes, First Sergeant hikes. It was just nonstop. It was just a matter of making sure the conditioning was there. We're just going to do fire team rushes while we're waiting to do something else. Just anything that can help prepare. I remember seeing fire teams huddled up in one of the barracks rooms and they're going over plans. This is what we do. This is how we respond. There's just nothing but training and preparation. The details are a blur because I did three deployments with 2/4. My first deployment was Ramadi. Second appointment when we got back in 2005 was Okinawa. We were just kind of on standby because we never really knew if we were going to get sent to Iraq or Afghanistan. Then, the last deployment, we went back to Iraq in 2006-07.

When we deployed, we flew into Kuwait first. We went to Camp Wolverine in Kuwait, then to a place called Victory. We spent some time there before pushing up north to Iraq. Some Marines didn't have proper body armor. It was our job to make sure every Marine had SAPI plates. There were two versions: the original, heavier version and a lighter version. Some Marines only had Kevlar, which wouldn't stop much. Later on, we got better armor and ballistic gear. We arrived in Ramadi around early March. Our day-to-day experiences involved living in the palace on Hurricane Point. I was fortunate enough to be living inside the palace on H.P. Day-to-day for me, I remember there was a lot of filling up sandbags for fortification. I don't believe the previous unit did a good job, or maybe it's the standard the Marine Corps has, to make sure that the building was fortified, the chow hall was fortified, and anything that needed to be fortified was completed. So, there were a lot of sandbags that needed to be filled. As a PFC, I had the joy of doing that for hours on end, dealing with that with gear on. It was an adjustment to kind of ease off as time went on.

A lot of it was on guard duty. Being H&S, I would do my job, my logistical portion of my job during the day. I did a lot of convoys between the line companies and Blue Diamond, making sure that they got what they needed. Making ice runs towards the government center because in hot, 100-degree weather, guys sitting on top of the roof of the government center needed to cool off or at least needed some cold water. Picking up supplies, dropping off supplies, offloading the 7-ton trucks, because we didn't have a forklift for the longest time, so everything was just working parties. If you found a Marine who wasn't doing anything, you picked them up and said, "Hey, guess what? You're helping us offload this convoy of 7-ton trucks." A lot of working parties. I'm sure that's where a lot of bad backs are coming from. In the beginning, it was really scary when you heard the incoming mortars. Sadly, we kind of got numb to it. The guys got complacent. But, you would hear these mortars, and you'd do the routine, right? Hey, you hunker down in place. It's over. You've been hit, and things go back to normal. There were a couple of times that they hit the compound. Thankfully for us, no one got hurt. We had a couple of wounded, but nothing severe - people recovered from those inbound rockets and mortar rounds.

Day-to-day life was watch, guard, doing my job as a supply Marine, and being with the convoys. I did spend time with the non-convoys either as an A driver or gunner or just as the person that had to go represent my section so that I can get the signatures for making sure that we got what we needed and then running it to the right line Company and having Gunny or an NCO sign for it. You know, it's weird. You've been out there - you're either waiting for something to happen or you're just thankful nothing's happening. It was crazy. Convoys broke you out of the routine, right? Sometimes, volunteering for a convoy, that's when I knew I would get to go to Blue Diamond. People would be excited. They had a PX. They had the internet. It meant the ability to connect back to a little bit of home. Reality of the real world.

Marines who had volunteered told me, "Shit, I'll do a run. It's better than filling sandbags. It's better than being on post for six hours. Let's do a run, and let's see what happens." It was a little scary.

I wasn't a trained driver, but I remember being told to drive right up to the vehicle in front of you, and that was the scariest thing that I remember at that time. Because you wanted to keep the distance between you and the vehicle, but you obviously didn't want to keep too much distance, because then you're holding up the entire convoy, or somebody's going to cut in and get in between you. Driving through the market was to me, "Shit, you gotta be kidding me?" You got these kids, and people that feel like they're invincible because they can jaywalk right in front of a 7-ton or Humvee and think they're not going to get hit. But yeah, it was crazy. The convoys were stressful, but they were also a way to have purpose. I know that I was always being tasked out. I was the most junior Marine in my platoon, so I was always tasked out to go. Luckily, one of my buddies would always go with me and we would always watch our backs. The last Corporal looked out for me and kind of taught me the ropes. But, it was kind of a way to break out of the routine, and have purpose. Being able to make a call home, or maybe being able to do video or maybe even a webcam with your significant other, your girlfriend, and family members. Probably crazy, definitely scary and stressful. But I mean, sometimes it was the thing that you had to do or the thing that was necessary to make sure that everyone got what they needed.

I mean, going to Echo to drop off stuff was scary shit because you were going across town on Nova or Michigan. Like, shit, here we go. You pray to God that nothing happens. Just a few times shit happened, but nothing crazy. At least, I was fortunate, nothing crazy during one of my convoys. So, it was a blessing that way. I was just thinking about something like that just recently. I saw roadkill. It was like a cat or a coyote, something. It was kind of a larger animal and it was just in the middle of the road. We have three lane highways out here. I just saw the middle road and I completely not only dodged it, but I completely went out of my way to go into the emergency lane in front of things. I kind of overreacted, probably to the average person, or average driver. But it just kind of came second nature to me. You have no idea what that thing is. For a year after I got home, my ass would tighten up every time I saw trash on the side of the road or roadkill. We never got hit, but it was an ever-present threat.

Nothing really sunk in until we started seeing people getting wounded. We started seeing stuff happening and we started seeing people get wounded. You can see some of the vehicles because Motor T would pull these damaged vehicles right in front of our shop where we worked. There were no KIA up to that point. I think it was late March when we got that first KIA. Then, it was like, holy shit, this is real. It started sinking in. You start realizing that it gives you purpose, but do I really want to go? Do I really have to go? Shit, fine. I'm being told that I gotta go, so I'll follow orders. But yeah, on that first day, it really sunk in for everyone. This is not a game. You can't hit reset. Got to take it a lot more seriously. It was very serious before, but that just took it to another level. Now, you're hyper-vigilant. Trying to look at everything and anyone, the kids especially. Those are the ones that you need to know if you can trust them or not because they would get so close to you. You would walk to the market, or you would escort the convoy to the market.

The first KIA we had was a Marine from the Bay area, LCpl Andrew Dang. It was close to home for me because we're from the same area. He was killed on 22nd March 2004. A few days after he passed away, we processed his personal belongings in my shop. My buddy, Lance Corporal "Chain" was the guy in charge of processing everyone's personal equipment. We had to go through his personal property. So, we're going through his personal belongings – photos, journals, they have things like that. We're preparing to ship it back home. I was selected for his ceremony duty. We were going to do his 21 gun salute and I was selected as one of the Marines to be a part of that. It was just crazy because even though we were in the middle of combat, everything paused. All the senior officers and senior enlisted came by and they were part of a really quick ceremony. I knew people convoyed to Hurricane Point and paid their respects. Right after that, they said, it was back to work. We got shit to do and we got to take care of things.

I was a part of what happened in Ramadi, but didn't have the front line combat experiences that the line Companies had. I remember the attacks at the end of March and the beginning of April. I remember there being Bug Hunt ops when we would go after certain guys. The patrol people talked about capturing guys. If someone was detained and questioned, they were brought up to Hurricane Point. I remember that happening towards the end of March. We also had a couple of KIAs at the end of March. When people were captured, detained and brought in, there was a smaller room that would share a wall with us. There would be some interrogation that would happen back there. I know there were uniformed Marines and there were other guys that were civilians or just not in uniform that they would take back there and they would question them. You could hear the intensity in the room. And even though this guy was a bad guy, even though this guy was trying to hurt you or a fellow Marine, at some point, you thought this poor stiff is getting the shitty end of the stick and he deserves it. It was just crazy to hear some of the stuff that was happening, the questioning or the muffled sounds that would come out of that room.

I remember hearing about Sergeant Major's confirmed kill when he went out on patrol in the beginning of April. Very motivating to hear the fact that he's out there on the front lines in the trenches with us and doing everything that everyone else is out there doing. Not hiding behind his rank. Yeah, Sergeant Major Booker. He was a Recon Marine and a Scout Sniper, too. It's the same thing with all the special Marines to be able to do all that type of stuff. He was always looking for a fight. If you got a Sergeant Major on a convoy you knew shit was going to happen. But, at the same time, it was probably one of the most motivating things. It was kind of a double-edged sword because some people would roll their eyes and say, "Shit's going to happen." But, at the same time, with those same Marines, the Marines were also excited to see that high-ranking Marine join them in the same thing that they're doing.

Slowly things kind of led up to that first week in April where shit hit the fan. April 6th, April 7th. I was just craving it. Shit got bad. You had the Cobras and jets flying over. You had all the radios just going crazy with all the back and forth, the small arms fire. It was just nonstop. We were at Hurricane Point, and I remember just coming off a post. I had early guard duty that day. I was trying to fall asleep, and they woke me up, and I'm told, "Hey, did you hear? You're on QRF." We're about to go out, and we're going to figure out how to help. They didn't know what the hell was going on. Corporal was trying to keep me up to date with what was going on – Golf's out here, Weapons already went out. Weapons Company was our QRF, and they're asking for another QRF after they already went out. It's like, "Fuck, what's going on?" It was crazy. There was a lot of communication, but at my level as a Lance Corporal, it was just all rumors - he said, she said type situations. We were set to go when everything went crazy and we were just always on standby for those couple of days. We didn't really get sent out. There was an emergency group of a hundred at Hurricane Point. It was made up of the H&S Marines, but we weren't really ever sent out.

Because of everything crazy that happened, they needed Marines to man the government center. So, a few of those Marines that were in that QRF got pulled and were told, "Hey, you're going on a convoy. You're going to go and stand watch at the government center." Half the Marines got put on the rooftop. The other Marines were put downstairs and were hovering, making sure that we were watching the Iraqi police and making sure they were behaving. It was a few long days of little sleep, with lots of crazy things happening. Then after the fact, I remember Lieutenant Stafford was telling us that we had 12 KIAs that day. No one really knew everything. I think the senior enlisted officers told us the facts, but no one really told us how everything happened. None of that really got to us. We were running convoys to Echo Marines, Golf Marines, Fox Marines, and Weapons, and then we were told the craziness of what happened. How people were trapped, and people went in to try to help, and it just got worse and worse and worse. I just remember all the noises, and the visuals of the things that were happening, seeing smoke out in the distance over the wall. You think, "Holy shit, what's going on?"

I remember we had a sniper that got hurt and he was going back out. I think this was 7th April. He had his hand bandaged. He was brought in and got taken care of by a doctor or medical, and then he was on his way back to the fight. You could just tell that the snipers were a different breed. They had a different swagger than every other Marine. I remember the injured sniper was trying to catch a ride back into the action that day. I never saw him again. We always looked up to the snipers. They were definitely motivated. We would see them leave the wall, no convoy. Just two at a time, walking out into Ramadi. We thought, "Holy shit, you guys are fucking crazy." But they obviously were trained, knew what they were doing, and they were extremely good at it. I remember being on patrol and my First Sergeant Ellis was a prime example. He later became Sergeant Major Ellis. He led a patrol. He had us going around. We were at the government center, and I think he wanted to get a view of something, and we had to leave the protected walls that were outside the government center. He said, "I have to go here. Let's go." I thought, "Shit, man. This is serious shit. We're out in the open. You've got a bunch of rooftops, a bunch of windows. You've got Iraqis all over the place and they're clearly seeing what we're doing." That was scary. In the heart of the city, trying to keep up with First Sergeant Ellis as we're moving across the compound to get a video of something he wanted to see. I can't imagine these fights going out there in the entire community in a city of unknowns.

I'll never forget Sergeant Major Ellis. Best friend ever. Sergeant Major and I checked in at the 2/4 at the same time, or similar times, I guess you could say. He was fresh to H&S when I got there. I remember that because he was settling in when my Staff Sergeant took over to check in with the 1st Sergeant and met the 1st Sergeant. He took the time to kind of find out who I was, find out a little bit about me. Looking back at it now, it just seemed normal. You're a PFC checking in. You don't really question anything. You just do as you're told. But hearing other Marines and how they checked into the other units or they check in, it's just a process. You check in and you leave. You go do what you gotta do. But, First Sergeant Ellis just asked, "Where are you from? What do you do? Where's your family? Do they know you're okay?" Our other First Sergeant was different. Instead of going into his office, he always seemed grumpy, frustrated, irritated. He always had a comment to make sure that you were squared away. He would always give me shit about my posture or my uniform. It was just little nitpicky things. It always frustrated me. It always pissed me off because he would always give me shit. It was never shit in a bad way. I think it was just his way.

As time went on, especially being in H&S, I ended up dealing with a lot of senior enlisted. I worked with a lot of First Sergeants because we always had to meet up to sign some type of paperwork. Some would give me shit without me interrupting them. They always had these mannerisms. Their presence was very demanding. You always knew they were in the room and it wasn't just because of their rank. It was who they were and their posture. On our family night before deployment, Ellis ended up getting to meet my wife at the time and met my son. He only met him once. I introduced them, "Hey, this is my wife, uh, Martha. This is my son Javier." But, he didn't remember me yet. As time went on, he would always remember that. Ellis would always ask me, "How's your wife, Martha? How's your boy? Is your boy getting big?"

I got in trouble once or twice. Ended up going to his office. I lost my libo in the Indian port because I was wearing a white t-shirt. When we were on libo, I took off my polo that we were required to wear. We were required to wear a collared shirt. Sergeant Major caught me and told me to bring my Staff Sergeant and CO and see him in his office. He was extremely disappointed in me. This was almost a year and a half later after I had introduced my son and my wife at the time, and he still remembered me. He said, "I expect better from you. I would expect better for you and your family, your boy, Javier." I'm thinking, "Oh, shit, man. How does he remember all this? You don't remember every name of every Marine in his command, right?" I think one of the things that I will always remember that he told me was, and it was when I got in trouble that he told me, "You're an NCO. I expect better out of you. You should always be at your best possible. You should be identified as a Marine in or out of uniform because it's not the uniform that makes you, it's your character. It's your morals. You should walk into a room and people should be able to see you and know you're a Marine. It shouldn't be the haircut. It should be your demeanor, your mannerisms, the posture should be able to tell everyone in that room that you're a Marine. If you're a Marine under my command, I expect better." That part of what makes a Marine a Marine stuck with me. It was just how you carry yourself. That's what he expected out of his Marines. He was always hard on us, but he was always there. He gave us a lot of shit and he always wanted better out of us. Then, when he got promoted to Sergeant Major, it just made sense. He's the closest NCO to me because I knew him. He spoke and gave me attention, gave me energy. There were very few times I ever caught that he was human and wasn't just Marine Corps 24/7. He would give you some advice and he would tell you, "Don't fuck up your marriage. Don't let the Marine Corps fuck up your marriage. This is my second marriage. Don't let libo be the reason why you have to separate." That was really hard to hear. As I'm leaving for libo, he would pull me aside and say, "How's your family doing?" You're standing at attention and paying attention. He would say, "How's the boy? Do you guys communicate? You guys stay in touch? Write letters? Yeah? Okay. Don't be stupid. Don't go catch something and get divorced over some broad on Libo." I'd say, "Yes, 1st Sergeant. Understand, 1st Sergeant." He would just give you a glimpse of his wisdom, just a great exchange.

On my second deployment to Ramadi, we were definitely pissed to hear that shit after everything that we went through in Ramadi with 2/4. The way it was when we got there was frustrating. When I got deployed the second time to Iraq, I thought, "What the fuck? Are you shitting me? Are you serious? Why did we go through all that?" Because for it to be this way, the idea was that it got better. So, it was a little frustrating to hear the kind of stories that were coming out of there. We knew that we were going to get sent back to Iraq, but what are the odds that we get sent back to that same hellhole all over again? That would be the cruelest joke ever. Some Marines ended up going back and it was just, "Not this again." Especially, when we were getting shot at on post the second time. At least they had the ballistic windows to protect themselves and still be able to see, and armored Humvees on my second deployment. But, they were still getting pop shots at them. Very scary times.

You asked me about supplies in 2004. We didn't have what we needed when we needed it, but that's just logistics. So, we made do the best we could. My Lieutenant, Lieutenant Rodriguez, did an amazing job with everybody that he could to partner with like the Captain at S4 and with the board on Blue Diamond. Then, obviously trying to get things through Kuwait and from back stateside. At the time, being a junior Marine and hearing some of these requests was just crazy because we were going through wish lists for logistical supplies back home before we left. We would get requests for high end water filtration systems, but that was high end in 2004. Now, in 2020, you can get the tubes with filters built in and you can suck water through that, like a big straw. Back then, cost and availability were huge issues. It was managing the cost versus the request, and trying to get everyone everything that they needed. We did the best we could. I think we did a great job stateside. Once we got to Iraq, the supply chain was a lot more difficult to manage. We tried to get everyone everything they needed, but it was definitely challenging. It was a matter of trying to requisition things through the DOD systems and trying to get the lowest bidder. That wasn't always easy or fun.

We ordered the Marines some of those Under Armour shirts in 2004. This was before all this fancy workout gear was really popular. But, we got the Marines the Under Armour shirts, the protective goggles, and Nomex gloves. I remember the armor situation. That was crazy because I don't remember what exactly happened with the armor situation, but I remember we used to have more connections that we made with an Iraqi that we would partner with, and he would get us supplies from within Iraq, and from Ramadi when possible. So, at first, we were getting metal to be able to up armor our vehicles through a local Iraqi. It took just a briefcase full of cash to try to get this stuff purchased. I think we got investigated, and my superiors had to answer the questions from higher-end officers, from senior enlisted officers, or senior ranking officers, in regards to how much money they spent and why they spent it the way they did. I was never part of those questions, those meetings, those conversations. But, I remember my new Lt. stressing because they had to justify it. He said, "Hey, do we have documentation of paperwork or any kind of paper trail to explain this? This is the date that we made this purchase for this amount." It was kind of like he was making decisions by himself, but it was things that he had to make sure that he would be able to track and answer any questions he was asked about. The armor was the hardest because we were trying to make sure that as the deployment went on from March to April, May, June, we were getting some of our vehicles up armored. Some of these vehicles coming in were just mangled. Our Humvees were aluminum with fiberglass doors, so bullets just passed right through them. We would wash the freaking blood off the truck, try to repair it and get it back on the road. It was crazy, sad, and frustrating.

Another very challenging issue seemed simple, but was practically difficult to accomplish. We were told we had to issue Marine Corps boots to the Iraqi Army attached to the line Companies. So, our USMC boots had the EGA (Eagle, Globe, and Anchor) on the side of them. My Lieutenant said, "I don't give a damn who gets those boots, but they're not getting the EGA. Carve each one out of all boots. I don't want them with the EGA on the side. So, we would have to scuff them up with a K-Bar knife or cut out a little patch of leather that had the EGA on it. That was one of the crazier requisitions that we had to meet. Apparently, we had Iraqi volunteers, and we had to comply with the Kevlar and boots and things like that, try to teach them what we do and how we do things. I remember that was very frustrating for a lot of Marines, putting these guys in our uniforms or in our gear that they shouldn't just have. I think definitely the ballistic armor was probably one of the more difficult items to try to get, and then the race to have it put on the vehicles, and then try to salvage whatever vehicles we could. Once something happens to them, what parts can we take out of them to use again? Kind of make a junkyard and use any parts that were still salvageable.

As far as supplying Iraqi, we didn't give them Marine uniforms for security reasons, but I definitely remember the boots. I remember having to sit there and try to either burn the brand, the EPA off, or cut it off, or scratch it off. I do think that we did have to issue them cami uniforms, but not USMC issue. I don't recall, to be honest. But it was one of those situations: I don't want to do this, they don't deserve this. I've gone through all of this to wear what I wear. You guys are just volunteering. Plus, no one trusted them. No one believed that they were going to do what they were expected to do. There were a few good Iraqis, but not all of them.

We used to have people that would do our laundry on Hurricane Point. On Hurricane Point, we were fortunate. We had the phone center that we would be able to use to communicate from time to time as long as things weren't crazy. We had one or two laptops where you could send out emails. But, we also had like a small local PX, whether it be an Iraqi that would come over and sell DVDs or whatever or do laundry for us. It was really nice because we were hand washing our stuff in Kuwait. We'd get a bag and then basically turn in your old camis and they would get washed and they would bring it back to you. It was amazing being on Hurricane Point, having that blessing. It didn't last very long because apparently, I think this guy either got killed or got threatened. For some of the few that were doing some good for the Marine Corps, there were always some consequences for them. I remember hearing a lot of line Company guys talking about how some of the guys that were attached didn't know how to do much of anything. They would just flag someone with a rifle. Just do something stupid that could possibly get someone hurt. Actually, if I remember correctly, I think someone did get hurt from an Iraqi volunteer in Fox Company from a negligent discharge. After that happened, things were really tense after that, not to mention all the IEDs.

You look back at all some of the close calls that you've had where you went a step left or right and something happens. There were some cases where a Marine got out of the shitter, getting ready, and walked away. You hear the incoming, you go for cover, and then the whole shitter gets blown up or slightly destroyed. One of the Marines in my platoon of supply Marines was Lance Corporal Cadena, called Change. He ended up getting hit with shrapnel on a rooftop. A rocket went off, hit the rooftop and he got some shrapnel in his arm. We always give him a hard time because it was a minor cut. It wasn't anything to joke about, but it was a minor cut and he survived. He still has shrapnel in him now. He's got a Purple Heart. But, it goes to show you how a small decision can change your life in the blink of an eye.

I'm a very proud military veteran. I work for a major cell phone company and they just asked me to share something for Military Appreciation Month in May. I'm that token military guy in my company in my area, Northern California. I'm always volunteering, and say, "Hey, yeah, count on me. I'll share my thoughts on this." Some Americans have short memories and we need to share our stories. People are always appreciated in a crisis. Right now, things are crazy with Covid starting up. Now, doctors and nurses are heroes. But, give it a few months, things kind of settle down, and people are going to take it for granted. It's just another doctor or nurse, right? The fact is that we need to make sure that we do everything we can to keep people aware of our service so they don't forget. Just ongoing appreciation and recognition. Because that's what makes it rewarding to me.

HQ, H&S, INTEL MARINES

SSGT. JAMIE BUNETTO, HQ-INTEL

Janney: Today is the 5th of May, 2020. Tell me why you decided to enlist in the Marine Corps?

Bunetto: I was 17, the youngest of four kids. The rule was you had to be out of the house at age 18. That was just the way it was with my parents. We didn't have money for college. I needed a place to live. I joined the military so I'd have a place to live and make money, a decent paycheck.

Janney: Did you ever have anybody else in your family that served in the military?

Bunetto: My father did about four years in the Air Force. I think my grandfather was in the Navy. But, no one had made a career out of it. That's the way it was. I've been working since I was 13 years old, and I knew when I was 18, I'd leave the house. I didn't have a wife or girlfriend at the time, didn't have a house, and didn't have a full-time job. So, the smartest thing to do was join the military.

Janney: Tell me when you enlisted. What month and year was that and what was your MOS?

Bunetto: It was August 20th of 1989. My original job was a diesel mechanic. I only did that for the first nine years. I had a break in service when my stepson at the time was going to a very, very good high school in Lubbock, Texas. He was on the magnet program and my father-in-law had just been diagnosed with cancer and he lived right down the street from us. I just had a reconstruction surgery done. So at my nine year mark, I had been Staff Sergeant Select and I had to make a decision whether I was going to try to press on with my physical condition or put my son out of school at the time or leave my in-laws. So, it turns out I ended up getting out of the service and I became a cop in Lubbock, Texas. Then, over the next four and a half years, all four of my parents ended up dying of cancer, right after the other. So it was my father-in-law, then my mother, then my father, and then my mother-in-law in a four and a half year time period. So, by that time, about two and a half years into the police department, which was about 2001, I had started getting the itch to go back to the military. My kid was getting ready to graduate high school. I went ahead and enlisted back up in 2001, but it took two years for my paperwork to get through, and I went back in in 2003.

That's when I changed to the Intelligence MOS. It was tough getting back in the Corps. I had reached out to quite a few people that I knew, and there was a First Sergeant I knew that was still in that assisted me. It wasn't just the paperwork processing. It just took time. They originally came back and said that I was not going to be able to enlist. I made peace with it and said, at least I tried. I had stayed as a police officer. I joined the SWAT team and I started getting on to that. Then, a message came out of Headquarters Marine Corps saying my enlistment was approved. In my first enlistment, I was in the Gulf War, so I got to experience a little bit of the process. Yeah, Desert Storm, Desert Shield. Yeah, we went all the way up into Kuwait and were there for about three and a half months. We were the first motorized unit in support of the troops that pushed straight through the minefields, went right up into Kuwait. The only real combat we saw was some artillery fire. I was 19 years old.

I think the biggest impact that experience had on me was at 19 years old, having to fill out a Will. That right there was a big changing point in my life as far as perspective goes. Oh, sure, you're basically pledging your life to when you sign up in the first place. Now, you're sitting down with the legal side of the house and we had to fill out a Last Will & Testament. I had nothing to leave behind. I had never even thought about dying before really. I was just 19 years old. Just that perspective, it changed a lot of my viewpoints. I was very fortunate. I was a Staff Sergeant Select in November of '98, and I got out in January of '99. So when I came back in, I didn't know what it was going to be. Most people that came back in, they started back out as a Lance Corporal.

But, in October of 2003, which was like a month or two before I was accepted back in, Headquarters Marine Corps came out with a message saying that whatever rank you get out as, you can come back in as. So, since I was a Staff Sergeant Select, I came right back in and I was Staff Sergeant. I was very fortunate and then I was fast-tracked all the way there. Less than 1% of Marines during their promotions get pulled out of a below zone. In the Marine Corps, there's a below zone, an end zone, and an above zone for promotions. They base it off your MOS, how long you've been in, your experience and all the different points systems. You have to fall within your time of grade; what your previous grade was and the amount of time that you have in the service based on a certain number. That puts you at a time when the Marine Corps says, Hey, we're going to look at all Staff Sergeants from this data rank to this data rank with this many years in the military, and only those are eligible, and they will fall in these three zones, below, end zone, and above zone. You can only get promoted out of the end zone unless you get pulled out of the below zone. I was pulled out of the below zone for both Gunny and Master Sergeant, which, not to toot my own horn, but it's unheard of. Most of them don't even get pulled out of below zone. But, just because of my career path that I chose with Intel, I think that was a lot of the stuff I did. I think that was a huge part of why. But yeah, I came in as a Staff Sergeant, and by 12 years, I was a Gunny.

In 2003, I started out as a Staff Sergeant, and I stayed a Staff Sergeant for three years before I picked up Gunny. I actually had to go to MOS school again, because I went to Intel then. Then, I went to VMA 214, the Black Sheep. I'm sure you've heard of the pilot in the show "Baba Black Sheep." That squadron was my first unit in Yuma, Arizona. I was only there for a couple months until the spot in Ramadi opened up and I volunteered for that deployment. 2/4 was in California, so I was not part of 2/4 when they were there. They were looking for Intel Marines. So, I volunteered and I was sent over to them in January of 2004. This was within two and a half, three weeks of their Ramadi deployment. They beefed up the Intel shop, but they didn't tell us what they wanted us to do with them. There was a big push for training 0311s, specific 0311s, in different intel capabilities, teaching them how to ask questions, what to look for, how to analyze stuff. So, during that big push is also when they were beefing up the Intel shops. I just happened to be part of that group when they were beefing them up. We put one down in each Company, which brought its own pros and cons. Because there was no internet connectivity at that time, the only way we could pass information was through convoys. Then, the material itself was secret, so it had to be done by the Intel Marines. Because at the time, not all 0311s were eligible for classified information. The only thing that I could really do when I got assigned to them was just assist with the embarkation, which was inventory and everything, packing everything up, getting all the security rosters, getting our gear.

I had no interaction really with the Marines other than daily when we showed up and did a little meeting. I didn't know any of the 03s. I didn't know the Company Gunnys. I didn't know anybody. And again, I had a brand-new MOS. So I was in a unique and admittedly a very scary situation. As a Staff Sergeant in the Marine Corps, you're supposed to be the expert in your job. I was a very humble person, and I used my Intel Marines to teach me. It worked. They were great kids. But, it was a whole new thing as far as having Intel Marines in the Companies. We obviously had the Intel Marines at the shop. We had enough to push down one per Company, so I had three or four in my shop there in the command operations center which is headquarters. Intel is a nonstop 24 hour process. You have to get it as soon as you can, get the information, process it, analyze it, make your assessments, and you have to disseminate it. That process was hindered quite a bit when you had to wait for the next convoy, which is usually the next day. The Intel Marines are so dedicated. The only way they could really get the information, the true information, was going out on foot patrols with the 0311s. So, the 03s at the Company level would send a squad out to a patrol for three or four hours. I'd come back. The next squad would then go out. My Intel Marines would go around the clock. They would go out on patrol, come back, try to get some notes done and get on the radio and pass the information up, and then they'd go right back out on the next patrol. So, there were a lot of headaches. What we started doing was, even though we had the Intel Marines down there at each Company, when there's a big operation, we initiated a collection for processing detainees and collecting wounded and gathering information. We would set up out in the field somewhere, in a designated area with some Intel Marines, some Corpsmen, and some headquarters element guards. We would have a vehicle or two, one for getting detainees and one for throwing the wounded in. Then, we would be the ones transporting them to the sick bays. So, then both 03s could stay out there and fight. That was something my shop did. It was something I pushed for, trying to form a collection point and that ended up working. A lot of time was wasted on this though.

You can't just send a wounded Marine back alone. You had to send the whole Humvee with guards. So, you take four or five people off the streets every time you have one wounded. Then, when you have 12 to 15 detainees, you're looking at almost a squad of Marines to bring them back to process them. So we started processing them out there in the field, if you will. My guys were able to do that. It was really weird because when you process a detainee, it's almost just like a civilian in America. We ask them questions. We have them fill out their names, dates of birth, their ages, where they come from. We have to get a statement from the Marines that actually captured them, just like an affidavit. Just like processing a detainee here in a local jail. It's very time consuming.

We usually had one interpreter if we were lucky. We had one per group. It was a process. We weren't the priority. But, if we went out there and set up somewhere, we usually had an interpreter with us. Intel is just so time critical. They would get Intel that this guy is a bad guy from some other person. You had to act on that quickly. One of the things that I pushed on was, without saying I was violating the laws, I was very open when I had a patrol come up and we had intelligence to pass to the Intel Marines and the Intel Marine wasn't there. I would find the squad leader and presume he had a clearance, give him the information before they hit the pit and went on patrol. So, things like that were done. We have a caveat, kind of like a need to know deal. I thought it was ridiculous that the Marines going out on patrols didn't have their security clearances, so they couldn't get all the information we had. So a lot of fine lines were probably crossed. But, it was that cooperation, and this human exploitation. The Human Exploitation Team (HET) - the best way to describe it is when you watch those movies with Jack Ryan, and those guys are dressed as civilians and they mingle amongst the crowd. Those are the human exploitation teams, the HET team. Those are a part of the Intel shop, and those are the ones who do interrogations. They're the ones who have an interpreter with them at all times. Most of them know the local language. They're the ones that go out there. They're kind of like the narcotics division of a police department. The narcotics and gang teams, if you will. They know who's who. At least they know the neighborhoods. They know the key players. They're the ones who work the sources, just like you would a CI in civilian life. So they had quick access, and they had ways to get information to and from local areas. So, that's a huge part of it.

The other part was the Intel Marines, we would stay up 24, 36 hours at a time to make sure the information was getting out. Across the river from us, the Headquarters Marine Corps put up what's called a TSC. This was a new thing. It was a tactical fusion center. It was a group of Intel Marines who assisted with processing Intelligence or information from all the regions in the area. They would try to send any information down to the local Companies. So, there was kind of a hybrid headquarters of Intel, but again, very little internet access. There was a lot of communication. We've had to take a convoy over to them to get them the information. There was something on the internet, but it was just limited. But, another big part of a big change in the Intelligence doctrine was having a tactical fusion center set up. So, that assisted as well, because you understand the AO that we worked in is the only Intel that we had.

All the Marine units and Army units had their own areas of operations. They had their own information and Intelligence process. But, there was no link between each other when I was there in 2004. It hadn't gotten to that level yet. So, we were getting things on the line with that, too. So, it made a whole new era for Intelligence in the Marine Corps. It's a lot smoother now, obviously. Comms were always an issue. The problem with the radios is you got that one who's on the Company level. He'll go on patrol for three hours and then he'll come back and he's only got 10 to 10 minutes for the next patrol or he's gonna crash out for a couple hours, so he can make it to the next patrol. So, unless he had time to get that information up on the radio, it may have gone eight to ten hours before we got information from one of the patrols.

God bless my Marines, my Intel Marines. I had to tell them they need to limit the patrols they go on and start training the 0311s that go on the patrols, what to look for, ask the questions, and do debriefings with them. Because the information that comes up from the Intel Marine is just rapid notes from a patrol that he took when he's in the middle of a firefight. He's not assessing much. He's trying to survive a firefight. So, when he comes back, his jotted notes are going off of memory. It's not very good information getting up to us. So, it was a hard balance for them to learn. To sort out false information from one sheik about another sheik or person, we had to concern ourselves with the Iraqi police. We knew that eventually we were going to have to use the Iraqi police to police their own. That was a decision we had to make. It was assessed and pushed forward. Colonel Kennedy, at the time, was forward as well. I want to say there was about 3,700 Iraqi police officers in the area. To begin working with them, the push that I did was, there's something called the BAT system which was a biometrics system. It's iris scanning of the eyes, fingerprinting, and a photograph.

You put notes in and that goes into a database. That was something that was coming out during that time from the Marine Corps, so we can start getting a database. I think it started through the identity information through what Homeland Security was putting together. What I suggested is we go to the police departments and we tell everybody who's a police officer that if they want to remain a police officer, they had to go through this biometrics database. I want to say we cut the line because over 50% of the officers never showed up. So, we started with the pool of half of the crooked officers that were out there. We just got rid of them because they never showed up because they didn't want to get a database done on them, so obviously they were crooked. From there, it was our HET team that did a lot of the vetting with the local sheiks. They would check their sources, they would do their thing, and they would make assessments on who would be the best to stick around or who they could work with. Who would benefit us and still be able to maintain control of the Iraqi police. So, that was another way of weeding them out. Then, through other intelligence, when you detain someone and you find out who they're working for. If we did a raid and there were six or seven Iraqi police officers in that house and there was a weapons cache found or anything like that, then those Iraqi police officers would be fired. So, it was a nonstop process of filtering information from the sheiks and the police department.

I still remember the very first Marine that was killed. I had to wake up SgtMaj Booker at 0230 and let him know. PFC Lee was the first one. I wanted to be perfect with intel because it was my job to get the information out there to protect them. So, I lost 36 pounds on that deployment. That was a rough one. Weeding out false intelligence was always tough in the intel world. If you get good intelligence out there and the 0311s kick down a door and they catch a bad guy, then the 0311s are the heroes. If we give the same information and there's no bad guy caught, the Intel Marines are the bad guys - we suck. So, it's a tough game. You do your best to try to filter out and assess with credibility what sources are accurate. Even if it was accurate, it doesn't mean it's going to be 100% foolproof because things change so rapidly. That was hard on the Intel Marines. It was hard on us because when there were dry holes, we knew we put the 03s in danger, sometimes for nothing. It was hard on them because they would go through the stress of the whole patrol thinking they were going to catch this bad guy, and when there's no one there, we just tried to minimize that the best we could. It's the intel game. You use what information you have and try to make the best of it. We knew that when we gave intel to the Company Commanders and Sergeant Majors, they were going to take it as valuable. So, we had to strive to make sure the intel was as perfect as possible.

Another issue we had to factor in was that even the "good" Iraqis could turn on us. They had good intentions and wanted to work with us, but their families are being held hostage, kidnapped, with a gun to their head. The "bad" guys would tell the "good" guys that if you don't give us this information or if you don't assist us, we're going to kill your family. Even the best intentions of the Iraqi soldiers or Iraqi police that wanted to help us couldn't be counted on because you never knew. FYI, I was stationed at Hurricane Point. I lived in the electric room outside. Strangely enough, there was a 120 millimeter mortar that hit the wall right behind me when I was sitting there after I got off work. Smoke came from that mortar and there was yellow powder in it. Those of us that were hit by the smoke actually had some effects. Our skin started burning and we were coughing and couldn't breathe. It had the same effects of CS gas. We put our gas masks on. Our NBC (nuclear biological chemical) warfare representative did a testing of it. After about a few months, it came back and they said it was just a CS round. But, I still got my medical records that show I was hit with a gas that causes burning sensations on the skin and coughing. It's the anticipation. That's what I learned in the Gulf War - the anticipation of death is worse than death itself.

We also had intel that foreign fighters were fighting us in Ramadi. We had vehicle-borne IEDs. Some of the materials that were used or the types of IEDs that were used, were able to link to different regions. As far as EFPs (explosively formed penetrators), we had reports that those came from either Iran or other areas. We didn't see a lot of that in Anbar in 2004. One thing we discovered is that there's a "signature" of the bomb maker or type of IED in that area. One of the processes that we used is to find out the type of wiring they used, and the type of explosives they used. Then, a month later, we see that same type a good 20, 30 miles west of that area. Then another month and you see that same type 20, 30 miles to the west of that area. So it's either: A) One person is going through the whole country and setting these IEDs. B) One person or group is going to this area and instructing and teaching and training insurgents in that area. Then they move on, you know. You can watch the trends. That's one of the things we did is we would say, two months ago there was a new type of IED, let's say an EFP. Based on trends, we can assess that it's going to be hitting our AO within the next 30 days. So, we can prepare ourselves and change our tactics on what to do, whether we're driving slow down the streets or whether we're just going to haul ass. Because if it's a remote detonated one, then you haul ass. You drive as fast as you can from point A to point B and hope that the spotter or the guy watching the convoy misjudges your speed. When it's a pressure plate, you're going to have slow convoys because you want to keep your eyes open and look for signs and signatures of an IED on the road. So, those tactics that the enemy were using is what we used to assess and tell our commanding officers, "Hey, when they're doing convoys, they need to start slowing down now because we're starting to see a trend of pressure plate IEDs." You know, command detonated versus pressure plate. You got hard wire and you got telephone. With the telephone ones, there's no wires there. With the wired ones, they push a button. And in other words, there's got to be someone there watching it close by, depending on how long those wires are going to be on a real open road. You're not going to see someone sitting off the side of the road. They're going to have a telephone most likely. So, if it's in a huge town like Ramadi, you could imagine a black plastic bag on the side of the road could have explosives in it. Those wires could be running right up the post and into the house and you'd never know it. But, those are the things, the trends we watch for, the types of things we do when you gather information on explosives or from detainees. If I find materials at a house that match the description of IEDs that we've captured or we've located or have been hit with, then we could use that for processing that detainee. We probably processed 2,000 or 3,000 detainees. My Intel shop did that.

Colonel Oliver North came to visit us in Ramadi. I actually kicked him out of my operations center. I got my ass chewed by Major Wiley because of that. I knew how North was, and he came walking in there with a camera and I said, "We have secret material out here." I knew that North knows what not to put out. But, he just walked in and was just looking for a story or something. It wasn't anything personal, but as soon as he walked in, I said, "Wait a minute, sir. Get out of here." But yeah, I got my ass chewed for that. We talked quite a bit because he was Intel as well. There was another author that was embedded with us for a while, too. Ben King? I can't remember his name. I think his son was one of the Company Commanders.

You and I talked a little about foreign fighters coming to Ramadi from Fallujah. It was just natural that if there's going to be a huge push over there, they're going to come our way. We knew that anytime there's something big happening in Fallujah, one of our pushes from the Intel shop would be getting information out there saying, "Hey, start watching local traffic and see what they do, because local traffic will clear out of the way for foreign people. The streets will start getting quiet as far as going indoors or there'll be an increase in traffic as people try to leave the city. air that, we're just watching lines of communication, which was the roads and whatnot. That was something that we assessed. A lot of it was from that. Just the detainees that we captured. Some of them were still alive, but there were a lot of dead. One of the things Intel Shop did is we processed the dead bodies, too, when we could. You can tell by looking at someone whether or not they're from the area or not. So, there were definitely different foreigners in our AO. There's different ethnic types other than just the people in Anbar.

They had their own ideologies, and there's some that just straight up needed the money. They got paid by the bad guys to cause shit. Not everyone we fought were necessarily terrorists, mujahideen or Al-Qaeda. Some of them were just local people that were paid money to do stuff. As far as combat strategy, coordination and control, there were definitely people in charge that were using their own intel. There was planning and foresight involved. One of the mistakes the U.S. made was that we presumed there was going to be no resistance, we're going to say how stupid they are and how weak they are, and our forces were superior, so there was no need to worry about an insurgency. Not to say that that was our fault when we walked into an ambush, but I think we underestimated the tactics and capabilities of the enemy. They just weren't very enhanced, and it wasn't very broad. But, there were groups out there that had tactics, and that's the other part that I thought was there was some foreign influence because of the tactics that they used. I think they had the foresight to set up out there and they knew what our reactions were going to be in the situation with the Gypsum and Nova ambush. We tried to change our reactions. We tried to do that all the time. But, you can't do it every time, especially when it comes to a QRF, because you're going to take the fastest route there. It just could have been one of those days where they had enough people in the right place at the right time. They drew us out and were waiting on us. There was another time where they did this at the school down south. They shot up the school and the kids out in the front because they knew that we'd respond and then they ambushed us.

They'd use donkeys and bicycles for IEDs. Anything their imagination could come up with, they would do. You mentioned that IED on Nova being discovered by the Echo squad on 6th April. It might have been a time constraint or whatever, but it was a well concealed IED. However, they left the wires visible. I don't remember the details of that IED, because sometimes the wires had to be exposed. If it was going to be remotely detonated, it still had wires sometimes, like an antenna. It could have been one of those, or it could have just been it wasn't very well buried because of time constraints. Our battalion initiated something called satellite patrols, not just going out from local units. The guys would tell me when they went on the patrols that they would take off running. The whole patrol would take off running one direction and then stop and go a different route just to stir up or make someone think something was going on in one area. If there was anything over there, they might move or adjust their direction. So, the patrols weren't random, but they had tactics to make them less predictable. The main thing is, I don't think it would have been above them to set up an ambush. After we left there, or it might have been while we were still there, they found either a school or a hospital that was being constructed while Marines were there and they found IEDs inside the walls of the building. While they were building it, they emplaced them.

A big thing that I remember when we first started getting there and took over the AO from the Army was the Army's tactics. I remember going on some convoys with the Army. The soldiers were telling us about how at that time there was a curfew, what their use of force would be, and their no warning shot policy - they wouldn't warn. As soon as they saw someone outside of the curfew, they'd just shoot at them. So, when we took over that AO, the people did not like the US. Maybe it was a necessity at the time by the Army because it was the early months, but it was a very harsh time for the civilians, even the good ones, because the Army used such heavy handed tactics. We had to learn the whole hearts and minds deal, which a lot of people laughed at and joked about, but there's a lot to be said about winning the hearts and minds. It's been proven, but we had to get the people's trust. We had to work with the people in order to get more information and to settle in and make a more peaceful environment. Eventually, it worked out. Not so much for our Battalion, but I think we laid the groundwork for it. The atmosphere was very charged, I guess. It felt like we were an occupying force. If I was a civilian there, I would get the sense that the Army was occupying, the US was occupying us. I don't think that was necessarily the intent of the Army, but I think because of the way they did things and that there wasn't anything established yet, it made the locals dislike and distrust us. It was all brand new to everyone. Because the Army had just gotten that AO and it was a bad AO. It was harsh. So, we spent our first few months trying to calm that aspect.

I think the overall mindset was to rebuild. I think when we got there, we had that intent and we knew the long run was going to be to take charge, weed out the bad guys, and set up local law enforcement. This is one of the reasons Colonel Kennedy had chosen me when we were back in Pendleton. There were several Intel Marines that had volunteered for different deployments. When he found out that I was a police officer, he picked me by hand. I think he had that foresight of knowing that we were going to eventually have to work with the law enforcement and help set up a local force. So even when we got there, it was unsure how we were going to handle things. I think the intent was always going to be to rebuild and work with the community. In the early months, it was working. It's just that it's hard to stay that way when you're getting big attacks, when you're losing six or seven Marines in one ambush. It's a lot of discipline for those young Marines to understand at that time that you gotta fight when you need to, but there was actually a bigger picture. I really think they exercised amazing restraint. I really do. They were totally following ROE (rules of engagement) that were pretty tight. It changed quite a bit depending on what was happening, too. It wasn't largely changed, but it changed often.

It was little things, but they made a big difference. We were just talking about restraint. I remember the Marines bringing in some detainees. A Corpsman was killed, and the Marines raided a house and found the Corpsman's materials, scissors, and his medic bag, on this guy. They arrested him and brought him into us, and that in itself showed restraint. He's innocent until proven guilty. But, on the battlefield, when your Corpsman gets killed, and then hours later, you find his property in the possession of somebody, you wouldn't expect that person to still be alive. So, a combatant had Doc Mendez's bag that they stole on Route Gypsum, and yet the Marines brought him in alive.

We also spoke earlier about Parker's team being killed on the roof of the OP house on 21st June. The scout snipers fall into the S2 shop, the Intel shop's purview. They use them for reconnaissance and that intel is invaluable to us – recon is actually the biggest part of a scout sniper's job. They can use their scopes while doing over watch from a high position so they can give us information. They're actually a collection tool. After the team was killed, there was some discussion about using that house as a fixed OP 24/7 and if anyone was to blame. There's two sides to using the same place. The bad part of it, yeah, someone's gonna know you're there. The good side is the defenses are already set up and it's a familiar place to the friendlies. Typically, it's a more fortified and in a lot of ways, a secure place. I'm not gonna second guess them because I don't know exactly what happened. When the reports came in, I went and looked at the scene. It looked like they were all sleeping except for possibly one of them. So, if they were set up as a regular watch and there was always somebody there awake and ready to go, maybe it wouldn't have been as devastating. Also, some communication check-ins were missed. I think the radio check-ins were supposed to be every 30 minutes, and it went like almost 45, 50 minutes before they did a radio check that was missed or something. Again, we're guessing if that would have stopped it or not. I don't know. Was the one person that was staying awake and doing the watch, did he get up to take a piss first or something? I don't know. As far as it being the same place, there are some pros and cons to that. But, we also set up at the government center pretty much the whole time we were there - we had an over watch there, too. But, the government center was not an isolated position. It's fortified. There's a fence around it and all that stuff. Were the snipers using claymores for protection? I want to say they had some claymores set up for protection, but then a message came down from the national level saying that we could not use claymores or any type of explosive device in that situation. But, that was something that originally was used for protecting our snipers. They would set up like booby traps. You mentioned one of the Marines had an interaction on 20 June with one of the Iraqis on the roof of the OP house – the Iraqi brought them ice, you said. Allowing them up on the roof is complacency of the Marines and not an issue of being in the same place. You know, if they had set downstairs or claymores or something set, then yeah. I mean, you know, they would have had a warning. They would not have been surprised because I think those guys were just so used to the guys popping out on the rooftop, "Hey, how are you doing?" On 21 June, they didn't say, "Hey!" They just murdered the Marines. These brothers could have been super U.S. supporters, but the night before, their wife could have been kidnapped. You know? There's nothing that's going to change that, you know? The things I've read said that those guys were kind of skeevy anyway, but who knows? Booker said that we got solid intel that the brothers "okayed" that. I remember hearing something about that. You know, the brothers "okayed" that, but basically said, you can kill them, but don't tear up my house. They left the house. Booker said that they gave them an opportunity to turn themselves in and they didn't. They fled. I think I was in Iraq in '06 when a 3/5 Marine sniper team recovered that sniper rifle. That's an amazing thing, and I'm so glad that they did. I wasn't part of the ones that caught them, but I was in Iraq in '06 that year when it was found. I was with the military instruction team. I was a mid-eight team. We were just north of the Euphrates, north of Habbaniyah. There were about seven houses surrounded by some Hesco barriers. We just lived out there. We trained Iraqi soldiers, went out and did patrols with Iraqi soldiers, taught them tactics and all that good stuff. So, you heard there were some 3/5 snipers who saw a couple guys in a car glassing the base. They saw a stocked rifle and killed the guys in the car and recovered the rifle. But anyway, I'm glad they did. That's kind of crazy because one of the guys you interviewed, that was actually his rifle because he swapped it out with Tommy Parker every day because Tommy had a M40A1 and this guy had a M40A3 with a PVS-10, the day-night scope, and he wanted Parker to have the best advantage. He got a lot of crap for swapping his rifle out and losing it in that incident. I don't think he lost rank or anything, but it was a big, hairy deal, you know? When it comes to weapons, there's investigations all the time. That's just one of those deals where it's not the time for it.

There were a couple of times where I remember there were some big battles where the headquarters units had to leave Hurricane Point, go out and set up perimeters. We went to the government center and replaced the 03s. The one that we got, we set up at a detention, a collection point at the soccer stadium. We had set up a collection point there. We had seven or eight 120 millimeter mortars hit us. There were 18 of us out there. Right after that, there was a firefight. I think that Company was right to our north. No one got seriously injured. But that was another big one. I think it was Wagon Wheels, the operation where they were doing.

I'll talk about the Marines that died crossing the river - Green and Schrage. I didn't agree with doing it like that because I knew that the water temperature was very low. At Hurricane Point, obviously, we had access to the water. I had checked the temperature of the water and it was at 48 or 49 degrees F. I researched it and found out that hypothermia sets in within 3 to 5 minutes at that temperature. So, I was against swimming across the river in the first place, but that's what they were trying to do. I'm not sure if they had wetsuits or not. It was a weird situation because they weren't found until later. Either the next day or the day after, there was actually writing found in the sand. I can't remember what it was, but it was letters ABC or something like that were written down. So, we thought that we found evidence that they were still alive because the search went on for a couple of days. They found out that they never even made it out of the water. So, it wasn't them, but it was just a weird situation. I don't remember the tactical issue, or why they even were trying to get out to the island. I can't remember if it was to just set up an OP or what they were trying to do. I think it was eventually going to be a raid, but I don't remember all the details of why they were doing it. It was the Marines that were out there, and that was just one of the things that Fox Company did as a boat Company and as raiders. They were trained in the water. So, I hold no blame for the commanding officer. It's an asset that they had. You use your assets. I read somewhere that they suspected they were hiding weapons on the island or something like that. I just know that Booker said they actually dropped the level of the river ridiculously low just so they could try to find those guys, make sure that they've recovered their bodies, and they did recover them.

As far as knowing any of the guys that were killed during our deployment, it's a unique deal for me and I had to go through this twice. As I said, I was only there for two weeks before we deployed. The day after I got back, I went to help set up the shop. I was there with one of my Intel Marines for about two weeks prior to the rest of the battalion getting home. As soon as they got home, I was gone. So, I never saw any of them again. I recently linked up with one of my Intel Marines from out there who's in the Dallas Fort Worth area. He went out on a lot of the patrols with Echo Company. He was wounded out there in that big battle. Other than working with them side-by-side, I really didn't know many of them - that was hard.

After 2/4, I got attached to a unit out of California, went and did a year-long assignment, came back and went to Union Arizona to an air wing. It's not a grunt unit, so there's very few combat Marines on that station. So, it was a tough battle for me when I returned home because I had no one to share with, no one to talk to. I just found out this past year they were doing reunions. But, they didn't know me either. I'm going to attend some reunions and try to get to know them better, and talk about those shared experiences. I just had a buddy of mine, one of my Intel Marines out here who was telling me about it. I linked up with a couple of them on Facebook. The Marines who were actually in my shop I've hooked up with. But as far as the 03s and all them, I didn't get to know them. A couple of Staff Sergeants and Gunnys knew me because I was the one that was going out and giving them intel briefs before the patrols. But, I didn't get to know them personally.

My hit team is what got me out in the field on another Iraq deployment in January 06 or 07. I got over 500 miles on foot patrols that year. I went on over 298 patrols. I saw a lot more action during that deployment. It was in the same area, just west of Ramadi.

You mentioned that Major General Kelly escorted you and John Wroblewski to Rt Gypsum on 6 Mar 2008. Back in 1995, my wife met (then) LtCol Kelly at the Subway on base. She was waiting in line and she had a black eye. The lady behind the counter knew her from being a regular customer and said, "What happened? My wife said, "Oh, my husband did it." You know, joking around. I did it, but we were sleeping. I rolled over and my elbow hit her in the eye and it gave her black eye. After she explained what happened, they laughed. Well, a few minutes later, my wife gets her sandwich. A gentleman approaches her and taps her on the shoulder and says, "Hey, I'm Lieutenant Colonel Kelly. Can I talk to you for a minute?" She said, "Well, sure." They're talking and he said, "I understand your husband did that to your face. She said, "Oh my God, not like that." She told him the whole story. Nothing happened out of it, obviously, but that's how she met him. She had a black eye and Kelly was thinking that I had punched her in the eye. But yeah, it was kind of funny. Later, he promoted me to Sergeant.

HQ, H&S, Intel Marines

Cpl Ed Hines, H&S Co (promoted to Sgt. while deployed)

Janney: It's 12th May, 2022. Why and when did you decide to enlist in the Marine Corps?

Hines: My father was in the Marines. He was a door gunner on a Huey during Vietnam. Like most kids, I wanted to be like my father. So, I think that was probably the biggest influence. I came in quite a bit earlier than most of the guys in the unit. I'd been with 2/4 probably the longest out of everybody there. I had a broken time, which meant that I got out and then I came back in. I enlisted in May 1996 right after high school. Then I went to bootcamp and then my first year in the fleet, stationed in Okinawa beginning December '96. I came to 2/4 in late '97. I went to Camp Pendleton, Minneapolis in late' 97. I checked into 2/4 and I was assigned to Fox Company. At the time, Fox Company was what's called a boat Company. Our job was to conduct amphibious raids. This is before MARSOC got stood up. MARSOC did not exist yet. So, the only Marine Raiders were the ones assigned to 2/4. I got out around 2000, and came back in the same year, about six months before 9/11, if memory serves me right. Then, 9/11 happened when I was at division school. After division school, we deployed to Korea.

Then, the following year, we deployed to Ramadi, Iraq in February 2004. We landed in Kuwait and were there for about three weeks for acclimatization. And basically, acclimatization was the reason we were given. About three days before we actually got into Iraq, they herded us all into a tent and said, "You'll be in Ramadi, Iraq for the next six or seven months. I had never heard of Ramadi, but I didn't know much about Iraq. I was a Corporal, an E-4 when we deployed. I left Iraq a Sergeant. When I actually went to Iraq, I was in H&S Company. They sent us in basically two parts. The first part was by air and the second part was by land. I went in the part that went by air. We got in and then the ground team linked up with us.

I was stationed at Hurricane Point. We had Weapons Company with us. We had H&S and a bunch of attachments that were associated with H&S as well. Hurricane Point was a palace that Saddam Hussein used for guests and was located on a tributary of the Euphrates River and the river itself on a peninsula that formed a point, hence the name. There were a couple of cement factories behind us. There were cement factories. Then, there was a bridge behind us. Across the river was Saddam's big main palace for Ramadi, and that was where he stayed.

The first impression I got of Ramadi wasn't clear because we got there at night around 0100. When we arrived, there were still some guys there from the Army, 82nd Airborne. There was a National Guard unit out of Puerto Rico also there, I think. Someone said they were from Miami, so I'm not sure. We were going to be replacing them, so we did the leftsy-rightsy familiarization. When we got there, I was not impressed with what had been done to beef up security. On the FOB, I think there was one medium machine gun on the front gate and that was it. They had the guards, and they all had their M16s. And the interesting thing was that as soon as we took over, we beefed up defenses on the entrance, the ECP to Hurricane Point quite a bit. There was a lot that needed to be done. I understand that they probably had more access to units like Green Berets than we did simply because they were the Army. They were operating in a different way. They weren't going really after the enemy. They were trying to work with the locals to find the enemy, or they just had a different approach. But when we pulled in, I remember looking up in the guard tower that led to the entrance of Hurricane Point. At the top, there's one of the Army soldiers just sitting in the window with his leg lying on the edge of the windowsill, and his other leg was hanging out the window. So, that said to me this guy's not really ready for action. That was my first impression. They definitely had a different mindset. They did mainly mounted patrols. It had a laid back feel to it when we first got there, but I'm thinking this is exactly why this isn't a good idea.

If you're going to be in a city, urban warfare is extremely lethal. Warfare is obviously lethal, but when you're in a city, there's not a whole lot of places you can go. At first, it was a very calm feeling. Nothing was really going on. We were doing patrols. There were some meetings with locals like the sheiks, and the police chief who was shady as shit. It was pretty obvious the police chief was playing both sides of the fence, so the impression I kind of walked away with was that the city was holding its breath. They're waiting to see who's gonna come out on top before they decide to cast their lot. There was a feeling out process. Soon, there'd be little ambushes here and there.

About a week into our time, we had a guy killed from H&S. His name was Lance Corporal Andy Dang. Andy was from California. I think he was Filipino. He died from an IED or an RPG and it was terrible. I'm not going to go into it. I was there when it happened. He was young, around 19. He was a really cool kid and he actually introduced me to "Family Guy." I'd never watched it before and he introduced me to it. That kind of hit me hard, you know, one weekend and we already have a Marine gone. Dang was a combat engineer who was assigned to us.

Everybody knows what happened April 5th through the 10th, but there were events that were leading up to the battle. I remember what happened after it, what the Battalion Commander did. He got fed up. He went to the CIA and said, "Hey, give us all your high value target lists. We're going to clean house." And that's what we did. It got real quiet in that city for the remainder of the tour. I remember that for one raid we did, we were going after a group. Before the raid, there were a couple of guys that were in the Battalion CP, in the palace, that went with us on the raid. They were two really tall dudes, really tall, like frigging football player and NBA tall. Come to find out later, they were Delta Force guys. But there was another guy that was standing there. I didn't know who the heck he was either, but I found out later it was the author Bing West. He was in the CP sitting there just taking notes and stuff.

I had a team with me. The team basically said, "Hey, sir, if you need us, we got QRF on standby. We got the Humvees lined up, you know, ammoed up and everything ready to go. You just let us know when you get there or just let us know if you need us to do anything." I do remember that after the Route Gypsum ambush, that one guy that survived the ambush where the entire squad was killed came back to Hurricane Point (author note: This description sounds like LCpl Deshon Otey, and Hines thought it was Otey.) He was in the chow hall. One of our commanders walked in the chow hall to go see him. Kennedy was there. Of course, somebody called, "Attention on deck." As soon as they called attention on deck, the guy kept eating and didn't stand up. Someone said, "Marine, attention on deck!" Stand up or whatever they said. The Marine kept eating and said, "You want to leave me the fuck alone?" That kind of caused a ruckus there, but I understood. I thought it was funnier shit that he said that. The guy just lost his entire fucking squad, and this is probably the first hot meal he's had all damn day. You might want to lay the fuck off. Anyway, the stance I've taken in recent years is the stance that I got from Alfred of "Batman" fame who said, "I have no desire to spend the remaining years of my life mourning for the loss of old friends." That's kind of where I'm at. I did my crying. That's all the crying I intend to do. It's the survivor's guilt thing. We all know about that. We've all been over that terrain repeatedly. I have no desire to beat that dead horse. You can do "what if this, what if that" thing all day, as much as you want. It's not going to change anything because even if you get your wish, even if you did get to act out and redo that whole scene, doing something differently, there's no guarantee that it would have changed the outcome. There's no guarantee. I've got a friend of mine, John Hancock, he's a Fox guy. He was a squad leader. He was Raider 3/2. He said "The firefights are all a crapshoot. You're either going to get hit or you're not. It's just that simple. It doesn't matter if you take a step left or right, back, forward, duck or don't duck. It doesn't matter."

There are guys that are drinking themselves to death or they're thinking about it endlessly. It's not healthy. I don't think there's a lot of trust with the VA to be able to resolve that either. They'll push pills on you if they can, but that's not a fix. No, it's not. Even if they try to say, "Look, we like to use pills in conjunction with therapy." Because that's how the VA works, that's the VA's policy is to address it. But, the problem with that is there's a huge trust deficiency. That trust deficiency is not surmountable by a guy who spent most of his time sitting in an office with a suit on who's probably never seen combat but has read a whole lot about it. So, here you got a guy who's trying to dig his way out of a hole and you're expecting him to be able to do that with a guy who probably doesn't know what the fuck he's talking about. To me, that's the short of it. There are very few people that I'll talk to about some of the shit that went on over there. And the funny thing of it is that the people that I would like to talk to most about it are people I don't need to talk to about it because they were fucking there. I'm not going to go over that terrain anymore. It's been beaten to death, and I have no desire to open up old wounds.

But yeah, that was our job. I was in the security platoon in H&S. Our job was to provide security and basically just do whatever the hell was needed. Combat patrols, whatever was required. Saw some things, saw some people - a lot of weird stuff. A lot of the news media people that were there - they were there and then they were gone. Okay, we got what we came for. Let's get back on the plane, get out of this shithole - that kind of a mentality. I met Oliver North. He was really cool. He was about the only one I got to meet that I really cared to meet. I mean, this is a guy that had been fighting and was in the shit since before I was born. You know, Sergeant Major Booker met him and he was so excited when Booker said, "Oh, we'll take you on a patrol." North was saying, "Yeah, that's great. I want to get into the shit." He was the only guy that was there on a consistent basis out of every freaking reporter. He was the only one that actually stayed out of all the reporters that popped in. He was a Marine to start with, of course. It was weird because you think he'd be a lot taller. There are people that try to attack his reputation, and I tell them, "Look, you don't know what the fuck you're talking about. That guy's a friggin' hero, period, point blank. He's the real deal." I will never ever believe that Oliver North is not a hero. I spoke to some folks about him afterwards, a Lieutenant General who said one of the reasons why Oliver North was not viewed favorably amongst a lot of Marine officers was because of the way that he was promoted. He didn't work his way up through the ranks in the infantry. He was in Intelligence and was a spy, essentially. They didn't like that as compared to the traditional way.

Another Marine that is really cool, is actually a cool cat, and he's a good dude too is Marc Coiner. Marc is from South Africa, and he'd actually been a diver in the South African Navy before he came to us. Marc's got a good life. I talk to him now and again. I believe you've actually interviewed him, too. Yeah, he lives in Tennessee with his wife, Kathy. Super, super nice people. I like to lie to myself and think that I turned Marc on to blacksmithing, because I still read a lot about it and stuff. So I'd like to get into that one day when I have the money for it. But he's a really cool dude. I like Marc a lot.

      Sergeant Major Ellis, he's one that I have a ton of funny stories about. I used to fuck with Sergeant Major Ellis a lot when he was a First Sergeant. To no end. One of the things I liked doing was impressions, like voice impressions, right? So, Sergeant Major Ellis would always, he'd always fricking talk like this (does a gruff voice impression.) That was how he talked. We'd already come back from Iraq and he was standing in the battalion CP. I think he was frocked at this point for either First Sergeant or Sergeant Major. Frocking is when you pin on the rank, but you're still getting paid at the previous rank until everything goes through. Ellis and I were in offices near one another. I made a prank call to a Staff Sergeant I knew. He's a dark green Marine up in S4. Pretending to be Ellis, I said, "Hey, devil dog. I'm downstairs. Why don't you come on down here? I want to have a word with you." So, he comes down and said, "You want to see me, Sergeant Major?" Sergeant Major looks at him and said, "I don't need to see you." I guess Sergeant Major had fucking ESP because I started laughing in the office. Then I hear him bellowing from down the hall, "Sergeant Hines!" I still laugh about that to this day. Knocked the wind out of me when I heard he died. I thought the guy was freaking immortal. Everybody thought that. I still, to this day, think that SgtMaj Ellis was a real life John Wayne. He talked like him, walked like him, the whole fucking thing, man. I paid my respects to him and reached out to his family when I read the obituary.

Combat Replacements and Delayed Deployment Marines

Cpl Jason Adams, Weapons Co

Janney: Today is 27 March 2020. Why did you decide to enlist in the Marine Corps?

Adams: I was 17, and I came from a pretty right-leaning family. My old man always said that when I turned 18, I had to get out of the house. I was not doing well in high school, no plans to go to college, didn't have a job, and I didn't know what to do. So, I considered joining the Army. I had already turned the Marines down earlier a year before that. So, this Army guy said, "I'm telling you I'm from the Marines. They don't even have guns, they don't have helicopters." I said, "What?" That night, the Marine Corps recruiter called me just randomly, a year later. I said, "Yeah, absolutely. But this Army guy said you guys don't even have guns or helicopters. He said, "What in the hell? Who said that?" I said, "I don't know, some big black guy." He said, "Give me a second. I'm going to call him." The guy in the Army called me back. He said, "Listen, man, I'm so sorry. I was way out of line. I apologize. The Marine Corps is amazing." The Marine called me back. He said, "Yeah, put him in his place." That's how I picked the Marine Corps.

I went to California for boot camp after enlisting in the infantry. I went to the School of Infantry, Camp Pendleton, as well. I got to my unit probably in March of 2001. That was with 3rd Battalion, 5th Marines on Camp San Mateo, Camp Pendleton. It was interesting how I got from 3/5 to 2/4. They told us there was an emergency formation at 0530. They said, "Listen, 2/4 is over there right now. They lost a bunch of dudes. They need help. There's 400 of you. 200 of you are leaving next week. 200 are leaving two weeks after that. If you volunteer to go to the first group, we'll let you pick four of your friends and we promise you'll stay together. So, that's how I got to 2/4. Me and three other buddies decided to go first, just so we could have somebody we knew.

I was a Corporal when I got there, and 2/4 made it clear they didn't care that there were PFCs in charge of us. We said, "Listen, we've already been through a year of this." They said, "Yeah, it's nothing like that now, dude. Just shut up and hang on." We took a CH-46 straight on to Hurricane Point. It was the day that Lance Corporal Jeremiah Savage was killed. We got there and they told us to take a day off to acclimate and tomorrow you'll get on your feet and go patrol. My best friend Mike and I thought, "Oh my God. This is gonna suck." The patrol came in just crying. They said, "Listen, man, one of the dudes died, but you got to go through his stuff." So, now my friend Mike and I were going through his belongings and they just hated us for it. You know what I mean? Not stoked, you know? They said, "Fuck you guys, get the fuck out of here." They were really shitty to us for a while. Talking to them later, they felt like we were trying to take their friend's place. They weren't having any of that. It was awful. They'd been there for a couple months and they felt like we didn't hack it because we weren't there for the Battle of Ramadi. In the end, we were all in the post, and it worked out. But it was touch and go for a month or so.

It was crazy. I was there for the push in 2003 when we crossed the border, the line of departure into Iraq. I did six, seven months of steady combat. But, it was a different combat. It wasn't like Ramadi. I came back, and it was just a different ballgame. The next morning, we're 10 minutes into a vehicle patrol, and it took a fat IED. Another best friend, Brian, got hit in the face. I had a little headset on, so I could hear them talking. They said, "Who the heck is it? I don't know. It's one of those fucking 3/5 guys." I thought, "Jesus Christ. 12 hours into the situation." It was like that every day. It's pretty much the rest of the tour. I was with Weapons Company. Justin Weaver became a new friend and he had my back the whole time. He always said, "Chill the fuck out, guys. They're Marines, too." It was break-in time for us because the ROEs were different than in '03 when we were leading the war. We were the aggressors then. In '04, you couldn't do anything unless you saw a weapon first, like Vietnam, you know? So, the particular dude who set that IED off, myself and my buddy Mike chased that guy down. We said, "Oh, it's ass-beating time." We took him back and are about to get after it and they said, "Whoa! He's innocent until proven guilty." I said, "What do you mean he's innocent until proven guilty? I saw him do it though, so we're good." They said, "No, it doesn't count. You gotta go to court." The combatants knew that. So, they just played that angle the whole time. You could see guys walking around with AK47s down the street just mean mugging you and you couldn't do anything about it.

We'd do half and half on our patrols. We'd be in vehicles when we dismount and walk foot patrol next to the vehicle. We were in Hurricane Point with Lieutenant Colonel Kennedy and Sergeant Major Booker and all those dudes. I gotta tell you a story real quick about SgtMaj Booker. When those four snipers were killed on that rooftop OP on 21 June, we were the quick reaction force. I've never seen anything like that in my life. There was media over there by that point with cameras. My mom told me she saw that on the news, saw the feet and all the blood. She thought, "Thank God it's not Jason. He has really small feet." It's funny in a sick way, but also horrible that she's trying to match up the pictures of the dead Marines' feet to make sure it's not me. We had to put those guys in body bags and take them to the morgue on the Army base. We took the guys and then said, "Fuck, we need to get a soda." We didn't actually have that shit on our base. So, we started making Cokes. We're in our soft covers and we're covered in blood. This Army Sergeant Major yelled, "Come here, motherfuckers!" Me and my friend said, "What's up?" He said, "You undisciplined pieces of shit. That's why you guys get shot in the fucking head because you're fucking lazy, and you don't do shit. You're fucking worthless." We were so mad that we're shaking. He said, "Oh, does that make you mad? Listen to me. Go back to your base, tell your fucking Sergeant Major to call the Sergeant Major of the base because that's who the fuck I am, and I'll put his ass in his place. Get off my fucking base." We said, "Okay." So, we went back and told Sergeant Major Booker and he said, "What the fuck? That was real?" He had this old phone and he called that base and spoke with the Army SgtMaj. He said, "What's the matter, motherfucker?" He just lays this dude out at volume level 10, right? He's screaming at him, "I'm gonna fucking murder you motherfucker, you fucking piece of shit." He slammed the phone onto the desk. We looked at each other and smiled. He said, "Was that good?" We said, "Absolutely, SgtMaj!" Booker said, "Okay. Get the fuck out of my office." People that never met him, don't understand. He's a damn fireball.

Toward the end of 2/4's deployment, one of 3/5's senior NCOs called me and said, "3/5 is about to go back to the battle. Why don't you extend for just that pump?" I said, "There's no way on God's earth that I'm willing to do that. Go to hell." He was really mad. We argued for a while. He said, "Okay, well, when one of your friends dies because you were too cheeky to do the pump, you can think about that every day for the rest of your life." Then on December 23rd, that's when four of my buds were killed in an ambush in the Battle of Fallujah. Hillenburg and Lance Corporal Raleigh Smith were the two that I knew. Raleigh came into 3/5 when I was about to pick up Corporal, so he was my boot. I was his teammate for the whole first tour in Iraq. Super close, really good guy. In '03, we got to Kuwait in early February. I think we crossed the line of departure on the 21st of March. We were there for a month, living in a hole, a worm pit - just waiting to go. We just got up one day, and they said, "We're going at midnight." We were in the armored personnel carriers, the tracks. So just 150 of us in the back of these things with no air conditioning, peeing in bottles. It was crazy. We saw a lot of action on that tour. But we were driving it. I think that's what the difference was. In '04, we just had to wait for it. I don't know what the hell we were thinking the whole time. No armor, and we would weld broken pieces of other blown-up Humvees onto our own vehicles. We'd call them Batmobiles. I was just hoping it would do something, but it never did. You know, we never could get any armor.

The second week I was there, me and all my friends threw our cameras out. We didn't ever want to think about this ever again. It's the worst thing that ever happened to us. You mentioned Musser earlier. In fact, when I was about to beat up that dude who set off that IED, Musser was the guy that broke it up. I was so mad at him, I said, "I fucking hate you, Musser." He said, "You'll get over it." I yelled, "Fuck you!" I did get over it. I just talked to the guy last night. He's a good guy. We still stay in contact. May, June, if I had to describe what I remember, it was just IED after IED every day. We were out missing them, so we would leave town. It would be like New York City, just bustling. They'd be cutting goats up and stuff. We'd come back, and it would just be like a ghost town, with signs just swinging like an old Western movie. We'd stop and say, "All right, we'll walk a little." We'd just stroll through the town, praying we wouldn't get blown up.

Ramadi was combat like you think combat's going to be like. You're just kicking doors and just hoping for the best, trying to accomplish the mission. But it's tough. It's hard. We were out there for 18, 20 hours a day sometimes, on our feet, and that also caused situations, right. Because people would start slipping at that point, in that kind of sun with no hydration or food. It just gets weird. Being the guys on the ground, we always felt like it was a very unorganized situation. No one really knew what the purpose was, what the mission was. We were just kicking in doors, hoping for the best. I do remember a lot of coordinated ambushes, ambushes and then a secondary by their teams. They weren't doing that in '03. I don't know what happened between 2003 and 2004. In '03, I met an Iraqi family. They taught me how to speak Arabic. I'd have dinner with them and stuff, but in '04, that's a hard no. I never really hated that culture. So, it was difficult for me to fight that.

It took a long time for the 2/4 guys to accept us. I brought an acoustic guitar. You know, music is always something that was a support for me. I would play guitar. We played music at night in the smoke shack. And all the dudes came out, 20, 25 guys. We'd just be singing, you know, off-key voices. Wheezer, Foo Fighters or whatever. I believe that's probably when we were accepted.

Like Combat Outpost - those guys got their ass kicked every day. I don't know how the hell they did that. They were just getting rocked. There's other shit that would just make no sense. There was a vehicle that we suspected was a VBIED that we had come across in our travels. We stopped, and whoever was in the front said, "Let's hit this with a TOW missile, and get it over with because it's sideways to the road, carpet over the front." The Lieutenant said, "No, drive past it." We said, "Sir, listen. No, no, no, just blow it up." Lt. said, "Those things are really expensive. We're not going to waste it on what could-be." He starts yelling, "You signed the fucking contract, boys. This is what you do. Sack up and go." So, we went past it, and it didn't blow up. He said, "Fuck, I told you, motherfuckers." We got back to base, dropped our tailgates and heard, "QRF, QRF, vehicle borne IED, Echo 3 KIA, up to 10 wounded." They hit the platoon that was right behind us. It rocked you for a while. I would say the consistent average was just an IED a day, a small firefight, you know. It might be five or 6 guys that would do that. They'd be like holed up in the fourth floor of a hotel or something with a stockpile of ammunition. You could never really kill them all. I remember when Oliver North came and we got in this firefight. We just couldn't get it done, even with Mark 19s and grenades and shit to get in there.

You know, for what it's worth, 2/4 had probably the best leadership that I've ever worked for in my career in the Marine Corps. Sergeant Major Booker was out there on foot with us every day, never missing a beat. Tracking blood trails and stuff, he was an animal. Colonel Kennedy, that guy was solid too, you know. The day before we all went home, he sat down next to me and my friend and he said, "Marine, what are you going to do when you get out of the Marine Corps?" I said, "I'm going to be a fireman, sir." He said, "What about you?" He said, "I want to be a postman, sir." He said, "A postman? Why?" He said, "I'm fucking tired of getting shot at, sir." dude. We were having memorial services on our base and we're getting shot at the same time. So, Kennedy changed the rules and we had to wear full combat gear, weapons locked and loaded just to go to the phones. I was pretty well known for being a smart ass. I had a big mouth and was always getting in trouble. I walked over there into the phone center in PT gear, and as I turned around, Kennedy said, "Hey, you come here." I didn't move. He said, "Get over here, motherfucker." I just stood there. He started walking towards me. I just turned around and sprinted. Jumped into my rack and rolled up in the covers. I heard Kennedy yelling. I said, "Everybody shut up." About a minute later, he comes in. He said, "You guys seen a Marine runner down here?" I said, "No, sir." He said, "God damn it." Then, he walked out. He never did figure out that it was me. I got away with murder that day.

Jason Reeves was a platoon NCO and he was probably the only dude we just never got along with the whole time. I just don't know if he just didn't like us for where we came from, but he was always at odds with us. But even with all that, he was a good leader. He was good in the field when it counted. I'm sure he saved countless lives. That's the important thing.

When Colonel North went out on our patrol with us, we got into a little firefight at the very end of the day. North was hooting and hollering. He's on his stomach, yelling, "Hell, yeah! Kill him, motherfucker!" Just getting loud about it. When it was over, he was just ecstatic. I don't want to say it was weird, but it was kind of weird. I don't know what I expected, but I didn't expect that from a 75-year-old guy, even if he's a retired Marine Colonel.

One of my buddies, Royce, was with us the day we had a bad rocket attack. I'll never forget it. He comes out of the port-a-potty with his pants around his ankles and he had American flag boxer shorts on. He yelled, "Adams, holy shit, man, we're taking fire!" I said, "Yeah!" I don't know what he's doing, but he starts trying to run and he just falls on his face. He said, "Son of a bitch!" He's just trying to get up and pull his pants up. I don't know why, but I'll remember that till the day I die. It was hilarious.

My time in the Marine Corps really affected me, but it also made me want to help others that had been through what I'd experienced. I've accepted that, and I'm just trying to use that for a better purpose. I'm almost done with my degree. I'm going to transfer to the University of Cincinnati for social work in September. The town where I live in Washington is a pretty small town, around 30,000 people, but there's a VA hospital here. There's a lot of vets that are in recovery here, but some don't know how to get help. I try really hard to just be transparent. I'd say, "Dude, this is what happened to me. I get where you're at right now, but it could get better." Before this, I did the same thing but with veterans for a year and a half. I loved nothing more than hearing the dudes say, "What the fuck do you know about the service?" I'd say,

"Let me tell you a story." They'd say, "Oh, my God." I switched though because part of it wasn't the population I wanted to serve. It was kind of like people that were entitled or were looking for a housing voucher. Our program would take guys who were on active duty for at least a day. So, a ton of dudes got just out of the camp week one, you know? That's a little secondary problem. The combat vets weren't coming out to see us. They didn't want help. I think part of the problem was they were too proud. You could go in the alleys, find them, and drag them out. But they weren't trying to get help. We're all too proud. I was too proud for years. I think also they feel like their problems really aren't that bad. I've heard this a million times, "There's somebody else out there who could use the help more than me. I'll stay in the alley. I'll just sleep in my tent." It's an honor thing. Like integrity - they don't want to be that guy that breaks it off with somebody else. I just want everyone to know that there's people that care and there's help available. I don't want anyone to be too proud to ask for help. It's not failure to need and ask for help.

Now, I'm a different person than I was back then. I was kind of cocky and arrogant. Pretty sure I was running the show. I think I have more empathy now. For most of my life, I claimed that I was an atheist. I would yell at God up until a few years ago. I went to visit the grave of my therapist in Montana. His brother was a Marine and was in Iraq when Raleigh was killed. The cemetery wasn't on the map, so I knocked on doors until someone took me there. I was looking at his grave, and I was just thinking to myself about God, "If you're real, prove yourself. Show me something." I'm also thinking how cool it is that 14 years later, people are still connecting over my friend's presence. Suddenly, my whole body started tingling and every hair on my body stood up. I can't really describe it. It was like a cat was poking at my chest. It just felt perfect. It was the most amazing feeling. I could feel like transmissions in my chest. I understood what it meant, right? It was like a language. I felt these words, "You're a good person. You made the right decision for you and your family. This is the game plan. Let the chips fall. They fall. This is my plan. We love you. You're an amazing person. You're doing an amazing job. You're right where I want you to be. This is it. Keep doing your thing. You're changing lives. I love you. He loves you." I have a different understanding of what God is now. I used to think you had to give money at church, pray every night or whatever. It's more like a connection in a relationship, you know? Like talking about what's going on with today. It doesn't have to be, for me at least, a big formality. I can just say, "God, this is tough. Let's talk about it." It's a personal relationship. It doesn't have to be in a formal setting like a church. You can connect with him anywhere you go, anytime of the day. I'm just so much happier now. These last 2 years have been the easiest years of my life.

Combat Replacements and Delayed Deployment Marines

Cpl Mike Martinez

Janney: You pushed for the MOUT (military operations in urban terrain) course.

Martinez: Doing door-to-door clearing was what I was trained in. So, I knew how to do it. So even though I was the mountain instructor for 81s for Weapons Company, working with India Company, I was able to train these guys who didn't have an instructor. So, we're kind of training on the fly. I was still just with the fifth team, but I would work with the individual platoons, especially in Kuwait prior to going into the, or crossing LOD (line of departure) in 2003 Desert Storm. So, I was training them how to do house to house clearing. So, unluckily for me, I'm still struggling with the injury now, but two days prior to the course, the mountain instructors course, I broke my ankle playing basketball for PT. I was not supposed to go to the course, but I didn't go to BAS. I just sucked it up, put an ACE bandage on and I went to the course anyway. So, I was able to get the training when I probably shouldn't have. After Baghdad, we went south and we went to Diwaniyah. And we stayed in Diwaniyah for the majority of our deployment. So, the beginning of September of 2003 is when we came home. During this time we were traveling around Iraq, just around the countryside, and, you know, we were essentially the first police department at Diwaniyah. We were doing patrols on foot and in vehicles and making arrests for various things. We were also teaching the Iraqis how to build their own police departments and how to stop crime from happening.

So, we get home in September. My son was born in July of 2003. I missed his birth. We were home in September and we're told you guys are done. We don't need you guys anymore. We're going to put you on guard duty. We didn't really do a buildup. We had a cool down phase where we were just kind of hanging out around base. We're put on guard duty, guarding the armory, guarding just different little areas around our camp in San Mateo. And yeah, so we're really on fat and happy days. We're just relaxing, enjoying the rest of our enlistment, and just having a good time with our friends basically is what happened. So Easter of 2004, so around April, we got called into the gym. Everybody who's getting ready to EAS gets called into the gym, and we're told that, so people who are going to EAS from September and October of 2004, the beginning of October, so it's like October 10th, which I think was the cutoff, some of those are early in October. These guys are going to be sent back to Iraq again, and not just for 2/4, but for three different units that needed combat replacements. So, these guys, our friends, are being told that you guys are going, so people like myself and Jason Adams weren't forced to extend - We joined October 24th. So, I was not in this group that was being mandated to return in the beginning of November. Jason wasn't either. But, all of our best friends are being told now that they are being recalled and forced to extend. This is when Jason and I had our talk and we still weren't sure again. My son has just been born and he's less than a year old at this point. My older son was already almost four. Jason and I have to decide what to do. I can stay home with my family or I can get redeployed and you know take this monster head on. So,

Jason and I made a decision - Screw it. Let's go. I talked to my wife and she understood. I mean it's hard for a young wife to take this kind of Information, but I explained to her that I couldn't stay home while my buddies re-deployed to Iraq. So that's when it was the beginning of May when we actually got sent back. I remember missing Mother's Day. We were all together and it was a ragtag group of guys. All of the guys from 3/5 are getting sent to different units in Iraq at this point. These guys are all getting sent back also. We're going to just random units throughout Iraq that were hit hard. And so the 81's guys, we got sent. Most of the 81's guys got sent too. 2/4 81s are their Weapons Company. And then some of our guys got sent to other units. Some of them got sent to 2/4 Echo Company and Golf Company. We were just lucky enough to stay with our core group of guys. It was Lee, Jason, Fox, Clark, and Larson. So, five of us went to 2/4's 81's platoon. And then I believe there were another five sent to their 2/4's CAT platoon. And then, again, the other guys just got sent around everywhere else. Beginning of May of 2004.

Even though we were combat Marines, 2/4 really didn't welcome us with open arms when we got there. With 3/5, I got a NAM with a combat V. I got a couple of awards for fighting with 3/5. When we got to 2/4, they knew about this. They were not very happy to have combat experience veterans with them. We were there to replace their brothers that had been killed, and they still weren't happy to see us. Right, and that really was how it was. The only ones who really accepted us were their young guys because we were Corporals at the time, we were E4. Their Corporals, their Sergeants, they did not welcome us. I thought these are going to be the guys I'll take care of. So instead of making it easier for the NCOs, I figured I'll make it easier for the junior enlisted. Most of us were at Hurricane Point. I remember getting to Hurricane Point the day that Jeremiah Savage died, so we arrived in their camp while they were still bloody from that battle. They gave us Jeremiah's bed to sleep in. That's the kind of stuff that happened. You're moving into this spot, and moving this guy's stuff out of the way.

We were doing a lot of mounted patrols since we had the heavy guns with Weapons Company. Since I was not the favorite, I actually got left behind a lot. I was more of a gear guard where they would use me for your radio watch or guarding our barracks. They would refuse to take me on some missions because I already had the combat experience that they were trying to catch up on. When we got there, they had no armor. I remember going to, I believe it was the Air Force base, and they gave us a little bit of a shack, I guess would be the best way to describe it, and some welding materials. The guys who knew how to weld welded pieces of metal onto the side of our Humvees. So that was our first armor. Corporal Musser had a little bit of experience with welding, but not a lot. He burned himself through his uniform just from using the welder. It was basically a sunburn on his skin that was just blistered from welding. He didn't have a lot of experience with welding, he just kind of knew what he was doing.

We would still do some door-to-door clearing in the city. But most of the time, we're traveling down certain routes kicking over trash piles, trying to find IEDs. We would look for freshly dug up areas and see if there was an IED there. We don't have an EOD guy with us. This is just a bunch of grunts kicking over trash piles and looking at freshly dug holes. We ended up getting some EOD guys later on who had the metal detectors. On Route Nova, one of our other buddies Lopez was with a platoon on a patrol. We're on one end, he was on the other and I remember the explosion going off, a loud boom in their direction. Apparently, the EOD guy stepped across an IED with a metal detector and it went off. Lopez was the first on scene and was the first one to try to assist him. That was pretty much our daily routine where we would head into the city in vehicles and wait for contact. We were taking a lot of IEDs, so it was pretty much daily. We would get hit with an IED, or we would see an IED and stop and shoot at it and try to make it explode, because we didn't really have the EOD to come and disarm IEDs at the time, so we were shooting at them to try to get them to explode without us being close. If we got hit by an IED, it was usually 4 to 6 guys that would follow up with small arms fire.

I think it was Operation Bug Hunt where I was guarding an alleyway and Adams was on a rooftop. I could see him, and he was maybe two buildings away. I could see him standing on top of a building with Jamie Rocha. I'm down on street level. They were overlooking traffic - just normal Iraqi everyday traffic. A group of five guys in a vehicle just started shooting at them. I just remember hearing the gunshots going off, "Pow, pow, pow, pow, pow." I see both of them drop on the ground. I still don't know how they survived. In my mind's eye, I see they both got hit and they're on the ground and now we're trying to find out how to get to them and they pop up. Neither of them got hit and they return fire and the vehicle gets away, so that's the kind of experiences we're having.

I was there when Rocha got wounded during that. I was in the back of the Humvee with him when he got wounded. I can't remember his exact injury and I don't know the circumstances either. It calmed down at one point when the ROEs changed to anybody with a video camera is now an enemy and we can shoot at them. Because that was what they were doing. They were using recordings as well as we're driving down different routes. Then they would blow us up and they would use the video for their propaganda films. Once the ROE changed where we could start shooting at those guys, they stopped doing that and so it kind of calmed down for a short period of time. But, then they would just find another way to mess with us. We would adapt to it and they'd attack again.

Jason and I were on the QRF to go check on Parker, Contreras, Lopez, and Otey. We were told that a few of our snipers had been killed. By the time we got there, there was no fight. There was nothing. It was just basically collecting bodies. We put two bodies into the back of our Humvee, and we drove them to the Air Force base. I just remember sitting in the back of the Humvee with these bodies, our brothers, laying there. I mean, it's Humvees, its Iraqi roads, and their heads are just hitting the bottom of the Humvee. You can hear the "dum-dum-dum-dum-dum" sounds of their heads hitting the floor of the Humvee as we're bouncing down the road. Jason and I sat on the deck and cradled their heads as we took them to the base. It was the worst memory I have of Iraq.

The rest of the deployment was more of the same hit and run stuff. I remember one patrol where I was with Staff Sergeant Cook, I guess he became Gunny Cook later after I left. But I was in his vehicle and we were driving and we saw a suspicious vehicle up ahead. He calls in a suspicious vehicle, and we drive past it. We're told there's no one available to check it out. So, we drive past it and as we're driving past, I just remember clenching, just getting ready for an explosion. We get past it. Our whole convoy gets past it. By the time we get back to the base five minutes later, we hear the explosion. One of our five guys, Benny "Benji" Gonzalez, gets killed in that VBIED attack. Jason said that they tried to get whoever was in charge of that patrol to fire a TOW missile at the vehicle, because they clearly suspected that it was a VBIED. But, he was told, "No, we can't waste a TOW on what "might be" a VBIED. We were just told, "Leave it alone, just go around it." Again, Gunny Cook was not happy about this. He was pissed that we had to leave a suspected weapon. Then, our next convoy that goes through gets hit. You know, all for the cost of a TOW missile. We lose American lives because we can't afford one missile.

One funny story occurred after we dropped off the snipers' bodies at the Air Force base. This is Jason and I just being the pains in the ass that we are. After dropping off the bodies, we headed back to the Army base across from Hurricane Point because they had better chow than we did, plus they had sodas we didn't have. We figured, let's go over there, let's get something to eat, and then we'll go back to it at Hurricane Point. Since we were on a QRF, Jason and I left our covers (hats) back at Hurricane Point. We decided that, who cares, we'll be fine. At the Army base, Jason and I are walking around, and just goofing off, we made up an op we called Operation Drawfire. We're walking on the Army base and some army PFC walks up to us and says, "Hey Marines, where are your hats?" We just said, "Shut up. Leave us alone." We walked away from him. We see an Army Specialist and he walks up and says, "Hey Marines, where are your covers?" We just said, "Leave us alone." Then, we came across the Army base Sergeant Major. He walks up to us and starts yelling. I mean, it's just full Army ass-chewing. He's going off about how Marines are lazy, we're pieces of shit and that is the reason our snipers got killed. Jason and I decide to go back to Hurricane Point. When we get back, we automatically go straight to Sergeant Major Booker. And SgtMaj Booker is definitely a badass. To this day, he's one of the most badass people I've ever met. We tell him what happened. We told him about our Operation Drawfire, just to see how many people could chew our ass. Booker said, "Okay, just give me one second." He gets on a regular telephone and calls the Army Base Sergeant Major. Booker said, "Hey, Sergeant Major whatever-your-name-is. This is Sergeant Major Booker with 2/4. I have two of my Marines here who said that you had a confrontation with them." The other Sergeant Major has some words back, but I can't hear him. Sergeant Major Booker replies, "Well, if you ever speak to my Marines again, I will make sure you never come back from another patrol." Jason and I, we were dumbfounded. I was just in complete awe of the badassery of this one Marine. It was just the coolest thing I've ever seen. So, that was probably my favorite story of 2/4 and their chain of command being willing to fight for their Marines.

So, Sergeant Major Booker is definitely one of my favorite people I've ever met in my life. That's something I do have to say about 2/4. Their chain of command was very welcoming, very accepting of the 3/5 guys. They were extremely grateful that we showed up and were willing to fight, even for a unit that we didn't know anything about. Marine Corps tradition, you get to your unit and this is your family, this is who you are. Being transferred over to another battalion was definitely different. As far as being welcomed into the unit by the other Marines, we really weren't at first. I didn't know Jeremiah Savage, I took his rack. There were so many injuries and KIAs throughout, but I don't think I knew anybody who actually got KIA. There were guys that I was actually close with that were injured by shrapnel from IEDs, but we were pretty lucky while I was there. I've always considered myself pretty lucky. Even in 3/5, none of my close friends got injured. When we went to 2/4, only one of our guys, Corporal Fox, took some shrapnel to the face during an IED. I think we were blessed.

Fox was in a vehicle that was known to get hit. I'm notoriously known for giving people nicknames and nicknaming things, so the Humvee that I was always in was a vehicle that we appropriated from the Air Force, so we called it Air Force One. There was another Humvee that was put together with just random pieces of other Humvees, so we called that one Frankenstein. Corporal Brian Fox was in a vehicle that Marines had recently been killed in before we got there. That vehicle was known to get hit by IEDs, so we called it White Lightning, because lightning never strikes twice, but that vehicle always got struck. But, that was the vehicle that Fox was in daily. Another good story is about Jamie Rocha. We're driving down a road and Rocha was up on the 50 cal M2 machine gun. This is a high-back Humvee with no top on it. We don't have enough vehicle seats. Again, I was usually left behind, but on this mission, I forced myself to go. Instead of sitting on the vehicle seat, I'm sitting on the ice chest in the back of the Humvee. As we're driving down the road, we get hit with an IED. I think Hansen was the driver and he just stopped abruptly. I fell backwards because I'm sitting facing the rear of the vehicle while on an ice chest, so I'm not secured. I sit back up and then I fall back down again. That's when I hear the boom. That's when I see the smoke. I'm looking in the sky and I'm not understanding or comprehending what's going on and that's when the debris starts falling. We're trying to gather our senses about what's going on. We see the guy who set off the IED running from us. Rocha was so angry because of these makeshift gun mounts that were in the back of the Humvee. It was just pieces of metal welded together with a machine gun mount. His foot got caught into it. Everybody got out of the vehicle and Rocha couldn't get out. His foot was stuck in it. He was just so angry because everybody's running and there he is stuck in a vehicle, screaming, "Get me out of this thing!" He was not happy about that. So, it was pretty hilarious. We ended up catching the guy who set off the IED, and this guy was special needs. He had a crooked foot so he couldn't run. He was disabled, physically and mentally.

The enemy, I can't even say Iraqis, because I don't know where they're from, just used anybody. They got people to do their bidding. That IED triggerman was trying to run from us and our guys easily tackled him and got him. Because of Marine Corp policy, combatants were innocent until proven guilty, even if we caught them in the act of setting off an IED or shooting at us. You and I talked a little about foreign fighters before the interview. The second tour with 2/4, I didn't really get a lot of up-close visuals of enemy KIA. But, when I was with 3/5, a lot of our enemies were from Syria, Libya, and Egypt. Most of our ambushes in 2003 were not Iraqi people. It wasn't the Republican Guard. It was Syrians who were being transported in to fight us. Even with 2/4, we had to be careful because of that sniper that was running around competing with Chris Kyle. I remember seeing some of the Navy SEALs there at the time. We were there surrounded by legends, but we also knew that there was a sniper that was on buildings shooting at us. We just had to always keep our head down and always stay vigilant. I just remember seeing those SEALs because they looked cooler than shit. We're wearing basic Marine camis, and these guys have beards, have cool gear on like plate carriers instead of our IBA, carried tricked out M4s, have pistols strapped to their legs, and were wearing baseball caps. So, it was definitely cool to see these guys.

When we got to 2/4, we were issued M16A4s that had ACOG scopes. Never used them prior. We got there and we carried them for a couple weeks without zeroing them in. Then, we were taken to a range somewhere and we got, quote unquote, qualified on them. Basically, you hit the target and you're qualified. The range is funny because the normal Marine Corps range is 200, 300, and 500 meters or yards. At this range, it was probably 50 yards - hit the target at 50 yards using an ACOG scope. I understand because we're shooting down alleyways and aren't shooting 500 yards at this point. We're going door to door, 30 feet, you know? I liked the ACOG, but iron sights never fail. Because scopes fail. No matter what kind of scope it is - it could get shot out, it could get broken, and it could lose battery, whatever. You must always have a contingency for failure. Honestly, the ACOG is probably one of the best scopes I've ever seen. It's still an amazing scope.

As we're approaching September and getting ready to go home, we were still under a lot of sniper fire. We're still getting a lot of IEDs, but I just remember at this point, I'm about a month out from my EAS. We didn't really want to get sent out anymore at this point. Because with a normal EAS or a normal person just returning from combat, they have a cool down period prior to meeting back up with family and going home. For us, it was combat, combat, combat. So, we're fighting, whether IEDs or clearing houses pretty much until they said, "You're going home tomorrow." Even though we did bitch about it. We did ask if we could stay out of some patrols, but it didn't work out. Adams said basically he's in combat and then 10 days later, he's sitting there with his mom and dad. It was just, boom, you're home. He said 10 days, but I think he's being generous. We're in combat. Then two days later, we're in Kuwait. Then the next day, 24 hours later, we're flying and then are at home again. It was definitely a weird experience where you're searching for IEDs and then two days later, you're in California.

Combat Replacements and Delayed Deployment Marines

Sgt. James (Buttrey) Baum

Janney: It's 21 April 2020. Tell me why you decided to enlist in the Marine Corps.

Baum: Well, that's a pretty easy one. I had never thought about service in the military. I'd always planned to be a cop like my folks. But, I was kind of a dirtbag in high school. I was in college, but I wasn't really going if that makes sense. I think in the one semester I attended college, I might have gone to class, out of all my classes, I might have gone to six classes the entire semester. When my parents found out about that, they were less than pleased. My dad told me I had two weeks to be out of the house - I was on my own. I said, "I'll show you." A week later, I was in boot camp. I went to the recruiter the next day to take a friend of mine. I wasn't even going for myself. I liked what the guy had to say. The very next day, I went to MEPS, took the ASVAB, and got screened. Then, a week and a half after that, I was in boot camp. I enlisted in February of 98. I came in as security forces. I was 8152. I was stationed in Bangor, Washington for the first couple years in the Corps in Marine Corps Security Force Company, Banger.

I felt like I was extremely motivated. I spent two years in Bangor and was motivated to come down to the fleet, but with the plan that I was going to get out in February 2002. I had a job lined up with the Department of Energy back in Washington. Then, 9/11 happened. I dropped reenlistment papers the next week and I ended up staying in. That's when I went to 2/4 in February of 2002. I wasn't there for the beginning of the Ramadi deployment, but I was still in that unit. After 11 and a half months of deployment to Okinawa, we got home from that in July of 2003. In August of 2003, I volunteered to go with the coalition, me and one other Sergeant from the 1st Marine Division. I left in August and deployed and didn't come home until after the Battalion had already deployed. When I deployed with the coalition forces, I spent most of my time in Baghdad, but I was all over Basra, Mosul, Sulaymaniyah, and Kirkuk. A lot of my time was in Kirkuk setting up the Iraqi Army's boot camp. It was a very interesting deployment. It was just the same as the beginning of the 2/4 deployment. We didn't have armored vehicles. I drove around in an unarmored Yukon. The whole time I was there, I got hit with one IED, and I think there were only two that the coalition got hit with in the whole seven months I was there.

I came back to Camp Pendleton at the end of February. It was just after the Battalion left. I was supposed to get married in March, so the Battalion let me stay in the rear. At the time, obviously, there wasn't really anything going on. Then when April 6 started and we started taking heavy casualties, the Battalion was sending me to the funerals, me and a couple other Marines, to be a representative and present the Purple Heart to the Gold Star families. After either the fifth or sixth one of those, I talked to Sergeant Major Booker and said that I wanted to come back out and rejoin the Battalion. It was excruciating. Plus, I was picking up all the guys that were getting sent back wounded. I had to go and pick up Zimmerman from a hospital in San Francisco. It was very rough. I went and visited Stayskal when he was in the hospital at Camp Pendleton. It was a rough time. When I finally returned to 2/4, I flew into Al-Assad, and then got transported to the Battalion and went straight out to the Combat Outpost and rejoined Echo Company.

I was a Sergeant and I had a different last name back then. My last name back then was Buttrey. I changed my name in 2005 when I took my step dad's last name for Father's Day. When I first got back, I rejoined 2nd platoon, which was my platoon prior to me going with the coalition. At the time, Corporal DeGutti had taken over my squad. He was one of my team leaders when I left. He took over the squad, so I was like the platoon guide. Sgt. Damien Coan would go with one squad on patrols. Lieutenant would go with another squad, and then I would go with Takeda's squad. Then at the end of July, they fired one of the squad leaders from 1st platoon, and I went and took his squad over. It was constant patrols and standing watch and men in the COC. By the time I had rejoined the Battalion, it wasn't like the beginning of their deployment in April. I think probably about once a week, we were getting into firefights at that point. From the time I got there in July, I don't think we took any KIA in Echo. I don't think we took any more casualties. I'd say that somebody got shot in the butt, but that wasn't in my platoon.

We did launch a few rounds at a few suspected IEDs. I think by that point, we had gotten some armor or at least makeshift armor for all of our vehicles. We were trying to take better steps, maintain distance when we could. But, we still had to do route clearance. I know Sgt. Valerio worked to cobble together makeshift armor. Valerio and I were actually drill instructors together after that. We also used the heck out of counter battery fire for mortar attacks. So anytime we would get a direction, anytime we'd hear launches, we'd immediately work to get an azimuth to the direction that it was coming from, estimate the distance, call it into a counter battery and then let them deal with it. I think at that point they were pretty mobile, so they would launch a few rounds and then they'd leave and go launch from somewhere else. We had been getting briefed about foreign fighters, but I didn't know the difference. Just a guy in a tracksuit. Just a dude running around shooting at me that I needed to shoot in the face.

You asked me when we met in Arlington to think of funny stories about my brothers. Contreras was one of the funniest guys I've ever met. It didn't matter. Even if you were trying to yell at him for doing something stupid, he would say something that would just make you laugh. He was just always in a good mood. Always a stupid smile on his face. I can remember being in Okinawa and him mouthing off to Sgt. Damien Coan and we had all been drinking. We did kind of an Echo thing when we were at a resort in Oki. He keeps mouthing off being stupid, being drunk when we're telling him to go to bed. He tried to fight Damien, and they rolled through the fire pit. It was hilarious. I don't think I ever saw that guy in a bad mood. Even if I was furious with him, I could never, ever stay mad at him. Because he would just say something else that would make me laugh. That guy was great. I went to a memorial when we got back from that deployment with his family in Texas, and it was just a very emotional time.

I'm trying to think of other stories. A lot of them got to the Battalion while I was gone. I knew Otey. He was on that first deployment to Oki. We didn't have a lot of interactions because we were in different platoons. I mostly hung out with the second platoon guys. He was a good kid. I don't know if anyone has shared with you the after action report (AAR) that Otey had written before he was killed. That thing one hundred percent sums him up in every aspect. The way he wrote, that's the way he talked. He was a good Marine. He didn't do a bunch of dumb stuff. He mouthed off a little bit, but only if he felt like he was being disrespected. But, he did his duty, and he took care of Marines. Ben Musser was in my squad. Ben was one of my SAW gunners for a long time. I tell you what, he's a smart ass, but that kid is loyal like no other. He puts off a persona that he's not very intelligent, but that dude is a lot smarter than people give him credit for. I think he does it on purpose, if that makes sense. If it doesn't seem like you can handle the responsibility, you're not going to get it. I trusted him more than most. He was just a loyal kid. He was loyal to his squad. He was loyal to the Marines. He was loyal to the platoon. He was loyal to the Marine Corps. I knew John Sims. Sims was just a goofy kid. But again, like Musser, he was loyal. He took care of his Marines.

I obviously knew the men in the second platoon. All of those kids, especially the third squad, when I was there - half of those guys, Morel Cruz, Degutti, all those guys were in my squad for a year and a half, almost two years before that Ramadi deployment. They were just good kids. When I left to go be with the coalition in Iraq, they were kids. When I rejoined them back in Iraq, they were men. They had been through the shit. I was proud of them because they trained hard even when we got extended on that Okinawa deployment and we didn't want to train. I told them we still had to and they did it. They pushed forward, and they excelled, and they did exactly what they were supposed to do in combat. I was just damn proud to know I'm in the circle.

So, I almost crossed decked to 2/5 towards the end of that deployment. The last couple of weeks when their leadership was on deck, they kept all the squad leaders and platoon commanders at Combat Outpost while everybody else went to the main station to get ready to fly out. We did a leftsy-rightsy with them. That Company commander had asked me if I would consider staying on with them and I almost did. So, it was me and the two other squad leaders and our Lieutenant Valdez who a week before had gotten shot in the elbow during a raid with some Shawanis. They had a negligent discharge and shot him. We operated all the way up to the time we left. In fact, we got in a firefight alongside the guys that were taking our place during a leftsy-rightsy patrol. The next day, we're supposed to be getting out of that area, out of Combat Outpost, and leaving everything to them.

I think some of the 2/5 guys that replaced us came in with a bad attitude, "Oh, we've already been to Iraq. We got this." We said, "Okay, man, this ain't the same as when you came and rolled through in March of 2003. This is a different world." We tried to tell them that. One of the platoon commanders I knew were in 2/4 together. Now, I think he's a Major, or maybe a Lieutenant Colonel by now. I took him out and got him in his first firefight, which I thought was pretty funny. So, while we're leftsy-rightsy, I'm with his platoon showing them our area. Some people start shooting at us for a couple hours and this is his first day there. He was a solid Sergeant when I served alongside him. As a Lieutenant, there was no difference. He's always been a force of nature. Even now, I think he's slated to take over a Battalion as a Battalion CO. He performed well and he performed well under pressure.

Battalion Sergeant Major James "Jim" Booker – Deployment Overview

Janney: It's 2nd May 2020. Why did you decide to enlist in the Marine Corps?

Booker: I lived in Ft. Wayne, Indiana, and the snow was really bad. I was pushing a lot of snow clearing streets. I said, "Okay, if I don't do it now I'm never going to" so I went to the Navy and they told me that I had to wait six months and they would put me in a 52 week turbine school in Great Lakes. I said, "Hey dude, that's colder than I thought" so I went to the Marine Corps and enlisted on January 18, 1983. Two days later, I was in boot camp. I turned 21 in boot camp, so I was 20 at the time. My dad had been in submarines from 1944 to 1967. I believe that everybody should serve in some capacity.

Janney: Once you got out of boot camp and SOI, were you 0311 (infantry) or what did you do?

Booker: Yes, I came in with an open contract and wanted to get into some kind of electronics field, but next thing you know, I'm a grunt and I'm thinking, "Holy cow, this boot camp kind of killed me." I'm really glad that I did it. They hung up a poster and said this is what your MOS is, but we still had the last week of boot camp, and then I went on to be a 0311. The first school they sent me to right after that was called SCAMP, Sensor Control Management Platoon, which is a division level asset of HOW, and it was basically the electronic sensors that they would put out in the Vietnam War in Ft. Huachuca, Arizona and went to about a 6- or 10-week sensor control platoon training. After that, I went to the Sensor Control Management Platoon for my first two years in Okinawa. I was a 0311, however that was deemed secondary to being a sensor guy. I thought that I was going to go to Beirut, but that never happened.

Janney: At some point after that you volunteered for ENDOC and went to the scout sniper school or were you selected for that?

Booker: I got sent to Okinawa in about August and it was right when that Korean Airlines got shot down in 1983 and when I landed, they were all in a big tizzy, "Hey, we thought you guys got shot down" and I was like, "WHAT?" We were on the same flight pattern as that god dang thing. But anyway, we landed, and half of the platoon were guys that read the sensors and stayed away in the Intel field because the SCAMP platoon comes under the Intelligence field controlling sensors and the other half have to go out with a reconnaissance unit and embed these things.

I really liked reconnaissance. In 1983, I went into the reconnaissance indoctrination program at 3rd Marine division under Colonel Teahan. That's when the bomb blew up in October of 1983. They called us in, and we thought they were screwing with us. But, we were in NTA and they were screwing with us pretty hard, but that was in RIP school. Because of that I always went with reconnaissance units and laid out the sensors and they all really liked that because they didn't know how to set them up and didn't want to do it. They would always use me as a pack mule. They would drop me off and I would set out the sensors and it worked very well because we got them placed correctly after that.

So, after those two years, I went to many schools. Combat Squad Leader school, TACP, all these schools because they were school intensive back then. When I came back in 1985 and '86, I went to 1st Battalion, 4th Marines. They were the only infantry unit in 29 Palms. Now they've got the 7th Marine regiment. They sent me to sniper school. After that I went to the drill field and after that I went to Force Recon, so it all was a progression of things.

Janney: Did you pass scout sniper school the first time?

Booker: Yes sir. I went to school with a kid named Ronk, a grizzly son of a gun. They did not prepare me and I'm glad they didn't because I thought, "Man, that would be pretty cool to go to scout sniper school." I went in there without an idea of how difficult it was. He told the rest of the unit that he didn't think I was going to stay with 1/4, "I don't think Booker's gonna make it" but he was the one that came back and I graduated.

Janney: That's quite an accomplishment in itself. Several of your guys didn't pass on the first go round and had to hope for another slot.

Booker: Yeah, like Longoria, three times. I know Longoria failed because of (redacted.) I ran the school in 1997 and 1998 and we revamped the whole school. I was so proud. We didn't haze them. We transformed the school and made a whole different version of how we trained them for stalking because we thought there was too much dropping them for stalking. Hell, now I don't even think they teach it. But at one time you were supposed to have partners and I thought my partner was going to take my sea bag with my ghillie suit in it and throw it on the truck. I had to go do something else like carry water cans or get ammo. When we got to the stalk site, I didn't have my ghillie suit. The instructor said, "Okay Booker. We're gonna make you stalk without it." I was glad they did because normally if you didn't come with your equipment, you got a zero, and a zero would be hard to get back from. So, I got a 10 without a ghillie suit. There were people that could just not believe it. They said, "Okay, but we're only going to give you a 9." I did all the stuff, the shot, and everything, but they ended up giving me a 9. But, I got a 10 without a ghillie suit. In 1986, I went to DI school and was a DI for 2 years.

Janney: Once you transitioned out of that, what unit did you go to and what were you doing?

Booker: I went to First Force Reconnaissance. I was there from '87 and '88. I was a Drill Instructor and then in '88, I got meritoriously promoted to Staff Sgt. As you know, it was a 0369 and that changes everything, because now you're a staff NCO and get a different monitor. But, they allowed me, even though I ran my ENDOC when I was a Sgt., they allowed me to go do Force Recon. SgtMaj Jerrold really made that happen. He was the Recon guy that was on the posters. 3rd Recruit Training Battalion even though I was in 2nd Battalion. I became a platoon Sgt. I had 5th platoon and then transitioned to 6th platoon during the Gulf War, but I was there for four years, '88-'91.

Janney: Did you deploy during Desert Storm?

Booker: Yeah. I was in 5th platoon, but they changed our name to 6th platoon as we got into Desert Storm, so I had 6th platoon and Force Recon. Then we got into a gunfight down there and I received a Bronze Star for that one. There's a town there called Khafji and there was a big gun fight there. Colonel Berry who originally had me at Force Recon was at G level staff. He invited all of us that were on the borders where there were six or seven checkpoints about every 10 miles. We had OP 4, 5 and 6 in Hamil Tiad where I was at. Another unit ended up taking OP 4 and they were shot up pretty bad. It was a TOW unit, and they lost a lot of people there. We got in a gun fight on April 20th in 1991 and it's in that book, Moskin's "The U.S. Marine Corps Story." There's a brief paragraph of my platoon getting hit by about 115 Iraqis. We were supposed to be processing them back because we were one of the first ones to tell them, "Hey, there's a whole butt load of these guys that are going to defect" because they got all the Kurdish guys in the front and they don't want to fight and they don't have shoes and stuff. They kept coming across, but we would get them and interrogate them as a friendly.

The Saudis were monitoring us because they didn't want any of us Christian infidels to be wandering around. In fact, they wanted us to shit in plastic bags because a goat can shit there, but they didn't want any Christians to do so. They were monitoring us and every time we would get them, those guys would find their tracks and come to our door and say, "We know you have them" and we would turn them over to them. So, we developed this plan. One of these guys came by and he was dressed better than the rest. He had a compass, some kind of shoes, and a beret. We set up a plan that we were going to go back into the town back to our supplementary positions and we were going to take in these guys.

Well, it's so damn dark and you can hardly see anything, and they surrounded us down around Hamil Tiad, about 115 of them and there were 19 of us. We were in a building where me and the platoon commander hunkered down and they were spilling a whole lot of machine gun fire into us. We had an antenna on our radio that had a chemlight on it, and after that I will never use a chemlight again because they could see that globe glow. They had us pinned bad, and we were scrambling for that radio and that antenna like a god dang greased pig contest trying to get that thing covered up. Then, they were on top of us with RPGs and then they shot artillery, but were too far to the right or left. They shot another one and had us bracketed.

So, me and my buddy took off to the main building because we had a team there trying to see where these defectors were going to come from. We were right on the border, so we could see the closest place to Kuwait City. We didn't have any more radio communication and the noise was so deafening that we thought that building would collapse like that building in Beirut. I'm thinking, "Man, they're done. We're just going to go out there and see if we can find any body parts and get back." Me and my buddy Tilly ran in and cleared the building because it was still standing. It was a three-story building. We got on top of it and found our guys. They were hunkered down, but they couldn't reach their radio. I called in to my platoon commander and said, "Hey, these guys are still here. We're all alive. I'm going to get in a Humvee and we're going to take off."

They called the reaction force under Colonel Powers in the 3rd LAR. Colonel Powers and them were supposed to come and get us within twenty minutes. Well, as they were coming in, their guy had restricted breathing and so they thought they got gassed. So, they went into MOPP mode because they thought they were getting gassed, so it took them forever to get to us. I got my guys down and we got into this god dang Humvee and I started taking off. Well, the Humvee had been shot through the radiator and all the fluids in that radiator keep it cool. So, as I was going, it slowly got slower and slower to the point that the engine was down. I can't see because it is so dark out. Somebody reached up and touched my shoulder through the driver's side and one of my guys said, "Hey man, this thing is not going anywhere." It scared the shit out of me. We left that Humvee there and transitioned on to another place and got back under this bridge and waited for reinforcements to show up. Then I took them back in and cleared the town again.

After that, we took off a day or two later. As we were going down, we hit 5 and went down to 4 and then we transitioned down to Khafji. Right as we got that last OP, there were a bunch of guys there and we dropped off some MREs to those guys and kept on going. As soon as we got a couple of hundred yards away from them, A-10s started circling around and they shot my whole platoon up. Three of them got wounded. One of them was a SEAL Team 6 guy that was with us. He got shot in the arm, but he lived. We were towing that Humvee that had failed before; we unbolted that thing and said the hell with it. It's completely flat there, like a pool table, so there's no place to hide. Those two A-10s circled around us and shot the hell out of us, but none of us got killed.

We already had the airborne frequency, so we called them up and said, "Hey man, get these things off of us." They go, "Hey, you're not supposed to be calling on this net." I said, "How in the hell do you think we got this?" It's a top-secret thing, you can't just get the frequency by dialing it up. They ended up calling them off, but our platoon had been shot up three different times. It was on the front page of the Marine Corps Times of the three times that we got shot at while we were up there on OP. So, this time, it's been since Vietnam, nobody realizes what it's like up here, so I'm going to get a piece of this depleted uranium round and I'm going to put it in my pocket and I'm going to slam it down and say, "Hey, look guys. We had our gun belts and our restricted air, we were south of the border. There's no reason these guys should have shot us up." So, I got that depleted uranium round and gave it back to them. Then a couple of months later after we got back, somebody called down to 1st Force Recon and they were doing congressional hearings and wanted our guys that got injured from that attack to go to the hospital to see if they have uranium exposure. So, Sharper had to go because he was sitting right next to me and the depleted uranium round cracked his rifle. I mean, it was that close. What was funny was Colonel Brooks, the new CO, Mike Brooks, was there at the time and he saw that happen. To tell you the truth, nobody was listening to us. The Air Force never believed that they ever hit anybody and they probably got medals for it.

After that, we pulled back because we did the reconnaissance for the breach. We were the first ones to say, "Hey, when this thing goes down, there's going to be a whole lot of people that will flee out of this place." And as you know, later on, there were thousands of people that gave up right away. We were the ones doing that, but the reason we pulled back is we were there when they started the bombing - we were just sitting there running OP. Our Colonel came up about 100 miles and we said, "What the heck is our Colonel doing up here?" Then he whispers to my platoon commander and says, "Fall back to your supplementary positions." We were like, "What?"

We went down there, and within an hour's notice, that's when all the air strikes started happening. You could see all the anti-weapon fire. It was nighttime so it looked like they were spraying hoses around. We just said, "Man, we are the closest spot to the city." It was incredible. Those guys were screaming with their birds only 10 or 15 feet above ground level and then they would pull up. It was all kinds of junk aircraft, Italians and British and all kinds of shit. During that time, we did get bombed a couple of times and hit us instead of where they were supposed to. They said, "Okay, we'll start the ground war." We were up there with all the oil fires. Within a day, they liberated Kuwait and then we went home.

After that, I was the first platoon Sergeant to make two deployments. Usually, you make one deployment, and they send you somewhere. After I was in Desert Storm, we started training for the 11th MEU with Colonel Hagee who ended up being the Commandant. I did a six month deployment in the Persian Gulf with the 11th MEU and then we came back. Now, I have two deployments which is unheard of for a platoon Sergeant to have. They were going to send me to Alaska to an INI unit and I was looking forward to it, but then they switched my orders and I ended up going to the Special Operations Training

Group in Camp Pendleton. I did that for three years, '93 to '96. After that, General Libutti, the 1st Marine Division CO, fired all of the division scout sniper instructors because they were hazing people. They called over to me and said, "Hey, would you want the job?" I went over to Colonel Powers who didn't want me to go. But I said, "Hey, I would love to go back to Division and unscrew this school." So, in '96 and '97, I went over there and unscrewed the school. That was a great two years there. Then in '98, I got promoted to First Sergeant, so I left the school and became a First Sergeant in 3/1. Immediately, we went back on deployment again with the 13th MEU. While I was in the Persian Gulf, I thought, "I'm going to be here for two or three deployments and it's going to be great." So, I was in India 3/1, we were a boat Company, and they called me about halfway through deployment and said you got orders to Waco, Texas. I thought, "What the heck is in Waco, Texas?" My three Sgt. Majors that were above me called me over to the big ship and they said, "How in the heck did you make communication and get orders there?" I said, "I don't want orders there." They said, "Well, we all want to take orders. Why don't you take them?" I said, "I want to stay here with the grunts." They said, "Well, you've got to take them." They calmed down, but in April of '99, I came to Waco, Texas. I did almost three years here and as a First Sergeant, I thought I'd probably retire. On 9/11, me and the boss saw airplanes hitting the buildings in New York. Right after that, I got orders to Headquarters Company, 5th Marines and that's where I met General Dunford. He was the Colonel of 5th Marines and he said, "Hey, you're in charge of Headquarters Company of my 5th Marine Regiment." Then we went and did the Baghdad thing.

From the time we deployed to the time they sent me back, I think it was 90 days. I was there in February, March. In Diwaniyah, Iraq on May 1st of 2003, they promoted me to Sgt. Major. General Dunford said, "Hey, we got this screwed up Battalion. You and Colonel Kennedy are going to go take it over." So, they sent me back there and that's when I linked up with 2/4. After Kuwait, they sent me and General Kennedy back on the same flight. I got to Camp Pendleton where I'd been before. I stayed in the same barracks as before. I didn't move my family up. I was in the same Regiment as before. I think it was about March 3rd or 5th, I saw 2/4 come back from their flight.

I didn't have any official title. They still had a Sgt. Major. But whatever that date was when they got back, I went up to Travis Air Force Base and watched them get off the plane. It was incredible because most of us had been training for OF1 and we were only there for 90 days. We were whittled down pretty thin, and we'd been training hard and all that. Well, these guys had been in Okinawa for well over a year because the Commandant said, "Well, some people have to watch the rest of the war sitting on the bench." That really pissed those guys off. When they came off that flight, they looked like they had just been let out of prison. These guys were buffed up and all of them knew immediately. I said, "Holy cow! These guys are huge." I was thinking, "Man, when these guys get back from Baghdad, there's going to be a fight." Because we were in a war and you guys weren't and these guys are going to pummel them. But I was there, and I watched them come down. I don't know exactly what day I received the sword for Sgt. Major.

We started training in the middle of the summer. We were Billy goats. We were all over just humping and humping and humping. We'd do Company humps and Battalion humps. Man, I was in great shape. As soon as I met them at the Air Force base a couple of weeks later, I got word back to General Dunford that, "Sir, these guys aren't messed up. The problem is that each one of the Companies is incredibly tight. They've been put through so much crap and mistreated so badly that at the Company level, they are incredible. We just need somebody to pull all five of these Companies together and we're going to dominate." So, that's what General Kennedy did. Then, I took leave, came back, and we shoved off in February.

Janney: At which point did you find out that you were going to Ramadi?

Booker: It might have been when we were getting on the plane. We were still debating where we were going to go. My Colonel, Kennedy, had been in the plans department before. He wanted us to go somewhere near TQ. It was an old fort back in the day when the Brits used to be there in WWII. He wanted us to go there because he thought that was where they were going to put us. But all of a sudden, when we were on the plane, he said, "God dang it, we're going to Ramadi. There's nothing going on there." He really wanted to go somewhere where he could make a difference and he didn't think that Ramadi was the place.

While we were training for Iraq, we learned that all the Companies needed individual Intel nodes, plus we did MOUT with the Marines. If you were a team leader or a squad leader everything was up to you to make the decisions and we had to push everything down to you. We had a big transition to make and said, "Okay, now we're so spread out, we've got to create our own little intel on each one of them; we have to stress the importance of being a squad leader." We had already done all that and I thought it was a beautiful transition and these guys were already set for that. We had them set up for success.

We were taking over from 82nd Airborne and 82nd had some problems with IEDs, but the Division hadn't had any casualties for 9 months. Well, we came in there and we thought that we were going to have them hook up generators and electricity; that we were going to get them some security. We were going to have water and they would all be happy. Well, that didn't happen. I guess my point is, I spent a lot of time, probably about a month, charting all of the things that you wouldn't think of. For example, you go down into Fox platoon and say, "Hey, do any of you have any experience in agriculture?" "Well, sir, I worked on a farm. I did all the agriculture for a million acres." I'd say, "Wow, okay. Do you have any generator experience?" "Yes, sir. I did all the generators for blah, blah, blah." We documented all of that and we ended up not doing it. I really felt that I wasted a lot of time documenting people that had certain capabilities that we never bothered with.

Janney: It's mid-February. What was your perception at that point?

Booker: We took the whole Battalion up except for Echo Company. The rest of us were already up there. During that time, we had guys come down that were from the Army unit that we were going to relieve. I think most of them were Miami cops, all reservists, and they all had a cop mentality. They threaded themselves with the Iraqi police and all their meetings were with the Iraqi police. Because they were cops, they acted like cops and some of them were dirty. They had 10 or 19 people injured. I'm thinking, "That's really something" not knowing what we were getting into. I thought that meant I was going to have to do certain things as a Sgt. Major, so I took those notes. Then they told us some of the stuff that they were doing and we realized that these were dirty cops. The reason that they didn't have any problems was that they were basically in with the Ba'ath party. They were in with the tribes and Islamists; they were bad cops and they were going to go along. We cut that off. I don't know that they were involved with running guns, but I guarantee you that they were into prostitution. I guarantee that they were into drugs. They let this shit go on and everybody went along with it. When we got there, we said, "We're an active duty unit. We're not involved with drugs and prostitution." They let us know that this cop let us do this, this, and this, and we said, "Nope, we're not doing any of that crap." They knew there was a new game in town.

Janney: I heard a lot of the Iraqi cops were dirty and they were actually attacking you guys, too.

Booker: Yes. But some of those reservist cops were dirty, too. They had a cop version of how to keep the

449

tribes satisfied. There was one guy that got wounded and General Mattis gave him an award for heroism. Come to find out he was fighting against us. It was one of our bullets that was in his ass.
Eventually, I called him that so much that, eventually on the intel reports, they started calling him by that name. He was related to this cop and he started mean mugging me when he saw me. He started looking at me and I was looking at him and I thought he was going to shoot me. He said something to me, "You shot my cousin." Or some shit. I said, "I shot him twice. I should have shot him a third time."

Janney: I have read that story, sir. That was a pretty amazing shot that you made. Two shots.

Booker: Yeah.

Janney: I heard you found him underneath a car or something where he bled out.

Booker: No, I thought he was underneath a car because in that area there were a bunch of tractor trailers. I thought I was going to get shot in the legs. General Kennedy called me back because he was behind me, and I got about half ahead of him. He called me back and he pointed down and sure enough the blood was coming out from a carport right there. Weapons Company busted in there and he was lying there. He died a miserable death. We used to call that cop Steve Garvey. Then later on the Intel guys started calling him Steve Garvey.

Janney: Wasn't his name Jibani or something like that? So, that was Jibani or Garvey's cousin. What I read in Bing West's "No True Glory" was that you and Kennedy and some other guys were going somewhere and that an RPG team fired a couple of RPGs at you and you jumped out and made a snap shot and took that guy out.

Booker: Yeah, the situation was that we were at a late-night meeting at the police department. There was a Four Star there. They were trying to tell the Chief of Police what my position was and he goes, "Oh, you look just like my Sheik so and so." They said, "Yes, same relationship." It was April Fool's Day.

    We finally decided I'm going to strip down, put on a man dress and have a light radio. I got in the back seat of this vehicle, and I had my NODs on and I went out there. At that time, they had neighborhood watches and it was cold out and they had burn barrels at the boundaries. There were little gangs all over the place. So, these guys would stop and get them to let us go through these places. I'm getting farther and farther back in these areas that I don't even know about. We got to this house, and I figured out what house it was and we came back. We said, "Okay, we're going to hit the house." I got redressed.

    We got into four or five Humvees, but this time we had shut down Route Michigan which is the center of the city. This time it was a straight line to get there and we said, "We're going to violate it and this time we're going to go down Route Michigan." About half way between the government center and Hurricane Point, a whole blanket of RPG and gunfire just wreaked the right side of our Humvees. At that time, we only had plastic doors or half doors. Kennedy pulled over and said, "Let's kill these son of a bitches." We all hopped out. I call it the AARP fire team because it was me, the Master Gunnery Sergeant, and Colonel Harrell the operations officer, and Kennedy. We immediately got out of the right side of the vehicles and we were pointed towards where the weapons fire was at, but it stopped for a second. I had a M14 which sounds completely different from any other guns, and I could vaguely see about a half a block down. There was a gap and I saw a human being darting into that gap. I don't even know how I could see because it was so dark out, but I took a shot at him. Immediately, Kennedy said, "Who in the hell shot that gun?" because it sounds like one of their weapons. This red headed kid said, "Sergeant Major shot, sir." I said, "I just shot over here and I think they're coming out over here."

So I went down this road as the lead guy and Kennedy and all them were behind me. We're doing the SAS waltz and the guy came around the corner and I could see a faint light above his head. He was standing right there, but he could not see me and he had an RPG. I shot and I think it was a good shot and I think I hit him in the hip because he flinched. Now, he could see me. He was getting ready to shoot me, so I shot again and this time I shot him in the neck. He was already aiming at me and he lunged forward. He dropped his RPG and went running down the street. He went down the same direction on Michigan back down to where he had come from.

So, I went up to the corner and said, "Okay, he's lost his weapon, so I'm going to pursue him." I came around the corner and I kept going and that's when I said, "Man, I'm going to get shot here near these trucks" and that's when Kennedy says, "Come on back. He's right here." We broke in and Weapons Company secured and we found him. There were seven of those guys. They were the same guys that had killed PFC Dane, the first guy that got killed. They probably were the same guys that got Corporal McPherson because he was lost in the same area. We took those seven guys into custody and then we made a plan to go get this guy's house. Well, now I think the sun was about to come up, so we went back and got this guy's house. I can remember specifically that when we raided that house, we went into his bedroom and he had tae kwon do belts in his bedroom all arranged real nicely and it said something like Dallas Karate Association. We got the guy and then came to find out that technically the cops were using us, because they wanted him eliminated and they were using us to get him.

Janney: Wow. That's a level of deception you wouldn't even think about as far as using you guys to do their dirty business. I know there was another story that I heard about a mortar team. You guys got mortared around chow time and it was a pretty regular occurrence. Then on the 25th of March, they got some radar information from the Army and they went out and killed that mortar team and captured one of the guys. Tell me what was happening in early April after that incident.

Booker: I don't have much to say other than each one of our Companies had a different area and were patrolling their asses off. It had started to become apparent that we were having a little bit of a problem with the Companies themselves. Their strengths and weaknesses are coming out and that we had a problem with Echo Company because they always seemed to be dirtier, more tired, and disheveled as compared to others. Echo Company was patrolling their asses off. They had farther places to go, more distances to go, but that's all I remember. I was always going out to check on or patrol with a lot of people just seeing what the strengths and weaknesses were of all the Companies.

Janney: My theory, and I'm sure that you know a lot better than I do, is that the events of early April kicked off because of the Marine battalions that were pushing surrounding Fallujah and some of the fighters just squirted out and headed west to Ramadi. Do you think that kicked off the events of early April?

Booker: Well, you might be right. But, my own question was anybody that was in Fallujah who wanted to fight, we had intel that they were coming from Syria traveling down the Euphrates, and on the way by, they would let these guys stay there. They would take pot shots at us and we were always thinking, "Why in the hell would you allow one of these foreign fighters who's working his way down to Fallujah stop by and fire, knowing that we're going to fire back into it and put your family at risk? We're protecting your ass, but this guy is going to cause you to get shot." So, I don't know that any bad guys fled. I think they were heading towards there to fight.

Janney: So, they were coming in on boats down the Euphrates?

Booker: Not boats - hiking or whatever. Whatever method they could get there. I don't know that they took boats necessarily. But, there were always lots of them walking the highways and walking the desert. You don't know where in the hell they are going.

Janney: That was something else too that you mentioned that I had a lot of thoughts about. The lower-level enlisted guys that I talked to said that they didn't really have a lot of information on this. I was curious about how many of the guys that you encountered there were foreign fighters and you said that your experience was quite a lot.

Booker: I guess this was happening here too. We had an Intel cell in each node I wanted to capture because every time we'd take a patrol, they're going to counter it and you don't have time to write documentation and put it up to the staff. Reporting has to be quick. So, I got with the Colonel and said, "Can we get a camera and each one of the Companies has it? They can come in and they don't have to write anything down because they are tired or they have other things that they have to do. Can we just have them come in and do a quick interview?" Well, obviously that worked a little bit, but it didn't work as well as I thought it would.

However, I have a six or seven minute video of one of our guys in Echo Company that died. He was PFC Pedro Gonzalez. It's interesting because he starts getting into this story and it's a very telling story about how they can dress like this and that, but we can't do it. He starts grabbing his rifle while he's telling his story. It's pretty impressive, but he ended up dying. Another thing was we ended up taking some pictures. I would say, "Hey man, these guys can't figure out the five or six types of mujahideen we have. You got foreign fighters, you've got guys from Syria, you've got this guy and this guy" and we were all looking for Zarqawi.

I thought I saw Zarqawi at the government center one time. That's all we were looking for; we were looking for people like that. But, we were not educated enough to realize all that stuff. We knew a little bit of it, but we needed to get some people that really knew the culture. So, I had a couple of guys take pictures of the guys that they killed. You can get in trouble for that because you are taking a picture of a corpse, but I turned a couple of those in. That didn't work out real well because there are so many different types and they're so nomadic that it's really hard to tell.

Janney: Jonathan Wood told me the snipers would take photos of people for intel.

Booker: I don't know a whole lot about that, but photographs and electronics are important. For example, one time at Hurricane Point, there's the river, but the real part of the river is between us and Blue Diamond. There's a dam right there that everybody would travel across to go to Blue Diamond. On the other side is just a canal, but it's an offshoot of the actual river. I was on top of that thing one time, and I saw this guy that I knew had to be a bad dude. I took pictures of him and he looked like somebody we were looking for. I took pictures of him and took them back to the CP and they rounded the guy up. Come to find out, he looked somewhat smaller when they got him and he wasn't the guy we were looking for. But, that did work sometime. I'm sure there was a lot more of it that I did not know was going on.

Janney: Let me ask you some questions about the 6th of April. I talked to Santiago and (redacted.)

Booker: I love him. Man, what a nut!

Janney: Yeah, he's a great guy. We had a long talk. He's quite a character. He said that they were doing route watch on Nova on those pump houses by the river, and that at some point, there was an Echo Company squad with Lt. Valdez, Katz, Lund, Roy Thomas, and Barron, that patrolled by them. That squad found an IED and they got some EOD guys out there, but they got attacked and took shelter in a house. During this time, Santiago's team was surrounded by up to thirty guys. Basically, Barron's decision to take the Humvee and push down towards Santiago with Benjamin Carmen and Marcus Cherry really saved the day because Santiago said they were about to get overrun.

Booker: I don't know much about that. I just know that Woodall was one of the guys that they left behind in a pump house. Woodall and his squad got into a pump house, and they were taking so much fire that they had to lie down. A guy was trying to coax him out, and later he heard tracks. The Army was trying to get them. Woodall was really pissed when I was with him at the PX. He said, "Those sons of a bitches left me." Apparently, his team had all left the building and Woodall could not hear because it was so deafening and he did not know that he was all by himself. It really pissed him off, "How in the hell could those guys leave me?" The Army guys came and got him, but he was going off on the Army because he was confused. But they got him, and he was all banged up. He had shrapnel all over him. He got shot in the arm and had to use his other arm to shoot with his left hand.

Janney: Let me ask you another question. This is just a wild ass theory that I have, but that Echo Company squad that I just mentioned found an IED on Nova further east than Santiago and those guys were, but the ambush at Nova and Gypsum had been set up for hours waiting on somebody to come up Gypsum. They had a DShK and RPG teams plus RPKs and other machine guns. Do you think that the bad guys had developed enough strategy to try to draw QRF up Gypsum to Nova to walk into that kill zone? Or do you think that they were just waiting because they knew at some point a squad would come up that road?

Booker: Well, I really don't have much intel on that. However, I would not put them in the stupid category. Whatever you said might be a fact because they are not stupid and they didn't waste any time. We should never underestimate them. I believe it's possible that they probably did that.

Janney: Like I said, it's just a theory. Whoever entered that intersection at Nova and Gypsum was pretty much doomed because of the way they had that ambush set up.

Booker: Also, that led me to believe that those guys had stolen armor on them. They just weren't waiting around forever, they knew something was going to happen. They had time to get body armor. I know that one guy that they shot there had a helmet, too. You've seen that picture that David Swanson took, right? So, somebody wasn't just loitering around. They had a reason to believe that somebody was going to come up there.

Janney: Yeah, a DShK is not something that is casually set in place.

Booker: Yeah, you're right.

Janney: So, anyway, that's kind of my thought is they had that planned out. At least they knew that somebody eventually was going to come up Gypsum and walk into that kill box.

Booker: I'll just go with it. I don't have any real intel on it, but I would tend to believe that is probably right. When they got hit, I was on the other side of town with Holt's team and all those guys. Then, I heard that Lt. Ski got hit. I never saw him. The other twelve guys or so, at the end of the day, I unzipped all of their body bags and put them all in one area and inspected them to see what it looked like they died of, and inventory a little bit. When I passed them off, I never wanted to go, "Whoa." The point is that I never saw the Lt. when he got hit. They already had him evacuated.

Janney: Yes, sir. So, did you know Lt. Wroblewski? Tell me what you remember about him.

Booker: Oh, yeah. Well, were you there when they dedicated that building to Lt. Ski?

Janney: No sir. I never got to go to the gym that they dedicated to him.

Booker: I just happened to be on the East coast, and someone said, "Hey, they're going to dedicate a building to Lt. Ski. We went there and we stayed in a hotel, I believe on base, and we took a small ride. It was beautiful. I didn't know that New Jersey was so beautiful. His whole family was there, and one guy was asking me all of these questions and it was his best friend. I couldn't answer him. I really didn't know him that well. I personally was not around Wroblewski that much, but I thought he's an up and comer. That guy is physically fit, he looks like a Marine. He's chiseled. He is nothing but a "state the facts, sir" kind of guy. Info out, info in, but in a good way, not one of those key punch kind of guys. I think he did care about his people, and his people really cared about him.

There was a professional football player there from the '70s, a huge man. I think he was in politics at some time. He was one of their friends and he was there. I wish I remembered who he was. They dedicated the building to Lt. Wroblewski, and when they were playing taps, an eagle flew over right as they were doing it. I still get the hee bee gee bees over remembering that.

You know, on that day (6 April), I had one Lt. and a Corpsman die. That Lt. and the Corpsman really hit a lot of people hard. A lot of people think, "The Lt. is not supposed to die. The Corpsman is not supposed to die" and it really had an effect on them because they think they are invincible. They know that PFCs and Lance Corporals might die.

Janney: Sure, Mendes-Aceves was the Corpsman that you are talking about. I heard that rendering aid to Staff Sgt. Walker when he was shot.

Booker: When we got back there, yeah, he was one of the ones that I had to inspect and same with the Staff NCO Walker. I remember when I opened his bag up, he had his DI t-shirt on.

Janney: Did you know Staff Sgt. Walker very well? Tell me what Staff Sgt. Walker was like?

Booker: Well, when I first met him, he was a monster of a man. He was huge. He had a skateboard. I'm like, "Why the heck does this guy have a skateboard?" He's kind of old to be a skateboarder. He had these crazy tattoos up and down his legs and arms. He was tough and you wouldn't want to screw with him, but at the same time he was the real deal. Later on, when I worked for General Hummer, they dedicated a building to him at MCRD San Diego.

I had a guy that received the Silver Star that wasn't from 2/4, but Walker was his DI. We had to find a place to do this thing because we had held it up for so long it was going to look bad. I said, "Can you get down to San Diego because Walker was his DI?" So, this guy got a Silver Star, a Hispanic kid from Mexico, next to Walker's building. Walker was big and tough, and kind of like Wroblewski, too. He just took information and channeled it back in, a no nonsense guy. He loved his Marines and he knew that they loved him. Most of our guys that were Drill Instructors were in Golf Company, but they had one in

Echo Company that some didn't like because he used to always screw with them. Walker wasn't like that. I think they really appreciated the fact that he could kick ass and really cared about them, too. I had several conversations with him about his Marines. He cared for them a lot. They called him "the Beast."

Janney: I want to ask you about a story that I heard about you. Mike Martinez and Jason Adams were 3/5 guys that were dropped to you as combat replacements. They said that they had just returned from collecting Parker and his guys on that rooftop OP on 21 June. I don't know if they went to Blue Diamond or Hurricane Point, but they were just trying to get some chow. They had just come in from a combat assignment, so they had on their IBA, were bloody, and didn't have their covers. One of the Sergeants there started yelling at them that they weren't dressed properly, were dirty and just gave them a hard time. So, when they got back, they saw you and told you what had happened. You called and gave that gentleman a piece of your mind and told him not to mess with your Marines anymore. They remember that still to this day and love you for standing up for them.

Booker: You know what was crazy about that is there's two parts to the story. First of all, when we came in, we had an Army unit ahead of us that was called the Big Red One. You would think that we would get screwed over more as an infantry Battalion falling under an Army Brigade. Actually, they treated us pretty good and our Battalion Commander probably got away with more stuff and spoke his mind better than their knuckleheads with their, "Yes sir, no sir" bullshit.

They had a guy who worked for a 4 star, at the time, the CSM for the Army at the Brigade level. He was a guy that I'd known when I went to Army Pathfinders School with him in 1990. He was the one that said, "Jimmy, I think I know you. You were in third squad and I was in…" I thought, "How in the hell does this guy remember this stuff?" He was an Army CSM and he was incredible. He ended up getting a Silver Star, too. He left about halfway through that deployment. They transferred him out of there and gave me another CSM. I didn't know that was the guy that ripped my guys because he said, "Tell them when you get across the river that the Post Sgt. Major said so." I said, "Who in the hell is the Post Sgt. Major?" So, I get ahold of my guy who is the Brigade Sgt. Major. I didn't know that he was calling himself those two things: Post Sgt. Major and the Brigade Sgt. Major. Usually, you would refer to yourself as the Brigade Sgt. Major because that's a higher calling than a Post Sgt. Major. So, I called the guy up and I said, "Hey man, I hadn't met you," but I had met him, and he said, "What's the problem?" I say, "Hey, do you know who the Post Sgt. Major is?" He said, "Yeah." And I said, "Can I get his number?" Well, I thought he was looking for the number and rifling through his papers and I say, "Yeah, that son of a bitch" and then I told him the story about my guys going there and somebody giving them shit. He says, "Really?" and I said, "Yeah, I just want to wipe that motherfucker up." And he said, "Well, this is him." I didn't know what he was saying. He said, "I'm the guy." I said, "You motherfucker. First time I met you, you said you wanted to come over here and patrol with my guys." And I can't believe I said this to him, "You will never, ever come on this side of the river. Because you know what? My guys will probably shoot your fucking dumb ass." He fucked them over. Then, he went off and I went off. He said, "Are you serious?" He thought I was joking with him. I said, "I'm goddamn serious, buddy. Don't ever come back here." So, that was kind of cool.

Janney: They cherish you for that, sir. They didn't expect that. "We just wanted to let Sgt. Major know what happened, but he tore that guy a new ass." Laughs. They love you for that, Sgt. Major.

Booker: Laughs. There were two other times that happened. Once at Blue Diamond, a friend of mine, who I thought was a friend, he had served time at Recon and everybody thought he was a badass back in the day. He was working for a one star and he had a word with some of my guys that were going to get chow that didn't have a cover. I called him up and said, "Hey, Sgt. Major, this motherfucking knucklehead at Blue Diamond is telling my guys at 4:30 or 5:30 in the morning they've got to have a cover on. These guys don't have covers, man." Maybe he could have just let them have chow and say next time have a cover. But, he did not allow them to eat and they'd been out all goddamn night long patrolling. He says, "Well Jim, there's gotta be rules." I say, "Motherfucker, you are the most undisciplined motherfucker I've ever met and you're going to tell me some goddamn Master Sergeant is giving them shit." I've never been friends with him since then.

Janney: What else do you want to talk about as far as the deployment? The events of the 6th, we've kind of touched on a little bit and then you guys had a big operation, Bug Hunt, on the 10th.

Booker: Yeah, we had three of them. I just know that we had three bug hunts. The reason that they called it a bug hunt was because we knew that there were certain people, certain "bugs" we were looking for, but we called it an operation and we swept the whole city. We knew damn well we needed to sweep the whole city, but there were certain people that we were looking for blanketing underneath it. I don't remember who they were. I thought maybe it was IED guys. I thought the original bug hunt started at places that were making IEDs. But, I just know that it was incredible to see these guys in the gear that they are wearing all over the city. It was just a masterful thing of professionalism, all the Companies working together. That's what I really took out of it. We can blanket this city and they're so effective at it and they're so good that these squad leaders are out there on their own. They don't have Lieutenants with them personally full time. That's the only thing I have there. I thought we were going after IED guys at that time.

Janney: I know Sgt. Majors don't hang out with Lance Corporals and Privates, but do you have any other remembrances of the guys that didn't come home?

Booker: I've got bunches of them. For example, I was walking down a base road in CA toward the PX and I see this truck and some fucker had painted it all black with a spray can. I looked at it and I thought, "What the fuck is this?" Above me to my right, there's a bunch of these guys on the third deck looking down at me asking, "Sgt. Major, doesn't it look bad? Doesn't it look mean?" I look up and it's Contreras and I'm like, "You son of a bitch." Other people might have got pissed off at stuff like that, but I just looked at it and said, "Hey man, that's just the way grunts go."

Another one is about Otey that died on the roof there. One time when we were getting ready, I think it was for Bug Hunt to tell you the truth, we stationed ourselves in Echo and Golf Companies' area for the night at Combat Outpost and we were going to push out in the morning. The CO, once again, kind of pissed everybody off when he said, "Sleep has priority." Sleep never has priority. Usually, you work and maybe you don't sleep for a whole night before you go. But the Battalion Commander said, "At 2300, all of them need to be in the rack." They had never heard that before. They asked, "Sgt. Major, what the hell is sleep has priority?" I said, "Hey motherfucker, we might patrol for three days. They are going to sleep." The guys were pissed. So, I was going through the area, and I saw these lights towards the outside of the base where the barracks were at. I go over there and I see this light go on. So, I go in and there's two guys in there. One of them was this big, tall kid and one of them is Otey and Otey is cleaning his rifle. I thought maybe their team leaders were screwing with them, making them clean rifles. I said, "Why in the hell is this light on?" I was getting ready to jump their squad leader and say, "Why are you screwing with these guys?" because I thought they were trying to rebel against us a little bit. Otey started to step up and I said, "No, no, sit down." I said to the other kid, "Sit down, what the hell are you doing, man?" Otey said, "Well, I heard that sleep was a priority and I know we're not supposed to do it, but my rifle got shot yesterday. I just got this new rifle and I just want to make sure it's clean for tomorrow's operation." I looked at him like, "Oh my God, I'm glad I didn't just rip their ass." I mean it made me feel so bad, and later on he died on that rooftop.

I've got so many stories about these guys. These burn pits that we have, I spent a lot of time in burn pits and looked in on these fuckers. They know that they are telling a story that I can hear. First of all, you can listen to the music that they have because if they are burning shit, they've probably got some music on or you're listening to these crazy fucking stories that they tell. Or you sit there and listen to them and maybe you'll chime in on something. I learned so much shit in burn pits. Some of it, I took actions because, "Okay I know somebody is fucking with them right now" and I'll go fucking use it for intel.

I had a great relationship with those guys. I would go on patrols with them. For example, when Savage died, I went on the next patrol with them because of how I felt when they took his gear out and were inventorying it on that squad bay. I told the CO, "Sir, I'm going out with them." He goes, "No, I want you to come back." I said, "Sir, I think something bad is going to happen. These guys are going out for blood." It reminded me of the My Lai massacre, reminded me that these guys are savage, blood thirsty sons of a bitches and if you ever give a wink to them like, "Yeah, it's okay," they'll go out and kill somebody. Then later on, they're going to have court marshals and regret it. It's just not the way we did it. I've been there. I wanted to kill these fuckers too. So, as they were leaving the patrol base, they all were touching his chair. He had a little chair that he sat in. I could tell in their eyes what they might do. I had a handful of mints or candy and I would give them to some of the kids. They were looking at me with them on a squad sized patrol and it calmed them down. It really made me feel good that there are so many areas like that. If I had just gone around and said, "Hey, dude, light them up" then they would have killed somebody and it would cause all kinds of problems. I have so many of these stories and I love them so much. Being an Infantry Battalion Sgt. Major is the best job ever. You know, there's other guys that I really wish you'd talk to. Do you know Hodges?

Janney: Yes, Reagan and I are supposed to talk soon.

Booker: I feel really bad about Hodges because of all the shit that he had going on. Hodges is a tough son of a gun. He was never afraid of a fight. He had a wife that was raking him across the coals, seeing another guy. I knew that he was getting fucked over by her, but he would never listen to me. By the time I came in, I would have to get him up there and say, "Hey look dude, you can do this, this and this." At one point, I thought he was going to kick my ass in my office. He was just getting ready to explode. I said, "Look Hodges, you just can't do that" and he kept violating every one of the orders that we gave him to stay in the barracks and not make contact with her boyfriend; the guy had a protective order against Hodges. I said, "Dude, you're pushing it." It came to a point where I told Captain Weiler, "Hey, this guy is going to get burned, man. He's violating and he wants to kick my ass, so he needs to get burned." Well, come to find out because he got reduced to Lance Corporal, he couldn't stay in the Marine Corps because he already had eight years in the Corps. I met him again at one of our memorial days in Houston and he was sitting there looking at me. Everybody kind of melted away and were drinking somewhere. We were just sitting there and I said, "Well, buddy, what's it going to be?" Hodges said, "Hey brother, I'm sorry for the way I acted." I said, "Brother, I think about you all the time. I wish that we never had to do what we did." It really bothers me. Now, he's involved with football programs, powerlifting. He's really done a great job, but it really bothers me that I couldn't do anything for him at the time. But, now he's doing great. He's doing incredible stuff.

Here's another thing too that I feel good about. When we got those 3/5 guys, we got them in

Hurricane Point in a big conference room and I said, "Look, guys, I know General Mattis really well and I know what he planned. If they're going to train together, they're going to fight together. When I send them to you, they're going to stay in the same squad." Well, I violated General Mattis' order and I didn't realize that it was such a big deal. I knew what his intent was, so I thought I was following his intent. His personnel officer to our Battalion one time called and I said, "What I did is I sent these guys to different units." She said, "What! You can't do that. General Mattis said blah, blah, blah." I said, "He's got machine gunners with machine gunners and you can't deploy them like that. It just doesn't work."

So, what I did was I said, "Weapons Company is doing this. Most of you guys are into machine guns. They're mobile assault and they do this kind of activity. Then you've got Echo Company. They hump farther than ever. They go way out in canal zones, mixed rural/city. Golf Company patrols in the city – they're good at that stuff and we need this kind of work. Fox Company generally is sitting in one spot, and they patrol a little bit." I said, "You've got a corner over here for all of you guys that want to go to that Company. You've got another corner over here and you guys go to that Company." And you know what, all of where they went was the exact number of people that I needed in each Company. Just by me dictating what their performance would have to be. I said, "Man, this is cool." I let them decide what they had to do. A lot of the 3/5 guys hated their Battalion Sgt. Major. I know who he is, he's a hard-core Drill Instructor and not much on tactics, but he's a drill connoisseur. They hated him. We were NCOs, we were looked after and took care of our people. When 3/5 said that they were going to deploy again, they just took all of us and shoved us all to the side and said, "You're not going to train with these guys anymore. Have a good time and sooner or later you're going to get out." They really felt that they had something to offer, but they were whisked away and now all of a sudden, we're going to say, "Okay guys, now you're going to 2/4." They probably thought that 2/4 was fucked up. They probably said, "Goddamn, these guys are going to get a bunch of people killed." But, then they realized that this was a good Battalion.

One of the guys that died, I called him our 36th guy. His name was Corporal Ryan. He was a little out of shape, good haircut, squared away. He was a machine gunner and I sent him over to Weapons Company. He decided he was going to stay with 2/5, but 2/5 didn't think they were going to have any casualties. They came in, "Okay, you guys probably screwed up" and sure enough they started having casualties right away. Well, the operations officer, they called him Frenchie; he was a Major, born in France and had kind of a French accent, but he's a tough dude. He was their S3 guy. When he came over to visit us, he got a Purple Heart because a bomb blew up and engulfed his whole face. His whole face was burnt up. Most of the time, it takes a while for the face to turn charcoal, but his face on the second day was charcoaled up and I was thinking, "Holy shit, these guys are going to think what the fuck happened to Frenchie?" The point is, we left a machine gunner with Weapons Company in 2/5 after we left.

Corporal Ryan was always cleaning his machine gun and doing what he had to do. One time he had his machine gun at the government center, and it was tilted up. We were just starting to move out and there were buildings to the right of us in the compound and this electric wire hooks on that goddamn thing and explodes. I thought it was an RPG coming at us. All of a sudden, I see my guys running backwards, not towards the fight. Within a split second, I'm thinking, "What's up? Our guys always run to the fight. Why are they retreating?" I got out there and it was fucking Corporal Ryan laying there and his mustache was burnt off because he caught that electric wire. We were laughing like, "You son of a bitch. You fucking burnt off your mustache. I knew it was something having that fucker, but now you don't have one." He was shaken up. Anyway, when we left, we left him there.

The 2/5 Weapons Company stayed in the same place that our Weapons Company did. They did the same kind of mission. Their Company Commander decided that he didn't want to have anything more to do with this war thing and refused to go out anymore. The Battalion Commander, Randy Newman, had to relieve him. He needed to get Frenchie, who was his operations officer, to go take over the Company. Both of them got killed right outside Hurricane Point. I always felt bad that we left Cpl Ryan behind. His parents do come to our anniversaries. We were at Conde Hall when he got promoted. They got that building at Quantico dedicated to him. So, Ryan's still part of the family, but I have a whole bunch of those heartbreaking stories.

I remember seeing Sgt. Conde in the chow hall. Sgt. Conde who was killed in Weapons Company. Word came out towards the end of this thing about my transfer. I guess the word was out that I was going to go somewhere else to get another unit. I didn't want to, but SgtMaj Estrada who came down to visit us thought that I'd seen too much shit and that maybe I needed to go somewhere else. I went to SgtMaj Kent and I said, "Sir, why are they going to make me leave? Am I fucked up?" He said, "No, they just think that you've seen too much stuff. When you leave here, you go to another unit." I said, "Sir, I want to stay here." Well, it didn't work and so word got out and Sgt. Conde was in the chow hall right in front of me. He just sat down and said, "Well, SgtMaj, I heard you're leaving us." I liked that these guys could stand up to you and they didn't cower down. They gave me respect, but I liked when they stood up like that and said what they had to say. I remember when Conde and his men were fighting down Easy street and he got shot. Conde had been shot in the back on his shoulder and didn't want anybody to hear about it. I felt bad that he hid the wound he had on his back. It fucking laid him open for quite a while, and later he was killed by an IED.

Janney: Sgt. Major, I'd love to hear whatever you want to tell me. One thing that I did hear about Royer was that he took Lt. "Ski" Wroblewski's RO on 6th April and that was why Lt. Ski had to operate the radio during the ambush instead of fighting his men. Royer just wanted 3 ROs for some reason. (Author note: Based on my interviews of the QRF Marines and Lt. Ski's RO, this severely hampered their effectiveness on 6th April because Cpl Smith and the rest of the uninjured QRF had no comms after 2ndLt Ski's radio handset was shot when 2ndLt Ski was mortally wounded.

(At this point, Sergeant Major Booker proceeded to tell me the backstory of Captain Royer, how he left 2/4, and the reasons why Royer left. As some people know, Captain Royer was a controversial figure. For many reasons, including my attempts to avoid dishonoring the Corps or any Marines, I have elected not to include this portion of Sgt. Major Booker's interview. I have left in the portions of the Sgt. Major's comments about the Marine awards process, because Sgt. Major Booker and several Marines expressed concern at the lack of awards for Echo Company Marines and I realize how important it is to recognize the valor and sacrifices of your men.)

Booker: At that time, we were handing in awards. I had worked three or four times with General Dunford, Colonel Dunford on awards boards before we left and I didn't know anything about it and now I'm in the know and I said, "Hey, we have General Dunford over here from division CP and is talking to our awards boards guys and tell us how he looks at awards." So, therefore come up to his position and it will be greased lightning because I already know how he thinks. So, he came down, and as only Dunford can do, he said, "I think this is a Navy achievement, I think this is a Navy commendation, I think this is a Bronze Star, I think this is a Silver Star" and so on. We all said, "Wow, he makes things sound so easy." Royer had not submitted any awards. I went to his First Sergeant and said, "Hey dude, we got all enlisted guys here and your clerks. We're going to screen your awards and later on we're going to bring your officers in and we're going to tell them the things that Gen Dunford suggested and if they agree to it, then they'll pump it up to their CO and it will be perfect." Well, some of the officers were asking very legitimate questions like, "Sgt. Major, why are we not involved in this?" I said, "The time will come. We're going to do the rough work now. We'll get General Dunford to explain it and you know what, you guys don't need to worry about this shit right now. You guys are going to have a key part in it. We're going to rough it out and it will be smooth as silk." Colonel Kennedy didn't want to have anything to do with it. At one time, he had already said to me, "I don't want you to be involved in these awards boards." When I was first printing them up in California I said, "Hey, Colonel Dunford wants me to come up to headquarters and do this shit." But he didn't like it. But now it all paid off. Kennedy did not trust that I was going to do it right and I felt so good that he didn't have to worry about this shit because he has too much to worry about. Well, Royer didn't have any awards. I was like, "What the fuck is going on?" It was really disgusting and after that, things started coming out that were so pitiful and I felt so bad for those guys. There was a Company Gunny that left early, Coleman, who I respected, and he ended up becoming a Sgt. Major. He had to leave early for some reason, but when he passed through my office, he said, "Hey, I didn't say anything to you, but all that shit that you've been saying about Royer, it's true." I felt so bad that I didn't push it farther. I think that's why those officers and staff NCOs, their troops loved them so much, because they stood in between them and Royer. So, even though Royer was doing some stuff, those guys protected them to the point that they didn't let it out.

Janney: Let me ask you this, going back to Parker, Contreras, Lopez and Deshon Otey, on that OP, that house that was under construction by the Iraqis. I was told that Royer wanted that as a fixed position, basically as a projection of force. Somebody on that rooftop seven days a week, twenty four hours a day. They did it with two rotating four man teams. But even as a civilian, I always thought that's not how you utilize snipers. You certainly don't put them on a rooftop without security downstairs on a two-story building.

Booker: I believe that it was the team leader Parker's fault that they got killed. Here's the deal on that. First of all, Parker was the only one of them that was a sniper. The rest of those guys maybe wanted to be a sniper one day. They were going to deploy those guys to learn how it is. Well, when we received the word that went down, I was visiting somebody and Kennedy at the hospital. By the way, the woman that treated all of our wounded, I called her a doctor, but she was a nurse, I know her and she's stationed in San Antonio at the burn unit.

They came in and started working on that house every single day at sunrise. I believe the people that killed them intercepted the workers and told them to stay away. They dressed up like they were workers, came up there like they thought they were going to. They were probably a little friendly with the guys because they had been there so many times and they shot them.

Janney: I've actually seen that NCIS report that you helped put together. I have no idea why they didn't have security set up downstairs because, like you said, those workers were in and out of that house all throughout the day during daylight hours.

Booker: Yeah, and not only downstairs. There's only one way to get up on that roof. They could have had tape or cans alerting them that they were coming up. I don't fault them. If it's daylight and we're getting ready to extract, some of the guys could take a nap then. There are less important times to be on alert. Whoever was there on that team that was supposed to be up did not alert them.

Janney: I talked to one of the guys, Ben Musser who was part of the other four man team that was rotating out with Parker and his guys. Musser said that on 20 June, the day before the team was killed, those Iraqi guys brought them ice which he thought was weird as hell because where does an Iraqi get ice in the middle of the summertime. But, they brought them ice. He thinks that they were basically scoping out where everybody was positioned to get ready to murder them the next day.

Booker: If you shut off the recorder, I can tell you this part.

Booker: His mother and family members were there in the middle of the night and we woke them all up. We told his mother what happened and she said, "Oh, no. He's not involved." We said, "No, he just killed a bunch of our guys." A couple of his brothers and sisters were sitting there and we told them, "Look if doesn't come and give himself up, we're going to blow up the house." So, we ended up blowing up the house and it took 400 lbs. of TNT.

You know, I went back in '09 as a Regimental Sgt. Major and I had everything from Fallujah to Hit to Lake Tharthar and down. That whole thing. Colonel Caporia is the Colonel that I said that I was going to fly back with when I made First Sergeant during OIF1. What a good guy. He ended up making Three Star General. Well anyway, when I went back, he had 1st Marine Regiment. 6th Marines in '09 and we took charge of their area. 1st Marines went away and we had 6th Marines. We were based out of Camp Ramadi which is the place where my guys got into it with that Sgt. Major. I saw this guy named Otey and I almost belted out, "Hey, I had this guy named Otey." But I didn't do it. I went back to the Sgt. Major, who was the outgoing Sgt. Major with 1st Marines and said, "Hey dude, I lost a guy named Otey." He said, "Yeah, I think his cousin was that guy. I heard something about that." So, I took him out there, he was a supply guy, and I said, "Do you want to know more about what happened to your cousin?" He said, "Holy shit. Yeah, I would." So, I took him out there. At that time, they didn't have a base there, but right across the street from Route Michigan, they had a new base there and the gun target line for the machine gun is right there on the south side and it's facing right where that house was. I put the kid on the machine gun target line. "Your cousin died right there." I told him the whole story. It was an emotional moment because now he could go back and say, "Hey, I was there."

Janney: Yeah, I'm glad that you got a chance to share that with him if he wanted to hear it.

Booker: Oh yeah. He said, "I want to hear it. We don't know." I said "Well, I'm going to take you to exactly where it happened." That shit doesn't happen very often, you know.

Janney: Ironically, I'm sure you know this, Deshon Otey was the only person that escaped alive out of lead Humvee on Gypsum and Nova on the 6th of April. Everybody else in that vehicle was killed, and then for him to get killed on the rooftop several months later was beyond tragic.

Booker: Yeah, he was a good goddamn kid.

Janney: I've got a copy of his AAR from the 6th of April.

Booker: Well, you know, what's funny, sir, I think it was the New York Times, maybe I'm wrong. Okay, I'll start backwards. After I left 2/4, I went to 3rd Tracks which was, ironically, in 1st Division, but its right down by the beach. But, I remember going to chow and for some reason, at that time, I remember every one of our dates exactly that somebody died, it stuck with me. That day I was going, "Man, this is the day that that happened." Then, sure enough, NCIS or CID, somebody called me up and said, "Hey, we want you to come down here and answer some questions." I went down there and introduced myself and this woman has all the doors open, but she's in her office. She said, "I want to ask you something about this incident." I started answering questions and she was acting like this was new information and I said, "Lady, this thing was completely investigated by General Mattis. You know how he is. He has a determination of guilt or innocence within 24 hours. This is a year old. I remember the day." She said, "Well, we've got these questions that we've got to answer." I said, "Lady, it's already been taken care of. I'll tell you whatever you want to hear, but this has already been documented." So, I started to tell her and all these people, and I was pretty raw. They were kind of leaning back going, "Holy shit, I never heard anything like this." I told them the whole story about how we took them down to TQ and said they were shot.

I then was transferred. I was in 2/4, so there's a Regiment there and everything you pass up goes up through a higher level. Well, there are separate Battalions. A separate Battalion is AmTracs, CID and other units. You do not have a Regiment, so you've got to be a little more on your game as a Battalion Commander because you don't have somebody to protect you or clean up your work as you're going up the Division. So, I ended up at 3rd Tracks and then after that, the natural progression for a Sgt. Major is for you to work for a Colonel. So, they sent me to Ft. Sill Oklahoma and from 2006 to 2008, I was the Sgt. Major of an artillery detachment. I had an office there and the phone rang. I answer this phone and this reporter starts asking me questions. I think he was the guy that was trying to say that Royer was a whistleblower, that this guy has written documents on how screwed up the body armor was and how screwed up the IED deflection equipment on the Humvees was. So, he was going that route and he was leading me to the point that Royer was railroaded by Kennedy and da, da, da. I said, "Whoa." I'll tell you what, I had a little PTSD back in those days and I just went off. I said, "You motherfucker, I can't believe you're saying that. What documents are you talking about?" I didn't know what document he was talking about. He said, "I'll fax it to you." Sure enough, it was that document that I said the NCIS or CID was that day. All they did was block out little words, but at the end of it says, "Now, the Sgt. Major is currently stationed at Ft. Sill, Oklahoma." I'm like, "What the fuck good does it do if you block out the shit that I said that you think is classified, but you can still call me?" I went crazy. I called up Kennedy. He was in charge of PAO at the time and he put me on the phone with his Colonel at the time and I said, "I'll kill this motherfucker. This sonofabitch is saying things. He's accusing me of covering up this deal." They said, "Settle down." Sure enough, they called the guy off. But I couldn't believe that shit.

Janney: You know, my biggest thing is that I don't want to write anything that is going to dishonor the Corps. I'm not trying to have a "gotcha" moment or anything like that. I just want you guys to tell your stories as they occurred and kind of let the pieces fall where they may.

Booker: I appreciate that. I'm really proud of what we did. Obviously I've not spoken to anybody about a lot of this stuff. I don't know what angle you're taking on it, but I'm satisfied with what we did and I'm glad you're telling the story. If I did something wrong or something like that, I think history will tell it. A lot of times, some of our guys would say, "How about Fallujah? Why are they putting so much effort into Fallujah? We've got a harder job here." I've heard General Mattis say this, "Hey, if we put one Battalion there of 2/4, we could hold it. We're some fighting sons of bitches. If that capital would have fallen, all kinds of stupid shit would have happened."

Hell, they had, what, five or seven Battalions in Fallujah which is a dinky in the middle of nowhere town, and our guys had to patrol through all this brush and trees. It's alongside a river, so the terrain is much different. Whereas Fallujah, hell, I went there again in '09, and of course, we only had two guys die, but it wasn't the same. They were starting the general elections then and we pulled out in July, but I spent a lot of time there. You know, they don't have a lot of trees. Everything is a Euclid system. They go block by block. There's no river running through it. It's square and you can navigate it. We had a hard job and a lot of our guys didn't realize it.

You know, we had two guys drown in Fox Company and that was the time that General Mattis said, "2/4 is our focus of effort." When they said that, I've never felt like that, but when we're the focus of effort for the Division, there's more shit coming to you and you've got to hold more shit back. We found these two guys, Green and Schrage, who were drowned in the river. That bridge that I described, upriver from Ramadi. I told General Kennedy, "We need to regulate this thing and bring the water down because there would be less space for us to look for these guys if they get washed down the river." We found the bodies and we thought we were drowned there, but we didn't know if they got captured. Also, they found a hat later on. I thought, "Man, what if these guys were captured?" We just didn't want that to happen. Finding them dead would be better than seeing them get their heads cut off. So, we lowered the river down. We lowered it down to the point that all those irrigation pipes that they have that come off the banks that feed the irrigation systems couldn't draw in any water. All the farmers said, "Hey, we can't get water to irrigate our stuff." They started pleading to their tribes to give it up. I was thinking, "You know what, if my catechism teacher could ever know that I took the Euphrates and let it down it would be like…"

Janney: Yeah, I heard you guys practically drained it and almost shut the river down.

Booker: So then, they'd think I was the antichrist. Man, who in the hell ever thought that I would be the man that drained the Euphrates River?

Janney: Draining the Euphrates. It's kind of like Moses parting the Red Sea.

Booker: That's right.

Janney: Sgt. Major, you guys did a hell of a job with one Battalion where they had the use of four or five Battalions in Fallujah. I don't know how you did what you did. I have nothing but the utmost respect for you and your men. It's amazing that they held the city. It didn't fall like they thought it would. It was just an incredible effort and the bravery of those young men never ceases to amaze me every time I talk to another Marine and hear their story.

Booker: Well, you know, maybe this isn't true, but my opinion is, it started when Kennedy said, "We're not having these old men's pictures on the walls. If you're a squad leader, you are held to a different standard. You can't lie, you can't cheat, and you can't steal. If we catch you at any of that stuff, you are not going to be a squad leader." He held them to such a standard they didn't have to go higher up if we couldn't handle the situation. Because the Army would have loved to get their armored personnel vehicles and just rip that place up. I've got a picture of them clearing a building with one of those M113s. It was the OP Hotel that we had and those sons of a bitches had a 50 cal just raking the hell out of it. They called that clearing. They didn't even get out of the vehicle. Then, when we left, they brought tanks in. Our guys didn't need all of that shit.

Janney: You didn't even have armor. You had plastic doored Humvees.

Booker: I don't know what else you've got for me, sir. If there are any times that you've got follow up on this stuff or I can remember something, there's a lot of crazy stories of what my guys did.
Him and another guy, they knew this guy was bad, but they couldn't arrest him at the time. So, these fuckers took a bag of flour and poured it on him, so he would be white so everybody could see him. Oh, God, who in the hell thinks of shit like that.

Janney: I want to ask you one more question before I let you get off the phone. I heard a story from one of the guys that I interviewed. Maybe you already told me this story, but I think it's a different version of it. You grew your beard out for a couple of days, then you put on a dishdash, slung an AK over your shoulder and you and Contreras just walked out the gate into the city to see what was going on.

Booker: Well, it's a version of that. We couldn't get enough secret squirrels to come in and get intel for us. We did by squad, but we didn't have a whole bunch of people doing intel for us. If we did, they reported to something else; Delta guys or someone else. That Chief's name, the 4 star, was Jordan. Chief Jordan was the Chief of Police there and we ended up arresting him and he was dirty as hell. I kind of felt bad that they put him in a prison there.

    We were relying on the Army's counter battery fire and it blew our minds that sometimes the Army would call up our CPs down to the Company level, bypass the damn Battalion and say, "Are we clear to fire?" They said, "What are you talking about, clear to fire?" "You have troops in our area." We said, "Whoa, whoa, we're not going to answer that. What are you talking about? You're going to fire in our area?" But the Army's stance, in their mind, is that if you don't have troops in the area and the CO wants to do harass and interdiction fire, if you don't have troops there, he's going to do it. We said, "Whoa, whoa, we don't operate that way." So, all the time from the glass factory, we would hear mortar rounds coming in. They always came from the same direction. On a moonlit night, on a Thursday, you knew damn well they were coming in from there. So, I said, "Hey, when we have Ops, we don't have enough troops. They can sneak up on us. If you know there is an OP there, you can sneak around and get to all the hide spots - get right up on them and shoot them.

    So, I proposed an idea that I would get some guys that were darker skinned like me. Let them grow their beards out, dress in the garb, get a couple of cars. Outfit with light radios in the front car and a heavy radio in the back car, and arm them. We'll take the cars from the prisoners or the people that we put away. In some cases, we painted them a little differently. We kept them under cover. So, we went out. We would always tell the Weapons or Golf Company that we were in their area and it was a very precise thing. We found about 5 or 6 of these guys and a couple of them acted like they were tough, but they melted away real quick and they were uncomfortable with this. I said, "Hey buddy, no harm, this is all a volunteer thing." So, I got Santiago, me, Hekmati, and Amir. Hekmati and the other guy both have the same first name, Amir. Amir lives in Michigan and the other Amir, Hekmati, when President Obama gave that money to Iran and they got four hostages out, Amir Hekmati was the Marine. He got transferred to us because of that money that he gave that Trump thought was the worst deal ever. Amir is a good dude.

    So, I got all four of these guys. A couple of them didn't pan out and that was no problem. There was a lot of discipline that these guys didn't think about that Recon guys would do. I said, "Hey man, if you ever lose communications for a second, we stop what we are doing and we go back." There were times when I would give a comm check from the first car to the second car and back to the Company. If it ever broke down at any time, we stopped where we were at and went back or fixed it. Sometimes, I would stop a mission and say, "Let's go back and do it again." These guys were looking at me and were thinking, "Damn, I think Sgt. Major is a pussy. WTF?" I said, "Hey dude, if you don't have comm, they are going to annihilate us." If you don't have comm, there is no reason for you to be out there doing reconnaissance. So, what we would do is we go out there and we would go around these outposts and see if anybody was coming up on them. We also had another mission to go out ahead of vehicles. There was another mission to go to these places where these mortar teams would go, and at that time, on a Thursday night, we would go where we thought they were planting these mortar teams. What you would find there is a whole bunch of these homosexual motherfuckers all getting freaky deaky. Because that's where they hang out. So, I didn't wear the body armor, but all the other guys wore body armor so they were protected. They had the full deal. I had pictures of them. I would wear the sandals. But, the first time that we did it, we had to delay the op because my feet were so white when I put the sandals on. So, I had to suntan my feet and get them dirty. The reason that we stopped it, there was a group of people at Special Forces that the Iraqis had called the Shishwanis or something like that. They killed the guy that was in charge of that, but they were fierce fighters and they came to live with us. As soon as they came in, I told Kennedy, "I'd like to do this again, but this could go bad. If we have the Shishwani guys here, there's no reason for us to do it." I brought the guys in, told them to shave their beards and we're not going to do it anymore.

Janney: I can't even imagine Sgt. Major. Like I said, that took some giant balls to do stuff like that.

Booker: Yeah, well, you want to support these guys. I had a guy named Sgt. Major Ellis, who I brought over. He was with me in the first Gulf War. He was my communications guy when I got in that gunfight in Hamil Tiat. He was a Sgt. and got meritoriously promoted to Staff Sgt. by the Commandant of the Marine Corps in about '89 or so. Then, he became a First Sgt. on the east coast and wanted to be the First Sgt. in my Battalion, 2/4. So, I got orders for him and he became my headquarters First Sgt. because he was the most senior guy and was more suited for that. When I left, First Sergeant Joe Ellis got the office and he was promoted to Sgt. Major. He ended up dying on December 7 of 2007 with Cpl Jennifer Parcell in a suicide bombing. He was a great man and very loved.

Janney: Yeah, you guys were incredible. As I said, it's just an honor just to speak with you today. I've heard so many stories about you.

Booker: Well sir, you know, on the other side of the city, you know the guy that President Bush met with and called him the hero of Ramadi and all that stuff? The guy that did the Awakening, they called it?

Janney: I don't know his name, but I heard a lot about him when I was over there in '08. That was the buzz word when I was there with 2/8. The 2/8 CO said, "Now they do their own security. They got tired of having their women and children blown up or getting caught in the crossfire between foreign fighters and the Marine Corps."

Booker: Yeah. We knew that family very well. I had been to their house several times. It's kind of funny because when we first got there, his brother, who was now the Abdullah, he was the one that lived. All of the other ones died. I went there in '08 and I saw him and they remember everybody. If they see your face, they'll know you. They came in with General Tryon in '08 when I had 5th Marines and they had this big meeting. This guy looked at me and he came up talking to me. I'm sitting down at the end and I'm going, "Oh my god." He said, "I know you, I know you." I said, "Oh, yeah, yeah." He wanted me to sit with him and said, "You must come to my house." I went to his house. General Tryon hated that kind of shit because if you were enlisted, he had no respect for you. They had a very tight relationship where only certain people could talk to certain people. They were always talking about firing the Chief of Police in Ramadi in '08 because he was dirty.

Well, I got invited to the Sheik's house. I went out there and they had built this place up. They called it something like the Orange Dome - some type of football analogy thing. But anyway, they built this house up to something incredible after Sheik Sattar, the one that was killed. When we were there in '04, the big guy in '09, Sheik Sattar's brother, was our plumbing contractor for plumbing. That's how tight these guys were. We build them up and build them up to this thing and now his brother is working for us. They took Abdullah and sent him to Syria because he was the heir apparent. So, they got him out of town.

Well, the guy, who now became Sheik Sattar's guy, he was always overdressed with all the bling, and all this Arab looking shit. It's funny, I was at their house one time and they had this horse out there, and his Dad had a 45 like Clint Eastwood, the golden one, the silver one, and I knew him very well. It really bothered me when they killed him because they wouldn't give his dad his head back. So, they killed his ass. But anyway, they were plumbers when we were over there. They were dirty as everybody else was at one time. They called it the Georgia dome. That's what it was. It was all orange.

Janney: The Georgia Dome?

Booker: It was the Georgia Dome or the Peach Dome. It was all orange. My point was, when I met him during '09, General Tryon said, "Oh my God, he's stalking the Sgt. Major." The Sheik said, "You must come to my house." He took a bunch of us to the house, but General Tryon was not allowed to go to the inner sanctum where I was at. Me and my Colonel were there. I knew General Tryon probably thought, "You son of a bitch. How come he's going to the inner sanctum of this place?" So, I went there and took all the pictures with Sheik Sattar and all the people, they always had a picture of the Chief of Police they wanted to fire. General Tryon was always saying, "We're going to fire this guy." I told my Colonel at the time, Colonel Lopez, 6th Marines in "09, "Sir, they are never going to fire this guy. He's in every one of the pictures. He's as dirty as they are. He fought with them. There's no way that they are going to say okay, you white bastards, you can come in here and fire my chief of police." He shook his head and said, "God damn it, man." So, it was kind of neat when you meet these guys and have a relationship with them.

Janney: That had to have helped. As I said, I met a couple of Sheiks while I was there with Captain Martin who was in charge of the JSS Karama where we stayed. Just inside the east gates of Ramadi inside the arches.

Booker: That's right. You know, I've seen those pictures with Mr. Wroblewski and General Kelly. You've seen that picture where, at one time we had six 4 stars? I worked for all those guys except for General Allen. I did meet him two times, but one time when General Powell was giving me an award for General Dunford, I had to fly back and get it and he was standing on the stage. I worked for all those guys and it's kind of funny that Kelly took you guys - he's a son of a gun. I was in his house one time when both of his sons were alive in Washington DC in the Navy yard. His daughter is gorgeous. All of his Irish buddies were drinking beer. I was thinking, "I'm not in a beer drinking mood." So, I saw her with a tumbler glass and I said, "Where did you get that shit?" She got me a glass and this whiskey was incredible. General Kelly came into the kitchen, looked at me and at that glass and said, "You son of a bitch. You got into my shit." Laughs.

Janney: That's funny. Yeah, he was very good to us. You know, when he found out about the mission, he supposedly said, "Well, if Janney's got him this far, then I'm going to get him the rest of the way." So, General Kelly and his PSD escorted us to Route Gypsum. We were about 200 meters south of the Nova-Gypsum intersection. It was as close as we could get and maintain some semblance of security with General Kelly being there, too.

Booker: Yeah, that was something. I studied those pictures a little bit. It was incredible.

Janney: It was an honor to do it, sir.

Booker: Yeah, yeah.

Janney: I know I kept you way longer than I intended to, but I really appreciate you taking the time.

Booker: I'll tell you what. That's the reason that I got Facebook because I realized when I got out of the Marine Corps, I needed to keep in touch with those guys. I had a bunch of jobs that I could have gone overseas, worked at embassies, and done some stuff, but when I had that stroke, I realized my brain wasn't gonna work as well. I didn't want to work for some letter agencies after that. So, I got Facebook just to keep in contact with these guys. At first, it was overwhelming because of the flood of information, but now I just go back to it when I can. It's pretty cool to see what these guys have done now that they are getting older.

Janney: A lot of them are still serving in some capacity or another, either working with veterans or law enforcement or whatever. It's a great bunch of guys. I've talked to about thirty of them so far, and I just can't get over their selfless sacrifice and bravery over there.

Gold Star Family Interviews

Gold Star Mom Shelia Cobb (PFC Christopher Cobb)

Cobb: Christopher was born New Year's Day 1985. He was precious to me because his father had passed away. His father was sick in the hospital with cancer, his real father, and so my sister was there with me when he was born. Christopher grew up, he lived with me when he was little and everything. He didn't know trouble; he was a mama's boy. He was a New Year's boy.

Janney: As was my Mom. My Mom was born on New Year's Day.

Cobb: From the time he was little until the time he was grown, he was a very, very good boy, but he was a momma's boy. He was attached to me. The last time Christopher was with his Dad was when Christopher was three months old because his Dad died when Christopher was three months old. From Agent Orange from when he was in Vietnam.

Janney: So, he passed away from cancer?

Cobb: Yeah, passed away from cancer from Agent Orange. He was only thirty-five when he passed away.

Janney: Was his Dad a soldier? Was he a Marine or what?

Cobb: No, his Dad was in the Air Force. He was a veteran; he was in the Air Force years ago.

Janney: Yes, they definitely got exposed to it to a large degree since they were dispersing it.

Cobb: Well, Christopher got attached to the Marines because the man that raised him, Howard, and his grandkids were all Marines. So, that's how Christopher got attached to the Marines. He had been at Parris Island several times to see the grand boys graduate from boot camp. He just came up one day and told us that he wanted to join the service. He was very, very fanatic on the computer. Oh, my, he played a lot of video games. He was a video fanatic.

Janney: And as far as video games, were there certain types of video games that he liked to play? Were they role playing games?

Cobb: Well, he was in karate school when he was younger. He used to dress up as a black ninja. I've still got the outfit that they wore. I've still got his weapons that he had. He was just into doing all this kind of stuff. I think that's why he went into the Marine Corps, into the infantry.

Janney: And so, you said that your second husband's family had a lot of influence on him. Tell me about their service.

Cobb: Well, he had three grandsons that were in the Marines. They all went to Parris Island and they all went to Camp Lejeune. Christopher used to go up there to their house and went to Parris Island for three of their graduations and it was a big influence on Christopher. He grew up with all these kids. They were from North Carolina.

Janney: At what point during his youth did he express an interest in the Marines, do you know? Or serve in the military?

Cobb: He was about 17 years old and he called me up one day, telling me that he wanted to enlist in the Marines and said, "Mom, you have to come down and sign some papers." I said, "What are you telling me?" He said, "I want to join the Marine Corps." He was 17 years old when he joined but he had to finish high school. They called it the delayed entry program. He joined with four other boys from here; his buddies.

Janney: The buddies that he signed up with; did he actually serve with them? Or did they go on to different units? What happened with that?

Cobb: They were in Iraq, but two of them were in Baghdad when Chris got killed. One of them was in Baghdad, one of his friends. The other one was stationed somewhere out there. I can't remember exactly where they were stationed. Christopher graduated from high school and then they took him up to Tampa and they swore him in. That's when he left for boot camp. Parris Island.

Janney: Were you there for that?

Cobb: I was there for his graduation from boot camp. Sure was. He was killed in 2004 and he graduated from boot camp in 2003 because he wasn't even over there a year when he got killed.

Janney: Let me backup just a little bit. I know that we talked a little bit about Christopher when he was a young man growing up. What did he like to do when he was a kid besides the martial arts? Did he participate in any sports? Did he have any hobbies?

Cobb: He played the violin from middle school to high school. He liked girls. He wasn't into any sports because he had hurt himself in elementary school, so he wasn't into the sports thing. He had broken his leg.

Janney: I know you told me once before when we talked that he had broken his leg. How did that happen?

Cobb: It happened when he was walking home from elementary school one day and one of the kids ran his bike over Chris. I had to go up there and take him to the hospital.

Janney: He played the violin. Did he have a favorite piece that he liked to play?

Cobb: No, he was just in concerts. He played in concerts whenever they went somewhere. He just liked it. I don't know how he got into the violin.

Janney: Was he pretty good at that? It sounds like he was if he played it through middle school and high school.

Cobb: Oh, yeah. He played from middle through high school. In fact, the high school orchestra played at his funeral.

Janney: What types of music did he like to play? Did he play country music or classical music?

Cobb: Classical. But, he liked rap music. He got himself a gift for his car and that's all I heard in his car was rap music. He had three jobs before he enlisted in the service and worked those jobs and saved some money for when he graduated from high school.

Janney: Let me backup just a little bit. What was his first car? What did he get?

Cobb: Oh, an old Chevy. I can't remember the year or the make or whatever. It was an old Chevy.

Janney: What kind of jobs did he work?

Cobb: He worked at Walgreens Drug Store. He worked as a clerk. He worked at Winn Dixie as a bag boy - that was his first job. He also worked in a nursing home in food service.

Janney: What was his favorite job? Did he ever express that to you?

Cobb: Walgreens. He liked waiting on people.

Janney: After he got his three jobs and started saving money, what did he do?

Cobb: Then he decided to go into the service. So, he went through boot camp and when he graduated from boot camp, they had changed his orders from Camp Lejeune to Camp Pendleton, California.

Janney: You saw him graduate from boot camp. Did you ever go out to Camp Pendleton and see him out there?

Cobb: Yes, I did. I went out there in February 2004. I went out there twice. I went out there for his SOI school in November 2003. Then I went out there in February just before he went overseas to Iraq. He came home in January for his 19th birthday. And then they went back, and he was home for Christmas and his birthday. They were here with him and they had a wild party.

Janney: I don't blame them for that a bit. When was the last time that you saw him before he went overseas?

Cobb: I saw him in February of 2004. Over at Camp Pendleton, I saw him the day he left for overseas. When Christopher went out to California, he used to go see my Dad's brother who lived out there and so Chris had family out there in California. Cousins were out there and my uncle. So, Chris got to go up there and stay with them and go see his cousin and everything. They always picked him up at the airport when he came back home from Florida.

Janney: After you saw him in February 2004, what communication did you have with him during the time he was going to be deployed or while he was deployed?

Cobb: That was it. When he was in Kuwait for six weeks, I would get letters from him and he called me on the phone twice, I think, while he was in Kuwait. He called me up one morning and woke me up and he said, "Mom, do you hear those bombs going off?" and I said, "Christopher, what are you talking about?" I sent two packages and they came back right after he was killed.

Janney: So, is that the last time you got a chance to speak with him, when he was in Kuwait?

Cobb: Yeah, I talked to him in March when he was in Kuwait. That was the last time that I talked to him on the phone.

Janney: Can you relay to me what he said to you on the phone?

Cobb: He said, "Mom" and told me to send him cookies and candy because all the kids called him Mr. Marine and he wanted some candy for the Iraqi kids so he said, "Mom, please send us a lot of candy and cookies for the kids, so we can give it to the kids." And that's what he told me. Just send him some cookies and candy. And he said, "If anything ever happens to me Mom, then you'll be taken care of." I said, "Christopher, don't tell me that."

Janney: I'm sorry. Repeat that again.

Cobb: The last thing he said to me was when I was out there; the thing was if anything happens to me Mom, you will be taken care of. Because he was always raised in apartments. He grew up in apartments and duplexes and he told me that he would have enough money when he got out of the Marine Corps to buy me a house.

Janney: Awww. That's very sweet.

Cobb: I took the money that we got from my insurance and we bought my mobile home.

Janney: So, the money from his death benefits you bought you a place.

Cobb: Yes. He had a girlfriend, but they weren't too close.

Cobb: You know, but when I find Jared's number, I'm going to text you Jared's phone number.

Janney: Yeah, I would love to speak to him. I know you said that you were his second Mom and that he had talked to JT, I guess, about 10 minutes before the ambush.

Cobb: Yeah, he was one of Christopher's best friends too. He was on the plane when Marcus Cherry's body came back too. I don't know - it's hard to do this. It's been ten years coming up here and it's going to be hard to go out there.

Janney: Yeah. Are you planning on going out to California for the reunion?

Cobb: Yeah, I am. I don't know what day it's going to be on.

Janney: Neither do I. Dianne extended an invitation to me and I would love to go if nobody has any objections to that.

Cobb: No, that's fine. I hope she gets a lot of the parents that'll show up. You know, she's working on it.

She'll get it done.

Janney: Oh, yeah. Like I said, she was gracious enough to send an invitation to the other Gold Star parents to ask them if they would want to be included in the book as far as the interview goes.

Cobb: Oh, did she? Have you heard from any of the other ones?

Janney: Yes, she did. I did. I heard from Nancy Walker.

Cobb: Oh, wow, yeah, well that's good that you've heard from her.

Janney: Yeah, matter of fact, when you and I first started talking; she called me on the other line, but I didn't take the call.

Janney: Let me ask you another question or two. I know Jim Dougherty probably is the first one to mention my name to you. Did you know about my trip with John Wroblewski to Iraq? How did you find out about it or did you find out about it?

Cobb: Jim told me about it. He knew about it somehow. I don't know how he heard about it.

Janney: How did you feel about that? I mean, about my trip over there with John Wroblewski.

Cobb: I felt that that's what John needed to do. That's what you guys needed to do. You had to do what you had to do; you know. I understand it. The memorial service was very good from what I saw on the computer.

Janney: You saw the memorial service that I posted on the blog? How did you see it?

Cobb: Yeah, you put it on the blog.

Janney: Okay. The main thing that I want to emphasize, and I'm sure you know, is that I didn't go over there just for JT, I went over there for all those men including Christopher.

Cobb: Right. I understand that. I'm glad somebody is trying to get a book started about these kids because they need it. I'm sure Jared will talk to you too. In fact, once I get off the phone, I'll find his phone number and I will text it to you.

Janney: All right. That would be cool.

Cobb: Then you can get a hold of him. His name is Jared Cole.

Janney: Okay, great. I've got that written down I think from our conversation back in March.

Cobb: Yeah.

Janney: And so, Christopher didn't have any children. Is that correct?

Cobb: No, no children. He was single. He was nineteen years old when he was killed.

Janney: As far as you know, was he in the truck with Kyle Crowley and Staff Sgt. Walker?

Cobb: Yeah. I heard Christopher was on the back of the Humvee. He was the first one killed. That's what I heard. That's what they told me. In fact, you know, my sister has…when I got the death certificate, the department from Quantico; they sent my sister pictures of when Christopher was killed. They swore they were to show me those pictures of Chris. I was never supposed to see those pictures at all. My sister has them.

Janney: Well, I know you and I talked a little bit about this before back in March when we spoke last; the first time we spoke. That it was a well-planned ambush and no matter what had happened there was really nothing that could have precluded the results.

Cobb: No, I know. Do you know Colonel Kennedy? The Brigadier General?

Janney: No, ma'am. I do not know him personally.

Cobb: Yeah, I'm trying to remember who told me that. Somebody told me that and I can't remember. I don't know, maybe it was Dianne, but somebody told me that it was a Brigadier General.

Janney: Yeah, I did not know that. I know they relieved Captain Royer of his command.

Cobb: Royer, Captain Royer. I didn't know what happened to him.

Janney: Did you know that the General that escorted John and I to the ambush site, at least part of the ambush site where JT got killed, that his son was killed in Afghanistan a few months later?

Cobb: No, what was his name?

Janney: Major General John Kelly.

Cobb: Oh, I know who you're talking about now. Yeah, his son was killed in Afghanistan.

Janney: Yes, ma'am. Just a year, not quite a year after we did the memorial service in Ramadi; his son was killed in Afghanistan in Helmand province.

Cobb: Wow. How old was his son?

Janney: I want to say that his son was 21 or 22.

Cobb: Yeah. That's sad. But I will find the pictures this weekend and get them to you.

Janney: That would be great. Like I said, once you send them to me, I will scan them. As long as they are pictures that you took, that's the whole deal. Mark Crowley sent me pictures that other people took of his son and I can't use those because there's a copyright issue.

Cobb: Yeah. No, I got ones that I took.

Janney: That would be cool. You know the picture of his Dad holding him and you and Christopher and his stepfather. Did you take any pictures when you went out to Pendleton?

Cobb: Yeah, I've got pictures of me and him.

Janney: That would be great. If you could send me one of those if you have one.

Gold Star Family Interviews

Gold Star Dad Mark Crowley (PFC Kyle Crowley)

Janney: It's 18 June 2013 and this is an interview with Mark Crowley who is the Gold Star father of Lance Corporal Kyle Crowley of 2/4 Marines. Mark, let's start at the beginning on 13th of May 1985 when Kyle was born. If you could just tell me about Kyle from the beginning. I know he had a twin brother Shane.

Crowley: Kyle was born on May 13, 1985. He was the smallest of two twins: Shane Arthur, and, of course, Kyle Dwayne Crowley. Shane was the biggest of the two twins and the healthiest of the two. Unfortunately, Shane died of crib death four months later after birth. Kyle and his sister lived with his mom. Not long after Shane's death, Kyle's mother and I separated without going through a divorce, at which time, Kyle and his older sister Nicole lived with their mother for about eight or nine months. I had left Reno, Nevada where they were living and I moved away with Kyle to California where I landed a produce job with a supermarket there in North Shore, Lake Tahoe. On the weekends, I would go get the kids and bring them down. Bring them up there to Lake Tahoe where I lived and where we would do things like go to the beach, go fishing there for trout there in the lake or on the Truckee River. When it would snow, we would play in the snow. In the summertime, we would do things in the water. I had a little 16 foot boat and we would go out and swim. I taught them how to hydro swim and water ski from that boat. On the weekends, I would take them back to Reno and every weekend I would get to see my kids. This went on for less than three years.

At that time, I had moved from Lake Tahoe and had moved down to Concord California right around my Dad. A couple of months after that, Kyle's mother called me and said that she couldn't raise Kyle - Did I want to take him? Of course, I jumped at that opportunity and so that day, I got Kyle. As a single parent we were together, and I raised him until he was eight years old. Then, there was a lady and I moved in with her for a while. That lasted for a year or so and then we moved out again. For a couple of years, it was just Kyle and I and we lived like bachelors. It was pretty tough to have day care and take care of that, working and going to school. I was trying to better my life to earn more money to provide for myself and Kyle. Day care was through the roof; very costly, but if I hadn't had it, I couldn't work and stay home at the same time. So, Kyle got to the age where he entered first grade and I had a pretty good day care that was close to the school that Kyle went to. I would drop him off in the morning and I would go to work for eight hours. At the end of the day, I would pick Kyle up and we would go home and do our laundry, have dinner, sit down and watch TV. On my days off, although sometimes I had to work seven days a week, but as time went by, I would get a day or two off and we would do stuff like fish. It was kind of lonely. It was kind of hard to juggle work and parenting and everything and still have a social life.

So, I entered this club called PWP. It was Parents without Partners. A lot of single parents there, so they not only did stuff for the parents, they did stuff for the kids as well. So, I joined that, and we made a lot of friends there. Good friends, you know. Not only would they take care of each other, they would have dances where some of the older children of the other parents would watch Kyle while I would go out and dance or do something where Kyle wasn't involved. Most of the stuff through that outfit was picnics, baseball games, ping pong, and camping. We would go to the lakes and camp out for the weekend and play in the water and have fun. We did that until Kyle was about twelve. Then I met another lady and that was the first time I ended up getting married. Life was pretty good for us. Mary was really good for me and Kyle. We lived in Hayward for some time. We got to the point some years later that we had moved again and sold the house in Hayward. Kyle did an elementary year and then junior high and then of course, California High at which time Kyle had friends. He had a girlfriend. He had stuck with this one girl from the age of fourteen through the age of eighteen. Kyle pre-entered the Marine Corps. He had entered and wanted me to sign him over to them and then when he graduated, he would go on and be a Marine. I didn't agree with that. I didn't want Kyle to be a Marine. I wanted him to go to school and get educated like all the other kids his age. Kyle wanted to be a Marine and he pretty much stuck to his guns. He was unhappy when I told him no. So, after four, five, six days of this back and forth, bickering at each other, I decided as a parent that we are obligated to give our kids the right to make decisions on their own and this was a decision that he was adamant about. This is what he wanted to do; this was his goal in life to finish high school and join the Marines. This was something that he was really into.

So, I went ahead and met the Marine recruiter. He came by the house and we sat down with him and filled out his entrance form into the Marine Corps. Of course, I had to sign him over and we threw the pages together and I helped to fill it out. At the end, we came to the beneficiary. Kyle asked me about that because I was a vet and I had been through that before myself. I told him that in case of an accident, injury and/or his death that he would have life insurance that would go to whomever he had chosen for his beneficiary. It was decided that he wanted to leave whatever it was to me and to his grandmother. He loved his grandmother. He spent a couple of summers, three or four summers throughout his growing up with her in Florida where she lived, and he stayed throughout the summer with her. He really loved her a lot. Of course, she loved him, so he decided to leave whatever the death benefit would have been to myself and to his grandmother. When it came time, Kyle kind of struggled through high school. He was an educated kid and all these tests that they give kids throughout their schooling, Kyle's test results would come back very high; very intelligent. Scoring up to third year in college. So, he was very intelligent, smart, and he just didn't want to apply it. At this point, between fifteen and eighteen, he kind of bucked me and bucked the system and just wanted to get through it and get into the Corps and drive on from there. Before he graduated, he actually just wouldn't do the work. He wasted his last year as a senior. He finished; he got out.

Throughout the years, the last three or four years, this girlfriend that he had, it was just him and her. He didn't cheat on her or anything or go with other girls. I'm not sure what she did not really being around her that much. She was supposed to go to San Diego State which is southern California. Of course, Camp Pendleton is not very far from that University of San Diego, probably within a forty-five to fifty-minute drive. You know Kyle wanted to be a Marine and be down there close to her. She was not going to a four to six-year college term down there at San Diego State. Kyle wanted to be as close to her as he could regardless of deploying or wherever the Corps sent him. The last so many months of Kyle's high school years, he and I were at odds with each other and he was becoming further and further distant from me and kind of aggressively separated from me. This girl and his friends were most important to him. Just friends and this girlfriend. Before he had finished school, this girl had broken up with him. Her parents didn't want her to have anything to do with Kyle. They broke up and got back together and were secretly seeing each other. Her parents wished that they would stay apart. Me and the dad kind of avoided it one way or the other because that was his personal business and that was the way he liked it. Keeping things personal. The father of this girl absolutely forbade his daughter to see Kyle or to be with Kyle. They hid that for quite some time. But in the end, they broke up. She was going one way and Kyle was going the other. The last couple of weeks, after he had gotten out of high school, Kyle left the house and approached the girl's father over there. Telling that father that I had kicked him out of the house which wasn't true. I think Kyle was just trying to get close to this girl because he felt that relationship melting away. He told the father that I had kicked him out of the house, and he didn't have any place to stay, and this guy took him in. Kyle was to enter the Marine Corps sometime later. I don't remember exactly how many weeks later.

It was only a matter of like four or five days after he left the house that his beneficiary was changed to this man and his daughter. I didn't find this out until after I had buried Kyle. This guy approached me saying that Kyle had left his death benefits to him and his daughter, Kyle's girlfriend. That's when I found out that Kyle's death benefit was going to him. I wanted to bury Kyle before I approached that. After burying Kyle, I approached this man. He told me that Kyle had sat down with him and discussed that he wanted to leave his death benefits to the daughter. Her father told me that he was told that I wouldn't allow Kyle to do that. But, Kyle had left his death benefits to him as the number 1, with the daughter as the number 2, so he would allow that. I struggled from Kyle's death tremendously and I struggled with that tremendously. I couldn't figure out why another man who didn't want his daughter to have anything to do with my son could sit down and discuss life and death with my son.

Dianne and I and all the Gold Star parents, those that were Gold Star parents, went down for the homecoming. We met Colonel Kennedy and some of his administration. Kennedy's wife had put together a reception and a dinner for us and we went through a memorial deal type thing with that. Of course, the next day everybody was in the barracks parking lot waiting for them to come up. They drop them off a quarter of a mile away and they muster and get in formation and they march up the road with all their gear in tow. These kids came up and met all of us - me and Dianne and Sandra, whose son, a medic, was killed. We hadn't met some of these other parents yet. We finally met Marcus Cherry's mother, and Corporal Young's parents there that day. Seeing these Marines march up that road was extremely tough. All of us had broken down when they finally got up there in formation. The Battalion Commander and the Company Commander released them to the Company platoon leader which is usually a Lieutenant. In this case, some of the Lieutenants were killed over there, and so they were replaced with Majors. There were some pretty cool guys. Once they were released from formation, everybody's loved ones went to their Marine. Once they met with their family, then they started seeing that we were there. Of course, we were approached, hugged, and cried a lot of tears. It was tough for them too.

People usually don't know how to act when a child is killed. I had a lot of friends from work and when I tried to go back to work, none of my friends really knew what to say. Before Kyle's death, they were all cool - fishing, hunting, coming to my house for barbeque, but after Kyle's death, they didn't know how to act. The Marines were similar. These kids are tough as nails. They've been through battle and at the same time, they've seen death. You don't know which one of them had picked your son up and carried him off the field unless they come up and tell you. Dianne knew who picked up Travis, but I don't know who picked Kyle up. They were stripped of all their weapons and clothing and the Humvees that they were driving were stripped clean. These kids, they came up to us. It was tough, really hard seeing them come up there knowing that your son is not with them. Most of the fathers were pretty quiet, but most of the mothers were bragging about their sons. I don't know if it was selfishness or what, but when I heard all these ladies bragging about their sons, I thought why couldn't my son have come up that road and not yours. I had a real bad feeling that those twelve young people were gone. Not just Kyle's death, but the death of Travis and the death of each and every one of those young men that they lost over there affected me the same as it would have been Kyle. Somebody had lost a son. Somebody they raised. Somebody they tutored, somebody they fished or hunted or golfed or played football with. It bothered me. It bothered me that Kyle was driving a Humvee that had no ballistic glass in it and no armor on it that day. I don't know if you got a hold of that picture. Swanson will probably send it to you if you get a hold of him.

Janney: Yes sir. I've seen it.

Crowley: Otey was the only kid that jumped out of that Humvee that day. Got behind a concrete wall and he survived that day, but later on he was killed in a sniper tower. But, he was behind that windshield when Kyle was driving. There's about twelve or thirteen bullet holes in that glass which was fractured and cracked. You look at Otey's face behind that and you think that your son was there and that's where he was shot. Those bullet holes hit all of them in the neck and in the face. I wasn't able to view Kyle. He was wrapped up like a mummy. He was viewed by the mortician and he was viewed by the Marine that escorted him home from Dover Air Force base and I was told that I wouldn't be able to view him. Travis Layfield had arrived in California and I went to his funeral and I met his family. Dianne and everybody were able to view him. It tore me up terribly to see that young man in that coffin and what it did to his family. Kyle hadn't arrived yet because they were trying to get him presentable, and they were unable to. He was cut to pieces, and I was able to touch his body, touch his face, his arms, his stomach, his legs and I could feel there was a lot of him that wasn't there.

Janney: I'm so sorry, Mark. I'm so sorry that Kyle was killed.

Crowley: I struggled with the loss of each and every one of them. Of course, my son more than anybody, but it felt good to be held by those kids that survived it. But, each and every one of them that was in that battle had that thousand-yard stare. They were all glad to be back home to their wives, girlfriends or families and they were all glad to know that they were going to get some time away from the Marine Corps. They were also there to meet the parents of those that had fallen. I was with Dianne and we had gone to a dinner the night before and we had only met two other Gold Star parents at that dinner. Everybody else there were Blue Star mothers and Mrs. Kennedy, Commander Kennedy's wife. I'm sure it was hard for others to meet the other Gold Stars. Everybody was severely torn up from the loss. Dianne and I probably went to eight or nine funerals throughout our becoming a member of the Gold Star parents' family. We did this to kind of give back to some of these newly made Gold Star parents to assure them that their sons or daughters were in a better place. To give back support to them and what to expect in the months to come, years to come. It doesn't ever get easier, but it loosens up on you. The dread and the depression, the disbelief that what your son believed in was the right thing to do. We didn't all believe in that. To hear all the stuff about Halliburton and the billions and billions of dollars that they made over there. Rumsfeld did a town hall meeting here in San Francisco not long after we buried Kyle. Dianne's daughter was there. There was a kid that came home wounded that asked Rumsfeld a question. He asked, "Why are so many military personnel going to war with vehicles that aren't armored and don't have ballistic glass. Why are they being cut to shit, shredded, and killed?" Rumsfeld replied, "When you go to war, you don't go to war with what you want, you go to war with what you've got." At that point, I could have killed the guy with my bare hands.

You might hear this later because they investigated it, but their Company commander did not order flak vests, did not have body armor for them as they were leaving or going to leave for Kuwait. He was relieved of duty before they even deployed. Then, Captain Royer was relieved in Iraq. So, when these companies trained together with an established chain of command, and then the Company commander is relieved of duty before they deploy to a combat zone - that softens up the Company. They have started out on the wrong foot already. Then, to have a second Company commander relieved, I'm sure that 2nd Battalion, 4th Marines Echo Company was probably the only unit that went through that.

It was good to be around those guys. I went down probably the first three years, two or three times, whenever they were around. I went down with the guys that were barracks rats meaning they couldn't redeploy to active duty, but they're waiting for their discharge from the Corps. They were shot at, blown up, beat up pretty bad before they could return. Being with those guys away from active duty and being able to put their civilian clothes on and get away from that military environment for a time was good for me. I'd been down and also met Nancy Walker, whose son was Staff Sgt. Walker who was sitting next to Kyle when they were ambushed. He died right next to Kyle. He actually was able to get out of the Humvee, took three or four steps and survived outside of the Humvee. He was still alive when the battle was going on. Fernando Mendez-Aceves was the Navy Corpsman there. He was under cover and he went out to try to help Staff Sgt. Walker who was still alive, and he was cut down right there beside Walker. I was told that Kyle had gotten out of the Humvee, mortally wounded and had taken two or three steps before he fell. He was dead, but his body was still moving. I was told that Kyle was found outside of the Humvee and that he was viewed from a distance. These guys were a distance away watching all of this unfold, but they couldn't get there. They were pinned down by RPGs, machine gun fire, and mortar rounds. The last kid that was still moving was Ryan Jerabek. He was the machine gunner. He was in a gun turret on the back of the Humvee that had an inch and a quarter thick steel gun turret with a 30-caliber machine gun belted inside of it. Jerabek was killing them wholesale and he was the last one to survive until a sniper took him out. All this was seen from a distance, but they weren't able to get there. So, by the time they did get to them, everything was gone and stripped. Swanson was with Captain Royer at that point. They were flat on the ground being shot at. If you've seen the video, they were being shot at. That's when they were trying to get to Kyle, but they couldn't move. They were pinned down. I guess Swanson had got some shrapnel or a bullet or something. He actually was bleeding at that point. Captain Royer had a bullet crease in his helmet where he got hit in the head, but it bounced off. He was a lucky sonofabitch too, I'll tell you. (Author note: Swanson had a grazing wound to his arm and Royer's Kevlar helmet was hit by a round on 10 April, not 6 April during the Gypsum ambush.)

Janney: I have actually stood there. This is hard for me to share with you, but I walked that whole intersection where your son got killed at Route Gypsum and Route Nova. I walked through that intersection and took numerous pictures throughout that area.

Crowley: You were there, huh?

Janney: I was there. I walked that whole route, you know.

Crowley: What I would give to go there! I have the coordinates, but I don't know if I will ever get there. I too, like John Wroblewski, would like to go there someday and lay a wreath. Just touch the ground where Kyle bled out, you know, where his last breath left him.

Janney: I would like to promise you that I could get you there, but I can only say that I could try to get you there one day.

Crowley: I wouldn't expect you to promise that. I would probably be beside myself if you said one day, "Hey, Mark. I've got a trip to this area." Man, I would have to plan what I was gonna do. I would definitely lay a wreath and just something else, some type of ceremony that the warriors would look down on. I say to myself so many times that I would give my life a thousand times over so that they could live a normal life. They could come back and serve their country or come back and go to school or do something with their lives. Some of these kids have tried to come back here to civilian life and they can't do it. I don't know if you remember Vietnam. I'm sure you do.

Janney: I do. I remember hearing the body counts on the radio. I'm 50 years old, so I'm probably older than you as far as that goes. I remember listening to WSB 750 AM in Atlanta and hearing body count reports on the AM radio. So, I do remember that very clearly. My Dad had a draft number and for whatever reason, he did not get called up. I'm very grateful to the men that went in his place.

Crowley: Yeah, I'm 54. I missed Vietnam by two years. I graduated in '75 and I went into the military in '76 and I got out in '80 with two years in active duty; reserve duty. These Vietnam vets that came back; they had survived a tour over there and a few of them started picking people off from towers, killing people, and other crimes - you know they were insane. Post traumatic syndrome. America wasn't ready for that, even though they're trying now to help these guys. I was told this by a Marine that survived three tours over there, if you go to the hospital for post traumatic syndrome, and your platoon or your squad finds out about it, they think you're weak and cowardly. Get a straw and suck it up. An 18-year-old kid telling a 24-year-old kid that was on their team to suck it up. Quit being a punk. What they tell me is if you're found out that you've gone to that early in their tours, you're a weakling and cowardly. Nowadays they don't do that, but back then, you were a weakling if you couldn't handle that.…you're the one that enlisted; you're the one that wanted to be a Marine; you're the one that trained for it; Marines are supposed to be killers and this, this and the other thing and that's the kind of shit that they would get from their peers. They were supposed to suck it up. Take it like a man; swallow it down; whatever and move on. That's what they were trained and taught to do. So for those that didn't suck it up and move on with life and be a man; that they were supposed to be whether a Marine or a Ranger or whatever; the anticipation of that; it's a shame when you're standing next to a guy bullshit fucking Iraqis can't shoot worth a shit and they've got bullets passing by them three inches from their head and they're all laughing about it. These guys can't shoot worth a shit and then all of a sudden you hear something and your ears are ringing and you look at that guy that you were just laughing with and he's dead and he got hit in the torso or he's blown in half or whatever. It's not a video game. It's not a game anymore.

So many times I've heard these guys laughing and brushing it off. The Iraqis can't shoot, but the day Kyle died they weren't Iraqis. All these dudes, from what I was told, the body count was over 900 within the first three days of April 4th. So, the body count wasn't Iraqi. They were from Syria, Jordan, and Egypt. Their Intelligence wasn't there. They didn't know that they were fighting foreigners. They thought they were fighting Iraqis. It's not a video game anymore buddy. You're dead and your family is severely wounded from the loss of your life. Oh, Lord, you know. I was told that they got $500 for every American confirmed kill.

Janney: All I know is that the guys that I interviewed there in 2007 and then the guys that I interviewed in 2008 were saying that about 70% of the guys they were killing were foreign fighters, based on the papers they were pulling off their bodies.

Crowley: Yeah. They didn't have the intelligence enough to stop it, you know. That was the first thing they should have done was close the borders. If they had got bin Laden earlier in the game, they would have closed the borders. I'm sure that the border is so big that it's probably impossible to do. But I mean, 10 times; 20 times the Mexican border; it's that big. They'd have a hard time closing that but they should have. Hey, that's a question I've got. I don't know if you can answer it. All this ammunition, these mortar rounds, and artillery shells they found that they blew up; all these mines; all these explosives; why didn't our military use that shit against the enemy? I don't get it. They spend millions of dollars on ammunition and explosives and rounds and everything for our military; wouldn't it be logical or financially reasonable to use their own shit against them? I could never understand that.

Janney: It makes sense to do that, I guess. Their issue is, I guess, that it's been buried in the ground and so they're not sure about the safety of using them. The Soviets were not shy about using anything that they captured in Afghanistan against the Taliban or whoever they were fighting there.

Crowley: It was the Taliban, mostly the Taliban. There were a few foreigners in there against the Russians, but not like there was against us.

Janney: There were a lot of foreign fighters there too, but I think that Iraq was a big call to jihad and a lot of foreign fighters came in. I interviewed dozens and dozens of soldiers and Marines that talked about how many foreign fighters they'd killed. At least in my interviews, it was very common knowledge that 60 to 70% of the guys they were killing were foreign fighters and not Iraqis.

Crowley: Yeah. John Wroblewski and I had talked about this before when I met him in Crawford, Texas. Not long after April 4, 2004, I lived in California and there was a little town about 15 miles from me. They had busted this house; there were Syrian, Yemenis, and some Afghanistan people who lived there in this little place. There were about fourteen people living in a four bedroom house. This place had bomb making materials, sensitive and secret material, computers, and these people were making bombs and they were getting ready to set this shit off when they were busted. My theory on that is that these people are already here. John and I talked about it and he agreed with me that they're just waiting. There's gonna come a time when the next hit on this country is gonna make 9/11 look like a bar fight. It's all gonna go down at the same time, different states, different cities, and it's all going to be coordinated because there's so much money flowing from these hardliners, they're all around us and it's just a matter of time.

Janney: The crazy thing about it is, if you look at the Boston bombers, those guys had government assistance for eighteen months, or twenty-four months. We're paying some of them to live here.

Crowley: At one point, I took off cross country. Kyle was dead - I don't know what the hell I was searching for. I just went across the country. I went to Wisconsin to the grave of Ryan Jerabek. I went to Ohio to the grave of PFC Young. Came back down close to the Mexican border to Marcus Cherry's grave site. Visited Fernando Mendez-Aceves' grave. I was just going to these grave sites leaving pictures of Kyle and Travis and praying. I was just lost. I was a lost soul. After my trek across the country, I had a heart attack and wound up in the hospital - almost died.

Janney: Did the attacks on 9/11 have any effect on Kyle's decision to join the Marines or was that something that he had already decided to do?

Crowley: It had a great impact on him on his decision to be a Marine. It devastated him. One of the things that he and I both touched on at that point was all the children that died in both the towers. They had daycares in the towers and every child in the daycares perished that day. So, that impacted him. It affected me. He approached me about it after it had already affected me, and he had asked me about it. I said, "Kyle, I've been struggling with that for three days now." He said, "Dad, this is horrible." We talked about Pearl Harbor and all the devastation during that point in history and some of the other things where there was a massive loss of life. You know, 9/11 was his Pearl Harbor. I wasn't alive during Pearl Harbor, but it affected me for years and years and years that so many perished, as well as in Germany and all the other countries we fought in. So, yeah, it had a great impact on him. He wanted to fight. We had discussed that. When he took his tests, they had these mechanical tests, aptitude tests, and other tests to see where he would best fit in. He came in best in the intelligence sector. They wanted him to go through that schooling and into Intelligence because the military needed some smart people to go into the Intelligence sector of combat and war and antiterrorism. He scored very high in that, but you know what he told me? "I don't want to do that. I want to be a rifleman. I want to kill bad guys." That was his answer to me. I could never fathom that. He was so intelligent, but he wouldn't use it. What do you think about that?

Janney: Well, he did what he wanted to do, I guess.

Crowley: Yeah. You sound like my Dad now. That's what my Dad said as well as my grandfather Lloyd. Lloyd had tried to persuade him to go into the Intelligence sector because Lloyd knew being a retired Marine; being in combat all those years. Lloyd got out a Gunnery Sergeant after all those years and he had tried to convince Kyle to do that, but Kyle wouldn't have it. He wanted to fight.

I think I told you about the one Marine that I talked to, Steiff. Steiff had told me that Kyle took his place that day because he was in the clinic. He told me that he probably would have been killed if it hadn't been for Kyle because Kyle took his place that day. I can't even imagine what these kids experience during combat. What they feel, how they feel, and how they deal with it. I went into the military thinking that I was going to make a career out of it, but I'll tell you, after four years I couldn't wait to get out. I could not wait to get out and go to school. I wanted to be a forest ranger when I got out and I tried and tried but couldn't get into the Forest Service or Park Service.

Well, we've all struggled. Shawn and John Wroblewski, they're both teachers and I think they've lived a pretty good life, both of them. Their kids all have good lives. Every son they've got, they're really great kids. Very respectful, intelligent, super smart. Very respectful of their parents. They've done great jobs with their sons, you know. I mean, look at JT. Everybody that I know that knew JT really admired him. They say bad officers get killed, good officers die with you.

You know my last name is Crowley, but everybody always wanted to sing that song "Mr. Crowley" so one of the officers, I believe it was a Major, and there was another Lt., who said that JT and Kyle did this; JT would go up to Kyle and would start singing that song, but he wouldn't sing it all the way through. He would stop midway through the song and Kyle would go "dunh, dunh, dunh, SIR!" Laughs.

Janney: Laughs.

Crowley: It was a sign of respect for the officer too. JT would sing the song and then cut it short and Kyle would finish it off or whatever. I thought that was pretty funny. He and Kyle were characters with these guys. JT was a character too, I guess. They all are, aren't they?

Janney: Oh, yeah. They're all larger than life.

Crowley: I don't know how many of those guys have gotten out now. The officers, the guys that really knew Kyle. The guys would approach you and tell me something about my son. I got letters from a lot of them about Kyle. They called him a good piece of equipment. I guess that's Marine jargon for he was a good soldier.

Janney: I know you said a lot of them showed up at his memorial service at Golden Gate.

Crowley: Yeah. There were shit tons of them. And they called me in the second year. They were trying for both me and for Dianne at the same time for her family and for Kyle. Dianne had a lot of stuff that she was involved in and that was my second year. I was going through a lot of medical stuff and I just couldn't do it. Every time Dianne called me and wanted me to go to a funeral or something and I did that. With their memorials here, there's quite a few bills where they were paying for our airfare and our hotel rooms. They wanted us to go to Houston, TX. They wanted us to go to Florida and to Ft. Benning, GA. They sent us to Washington three or four times. They sent us to Quantico, Virginia. Every time they paid our way. They paid for our motel, our food and they did these memorials, and you know, after about seven or eight of those things, it was just difficult.

I just can't do that stuff anymore. I've given of myself and given back as much support to those that I can. Sometimes I want to go, like the other day, we went up to a lake up here. We had twenty-five boats up here at New Melones Lake. We had 118 vets, Purple Heart vets from WWII, Vietnam, some of the Korean guys, from Iraq and Afghanistan and every boat out there, these guys were pros. Each boat took six vets and we went out fishing on these lakes. Stuff like that, I enjoy. Because these guys get out there; they fish, the pro guides get them on the fish. They catch a shit ton of fish or we take them out hunting, bird hunting or deer hunting or pig hunting, archery hunting, or gun hunting. That's something I can enjoy. There's not too much depression involved in that. I don't know if I told you but the local high school football team; the football team will drag these wheelchair guys into the field. We'll get the dogs pointing on the birds and roll them up and these guys will shoot pheasants and stuff. That kind of thing is what I like to do. But to go to funerals and cry about the loss of this or that, I can't do that anymore.

Gold Star Family Interviews

Gold Star Mom Margaret Kellum (LCpl John "J.T." Sims)

Janney: The basic purpose of the book is to let the readers know about these young men that lost their lives in defense of their country, fighting alongside their brothers in the Marines. Please tell me about JT. Just start at the beginning from the time he was born, and share as much as you like.

Kellum: Well, he was born on August 10th of 1982 and he was a little preemie. He was 4.6 when he was born.

Janney: I saw that he was the smallest of your four kids, but he had the biggest heart and took up for everybody else.

Kellum: He did. He was the one child; he was the only boy, and the baby. He was the one that I knew for a fact when I got old that he was going to be the one here to take care of me. He wasn't going to let nothing happen because when he was here, he didn't let nothing happen to me anyway. That's what I worried about him going into the military because you just didn't talk about his momma. I've seen these movies where they call "your momma is sorry" and "your momma is a maggot" or whatever, you know, and I just knew he was going to get kicked out of boot camp for flooring a Drill Sergeant. Laughs.

Janney: He was a good Southern boy, in other words, just like I call myself. I was born and raised here in Georgia, so I consider myself a well raised Southern boy. It sounds like he was also.

Kellum: He was. He was a Southern redneck boy. He only had one job when he turned sixteen. He went to work at a local Arby's up here. He could basically get by with murder up there. He had a former Marine, his boss that hired him first. I can't think of his name, but he was a district manager. Then he left. He would get on the intercom and tell his friends, "I'm sorry we don't serve burgers here" or tell some of the girls that he knew; they would get on there and say, "You're hot." Just cutting up. I would walk in and he would jump over the front counter just to come over there and hug my neck or just act silly. I asked his boss lady one day and she says, this was after he got killed, she says, "You know, John is the only one that I didn't mind cutting up because when we get really busy, I knew that he was going to be right by my side just working just as hard as I am." She said, "So, yeah, we let him cut up when he wanted to because when we got busy, he was there." All of his friends admired him.

Janney: Now, I did hear several stories about how one of his favorite expressions was "that didn't hurt" and it started off with some incident about running in and out the door and somebody locked the screen door and he ran into it and it knocked him down and he jumped up and said, "That didn't hurt."

Kellum: Yeah, that was his favorite saying. Even when you would whip him, he'd say, "That didn't hurt." He got the nickname Hammerhead from us because when he was about 10 or 11, he got mad and went behind the house and used my hammer to hit the post. The hammer jerked back and hit him in the back of the head. The claw part of it. And he came out from behind the house and was drooping his head and my husband asked him, "What did you do?" He said, "JT, what's wrong?" JT said, "I hit my head with a hammer." My husband said, "Son, how'd you do that?" JT said, "Well, come and I'll show you." He went back out there and hit the exact same spot. Laughs.

Janney: Laughs. Oh, my goodness.

Kellum: I said, "I can't believe you did that." But, he's the type that if he saw anybody else's blood it would bother him. If he saw his own, he went ballistic. He thought he would just die. He used to get in Walmart on those little bicycles and go and get the largest pair of panties they had and ride around Walmart with a pair of panties on his head, calling out his friend's name, "Hey, Mickey!" Laughs.

Janney: Laughs. Well, did he get interested in the Marine Corps from his boss that he worked with at Arby's, or how did he get interested?

Kellum: No, he always wanted to go into the Marines. That was his first choice. When he got in the eleventh grade, the Marines recruiter never did come around. He waited and waited and just finally went ahead. The Army recruiter came by and so he joined the Army. When he went to MEPS up there in Montgomery, he failed the physical by seven pounds. He was seven pounds overweight, so that kicked him out. They told him to come back next month after he lost seven pounds. Well, before he could go back, the Marine recruiter came, so he joined the Marines. The Army recruiter asked him, "Well, what did he tell you that I didn't?" JT said, "The truth." So, he went ahead and joined the Marines. I'll tell you, I think he was born to be a Marine because he was fearless. He'd give you the shirt off his back and tough. He loved to wrestle.

Janney: I heard he was a great wrestler.

Kellum: He was in school. He was my little hellion when he was little. He was mischievous as all get out. He liked to drove me crazy when he was little. He was hanging around some rough crowds in middle school and all. When he went to work for the Arby's, he sort of faded away from that group and got with this group that he was with when he went into the Marines, which was an awesome group. I mean, you know, I love them to death. That's what helped turn him around. That other group that he was hanging around with stayed in trouble all of the time. But the group that he wound up with when he got over into high school, they were just an awesome bunch. And one of their parents owned Domino's in Alex City. He'd go up there and at the time, if he wasn't over there, he was at their house because they lived on the lake. He loved to ski. He loved to boat ride. He loved to swim. Jump off Chimney Rock. He was just fearless and it scared me to death.

    When the Marine recruiter came to the house, he was marking off in the book what all he could do and I kept looking at him and I said, "It would be easier to mark off what he can't do, won't it?" He said, "Yeah, I believe you're right." Of course, he wasn't but seventeen at the time. The Marine recruiter said, "You know, next year he's not going to have your signature to get into the Marines." And he said, "It sure would be nice if he had your support now." I said, "Whoa, you wait a minute, buddy. You back up." I said, "This is my child. I might not agree with his decisions. I may not like his decision. But he made the decision and whatever decision he decided to make I am behind him 200%. He's got my support. Like I said, I might not like it. I might not agree with it, but he's the one that's got to do it."

Kellum: He went to Parris Island for training. He left there, finally he came home for two weeks and then he went to Camp Lejeune or Camp Geiger.

Janney: Before we get too far along in his career with the Marines, is there anything that you want to share, anything else you want to share about him growing up? Did he have a car that he was particularly fond of? What were his hobbies? What did he like to do besides wrestle?

Kellum: He had a '74 Maverick.

Janney: I did read about that. I heard a story about some friend of his asking him, "When are you going to let me drive it?" JT jumped in the back and said, "Right now." Laughs.

Kellum: Laughs. He loved that car. Then we went and got a Pontiac Grand Am. Bless his heart, he got so many speeding tickets. It wasn't even funny. One day, he was going through a green light and a lady ran the red light and T-boned him.

Kellum: So, he totaled that car out. Or she did. So, every time he came home, he borrowed my car. And then when he got stationed at Camp Pendleton, he got that Ford Ranger. When he got stationed at Camp Pendleton, I was talking to him one day and I said, "Son, what have you done now?" He said, "I don't know Mom. Why do you ask?" I said, "Well, the police were just out here." He said, "The police? What did they want?" I said, "They just wanted to know if you were all right. Were you sick or anything?" He said, "Why would they want to know that?" and I said, "Well, they're sure missing your money." Laughs.

Janney: Oh, my goodness gracious. That's hilarious. That is too funny.

Kellum: He got so many speeding tickets that he got his license suspended for 90 days. They were just wondering if he was all right because they're sure missing his money.

Janney: That is too funny, Margaret, that is so funny.

Kellum: He said, "All right, Mom." Laughs.

Janney: Well, he sounds like he was a joker, kind of like his momma.

Kellum: Yeah. Laughs. Sounds like that's where he got it from.

Janney: That is just too funny. Well anyway, he went into the Marines and he got assigned to 1st Division. You said he came home from boot camp before he went to Camp Pendleton for a couple of weeks. Tell me what happened after that.

Kellum: Well, after he got stationed at Camp Pendleton, we always talked on the phone and emailed and stuff. He had one buddy that he met when he first got out there. It was Jeremy Weber. This guy is so tall; I think he's about six foot. And poor John, JT's about 5'3" or 5'4". You know, they looked like Mutt and Jeff. But he had the deepest voice and he was from Michigan. So, I always called him a Michigan cowboy because he sounded like a Texan. Well, then he got up with this other guy named Joe McGee from Boston. So, every time I would call him, he would say, "Well, hold on, I've got two here that want to talk to you" and they'd say, "Hey, Mom" so I went around telling everybody that I only gave birth to one son but when he joined the Marines, I got 230,000.

Janney: Exactly.

Kellum: Every time I called out there one of them would be with him and they were side by side every time. He really liked it out there. One time he said, "I can get from Camp Pendleton to the guard station on the other side of the base in 27 minutes." You know, on the base it's just one straight road down to San Diego or Oceanside. I said, "No, you can't" and he said, "Yes, I can. Ask my Ranger." Laughs. When he went to Camp Pendleton, he called out here one day and he talked to my husband and he said, "Can I use your CB handle out here on base, right here in California?" My husband says, "I don't care" because my husband goes by Alabama redneck, so he wanted to use that same CB handle out there. That just puffed up my husband's chest way out, "He's going to use MY CB handle."

Janney: Did you get to see him before he deployed at all?

Kellum: No, me and my husband were in a bad car wreck October 10, '03 and so the American Red Cross flew him to Birmingham where I was in intensive care. They weren't looking for me to live. They flew him in, but they had me so heavily sedated that I really don't remember him being there until about two days later. They didn't have anywhere in trauma to put me in Birmingham, so they had to put me in a burn unit, and he brought me a little dog, a stuffed dog. He handed it to me and the nurse came in there and said she can't have that in here. Well, they went to get it from me, and I was still unconscious, and they said I had a death grip on that dog.

Janney: Aww, bless your heart.

Kellum: I wouldn't turn it loose. And he says, "Well, I guess she can keep it." But I did get to see him when they put me in step down. He came in there. He left on the day that I got out of the hospital. The next day after I got out of the hospital. Then when he got back to Camp Pendleton, he told me that they were deploying in February. I was wanting to go out there then, but I was still under doctor care and I couldn't leave. But they deployed on February 9th and he wasn't even supposed to go because he had done something to his knees, and he had a doctor's excuse. He wasn't even supposed to stand in formation for over 10 minutes. So, he wasn't even supposed to leave the base and go to Iraq or anywhere. He hid the orders from his Sergeant, and he told me, he says, "Momma, I wouldn't be able to get a paycheck if I didn't go with my unit". He said, "I wouldn't be able to back up to get my paycheck." He said, "That's my unit and they're my brothers. I've got to go." So, he got deployed on February 9th and was killed on April 10th.

Janney: Let me ask you. The book that I've written is obviously about, and you may not know, I'm not sure how much Dianne shared with you, but the book that I've written was a trip with Gold Star Dad John Wroblewski to honor the twelve guys that were killed on the 6th of April a few days before your son passed away.

Kellum: Yeah, that was John Wroblewski. Lord, my mind just went blank as all get out. I'm sorry, I'm a blonde you know.

Janney: Laughs. It was Lt. Ski.

Kellum: I have a legal right to go blank. Laughs.

Janney: No, that's fine Margaret. I appreciate that. But that's Lt. Ski and your son was in 2/4, was he in Echo Company or Golf?

Kellum: Echo.

Janney: Echo. So, he was in that same Company with Travis and those guys?

Kellum: Yeah, he was in that with Travis Layfield and Ayers and Wroblewski. All of them. They were killed on the 6th and he was killed on the 10th.

Janney: As far as you know, was your son involved in the action in the market in Ramadi on the 6th of April? Did he go with those guys to rescue those guys that got caught in that ambush or do you have any idea about that?

Kellum: Yes, I think he was coming up from behind when they got ambushed.

Janney: I had no idea about the actual action itself since obviously I wasn't there. I know there was a journalist, Dave Swanson, that was with the Philadelphia Enquirer. I don't know if you've ever seen his pictures.

Kellum: Oh, yeah, I've met him. Oh, yeah, when we went out there when they were dedicating the plaque, I flew out there and we had a luncheon and David was there. So, I got to meet him. Along with Captain Royer, but I have mixed feelings about him. Because from what Virginia and all them told me that's the reason that they lost so many in that ambush in the month of April is because he wouldn't listen to his Corporals or anyone. He kept using the same route at the same time every day. So, that was the reason that they lost so many in that ambush. The Corporals and all kept telling him, "You can't do that. You have to change it up. You've got to go at different times." They said that he said he was the Captain and the leader, they did what he told them to do. He was there at that ceremony, but like I said, I had so many mixed feelings about that, so I really didn't talk to him much.

Janney: And obviously, you probably know that he got relieved of his command not too long after that, right?

Kellum: Yeah, I had heard that they had escorted him back to Camp Pendleton and escorted him into his office, made sure he got everything, was escorted off Camp Pendleton and told not to come back. But they just moved him to another unit. Yeah, but you know, anybody that's in authority, in a situation like that, you can't be so self-righteous that you can't listen to other reasoning. I really do want to forgive him, but I've got mixed feelings about that too because look at all the kids he killed just by him not listening.

Janney: Did you ever have a chance to speak to JT when he was in Iraq at all?

Kellum: Yeah, I talked to him a couple of times. He used to call, and he would say, "Yeah, I'm fine." I would tell him, "Don't you go over there and get hurt." "I ain't, Momma." He always called me "Lady." When he called me Lady that's when I knew he wanted to talk. When he called me Momma, I knew he wanted money.

Janney: Aren't they all like that? I've got three boys and a girl myself, so I know where you're coming from.

Kellum: I had two girls and a boy and JT was the baby. Like I said, he weighed 4 lbs. 6 oz. My daughter, who was born in '80 weighed 2 lbs. 5 oz. She was a six-month baby. Then I've got my oldest daughter. She was born in '74. Now I've got one left. My youngest daughter died in February of last year.

Janney: Oh, I'm so sorry. I didn't know that. What happened to your daughter?

Kellum: Her husband allegedly overdosed her.

Janney: I'm so sorry to hear that.

Kellum: We go for his sentencing this Friday. It's not for that; it's for all the things that have gone wrong. They haven't proved murder yet.

Janney: I'll be praying that justice is served. That's terrible.

Kellum: They say today that even though we're knocking on ten years, JT's grave is the most visited. I wear a black bracelet with his name and all on it; and, of course, my dog tags and I do the pictures of dog tags that you wear around your neck. I've got a picture of him on my neck. They say, "Was John your son?" I say, "Yeah." Many say, "I went to school with him. I thought the world of him".

Janney: Like I said, everything that I've read about him online; he was an amazing young man and I'm so terribly sorry for your loss. I really am.

Kellum: He was. He was always right there when I called him. One day I had a flat tire and he was at work at Arby's and I told him. I called him and said, "Baby, I've got a flat tire." He said, "Where are you at Momma?" I said, "Over here at the Dollar General." He said, "I'll be right there." He told them, "I'll be back in a minute. My momma needs me."

Janney: He was a good son.

Kellum: We used to drive a truck over the road; me and my husband. He went with us one time and we stopped in Lincoln, GA at a little old truck stop restaurant there. Usually, he made me order whatever food he wanted and he would always drink Cokes with no ice and this pretty little waitress came over there and he stuck his chest out and she said, "What would you like to have?" I started to say something. He said, "I'll have a Coke with no ice, a cheeseburger with fries, and I'm sitting over there." I said "Huh?" Laughs.

Janney: Laughs

Kellum: Every time that phone rung when he was in high school it was a different girl that wanted to talk to JT.

Janney: Yes. He sounds like an amazing young man. I wish I had gotten a chance to meet him.

Kellum: You would have fallen in love with him.

Janney: I know I would have. I do want to ask you one other question about a picture that I saw on a website that was dedicated to him. It's kind of a situation room picture, you know, with a bunch of Marines standing around. Do you know the picture that I'm talking about?

Kellum: Yes, sir.

Janney: Is he the young man that's kind of in the middle of the picture?

Kellum: Yes, sir. He's sort of facing the camera, but he's turned a little bit.

Janney: So, do you know anything about that picture? I don't know where you got that or who sent it to you. Can you tell me?

Kellum: That was the last picture that he had taken before he got killed and it was in Ramadi, Iraq.

Janney: Do you know if he was at the Combat Outpost there on the outskirts of town? Do you have any ideas?

Kellum: I really don't know. I don't know where they were.

Janney: You know I was there, right? I went there and spent a couple of weeks in that town and stayed at that Combat Outpost where JT was stationed so I'm sure that your son was there.

Kellum: More than likely.

Janney: But anyway, that picture is very intriguing. He looks like a very serious young man in that picture and looks incredibly focused on whatever was taking place at the time.

Kellum: When he did anything, he was 100%. When we went out there when his unit came back from Iraq, his Sergeant told me, "Mrs. Kellum, some of these men I had to worry about. If they had everything they needed. I never had to worry about John. I could walk in there and tell them let's go and the only thing he had to do was grab his bag and he was set." They called JT "Lunchbox."

Janney: Lunchbox? Why was that?

Kellum: Well, no matter where he went, he had food stuck in his pockets. He'd have snacks in every free pocket - crackers and cheese, you name it. So, they got to call him lunchbox because they knew if they needed anything to eat, they just had to go talk to John. If he was sitting in his barracks at Camp Pendleton playing his video games and stuff; and they come in, "John, are you going to eat that hamburger?" because he'd buy two or three, you know, and he'd say, "No, go ahead and take it." He said, "I'd only eat one or two." He never gained any weight. They'd be out there in the field or something and they'd say, "Oh, I'm hungry." JT would just pull something out of his pocket and say, "Here." Laughs.

Janney: So, that's why they called him Lunchbox because he always had something ready.

Kellum: Yeah. That's why they called him Lunchbox.

Janney: Did you ever hear anything from any of his Marine buddies; his Marine brothers about his role in the support for the ambush? Or anything that happened afterwards?

Kellum: One of his buddies, John Huercamp still calls. He still feels bad about what happened to JT. I got a call from one of his friends from Iraq. He kept saying, "Sims didn't make it. Sims didn't make it." I said

"Well, slow down. I can't understand you." He finally calmed down a little bit and he said, "Sims didn't make it. He got shot." He took it real bad and is still taking it bad. His parents came down for the viewing and all. Let me tell you about his crazy friends. The night of the viewing I had a Lincoln Continental car and, of course, I was in there with JT and all, Jaime, the one that owned the Domino's. She comes in and says, "Margaret, let's go outside. You need some air." I said, "No, I'm fine" and she said, "No, let's go outside" and so she gets me by the arm and she starts leading me out and when I get out on the porch I see my preacher and his wife over to my right and they're all white faced. I mean, all the color has left out of them. I'm thinking what in the world. So, she led me down to the parking lot. I didn't notice that my car had been moved. Everybody is stopping and looking at me like a deer in the headlights that won't move, you know, and I said, "What is going on?" So, she leads me up the driveway. There was a policeman, an off-duty policeman and a friend of ours looking at me real weird. My sister is looking at me. I said, "What's going on?" So, I get there and I'm looking around. I turned back around and then looked toward the right and I said, "Whoa, wait a minute." I look back to the left and six of his friends are in my car and they've got my car saran wrapped.

Janney: What?

Kellum: Yeah, they had Saran wrapped my car at the viewing. Laughs.

Janney: Oh my word!

Kellum: Then, when I got up there, they handed me a teddy bear dressed in dress blues. One of those talking ones. They said, "Mom, mash the button." So, I looked at the teddy bear and I mashed the button and it said, "Hey Mom, this is Mickey, Joe and friends. We're in the bathroom and we just want to tell you that we loooove you." Laughs. But, they almost got arrested because the police friend of mine kept telling them, "It doesn't pay to be doing all this. I could get my friends out here. It won't take but five minutes." The Daddy of the one that owned the Pizza Hut, he kept the saran wrap and gave it to his other friend and they kept wrapping up my car. He says, "If you'll just wait five minutes, we'll be through." My preacher, Don, is the one that worked at the funeral home, and the owner of the funeral home is going to make sure nothing goes wrong. My casualty officer already told them that nothing had better go wrong with this viewing. So, I go walking back down toward the funeral home and my preacher's son is walking around in circles out there in the yard talking to his boss, I look over at him and I said, "Russell, Russell" and he says, "Yes ma'am, Ms. Margaret." I said, "It's okay." He's on the phone and he tells his boss, "Oh, she says it's okay." Laughs. He just knew he was fixing to be fired.

Janney: Well, that's funny that they tried to brighten your day. That sounds like something that JT would have done had he been there.

Kellum: Yeah. How that all started is that his buddy that he went into the Marines with, and went to high school and all with, when he and his wife got married, JT came home for the wedding and he went out there and put shaving cream up under the door handle and saran wrapped their car.

Janney: Oh, my goodness. Okay, so there was some history of that.

Kellum: So, he put saran wrap on my car at JTs viewing. It was hilarious. I mean, it just showed me how much they loved him and me.

Janney: Do you remember the name of your casualty officer?

Kellum: Getry. He was stationed out of Montgomery. I think he's retired now. I think he went to Mississippi or somewhere. But he was the best casualty officer. I mean just anything we needed he was right there. He went to the funeral home and got with the funeral home owner and he said, "All right. Tell me where you're going to put him when he gets here." They showed him and he just said, "The casket over here. The flowers over here." And he said, "And you know, nothing goes on that casket." The funeral director, Mr. Anderson, said, "Yeah. All right." Getry said, "Now take me to the church." He went over to the church and he said, "This is where the family is going to be coming in. And this is where the family is going out. This is where the family is going to be sitting." And then he said, "Now, take me to the cemetery." So, they took him up there and said, "This is how it's going to go over here." Then he went back to the funeral home and he said, "This family better not have to ask me for anything. Nothing had better go wrong." So, that put all of them on pins and needles right there and then they went out there and Saran wrapped my car. Laughs.

Janney: Yeah. I'm sure that gave everybody fits.

Kellum: It was hilarious.

Janney: That is hilarious. Can you tell me where JT is buried?

Kellum: At Hillview in Alex City. We even had one lady that was bringing up the rear of the procession who stopped along the side of the road and picked up a hay bale and put it in her trunk and went on to the funeral; to the graveside. If that ain't redneck, I don't know what is.

Janney: Oh, no ma'am, that's fine. I heard there was quite a turn out for him; a homecoming celebration for him when he came home.

Kellum: Oh, yes. I have never been so proud of a town in all my life as I was of Alex City when he came home. There were 152 cars in the funeral procession.

Janney: I'd be lucky to have 10 at mine.

Kellum: No, you won't. You know, when we got JT's truck back from California, it was all banged up, and I took it to the Ford place and got it all fixed back the way it was and I took pictures of it. So, when I went out there for the ceremony, I was showing Joe McGee pictures of it and I said, "Remember this truck?" He said, "Yeah, it doesn't look the same." I said, "It had $4,500 damage." He said, "Yeah, I was with him when he put $4,200 of it on there." Laughs.

Janney: Laughs.

Kellum: I said, "I don't doubt you were, son." Somebody had seen the truck up here in Alex City one day last week or last weekend. She remembered who JT was and she called me and told me that, "I just saw that truck with your son's name on the back. I just wanted to call and tell you that I have not forgotten, and I still remember when this happened. I still remember JT. I just wanted to let you know that I have not forgotten and you're still in my prayers." That just meant the whole world to me. I was feeling really down that day, but she lifted my spirits a little bit.

Kellum: We're supposed to be planning a 10-year reunion in San Mateo. Because I've already asked my boss, who is a former Marine. He served 21 years in the Marines. I asked him, "Can I be off from like the

3rd or the 4th to maybe the 10th?" He said, "Take the whole week off. You deserve it. You haven't had a vacation since you've been here in two years. Take that whole week off and relax." I told him "Well, I want to go to Camp Pendleton because they are having their 10-year anniversary." He said, "Take as much time as you need."

Kellum: I bought that book "No True Glory" by Bing West and it's got chapters that explains what happened with 2/4 and all. I managed to get through two or three pages, but JT's part is later and I just hadn't gotten that far yet. But I will get around to reading the whole book.

Janney: I just really appreciate your time telling me about JT.

Kellum: I appreciate you doing this. I admire you for it.

Janney: Like I said, it's the least I can do. I got hired as a photographer by this journalist by the name of Martha Zoller. She actually came up with the idea to take John Wroblewski to Ramadi back in 2007. She hired me as her photographer to do that, but we couldn't get there because of all the sniper activity in January 2007. I just felt so bad for this Dad because I had spent twenty-four hours a day for ten days with him that I promised him on the plane ride back that I would get him to the site where his son got killed, and fourteen months later we actually made it there. I organized and planned the trip, raised a lot of donations from different individuals and the National Christian Foundation. I took him there and did a memorial service for those twelve guys and your son. I was honored to be a part of that. I really was.

Kellum: Yeah, I was asked if I wanted to go to Iraq and visit where my son was at and I told them, "No." Then, Dianne Layfield asked me, "Do you want to go to Iraq?" I said, "No." She said, "I don't either. It's not going to bring Travis back." I said, "No, it ain't going to bring JT back either." I said, "If I get over there, I'm likely to go a little bit stupid and I won't ever come back because I'll be in heaven, or something."

Janney: Well, the sad thing about that Margaret is that Mark Crowley told me that there is an organization that made a lot of promises to take people over there and they used them to raise a lot of donations, but they failed to get them to Ramadi. I just felt very honored and very blessed to get there, just John and I, without any kind of support at all, other than through donations to actually accomplish that. Mark Crowley is very upset that he spent so much time raising money for this organization that he told me about that promised to get him there, but never followed through on their promises.

Kellum: I was really blessed that we had such a close relationship, Mark and Kyle had gotten in a big argument before he left and there were some real harsh words said and words that can never be taken back. They never did speak again even when he was over in Iraq.

Janney: Right.

Kellum: None of that got resolved. It was just too late. Mark's got to live with that for the rest of his life, you know.

Janney: Right. He and I have talked about that. It breaks my heart.

Kellum: I can't imagine. I am so thankful that we had such a close relationship that nothing like that took place.

Janney: Yeah, and I agree. As I said, Mark and I have talked about that a number of times and I can't fix that, but I can let him know that his son is not forgotten. I stood at the spot where his son passed away and did a memorial service for him. I've written this book so that people will remember his son and what he fought and died for and that he and his sacrifice are not forgotten.

Kellum: I just can't imagine going through the rest of my life with that laying heavy on my heart. I feel so bad for him.

Janney: I do, too. He and I talked and we cried together on the phone several times.

Kellum: Like I said, I admire you for stepping up and doing it.

Janney: Somebody needs to remember these young men because they were a lot better Americans and people than a lot of us.

Kellum: I've got a big old picture of him on the back of my truck that I printed off. You know, because I've got him on one side and my daughter on the other side.

Janney: Well, I just appreciate your time speaking with me. I didn't mean to dredge up the sadness, but I want you to know that I do think a lot of those young men and especially your son and I promise that I will not forget them as long as I'm alive.

Kellum: Thank you. I don't mind talking about him; that means he's still alive.

Janney: Exactly. He is alive. As I said, I've read so many stories about him on the internet. I spent a lot of time trying to learn a little bit about him before I had a chance to speak to you and I promise I won't forget him.

Gold Star Family Interviews

Gold Star Mother Dianne Layfield (LCpl Travis Layfield)

Janney: Dianne, obviously I've watched the film; the memorial film that you put up about Travis. That was an amazing recount of his life, but I wanted to go into more detail to learn more about him. I know you said that he was born Memorial Day weekend, but tell me a little bit about Travis as a baby and growing up.

Layfield: I had a daughter who was twelve years older than Travis, so she was like a second Mom to Travis. Her name was Tiffany and between us, I don't think Travis was put down very often growing up. He had a good family life. Just an amazing young man. He grew up like any normal child playing ball. He didn't have many problems - we worked on those. He did have a heart murmur which was a BSD - a defect in his heart. Growing up, I kept a pretty good eye on that to see if they had to do surgery on that, but they never did at any point. It just never got any worse. He was able to do normal functions and play and do anything that a normal child would do. He loved to camp. My older son had a motorcycle and Travis' father, so they would go out camping and they would dirt bike and, of course, Travis had a little dirt bike, so he had some camping growing up as a kid, fishing and all that kind of stuff. Just a normal young man; always had a smile on his face. Always very loving. Never gave us a lot of problems.

Janney: I know you have mentioned that he decided to join the Navy SEAL program.

Layfield: He joined that.

Janney: I think you said he had joined that when he was 10 in school?

Layfield: Yes, two of his buddies in junior high were in that program and they hung out all the time together and they were in it so they wanted Travis to join so I said okay. We did and it was probably one of the best things that I did for Travis at that time. He was never a problem child, but it was just a good way for him to learn respect and discipline at a young age. I don't think they could ever hurt any child and he loved it and went to boot camp after being admitted to that program in June that year and graduated. At that time, he said, "Mom, one day I'm going to be an officer. You'll be really proud of me. I'll make this my career." So, a few years later when he went into high school and the Marines were always coming around, he decided that he would rather be on land than in the water. He didn't think he could stay in the water all that time on a boat. So, he decided to go into the Marines, and he did the pre-entry program which was in his junior year. He was always down at the recruiting office hanging out with them and going on field trips with them. That's just what he wanted to do. Then, in his senior year he knew we were at war, but he said, "I want to serve my country Mom. This is what I want to do" so there was no stopping him or changing his mind. It was what his heart wanted to do.

Janney: As far as his interest in the Navy initially, was it just because his friends were doing that and it was just something he wanted to do or was there a family member that was in the Navy or Marine Corps that got him interested in those branches of service?

Layfield: Yeah, my father was a Navy Seabee in WWII and he idolized my father and I think that that had a big impact on him that he wanted to be Navy like his grandfather. I'm not sure what happened later on that made him want to go into the Marines, but I think my father being in the Navy absolutely had some impact on Travis.

Janney: As far as 9/11, did he have any reaction to that attack on our country?

Layfield: Not a whole lot. Of course, it added to it; escalated it, but he'd already had his mind set even before that he was going to become a Marine, the best of the best he would say. I'm sure that escalated after it all happened, but he had already knew that he was going into the military and he was going to become a Marine one way or another. With his heart murmur, on this end, he had to have the doctors check him and everything and give their okay. His cardiologist gave his okay and cleared Travis and said he was perfectly normal to go. Then when he got down to boot camp in Camp Pendleton, of course they did a full physical there and they were not going to let Travis become a Marine. Now, I'll never forget the day he called me just crying so hard and saying, "Mom, I'm in the hospital." I said, "Oh, gosh, what?" and he said, "I just had my physical and they're not probably going to let me go into the Marines because of my heart defect." But not long after, the doctors cleared him, and they said that there should be no reason that he cannot go. So, that was a little bit of a to-do. Travis was devastated that he might not be able to become a Marine and it delayed him going into the first group because he wanted to be with some of his buddies. But, he was in the hospital and they had to prolong it because they had to find out for sure if they would release him. They finally did so, but he was devastated. I don't think there would have been anything else in this lifetime that Travis would have wanted to do other than become a Marine and serve his country.

Janney: As far as his buddies that you mentioned, the group that he was not able to link up with; did they in fact complete basic training and deploy with the Marines at any point?

Layfield: As far as I know they did. I kind of lost track of that first group. I didn't really know much about them because he had only been there a couple of days when all of this happened, so I'm honestly not sure but I know he was postponed and went to Iraq a little bit later with a group that he was involved with.

Janney: Okay, so it wasn't his high school buddies that he joined with. It was some guys in basic?

Layfield: Just some ones that he had gone through the recruiting process with here in Fremont and had transferred down to Camp Pendleton. They all went on a bus together to boot camp. Those were a couple of other buddies that he had just met through the recruiting office and kind of hung out with. That just kind of pained him from being with that group of friends that he had just met.

Janney: Let me backup a little bit. I know you said that Travis was devastated when his cousin Raymond died. He had always admired Raymond's '62 Ford Galaxy and he was able to buy that, I guess, when his aunt kept the car for him and allowed him to buy it when he was sixteen. Can you tell me a little bit more about that? Obviously he had the car for several years before joined the Marines, but was that car his baby?

Layfield: That was his baby. He loved that car and he worked really hard to pay my sister the money for the car and told her that he would always take care of it and he did. He loved it and his nephew would go riding with him and hanging out. Again, it wasn't too long because then he was going into the military. His father still has the car. He lives in Las Colinas in northern California and his father still has the car. At his memorial, when they had it at the high school here, Travis's father brought that car and they had blocked off Fremont Boulevard which is a huge strip in Fremont, you know, entryway type thing through the whole city and they had blocked it off and at the high school did a memorial for Travis and there were just people and fire trucks and cops and everything everywhere. Travis' Dad drove Travis's car up front so everybody would see it by the podium. The police said, "Oh, no, no, you've got to get that car out of here" and his Day said, "Oh, no, no, no this car was Travis' thing" and they let him stay there.

Janney: Aww, that's very cool.

Layfield: It is. I'm hoping his Dad, because, you know, we found out that Travis has a little boy, so we're hoping that Johnny, who is Travis' father will definitely hand the car down to Dylan, Travis' son, if anything should happen to him and Dylan should be next in line to get the car, we're hoping.

Janney: Travis had a son, Dylan. How old is Dylan now?

Layfield: Dylan will be twelve next week. But, we didn't find out about Dylan until two and a half years after Travis was killed. Yeah, I didn't know if you knew that part of the story or not. So, Travis never knew that he had a son. He thought it might have been his, but then he left right away and she never told him any different that it was their roommate's. But it ended up being Travis' little boy and he's a part of our life now, so we're blessed.

Janney: Oh, that's very good.

Layfield: It is, yeah. His legacy lives on. We didn't know about it until two and a half years later. She had kicked Travis out. They had moved down to Atwater which is about three hours away at the end of his junior year . Actually at the end of his sophomore year almost, he lived down there almost his whole junior year and then she kicked him out. She was quite a few years older than Travis and so she kicked him out because she decided that she liked the roommate better . Long story short, I guess she was pregnant at that time. But Travis did not know that, so she married the roommate and led him to believe for two and a half years that Dylan was his child, but he wasn't. My daughter, like I said, who had a twelve year age difference between Travis and her, and kind of was like a second mom to Travis, she kept having dreams that Travis was still alive. He kept coming to her in her dreams saying that he was still alive. He used to call her Mimi, and so she decided to go on this mission to find this lady. She found out that she still lived where they had lived, so she left a note because the lady wasn't home. The neighbor later said that she did have a little boy, so my daughter left her a note and three weeks later, Catana is her name, she finally contacted my Tiffany and said that she was sure he was Travis' son. So, we did the DNA, and that year, a week before Christmas, we found out that Dylan was my grandson and Travis' son. So, ever since that, he has not come back to Tiffany's dreams at all. He's at peace now; he knows we know Dylan.

Janney: What a neat Christmas present.

Layfield: Yeah, yeah it was. It was pretty exciting. She and him actually came up and did a big story about it. I can get you a copy of that in the mail if you would like one.

Janney: Yeah, that would be amazing. Who was the reporter that did the story?

Layfield: I don't know offhand. I'm not sure what she said there. I think she does. I probably have it written down somewhere. I can research that and see if I can find it. Yeah, it was a lady that came out and did the whole story, but I don't remember her name. It's been a lot of years.

Janney: No, ma'am. I understand.

Layfield: But if you email me or you just want to give me your address, I'll send you a copy.

Janney: Yeah, I'll send you an email.

Layfield: Oh, what I meant was I am going to send you a hard copy. You can send me the email with your address in it. That'll be fine. Send me your home address and I'll snail mail you a copy.

Janney: Yeah, of course. I appreciate you doing that. Now, I've got a piece that Brigham Young University did on John Wroblewski's trip overseas. I'll be glad to send you a copy of that back too if you would like to see it if you haven't already seen it.

Layfield: I'd love to. I might have but I'm not sure. If you'd send me that, that would be great.

Janney: What about the Marine Corps video that combat cameraman Corporal Angel photographed? I know that I've got a copy on my blog, but have you seen that?

Layfield: I think you sent me that, right? I think when we first contacted each other or something. If you want to send it again, that's fine too.

Janney: I just didn't know if you wanted a hard copy of that video?

Layfield: I think that would be great. I'd love it. That would be great for his library. My house is a shrine of Travis and even though I have two other children and I try not to do that but still today, like I said, I've got two other children who do wonderful things, still today, people still recognize Travis and remember him and he did some beautiful things and I can't just shove them in a drawer.

Janney: Oh, no, you shouldn't. As long as we remember our loved ones, they will never die, right?

Layfield: Absolutely true, yes. So, anything you want to send me, absolutely, I would love to have it.

Janney: I know you mentioned in the video, and of course, I've seen photographs and the video of Travis' feather tattoo, to honor his Lakota Sioux heritage, can you tell me a little bit about that? How closely was he connected with his background and his ancestors?

Layfield: Well, his grandmother was 100% Lakota Sioux and he was half Lakota Sioux and he grew up with his grandparents all around us all the time. He knew all his ancestors on that side of the family that he grew up with and knew of their Native American heritage. He had always said in his high school years,

"Mom, when I get home from the service, I want to learn more about my heritage. I really want to get involved and do a lot more research." When he got the feather, he always thought it meant one thing - it ended up meaning "fallen warrior" so that today still gives me chills when I think of it because, it's funny how things go and how things work out. But, so many signs lead up to Travis just not coming home. He knew down the road what was going to be in store for him. We had a hard time getting anybody to tattoo him because he was only sixteen and he had to be eighteen with a parent's permission. But we finally called and talked one of those guys that does these tattoos into doing the feather. He did it and that's what became of it.

Janney: You said you found out about it after the fact when his grandmother told you? No, it was the medicine man that was at his memorial service? Is that correct?

Layfield: Yes, that was it. You are right. Because his grandmother told me it meant the first-born son which would have been Travis' father, my second husband, and Travis was his first-born son. So, that's what we thought it meant for a few years. Then the medicine man saw a picture of him at his services and he came over to us afterward and asked us if we knew what it meant. Of course, we said first-born son and he said, "No, it means fallen warrior." So, quite a shock. It just really blew us away.

Janney: Yeah, and that definitely does give me chills as well.

Layfield: Yeah, it really does. I'm going down to see Dylan actually this Saturday. He's doing a run for the fallen. We do it in Merced County; a big event so Dylan has been running and walking for his Dad for the last few years and my younger son, Travis' brother and I and his wife, we go down and we run and walk with Dylan in this 10K run. So, we're kind of excited to see him this weekend.

Janney: Oh, absolutely. And you said his brother's name is Tyler. And you said that they were inseparable, I believe, is what you had mentioned?

Layfield: Yes. They did everything together. They did sports together and hung out together and were the best of friends and the best of brothers. They had a few spats here and there, but for the most part they were just inseparable. They got along really well.

Janney: I know another thing from the video that had touched me was the fact that you had gone down to Camp Pendleton for three days.

Layfield: Yes. For three days. Ten days after he graduated, I got to spend some time with him.

Janney: Can you share a little bit about your time when you were there with him?

Layfield: It was because my ex-husband had gotten a couple of rooms for us and so I stayed in one, I was going to stay with Tyler, and my youngest and Travis, I thought, would stay with his Dad. But Travis said, "No, Mom, I want to stay with you. I want to be with you as much as I can and spend as much time with you as I can." That meant a lot to me because I always thought boys are closer with their Dads, but my boys and I had a really good rapport and we had the best time ever. We went out to dinner and he showed us around the camp, the whole Camp Pendleton area and told us little stories and things and he just loved it. He loved going to the church there and he said, "Mom, it sounds like the Tabernacle Choir in here when the guys all sing on Sunday morning." There were little things like that, and he loved what he was doing. He really did. He really loved being a Marine. He was very proud, and I was very proud of him.

Janney: It sounds like he went up through the ranks pretty quickly if he was a Lance Corporal when he deployed to Iraq.

Layfield: Yeah, he did.

Janney: So, he must have been pretty good at being a Marine.

Layfield: He really was. He was devoted and very disciplined. He learned all that with the Navy program in the beginning and he was just dedicated. Doing something worthwhile was what he really wanted to do. He just had his mind set and that was his goal. He loved being a Marine.

Janney: So, when he flew to Kuwait on 16 February were you able to see him off at all?

Layfield: We weren't. We went down a couple of weeks before, and then got to say goodbye to him, but he didn't want us to come down and be there when he actually left. We didn't go down to actually see him and I really regret that to this day, but that's 20/20 hindsight. I really regret not going down and saying goodbye to him, but that's how he wanted it so we honored his wishes. I knew it would be hard for him and I think that's why he didn't want us there.

Janney: That makes perfectly good sense. Saying goodbye is always the hardest thing, I think.

Layfield: Yes, it is. Even in my little video I said, "I wish I could have jumped back out of that car and run back and hug him again, but he was already walking to go see his buddies." When we saw him the last time, there was a little party, so it's like all right, time for Mom and Dad and brother to go, so it was really hard turning around. Of course, had I known that I would never see him again, I would have never let him go. That's life, right.

Janney: Yes. As far as the Marines go, was he always a 2/4 Marine?

Layfield: Yes. As far as I know, yes. He was in such a short time that, yeah, it was pretty much 2/4. Now, do you know of David Swanson?

Janney: I do know David. I know that he was with them on the days leading up to the ambush, that day, and of course, for a period of time afterward.

Layfield: Yeah. I didn't know if you had contacted him to discuss notes or anything like that, but he'd also be a good contact for you if you wanted to.

Janney: I'm hoping that he agrees to, with permission, to use some of his photographs in the book. Because, you know, they obviously tell another facet of the story.

Layfield: Yes, absolutely. And he's a great man. A great man.

Janney: We've talked on the phone many times and conversed back and forth via Facebook and email and I think the world of him. Like I said, he, unfortunately, it changed him in ways that it's been tough for him, too. I think it was tough seeing the guys that he had gotten to know and spend time with not be able to make it home.

Layfield: Yeah, very true. I can't imagine what any of them went through. It's just unbelievable but…Did you have all the family's names and addresses and things or contacts that you need to talk to any of the other families?

Janney: No, ma'am. I do not. I really only have contact information for yourself and Sheila Cobb and John Wroblewski and Shawn so if you want to share that, I think you had mentioned that you were going to speak with them and see if they would be willing to do an interview. Obviously, I'd like to talk to as many parents as want to talk to me.

Layfield: Okay, okay. I will send that out after we get done and everything and see if anybody, I can give them your email, would that be how you would want them to contact you?

Janney: Sure. My phone number is fine. I want to give them as easy an access as they can have.

Layfield: I will do that then. I was just kind of waiting to hear from you and see how the interview went and all that. And talk to you about it, so that would be great, and I will do that.

Layfield: Mark Crowley has spent lots of time with us and I actually think I kept Mark Crowley alive. You'll hopefully do an interview with him or hear a lot about him, but literally he calls me his big sister because I literally kept him from going over the edge. He's been a great resource to me too, but I was such a friend to him, and we became really bonded after that. It's funny how things work out sometimes.

Janney: Yeah. I'd love to find out more about Kyle. Like I said, I obviously have never met any of these guys, but this is how I'm trying to learn about these fellows is by talking to their parents so, I mean, I would appreciate an introduction to Mark.

Layfield: I will do that. Hopefully he will be open to it and that will be great. I know that there's a couple of Moms that don't have emails, so I have their snail mail addresses. I will write them a note and ask them and forward your information and hopefully they might also contact you. But, that might take a little longer for them because I have to get to them so just, I'm just letting you know. Everybody else will.

Janney: I appreciate that. There's no urgency and I sure don't want to force anybody to do it, but I'd like to. The trip that we made over there was not only to honor JT, but to honor all those guys and it'd be great if I could share with the readers about their sons as well.

Layfield: I'm really honored and thank you for doing this. It's wonderful.

Janney: It's my honor, Dianne. Like I said, my heart just broke for John on that 1st trip back after Martha's mission was unsuccessful. For him to have gotten within 60 miles of our goal and to have been turned away, it just broke my heart. I felt compelled to promise him somehow or another that I would get him back to Iraq.

Layfield: Wow, bless your heart. That's one thing I never had any desire for. I had a desire to go when the Battalion came home, I wanted to go see the guys that served with Travis. I wanted to welcome them home and let them know none of us held any grudges. It wasn't their fault because they were all doing that guilt thing and I wanted to be there. So, that was harder than anything, almost as hard as losing Travis, to watch them all come up the hill without my son, but it's something that I'm so glad I did and greeted his buddies and to still be in touch with a few of them. It was worth going to, but going to Iraq is just something that I didn't think I would ever do. To see where he was actually killed, that I honor JT and another friend of mine, Michael Anderson has gone. It's something that I didn't have the guts to do, I guess. I did everything else that I possibly could to keep Travis with me and carry his legacy on and I know the guys that were with him when he was killed.

Janney: You mentioned Michael Anderson. Was his son in that unit also?

Layfield: No, his son wasn't. He was with 1/5 or 3/5, I think it was 1/5. He was killed in Fallujah in December of 2004, so he was not with Travis at all, but his father and I have become really good friends just because we both buried them. We're real good friends and he lives about thirty or forty miles away, but we see each other quite a bit at different events and his family and I have become really close. That's kind of how I know Michael and I know Mike, his father had gone to Iraq as well. He wanted to see where his son was killed.

Janney: Did he actually make it to the site where his son was killed?

Layfield: I believe he did. Yes, yes. His mind is now at peace as well; he's gotten to do that.

Janney: Getting back to Travis, I know you had said you had talked to him a couple of times when he was in Iraq, or I don't know if one of the times you spoke to him was when he was in Kuwait, but do you mind sharing what you were able to discuss with him or what he said to you the first time that you spoke with him?

Layfield: He really had a hard time back then getting hold of a phone. He hardly got to call us at all, so everything was really brief and just, "Hi, Mom, I'm doing fine. How's everybody there? I love you and I miss you." Then when we finally got to talk to him the last time, I talked to him on March 17th, St. Patrick's Day. That was the morning he called me at work and he just wanted to talk and talk. He got the phone and, in fact, I was on my lunch break when he called. I didn't even know he would remember the number or anything. He had it written down and he called, and of course I found out later that day that he had called his sister and his Dad, so it was like he was just trying the final goodbye because that was just the last time we heard from him. It was just a sweet conversation. We laughed, we joked, and he told me that he was seeing a lot of bad things. Some of his friends were getting blown up and bad things were happening, but he was okay and said not to worry about him because he knew what he was doing and he knew what he was supposed to be looking for. So, then my boss walked by and he said, "take your time" because he knew I was talking with Travis and of course he knew he was in Iraq. So, we just talked; probably about fifteen minutes and it was just wonderful, you know, to hear that laughter and he just wanted to make sure that I knew that he was okay and he was doing fine and not to worry about him. So, that was the last time that I got to hear from him.

Janney: You said that he was seeing some bad things. Do you want to share any of that with me? Other than his friends being killed or injured with IED blasts. At that point, based on my understanding, there was not a lot of combat in Ramadi. I think it kicked off toward the end of that month not long after you spoke with him.

Layfield: He didn't go into a lot of detail because I'm sure he didn't want me to hear a lot of it. But he just said, "Mom, I've been seeing guys getting blown up" and he said, "It's pretty bad over here" and that's about all he said. He didn't go into any gory details or anything. Just that he was seeing some of his buddies getting blown up; not to worry about him that he was going to be fine. He didn't go into any details on anything. I don't think he wanted me to know how really bad it was because I definitely couldn't fathom any of it because I wasn't watching the news. That's what he told me, "Stay away from the TV, Mom. You don't need to be watching the news. You can't believe everything you hear." And I didn't want to. I didn't want to hear or see what was going on. I was scared to death as it was. I sure didn't want to have any reminders of all of that by watching it on the TV or the news.

Janney: Yeah, and I can understand that. You know, operational security, he probably couldn't say a lot too. I wasn't fishing for gory details; I wanted to know if he said anything about the general climate, but I mean in general.

Layfield: Not really. He had said that they had given candy out the day before. They had given it out a couple of times. They had been in the marketplace a couple of different times. He loved the kids over there. He did mention that. He didn't get our care packages until after he had passed. It ended up coming back to me and my Mom's was also returned because he didn't even get the packages with our pictures and stuff in it. He never even got a chance to open all of that. But when I was on the phone with him, he was just telling me that he was having fun with the kids and giving candy to them and it was a whole different world over there. It was lovely he was asking how we were all doing here. Asking about his sister and his brothers and I think that's what convinced him to call his sister. He never got to talk to Tyler. He did get to talk to his sister Tiffany and his Dad March 17th.

Janney: I think you shared with me that several of the men in his unit had contacted you too, afterwards, have you been able to stay in touch with those guys, did you say?

Layfield: I have; a couple of them. One of them ended up going to his wedding in Texas. My daughter and I when they got home; a year later when he had gotten married, he invited us to his wedding. That was Travis' Sergeant. There is another one that has been in touch with me quite a bit. He's been in Okinawa and in Colorado and now he's back at Camp Pendleton, so I'm hoping to see him soon. I'd like to go down and visit with him.

Janney: What was the Sergeant's name whose wedding you attended?

Layfield: Marcus Waechter.

Janney: Oh, yes ma'am. I'm just honored to play a part in helping folks know who these guys were and who your son was and what they went through and the sacrifices that they experienced and suffered and that their families suffered.

Layfield: Yeah. And every day, it's been almost 10 years; and every day is like yesterday. I can still see his car coming into my carport and everything is so focused that it will be embedded in our lives forever. It's definitely changed our lives and left a big void. There's no doubt about it.

Janney: I know that you had shared a little bit about that in the video about how you saw them coming up and you said, "No, not my Travis."

Layfield: Yeah. Quite a shock; quite a shock. And you don't know what you can live through and what you can withstand until it actually hits you. I never thought I would have to face something like that. And then you don't know how you will deal with it and you try to think about how you would deal with it, but you honestly don't have a clue until it actually hits you. And then it just takes its course. But I felt good. I live in Travis' memory and I've been to, I've lost count of how many funerals I follow in the Bay area and I try to go to as many as I can just to support the families and to let them know I'm here and if there's anything I can help them with. The day God took Travis, I told Him, "You've given me a calling and a mission." Just lead me in the right direction and I've gotten pretty well known because of what I've tried to help the Gold Star families go through and through my experience just letting them know what's out there for them and what's here for them. And I can actually say "I'm walking in your shoes." And to me, because I am.

Janney: Yes, indeed. Like I said, I hope I never experience what you've gone through, but I do have three sons that are all military age. They've not professed a whole lot of interest in doing that but who knows, they might change their minds and I couldn't imagine what you've been through.

Layfield: Yeah. I sure hope you never do but, my grandson, you should see him. He just now joined the Navy cadet program about three months ago. So, he's traveling in his uncle's shoe steps. I'm very proud of him too. He just loves it. It's only been a very short while, but he loves it.

Janney: We talked a little bit about David Swanson. Has he shared all of his photographs with you? Have you seen the powerpoint presentation that is circulating online? "Echoes of War " is the name of it.

Layfield: Yes, I have seen it. I have a hard copy of "Echoes of War."

Janney: So, you basically saw the marketplace and the places that John and I walked through a couple of days when we were there on our trip.

Layfield: Yes, yes. I did, yes. And a lot of it was about John actually. Walking through that field and stuff too there. That was John I believe. Yeah, okay.

Janney: Yes, ma'am. You are right. I haven't communicated with David about this, but are there any photographs in that video or that he has shared with you of Travis in the field at all? I know that he was with the unit for several days before the ambush.

Layfield: I didn't see any of Travis. Only one that is real, to think in my mind, is the one where the men were carrying a body bag. Do you know that picture?

Janney: Yes, ma'am. I do.

Layfield: Yes, that is Travis. When I saw that picture, I told my daughter, "That's Travis" and she said, "Mom, how do you know?" I said, "I know that's Travis" and we didn't know at that time, but I just knew that was Travis. And when we went down to meet them when they came home, I met David Swanson and he told me that was Travis and he wanted me to meet the young man that carried Travis. And I got to meet Apple. His last name is Apple or Appel or something like that. So, I did get to meet him and then, like I said, I went with Swanson at that time but yeah, that's very much embedded in my mind.

Janney: I'm sorry. I didn't mean to bring up a sensitive thing, but I didn't know if David had any pictures of him in the field before.

Layfield: Oh, no that's okay. I don't think there's any of him in the field that I know of. Although one of his buddies did give me, or actually give them to Mark Crowley, some pictures of Travis and Kyle over in

Iraq. I didn't get any pictures back with Travis' stuff. But when his friend came to visit Mark Crowley, he had this whole binder of pictures and Mark said, "Hey, look here's some pictures of Travis." And he had a few of Travis, of course, over in Iraq. Like in a bunker and stuff like that. Not a lot, but there were a few that I didn't know existed. So that was nice to see Travis in a few of those pictures. I don't think they were ever together in any of the pictures but the friend of theirs, his name was Ryan Downing, he had taken a lot of pictures and he knew both the guys and stuff. Because Travis, when I did talk to him, he had told me about two guys that he met from the Bay area here and he never did mention their names but he just said that he was becoming friends with two guys that he met from the Bay area and that was Kyle Crowley and Vince. Vince had gotten wounded a few days before their ambush and he had gone home. So, he lived with that regret that he wasn't there to live through that with his buddies.

After we met Dylan, Tiffany told us about a dream she had about Travis coming to see her and Dylan. The light shone down, the sun came down right where we were at. There were people there with Travis, and Dylan was only 2 and a half years old and he reached up and said, "Look, Mommy! An angel!" I will never forget that. And ever since then, we have not heard from Travis. He has never come to see Tiffany in her dreams any more so he's really at peace and for that little boy to say, "Look, mommy, there's an angel" you know it was Travis.

GOLD STAR FAMILY INTERVIEWS

GOLD STAR GRANDMOTHER FRANCES MABRY (PFC CHRISTOPHER MABRY)

Mabry: Three years after Christopher was killed, I held a personal memorial for Chris and 18 of the men in 2/4 came to the memorial. Carlos Lopez, Staff Sgt. J.M. Riccardi, three other Sergeants, Jason Gatto and twelve other Marines came to the memorial. They were visiting families around the country to honor all seven men of the original group that Chris was with that had been killed. It was wonderful to meet all of them and I appreciated them coming to honor Chris and his sacrifice.

As I mentioned earlier, Chris played football. He played three years of junior basketball. The last year, his senior year, he ran track, and came in second place in a state meet. He lifted weights. Anything athletic and competitive, Chris was in it. That was one reason he joined the Marines. He thought they were the toughest, best branch of service, and he was determined to be a Marine. My late son-in-law George Woodall who got killed in a work accident had been in the 2/4 Marines "Magnificent Bastards" in the Vietnam War. He was a Corporal, a Marine Corps sniper, and he was on one of the last helicopters coming out of Vietnam and was wounded then. He didn't have any serious permanent injuries, but he had taught Chris about his service in the Marines. Chris decided that was the branch of service he wanted to join. George was killed in a work accident in September 2003 and six months later, Chris got killed. I raised Chris from the age of four. He was competitive. He wanted to do everything and he loved art. He drew pictures of vehicles and things and he wanted to come back when he finished his service and use his VA educational benefits to open his own automotive business. I don't know if he really intended to do prototypes as such, but he wanted to do the designs, painting, and things like that. So, as a young boy, he loved drawing, he loved animals, and even in school as he got older, he worked for a veterinarian in high school.

I worked as a nurse and Chris did maintenance at the nursing home where I worked. He was always active, always busy. As a young boy, his favorite thing was arm wrestling. He was so proud of those big biceps. He lifted weights. He would run for an hour or longer at a time, over and over. People here even got to where they wouldn't even stop and offer him a ride because they knew he was not going to get in the vehicle. He would run and would do that at school around the football field. He would run for as long as they didn't have class or anything. He just was that type of person. He was an honor student. He was in the "Who's Who" of high school students and honor roll students. He was in the National Honor Society. He just stayed so busy because there were so many things he wanted to do. He loved girls. He had girlfriends. I was told that five days before his death, he had asked his girlfriend, Jessica Gilbert, if she would marry him when he got back to the States and after his time was up in the Marine Corps and she had said yes. So, for at least five days, I know he was one happy guy.

Yeah, growing up, he was just competitive. He was quiet. You would think that he was just shy, but if there was something interesting to him, he liked to talk about it. But he was not one just to carry on all the time, even with his friends. When he was out just with a bunch of them playing ball or whatever, he'd laugh and joke. His favorite thing was to wrestle with them. He liked to watch movies and go skating. He was just always active, always wanting to be on the go and do things. Even as a little bitty kid, my first thought of him as a baby was crawling across the floor with his little diaper pulling off because he was trying to go as fast as he could across the floor on his little hands and knees. He was blonde-headed, pale-complected, and thin when he was really young. He was still thin when he grew up, but he was solid muscle. I don't think you could have pinched one bit of tissue up between your fingers. He was that solid.

I very seldom had to get on to him. He minded very well. He obeyed the rules at school. As far as I know, all of his teachers and principal talked like he was a model student. He worked in the office and helped out. And other than that, he, as I said, loved art and spent a lot of time in his room when he was at home with his earphones on listening to music and with his sketchbook and drawing pictures. Like most teens, he talked on the telephone with his friends the rest of the time. He had varied taste in music, liked some country songs like "Fast Car" by Toby Keith. He liked some rock music if it had a lot of rhythm like Led Zeppelin and Whitefish, but he didn't care for heavy metal and he didn't like rap.

He bought a 91 Chevy Corsica that he drove around while he worked and went to school. When he got out of the military, he intended to use his car to practice on, to do the designs and things like that he wanted to do. He had told one of my other grandsons that if he didn't make it back, he could have the car. He was going to trade it in after he got it done or sell it if he ever got back and got it fixed the way he wanted. I didn't really want to part with it, but I didn't want it to sit around and get ruined either. I had one and didn't need it. It just broke my heart to try to drive it. I never cranked it after he got killed. I knew that he had told his cousin that if he didn't come back, he could have a car because they hung around together all the time. His first cousin's name is Sam Hurt. I let Sam have the vehicle. Something had happened to it and it needed repairing. It didn't run well, so Sam finally traded it in to get him one that would run, so I don't know what happened to it.

I still have his basketball goal, basketball, and his football. The school where he went retired his football jersey number 5 and gave me his jersey. After his death, they dedicated the football field to him which is now Chris Mabry Field at his school. They also started an Achievement Award. At that time, they were trying to get a scholarship started, but that didn't pan out, but they still do the Achievement Award every year at the athletic banquet awarded to the best athlete of the year and the one that's improved the most in football. So, Chris still lives on in Clarkdale Attendance Center, at the football field anyway.

Chris had written a letter to one of his cousins. I don't have the letter, but he just said, "We've seen 10 men get killed in the last two days. Yesterday, I saw one who got his jaw blown off." Later, we saw Lieutenant Wroblewski's wife on TV when someone had interviewed her. They were just giving a little short talk, a short interview. She had said her husband got his lower jaw blown off.

Staff Sergeant J.M. Riccardi told me they got attacked. It was an ambush, and there were three groups. Chris and his team were in the middle group, fighting their way to and trying to get through to the first group. It was a long time, they said, before the last group could get to them to help them. All of the guys told me that they credited Chris with actually saving their lives, not physically with his gun, but when he was shot, they were running out of ammunition and they used all of his ammunition during the firefight. They said when the other group finally got to him, each one of them only had one magazine of bullets for each left over. So, if Chris had not been able to help them that way, had they not been able to use his ammunition, a lot more of them would have probably died. So, Riccardi told me that they credited Chris with saving their lives. Of course, he didn't run and drag anybody out or anything, so he's never been acknowledged by the military or anything. It's just enough for me to know that his friends and the other Marines respected him and honored him. Some of them apologized for not being able to save him and told me how sorry they were and hoped I didn't blame them for not being able to do anything. I told them, "No, I don't blame any of you in any way because you were all in the same harm, in the same danger." It could have been any of them. I was just thankful that they all made it back okay.

As I said, I was working as a nurse and our group of nurses and people that come in and out of Houston adopted Christopher's entire unit after his death. We sent care packages to the group. We just sent a huge box about every four to six weeks. We had pharmacists, paper companies, several different businesses, plus everyone coming in donating socks, foot powder, shampoo, you name it and they were giving it. It cost us $270 postage for the first box we sent. But, we sent everything from Gatorade to Bayer Aspirin. We did that until they all made it back home that next September.

I did know that Dianne Layfield and her husband got to lay the wreath at the Tomb of the Unknown Soldier. My sister and I went three years in a row for the Day of Remembrance, and if we had known ahead of time what time we would be there, we could have contacted Senator Trent Lott, and I could have placed the wreath myself. But, we didn't know that at the time, so we never did it. We did go to the tomb, but I never got to place any flowers on it because I didn't know that I could have done that. But, I did see Dianne, and she and her husband did. I signed Travis' guestbook, and kept up with different things that Dianne is doing. Dianne is a sweetheart, and every Christmas, she sends me a beautiful card and says, "Our boys are sitting around the table together. I'm still wondering what they are doing." I said, "Well, whatever it is, they're probably in mischief, but they're like they were when they were home."

It's hard to believe that it's actually been almost 10 years. It seems like maybe a few months. I can still turn around in this house and almost feel Chris behind me. His favorite thing was eating. You couldn't fill that boy up. No hound dog ever had anything on Chris. Chris could eat more than any man you'll ever meet and would still be ready to eat more. But he didn't gain weight. He ran it off. He worked it off. But, he was always either opening the refrigerator door, getting in a car, or on the telephone, or just something all the time. Some days I feel like if I could just turn around quick enough, I'd see him there behind me. I know he still comes around to check on me.

Matter of fact, I had a poem published in the paper, and I don't even remember where I read it somewhere a long time ago, so I guess I actually was guilty of plagiarism. But, it basically said our loved ones are always with us. We were given the option of burying him in Arlington Cemetery or the veteran's cemetery just a few miles down the road from me at Newton. Christopher's picture is on the granite wall at the memorial at the cemetery, but we had him buried in the church cemetery where he had always loved to attend with his cousins and where he'd grown up. I told them that I wanted him close to where we can go anytime and anyone else that knew Chris and wants to can go to his grave site. I go many times and I will find Marine Corps mementos hanging on his flowers. I've got his picture in his headstone, so no one ever puts anything really on it. But, in the vase at the base of the headstone, they leave items. I went one day and a Marine Corps utility cap was hanging on a stick in front of the American flag on his grave. I don't know who did that. I have no idea. People still go, still think of him, still honor him in their own way. So, he's certainly not forgotten around here.

His school teachers, when they mention him, still cry. They still talk about things he did at school and how he honored and respected them. They had a memorial ceremony for him after his death. His

National Honor Society teacher was in charge of it. She's the one who did the dedication and gave me a beautiful little book. They played music for him and different students made little testimonials. He was the type of person that if you ever met him, he touched you. He was just someone you would never forget. If you saw him one way this time, that's how he was the next time. As I said, he was quiet. He never got into any trouble at all. But he wasn't wishy-washy. He'd fight if the boys picked on him or whatever. But he just was not a troublemaker. He wasn't wild and mean. He was tender-hearted. He couldn't even be mean to a little animal. He had a little hamster that got loose and got in my closet and made a bed there. Sucker scared me, ran out on my foot, and I tried to stop him, but Chris got him, cupped him, and held him like a little bird. That little devil had been missing for more than five days. I thought he was gone. But that's how he was. He loved people. He loved animals. He wanted to be a Marine and to him, that was just the highest thing that he desired. He just thought if he attained that, he had reached the max.

The day he graduated at Parris Island, there were about 600 Marines, including the female Marines. And they were all in different groups, of course. That was such an awesome sight, looking out over that parade ground with about 600 people in uniforms standing there. I told a reporter from a Mississippi paper who called me one day, "Every one of the people wearing a uniform, every person in the world, or the United States, deserves to know something about that person before they put that uniform on. Because they had a life besides the military. People need to know that. They need to know what the troops gave up and what our people can do for them, to honor and remember them. So, that is another reason that I am so thankful for your book. It means the world to me. I know even with Christopher's death and the memorials at Camp Pendleton, they didn't provide a place to stay and transportation for the attendees. I know what a huge sacrifice it is for people to have to give up their time, their own families, and that's what I told the guys when they came for memorials because they were doing that for all seven of the guys in Chris's group, and they were going to do it each year.

Gabe Henderson was the one in charge, and he said when they had held the memorials for all seven that were going to start over. They had me draw the name for the next memorial for whoever would be next. I drew Derek Hallal in Indiana. That's just how close-knit that group was. They were just all together, all good. One reporter from the Chicago Tribune called me one night, months and months after Chris's death. He said he just wanted me to know personally that he had just returned from Iraq, and he had met with many of the groups, you know, not just Christopher's, but with many of the groups. But, he had talked with the group and with the ones that survived, and said he just wanted me to know how the guys felt about Chris. He said, "I just want you to know that everyone that knew him in Iraq is saying the same thing that the people in Marina, Mississippi are saying. I've talked to over 200 people." Now, I don't know that they all knew Chris, but he said he spoke to over 200. He said not one person had a bad word to say about Christopher, said they all agreed that he was easygoing, laid back, did his job, tried to be professional at all times and that they never heard him complain, never heard him call anybody a bad name, and never heard him lose his temper. He was just calm and they said Chris had a calming influence on a lot of the other guys.

Originally, I'm from Florida. I lived there as a child and my father was killed in a work accident. He was working when they were building the new runways at Pensacola Air Base. He got killed in a construction accident. I was only three years old. My brother was born the day daddy was buried. Dad was from Mississippi, from here at Meridian. So, they sent his body back here to his mother and father. As soon as mother was able to come back, she came back to Mississippi and let me stay with my grandparents until she was able to get settled and get a place. I've been here ever since. My husband, he was a Mississippi hillbilly, and he wouldn't leave. He didn't even like to go to the beach. He went beach sea fishing one time, and I think he was terrified. But, every time I had an opportunity, I headed back to the beach, either the coast or Florida. I like Mississippi, and I've always loved to hunt and fish. I've driven too fast many times, so I'm guilty of it too, but I'm no redneck country girl. When I was growing up, we sold milk to a dairy. We raised pigs, cattle, and chickens. My brother and I couldn't count the times we've had to either hook up a horse to a plow or a seed hopper or a fertilizer distributor. I never learned, really, to drive the tractor. My brother did. It's one of those old-timey big tall deals. I almost ran the chicken house down with it, and I never learned to drive it well. We grew up working, and Chris was like me.

Chris was not old enough to go to kindergarten when I first got custody of him. His brother had just started. He's almost four years older than Chris. Christopher and I would sit here during the day. I have two sets of encyclopedias and I have one of the world globes. We would look through the encyclopedia, talk about different things and whatever he wanted to do or talk about, that's what we did. We spent a lot of our time reading, writing, and drawing. I don't like to boast about myself, but I was an honors student. My mother, when she died in 2001, still had some of my report cards in her dresser drawer because they were all A's. When I finished nursing training, I made the dean's list with honors at the college. I scored the second highest on the state board that year out of the nurses group. So, Chris was like me. He just loved to study. He was interested in everything, and my biggest memory of him was the globe. We would spin the globe, and I would let him put his finger on a place that he'd want to study, and then we would look it up in the encyclopedia. One day I said, "There's really an Easter Island" because it was Easter time. We checked around on the globe for the Solomon Islands and then found Easter Island. Then, we looked it up in the encyclopedia and I read it to him. He said, "Grandma, someday when we get a bunch of money, can we take a vacation and go to Easter Island?" I told him, "Yes, someday, baby, if we get enough money, we'll go to Easter Island." Well, we never made it there unless he's visited it since then, but that's the kind of person he was. He wanted to know as much as he could about everything. He was just, you know, interested in everything. He was competitive even against himself. If he scored like a 92 or something on a paper, he immediately set out trying to make better and he wouldn't stop until he did. One reporter from the Associated Press looked at all the material here, and he said, "It's hard to imagine a 19-year-old kid accomplishing so much and being that astute and just wanting to learn and wanting to do as much as he did at his age. It's hard to believe that one person accomplished so much in 19 years, even with the proof right here in front of me."

Chris liked drama and plays at school. He played Commodus in one of the Greek dramas with the Romans. He'd put on a sheet and wrap it around him in a pair of sandals and he was just as much a Roman statesman as any you've ever met. Then, the next time you saw him, he'd have on blue jeans or a swimsuit to hit the creek down there. He just wasn't afraid to try new things, but he wasn't really a daredevil. He didn't go out and try to get into anything. I credit my husband most of the time because my husband loved to fish. He loved to hunt. Matter of fact, he won the world state record on the largest catfish caught in Mississippi. Then he won the world record in the Freshwater fishing Hall of Fame because he caught it on lightweight tackle. If Pop said it was okay and he went fishing or hunting, Chris said it was okay. He wanted to go, too. My husband was disabled for 14 years before he died in 1995. So, he was at home while I worked 12-hour shifts at the hospital at night. He could be with the children when I was gone, and then I was here with him in the daytime when they were gone. So, I credit him with being responsible for the boys the majority of the time because I couldn't have done it myself or by myself. Someone bragged about Chris and complimented him and what a good job I had done with him and I said, "Well, I really can't take out the credit for it. God gave me some good material to work with." Chris was a Christian and had been saved two years before his death. He'd been brought up in the church all of his life, so you know he was just a good guy all around. From what everyone that knew him as a Marine said, he was the same way in service. We'll get to see him again one day. I can think about him without getting too upset now. Don't ask me how I feel about terrorists Muslims because that's a whole different ball game. I know I'll be with him again sometime and he'll be my Chris, so I'm satisfied with that.

Well, I hope that you can contact Carlos Lopez. As I said, I told my daughter to send you pictures on her computer email. I know that she sent him a copy of your paper that Dianne had sent me, showing what you were doing and your contact number and all. They'd been living in Oceanside, California before he left. I don't know that he's still got an address out there or what, but he's been in Okinawa. He was sent to Okinawa for seven months, and I think he still has probably two, maybe three more months there to go. But any of the guys in the group, if they know it, would love to talk to you. Now, I know Jason Gatto got out of the service not too long after I saw him. I met him in 2010, I believe it was. He had already got out of service and was going to Quantico and he studied criminal law. He passed and he's working in the Justice Department now. And one of the other guys, Eli Abbott, is a troubleshooter for different groups and the last contact I had with him, he was in China. He speaks about five languages. I met him when I was in California and then he and Juan Munoz, who were in the group, got hurt with Timothy Prather. Timothy was one of Christopher's buddies and friends. Juan and Timothy were near a suicide bomber one day. Juan saw her before she actually pulled a string or whatever it was to blow herself up. She hollered at Timothy to get out of the way and tried to shove him out of the way. Timothy lost some of his fingers, injured his leg, his hand, and his face. Juan Munoz lost some of his hearing.

So, as I said, all of the guys, every one of them that I've met, they've been top notch. I told every one of them, "I love every one of you. I respect you. I'm glad you're all here. Even though you're not mine, I call you my Marines." And I do. When I sent an article to the paper here after their memorial that is how I described them. I said, "I'm proud to call these guys my Marines. I've got their cell phone numbers. I have their emails." Timothy puts pictures on the computer on Facebook. I have pictures of their children, and they try to let me know what they're doing and how they are. At least two of them have told me anything I needed, at any time, anything I wanted to know, just call them and ask them. Ricardi told me he didn't care what it was, if I needed anything, day or night, call him no matter what. That's just how all the guys are. They look out for every one of them that got killed. I very much appreciate that.

Well, they actually brought me a flag after Chris was killed. His group got me a red flag with a picture of the silhouette, you know, the memorial on it with the helmet and the gun and all. And they drew the street and put 20th Street, Ar Ramadi. They put Christopher's name, wrote it down across that street like a line. Then every one of the group signed that flag and they brought it to me when they came back from Iraq. I have it in a special shadow box flag case. Of course, I have his flag from the casket at the funeral. They gave his father one, me one, and they gave his mother one. They didn't even know he had a mother until the night after they came to tell us he was dead because Chris just left all her information blank on the paper. He had not been around her, and she had disappointed him so many times, but he didn't hate her. She just didn't have anything much to do with him, and he didn't like her husband, and so he just didn't go around her, but he left her address, phone number, everything just blank, and they didn't even know she was still around until they came here that night to tell us. But, she's changed. She regrets what happened many, many times over. I told them, I don't have any animosity towards her because I had the pleasure of raising Chris and seeing him grow up and become the man that he was. I've attended school more than half of the school kids have because I was there for conferences, for their programs, and for their taking part in whatever they were doing and all their football games. Grandma was in the bleachers screaming her lungs out and jumping up and down, hollering, run, run, run, or catch him, bring him down, you know. I enjoyed every day that I had him and I miss him every day that he's gone, but as you said, I know I'll be with him again. So, you know, it helps and it makes up for a lot.

It's still hard to believe it's 10 years. That's what Dianne and I and our Christmas cards this year said. We just found it almost impossible to believe that we were fixing to go through a 10-year anniversary from that first year we met in Washington. But with my health, I'm 74 years old and I've developed a multitude of health problems. Now I've had a heart attack and a heart blockage. It causes me some problems sometimes, so I don't know. The last time I wanted to go fly to California, the doctor told me I would have to use oxygen all the way out. It will depend upon several different things whether I'll be able to attend it. I'm hoping that if I can't go, that I can get some of them, Dianne, somebody, or maybe even one of the Marines themselves, to make me a copy or a CD or something of it and send it to me. I didn't go when they first came back. I regretted it, but I told Riccardi, I just didn't feel like I could face that group coming in, getting off the plane, getting off the bus without Chris knowing his last trip home, you know, in this plane. So, I just didn't feel like I was ready to handle that. So, they sent me a CD of that, and I'm hoping that if by some chance I'm not able to go, that maybe some of them will send me one this time. But, I hope that it'll be where I can go. I have a nephew who lives between Victorville and Barstow. I drove across the country a few years ago and stayed two weeks with him. So, you know, it might be that maybe they could take me down there or something, if I can get out there. But, that's still a few months down the road. Right now, being a 74-year-old widow and my daughter has had some health problems. Now, I'm living on Social Security and it's not easy. But my home, 12 acres of property and my vehicle, they're all paid for. I'm looking forward to when your book comes out. I did get to meet President Bush.

Carlos sent me one of their group actually in battle, in the battle for Ramadi. I forgot who, but somebody sent me another one. And one guy that worked with my nephew went into the music business, Frank Ortega. his family or their names even in the museum in Arizona for something. I don't remember what, but he had a communications business in Arizona. Then, he branched out and went to the southern states. My nephew was vice president and he was in Atlanta, Georgia. Their offices were on the ninth floor of the bank. He decided to go try to be a musician, you know, and get into that branch of business. He wrote a song and dedicated it to Chris entitled, "A Soldier's Prayer." He sent me two CDs, copies of the song. It's just the one song, it's a single. And I gave one to a D.K. here in Meridian that was a Marine that is always at the memorials. At that time, he was still at the radio station. But, he couldn't play it on the radio station because he didn't have a copyright release. So, I told him to keep it for his own enjoyment, you know. But that's how much people have done since Chris was killed. So many different ones that I know if I tried to remember and go back over everything, I would forget somebody and leave somebody out. As I said, Christopher was a proud Marine.

So, if you can get in touch with Carlos, I would love for him to talk with you, because when he talked to me, he was really upset. He had lost a close friend, and he didn't know Chris was shot four times, and well, none of the guys did. The medic thought he was going to live when they took him out and told him that they believed he was going to make it. But he didn't know how badly he was injured. As I said, he was shot four times. The bullet that killed him went through between the fifth and sixth rib on the right side, went through the diaphragm, liver, lung, and the apex of the heart. and exited between the 7th and 8th rib on the other side, on the left side. He still survived six hours, but of course he had blood and fluid in both lungs. They put in chest tubes and had to do a laparotomy and close off the large intestine because of the gut wound. He was also shot through the left thigh. His left arm from the shoulder to the elbow was just literally blown apart. And the guys didn't know that because he had his top on and his jacket. So, all they knew was that he was shot. Carlos tried to get the jacket off to help stop the bleeding. Chris was awake and speaking to him. I know Ricardi said that he spoke to him. The only thing Chris said was "It's hard to breathe." So, he didn't complain, he didn't whine, he didn't get in a bad mood about getting shot. I guess he figured that was a chance they were taking.

But I will get whatever information I can and pictures I can to you and good luck with getting in touch with some of them. I've talked to so many reporters and all since his death. My brother even asked me one time, joking, he said, "I don't know what office you're running for. Every time I turn on my TV or pick up a paper, you're in it." But I joined the latest Marine Corps League Auxiliary Group, and I worked with VFW. My group of nurses did the care packages and all. It was the least we could do to bring attention to the military and their needs, and especially now that people don't understand PTSD. I guess they figure the war's over, the younger ones don't even know or realize what all was going on. I guess they didn't appreciate the depth of what those guys, and not just them, but everyone that's ever been hurt or been killed in the military, the sacrifice they all made, so all of these whiners and complainers can get out there and do it. Because most of the families, a lot of them have resentment, but most of them are like us. They just have sadness, really. It's just more sadness that things can't be worked out without fighting. I'm a Christian and I study my Bible daily, and I know for a fact that He says it's going to be that way till He comes back. So, whatever it is, we just have to be prepared and roll with the punches.

Gold Star Family Interviews

Gold Star Parents John and Shawn Wroblewski
(2ndLt John "J.T." Wroblewski – "Lt. Ski")

Janney: It's February the 2nd, 2021. Tell me about JT. What kind of kid he was, from the time he was born until he was a man. Just tell me as much as you want to share about him.

John: John Thomas was born on April 16th, 1978. Everything went well with the pregnancy and the delivery was fine. We brought him home at that time. We lived in an apartment in Garfield, brought him home there and then started looking for a house. In order to afford something, we had to drive out more west - from Garfield to West Jersey. There the houses were more affordable. We were able to get a ranch - a small three-bedroom, one bath and purchased it. When we moved in, he was two months old. JT was the firstborn. So, we moved in and didn't have a lot of money. We had friends help us move in and did it in one really long day. At the end of the move, everybody had pizza and a couple of beers and people went on their way. We went to lay down, and the funny thing, and we'll always remember this, the previous owners had cats, and we never realized it until then. I said to Shawn, "What the heck is that smell?" So, we both got up and realized that it's cat urine. There's no way either one of us could sleep. I guess we started probably about 11 o'clock at night, ripping up all the carpet. It wasn't a big house, but here we are, ripping up carpeting. Then I take out the bleach and bleach under the radiators trying to kill that urine smell. We finished up at around four o'clock in the morning and then laid down. Now everything smelled like bleach, but it was a lot better than the urine smell. We were concerned about having a new baby though. JT slept right through it.

Five o'clock he was up and wanted to be fed and Shawn fed him. We just tried to rest as much as we could and put him down again. Being a newborn, he was up about almost every three or four hours for his feedings. I'll tell you, we lucked out. He really wasn't a crier and he never really was colicky. We're new parents and she would wake him up for a feeding, fearing that he was supposed to be fed. And then, when she spoke to the pediatrician, he said, "No. If they want to sleep, let them sleep." It wasn't long where he was finally sleeping through the night. Our second son, Mike, was only 19 months apart. It was short lived with JT. You know, being alone with him for 19 months. And then along comes Mike, and like I said, we had three bedrooms. So, we put JT in one room and Mike was in the other and Shawn and I in our bedroom. It wasn't too long after that, Dave came along and then finally our fourth son, Rich. In the one room, I built bunk beds and we put JT and Mike in one room and Dave and Rich shared the small room with the bunk beds. That's basically how we survived.

All our kids went to preschool and then to kindergarten and we found out that John was a gifted kid, academically and was always ahead of his class in reading and everything. Then he moved on from elementary school to middle school and then into high school. All during that time, he also was involved in athletics, played baseball at a younger level and also played football, and basketball too in the lower levels - was always good. Wasn't like the best kid on the team, but he was one of those kids that would be the starter. He just loved athletics, especially baseball and football. In middle school, we really realized what a good student he was. His academic prowess is sort of unique, along with being a really good athlete. Wound up playing baseball and football in high school. He got accolades as all-league honors in baseball. He was a second baseman. In football, he was a linebacker. The other thing about JT that everybody always remembered him for was his great sense of humor. He had a great sense of humor. He would spark up a party with his sense of humor. He always had nicknames for all his friends. A friend of his was David Mangold, and so he nicknamed him Lady Silver. He played around like that. It's amazing that those nicknames stuck with those kids for the rest of their lives. In fact, they're grown men now and some of them are still referred to by the nickname that JT gave them.

He wasn't boisterous. He was kind of on a quiet side. When he was having a good time, he was extremely funny and he could really break your horns. He was a jokester, but in a good, positive way. People enjoyed that. In fact, I remember one time when he was in middle school and they were having some type of program and kids were performing on the stage. And then, you know, after the performance, people clap. His clap was one of a kind. It just annoyed the heck out of you, and he knew it. After the students finished their performance, he was in the audience and he would clap. And then somebody else, then he would clap. Then finally, the teacher got up on stage on the mic and said, "Whoever is clapping like that, could they please stop?" His friends were laughing because everybody knew it was JT. That's just one story, but there's a million of them.

Also, as a kid, when he was younger, he would organize games in the backyard and they would play manhunt and things like that. They would play army and he would be back there taking charge of everybody telling them to do this, do that. We would hear it because we'd be inside and not think anything of it, not make any connections. We were the kind of family that everybody's running around trying to make a living, making sure you're paying the mortgage and the other bills. You just get busy with life. But, we would make an effort to sit down whenever we could as a family for dinner. It didn't happen every night, but it happened most nights. I worked late many nights, too. Sometimes, three, four nights in a row, I wouldn't be home until late. I'd get home and the kids would be in bed already. But on the weekends, if I was around and everybody was here, we really made an honest effort to sit down and have dinner together. We always relate to the kids the importance of some of our main holidays like Memorial Day. At dinner, we would make reference to Memorial Day, what the significance of it was, and give them a little history and thank our heroes. The other thing that we always tried to do is be at our kids' events, whether it be a baseball game or a back to school night at school, performing at a play or something in school. I would honestly say they were never there by themselves. Usually Shawn and I were both there. If one of us could not make it, obviously the other one was there, but they would never do something or play a game or be involved in a school activity where we wouldn't show up or support them because we felt that was important to show them that what they're doing is important to us also.

John went off to college and a couple of colleges recruited him for football. He decided to go to a college in Pennsylvania and went there to play football. He didn't like the school in all honesty and ended up coming home. We explained that it's important to get an education and couldn't just hang out at the house. So, for that summer, he actually worked for my son, Dave who was starting up a landscaping business, but Dave was also working for a landscaper. His business hadn't evolved enough where he could work it full time, but he did have enough customers that JT could go out and service customers, basically working for Dave. Then on weekends, they would work together. So that's what he did more or less for spending money. Then, he decided to go to a community college here in Morris County, Morris Community College and he liked it a lot. His classes were during the day and when he would get home, he was able to work with Dave to help him out with some spending money. He went there for two years and graduated from there. Morris Community College had a reciprocal agreement with Rutgers, so from there he went off to Rutgers, stayed there and graduated from Rutgers with the highest honors. Also, when he was at Rutgers, he helped all the time with the Special Olympics, running it as a support staff and enjoyed it immensely.

I know the one story that kind of touched our hearts was when he was staying in a house in the room upstairs. The people that owned the house had a son that was very disabled. He was bedridden and couldn't get out of bed, could maybe move his fingers to operate a thing here, a thing there. But the kid was very limited in what he was able to do. He couldn't talk. I don't know how he got in that condition. I don't know if it was a car accident or whatever. The parents told us after finding out that J.T. was killed, that JT would often visit with the son Michael when he got home from school. J.T. would go downstairs and talk to Michael. And the parents told us that Michael would react with John and John would spend time with him there, read to him, you know, do things like that. We had no idea that this was going on until after J.T. was killed when the parents shared that story with us. It's not like John calls us up and said, "Hey, you know what I did?" He wasn't that kind of person. He was not the braggadocious type of guy. He just did it because it's something that meant a lot to him.

So anyway, 9/11 happened when he was at Rutgers. He goes to the recruiting office of the United States Marines, and talks to a recruiter. Unbeknownst to us, he decides to go to OCS (Officer Candidate School) in Quantico. All my boys to this day have a tremendous relationship with their mom. I mean, they confide. They tell her things that I never knew. I mean, to this day, there are things that I'm just finding out - about a car accident or something that took place. I say, "What? When did that happen?" Well, we never told you because we knew you were going to get mad. That's how it was. Yeah. Again, all my sons, JT included, had that special relationship with their mom. So anyway, he confides in her and tells her what his plans are with the United States Marine Corps. Shawn's kind of waiting to pick her time when to tell me. So, there's one day when I'm home and doing something around the house. I come through the garage and I'm walking through my kitchen. John is sitting in the kitchen with his girlfriend. I hear his girlfriend say something about, "Oh, you didn't tell your Dad? Well, you won't be allowed to do that." And I stop. I said, "What do you mean? What are you talking about?" So, she said to John, "You didn't tell your Dad?" I said, "Tell your dad what?" John said, "I joined the Marines." I said, "Oh, okay." And I just continue out the kitchen and go out the back door because I think I was working on a deck or something. I go out there and finish what I was doing. I didn't see John until the next day. So, I saw him in the workout room and I walked in there and said, "Hey, John, I know you said you were joining the Marines. Just let me tell you this. You're very intelligent. You got that going for you. You have a great sense of humor. You're a good looking kid. You've got a lot to offer companies in whatever you want to do. You have so much to offer." Then he looked at me. I remember this like yesterday. He said, "Dad, what could be better and what could be more honorable than to serve the country?" I said this a number of times that sons always look up to their dad; their dad is kind of on a different level. That day, I can honestly say our roles kind of reversed. When he said that, I was just struck by it. And, you know, our roles are reversed. Now, I'm looking up to him. He's becoming my role model.

John built an obstacle course in our backyard in the yard and in the woods. He's got a PT course like the Marine Corps PT course. He had a pull-up bar. He had other things like that. What he would do is he would run through the woods there with a backpack full of rocks. He would do his thing and then do pull-ups, push-ups. He made me build him a dip bar and then he would lift weights in the back room. He was there with a real good friend of his, Ryan, and I would always know when they were in the back room, because I would pull down the road and they would have the music blasting. Bruce Springsteen's song, "Born to Run" would be blasting through the walls and I'd hear it down the block. So, they worked out to various other songs. John trained hard to get ready.

Then he went off to boot camp around October 2002. We went up there one time to see him. After boot camp, he did extremely well in OCS. I remember it was Thanksgiving and he was able to call and we're talking. When he went in, he wanted to go infantry, but his recruiter said that was pretty full. He was probably more suited academically to go to be a pilot. So, he took a test to do that and then he wanted to be a Marine pilot. Had no problems with the test, did very well. And so when he went to OCS, he was doing his thing. He gets called into the Commandant's office over there in Quantico. The General was talking to him and said, "I'm hearing good things about you. I know you want to be a Marine pilot. I know your MOS is a Marine pilot, but I feel you would be better suited leading men. I'm not putting down pilots, but we need leaders. We need guys that can lead men. And from all the reports I'm hearing, you would be a great leader." John said, "Yes, sir. That's really what I wanted to do. But I'm told that once you're here, you can't change your M.O.S. It's impossible." So The General said, "Listen, if you want to go that route, I don't do this often, but I'll make sure we change your M.O.S." So, John agreed to that and changed to infantry.

John's telling me the story on Thanksgiving. He would become a combat commander. He graduated OCS and TBS with the highest honors. He was meritoriously augmented because he did so well in TBS. This meant that he could stay in the Marines as long as he wanted. Meritoriously augmented basically meant that you didn't have to re-up and you wouldn't be re-evaluated like a lot of these other Marines would be re-evaluated and if you didn't hold up to the standards, then they could get rid of you. But, when you're meritoriously augmented, that eliminates that process. It's a very prestigious honor. When he graduated from TBS, he graduated at the top five in his class, and again, this is stuff we didn't hear anything about until we went to his graduation. We're reading all of these things about him, and then you kind of say to yourself, that's our son. He also became a marksman in the sidearm and the rifle, and also in martial arts, becoming a martial arts instructor. Even though he didn't have any martial arts background, he caught on to things extremely fast and he had a great mind for envisioning things, putting things together and things like that. Even as a kid, when he would work with me he was able to visualize what it was supposed to look like, how it's supposed to go together. In spite of his successes, JT was always a modest person and didn't talk much about himself. His philosophy was always that he was no better than anyone else.

So, we go to graduation, and we hear all of these accolades. Obviously, you're a proud dad and mom. We're on base with him and he's showing us different places. We get in the car and go off base for dinner. I'll never forget this. We come back onto the base and have to be cleared by the gate sentry. The sentry sees John who's now a 2ndLt and salutes him. J.T. returns his salute. I imagine that JT felt good about that. For Shawn and I, it was just a feeling of pride in what our son had accomplished, especially hearing, you know, a little while ago at the graduation, all those accolades.

After graduation, he came back home for a little while. Then, he gets stationed at Camp Pendleton, California and he's going to assume command of forty men in 2/4's Echo Company. We're keeping in touch with him and he tells us they are deploying to Iraq. I remember talking to him on the phone and I said, "Where?" He said, "We're just going to supply water, help them build schools and stuff like that. Don't worry. There's nothing really going on. It's a pretty safe place." I asked, "Where are you going?" He told me, "Anbar Province near Ramadi." I never heard the word Ramadi before in my life. We talk a little bit more on the phone, he talks to Shawn, and then we end the phone call. I immediately started to research Ramadi. I found out it's not as safe as he's saying and is a very dangerous place. In fact, what I'm reading is that it is one of the most dangerous places, at that time, on the face of the earth, as far as Americans are concerned. At this point, we're very concerned. We find out when he's getting deployed and we fly out to California and spent a week with him before he got deployed.

We also found out that when all the men get a holiday or pre-deployment leave, some of the guys have nowhere to go and stay in their barracks on post. They either don't have a home to go to because they've left home or have a difficult home life, or they can't afford to travel back home., When we were there with him, he said, "All right, I just got to go on base for a little while. I'll be back." But what he was doing, he was visiting those guys, seeing how they were doing, if they needed anything, just spending time with them. He was just being a leader. He didn't have to do that because he was on leave. It was his free time. Later on, the guys knew that and told us how much they appreciated John caring about them like that. They said, "Not all the lieutenants were doing that. Lt. Ski was one of the only ones that spent that time with us." That's why they referred to him as "my lieutenant." We found out from Colonel Kennedy that they referred to John as "my lieutenant." Colonel Kennedy told us that's an endearing term when a Marine refers to their commanding Lt. as "my lieutenant." So, there was a relationship there between his men and him. That was leadership.

Janney: I tell you, John, everybody that I've spoken to that knew JT, they just loved him. I mean, they just sing his praises and talk about what an amazing man that he was and what an incredible officer. I just never talked to anybody that has had anything but just wonderful things to say about him. Marines are pretty outspoken, as you know, and if somebody is not doing the right thing, they're going to say, "Well, he was okay." More than likely, if they don't like someone, they'll just come out and tell you that, "No, I didn't really care for him." But Lt. Ski (as everyone called him) was not one of those people. You have a right to be proud of him. He was an amazing young man.

John Wroblewski: You had mentioned that to us before and that makes us feel very proud. That makes us feel proud. I think they loved him. I know he loved his men. He said, "Listen, I treat all my men like they're my brothers. I meet them there like they're Mike, Dave and Rich. I love them like my brothers." We weren't aware of this until later we spent some time with his guys after J.T. was killed. We eventually had that opportunity. He went off to Iraq and we sent some letters and packages. I remember one of his letters, he was asking us for cigarettes. We sent him cigarettes and in my letter, I asked, "What the heck do you need cigarettes for? You're dodging bullets. You don't need to start smoking." So, he writes me back and tells us that he uses the cigarettes because he can get information from Iraqis, "Hey, here's a pack of cigarettes. You know, what do you got going on?"

Janney: They loved him. Sure. American cigarettes are just so far superior to those Miami's and some of the other terrible Iraqi cigarettes that I've smoked when we were there in 2008. Don't you remember the one time I tried to trade one Marlboro for one Miami cigarette? I just wanted to try a Miami cigarette, and so I took a Marlboro cigarette, and I handed one Marlboro to the Iraqi guy and he handed me a whole pack of Miami cigarettes. I said, "No! One for one trade." He's looking at me like I'm crazy, and gave me 5 Miami for one Marlboro. So, yeah, that makes sense that he could use American cigarettes to trade for information or favors.

John Wroblewski: I was glad he didn't start smoking. Shawn worked for American Home Products, but it's now Pfizer. When she worked for this Company, they found out that John was in Iraq. They said to her,

"Listen, what does he need?" John said, well, they could use Advil and other products they made. So, every week American Home Products and Pfizer were sending a box of their products – Advil, gloves, and all these other products. The Marines and Echo Company really loved the care packages. It was really above and beyond a call of duty for her company to do something like that. The president of the Company sought her out. When he found out about John, he sought her out and said, "This is what we're going to do."

The day that J.T. was killed, Shawn and I were at work. So, you know, obviously he goes off and the day that he was killed, Sean was at work. I was at work. I remembered I had to go check a field to see if it was playable; it was early April and we had started the baseball season and had a game the next day. So, I get in my car and I drive to the field and check it out. It was okay. I get back in my car and have the radio on. The news is talking about Marines and this tremendous battle that's going on in Ramadi. And I'm saying to myself, "Holy shit, that's where John is. I wonder if he's involved." I get back to the school and my computer skills were very limited, if any. But, I had a computer in my office, and a good friend of mine, John Regan came in. I said, "John, do me a favor. I heard on the news there's something going on in Ramadi. Can you look it up on a computer? Maybe there's a little bit more information." So, he looks and now he's reading it. Now I'm reading it, and it's talking about the battle, and it's talking about Marines being killed. They don't have any names. This was a fierce battle. I went to my principal and I said, "Listen, there's something going on in Ramadi. I don't know what it is, but I gotta go home." I get in my car driving home and I obviously have the radio on and I'm listening to the news every chance I get. And I'm just playing, constantly playing with the radio and I'm saying "Our Father" prayer. Shawn and I had spoken on the phone and she was also aware of what was going on, that something was going on in Ramadi. She got home before I did and I pulled in and then we turned on the TV and we saw the Chiron crawler thing, they're talking about Ramadi and they're saying 10 Marines and one Navy Corpsman killed in a fierce battle. No, they said at that time 10 Marines were killed and they were saying one Marine was taken by helicopter. And when he left, he gave the thumbs up. Shawn and I looked at each other and we kind of knew it was JT because whenever he would dance, he would put his thumbs up. I remember him dancing at his wedding putting his thumbs up that he was having a good time and acknowledging the people around him. We said, "Listen, that's gotta be him. There's no question in our mind. We kept listening to the news - very limited. We weren't getting any more information.

We called our Congressman and we didn't make any contact with him. We spoke to one of his aides, didn't get any information, and didn't hear anything. Got a call from a reporter, Rob Jennings. He had heard about the battle and he wanted to know if we knew anything. We said, "No. We don't." The day goes on and gets into nighttime and we get a call from Joanna's father. He tells us that John was wounded and that's all he knew. He thought he was going to be airlifted to Germany. We got off the phone and started to make phone calls, letting people know. People started to call us. Rob Jennings calls us again because I guess it goes out on the wire that John was wounded and he calls us and is talking to us to give an interview. He says, "I come by tomorrow and we'll have an interview and you can give me a picture of John." I said, "Yeah, no problem."

So, the next morning and now all my other sons are here. We're outside on the deck and we're trying to figure out how we're going to get to Germany, get passports and so forth. My kid is going to the post office to see how passports work. Everybody's doing their own little research. The president of Shawn's company hears that John was wounded and also gets word that we're looking to line up a flight that once we find out so we could head to Germany. He calls Shawn and says, "Don't worry about a thing. We will have our private jet take you. You just call me. And we'll have a limo pick you guys up. We'll bring you to Germany. We'll put you up. We'll make all the reservations. Don't worry about anything." So, that was a big thing to offer, a big relief for us. So, the day goes on and my sister comes over and a couple other people come over to hear about John being wounded.

It was 8:05 p.m. and we got a knock on the door. It was a knock unlike any other knock. I mean, to this day, it's a haunting knock. For some reason, you could tell it was different. I opened the door and we saw a Marine Major, an Army chaplain and a Marine Gunnery Sergeant. I just start shaking my head and saying, "No, no, no." It's just like you see it in the movies. You know what they're coming for, but you don't want to hear it. You just don't want to hear the news. I'm saying, "No, no, not my JT, please. Not my JT." They ask if they come in. Obviously, we let them in. Major Paulus proceeded to tell us, and gave us the news that it is with regret that we inform you that your son was killed in action protecting the country." People start crying in our house and I look at him and I say, "How was he killed? What happened?" Paulus looked at me and he said, "He was shot." I said, "Where was he shot?" Now I look at him and there's a tear coming down his cheek. Again, it's like you're not really there and you're kind of watching it. I watched his tear coming down his cheek in slow motion and it got to the end and I could see that the tear fell off his face. He looks at me square in my eyes and he says, "Sir, he was shot in the face." Now I hear Shawn say, "No, not his beautiful face, not his beautiful face. He's so handsome."

By this time, news got out that John was killed. Like I said, they knocked on the door at 805. By 915, our house was packed with people. People from the town. Strangers. Kids that I had coached in football. John's buddies - all of his buddies were here. Just an outpouring of support that you just never expect. The Marines, and the Army chaplain wound up staying till past midnight. Maj Paulus said, "We will stay here as long as you need us. If you need us to stay over, we will stay over. We will sleep on the floor and do whatever you need. We will do everything in our power to make sure you get whatever you need. We told him, "You don't have to sleep over. You go get a good night's sleep. If we need anything, we'll call you guys. He gave us their number. The next morning, we woke up. And I say wake up, I mean, we didn't even sleep all night, but we get out of bed and I see a cop car outside with his lights on. Shawn says to me, "Well, what the heck could this be?" I said, "Well, they can't give us any worse news than we just got. This can't be any worse." So, we open the door and the two cops come down the driveway. They tell us, "Every news outlet in the country is called asking where your street is; what direction is that? You're going to be inundated with news outlets in a little while. We just wanted to know what you wanted. If you don't want to be bothered, we'll put a roadblock up on this corner at the top of the street and nobody will get in. But, it's your call. What do you want?"

So, my three sons were there. I said to them, "What do you guys want? What do you want to do?" They said, "Dad, we want everybody to hear JT's story and we want everybody to hear what a hero he was." So, we told the cops, "Don't worry. Let everybody in and we'll talk to any news outlet." And it was unbelievable, Greg. I'm telling you, outside our house, the trucks, Channel 2, Channel 4, Channel 5, Channel 7, the cable companies just all parked on the street. They came to our door and they just said, "Can we interview you?" We would let them in and said, "Listen, if you want, you can come in and we'll do interviews." By this time, people brought so much food. Shawn's company had food catered and sent to us every single day. We invited the different news outlets in and they were talking to other people. They were here to interview Shawn and I. We would find a spot, usually downstairs in our family room where it was out of the way and quiet and they would carry on their interviews. That's the way we dealt with it. That was the way we dealt with our grief. Like my sons said, we wanted people to know his story.

We found out later other soldiers and Marines were killed from our area. We would go and visit those people once we found out. Sometimes they didn't want any visitors. They didn't want any news outlets. They just wanted to be completely left alone. That's fine, too. It depends on an individual how they decide how they're going to deal with the grief. That was the way we decided. Other people may decide differently. Shawn just reminded me, the cops were telling us that there were people just coming off of Route 80 and asking people where Lieutenant J.T. Wroblewski lives. Obviously people knew us in town, a lot of people knew us because of our four boys in athletics. Strangers, people that we didn't even know, driving to our house, asking to meet us and just wanting to shake our hand and extend their condolences. The cards, the letters, the emails. We got cards and letters, emails from all over the country, Russia, Germany, France. It was unbelievable. The outpouring of support just something you would never think could happen. It went on for days - the people would just come to our house. We would wake up in the morning and we would open our garage door and there'd be people there with coffee and donuts and cake.

Like I said, at around 11 o'clock or thereabouts, Shawn's company sent food over because they knew people were coming to our house. So, we always had food for everyone. They would send food here at 11 o'clock and then they would send us dinner at five or six o'clock. It was unbelievable. This was Easter time. For our Easter meal, her company had ham, turkey, soup to nuts, and seven course meals delivered here. It was just unbelievable what Shawn's company did for our family. Just an unbelievable outpouring of support and generosity on their behalf. When her parents came in from Florida, they flew them in from Florida. Shawn's company flew them in, paid for the flights, and had a limo there to pick them up, and then drove them here by limo. The same thing when they went back to Florida, paid for the flights, had a limo here to take them back to the airport. Then, we had the preparations for the funeral.

We had the funeral service with him here in April at the funeral home. John was buried on the 27th in Arlington. So, we had the funeral in the funeral home. They were going to have it on April 16th. Shawn said no because that was his birthday. So, we had the funeral a couple of days later. I'm going to say it was probably April 20th. So, we had the service up here - it was two days, and we had the Mass. And that's the other thing - well over a thousand people showed up at the church. It was packed, standing room only, and there were people standing outside. It was a scene. I never envisioned anything like that. There was not a seat in the church. There were people outside. The doors were open. Just very humbling that so many people cared about us and John.

Prior to that, we were setting up because there was going to be a repast after church. Obviously you have a little food or whatever. We were kind of expecting that there would be a large crowd. And again, Shawn's company heard about that and they said, "Don't worry about it. We will supply all the sandwiches, the beverages, the drinks, the water, the cookies, and the coffee. We'll have everything catered there so you guys don't have to worry about a thing." And I'll tell you, the amount of food that they sent at that repast, it looked like almost a thousand people came into the big gigantic hall at the church. And there was probably still food left over. Tables of food were like never ending.

Afterwards, we're saying goodbye to people and thank you. We're in the car and we're leaving to come home. And one of my sons said to me, "Dad, there was somebody there in the church dressed as a Marine, but I don't think he was a Marine. I think he may have been acting out or something. I saw the Marines by his car and I said, "Get out of here." So, we got home. I got a call from the priest that did John's service. The priest did a lot of different services for and funerals for cops and from 9/11; he's performed funerals for CIA; and he knew CIA officers and a real good friend of his was also an FBI agent. Cotone was the last name of this FBI agent. The priest said, "John, they arrested a guy at John's funeral because he was a Marine imposter." I said, "What? What are you talking about?" He said, "There was a guy there, a disheveled older guy in a Marine uniform." I remember now seeing him as I was up front and I looked over to my right, the way the church was set up, you know, you could see the pews. And there was this older Marine, and he had long hair hanging over his collar. He looked kind of disheveled. And I remember thinking, "Man, that guy must have been through a lot. He had all of these medals on his chest." And he was standing right up front and center. So, I remembered him. And so now the priest is telling me about him.

I said, "Well, what do we do? How did he get caught?" He said, "Well, remember the FBI agent?" And I said, "Yeah." The priest had said Cotone called him and said he was thinking about coming to the service and the priest said, "Yeah, that'd be good. You could help keep an eye on things." After the fact, I got to talk to Cotone, and he told me this story. He sees this Marine, and he sees all the medals he has and he sees the Medal of Honor. He sees the Medal of Honor on this Marine. So, this guy Cotone was very educated on Vietnam and all the battles that took place and he knew them inside and out. So, he goes up to the Marine and he says, "I just want to shake your hand. I see that you are a recipient of the Medal of Honor. That's few and far between. Where did you serve?" The guy said, "Well, I served in Vietnam." Cotone said, "All right, yeah, okay. But where, exactly where?" I can't remember where the guy said, but he told Cotone wherever, and Cotone knew that there were no battles or active battles going on there. It was quiet there, you know what I mean? There was no way he could have got a Medal of Honor in an area like that. So, Cotone kind of knew that this guy was full of shit. Cotone follows him. And now the guy, the fake Marine, goes to his car. Cotone follows him to his car in the parking lot and confronts him, now shows him his FBI ID, and said, "Listen, I know you're not a Medal of Honor recipient, and I know probably all these other medals are a lot of bullshit. I want you to open up your trunk." So, the guy opens up his trunk, and he has all of this other paraphernalia, Marine paraphernalia, this and that. Cotone gets out of him that he's an imposter. That's what he's doing.

And so now, Cotone sees up on the hill, there's about eight or 10 Marines and they're walking down towards Cotone. Cotone is telling me the story. He said, "John, I'm saying to myself, these Marines think that I'm harassing this Medal of Honor recipient. They probably want to smack me around." He tells the guy, "Get in your car." So, the guy gets in the car and he tells the guy, "Lock the doors." OK, so the guy gets in his car and locks the doors. These 10 Marines came down at John's funeral. Six of them were John's pallbearers. And so now Cotone is relaying the story to the Marines. They want Cotone to open up the door. They said, "Let us have him!" You know what I mean? They were livid. So, obviously he didn't open up the door and he calmed them down and got them on their way.

But, now back to the priest. The priest called and said to me, "What do you want to do? Do you want to file charges against this guy?" So, I said to the Priest, "Father, listen. I just buried my son. I really don't have a clear head. What would you do?" The priest said, "I'd burn the bastard." I said, "Yes, I get that." So, I filed charges. But, that was the story of the imposter. But the next day, Greg, in the newspaper, they have a picture of us coming out of the church. You know, the flag draped coffin, and Shawn and I were there at the coffin. And who's right behind Shawn and I, right in between us? It's the imposter. It's that guy. Front page. Front page of how many newspapers? He's right in between us. You just talk about an unbelievable situation. I mean, you hear about people like that, but you certainly don't think that they're going to be at JT's funeral, you know? It's just the nerves of folks that do that. All the guys in my son's unit would have torn him apart. I mean, they had the right. They were furious. Cotone said they were ready to just rip him apart. They would probably have beat him to death.

Now, the next morning, we've gotta get up. What they did is they took John's body to Arlington to where he was going to be buried. So, the next morning, we got up and drove to Arlington. Our town gives us a police escort. The different states that we go through, the state troopers, meet us and escort us through the state on the way to Arlington. We get to Arlington and we get to the rooms that we had reserved. Again, Shawn's company picked up the cost of the rooms. I think we needed four rooms for our family and a couple of John's friends wanted to come, so we had them with us too. We go to Arlington and show up in the morning at Arlington Cemetery and they tell us where we're going to assemble. So, we went there to assemble and this guy said, "All right, there's a room back there for everyone. Anybody that comes can wait in this room." So, that room got filled in a heartbeat and then, all of a sudden, he comes to me and he said, "How many people are you expecting?" And I said, "Listen, I have no idea." So he said, "All right, I'm going to open up these other two rooms." He opens up the other two rooms and they get full. He came up and said, "How many more people are coming? I got one more room." By then, people are going to have to wait outside. I said, "Well, I don't think they'll mind." He said, "Never have I seen a funeral that had this number of people show up since you guys live five hours, six hours away. Usually in that kind of scenario, we'll get 50 to 100 people. You've got 500, 600 people here, maybe more." And so, it was a tremendous outpouring of busloads of people from our town that went there. They had the teachers from my school that were there. People that Shawn worked with, not just co-workers, but the suits, the President of the company. Just an unbelievable turnout.

We go through the funeral. That was the first time in my life that I was ever at Arlington. First time in my life. I just could not believe the funeral. The horse drawn caisson, the first riderless horse with the boots turned around. And when the Marines would move, every step was like a picture. It was like a photo frame being shot. When they put John from the hearse onto the cassion, the way they moved the casket onto the cassion was perfect. Every move was like you were shooting a frame camera, as were their steps. We went from there to his grave site. We followed in a car. But, the caisson and the horses that were behind the caisson were in unison, watching every step of the horse is the same as the other horses. Just unbelievable.

At the burial itself, they set the casket up and they have chairs set up. Obviously, they didn't have enough chairs, so most of the people were standing. They go through the whole thing and they play taps and the Marine Corps band is there playing. Then, we had the 21-gun salute. When the Marines folded the flag that was draped over his casket, they held it so taut. It looked like it was a piece of plywood. Every fold was a picture of perfection. Just an unbelievable thing to see. To this day, I still speak to elementary school eighth graders that are going on the Washington field trip. I bring pictures of the funeral. I talk about John, you know, what I was telling you. And I tell them, if they get a chance, they should try to attend an Arlington funeral. I tell them to watch the horses, to notice the precision of the Marines or soldiers in the funeral ceremonies, to look at the caissons. I did some research and these caissons, if I have it correct and if my memory serves me correct, I believe there were only three caissons ever made in 1918. Ronald Reagan was on one, John Kennedy's been on one. Anytime a president's buried or you see them in a horse-drawn caisson in Washington, it's one of those. One of those three. So, obviously my son was on a caisson that Ronald Reagan, John Kennedy or famous generals like Patton could have been on. Just a tremendous thing to see.

I've been to Arlington quite a few times after, but it's a place I believe that every American should really go visit because once you get in those gates, you just have a feeling of reverence there. Just honor and reverence is what struck me when I was there this past summer. John's funeral was the first time I've been there since I was in the seventh grade, but I felt that same honor and reverence that I felt when I was a kid. It just makes you feel so proud to be American. The feeling is unbelievable. So anyway, we got through that service and then we had a little repast for the people that were there. And then, we had an escort that followed us home. It was the same thing going state-to-state with the state trooper escorts until we got home and our police gave us an escort again all the way home to our house.

The next part of the story is how I got to meet Mark Levin and Sean Hannity and how Sean

Hannity paid for Shawn and I to go to a rally that they were having for the troops in Crawford, Texas. That's where I met Martha Zoller, the journalist and radio show host. I'm driving home one day from work and it's before Memorial Day. I've always listened to Mark Levin's and Sean Hannity's radio or TV shows. They've always said such great things about Memorial Day and about our heroes, our fallen. So, I was listening to Sean Hannity on my drive home. I said, "Well, I'm going to try to give him a call." So I'm trying to call him, and I don't get through – busy signal, busy, busy. I get home and we're getting ready for dinner. Mark Levin comes on and he leads his show talking about Memorial Day and our fallen heroes. He always does a great job with that. I thought, "Well, let me try calling Mark Levin." So, I dial the number and all of a sudden, it's ringing. In all honesty, I'm getting scared. What am I going to say? I'm thinking about hanging up. So, his call screener answers, "Hello, it's the Mark Levin Show. Who's calling and what are you calling about?" I said, "I'm John Wroblewski. My son, Lieutenant JT Wroblewski, was killed in Iraq on April 6. I just want to thank Mark for all the great things that he says about our heroes on Memorial Day because now my son is one of those heroes that he's talking about. I'd like to just let him know that I'm so appreciative." The call screener said, "Listen, you're on next. I'm putting all other callers aside."

Now, I'm really getting nervous. Sure enough, click, boom, and it's Mark Levin on the phone. He couldn't be more compassionate, "Mr. Wroblewski, so sorry to hear about your son. Tell me your son's story." He goes on talking about how he's a hero. I said, "All right, I'll be brief." He said, "Don't worry about being brief. You got all the time that my show has got to offer." Just an amazing individual. To this day, Mark Levin and I are friends. We constantly text back and forth. We've maintained our friendship. On John's 10th anniversary of his death, we had a memorial at Arlington. It was something that was just going to be family, and it got out. Before you know it, we probably had 150, 200 people show up. I had invited Mark Levin and Mark Levin showed up. He gets to Arlington around 10 o'clock. After we had a little service at the gravesite, Mark stayed. We had sandwiches and soda. Mark was there until six o'clock, until everybody left. Then on his radio show a day or two later, he talked about the experience. He said, "The Wroblewski's invited me. What a tremendous place Arlington is."

And then from there, Sean Hannity and Levin are good friends. Hannity heard me on Mark Levin's show and a couple days later after doing Levin's show, I got a call from Hannity and he wanted to do a show. So, I do one with him and we become friends and has me on his show a number of times. Many times, he would have me as part of a panel. It would be me, Bob Beckel, and Melanie Morgan on one show. Melanie Morgan was the one who put together that rally and she started it in California. But anyway, it's December 2006 and I'm on a show with Hannity. He said, "Are you and your wife going to that rally in Texas?" I said, "I don't know. We were kind of thinking of it." But, I knew we couldn't go. I couldn't afford the flight and room. So, he said, "You and your wife should be there. That's an important thing for you to do. I'll tell you what, I'll pay for everything. I'll pay for your flight. I'll have a limo pick you guys up. I'll pay for your room. You won't have to pay a nickel." I said, "Sean, thank you, but I got to run this by my wife." I'm just shocked. Sean said, "All right, I'll have my assistant give you a call later." I said, "Okay, thank you. Goodbye." This is probably around 3.30, 4 o'clock. I drove home. Shawn comes home from work and I tell her the story. Now the phone rings, it's 5 o'clock, and it's Linda, his assistant that's from Jersey.

Linda said, "Have you had a chance to speak to your wife?" I haven't. We didn't give our answer yet. Shawn was saying, "There's no way we are going to that." So, Linda calls again later and says, "Look this is a blessing. Sean wants to do this for you. I just need to know if you and your wife are going. It's in a couple of days. I've got to make reservations." I said, "Listen Linda, I really don't feel comfortable taking handouts from somebody. I've never done that. I just feel uncomfortable." Linda said, "Sean really likes you guys (because we had met him too at the Freedom Concert) and you've been on the show a number of times. I'll be honest with you. He wouldn't do this for just anybody. If you guys turn him down, it's going to hurt his feelings." So, I started to laugh and said, "Really?" She said, "Yeah." I said, All right. Linda, we're going." Linda said, "Okay, great. I'll give you a call either later or tomorrow with all the specifics. We'll get this thing moving."

Linda called the next day. They picked us up in the limo, drove us to the airport. They had everything ready. You know, the flight, the room, everything was paid for. This is when I met Martha Zoller. She was helping Melanie Morgan and Sean with driving guests and other things. We were told, "Okay, we have a car, the driver is Martha Zoller, and she'll be with her assistant Autumn. So, it was Autumn, Martha Zoller and me and Shawn. The rest is history. We hit it off well with Martha and her assistant. Martha is a great person. She took a liking to us. She told me that she was in Iraq on an embed previously. I had said to her that if you ever go again, please include me. One day, I got that call. Martha told me that you (author Gregory Janney) were going as her photographer, and Lieutenant Colonel Quinn would accompany us as unofficial security. I should say that you and I know the rest of the story. It was an honor to be included on that January 2007 trip and to have met you.

Janney: I could see the disappointment in your face when that crew chief turned us back at the helicopter flight to Ramadi at LZ Washington. Remember, the Marine crew chief said, "I'm not putting them on this helicopter. I need those seats for Marines. I'm afraid I'll be bringing you guys back in body bags." That's why I promised you on the flight back that I would get you there one day. You were so funny, John. You said, "Okay, great. Thanks a lot." With that strong Jersey accent. As soon as I got back home, I contacted the former CO of the Ranger Training Battalion who was at the Pentagon that I knew. I said, "Colonel Flohr, who can I talk to at the Pentagon about doing a military embed? I'm particularly interested in going to Ramadi and embedding with a Marine unit." He sent me the contact information later that day. Then, it was just a matter of getting you on board with going back and raising the money from the corporate sponsors to pay for the trip. So, I thank God it worked out. It was a huge honor to have done that.

John Wroblewski: Greg, your ears must be burning so many times. I tell that story often. How you promised to get me back there. I told people about that promise, but I didn't really think it was going to happen. I thought maybe the guy's just being a nice guy. Then, finally it came to fruition and you gave me the call. I thought, "Holy Cow, he did it!" I tell people the story and what we went through and about meeting USMC MSgt Ellerbrock again in Fallujah in 2008 after Army Lieutenant Colonel Quinn kidnapped his Marine, PFC Eberle, 14 months earlier back in Jan 2007 on Martha Zoller's mission.

Janney: That was crazy. LtCol Quinn wanted to take you to the CSH and decided to commandeer a Marine to take with you, so he grabbed PFC Eberle from the Marines there waiting on a helo without asking his NCO if it was okay. MSgt Ellerbrock comes in and starts looking around, doing a headcount. He said, "Hey, where's Private Eberle?" Some young Marines said, "I don't know, sir. Some Army Lieutenant Colonel took him." Ellerbrock said, "What do you mean he took him?" He starts getting more and more frantic. "What do you mean he took him? Where did they go? We're waiting on a helicopter. You can't just take somebody's Marine." Martha Zoller starts trying to explain it, but he just cuts her off in the knees. He just said, "Look, lady, you don't understand. If something happens to that kid, I'm going to be court martialed. He's my responsibility. He should never have taken him without clearing it with me." Martha said, "Oh, you don't understand. 2ndLt JT was a Marine who died at the CSH and LtCol Quinn thought it would be appropriate to have a Marine escort John to the hospital where JT died." Ellerbrock said, "Lady, I don't care about any of that. What I need to know right now is where that private is and when he's going to be back, because he cannot miss that helicopter flight to Ramadi." Oh, my gosh, it was horrendous. You were there for the grand reunion, when Quinn and you and the Marine came in and the heated exchange between LtCol Quinn and MSgt Ellerbrock outside the LZ guard shack. What was Eberle like? Was he confused about why he was pulled aside by an Army LtCol to go on an unauthorized field trip? Laughs.

John Wroblewski: I never got much of a chance to speak with him. He was nice. I guess Quinn told him about my son and he was very respectful. He was saying, "I'm really sorry about your son. He was my brother." The kid was great. Quinn said, "Well, we got to go here." Eberle knew how to get there. I never really knew the story until after I got back. I thought he was one of Quinn's chauffeurs, so to speak. So that's what I was thinking. I'm saying, "Holy jeez, I guess he still has connections here." I had no idea what was going on. When we got back, MSgt Ellerbrock was yelling and screaming. I said, "Holy shit!" And he finally calmed down and apologized to Martha and to you, and offered his condolences. But, I thought he and Quinn were going to kill each other outside that little guard shack. I was sitting there thinking there's about to be some fisticuffs thrown, you know? I mean, they went at it. They were going at it. All of a sudden, here we are, right? We got approved for the 2008 embed. And, who's there? It was MSgt Ellerbrock!

Janney: I'll never forget that, John, because you and I walked into that little guard shack on the Fallujah airfield, following the little chemlights. It was pitch black out there that night. We walk in and this very familiar Marine Master Sergeant is standing there and I'm sitting there thinking, "No, no! It can't be him. It's the same guy. And he looked at me and said, "You! Yeah, you! Weren't you with some Army Lieutenant Colonel that stole my Marine about a year ago?" I'm standing there thinking, "Well, we're going to be sent home. This is so horrible. I can't even believe this could happen. How can it be the same guy? And I just started apologizing and falling over myself. He finally just started laughing and said, "No, no, you're okay. Let me get you settled in your quarters."

John Wroblewski: Yeah, you're right. He was great. He was telling us something about being promoted. I can't even remember the specifics of that other than I don't know why he was still in Iraq 14 months later. I have no idea. I don't remember what that story is. Yeah. But anyway, like I was saying, I often told the story and your name to a number of people. When I relay the story and I tell them how you were able to get us to Iraq for the second time and the experiences we had, it's unbelievable.

Janney: I lied to get you back there, John. I listed you as my photography assistant. You were my assistant until the public affairs officer, who was MSgt Ellerbrock, although I didn't know it at the time said, "Okay, get on the helicopter at LZ Washington and I'll meet you in Fallujah and we'll go get some pizza and do some stuff here. I said, "Okay, that's cool as long as I can get to Ramadi." And he says, "No, no, no, we're not going to Ramadi. You're going to do whatever you want to do here." And I said, "No, no, no. I have to get to Ramadi." And he said, "It's the same shithole that Fallujah is. There's no need for you to go to Ramadi. They're doing a unit changeover, so you can't go to Ramadi. And at that point, I had to confess. I said, "Well, no, I have to get there. I have grid coordinates that I have to get to." And he says, "Why?" And I said, "Because my assistant is not really my assistant. He's a Gold Star dad and his son was 2ndLt JT Wroblewski of 2/4 Marines. He and 11 other Marines were killed in an ambush in Ramadi on 6 April 2004. I'm trying to get to the ambush site to do a memorial service for those guys. There was this long silence on the phone. Then he says, "Wow. I don't know. Get on the helicopter and we'll talk about it when you get here."

John Wroblewski: I think this is the Major you were talking about. It was this public affairs guy and they were in the meeting. Major Peters, I think. I do remember meeting him later. Then when we got there, it was Ellerbrock, the same Master Sergeant Ellerbrock from fourteen months earlier. That was horrifying. I thought I was going to throw up when I saw him in that little guard shack at the Fallujah airfield.

Janney: When JT was doing that patrol with Dave Swanson, he and Dave got on the subject of fishing. Did he really like to fish or was it just something that interested him? Or was he just talking about fishing to pass the time with David Swanson? Tell me the backstory on the fishing thing.

John Wroblewski: He loved to fish. I did, too. I used to love fishing. There were times when I would get up extra early on my way to work. There was a stream on my way to work right off Route 23. I would stop and fish there for a half hour or so before I would go to work. All of my sons loved fishing because obviously I loved it. So, I used to take them fishing all the time. I taught JT how to fly fish in the Rockaway River. It was a little secluded spot, and that's what he was talking to Swanson about, about that spot. Because Dave spoke to me about it. In fact, I took Dave there. I said, "Yeah, I'll show you the spot that JT was talking about." They spoke about fishing because they were walking by the Euphrates River in Ramadi. Yeah, they were walking and they saw Euphrates and then they were talking about going fishing. John brought up fishing and Swanson said, "Yeah, I'd like to go fishing one day." And John said, "Listen, when we get home, we'll get together and I'll bring you to the spot where my father taught me how to fly fish. We'll do some fishing." My son used to go fishing a lot. I mean, his football coach would go fishing with him, and baseball coach, too. He had a couple of real good friends that also enjoyed it. And he fished a lot. He really did. I would have to say he probably loved fishing more than my other three sons. But, John really loved it. He enjoyed it tremendously. There was a little pond or a little lake at Camp Pendleton. When we got there, he was telling me that he wanted to fish that lake. It looked like you know it had some bass in it. I've never been there, so I don't know that place, but I hope to get there one day.

Janney: Now, did you know that he was a big Pantera fan. Pantera is a heavy metal band.

John Wroblewski: My wife knew it, yes. We both liked Led Zeppelin, too. JT's first choice was Led Zeppelin. When he worked with Shawn as a security guard, back when he was out of college in a summer job to make money, on the way there and back, he would pop in the Led Zeppelin tape. And, you know, he loved Led Zeppelin. One thing I did want to ask you, was it as unbelievable to you as it was to me that we actually made it to Route Gypsum? It seemed very surreal when the convoy stopped. The word came back, "Let's go. You're up. Let's go over here and do this thing." And so it just seemed very surreal to me. And Greg, after the first trip, and obviously you went the second time. I kind of figured that we probably weren't going to make it there, but let's give it our best shot. Let's just try. If the opportunity is there, let's try. Rather than saying we should have, we did it. I guess when it all kind of came together, I remember when we were in Camp Fallujah, and I believe a Sergeant came to our barracks and said that the Colonel wanted to see us. We went to his office and then he told us that he got an email from General Kelly. He started to read it and then he said, "Oh, you read it." Then he gave it to me and I'm reading it. Basically, General Kelly was saying he'd love to and he wants to meet with us. Then from there, head on to Ramadi. Now, at that point, I'm saying, wait a minute, if the general is involved I felt that we were definitely getting there. You know what I mean?

Janney: Yeah, I knew when Kelly got involved that we would get there. Because Ellerbrock said, "Look, it's gone up the chain of command and General Kelly is involved now. I think he's going to make sure that you get there." So, I mean, yeah. But like you said, when we got that, when you saw, read that email, then we knew then that it was really going to happen. But, on the actual day itself, it was very surreal. I was sitting there thinking, "Well, something's going to happen. We're not going to be able to get there. God forbid somebody gets hurt, you know, there's an IED or a sniper attack.

John Wroblewski: I never thought about getting hurt. The one time that went through my mind, we went out on patrol with Captain Martin. I was talking to Martin, and I remember saying to him, "Listen, my son took this trip and they were having gun battles here routinely." Capt. Martin said, "It's a lot safer now." I said, "What do you worry about the most?" He said, "Well, in all honesty, we really keep an eye out for snipers." That's going through my mind real quick. I'm saying, wait a minute, here we are with all of these Marines. Greg and I are the only ones in civilian clothes. We kind of stand out where a sniper may say maybe these guys are somebody important and look to pick us off. You know what I'm saying? That went through my mind and I realized it might be a possibility. Then when we went to the markets, Martin's telling me, "Look at those guys in the background. You notice there that they're keeping an eye on things. They're watching what's going on. Those are the guys that usually know the bad guys. Those are the guys that will have somebody tell us, "Oh, so-and-so is living at such-and-such. They're new. They're newbies here. They haven't been around long." Because it's very clannish in Iraq - family and clans and they know who comes and goes. When somebody new comes in, they know that they're new. Why are they here? This is what Martin was kind of relaying to me. Then they would give Martin the information or pass on the information to the Iraqi police. They would go to the guy's house and do what had to be done.

But yeah, it was something when the trip started, I didn't think we were gonna be successful. But then that email, it said a general's involved, so we're gonna get this done, and Greg's gonna be successful. So, I tell the story a lot of times. I just thank God that it worked out and that nobody got hurt because a lot of other people tried to do what we accomplished and we were the only successful mission that did that. That's the other thing I was going to say. There were other fathers and other family members that got to Iraq, but we were the only ones to get to the exact site where my son was shot. I mean, we're the first and only. And when you think about it, it's historic.

Janney: As I said, it's just a huge honor to me. That's why I had to make sure that it happened when I made you that promise on the plane. I knew I was going to do everything I could to keep that promise. I tell you though, John, when you showed up at the airport with a Vietnam-era flak, I almost shit myself. You know that, right? When you showed up at the airport in Atlanta, and I came to pick you up, and I had told you needed body armor like we had in 2007, and then you showed up with a Vietnam-era flak vest which is not even bullet proof. I'm sitting there thinking, "Oh, my word! Now it's too late for me to get body armor for him." So, I worried the whole time you were out there that you were going to get shot and it was going to be my fault that you got hurt.

John Wroblewski: You didn't say a word. You didn't say anything. You know when it was finally brought to my attention? When we were in Baghdad, Iraq at CPIC with all the other reporters. There was a room of bunk beds. We were getting our press credentials. We were staying there, and I had the flak jacket hanging on the corner of my bunk. One of the reporters said, "Whose is this?" I said, "Oh, that's mine." The reporter said, "That won't even stop a pistol bullet." Oh, my gosh, man. That's exactly when it was. I thought to myself, "Oh, shit."

Janney: That's exactly what I thought when you showed up with it. I thought, "Oh my word! I hope nobody shoots him. That's gonna be terrible."

John Wroblewski: Yeah, because the guy here, Mario Monaco, who is a former Marine, we became close friends through the death of JT. When he heard I was going to Iraq, he said, "Oh, I got it. You can use my flak jacket and helmet." I didn't know anything about him. I said, "Yeah, fine." When he gave it to me, I said, "Don't they have any plates in it or anything? I remember when I was there last time, the jacket was a lot different. This is very different." He said, "This is what we wore in Vietnam." And I said, "Oh, okay."

Janney: I should have made that clearer to you when I told you that you needed to get body armor, but I guess I didn't. I think you just said that you got it covered. So that was horrifying, man. I was really scared the whole time you were there that something was going to happen to you, and it was going to be awful.

John Wroblewski: I had no idea until that guy saw it hanging on my bed. The reporter said, "That can't stop anything. It won't stop a BB gun." I said, "Oh, shit." I didn't say anything, just kept my mouth shut because I knew there was nothing at that point that could be done. I just said some extra prayers that night.

Janney: When I saw your flak when you got to Atlanta, I just thought, "What are we going to do? There's nothing we can do at this point."

John Wroblewski: Right. There's one other story I'd like to relate. After John was killed and buried, a student I taught, Maria, became a teacher in a school district up here in Jersey in Rivervale which is a small town not too far from here. When she heard of John's death, she went to his funeral at Arlington. She had called and said to me that her class was going to Arlington, and would I mind if they visited his grave? I thought she was just being cordial. I said, "Maria, no problem. It'll be great. John is always looking for visitors. But Maria, I don't think his headstone is going to be ready yet because they told us it wouldn't be there until late June or July. It may be just a grave marker, but if you go to the office, they'll tell you exactly where it is. They leave to go to Arlington and, you know, these things stay in your mind. I'm in my office, the phone rings. I pick it up and it's Maria. She said, "Hey Coach, this is Maria. I just want to tell you the group left. But I'm standing here now at your son's gravesite." I said, "Oh, Maria, thank you." But you could tell that she was emotional. She said, "The reason why I'm calling is I just want to tell you that the headstone is here. Let me ask you a question. Did your son have anything to do with car racing?" I said, "No, not car racing. Why?" She said, "I found an Indiana state quarter on his grave in the grass with the Indianapolis 500 on the back of the coin." I started to get real emotional. She said, "Coach, what's the matter? I'm sorry." I said, "Maria, it's amazing because you don't know this story. Only a few people know this story. I'm going to tell you why it's emotional." So, let's go back a little bit. John's at Rutgers. I was an avid saver of the state quarters when they came out. I would go to the bank and I'd buy $10 worth. I'd put them in this drawer. At the end of the day, I would look in my pocket, and any state quarter I'd throw into this drawer. John comes home from Rutgers. I'm not home. Shawn's here. He said to Shawn, "Mom, do you have any quarters? I need quarters for the washer and dryer." She said, "Oh, yeah. Go in the top drawer over there." John said, "But those are Dad's." And she said, "Don't worry, he won't miss them." He grabs handfuls of them and he goes on his way. That night, I come home, empty out my pocket, and I have three or four state quarters. I open up the drawer and it's so much lighter. I threw the three quarters in there, and I said, "What the hell happened? Come on. Where the hell did my quarters go?" She said, "John needed them for the washer and dryer so he can do his laundry." I said, "Are you fucking kidding me? I'm saving them. I go to the bank to get them." So, I call John. I can't get through to him, but I get through the next day. I said, "You gotta be kidding me, taking my quarters." And he said, "Well, Dad, Mom told me I could take them. Don't worry. I'll make it up to you." So anyway, we forget about it.

So, now I go out to Camp Pendleton when we visited John. We come in and he's showing us where we're going to stay. We come down to his living room and he's got a water bottle there and it's almost full with coins. He said, "Dad, remember I told you I'd make it up to you? Those are state quarters. Take them." So, I empty the thing out and I'm going through and taking all the state quarters I want. These are great because all of them are Denver mints. All of the ones I'm getting are Philadelphia mints. I'm taking all of the Denver mints, and putting the rest of the coins back. So, he said, "All right. Are we square?" I said, "No, you're not getting off that easy." And he starts laughing. That was the end of our conversation.

So now, Maria has her third grade class at John's grave, and what's in the grass? An Indiana state quarter. You know what I mean? When I speak at the school for those eighth graders that take a trip to Arlington, I tell them that story, and I tell the teacher, Maria, to select one boy and one girl, and I will give them each a quarter to put on John's headstone. I guess I've been speaking at that school for 16, 17 years. It became an honor for these kids to have that quarter. And what one class did is they all brought quarters. When they went to Arlington, every kid went up to the headstone or another headstone and put a quarter on it.

But, you know, you talk about coincidence, the Arlington gravestones are made in Vermont. A newspaper in Vermont did a story showing how the headstones are made with pictures in their local newspaper there. A friend of mine vacationed in Vermont and she sees the newspaper and whose headstone is on the front page? JT's headstone! Another unbelievable story is from a baseball league up in Massachusetts. They would put the name of every fallen hero on a baseball, the date that they were born, and the date that they were killed in Iraq. They would put the baseball in a big, gigantic glass covered shadow box that they had mounted at the baseball field. They're doing a news report about the league putting fallen heroes' names on baseballs to honor them and their families. So, a friend of ours called us and told us about the show. We put it on and are watching. When the segment is over, they're zooming in on the gigantic shadow box - it starts from a distant shot and it shows how big the shadow box is. They zoom closer and closer until it focuses on just one baseball and it's Lieutenant JT Wroblewski. Phenomenal, just unbelievable.

When we were embedded there with Captain Martin, at the end when we were leaving, I spoke to him and I thanked him for everything that he and his Marines did. I said to him, "Listen, I just hope we weren't a burden and I hope we weren't the reason for anybody losing their focus. Captain Martin said, "No, just the opposite. You guys were a shot in the arm. These guys, when they heard why you guys were here because your son was killed and you're trying to get to his site to honor him. For these guys, it was like a shot in the arm. It really livened them up. It was a good thing. You guys were not a distraction. If anything, you boosted our morale."

Janney: You're talking about Captain Martin. We were walking around outside that joint security station. You remember that sketchy place where the Iraqi army guys were? That was really a sight. Had those bullet holes in the walls of our room. So, we're walking around outside. I got a headlamp with a green lens on it. We're being kind of quiet, you know, just kind of creeping around. Captain Martin didn't have a headlamp on, and I said, "Captain Martin, how come you don't have a headlamp on?" And he said, "Because the snipers shoot the guys with headlamps on." I just kind of chuckled, and then just reached up and turned it off. Remember when he took us up in the guard tower. The guy said, "At 20 meters, we fire a warning shot. At 15 meters, you just kill them." They had it marked that way outside the JSS.

John Wroblewski: I tell you what a trip. Unbelievable. It was quite an experience, man. I'm glad that we got to do route patrols and stuff like that with the 2/8 Marines. That was really cool because we got to see a lot of the stuff that J.T. saw. So, we were right on MSR Michigan and we walked the whole way up Route Gypsum to Route Nova and down Nova a good distance. MSR Michigan was where David Swanson took that photo of JT walking into the sunset. I have that on my living room wall.

A Father's Journey To Ramadi

Iraq Memorial Service Missions With Gold Star Dad John Wroblewski

2007 1st Iraq Embed and Memorial Mission Attempt

John Wroblewski, father of Marine 2ndLt J.T. Wroblewski, and I first met on January 12, 2007, as we set off with journalist Martha Zoller on our first trip to Iraq. Martha's plan was to get John to Route Gypsum in Ramadi where J.T. was mortally wounded. 10 other Marines, and a Navy Corpsman were also killed there or near there on 6 April 2004. We arrived in Baghdad on January 14, but despite numerous attempts to get to Ramadi, we were unsuccessful due to sniper activity there. Having spent seven days by John's side during which we heard more about J.T. and the Wroblewski family, I felt John's anguish as we risked so much in our unfulfilled attempt to honor his son's sacrifice. On the solemn flight home, I promised John Wroblewski that I would bring him back to Iraq to fulfill his dream.

Due to space limitations within this book, I will not describe the entire ten days of the 2007 trip. However, there are two incidents that I feel compelled to write about – one that could have had a direct consequence for the success of my 2008 mission and the second because it was just so weird.

Near the end of the ten days allocated for Zoller's mission, we had gotten clearance to fly our team from LZ Washington to Ramadi in a last-ditch effort to get Gold Star dad John Wroblewski to Ramadi to perform the memorial service for John's son, 2ndLt J.T. Wroblewski and the other 11 men of 2/4 Marines that were KIA on 6 April 2004. When we arrived at LZ Washington, the guard shack next to the LZ was full of Marines waiting on the same helicopter. Martha's friend, Army LtCol Quinn, decided that it would be a good idea to get a Marine to escort John to the CSH in Baghdad where 2ndLt Wroblewski passed away. So, Quinn asked PFC Eberle to come with him and John and they left to go to the CSH. A few minutes later, MSgt. Ellerbrock returned and did a quick head count of his Marines, finding PFC Eberle missing. He asked around and one of his Marines said, "We don't know where he went. Some Army LtCol took him." At this point, MSgt. Ellerbrock became visibly concerned about the safety of this PFC since he was responsible for this Marine. Martha Zoller began to explain why LtCol Quinn commandeered PFC Eberle, but MSgt. Ellerbrock cut her off, "Lady, I don't care about that. That Marine is my responsibility and if something happens to him, that's the end of my career." Later, LtCol Quinn returned from the hospital with PFC Eberle and was summoned outside by MSgt. Ellerbrock where there was a heated exchange. In a few minutes, MSgt. Ellerbrock returned and offered his condolences to John and apologized to Martha Zoller. A little later, the Marine crew chief refused to board us on the helo, saying, "I need these seats for Marines. There's too much sniper activity in Ramadi. If I take you, I'll end up bringing some of you back in body bags." At this point, our time was up and we headed home to the States.

The second incident occurred on our way out of Baghdad headed back to the States. Our team got on an armored bus called the Rhino for a ride to Camp Stryker. Just as we approached a checkpoint lit up by security lights, I saw an ocelot run in front of the bus, threading its way through the jersey barriers. Later, I learned that the area had been used by Saddam Hussein for a zoo, and the ocelot was an escapee from that zoo. It was the weirdest thing that I ever expected to see in Baghdad.

A Father's Journey To Ramadi

Iraq Memorial Service Missions With Gold Star Dad John Wroblewski

27 Feb – 13 Mar 2008 Our 2nd Journey to Iraq

27 Feb 2008 - 1720

John and I are about to board KLM flight #622 at Concourse E gate 26 from Atlanta to Amsterdam, and then on to Kuwait. This is our 2nd attempt to get Gold Star father John Wroblewski to the site in Ramadi, Iraq where his son, USMC 2ndLt John Thomas Wroblewski was mortally wounded in an April 6, 2004 ambush during a QRF mission by 2/4 Marines along Route Gypsum.

This plane's personal movie screens with remote on-seat controls rocks! I've got a window bulkhead seat 10J (assigned 12J), but the plane is almost empty so everyone moved to where they would have the most room. I had a little bit of stress just before and immediately after boarding, as I waited until the very last second to activate my Motorola Global phone. I almost didn't get it activated before we were told to turn off all electronic devices prior to taxiing out.

I'm watching "Over the Hedge" on my personal LCD monitor with remote control, boots off, cover on, pillow fluffed, and ready to fall asleep. Flight time is only 7+ hours instead of the normal 8 hours and 20 minutes due to a strong tail wind at this altitude. We were just served a very good meal of chicken, brown gravy, potatoes, corn, couscous-chickpea salad, and a fudge brownie for dessert. Now, I am going to watch "The Kingdom."

Later…WOW. That movie choice was seriously messed up. Wish I hadn't watched that. A FBI team goes to Saudi Arabia to investigate a terror attack and 1 agent is kidnapped by the terrorists, who were about to behead him on video. His FBI agent partners and Saudi police rescued him at the very last minute. This was a very poor choice of movie to watch before heading into a combat zone in the Middle East.

28 Feb 2008 – 0708

We touched down here in Amsterdam at the Schilpol airport at 0709 in foggy conditions with a temperature of 49 degrees. John and I looked around and went into a duty-free shop for a bit, before John decided to buy a chocolate muffin and bottled water for $8.00. I ended up getting a pint of Heineken and a spinach/mushroom pizza for 8.25 Euros. Not sure about the exchange rate, but I was hungry. It's now about 0915 local time and still very foggy. We got through another screening and passport control. Now we are just waiting to board KLM flight #459 to Kuwait.

It was a cramped, but pleasant 5-hour flight from Amsterdam to Kuwait. It is now 1740 local time, 21 Celsius, and we're to land in just a few minutes. We had a delicious in-flight meal of Bertolli chicken with red sauce, pasta/rice, cold salad, and tiramisu for dessert, and had a small slice of quiche with apple juice for breakfast. John and I are to meet our U.S. Army public affairs escorts in the Starbucks for transport to the military side of the airport. I hope they'll be there like they have said they would be! I have been shooting some videos during our trip. I am so thankful that Rick Albrecht loaned me his camera.

John and I got off the plane into a clean, but very busy terminal with Arabic (and very little English) signage. Almost everyone is dressed in traditional Arab dress and speaking Arabic, with many TCN's (Third Country Nationals) and Indians, too. We saw one particularly striking group of Indian girls dressed in identical sea foam colored saris. All of them were tall and looked to be about 18 to 22 years old. It was a little confusing getting through passport control, since in January 2007 we had flown World Airways on a military R&R flight and had bypassed this.

When John and I attempted to get through passport control to claim our checked bag (containing our body armor and Kevlar helmets), I was asked for more ID after I handed the officer my passport and plane ticket. I handed him my Georgia driver's license and Press ID, which he looked at and handed back. We finally understood when he handed our paperwork to a Kuwaiti soldier, and he told us to go upstairs to get a visa. John and I made our way upstairs, and took ticket #769 and waited for our turn. After 15 minutes, a man looked at our paperwork, and asked me for 3 "Kit" (Kuwaiti dinars - about $12.00) for each of our visas. I finally found the correct change in dollars and paid him. Twenty minutes later, they called John up to the counter and stamped his passport. A few minutes later, my name was called and my passport was stamped. John and I then went back downstairs and breezed through passport control with barely a glance.

Our bag was off to the left side of baggage claim. We quickly found my silver bag containing my IBA (ceramic body armor), Kevlar helmets, and John's Vietnam-era flak jacket and rolled it to the final x-ray checkpoint where all our bags were examined before we were allowed out the 2 huge sliding glass doors. The TCN's working the checkpoint repeatedly yelled, "Yellah! Yellah!" ("Hurry! Hurry!" in Arabic,) even though John and I were the only people in the building. It was a little surreal, but I guess it made them feel important.

At 1830, Sgt. Taylor met us in the Starbucks as promised. Standing 6'3" in a bright orange polo and khaki pants, Taylor was hard to miss. He just looked like the "right guy" as he seemed to be looking for someone – yes, us! We walked through the exit doors, into an underground garage, and across a very busy driveway to the parking lot where Sgt. Doug Veitch waited for us in a civilian SUV. After a bit of a traffic jam at the round-about, we got on the expressway and drove the 30 minutes to Ali Al Saleem, the U.S. military side of the airport. We came up to the 1st checkpoint for an incredibly thorough screening. A <u>very</u> intense black security contractor said to me, "You've gained some weight, haven't you?" I smiled, and said, "Yes, 6 years and 3 teenagers will do that to you!"

After security, we parked next to the PAO intake office at Building 2, and stowed our gear. A little later, Sgt. Veitch walked John and I over to Building 1, where we signed in to get our official orders, and submitted our passports for processing out of Kuwait. At this point, we had to pay another 3 KD, and will hopefully pick up our passports at 1630 tomorrow. If so, we report to Building 1 again to take our names off of passport hold in order to get our names on a list for a flight out on a C130 aircraft to BIAP (Baghdad International Airport.) Following this, John and I went to the billeting office where we were assigned bunks in tent P-5. John and I then walked about 1 mile across the camp to the area where food service vendors are located. We were the next-to-last people to place our order before they closed for the night. It was a pretty chilly dinner at our picnic table, with a temperature in the low 50's and a cold breeze blowing off the desert.

I spoke with a black female soldier who was just in from emergency leave to attend her brother's funeral. She said she had 2 more months left on her 15-month deployment. Both she and her husband were in the same engineering Battalion, although in different Companies, and had deployed at almost the same time. She said that being deployed together was a great opportunity to have some contact while overseas. After our meal, John and I hiked back to P5, took turns going to the latrine, and crashed about 2230.

29 Feb 2008 – 0730

John and I woke up late and hurried to DFACS (mess hall) because it closes at 0800. I got a breakfast of burrito, lots of bacon, 2 hash browns, biscuits with chipped beef, large Coke, 1 croissant, and 1 banana nut muffin. Since I had forgotten that you cannot bring a bag into the DFACS, John watched my camera bag while I went in and got a plate. John then went in and got his food. We ate outside in the oh-so-bright sunshine. It was at this point that I realized that I had forgotten my sunglasses! Afterwards, we made a cursory visit to the PX (post exchange store) and went to the finance office when they opened at 0900 to exchange $14.43 for 4 KD – 3 for our passport visa fees, and 1 extra as a souvenir.

I then walked over to the TMC (first aid station) and got some liquid tears for my case of pink eye, the PX for a towel (forgot mine) and Croakies eyeglass retainer. I then called my wife to let her know that John and I had made it to Kuwait safely. Then John and I took turns going to the latrine to shower, while the other watched our gear in P5. One thing that most people fail to realize is that Ali Al Saleem is basically a huge transient camp, with no storage facilities for expensive personal gear. Thousands of contractors, TCN's, and military personnel rotate through this base daily, and there is no security for your gear other than what you or your team provide. Therefore, with our team consisting of just John and I, we can never go anywhere together without bringing our gear with us. My camera bag contains 2 cameras, 2 flash units, multiple lenses, rechargeable batteries and chargers, laptop, clothes, and toiletries. This pack weighs about 50 pounds. My body armor and Kevlar helmet weigh an additional 30 pounds. The alternative to carrying it with us is for one of us to secure the gear while the other goes to the DFACS, latrine, internet café, or wherever. If any of my gear was lost or stolen, it would mean returning home immediately, as I cannot travel in a combat zone without armor, and a photojournalist without a camera is just a journalist.

Our accommodations in P5 are rather spartan. Each huge canvas tent houses up to twenty people, with 3 tiers of bunks along the sides, and 2 tiers along the back wall. The tent has an alcove entrance that has 2 doors to maintain black-out conditions, help maintain the minimal climate control, and to keep the dust and mud from entering the tent. While practical in theory, this is not possible in reality. The flow of people in and out is nearly constant, at all hours of the day and night, and the doors are often both open at the same time, allowing the heat or cold in and out, the dust to blow in, all accompanied by the constant sound of the plywood doors banging shut.

After a long nap, John and I walked to the internet café and checked email. I was unable to sign onto my blog, but resolved this problem later. It became much easier after I switched the computer language from Arabic to English! I received an email from my U.S. Marine Corp PAO (public affairs officer) MSgt. Willie Ellerbrock, who informed me that he had arranged helicopter transport for John and I to Fallujah on 2 Mar. I replied that we still needed transport to Ramadi, and that I could advise the specific details if he needed to know them. I also sent an email to my wife, Tom Dugan, and another to Tony McKinney, a 5th grade teacher at Riverbend Elementary and friend, before I ran out of internet time. Then, John and I walked to the DFACS for chow at 1700. I ate a huge meal of Salisbury steak, scalloped potatoes, noodles with beef gravy, salad, <u>excellent</u> turnip greens, and vanilla soft-serve ice cream with fudge and peanuts for dessert, all washed down with a giant Coca-Cola.

John and I exited the DFACS right at sundown and enjoyed the sunset view for a few minutes before the chill of the night crept in. We reported to Tent #1 for a 2030 roll call required in order to get on a C130 flight to BIAP. John and I are currently #4 and #5 on the list, so we should be able to get on the 1st flight available to Baghdad. Unfortunately, we will both have to be present for a 0230 roll call in order to keep our slot. Just like the military – hurry up and wait! Military personnel obviously have 1st priority for seats on military flights, so if we are unable to get seats on this flight, we will have to be present for a 0300 roll call for the embassy "Chrome" flight. The embassy usually holds 10 to 15 seats in reserve, and releases these at the last minute for journalists and other VIP's. During our last trip to Iraq in January 2007, we flew to Baghdad on an embassy flight. So, if experience repeats itself, we will probably board a C130 and take off for Baghdad at sunrise.

John and I then walked across camp to the internet café. After logging on to my email, I read MSgt. Ellerbrock's response. He advised that when John and I arrived in Fallujah, "We'll hang out, and go eat at McDonald's and Pizza Hut. Oh, by the way, tell me your pitch for Ramadi, so I can see if I can get it approved." This email was absolutely the moment of truth for the whole trip to Iraq with John. The Marine Corps and the Department of Defense do not generally look favorably on a Gold Star dad visiting the site in a combat zone where his son was killed. I had listed John Wroblewski as my assistant in order to get him back into Iraq. If the DOD or Ellerbrock follow strict military guidelines, we could very well be sent back to the States at this point, ending our 2nd attempt to get John to the site where J.T. and the other Marines were killed in 2004. If I fail to convince the USMC that our mission to perform the memorial service for the fallen of 2-4 Marines is important, then we will have failed again to accomplish what Martha Zoller had proposed in 2006 – the first visit by a Gold Star parent to the actual site in Iraq where their son or daughter died in combat. I had no choice but to lay it all out for Ellerbrock at this point. I explained about the 6 April 2004 ambush of 2-4 Marines, with the resulting loss of 12 men in that action, including John's son, 2nd Lt. J.T. Wroblewski, after his QRT responded to the initial ambush in the market. I asked him to help us accomplish what we had tried to do in January 2007 – to get John to the site to perform a simple memorial service for his son, J.T. and the fallen of 2/4 Marines.

After this email, I had time to post a single, short post to my blog before I ran out of internet time. "John and I made it safely to Kuwait. We are now waiting on a flight to Baghdad and on. More to follow tomorrow." Once we left the internet café, we just hung out in Tent 2 with the PAO's. Sgt. Taylor is going home in 72 hours, too. He felt sorry for me since I forgot my sunglasses, and gave me a nice set of Revision ballistic glasses, with a case containing both a smoke and clear lens.

1 Mar 2008 0230

John and I just attended the 0230 roll call for the military flight we had requested, and were told that no seats are available. We will stay here and wait for the 0300 roll call for the embassy flight. The internet is still down. I just paid for 1 hour of internet time, and used 8 minutes before it went down. I still have not received a response to my second email to MSgt. Ellerbrock requesting assistance in getting John to the ambush site in Ramadi, so I have no idea if we will even be allowed into Iraq, much less if we will be given an embed assignment with a Marine unit that can escort us to the site. I did receive an email response to my blog post from North Hall Middle School principal Barbara Jenkins, and Lana McGinnis, my nurse friend in Dothan, AL.

It is unbelievable to me how our troops do what they do, carrying all their gear with them throughout their travels and duties. Just my photography backpack, Kevlar helmet, and body armor weigh 60 pounds, and it is getting very tiresome carrying this gear everywhere I go, back and forth across this huge military camp. I wish I had a smaller, lighter laptop computer, but this is fine for now. At least I have a way to process images and post them to my blog while we are traveling. Now, it's time to sneak out and get a breath of fresh air before the 0300 roll call.

John and I were called at 0315 for the Chrome 21 embassy C130 flight to Baghdad. We will attend the final roll call with all our gear and board the bus at 0500. I am stoked! Later, John and I climbed on the bus, and although it wasn't long at all, it seemed like forever before we began the drive to the runway. We got to the tarmac, boarded the plane wearing our Kevlar, body armor, and ear plugs, carrying our personal gear in our backpacks. The plane didn't take off until 0700. It is approximately a 1 hour flight to Baghdad. I shot some photos and video while on-board the flight. The images look pretty good on the camera LCD, but the low light meant slow shutter speeds, so we'll see. At this point, I discovered the battery on my personal media device (PMD) is dead, so no Godsmack music for the flight. Sorry, Tony McKinney! My friend, Tony, had loaned me his Godsmack IV CD so that I would have some "inspirational" music to listen to while in Iraq, and now I was unable to do so on my 1st military flight of this trip. Apparently, I had forgotten to turn off the PMD, or it had been turned on while in my bag, running down the battery. Almost everyone on the plane is either reading or is asleep (mostly sleeping.) I just pray we can get a ride to the Central Press Information Center (CPIC) pretty quickly once we arrive at the Baghdad International Airport (BIAP.) On our last trip in 2007, we waited over 3 hours for a ride to Camp Victory before we boarded a Rhino (armored bus) for the trip to CPIC. All press must travel to CPIC to get their military press passes before being allowed to travel or embed with military units. Press passes are generally good for 1 year from date of issue, but mine has expired since we were last here 14 months ago.

Once we arrive at BIAP, we'll need to check in with the 3rd ID (3rd Infantry) to see if they can help, or ask CPIC or Catfish Air for assistance. Catfish Air is a military helicopter service that usually flies VIP's directly into LZ Washington inside the Green Zone. Otherwise, the standard procedure is to request a seat on the Rhino, which travels twice a day from BIAP to Camp Stryker. Once you arrive at Camp Stryker, you have to request another Rhino transport from Stryker to the Green Zone. This process normally takes 1 to 2 days, so it is rare to arrive at BIAP and get to CPIC inside the Green Zone on the same day. I am surprised to see several young girls and a reporter on the plane without Kevlar and IBA (body armor), which is normally required equipment in a combat zone. What's up with that?

A few minutes after our arrival at BIAP, Sgt. Kimber at Catfish Air hooked John and I up with a helo flight direct into LZ Washington. We lifted off at 1015 and enjoyed a very cool 15-minute flight over Baghdad, during which I shot over 100 images. The air is very hazy, but the images will look great after an adjustment in Photoshop. Many people are in the streets and markets below us, and many cars are on the roads. Some neighborhoods have satellite dishes and generators on almost every roof, especially near the Al Faw palace (Camp Victory), while other neighborhoods are flooded and look like deserted ghost towns. I shot some great images of the Crossed Swords monument, Al Faw palace, Tigris River, and the fertile fields and palm groves next to the river.

Once we landed at LZ Washington, I disembarked first and got a couple great shots of John as he exited the Blackhawk helicopter. After entering the office, I called CPIC and got Cpt. Hood on the phone (although I didn't realize it was him, and thought it was an enlisted man.) He said he didn't know anything about us, or our embed assignment. I replied that I had applied for an embed assignment in August 2007, and that Cpt. Signori and Specialist Deady were aware of our impending arrival and embed. I explained that beside all that, we were here, and desperately needed someone to come pick us up and transport us to CPIC so that we could renew our press credentials. John and I had to wait a couple hours for our ride since we had arrived a day ahead of schedule, having bypassed the Camp Stryker leg of the trip. U.S. Army personnel at CPIC are also very busy now that the local Iraqi businessmen use CPIC as a secure conference facility and meeting place, and they all need U.S.-issued identity cards, too. We rode to CPIC in an armored SUV, and once there, quickly dropped our gear on bunks to secure spots to sleep until we could get a helicopter flight to Fallujah.

Once we settled in, we met Sgt. Erik Burmeister, a military combat photographer and journalist at CPIC, E-5 Steven Hernandez, and Sgt. Rebekah Spencer, who cheerfully assisted us in completing the paperwork, photographs, and biometrics scans for our new military press identity cards. We also met Cpt. Hood, who graciously welcomed us to CPIC, and promised to assist us in any possible way. I then requested an escort to drive me to the outskirts of the Green Zone so I could interview Reverend Canon Andrew White about the ongoing genocide of Iraq's native Christians. Canon White is the only representative from the Church of England, sent to Iraq by the Archbishop of Canterbury, to minister to Baghdad's large Christian population. Cpt. Hood expressed grave concern about traveling to this area, and explained that although he could provide a driver in an armored SUV, he could not guarantee my safety once I was dropped off at Canon White's compound. I advised him that I accepted that risk, but this interview and the plight of the Christian's was also a major focus of my trip to Iraq, and that White and I had scheduled this interview weeks ago. Cpt. Hood spoke with his men, and assigned me a driver.

After donning my Kevlar helmet, body armor, and camera backpack, my driver and I climbed into the armored SUV and set out for the address I had been given. Although the drive was uneventful, it was very stressful for me. The driver advised me that my anxiety was probably needless, as security in the Green Zone had been good lately. After a short drive, we began looking for the compound, which was described as having a red gate.

After driving around the block 3 times and seeing no compounds with a red gate, I called Canon White. He laughed and said that he would go stand in the driveway and wait for me. After driving around again, I saw Canon White standing in front of a compound with a black gate. I climbed out of the SUV and bade my driver farewell. Canon White apologized for the confusion and told me that he didn't realize that the gate had been repainted black just yesterday. After close scrutiny by a very fierce looking Gurkha armed with an AK47, I was allowed to enter the compound. Canon White was a very gracious host, and offered me tea and biscuits upon entering his private quarters, which were served by Samir, his assistant and friend of many years. White and I made small talk for a few minutes until I began the interview.

Canon White told me that he loves his people (parishioners) so much, and would never leave, despite his illness and multiple assassination attempts, as well as having a price on his head. He said that when he looks out the window, "I look up and see the Glory of God. If you look down, you see the blast walls and barbwire." White then explained that the church he pastored has a membership of 1,500 Christians, and that he felt that he must be here for them, even though he has 2 sons back in the United Kingdom. He told me his sons were 10 and 13 years old, both with blue eyes, and that they were very cute.

The following is my article on my interview with Rev. Canon Andrew White:

If you are a Christian, and your son or daughter was killed because of your faith, what would you do? If your Muslim neighbor came to you and said, "Your daughters must convert to Islam and marry our sons, or we will kill your entire family," how would you respond? "Convert or die" is the message of choice for Islamic jihadists in Baghdad and Iraq who are working overtime to rid Iraq of "infidel" Christians.

On 1 March in Baghdad, I met a man who refuses to give up hope for the future of Iraq's Christians. This is in spite of repeated threats to his life; the kidnappings, extortion, torture, and murder of hundreds of his congregation; and the abductions and subsequent mutilations and murders of his friends who are the few remaining religious leaders in Iraq. On top of all these challenges, the church he pastors has little money to continue to feed and provide medical care and medicine to the 1500+ members of the church he pastors for the most despised, yet often poorest of Baghdad's citizens - Baghdad's Iraqi Christians. His name is Andrew White, and his official Anglican Church title is Canon White, but almost everyone he knows just calls him "Baba" ("Father" in Arabic.) He is a thoughtful, intelligent man who is a gentle giant, yet is terribly afflicted with the awful disease of multiple sclerosis.

Prior to my first military embed assignment in Iraq in January 2007, I was unaware that Christians even existed in Iraq. However, during that embed, I met a young Iraqi Christian woman, who had served as an interpreter for the U.S. for over three years, despite death threats to both she and her family. After she explained that there were over one million native Iraqi Christians, and that most were in great danger because of their faith, I was horrified and had to know more. After our meeting, I spent three months researching the story of the ongoing religious genocide of Iraqi's native Christians. Wait, did he say Iraq's native Christians? Yes, Iraq's Christian population pre-dated the Muslims in Iraq by almost seven hundred years!

Most of Iraq's native Christians call Nineveh (or the Nineveh Plains) their ancestral homeland. Those of you who have read the Old Testament will recall the story of Jonah and the whale in the Book of Jonah. Jonah was a reluctant prophet, who finally went to Nineveh where his testimony inspired the entire city of Nineveh to convert and repent. After Christ was crucified and was resurrected, Thomas the Apostle traveled to Nineveh, and his preaching facilitated many of the Ninehvites to convert to Christianity.

The Christian majority in Iraq gradually became a minority by the immigration of Muslims from the Ottoman Empire into Mesopotamia after 682 A.D. (Does this scenario sound somewhat vaguely familiar to my fellow Americans and our European allies?) The Christians and the Muslims had many differences, and much blood was spilled during the 1300+ years between then and now. The two religious groups generally settled into a tenuous co-existence throughout history, even through Saddam's reign. In fact, Saddam's right-hand man, Tariq Aziz, was a Christian (and an "enforced" acquaintance of Canon White's during the First Gulf War.) I have learned that many of Saddam's household staff and servants were Christians because he felt that he could trust them not to assassinate him!

After Saddam was deposed during the U.S. invasion in 2003, some Islamic jihadists took the opportunity to begin a program of discrimination, extortion (jizyah), and coercion to rid their country of the (infidel) Christians. This program eventually changed into more violent tactics such as kidnappings, torture, and murder. My friend's brother was kidnapped, and a ransom was paid, but he was shot in the leg before he was returned to his family. Unfortunately, this is one of the success stories.

Truth be known, much of Iraq's Christian population has either been forced into exile or has been murdered as a message to the remaining Christians. Of the estimated 1.4 million Iraqi Christians in Iraq in 2003, only an estimated 300,000 to 500,000 remain there. Many families are simply told, "Convert or die! Your daughters must convert to Islam and marry our sons or we will kill your entire family." There are eyewitness accounts of the kidnappings of priests who were subsequently decapitated, and had their bodies deposited at the door of their church with their heads placed upon their chests. Those young Christian men unwilling to convert or carry out suicide attacks are kidnapped and murdered. I heard accounts of both the crucifixion of a teenage boy, and of a boy who was roasted atop a pile of rice as a message to his parents and fellow Christians. Canon White related how he preaches at many funerals of the murdered, but the families seldom get the bodies back to bury them after they are kidnapped.

I know these stories are both shocking and horrifying. However, there are both recorded interviews and/or photographs to substantiate what I am telling you. As fellow Christians here in the U.S. and abroad, we must do something to help our fellow brothers and sisters in Christ who live in Iraqi, or are refugees before they are all driven out or killed.

In spite of their sufferings, there is hope. The U.S. Congress passed legislation in December of 2007 which acknowledged the plight of Iraq's Christians, and allotted $10 million to conduct a study to see what could be done to resolve the issue (too little and too late, but a good start.) My friend, Michael Youash, of www.democracyforiraq.com co-sponsored this bill, and I encourage everyone to view their website for updates. Canon White encouraged us to research this issue, but also to take action by writing our Congress and Senators, spreading the word to fellow Christians, and most importantly, to donate money for the direct support of these fellow Christians in Iraq, Syria, Jordan, and internally displaced persons within Iraq to those organizations listed below that will *make sure the aid is distributed properly.*

Please contact your fellow Christians within your congregation and have them write their Representatives and Senators within Congress, and pray for our fellow Christians in Iraq and elsewhere. You can find additional information by visiting www.rfcnet.org, www. csi-usa.org, and www.iraqdemocracyproject.org or by doing a Google search for Iraqi Christians, or by contacting me via a comment or email.

2 Mar 2008

Our day at CPIC was incredibly hectic, but productive. John and I crashed about 0430, and awakened at 0730 feeling pretty rested. Sunny weather was forecast, with a high of 85 degrees F – fine weather for the travel west that I hoped we would soon experience. After eating a banana nut muffin and drinking a Diet Coke, I used my laptop and a CPIC connection to check and send many emails, as well as updating my blog and posting 6 images. Shortly afterward, at 0815, the circus came to town.

Preceded by the cry, "Ahmadinejad is coming! Ahmadinejad is coming!" The Iraqi press trickled in at first, the volume quickly increasing until it became a veritable tidal wave! Of course, as has been my experience with the Iraqi press, they wanted coffee and chai, and began gobbling up every food item in sight. Strangely, many of the Iraqis would eat half an item only to place the uneaten half back on the tray or back in the package. After only a short while, the press living area was transformed into a sea of empty coffee cups, water bottles, and empty soda cans everywhere, with spilled sugar and coffee covering the coffeemaker table and surrounding floor. The press then rushed into the studio to set up for an interview with a deputy of Ahmadinejad, since the man himself is not allowed on U.S. soil (CPIC.) Since the interview was in Arabic, and didn't factor into my missions, I graciously declined to attend.

Soon after, I met Simon Klingert, a young independent photojournalist from Nuremberg, Germany, who was looking for transport to anyplace exciting that would yield a story. Klingert, who is also a student at university, is an accomplished combat photographer with an extensive curriculum vitae, and portfolio, including reporting from the Gaza Strip during the Uprising. We exchanged business cards, and I cautioned him to keep his head down so he would live to be a dad someday.

By lunchtime, the press living area was packed with Iraqi TV crews, who dumped their gear atop that of U.S. journalists, monopolizing all the chairs and both computers. Then, I recognized my old friend, Mohammad Fozi, an Iraqi I-TV cameraman that I had met at CPIC in January 2007. I was excited to see him, and told him so. We went outside to smoke, which was especially great as I was out of cigarettes, and Mohammad offered me a Gauloises, which I gratefully accepted. We sat, smoked, and talked for a while. From body language and the tones of their voices, I got the impression that his fellow Iraqis sitting near us were giving him a hard time for talking to an American photojournalist. Although I couldn't understand what they were saying, they were laughing in "that way." Mohammad responded to them in a somewhat defensive tone.

We then talked about our lives in general. Mohammad seemed a little depressed, but friendly. He told me he had a girlfriend who he wanted to marry. He had asked her parents for permission for them to marry, but they had said no, unless he built her a house. Mohammad explained that houses cost around $200,000 and he didn't have the money, so he was not going to be able to marry his sweetheart. He told me that he lived in an apartment upstairs from his parents in the Jihad neighborhood on the west side of Baghdad. Both his parents and girlfriend were fine with them living in the apartment, but that her parents would not accept this arrangement. Mohammad told me that he still worked for Eye Iraq TV as a TV cameraman, and also was working, filming a movie, doing final editing and adding music, although the project was running 2 months behind schedule. I promised him that I would pray for God's will about his wedding (Inshallah in Arabic), and that I would email him soon. Before Mohammad and I could finish our conversation, Specialist Deady came to call me to the phone to talk with Major Peters, the senior PAO with the Marine units in Fallujah.

I picked up the phone, and spoke with Maj. Peters, who wanted to discuss my request to get to Ramadi with John. MSgt. Ellerbrock had forwarded my email to him. I told Maj. Peters about the 6 April 2004 ambush of 2/4 Marines (which he already knew), and the positive aspect of performing a memorial service for the 12 men killed-in-action, including John's son, J.T. I reminded him that I had specifically requested a Ramadi embed in August 2007, and also emphasized the positive focus of my mission. I also mentioned the possibility of airing the story on Fox News via John Wroblewski's connection to Sean Hannity, and told Peters that we weren't spoiled, and would appreciate the chance to hang out with his cadre'. Maj. Peters sounded very excited about the mission, and stated that he would handle everything. Peters explained that we would get us on a helicopter flight to Fallujah at 2110 tonight, and that we would embed with the Marines in Fallujah from 2 to 5 Mar. Then, on 5 Mar, John and I would fly to Ramadi and embed with 2-8 Marines through 5 to 10 Mar. Then on 10 Mar, we could board a fixed-wing direct flight from Ar Ramadi to Kuwait. Major Peters said, "2/8 Marines own the city. We haven't had any violent incidents within the last 245 days in my AO (area of operations.)" He stated that the Marines in the Joint Security Station (JSS) near Route Gypsum and Ar Ramadi's famous Eastern entry arches on Route Michigan patrol the area where the 2004 ambush took place, and live within the city among the citizens. The Marines there had few facilities, so food is brought out to them every day. He explained that we could give them the coordinates and map of the ambush site that John had obtained, and that these Marines could escort us to the site to perform the memorial service. Peters said, "I will get the orders cut and make it happen!" After the call, I spoke with Cpt. Signori (who had been present during the phone call) and lavishly complimented his staff for their courteous, professional manner, and attention to detail. Signori thanked us for what we were trying to accomplish. I was relieved and ecstatic, and went to share the news with John that it looked like we would make it to Ramadi after all.

John and I slept from 1730 to 1915, when a reporter from the Atlanta Journal Constitution woke us up for chow – a weird meal of BBQ beef over rice, smoked pork, and boiled cauliflower. We ate it for the energy we would need for the trip ahead. There's an old saying about travel that you should never pass up food or a bathroom, and this is certainly true in a combat zone. We packed up for our 2110 "Showtime" (arrival for roll call to board the helicopter). Our 2110 deadline rolled by, and I asked Sgt. Spencer about it. She immediately got Sgt. Veitch to suit up and drive us to LZ Washington, the same LZ where our 2007 mission had been turned back. Sgt. Veitch got us to the LZ in an armored SUV, presented our orders, and we were listed as CF 233, which was written on the back of our right hands in permanent marker. I had just enough time to make a quick phone call home to speak to my children before John and I boarded a Marine CH-53 helicopter for our trip to Fallujah.

At 2205, we got the word to line up outside for a briefing, while wearing our full body armor and Kevlar helmet, carrying all our gear. Two CH-47 twin-rotor Chinook helicopters landed, and we were escorted to the lead helo. We climbed up the tail ramp, sat down, and buckled up. I turned on my PMD (personal media device) and jammed to some Godsmack, compliments of my friend, Tony McKinney. Our helo took off within just a couple of minutes. The ride was very loud, but smoother than I expected as compared to a Blackhawk. I also didn't realize that the tail ramp remains partially open during flight. Jutting out of both the port and starboard sides of the helicopter were Browning M2 .50 caliber machine guns that are each manned by a gunner wearing NOD's (night vision optical device), who scan the rooftops, road and fields below for any threats. As we flew over Baghdad, I was amazed by how many areas had power – large swaths lit up just like back home. I had heard some complaints from the Iraqis I spoke with yesterday, who said that power supplies now were less than they were before the war. But, even at 2320, lights at numerous houses, facilities, and streetlights dappled the dark landscape below. It was a great night for flying as it is a new moon with very little illumination to silhouette us to any possible ground fire. There were a couple tense moments when the gunners began swiveling their guns toward possible ground targets, but thankfully, we received no incoming fire. After a couple 5-minute stops, and another 5-minute stop for fuel, we landed at the LZ near Camp Fallujah.

John and I unbuckled, shouldered our gear, and marched down the ramp. We spied a line of glowing chemlights on the ground that led to the shack that served as the LZ gate. Following these, we entered the shack and were met by ebony-skinned Ugandan security guards armed with folding-stock AK-47's. The lead guard asked for our press identification cards and orders, and then refused to return our IDs. During our insistence that they be returned, a USMC Master Sergeant Willie Ellerbrock entered the shack. I began to feel a strange uneasiness, as MSgt. Ellerbrock looked strangely familiar.

This uneasy feeling turned to panic when Ellerbrock turned to me and said, "You! I know you! We met at LZ Washington a year or so ago." With a sinking feeling and dread in my heart, I realized that MSgt. Ellerbrock was the Marine NCO whose Marine PFC had been commandeered by US Army LtCol Quinn had commandeered back on our last embed in January 2007. Since LtCol Quinn didn't ask MSgt. Ellerbrock for permission prior to taking Private Eberle with John to the CASH (Combat Support Hospital), MSgt Ellerbrock was quite upset at finding his Marine missing-in-action while waiting for a helicopter flight into combat in Ramadi. It was hard to believe, but out of 155,000 U.S. personnel in Iraq, the chances that my PAO would be the same NCO are literally 1 in 155,000. Now, I guess my number is up as I'm in Fallujah, and the man who literally holds the success or failure of my mission and stories is my PAO, MSgt. Ellerbrock who is standing in front of me with a rather stern look on his face. I fell all over myself apologizing and explained that I was just the hired photographer in January 2007, and hoped he didn't hold that against me personally. Ellerbrock's face broke into a wry grin and he said, "Hey, it wasn't your fault." He shook my hand and welcomed us to Fallujah. Breathing a sigh of relief, I heard Ellerbrock ask us if we had our gear and IDs. I advised that we had our gear, but that the Ugandans had kept our IDs pending clearance from their Australian Defense Force boss. The Ugandans refused to hand over our IDs despite a vigorous ass-chewing from Ellerbrock, and insisted that he speak with the Sabre Security supervisor, whom they had called.

Ellerbrock met with the Sabre supervisor, a tall thin Australian, when he arrived shortly thereafter. The Aussie explained that the reason for the problem is that no one had informed his people anything about our visit. Ellerbrock advised him that our embed had been in the works for months, our photos were on our security IDs, and that our orders were signed. He explained that we were American citizens and photojournalists, and he would not stand for us to be treated this way. After getting our IDs back, Ellerbrock got the name of the supervisor who'd dropped the ball, and we left.

Ellerbrock drove us a short distance to Camp Fallujah where he gave us 2 keys to a very nice press trailer, complete with air conditioning, 2 beds with clean linen, a stocked mini-fridge, high-speed internet connection, desk, power converter, snacks and a table. Quite posh compared to what I'd previously experienced in Kuwait and Iraq! I worked on emails, called my wife ("I have a migraine and don't feel like talking!"), and posted to my blog until I crashed at 0430 on 3 Mar.

John and I woke up at 0600 when the alarm clock went off, but quickly fell back asleep after one of us hit the snooze button. We finally awakened at 0815, and Ellerbrock came to get us at 0845. We all went to the DFACS (chow hall) and had a delicious breakfast. Afterward, the MSgt took us to the USMC PAO office where we met SSgt Amy Forsythe and MSgt Evans. SSgt Forsythe was not only an accomplished combat photojournalist, she was one of the best looking Marines I had ever seen – petite, blond, with a beautiful smile. Forsythe informed us that she was in charge of TV relations and media PR, and told us that she could coordinate a SAT-link to do a TV broadcast of the memorial. She explained that she could mix my still images with the video for a TV broadcast. We discussed my plan to get John Wroblewski to the 2004 ambush site to do the memorial service, and what Marine assistance would be necessary to accomplish the mission. After we finished brain-storming, Ellerbrock sent us back to the trailer to rest for a couple hours, and advised that he would send Lt. Vickers to get us when it was time to take the next step. Once back at our trailer, John and I checked our email, and I updated the blog, being sure to follow military operational security (OPSEC) procedures to avoid compromising the security of the upcoming mission. Then, I called Lt. Vickers and left him a message. Shortly afterward, Lt. Vickers returned and we walked to RCT (Regimental Command Team), where I was to interview Major Matt Reid, COC Commander. Along the way, Lt. Vickers explained that the Sisters of Fallujah (female Iraqi police officers) did a great deal of dangerous police work here in Fallujah, searching women and houses when necessary, in order to better provide security in a society that forbids most male-female contact unless the parties are related. Once I arrived at RCT, I met and interviewed Maj. Reid. During our introductions, we learned that Maj. Reid was a close friend of Major Paulus, who had been the CACO (Casualty Assistant Officer) for the Wroblewski family following 2ndLt J.T. Wroblewski's death. Reid and Paulus had been Lieutenant's together many years ago.

USMC Major Matthew Reid explained that since October of 2007, the Marines AO had had no direct fire attacks, whereas there had been 100's in the spring of 2007. The recent focus in Fallujah had the Army's 1st Brigade of the 3rd Infantry Division pull out of the province, placing the RCT in charge of both Fallujah and Ramadi. Reid said, "The challenges are tribal divisions and strife, AQI (Al Qaeda in Iraq), and an occasional suicide bomber. The goals are to put the IP (Iraqi Police) in charge of security for both areas (estimated 10,000 IP in between 2 cities with a combined population of 1 million), while the PSF (Provisional Security Forces – local neighborhood watches) either convert to IP or transition back to civilian life. So far, the strategy has been fairly successful." Maj. Reid explained the strategy had helped build bonds between the provincial and city governments and the local sheiks. Also, the linkage had achieved a balance between the resources of Fallujah and Ramadi, as well as forming ties between the Anbar provincial government and the central government in Baghdad in a "National to Provincial" strategy. Maj. Reid advised that local power was all about "wasta" (local power or clout), and that all the tribal sheiks vied for this. When questioned about Marine resources, Maj. Reid said, "We're good to go!" He explained that he had no problems at all with the amount of military personnel, equipment, or supplies he needed to accomplish his mission in Anbar.

Following the interview, John and I walked back to our trailer, gathered laundry, and hiked over to the PX (Post Exchange store) to buy mesh laundry bags. We then headed to the laundry with our dirty clothes. Usually laundry takes 48 hours, but the attendants graciously agreed to do it in 24 hours since we were headed out soon. We then walked back to the PX, where I bought 6 non-alcohol Becks beers, a 12 pack of Diet Pepsi, some AAA batteries for my headlamp, and some bags of candy. By this time, John was getting hungry, so I dropped off my goodies at the trailer before we headed over to DFACS to eat.

After dinner, I went back to the PAO bunker to meet with Lt. Vickers and SSgt. Cazee to discuss tomorrow's plan. We would link up with SSgt. Cazee and RCT 1 at 0530 the next morning, and drive 1.5 hours to Al Faris to meet the mayor, inspect the mayor's prospective new jail, visit the new irrigation pumping facility and make a payment to the contractor for the finished work. Then, we would go the local market and walk around and talk to the local Iraqi shopkeepers and villagers.

After I returned to the press trailer, I met Tony Perry with the LA Times, who said he had been in Iraq for about a month, having just returned from Al Sahd. I was careful not to reveal too much information about my mission for fear that he would try to report the visit before it occurred, possibly compromising our security. I then sat down, finished the last 2 Beck's, typed and posted a blog entry, checked email again, read a little, and finally went to bed.

4 Mar 2008

0530 – Clear, 58 degrees

John and I awakened at 0530 and gathered our gear, just in time to hear Lt. Vickers knock on the door at 0540. We climbed into a white SUV and quickly drove to Inchon, where we linked up with SSgt. Cazee and the men of RCT 1. SSgt. Cazee explained that we would be driving in convoy about 25 km to the town of Faris. Once there, we would meet with the mayor, make payment to several contractors for completed work, inspect a building in which to locate the new jail, view a reconstructed irrigation pump facility, and possibly walk through the marketplace. Cazee then told me that Faris didn't see too many American troops, but that the town supported the Anbar Awakening, and helped force AQI out of their area. The Anbar Awakening was a sheik-led movement in Anbar to take back Anbar from AQI and foreign fighters in order to make their towns safe from VBIED (vehicle borne improvised explosive devices), suicide bombers, and the resulting combat with American troops. It seems the locals got tired of having their wives and children blown up in the market place when foreign fighters would detonate car bombs when the Americans passed through the market on patrol.

The NCOs of RCT1, along with John and I, waited around for a couple hours while the rest of the men assembled, readied their gear, mounted their weapons, loaded coolers and ammo, and prepped the Humvees for the trip. John and I climbed into an armored Humvee together, and our convoy rolled out at 0730. Since the Humvees are only authorized to travel at a maximum of 30 mph, it took quite a while, but we entered the outskirts of Fallujah, once known as "The Most Dangerous City in the World." John and I were teamed with a great Humvee crew, consisting of

Sam Michaels, VC (vehicle commander); .50 caliber turret gunner Crosby "CC" Carlsen; and driver Christian Martinez. We all spent the long ride to Faris getting to know one another. These Marine warriors were unsure of us at first, but quickly warmed to us once they heard about the 2004 ambush of 2/4 and John's son, 2ndLt Wroblewski, and our mission to get to Ramadi to do a memorial service for the fallen of 2/4 Marines.

Once we arrived in Faris, the line of Humvees drove to the end of the market place and turned around pointing back in the direction we came. We dismounted from the vehicles after the Marines fanned out to provide security for the convoy and each other. SSgt. Cazee invited John and I inside to attend the meeting with the mayor and the contractors. The meeting took place inside a mud brick building with a beautiful carpet covering the floor, and we sat in chairs along the walls. After the meeting was over, we were invited to eat a traditional lunch of roasted lamb and rice, which I graciously declined. I wanted to get back out on the street to take photographs. Before I could do that, John mentioned that he had to use the restroom, and headed around the corner. One of the Iraqis directed him to the bathroom, which consisted of a hole in the floor of one of the rooms. John was incredulous, and insisted that he'd rather piss against the wall the next time. I had to laugh at him.

I spent the first few minutes outside taking photographs of the children as they walked home from school, the boys and girls separating themselves into small like-gender groups that walked together, many of them holding hands. After taking quite a few shots, I broke out the Flix video camera again to shoot some video. One group of students came up to me and some of the Marines and chanted, "Choc-o-lot, Meestah?" I pulled out a Wrigley's Plen-T-Pak of peppermint gum, which I handed out in about 10 seconds. I shot video of the discarded foil gum wrappers tumbling down the street in the breeze. Based on the amount of trash everywhere in Iraq, littering is not a big deal here. SSgt. Cazee asked me if he could take my photo, and I handed him my Nikon. He snapped a couple images of me standing in the street with the local mosque in the background. By now, the wind was picking up and the sky was beginning to turn a weird shade of pink as the wind picked up the ultra-fine, red "Mars dust" that makes up the surrounding desert. The pink tinge of the sky lent a surreal look to the town and sky. Michaels and I walked through the marketplace, with me shooting images all along the way of the colorful shops and shopkeepers as they looked on us curiously. Inside one shop, I noticed packs of local Miami cigarettes for sale. The young man in front of the shop was smoking one of these cigarettes, and I walked up to him and smiled as I pulled out a pack of Marlboro menthol cigarettes. I gestured that I would like to trade 1 of my cigarettes for 1 of his. He looked at me thoughtfully, and then pulled out 3 of his cigarettes to exchange for 1 of mine. I shook my head no, and made hand gestures that indicated I wanted to swap even, 1 for 1. He shook his head, grinned, and we exchanged cigarettes. He had a big smile on his face like he thought he had pulled one over on me. I didn't care, as I just wanted to try one. I lit up and drew in the smoke from the strong, somewhat stale cigarette, and asked him how much for a pack. He held up 3 fingers and said, "Three dollars." I shook my head no, and offered him $2, which he quickly accepted.

About this time, the mayor and the Marines exited the government building, having finished lunch. The Marines formed a phalanx around the body-armor clad mayor and we walked together down the street to the building designated as a possible new jail. I ran ahead of the group and photographed them as they walked down the street with John Wroblewski. Once at the building, I entered and began extensive photography of the place. The building had bars on the windows, and lots of Arabic graffiti inside, but seemed relatively solid. The Army officers who had handled the payments to the contractors in the mayor's office walked around with the mayor, inspecting the many rooms. Later, I was very glad that I took so many pictures, as one of the officers asked me if I would share my images with him to send back to his superiors, which I gladly did. After the grand tour, we climbed back in the Humvees with the Marines, and headed off to view the reconstructed irrigation facility.

Strangely, the quickest way to the irrigation facility was up a side road near the center of Faris that had just been sprayed with tar so that it could be paved. The Humvees absolutely destroyed this freshly sprayed tar, churning it up and flinging it every which way with their huge all-terrain tires. I silently wondered how much money the U.S. had paid some contractor to spray the tar that was now ruined. After a short ride along the Tigris River, we reached the irrigation facility and dismounted. The Marines fanned out with M4's at the ready, pointed at the tree lines lining the surrounding fields, and scanned for possible threats. After much discussion and gesturing, someone flipped a switch and the pumps roared to life, flooding the irrigation ditches with water drawn from the nearby Tigris. Everyone in attendance shook hands and seemed very pleased. After shooting some more images, our convoy boarded the Humvees again and headed for our new home, Camp Fallujah. Along the way, "CC" Crosby, the gunner, asked for my video camera to film the highway and bullet-riddled signs with Fallujah in both Arabic and English. These 2 short videos did a great job of capturing the locals as they sat and waited as our convoy passed, as well as the blast walls and barbed wire along parts of the route. By now, it was almost dusk, and the mild sandstorm had turned the skies to something one might see on Mars.

Once back at Camp Fallujah, John and I were dropped off at our trailer, where I spent the next few hours backing up my images, checking email, and posting to my blog. Tomorrow is going to be a big day, as MSgt Ellerbrock informed me that we were to meet USMC Maj. Gen. John F. Kelly for lunch tomorrow to discuss our mission to get to Ramadi to perform a memorial service for the fallen of 2/4 Marines that were killed in the 6 April 2004 ambush in the market place along Route Gypsum on the east side of town.

5 Mar 2008

Unbelievably, John and I slept until 1100! We got up, showered, and then walked to DFACS to have lunch with Major General Kelly, General Mills, and LtCol Hughes at 1230. John and I met them outside the mess hall, and I shot several images of the Marines greeting John and shaking his hand as introductions were made. We got in line at the mess hall, got our food, and carried our trays over to an empty table. The conversation was pleasant and informal. Kelly explained that if we convoyed with him tomorrow, he could definitely get us to the ambush site based on the coordinates John had obtained. We planned to meet at 0820 the next day at the PAO, and parted ways. Both John and I are very excited, never having gotten this close to our goal during our last trip in 2007. It seemed like we might actually make it to the ambush site this time.

After lunch, we headed back to our trailer F3, checked email, and I posted an update to my blog to let everyone back home know that we were okay. I took a nap at 1500, and awoke at 1700 to check email again. I sent my wife Melissa an email to let her know that I was pretty disappointed in her lack of support during this embed. Despite numerous attempts to contact her via phone or email, she had been either uncommunicative or unavailable. After venting my frustration, John and I headed to DFACS for chow, which we brought back to the trailer. We stopped by the PAO office, where Sgt. Evans handed me a frosty Becks N.A. beer, which was quite tasty, but left me wanting more. I walked to the PX and picked up a six-pack of Becks N.A. to drink back at the trailer.

Once back at our home-away-from home, I sat outside while John wrote some emails, and read some of Rev. Canon White's book, "Iraq: Searching For Hope", about the persecution of Iraq's Christians by the Muslims. Afterward, I posted 19 images to my blog and crashed at 0350. Today's the big day, and only exhaustion allowed me to sleep before our planned convoy to the ambush site in east Ramadi.

6 Mar 2008

John and I awakened at 0600 to a beautiful dawn with a temperature of 60 degrees F. I am mostly packed, did a last check of email, and added a counter to my blog. We headed to the DFACS, where I had a big breakfast of an omelet, one burnt hash brown, breakfast burrito (disgusting!), 2 V-8's, and 2 Pepsis. John and I are to meet MajGen Kelly at the PAO office at 0820 to convoy to Ramadi. After a final gear check, and cleaning up our trailer, we headed to the PAO office, where I met USMC Cpl. Angel (combat cameraman) who is to videotape the memorial service.

At 0800, we walked outside into the bright sunshine, and linked up with MajGen Kelly, PSD Co., and 2/8 Marines that will be our escort during the convoy. John was assigned to Vehicle #3 with the General, and I was placed in Vehicle #7 with SSgt. Smith. The convoy formed up, and then Kelly gave a briefing on the mission. He went over emergency procedures, including a quick class on combat tourniquets in case someone lost a limb or was wounded during the mission. We all made sure that we had a handkerchief close-at-hand, so no time would be wasted looking for something to use for a tourniquet. I tied my camo handkerchief to the left shoulder strap of my body armor, just in case. This short class really made the danger we were facing very real, and I prayed silently that we would all make it there and back without incident.

We climbed into our armored Humvees, and headed west toward Ramadi at 0905. Maj. Gen. Kelly had explained that speed was the key to our security today, so the 30-mile trip went much faster than our 30 km/hour journey to Faris on 3 Mar. Our convoy arrived at the Government office in Ar Ramadi at 1100. With a mob of people, Iraqi Army and Iraqi Police outside, the white government building was surrounded by a 6-foot-high masonry fence topped with wrought-iron fencing. The building is flanked on either side by buildings heavily damaged by combat, and while clean, the office is not overly impressive. What is disturbing to me is that when our convoy of 7 Humvees pulls into the small courtyard, the heavy iron gates clang shut, essentially trapping us inside with no room to maneuver if we are attacked.

At this time, almost everyone dismounted from the Humvees, sheds their IBA (Interceptor body armor), and forms around the governor at the front of the building. I got out and asked Cpl. Angel if he thought I needed to do anything. Angel said no, and headed inside with John. I had already been told during the pre-briefing by the PAO staff that they did not really want me to take photos inside the building for the security of the Iraqis in attendance, and that of the facility. Somewhat relieved, I climbed back into the Humvee with the .50 turret gunner providing security, and began to watch the "show" outside. Our contingent headed inside, and shut the mirrored, bullet-proof glass door behind them. By now, a crowd had formed at the entrance gates, and the Iraqi security personnel began to admit a few business men and black burkha-clad women after thoroughly checking their id cards and searching both them and their belongings. Many of the women had children with them, and some of the kids began to climb the iron gates while they waited to be admitted.

The Humvee crew and I passed the time telling jokes and smoking, and I shared some candy with them that I had bought at the PX. One man outside the fence drew both my and the gunner's attention. He just stood there, staring intently at the front door, while ignoring all the people around him. I asked the gunner if he thought the man's behavior seemed strange, and then noticed that the gunner had his .50 machine gun pointed in the man's general direction. He said, "Yeah, but I'm watching him." I snapped a few photos of him for the intelligence officer traveling with us, just in case. The meeting lasted about an hour, and a call came in over the vehicle radio to start up the trucks as everyone was headed back out. As soon as the trucks fired up, the suspicious man outside the fence immediately got on his cell phone and made a call. I said, "Uh oh!" The gunner replied, "Yeah, that doesn't look good." During a 2007 anti-terrorism class, I had been taught that terrorists often got tunnel vision before carrying out an attack, and were often oblivious to everything around them except their intended target. This man's behavior certainly seemed to match that profile. After a short delay, the Marines and John walked out of the government building and climbed back in the Humvees.

The gates opened, and after much backing and maneuvering, our convoy headed out, only to park along the front of the compound fence. General Kelly and several other officers walked along the convoy, exchanging greetings and information along the way. After another 30 minutes, our convoy headed back east, toward the arches on Route Michigan (Hwy 10) that marked the entrance to Ar Ramadi, and our ultimate destination, the 6 April 2004 ambush site in the market on Route Gypsum. Our goal of performing a memorial service for the 12 fallen men of 2/4 Marines was almost within reach.

In a few minutes, our convoy reached the arches and turned left (north) onto Route Gypsum. The narrow street through the market place seemed almost too small for our Humvees to pass through, but we slowly picked our way up the street, dodging Iraqi passenger vehicles, bicycles, and pedestrians along the way. Along the way, I was filled with a mixture of excitement and fear, praying that the Marines and John would remain safe and that nothing would prevent us from reaching our goal. One good sign was that the market was full of people shopping. One journalist that had written about the 2004 ambush had stated that on the day of the 6 April ambush, insurgents had told the shopkeepers to close up because, "Today, we are going to kill Americans." Suddenly, the convoy stopped, and a call came over the radio. SSgt. Smith turned around and said, "We're here. They need you up front." I climbed out of the Humvee, and one of the Marines grabbed my camera bag out of the rear cargo area and handed it to me.

I hurriedly walked toward the front of the convoy, still praying that no shots would ring out, and that John would be able to reach his goal of seeing the spot where J.T. fell in combat, and that we all would remain safe during the memorial service. It all seemed very surreal as I turned left into a narrow alleyway and joined the Marines ahead of me. As we walked down the alley, I saw the Marines fan out along the cinder block walls lining the narrow street, providing security for us while making a minimal target for any possible snipers. MajGen Kelly urged John toward the left side of the alley as they walked along near the front of the PSD Company Marines, reminding him that the middle of the alley was not a safe place to stand or walk. Meanwhile, I hurried along, scanning the surrounding rooftops and open windows for any signs of danger, snapping images along the way. Walking ahead of John and MajGen Kelly, a Marine held a GPS unit, searching for the coordinates that John had provided. Reading the coordinates, the Marine directed John forward and then back, directing him to the location where his son, 2ⁿᵈLt J.T. Wroblewski was mortally wounded.

Once at this sacred spot, John withdrew a red T-shirt printed by another of his son's that had these simple words printed on the back:

In Honor and Memory of

Lt. J.T. Wroblewski

4-16-78   4-6-04

Operation Iraqi Freedom

"Earned – Never Given"

John tenderly draped the shirt atop the cinder block wall so that it hung down like a flag, and then pulled out a small piece of paper, and began the memorial service for the 12 men that had fallen during the 6 April 2004 ambush. His voice tense with emotion, John said, "Eleven Marines perished in and around this area, along with a Navy Corpsman. If I could read their names in their honor, all of 2/4, in that deployment, lost 34 heroes: LCpl Benjamin Carman, LCpl Marcus Cherry, PFC Christopher Cobb, LCpl Kyle Crowley, PFC Deryk Hallal, PFC Ryan Jerabek, PFC Moises Langhorst, LCpl Travis Layfield, HM3 Fernandez Mendez, LCpl Anthony Roberts, SSgt Allan Walker, and my hero, 2ndLt John Thomas Wroblewski."

John then asked for a moment of silence to honor these unforgotten heroes. We bowed our heads together, and I offered a prayer for John and his family, the fallen men and their families who had given everything to our country, and for the continued safety of the Marines that surrounded us. Afterward, John asked the Marines to offer a loud "Ooorah!" in honor of the memory of these men. Their emotional cheer rang throughout the alley, echoing off the walls of the houses behind the cinder block walls, finally fading away into the palm groves that surrounded us. In a voice cracking with emotion, John thanked Major General Kelly and our Marine escorts for their assistance in fulfilling his dream of visiting this now hallowed spot. Some of the Marines visibly struggled to control their own emotions, yet never relaxed their vigilance in their goal to provide security for the ceremony.

I truly felt J.T.'s presence there with us that day, especially when I observed the following - As John knelt to collect a handful of soil to bring home with him, the red T-shirt with J.T.'s name that John had hung on the wall above him "flew" off the wall, landing at John's feet. This incredible "coincidence" is even more amazing in light of the fact that I felt no gust of wind at that moment! I have posted a photograph of the shirt as it "flew" over John, and another as it landed at his feet to show you what I saw. Einstein said, "Coincidence is God's way of remaining anonymous." Was J.T. there with us that day? I prefer to believe that he was, and I know he was proud of his father for coming all this way to honor him and his Marine brothers that fell in combat.

7 March 2008

I finally got the blog updated with the story of the memorial service and posted 22 images this morning at 1100 along with a short write-up about the service. I really feel sorry for the men that are stationed here. Not only are they in a combat zone under less-than-ideal conditions, internet connectivity is spotty at best, and keeping in touch with their loved ones back home is tough. Unless you are stationed at a large camp like Ali Al Saleem in Kuwait with large internet trailers, there is almost always a wait just to get online. Prepaid phone cards sent by families and supporters of the troops are a much-valued commodity, as there is nothing like hearing the voice of a loved one while you are so far away from home. I have a Verizon global phone with me on this trip, but rarely find my wife receptive to my calls, despite the danger that I've placed myself in. This lack of support, empathy, and unfaithful "significant others" is something that the troops mention frequently. I cannot tell you how many men have told me horror stories about getting a "Dear John" email during a 9- or 12-month deployment, complaining that they are tired of waiting. Americans have truly forgotten their history. It wasn't uncommon for troops during WWII to be deployed for 4 years before coming home after the war, and I think that women on the Home Front had a lot more patience back then.

John and I got a surprise late this afternoon. I was napping at 1630 when a knock on the door of our trailer awakened me. John said, "They're here to take us out." I replied, "Who?" John said, "2/8 Weapons Company to take us to JSS Karama." We quickly packed, much to my later detriment. I ended up forgetting my bedding, laptop, charging cord for my phone, and my converter, as I had let John borrow it. John and I threw our gear into the MRAP (mine –resistant armored personnel carrier) and rode along with 1st Sgt. Rowe. Rowe advised us that we were to be out for a couple of days with Weapons Company at JSS (joint security station), and then on to Echo Company through Sunday. 1st Sgt. Rowe also advised me that they wanted John and I to spend some time at an Iraqi police checkpoint with USMC Cpl. Masters. When Rowe mentioned the IP checkpoint, I immediately thought, "Fuck that! No way am I going to put myself in a position to be blown up by a VBIED at an IP checkpoint! Not to mention, I don't trust IP as far as I can throw them."

Rowe explained that John and I would be doing both mounted and dismounted patrols tomorrow afternoon after visiting a wounded Iraqi girl with Dr. Fahid. USMC Capt. Martin and Weapons Company will be providing security for the Iraqi doctor who will be treating the 2-year-old girl, who was wounded by a stray round, possibly celebratory gunfire. Following the visit to the girl's home, we would be visiting a faux Sheikh that has a lot of local "wasta" (power) who was hosting a traditional luncheon for us. I can't wait to see how John reacts to local hospitality and dining customs, but will give him a couple pointers so he doesn't embarrass himself or us (i.e.: sitting Indian-style on the floor without showing your host the soles of your feet, and accepting food and drink and eating with the right hand only.) I am looking forward to the thought of some "home cooking", even if it's Iraqi cuisine, but it also makes me a little homesick. Right now, I miss my children, dry socks and boots, beer, and TV the most. I can only imagine how these Marines feel that are deployed here for 7 to 12 months – an almost unimaginable sacrifice to most Americans that have never been here.

JSS Karama (Joint Security Station) is very different from Camp Ramadi! No running water except for showers, 1 hot meal per day, no phone service, and limited internet (must sign-up for appointment to use 4 PC's available to 110 Marines.) The 2/8 Marines here are very cool, though. At the first meeting, Captain Martin seems to be a typical officer – very organized, but he likes to hear himself talk. However, after spending some time with him, I've come to understand that he cares deeply about his men, their security, and his mission. This unit has made amazing progress, though!

I am so glad that my daughter Ashley is not here. These young Marines are in peak physical condition, very confident, and most of them are extremely good looking. One Marine in particular, James Pizzillo from NJ, is definitely a young man that could break any woman's heart. He told me that his current girlfriend attends Rutgers (the same university from which J.T. Wroblewski graduated), and she works in public relations for fashion designer Hugo Boss. I wonder how she is handling his deployment in Iraq, but I am reluctant to raise this issue with him.

8 Mar 2008

JSS Karbala

0730 – Clear, 60 degrees F, Wind S 3

I awakened at 0600, and have been noshing on Famous Amos Chocolate Chip Cookies while drinking bottled water flavored with Wyler's single-serve flavor packets. Being at the JSS (Joint Security Station) is very different from any camp that John and I have experienced so far. The JSS is established around a huge Iraqi 2 story house that some people might consider a mansion. It has large rooms with high ceilings, large window openings without any glass or curtains, and is surrounded by high Hesco bastion walls. Hesco bastions are 6'x6' square wire baskets lined with plastic that are filled with sand, and then stacked like building blocks. They form bullet and blast-resistant perimeter walls around the compound to provide some protection for the Marines and IA (Iraqi Army) troops stationed together inside the JSS. The Marines conduct joint patrols and missions with the IA to give them additional training and support until they can operate on their own. Sandbagged machine gun nests/observation posts draped with camo netting and roofed with plywood and more sandbags are set atop the corners and midpoints of the walls to provide security for the compound. These posts are manned 24 hours a day by both Marines and IA.

The room John and I are placed in an exterior room that was probably a closet in its former life. The window opening is closed off by a sheet of plywood screwed to the wall, and the walls themselves are pockmarked with bullet holes. There are 2 beds with thin mattresses set in an L-shape in one corner of the room, with a bag of polyester fleece blankets to use as a sheet and cover. Once we are in the room, we toss our gear on the floor and flop down on the beds, exhausted. Also in the room are large care packages sent to the Marines from folks back home. These boxes contain all manner of things from batteries, phone cards, snacks, hand-written cards, to DVD movies. These care boxes are prized by the Marines, but they graciously offer John and I anything within them, explaining that sometimes they have more snacks than they can eat before they get stale. One of these boxes is the source of the Famous Amos cookies I was eating earlier. Both the Marines and I thank you, U.S.A., for your support of the troops.

The rest of the JSS is equally Spartan. There is no running water except for that fed by tanks to the 2 toilets and 2 showers. The gray water flows by pipe into a huge plastic tank that is buried in the sand. Unfortunately for the Marines, the tank is now overflowing, and the head (latrine) must be reached by walking atop boards placed over the gray water. The Marines have 1 hot meal brought to them per day, and have freezers that contain microwaveable food.

I spoke with SSgt. Lalchan at JSS Karama for a long time this morning. Originally from Trinidad, he has been a Marine for 11 years, and now lives in southern New Jersey. He joined the Marines in 1996, and both he and his wife discussed and agreed that he should re-up (re-enlist) in

1999. Lalchan told me that he originally joined the Marines for job security, and re-upped because of that, too. He had been promoted to Corporal in 1999, and Sergeant in 2000, and said that he plans on doing his 20 years before he retires at the age of 46. Sgt. Lalchan told me that he had served as a recruiter for 3 years, during which time he had a quota of 2 recruits per month. He said that he still maintains contact with almost all the families of his enlistees.

Lalchan explained that the situation in Iraq is very different now for a couple of reasons;

1) Local Iraqis are tired of the violence, especially the outside influence and brutality of AQI (Al Qaeda in Iraq), so they decided to take back their city. The citizens formed the Awakening Groups aka Sons of Iraq to provide security and fight back against foreign insurgents. Later, concerned women formed the Daughters of Iraq, an all-female security force that searches all burkha-clad females at checkpoints, and are also involved in house searches to deal with any females in the household. Per Sgt. Lalchan, the Iraqis also realized that the Marines were here to stay and backed them up in their endeavor. The locals turn in outsiders (as do IP because they are also locals) because they don't want anything to happen to the Marines. The Iraqis have come to realize that the Marines will go back to killing and destroying anything and everything if they are attacked or if any Marines are killed.

2) The Iraqis just want to be able to shop, travel, and live normal lives in peace, without fear of violence. Route Michigan, the main road from Baghdad through Ramadi and west to Syria, is open from 0700 to 1700 every day, interspersed only with IP checkpoints, manned by IP with USMC "advisors." Military-only lanes exist, but unless they have business in the city, the military doesn't enter the city after 0900 to avoid causing traffic jams. The locals are rebuilding, clearing the rubble and trash, and many have jobs paid for by the U.S. government to put money in their pockets to keep them from being out of a job, and therefore susceptible to AQI offers of money to plant IEDs or launch attacks. Sgt. Lalchan explained that the shops are open, with lots of vehicle and foot traffic.

The IP checks everyone and every vehicle coming through the checkpoints with fingerprints or retinal scans. The Marines note the license number, make, model, and color of every vehicle, including the time in and out, and issue a pass good for 1 entry into the city. The IP's do vehicle searches, and the Marines supervise individual searches using metal detecting wands. The female IP's search the females to avoid any cultural offenses. The IP have also set up a "trucks-only lane" and make every 5th truck dump their entire load (the order of the truck required to dump their load changes every day.) Per SSgt. Lalchan, a Turkish contractor performs this job for a purported price of $1 million for a 6-month contract. It is definitely not a contract I would want to have, as the chances of finding a VBIED (vehicle-borne improvised explosive device) are substantial, and if driven by a suicide bomber, there would be no chance of escape.

Lalchan explained that the people do not want their city to be destroyed again, and rebuilding is going on everywhere you look – fresh paint, median project in high gear with red and white tiles being laid in the center median, street lights are up, flower boxes being set along median and at traffic circles. The city is finally returning to normal, and the citizens want it to stay that way.

John and I have been briefed on our next mission. First, we are to depart with 2/8 Marines at 0845 to link up with Dr. Fahid to visit a young girl who had been accidentally shot about 3 weeks ago. Then, we will do a dismounted patrol along Rt. Gypsum, through the area where the 6 Apr

2004 ambush took place. Next, we will go to the local sheik for lunch until about 1530. Finally, we are to either transition to Echo Company or go back to Camp Ramadi. I am hoping that we will be back in Camp Ramadi tonight, as I am very tired. I just put on fresh socks for the first time in 4 days. Oh, it feels so nice!

Promptly at 0800, we gathered in the parking area of the JSS with Capt. Martin and the 2/8 Marines that we will be embedded with today. The Marines crewing our Humvee are as follows: VC (vehicle commander) - Cpl. Foxx, Driver – Cpl. Jonathan Yale (from Meherrin, Virginia, a 2nd year Marine on his 1st deployment with 7 months in-country); Gunner – LCpl. Rob Cummings (Stanford, Connecticut, 1.5 year Marine veteran on his 1st deployment); and "Rosco," our Iraqi "terp" (interpreter) (Diwaniyah, Iraq, 8 months with the USMC.) These young men seem very innocent, laughing and joking one minute, but are deadly serious once they step out of a vehicle in a combat zone. They are incredibly motivated, but tell me that they are a little bored. "We are used to killing people and blowing things up, so doing humanitarian missions is something we're just getting used to," said Corporal Yale.

Unbelievably, Cpl Jonathan Hale of 2/8 Marines along with LCpl Jordan Haerter of 1/9 Marines were killed by a VBIED on 22 April 2008, while guarding the Ramadi security station gate. While everyone else ran away, Yale and Haerter stood unflinchingly, firing their weapons at the fast-approaching suicide truck bomb until it slowed and exploded just feet from them. Both Yale and Haerter were posthumously awarded the Navy Cross. Their bravery in the face of certain death was also honored by MajGen John Kelly in his "Six Seconds to Live" speech[2] on 13 Nov 2010 given only four days after his own son USMC Lt. Robert Kelly was KIA in Afghanistan. Kelly's "Six Seconds to Live" is now read at Marine Corps boot camp training to all recruits. A short film "The 11th Order"[3] by Marine veteran Joshua DeFour also honored the bravery of these two young heroes who gave their own lives to save the lives of 50 of their fellow Marines and 100 Iraqi police at the security station they were guarding. The 11th Military General Order is "To be especially watchful at night, and during the time for challenging, to challenge all persons on or near my post, and to allow no one to pass without proper authority."

Capt. Martin personally checks on every Humvee crew to make sure that the Marines have everything they need, and are appropriately clad with Nomex gloves, Kevlar, and RBA. Martin seems to be a great officer that really cares about his Marines. When John and I were first escorted upstairs to our closet room, the Marine quarters we entered were unlocked and had unsecured weapons everywhere. Capt. Martin immediately got on the radio, and within seconds, several Marines appeared, secured the weapons and locked the door. I am comforted by the fact that if anyone with bad intentions comes in, they will have to pass through a room full of Marines with extensive combat experience, who are armed to the teeth. My 7 March briefing just confirmed to me that some of the IA and IP cannot be trusted, and I'll sleep like a baby tonight knowing the Marines are just outside my door.

Once everyone and the convoy are ready, our Humvees form up and exit the JSS. We turn left (east) on Rt. Michigan and drive the short distance to Rt. Gypsum. Our convoy turns left (north) just before the Ramadi Arches, and slowly drives into the marketplace. After a short drive, we pull over in an empty but garbage-strewn lot near a date palm grove, and deploy from the vehicles. All

---

[2] "Six Seconds to Live" by LtGen John Kelly: https://www.legion.org/magazine/101297/six-seconds-live

[3] :The 11th Order: (Short, 2019) by Joshua DeFour: https://www.youtube.com/watch?v=Q23gKyHWjjg

the while, the Marines are hyper-vigilant, and seem to be looking in every direction at once, weapons at the ready. We walk alongside a medium-sized house with a fenced backyard. Laundry flaps in the slight breeze, and one of the light purple head scarves draped over a stick frame ironically forms a cross-shape that I photograph as we walk by. Our squad, with John and I at the rear, passes a fat black & white cow just as we enter the palm grove.

After a short walk through the palm grove beautifully lit by the morning sun, we came to a small stucco house. Normally, we would remove our shoes before we enter a Muslim home, but the Marines explain that they do not do this for security reasons. I can imagine it would be difficult to respond to an attack if you had to put your lace-up boots on first, so this makes perfect sense to me. We enter the 1 room house, and greet the father, and meet his 2 young sons, and 2 daughters, one of whom is the wounded girl we have come to see. This is my first visit to an Iraqi home, and I am surprised to see all the furniture and bedding pushed against one side of the room. A worn rug covers the small floor area. Capt. Martin and Rosco immediately explain to the father that we are here to provide security and assistance to the doctor that is en route to treat his daughter. The little red-haired girl is lying atop a blanket on the floor. She has good color, but seems to be in pain. Capt. Martin offers candy and coconut milk to the children, which they quickly consume. He then pulls out some toy cars that he gives to the children. The boys shyly begin playing with the toy cars after their snack.

In a very short time, Dr. Fahid, the Iraqi doctor enters the home, and introductions are made all around. The doctor and Capt. Martin sit down on the carpet with the girl's father, and the doctor begins examining the little girl. She tolerates the exam until she begins to cry when Dr. Fahid examines her belly. More coconut milk soothed her back into silence, although huge tears glistened on her plump brown cheeks. Meanwhile, I am photographing everything in sight, from Capt. Martin watching the exam and trying to entertain the other children, to the boys playing with their new toy cars, to the worried father pacing the floor, to John Wroblewski offering the boys some candy. After the exam is complete, Dr. Fahid hands medicine to the father (which I believe was paid for out of Capt. Martin's pocket.) We exchange best wishes and exit the modest home back. We start back through the palm grove to our vehicles, while I shoot photos along the way.

On the path near our vehicles, we meet several Iraqi boys of school age, and they ply us for candy, "Meestah! Meestah! Choc-o-lot?" The boys are quite eager to pose for a few photographs, which I am only too glad to take. After we reach our vehicles, we walk a short distance further to a madrasa, a Muslim boys' school. 2/8 Marines have been sponsoring the school, and make frequent visits to give them supplies, and try to meet their needs. On this trip, Capt. Martin advises the headmasters that he has obtained more school desks for the students, which will be delivered by his unit later this week. We entered a classroom full of smiling Iraqi boys, and I began photographing everything I could, despite the poor lighting conditions. The students' most requested item is pencils! It saddens me to think that students anywhere do not have enough pencils to go around, and that they must be shared. Some students are obviously thin and are dressed in dirty clothes. Capt. Martin leads an impromptu English lesson and awards pencils and erasers to every student that even comes close to answering correctly. The remaining school supplies are given to the headmaster to be dispersed later.

We depart the dark classroom into the blazing sunlight in the courtyard. Several men and older students begin asking the Marines about helping them find work. One old man with a traditional Arab headdress begins to complain that there are no easy jobs for an old man like himself. The other Iraqis laugh, and tell him that there are no jobs for men who won't work. Shortly afterward, the school boys flood into the courtyard, surrounding the Marines and John Wroblewski. Several of them ask me to give them my camera, but I graciously decline their request. John is very much at home with the students, and fits right in, and poses for several group photos with the students. Many of the same students are appearing in all my images, so some of the boys are definitely "camera hogs." After handing out more school supplies, we depart the school and return to our Humvees.

After a short drive, we reached Sheik Mahmoud's house, parking along the road in front and beside his compound. We are graciously welcomed into the whitewashed home by one of the sheik's sons, who directed John to a cushion near the head of the spacious room. The room has large windows on one side, and the floor is covered by a beautiful hand-knotted rug. Along all sides of the room are thick green cushions upon which to sit. Again, we do not remove our boots due to security issues, although one would typically do this in any Arab home. John and I sit on the right side of the room, near the head of the room where the sheik will sit. John remembers my etiquette lesson and sits cross-legged to avoid displaying the soles of his feet to our hosts. As soon as we sit down, the sheik's sons bring silver trays with small crystal glasses of hot tea (chai) on little saucers with a tiny silver spoon. The crystal glasses contain as much sugar as tea, and following the lead of the Marines, we vigorously stir the chai using the silver spoon, which makes a delightful tinkling noise as others around the room do the same. We drink the tea quickly, and our glasses are refilled almost instantly. Sheik Mahmoud's brother enters the room, at which point we rise in deference to him. After greeting each of us with a hug and a kiss on the cheek, he sits on the cushion at the head of the room. After a couple of rounds of tea, one of the sons begins offering each of us a drink of espresso from a small china cup, beginning at the head of the room and working his way down the line. Each of us accepts the cup with our right hand, gulps down the espresso, and hands the cup back to the young man with our right hand. The boy uses a small embroidered towel to wipe the rim of the cup before filling it again from a china pitcher, and then offers it to the next person down the line. Meanwhile, the Camp Ramadi CO (commanding officer), LtCol Bargeron joins our gathering, and takes a seat alongside Sheik Mahmoud at the head of the room.

After several rounds of chai and espresso, one of Mahmoud's sons brings in a silver tray with packs of Kent cigarettes stacked in a little pyramid, and brings the tray around to each of us. I take a pack, open it, and light a cigarette, as does almost everyone else in the room. John Wroblewski graciously declines the proffered cigarettes when the son pauses in front of him.

After everyone has a cigarette or two, the luncheon begins in earnest. Mahmoud's son brings in a large vinyl sheet, and lays it over the carpet to protect it from any spilled food during our luncheon. Then, both sons bring out a 4' wide platter, piled high with rice, roasted chicken, and goat wrapped in grape leaves, and set it on the floor at the head of the room in front of the sheik, CO, John, and I. We scoot off of the cushions, and sit Indian-style on the floor. Using our right hands only, we dig into the pile of delicious food. John gives me a perplexed look, and whispers, "Where's the plates and silverware?" I explain that there are none, and to just reach in with his right hand, scoop up some food, and put it in his mouth, placing the chicken bones in front of him. He shakes his hand, and grimly proceeds. After everyone at the head of the room had eaten their fill, the platter is moved down the room, and the rest of the men (both Iraqis and Marines) take their turns in front of the platter.

Once everyone had eaten and smoked a cigarette (if so inclined), the platter is removed, and more rounds of chai and espresso are served. Now that the pleasantries and lunch are complete, the "business" part of the luncheon begins. Sheik Mahmoud begins by saying, "The civilians are tired of fighting. My son was killed, and my brother was wounded, and everyone is glad the violence is ending. AQI is an anti-Islamic organization, so the citizens along with the coalition forces have driven them out. They (AQI) have tried to kill me 5 times. Send a message to everyone in America that AQI is anti-Islamic and is a bad organization. Islam is a religion of peace and love for everyone, while AQI fights everyone. I like working with the Marines, Col. Bargeron, and coalition forces, and am trying to get more projects for this area, although we have gotten some. I know the Marines are doing good things for the people. Both my brother and I agree that Iran is also a bad influence in Iraq. Iran only wants power in Iraq, and wants to use nuclear technology for weapons, not for electricity and good science. I thank John (Wroblewski) for the sacrifice J.T. made for his country. John is a hero, because J.T. is a hero. It is great that they (Marines) are making Iraq free again."

Capt. Martin then stated, "The time to complete the College of Agriculture project is 3 months. I met with all nine contractors because some contractors do not like Hamid and Sheik Mahmoud because they work with Coalition forces. I will explain to them that they all have to work together for the common good."

Sheik Mahmoud then said, "The dean of the college took some stuff from the college and sold it in the market, and put the money in his pocket. The dean is not a good person." This issue was discussed by Capt. Martin and the sheik, and Capt. Martin promised to follow up with the dean if the money was not returned to the college.

After this, there was a heated discussion between Capt. Martin and an Iraqi Police LtCol who had entered the sheik's house late. The IP LtCol had "heard" that LtCol Bargeron was coming to visit the sheik, but was upset because he was not invited to attend the meeting. Capt. Martin explained that "fake" officers are the real problem, because the IP LtCol and his officers had promoted themselves, and the IP LtCol does not work most of the time, and had failed to fight the insurgents when they were in Anbar. "Rosco", the interpreter, translated Capt. Martin's comments to the IP LtCol and other Iraqis, and this caused general laughter around the room among the Iraqis, and the IP LtCol left in a huff, having lost "face" among his fellow Iraqis.

Col. Bargeron arrived at this point, and Sheik Mahmoud greeted him with a kiss on the cheek and thanked him and the coalition forces for improving security in the region – security that had allowed reconstruction and less violence. Bargeron sat down at the head of the room next to Mahmoud, and accepted chai. After a couple rounds of chai, Bargeron stated that there had been a couple of protests in Ramadi this week - one was in protest of the Danish cartoon about Mohammed the Prophet, and Bargeron asked the sheik who had supported this protest. Sheik Mahmoud said the protest had been organized by the old provincial party. Col. Bargeron continued, "The second protest was over the visit by Iranian President Ahmadinejad, and the Iraqi people don't like him because he has not been here in a long time and they suspect that something is wrong with him visiting now. Also, Iran is currently under sanctions from most U.N. members. Iran wants to open trade with Iraq, and is willing to loan Iraq $100 million. Iran is looking to control Iraqi oil supplies, because Iraq has no need to borrow money since Iraq has $20 million in reserve in the treasury. Therefore, Iran is just trying to exert influence in Iraq." Col. Bargeron continued, "Sheik Mahmoud is about to start a new project paving a road 12 km east of Ramadi, the airport project, while continuing the paving at the Al Assad base. Hamid also started a project at the Agricultural College

4 days ago."

After all the discussions, Col. Bargeron, Sheik Mahmoud, Hamid, and all the other Iraqi VIPs gathered in the courtyard with John Wroblewski for photo ops. Capt. Martin asked me to get close-ups of each Iraqi for their intelligence files, which I gladly did.

We exchanged email addresses and I gave the sheik my business card, and we bade them farewell.

Due to space limitations, I will not detail the final five days that Gold Star dad John Wroblewski and I spent in Iraq. When it came time to depart Ramadi, we reported to the small Ramadi airport where I thought we would board a CH-46 helicopter for our flight to Balad. I was happily surprised when we were told to board a V-22 Osprey tilt rotor aircraft. John and I saw next to a Marine LCpl and across from an Army Captain. During the short, smooth flight, the Captain began turning pale and hurriedly grabbed the plastic bag containing his shower shoes, which he proceeded to use as an air sickness bag. I couldn't resist elbowing the LCpl and said, "Obviously, the Captain isn't Airborne." The LCpl just laughed and shook his head.

THANK YOU!

First, I thank God for all my blessings and for keeping John Wroblewski and I safe during our two trips to Iraq and military embeds. I also thank the Marines and Corpsmen of the 2nd Battalion, 4th Marines for their heroism and sacrifices during the 2004 Ramadi deployment, and 2nd Battalion, 8th Marines for keeping John and I safe during our 2008 embed. I especially thank those 46 men of 2/4 Marines and the Gold Star family members that graciously completed interviews for the book because without you this book would not be possible. A special thank you to the Syfan Family and Syfan Logistics of Gainesville, Georgia for their generous donation that funded our 2008 Iraq memorial service mission expenses and another donation that helped a Christian Iraqi interpreter emigrate from Iraq to the U.S. to avoid being killed during Iraq's Christian genocide. Thank you to Canon Andrew White for his 2008 Baghdad interview during Iraq's Christian persecution.

I also thank Delta pilot Captain Russell Copeland for his donation of a 1st class round-trip ticket from Newark, NJ to Atlanta, GA for John Wroblewski for the 2008 mission. Another special thank you to journalist Martha Zoller of WDUN Radio in Gainesville, GA for introducing me to John Wroblewski and coming up with the idea to take John to Ramadi to do a memorial for 2/4's fallen heroes and hiring me as your 2007 mission photographer – you're one of the bravest and most positive people I know.

Thank you to David Swanson for your generous donation of one of your 2004 Ramadi photographs for the book. I also thank J.A. Lewis for your transcriptions of many of the Marine and Corpsmen interviews. A huge "thank you" to Major General John Kelly and his PSD for escorting us to Route Gypsum on 6 March 2008 so John could perform the memorial service for the fallen. I also thank MSgt. Willie Ellerbrock, PAO for helping John and I make our mission not only possible, but a success, especially in light of our ill-fated first meeting in 2007 at LZW.

A heartfelt thank you to my wife Lisa Aiello for your advice, patience, and understanding during this long process. Last, but not least, a huge thank you to my publisher Neal Minor at M. LiClar Publishing Co. for believing in this project and all your hard work in getting the book edited and published.

About The Author

Gregory developed a love of photography in 1978 when he was the photographer and Editor-in-Chief of his high school newspaper "Excalibur," and shot his first professional assignment in 1985 as a wedding photographer. His career includes photography for many Fortune 500 companies, a 16 year USDA Forest Service photography contract during which he photographed all the National Forests in Alabama, Mississippi, Louisiana, Chattahoochee National Forest in north Georgia, Sumter National Forest in South Carolina, and the Sabine National Forest in Texas (living in a tent for 3 months each year to do so), 2010 "Best Wedding Photographer" award by "The Knot," Georgia Veterinary Medical Association awards photographer for five years, as well as his two military embeds in Iraq in 2007 and 2008. Gregory graduated Magna Cum Laude from Piedmont University in 2010 with a B.A. in Early Childhood Education and escorted John Wroblewski to Iraq during his 2008 Piedmont Spring Break. His hobbies include backpacking, cooking, and spending his leisure time with his wife Lisa and their two dogs.